# LIVING BY
# CHEMISTRY

# LIVING BY
# CHEMISTRY

**SECOND EDITION**

**Angelica M. Stacy**
Professor of Chemistry
University of California at Berkeley

*with*

**Janice A. Coonrod**
Bishop O'Dowd High School
*Senior Writer and Developer*

**Jennifer Claesgens**
Weber State University
*Curriculum Developer*

W. H. Freeman and Company / BFW

W. H. Freeman and Company / BFW

*Publisher:* **Ann Heath**

*Sponsoring Editor:* **Jeffrey Dowling**

*Development Editor:* **Donald Gecewicz**

*Project Manager:* **Karen Misler**

*Marketing Manager:* **Janie Pierce-Bratcher**

*Editorial Assistant:* **Matthew Belford**

*Copyeditor:* **Kristin Ferraioli**

*Director of Editing, Design, and Media Production for the Sciences and Social Sciences:* **Tracey Kuehn**

*Managing Editor:* **Lisa Kinne**

*Project Editors:* **Enrico Bruno, Robert M. Errera**

*Production Manager:* **Paul Rohloff**

*Photo Editor:* **Robin Fadool**

*Associate Photo Editor:* **Eileen Liang**

*Art Director:* **Diana Blume**

*Text and Cover Designer:* **Marsha Cohen, Parallelogram Graphics**

*Art Manager:* **Matthew McAdams**

*Illustrator:* **Joe BelBruno**

*Technical Artist:* **Precision Graphics**

*Composition:* **Aptara®, Inc.**

*Printing and Binding:* **LSC Communications**

*Cover:* **Photo by C. Mane / Getty Images**

This material is based upon work supported by the National Science Foundation under award number 9730634. Any opinions, findings, and conclusions or recommendations expressed in this publication are those of the author and do not necessarily reflect the views of the National Science Foundation.

Companion Website
bcs.whfreeman.com/livingbychemistry2e
The *Living by Chemistry* Companion Website is available free of charge and provides all access to all media for this program.

Media and e-Book Offerings
For information about additional resources that are available to support *Living by Chemistry*, please visit the catalog page:
highschool.bfwpub.com/LBC2e

Library of Congress Preassigned Control Number: 2014953443

ISBN-10: 1-3192-1280-8

ISBN-13: 978-1-3192-1280-3

First Printing

Printed in the United States of America

W. H. Freeman and Company
One New York Plaza
Suite 4500
New York, NY 10004-1562
highschool.bfwpub.com

A number of individuals joined the project as developers for various periods of time along the way to complete this work. Thanks go to these individuals for their contributions to the unit development: Karen Chang, David Hodul, Rebecca Krystyniak, Tatiana Lim, Jennifer Loeser, Evy Kavaler, Sari Paikoff, Sally Rupert, Geoff Ruth, Nicci Nunes, Gabriela Waschewski, and Daniel Quach.

David R. Dudley contributed original ideas and sketches for some of the wonderful cartoons interspersed throughout the book. His sketches provided a rich foundation for the art manuscript.

This work would not have been possible without the thoughtful feedback and great ideas from numerous teachers who field-tested early versions of the curriculum. Thanks go to these teachers and their students: Carol de Boer, Wayne Brock, Susan Edgar-Lee, Melissa Getz, David Hodul, Richard Kassissieh, Tatiana Lim, Evy Kavaler, Geoff Ruth, Nicci Nunes, Gabriela Waschewski, and Daniel Quach.

Dr. Truman Schwartz provided a thorough and detailed review of the first edition manuscript. We appreciate his insights and chemistry expertise.

Thank you everyone at Bedford, Freeman & Worth who contributed to the development of this second edition. Special thanks to Jeffrey Dowling for masterfully guiding the development, to Donald Gecewicz for thoughtful development editing, feedback, and advice, and to Karen Misler for keeping us all on track with expert project management. Thank you, Sharon Sikora, for help with assessments and appendix material. Finally, thanks go to the publisher Ann Heath, who believed in this program and helped to assemble and guide the team along the way.

**Science Content Advisor**

Dr. A. Truman Schwartz, Macalaster College (emeritus), St. Paul, MN

**Teaching and Content Reviewers**

Scott Balicki, Boston Latin School, Boston, MA
Greg Banks, Urban Science Academy, West Roxbury, MA
Randy Cook, Tri County High School, Howard City, MI
Thomas Holme, University of Wisconsin, Milwaukee, Milwaukee, WI
Mark Klawiter, Deerfield High School, Deerfield, WI
Kathleen Markiewicz, Boston Latin School, Boston, MA
Nicole Nunes, Liberty High School, Benicia, CA
Carri Polizzotti, Marin Catholic High School, Larkspur, CA
Paul Thompson, Jr., High School for Math, Science, and Engineering at CCNY, New York, NY
Matthew Vaughn, Burlingame High School, Burlingame, CA
Rebecca Williams, Richland College, Dallas, TX

Molly Hodul

## About the Author

Angelica M. Stacy is a Professor of Chemistry at the University of California, Berkeley. At Berkeley, she teaches introductory chemistry and does research in materials chemistry and chemistry education. She has received numerous awards and honors for her teaching and held the President's Chair for Teaching at the University of California from 1993 to 1996. In 2005, the National Science Foundation named Dr. Stacy a Distinguished Teacher Scholar. As a part of her work to help more students succeed in science and gain a better understanding of the principles of chemistry, she has participated as a member of the Physical Sciences Design team that helped develop the Next Generation Science Standards. She has also participated on various AP® chemistry course committees as part of the College Board's efforts to redesign the AP® chemistry course and exam. Dr. Stacy has worked for many years in designing and refining the *Living by Chemistry* curriculum. She developed this curriculum to support students as they learn chemistry by offering them engaging challenges that relate to the world around them.

Dirk Wiersma/Science Source

## Unit 1 | Alchemy                                                          1

### Matter, Atomic Structure, and Bonding

Alex Cao/Getty Images

Stockbyte/Getty Images

Charles D. Winters/Science Source

Vadym Volodin/iStockphoto

## Unit 5  |  Fire                                                     481

### Energy, Thermodynamics, and Oxidation-Reduction

David Arky/Getty Images

# Next generation chemistry brought to life.

■ Chapter-opening pages introduce the main ideas and learning goals of the chapter.

■ **THINK ABOUT IT** prepares you to read about chemistry. Take a moment to reflect on the lesson question and topic outline before you read the lesson.

**CHAPTER 4**

## Moving Electrons

*In this chapter, you will study*
- the systematic arrangement of electrons in an atom
- how ionic compounds are formed
- valence electrons and ionic bonding

*When heated, different metal atoms produce flames of different colors. That is what causes the many brilliant colors of a fireworks display.*

While the nucleus of an atom can be difficult to change, the electrons are a different story. Electrons can be moved around within an atom or transferred between atoms. When atoms transfer electrons, they become ions with positive and negative

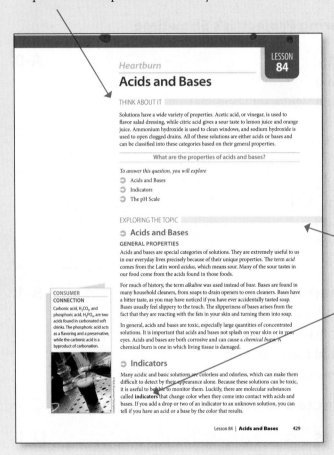

*Heartburn*

**LESSON 84**

## Acids and Bases

THINK ABOUT IT

Solutions have a wide variety of properties. Acetic acid, or vinegar, is used to flavor salad dressing, while citric acid gives a sour taste to lemon juice and orange juice. Ammonium hydroxide is used to clean windows, and sodium hydroxide is used to open clogged drains. All of these solutions are either acids or bases and can be classified into these categories based on their general properties.

**What are the properties of acids and bases?**

*To answer this question, you will explore*
- Acids and Bases
- Indicators
- The pH Scale

EXPLORING THE TOPIC

### Acids and Bases

GENERAL PROPERTIES

Acids and bases are special categories of solutions. They are extremely useful to us in our everyday lives precisely because of their unique properties. The term *acid* comes from the Latin word *acidus*, which means sour. Many of the sour tastes in our food come from the acids found in those foods.

For much of history, the term *alkaline* was used instead of *base*. Bases are found in many household cleaners, from soaps to drain openers to oven cleaners. Bases have a bitter taste, as you may have noticed if you have ever accidentally tasted soap. Bases usually feel slippery to the touch. The slipperiness of bases arises from the fact that they are reacting with the fats in your skin and turning them into soap.

In general, acids and bases are toxic, especially large quantities of concentrated solutions. It is important that acids and bases not splash on your skin or in your eyes. Acids and bases are both corrosive and can cause a *chemical burn*. A chemical burn is one in which living tissue is damaged.

### Indicators

Many acidic and basic solutions are colorless and odorless, which can make them difficult to detect by their appearance alone. Because these solutions can be toxic, it is useful to be able to monitor them. Luckily, there are molecular substances called **indicators** that change color when they come into contact with acids and bases. If you add a drop or two of an indicator to an unknown solution, you can tell if you have an acid or a base by the color that results.

**CONSUMER CONNECTION**
Carbonic acid, $H_2CO_3$, and phosphoric acid, $H_3PO_4$, are two acids found in carbonated soft drinks. The phosphoric acid acts as a flavoring and a preservative, while the carbonic acid is a byproduct of carbonation.

Lesson 84 | **Acids and Bases**   429

■ **EXPLORING THE TOPIC** allows you to review and practice the chemistry at the right pace. Each lesson follows in step with the daily hands-on, inquiry-based lesson from your class.

■ Look at the **key terms** in bold. Find definitions in English and Spanish at the end of the book.

■ Develop the science practice of using visual representations and models to explain chemistry by studying the **figures, diagrams**, and **illustrations** used throughout the book.

■ Pay attention to **Big Ideas,** especially when you are reviewing for a quiz or exam. These are fundamental concepts of chemistry.

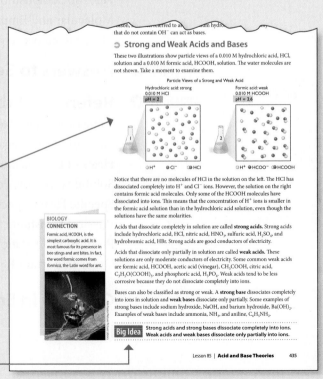

...that do not contain $OH^-$ can act as bases.

### Strong and Weak Acids and Bases

These two illustrations show particle views of a 0.010 M hydrochloric acid, HCl, solution and a 0.010 M formic acid, HCOOH, solution. The water molecules are not shown. Take a moment to examine them.

**Particle Views of a Strong and Weak Acid**

Hydrochloric acid: strong
0.010 M HCl
pH = 2

Formic acid: weak
0.010 M HCOOH
pH = 2.6

○ $OH^+$   ● $Cl^-$   ● HCl        ○ $H^+$   ○ $HCOO^-$   ◑ HCOOH

Notice that there are no molecules of HCl in the solution on the left. The HCl has dissociated completely into $H^+$ and $Cl^-$ ions. However, the solution on the right contains formic acid molecules. Only some of the HCOOH molecules have dissociated into ions. This means that the concentration of $H^+$ ions is smaller in the formic acid solution than in the hydrochloric acid solution, even though the solutions have the same molarities.

Acids that dissociate completely in solution are called **strong acids**. Strong acids include hydrochloric acid, HCl, nitric acid, $HNO_3$, sulfuric acid, $H_2SO_4$, and hydrobromic acid, HBr. Strong acids are good conductors of electricity.

Acids that dissociate only partially in solution are called **weak acids.** These solutions are only moderate conductors of electricity. Some common weak acids are formic acid, HCOOH, acetic acid (vinegar), $CH_3COOH$, citric acid, $C_3H_5O(COOH)_3$, and phosphoric acid, $H_3PO_4$. Weak acids tend to be less corrosive because they do not dissociate completely into ions.

Bases can also be classified as strong or weak. A **strong base** dissociates completely into ions in solution and **weak bases** dissociate only partially. Some examples of strong bases include sodium hydroxide, NaOH, and barium hydroxide, $Ba(OH)_2$. Examples of weak bases include ammonia, $NH_3$, and aniline, $C_6H_5NH_2$.

**BIOLOGY CONNECTION**
Formic acid, HCOOH, is the simplest carboxylic acid. It is most famous for its presence in bee stings and ant bites. In fact, the word formic comes from *formica*, the Latin word for ant.

**Big Idea** Strong acids and strong bases dissociate completely into ions. Weak acids and weak bases dissociate only partially into ions.

Lesson 85 | **Acid and Base Theories**   435

**Important to Know** highlights ideas that you should retain for further use throughout the course.

# Practice makes perfect.

So far you've been introduced to three of the gas laws. These gas laws apply when any two of the variables, $P$, $V$, or $T$, are changed and everything else stays the same.

Charles's law $\quad V = kT \quad$ $P$ and amount of gas do not change. $\quad k = \dfrac{V}{T}$

Gay-Lussac's law $\quad P = kT \quad$ $V$ and amount of gas do not change. $\quad k = \dfrac{P}{T}$

Boyle's law $\quad P = k\left(\dfrac{1}{V}\right) \quad T$ and amount of gas do not change. $\quad k = PV$

Note that the proportionality constant, $k$, is a generic symbol and is different for each gas sample and for each gas law.

**Example 2**

**Balloon with Air**

Imagine that you fill a balloon with air to a volume of 1.5 L. The air is at a temperature of 20 °C. You place the balloon in a refrigerator for half an hour until the air in the balloon is at a temperature of 10 °C. What is the new volume of the balloon?

*Solution*

First figure out which gas law to use. The external pressure on the balloon remains the same because the air pressure in the refrigerator is the same as the air pressure outside. Use Charles's law. You can predict that $V$ will decrease because $T$ decreases.

| Initial Conditions | Final Conditions |
|---|---|
| $P_1 = 1.0$ atm | $P_2 = 1.0$ atm |
| $T_1 = 20$ °C | $T_2 = 10$ °C |
| $V_1 = 1.5$ L | $V_2 = ?$ |

Convert temperature to the Kelvin scale.
$$K = 273 + C$$
$$T_1 = 273 + 20 = 293 \text{ K}$$
$$T_2 = 273 + 10 = 283 \text{ K}$$

Determine the value of $k$. $\quad k = \dfrac{V_1}{T_1} = \dfrac{1.5 \text{ L}}{293 \text{ K}} = 0.0050 \text{ L/K}$

Use $k$ to determine $V_2$. $\quad V_2 = k\,T_2 = 0.0050 \text{ L/K} \cdot 283 \text{ K} = 1.4 \text{ L}$

The balloon has shrunk to a volume of 1.4 L.

312   Chapter 11 | **Pressing Matter**

Lessons offer critical math support. The **examples** show you how to solve problems step by step. Try to answer the question yourself before reading the solution to check your understanding.

The **LESSON SUMMARY** recaps the main ideas of the lesson that answer the opening question. Reading the lesson summaries is a good way to review for a quiz or an exam.

Doing the **Exercises** will help you practice what you learn each day.

Answer the Reading Questions to make sure you understood the main ideas.

Answer the Reason and Apply questions to practice problem solving and using evidence-based reasoning to justify answers.

**LESSON SUMMARY**

How do meteorologists keep track of the amount of rainfall?

**KEY TERMS**
proportional
proportionality constant

Everywhere around the world, the amount of rainfall is measured by its height. If volume were used to report rainfall, each meteorologist would report a different number, depending on the area of the base of the rain gauge used. Height and volume for a specific rain gauge are related by a proportionality constant, $k$.

**Exercises**

**Reading Questions**

1. Suppose you have a cylindrical rain gauge with a base area of 2.0 cm². Explain two different ways you can determine the volume when the height of water is 3.0 cm.
2. Explain in your own words why meteorologists prefer to measure rain in inches or centimeters, not in milliliters or cubic centimeters.

**Reason and Apply**

3. If the amount of rainfall increases, do both the volume and height of water in the rain gauge keep track of this increase? Explain your thinking.
4. If you use a beaker for a rain gauge and the weather station uses a graduated cylinder, will both instruments give the same volume? The same height?
5. If a large washtub, a dog's water dish, and a graduated cylinder were left outside during a rainstorm, would the three containers all have the same volume of water in them after the storm? Explain why or why not.
6. Inches are used in the United States, but centimeters are used by scientists and in the rest of the world. Look at a ruler that is marked in both inches and centimeters.
   a. Make a graph of centimeters versus inches so that the $y$-axis is centimeters and the $x$-axis is inches.

   b. Convert 12 in. to centimeters.
   c. How many inches is 10 cm?
   d. How many inches is 1 cm?
7. A student placed the same empty rain gauge outside before five different rainstorms. She measured the height and volume of the water in the gauge after each storm. Her results are in the table.
   a. Which rainstorm dropped the most rain? Did you use height or volume to answer this question? Why?
   b. What pattern do you notice?
   c. Draw a graph of the data to show that the volume and height are proportional.
   d. Explain why the data points do not all lie exactly on a straight line.
   e. Predict the volume for a height of 6.0 cm. Show your work.

| Storm number | Height (cm) | Volume (cm³) |
|---|---|---|
| 1 | 1.0 cm | 2.5 cm³ |
| 2 | 2.5 cm | 6.3 cm³ |
| 3 | 0.5 cm | 1.1 cm³ |
| 4 | 8.0 cm | 20.0 cm³ |
| 5 | 5.0 cm | 12.5 cm³ |

266   Chapter 10 | **Physically Changing Matter**

# Review and prepare for quizzes and exams.

## Toxins Update

A toxic substance causes an undesirable chemical reaction, producing a harmful or unhealthy change in a living system. Chemical equations keep track of these changes and allow you to predict what you will observe when compounds combine. Becoming familiar with chemical equations is the first step in understanding these chemical changes.

**CHAPTER 13**

**Toxic Changes**

## SUMMARY

### KEY TERMS

physical change
coefficient
formula unit
combination reaction
decomposition reaction
single exchange reaction
(single replacement)
double exchange reaction
(double replacement)

### REVIEW EXERCISES

1. Why is it difficult to identify a physical or chemical change through observations alone?
2. Based on the chemical equation for a reaction, can you tell if any of the substances are toxic?
3. Consider the equation for the formation of a kidney stone.

   $Na_3PO_4(aq) + 3CaCl_2(aq) \longrightarrow Ca_3(PO_4)_2(s) + 6NaCl(aq)$

   a. Is each reactant bonded ionically or covalently? How do you know?
   b. Is this is a combination reaction, decomposition reaction, single exchange reaction, or double exchange reaction?
   c. Is this a chemical change or a physical change?
   d. How does a balanced reaction show that matter is conserved?
   e. What is the chemical name of the solid that makes up a kidney stone?

4. Copy and balance these chemical equations.
   a. $N_2(g) + H_2(g) \longrightarrow N_2H_4(g)$
   b. $KNO_3(s) + K(s) \longrightarrow K_2O(s) + N_2(g)$
   c. $H_2SO_4(aq) + NaCN(aq) \longrightarrow HCN(g) + Na_2SO_4(aq)$
   d. $H_3PO_4(aq) + Ca(OH)_2(aq) \longrightarrow Ca_3(PO_4)_2(aq) + H_2O(l)$
   e. $C_3H_8(g) + O_2(g) \longrightarrow CO_2(g) + H_2O(l)$
   f. $H_2S(g) + O_2(g) \longrightarrow SO_2(g) + H_2O(l)$
   g. $H_2(g) + O_2(g) \longrightarrow H_2O(l)$

### Toxins in the Environment

PROJECT Research a potentially toxic substance (your teacher may assign you one). Find out where in your environment you might find this substance and describe its effects on the body. Prepare a short report.

Chapter 13 | **Summary**    379

- The **Chapter Summary and Review Exercises** help you review after each group of lessons.

- **Projects** may be assigned by your teacher to give you a chance to do some research on your own.

- Two kinds of **Review Exercises** help you to review the unit, starting with **General Review** exercises covering the major concepts in the chapters.

---

- The **Unit Review** summarizes what you learned in that unit.

## Alchemy | REVIEW

**UNIT 1**

Matter is anything that has substance and takes up space. Mass is a measure of the substance of matter. Volume is a measure of the space that matter takes up. The density of a substance is the mass per unit of volume.

Everything on the planet is composed of elements or compounds made from combinations of these elements. The elements are organized according to their properties and atomic structure into a chart called the periodic table. The periodic table holds a wealth of information about the elements and their atoms. Elements that are located in the same column of the periodic table tend to have similar behavior and reactivity. The noble gases are very stable elements with extremely low reactivity.

| 79 |
| Au |
| Gold |
| 197.0 |

Matter is composed of individual building blocks called atoms. Atoms are much too small to be seen, so experimental evidence has led to various models of the atom. In the center of each atom is a dense nucleus consisting of protons and neutrons. Most of the mass of an atom is located in its nucleus. Most atoms have more neutrons than protons in their nuclei, which keeps the nucleus more stable.

Electrons are located at specific distances from the nucleus, called shells. The electrons in the outermost shell of an atom are called its valence electrons and are key to its behavior. Atoms will sometimes lose or gain valence electrons, forming ions with positive and negative charges, and achieving the electron arrangements of the noble gases.

- Negative electron
- Uncharged neutron
- Positive proton

A compound is a substance that contains atoms of more than one type of element, bonded together. There are millions of compounds in existence and many yet to be discovered.

140    Unit 1 | **Review**

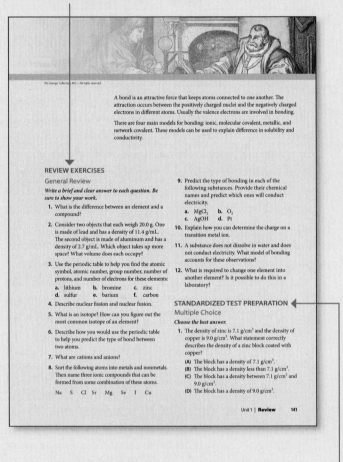

The Granger Collection, NYC—All rights reserved.

A bond is an attractive force that keeps atoms connected to one another. The attraction occurs between the positively charged nuclei and the negatively charged electrons in different atoms. Usually the valence electrons are involved in bonding.

There are four main models for bonding: ionic, molecular covalent, metallic, and network covalent. These models can be used to explain difference in solubility and conductivity.

### REVIEW EXERCISES

#### General Review

*Write a brief and clear answer to each question. Be sure to show your work.*

1. What is the difference between an element and a compound?
2. Consider two objects that each weigh 20.0 g. One is made of lead and has a density of 11.4 g/mL. The second object is made of aluminum and has a density of 2.7 g/mL. Which object takes up more space? What volume does each occupy?
3. Use the periodic table to help you find the atomic symbol, atomic number, group number, number of protons, and number of electrons for these elements:
   a. lithium    b. bromine    c. zinc
   d. sulfur    e. barium    f. carbon
4. Describe nuclear fission and nuclear fusion.
5. What is an isotope? How can you figure out the most common isotope of an element?
6. Describe how you would use the periodic table to help you predict the type of bond between two atoms.
7. What are cations and anions?
8. Sort the following atoms into metals and nonmetals. Then name three ionic compounds that can be formed from some combination of these atoms.

   Na    S    Cl    Sr    Mg    Se    I    Cu

9. Predict the type of bonding in each of the following substances. Provide their chemical names and predict which ones will conduct electricity.
   a. $MgCl_2$    b. $O_2$
   c. AgOH    d. Pt
10. Explain how you can determine the charge on a transition metal ion.
11. A substance does not dissolve in water and does not conduct electricity. What model of bonding accounts for these observations?
12. What is required to change one element into another element? Is it possible to do this in a laboratory?

### STANDARDIZED TEST PREPARATION

#### Multiple Choice

*Choose the best answer.*

1. The density of zinc is 7.1 g/cm³ and the density of copper is 9.0 g/cm³. What statement correctly describes the density of a zinc block coated with copper?
   (A) The block has a density of 7.1 g/cm³.
   (B) The block has a density less than 7.1 g/cm³.
   (C) The block has a density between 7.1 g/cm³ and 9.0 g/cm³.
   (D) The block has a density of 9.0 g/cm³.

Unit 1 | **Review**    141

- The **Standardized Test Preparation** section contains multiple-choice questions. Each Exercise has a stem and four possible answers.

# Extra support when you need it.

■ When you see a note about **Math Spotlights** in a lesson you can turn to the back of the book for a quick review of that math topic.

---

## Appendix A: Math Spotlights

### SI Units of Measure

| Quantity | Unit (abbreviation) |
|---|---|
| length | meter (m) |
| mass | kilogram (kg) |
| time | second (s) |
| temperature | Kelvin (K) |
| amount of substance | mole (mol) |

Scientists rely on repeatable measurements as they study the physical world. It is important that they use consistent units of measure worldwide. In 1960 an international council standardized the metric system, creating the *Système International d'Unités* (International System of Units), abbreviated as SI.

In the table at left, there are the basic SI units used in chemistry. Other units such as density and volume are combinations of these.

SI units are based on powers of 10. Larger and smaller units get their names by combining standard prefixes with these basic units. For example, the word *centimeter* is a combination of *centi-* and *meter*. A centimeter is one-hundredth of a meter. These are the prefixes used in the SI, along with their abbreviations and their meanings. (A kilogram is the only basic SI unit that has a prefix as part of its name.)

| Prefix | Multiple | Scientific Notation | Prefix | Multiple | Scientific Notation |
|---|---|---|---|---|---|
| tera- (T-) | 1,000,000,000,000 | $10^{12}$ | pico- (p-) | 0.000 000 000 001 | $10^{-12}$ |
| giga- (G-) | 1,000,000,000 | $10^{9}$ | nano- (n-) | 0.000 000 001 | $10^{-9}$ |
| mega- (M-) | 1,000,000 | $10^{6}$ | micro- ($\mu$-) | 0.000 001 | $10^{-6}$ |
| kilo- (k-) | 1,000 | $10^{3}$ | milli- (m-) | 0.001 | $10^{-3}$ |
| hecto- (h-) | 100 | $10^{2}$ | centi- (c-) | 0.01 | $10^{-2}$ |
| deka- (da-) | 10 | $10^{1}$ | deci- (d-) | 0.1 | $10^{-1}$ |

**Example 1**

#### Length Conversions

How many meters does each of these lengths represent?
**a.** 562 centimeters  **b.** 2.5 kilometers

*Solution*

**a.** $562 \text{ cm} \cdot \dfrac{1 \text{ m}}{100 \text{ cm}} = 5.62 \text{ m}$  **b.** $2.5 \text{ km} \cdot \dfrac{1000 \text{ m}}{1 \text{ km}} = 2500 \text{ m}$

**Example 2**

#### The Mass of One Liter

One milliliter, or cubic centimeter, of water at 4 °C weighs 1 g. How much does 1 L of water at 4 °C weigh?

A-0

---

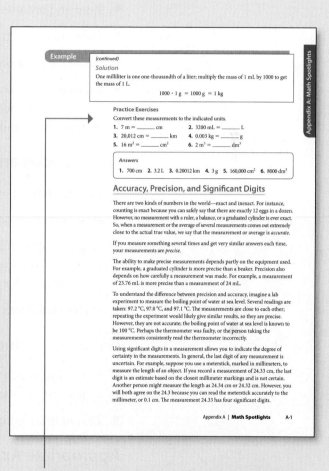

**Example**

(continued)

*Solution*

One milliliter is one one-thousandth of a liter; multiply the mass of 1 mL by 1000 to get the mass of 1 L.

$$1000 \cdot 1 \text{ g} = 1000 \text{ g} = 1 \text{ kg}$$

**Practice Exercises**

Convert these measurements to the indicated units.

1. $7 \text{ m} = \underline{\hspace{1cm}} \text{ cm}$   2. $3200 \text{ mL} = \underline{\hspace{1cm}} \text{ L}$.

3. $20{,}012 \text{ cm} = \underline{\hspace{1cm}} \text{ km}$   4. $0.003 \text{ kg} = \underline{\hspace{1cm}} \text{ g}$

5. $16 \text{ m}^2 = \underline{\hspace{1cm}} \text{ cm}^2$   6. $2 \text{ m}^3 = \underline{\hspace{1cm}} \text{ dm}^3$

*Answers*

1. 700 cm   2. 3.2 L   3. 0.20012 km   4. 3 g   5. 160,000 cm²   6. 8000 dm³

### Accuracy, Precision, and Significant Digits

There are two kinds of numbers in the world—exact and inexact. For instance, counting is exact because you can safely say that there are exactly 12 eggs in a dozen. However, no measurement with a ruler, a balance, or a graduated cylinder is ever exact. So, when a measurement or the average of several measurements comes out extremely close to the actual true value, we say that the measurement or average is *accurate*.

If you measure something several times and get very similar answers each time, your measurements are *precise*.

The ability to make precise measurements depends partly on the equipment used. For example, a graduated cylinder is more precise than a beaker. Precision also depends on how carefully a measurement was made. For example, a measurement of 23.76 mL is more precise than a measurement of 24 mL.

To understand the difference between precision and accuracy, imagine a lab experiment to measure the boiling point of water at sea level. Several readings are taken: 97.2 °C, 97.0 °C, and 97.1 °C. The measurements are close to each other; repeating the experiment would likely give similar results, so they are precise. However, they are not accurate; the boiling point of water at sea level is known to be 100 °C. Perhaps the thermometer was faulty, or the person taking the measurements consistently read the thermometer incorrectly.

Using significant digits in a measurement allows you to indicate the degree of certainty in the measurements. In general, the last digit of any measurement is uncertain. For example, suppose you use a meterstick, marked in millimeters, to measure the length of an object. If you record a measurement of 24.33 cm, the last digit is an estimate based on the closest millimeter markings and is not certain. Another person might measure the length as 24.34 cm or 24.32 cm. However, you will both agree on the 24.3 because you can read the meterstick accurately to the millimeter, or 0.1 cm. The measurement 24.33 has four significant digits.

Appendix A | **Math Spotlights**   A-1

---

■ Do the **Practice Exercises** for a quick refresher. Answers are provided directly below so you can check your understanding.

---

## Glossary

### English/Inglés

**A**

**absolute zero** The temperature defined as 0 K on the Kelvin scale and −273.15 °C on the Celsius scale. Considered to be the lowest possible temperature that matter can reach. (p. 279)

**absorption** Of light, not being transmitted because certain colors are removed by an object. Absorption transfers energy back to the object. (p. 579)

**acid** A substance that adds hydrogen ions, H⁺, to an aqueous solution; a substance that donates a proton to another substance in solution. (p. 430)

**actinides** A series of elements that follow actinium in Period 7 of the periodic table and that are typically placed separately at the bottom of the periodic table. (p. 44)

**activation energy** The minimum amount of energy required to initiate a chemical process or reaction. (p. 537)

**activity series** A table showing elements in order of their chemical activity, with the most easily oxidized at the top of the list. (p. 564)

**actual yield** The amount of a product obtained when a reaction is run (as opposed to the theoretical yield). (p. 473)

**air mass** A large volume of air that has consistent temperature and water content. (p. 288)

**alkali metals** The elements in Group 1A on the periodic table, except for hydrogen. (p. 44)

**alkaline earth metals** The elements in Group 2A on the periodic table. (p. 44)

### Spanish/Español

**cero absoluto** La temperatura definida como 0 K en la escala Kelvin y −273.15 °C en la escala Celsius. Se considera como la temperatura más baja que puede alcanzar la materia. (p. 279)

**absorción** De la luz: cuando esta no se transmite porque ciertos colores han sido eliminados (por un objeto). La absorción transfiere energía al objeto. (p. 579)

**ácido** Sustancia que cede iones de hidrógeno, H⁺, a una solución acuosa; una sustancia que dona un protón a otra sustancia en la solución. (p. 430)

**actínidos** Serie de elementos que están después del actinio en el séptimo período de la tabla periódica y que usualmente, aparecen en la parte de abajo de la tabla. (p. 44)

**energía de activación** La mínima cantidad de energía que se necesita para iniciar una reacción o un proceso químico. (p. 537)

**serie de actividad** Una tabla que muestra elementos ordenados de acuerdo con la actividad química de cada uno de dichos elementos, empezando por aquellos que se oxidan con mayor facilidad. (p. 564)

**rendimiento real** La cantidad de un producto que se obtiene cuando se ejecuta una reacción (al contrario del rendimiento teórico). (p. 473)

**masa de aire** Gran volumen de aire cuya temperatura y contenido de agua son constantes. (p. 288)

**metales alcalinos** Los elementos del grupo 1A de la tabla periódica, con excepción del hidrógeno. (p. 44)

**metales alcalinotérreos** Los elementos del grupo 2A de la tabla periódica. (p. 44)

G-1

---

■ The **Glossary** contains all the key terms, defined in English and Spanish.

# Laboratory Safety

Laboratory experiments are an important part of chemistry. Follow these safety precautions to avoid danger.

## Before Working in the Lab

- Read and become familiar with the entire procedure before starting.

- Listen to instructions. When you are in doubt, ask your teacher.

- Know the location of emergency exits and escape routes.

- Learn the location and operation of all safety equipment in your laboratory, including the safety shower, eye wash, first aid kit, fire extinguishers, and fire blanket.

## Emergencies and Accidents

- Immediately report any accident, however small, to your teacher.

- If you get chemicals on you, rinse the affected area with water.

- In case of chemicals on your face, wash off with plenty of water before removing your goggles. In case of chemicals in your eyes, remove contact lenses and wash eyes with water for at least 15 minutes.

- Minor skin burns should be held under cold, running water.

## General Conduct

- Clear your bench top of all unnecessary materials, such as books and jackets, before starting work.

- Do not bring gum, food, or drinks into the laboratory.

## Appropriate Apparel

- Always wear protective safety goggles when working in the laboratory.

- Avoid bulky, loose-fitting clothing; roll up long sleeves; and tie back loose hair. Lab coats or aprons may be required.

- Wear long pants and shoes that cover the whole foot (not sandals) so your feet are protected from accidental spills or broken glassware.

## Using Glassware and Equipment

- Do not use chipped or cracked glassware or damaged equipment.

- Be careful when handling hot glassware or apparatus. Remember, hot glassware looks like cold glassware.

- Place hot glassware or apparatus (such as a crucible) on an appropriate cooling surface, such as a wire gauze.

- Never point the open end of a test tube toward yourself or anyone else.

- Never fill a pipette using mouth suction. Always use a bulb.

- Keep electrical equipment away from sinks and faucets to minimize the risk of electrical shock.

## Using Chemicals

- Never taste substances in the laboratory and avoid touching them if possible.

- Check chemical labels twice to make sure you have the correct substance. Some chemical formulas and names differ by only a letter or a number.

- Read and follow all hazard classifications shown on the label.

- Never pour anything down the drain unless instructed to do so by your teacher.

- When transferring chemicals from a common container to your own test tube or beaker, take only what you need. Do not return any extra material to the original container, because this may contaminate the original.

- Mix all substances together slowly. Add concentrated solutions to dilute solutions. When working with acids and bases, always add concentrated acids and bases to water; never add water to a concentrated acid or base as this can cause dangerous spattering.

- If you are instructed to smell something, do so by wafting (fanning) some of the vapor toward your nose. Do not place your nose near the opening of the container.

## Before Leaving the Lab

- Clean your lab station and return equipment to its proper place.

- Make sure that gas lines and water faucets are shut off.

- When discarding used chemicals, carefully follow the instructions provided.

Welcome to *Living by Chemistry*. In this course, you will actively participate in uncovering the chemistry in the laboratory and in the world around you. Rather than simply writing "correct" answers to chemistry questions and problems, you will learn to support answers with evidence. Learning chemistry is a bit like learning a new sport or dance—you will get better with practice. The more you participate, the deeper your understanding will be.

This textbook is designed to be used for reading and reference. The readings focus on the concepts you investigate in class. There are two or three pages of reading to go along with each day's activity, followed by homework exercises. The readings provide real-life examples to further explain and clarify the chemistry concepts. After each day's lesson, the reading and exercises will help you understand and remember what you learned in class. We designed this textbook to be highly readable and sincerely believe it will make the study of high school chemistry enjoyable for you.

*Angelica M. Stacy*

# Unit 1 | Alchemy

Emperor Rudolf II (1552–1612) governed the Holy Roman Empire from 1576 to 1608. However, he often neglected his duties as ruler and is better known as an alchemist and patron of the arts.

## In this unit, you will learn

- what matter is composed of
- to use the language of chemistry
- to decode information contained in the periodic table
- how new substances with new properties are made
- what holds substances together

## Why Alchemy?

Chemistry has some of its roots in the ancient practice of alchemy. The alchemists experimented with trying to make gold out of ordinary substances. In the process, they learned a great deal about matter and about chemistry. When you understand the nature of matter and its composition, you will be able to answer the question, "Is it possible to turn ordinary substances into gold?"

# CHAPTER 1

# Defining Matter

*In this chapter, you will study*

- the tools of chemistry

- how matter is defined

- how to measure mass, volume, and density

- how types of matter differ from one another

*This photo shows gold. How can you find out if an object is made of solid gold or simply coated with gold? What makes gold different from other metals? How is it possible to identify a metal based on its properties?*

With a simple procedure, you can make a copper penny look like gold. But is it really gold, or has something else been made? A way to tell real gold from other substances is to compare the properties of those substances with the properties of gold. Chemists study various properties of matter and use the results to compare and identify substances.

# Lab Equipment and Safety

## THINK ABOUT IT

A chef depends on a wide variety of gadgets and kitchenware to create delicious meals—from whisks and mixers, to ovens and saucepans. An auto mechanic relies on a toolbox of wrenches. In every profession, it is important to have the right tool for the job. Chemists have their own special tools and equipment that allow them to study the world around them. They also have a set of guidelines for using the tools safely.

### What tools and equipment do chemists use?

*To answer this question, you will explore*

→ The Tools Chemists Use

→ Laboratory Safety

## EXPLORING THE TOPIC

### → The Tools Chemists Use

Chemistry often brings to mind a laboratory filled with unusual glassware and bubbling beakers. Chemists depend on a variety of tools in their explorations. In particular, chemists need tools that allow them to measure the mass and volume of substances, mix them, heat and cool them, and observe and separate them. Take a moment to examine these illustrations to see some of the tools that are used for these purposes.

*Tools for measuring mass: balance with weighing paper, spatula.*

*Tools for measuring volume: graduated cylinders, Erlenmeyer flask, burette attached to Erlenmeyer flask.*

*Tools for observing change: test-tube holder, test-tube rack, brush for cleaning.*

*Tools for mixing: stirring rod and beaker.*

Tools for separating: Funnel with filter paper, wash bottle, beaker.

Tools for heating: Hot plate with beaker, boiling chips, and stirring rod, a thermometer.

Tools for heating: Bunsen burner with striker, ring stand with utility clamp, and a triangle holding crucible.

Measuring accurate amounts is important to chemists. They weigh solids on electronic balances and measure volumes of liquids in special glassware. You might notice that many of the containers chemists use are made of glass. Glass is a material of choice because substances in a glass container are visible. Chemists use tempered glass containers, which can be heated over flames without shattering. Also, glass containers are relatively easy to clean and reuse. Finally, notice that chemists use ring stands and special clamps to keep glassware from toppling. Spills can be hazardous.

## ⮑ Laboratory Safety

The chemistry laboratory is a place for discovery. However, as in any workplace that uses specialized equipment, safety is always important. There are many situations in a lab that can become dangerous. Before participating in any chemistry activities, you should familiarize yourself with the safety equipment in your lab.

Take a moment to examine these illustrations. What safety equipment and precautions do you notice?

Know the location of the safety equipment and how to use it. Immediately report any laboratory accident, however small, to your teacher.

Never taste or touch chemicals. Never touch hot glassware. If you get chemicals on you, rinse with plenty of water.

**CONSUMER CONNECTION**

Like labware, bakeware is often made of tempered glass so that it can withstand rapid temperature changes and high temperatures without shattering.

A few do's and don'ts:

- When working in a lab, dress appropriately. Roll up your sleeves, tie back long hair, and wear closed-toe shoes.
- Be sure that you have read the instructions for the procedure carefully.
- Double-check that you are using the correct chemicals.
- Before you begin working with chemicals or glassware, put on safety goggles.
- Before leaving the lab, clean your lab station and return equipment to its proper place.
- Do not put chemicals back into the original bottle. Doing so might contaminate the chemicals in the bottle.
- Your teacher will provide waste containers; never put chemicals or solutions down the drain unless instructed to do so by your teacher.

## LESSON SUMMARY

### What tools and equipment do chemists use?

Chemists use their own specialized tools and equipment in the laboratory. These tools and equipment are designed to allow chemists to measure mass and volume, and to mix, heat, cool, observe, and separate substances. It is important to work safely and carefully. When working in a chemistry laboratory, always wear safety goggles. Always wear appropriate clothing and closed-toe shoes. Be prepared to know what to do in case of an accident.

## Exercises

### Reading Questions

1. Why are most chemistry containers made of glass?
2. Describe the appropriate clothing to wear in a chemistry lab.

### Reason and Apply

3. List three things you should do before beginning any laboratory procedure.
4. Describe what you would do in the case of an accidental spill in class.
5. List three things you should do before leaving the laboratory.
6. What is a fire blanket used for? If necessary, do some research to find out.
7. What is a hood used for in the chemistry laboratory? If necessary, do some research to find out.
8. Why do chemists use clamps and ring stands?

# A Penny for Your Thoughts

### Purpose

To observe a chemical transformation firsthand.

### Materials and Safety

1. List the equipment used in the demonstration.
2. Briefly describe your observations of each substance used in the demonstration.
3. Safety is extremely important in the chemistry lab. Write three important safety considerations for this demonstration.

### Procedure and Observations

Record your observations for each step of the demonstration.

1. Place a beaker containing zinc filings and sodium hydroxide on a hot plate set to 4.
2. Use tongs to pick up the penny and place it in the heated beaker.
3. While holding the beaker steady with tongs, remove the penny with the other tongs.
4. Put the hot penny in a beaker of cold water to cool and rinse it.
5. Use tongs to place the penny on the hot plate.
6. When the penny has changed color, use the tongs to place it in the beaker of cold water.

### Analysis

Working with the students at your table, spend a few minutes discussing what you observed during the demonstration. Then answer the questions individually on your own paper.

7. Describe what happened to the penny during the demonstration.
8. What do you think turned the penny silver?
9. What do you think turned the penny gold?
10. **Making Sense** Do you think you made real gold? Why or why not? How could you find out?

# Introduction to Chemistry

## THINK ABOUT IT

Gold is worth a lot more than copper. If you could turn pennies into gold, you would be very rich. Beginning in ancient times, people known as alchemists tried to transform substances into other substances. In particular, some of them tried to turn ordinary metals into gold. Today, we recognize these alchemists as early chemists. In fact, the word *chemistry* is derived from Arabic *alkīmiyā'* and from Greek *khēmia*, meaning "the art of transmuting metals."

### What is chemistry?

*To answer this question, you will explore*

⇨ The Roots of Chemistry

⇨ Chemistry: The Study of Matter and Change

## EXPLORING THE TOPIC

## ⇨ The Roots of Chemistry

While trying to make gold, alchemists developed some of the first laboratory tools and chemistry techniques. They classified substances into categories and experimented with mixing and heating different substances to create something new. When alchemists succeeded in creating a new substance, they faced the challenge of figuring out whether or not that new substance was really gold. Often, alchemists were fooled into thinking that a substance was gold just because it looked like gold.

In class, you watched a procedure to make a "golden" penny.

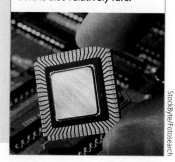
During the procedure, when the silver-colored penny was heated, it turned a gold color. You came up with a **hypothesis** to explain what happened. A hypothesis is a possible explanation for an observation. It can be tested by further investigation or experimentation. Suppose your hypothesis is that the penny turned to actual gold during the procedure. To test that hypothesis, you can compare your gold-colored penny to actual gold to see if it has all the **properties,** or characteristics, of gold. You can check the penny's physical properties, such as color, hardness, weight, and the temperature at which it melts. You can also test its chemical properties, such as whether it changes when you pour acid on it or rusts over time.

Over the course of this unit, you will explore whether it is possible to turn copper or any other substance into gold. But unlike the alchemists, you will have the advantage of hundreds of years of chemistry knowledge to help you answer this question.

# Chemistry: The Study of Matter and Change

Sergey Plakhotin/Shutterstock

*Chemistry at work—an iron train rusting*

Changes are constantly occurring all around you. Nails rust, colors fade, milk sours, and plants grow. You can mix ingredients and bake them in an oven to make cookies. You can bleach your hair to change its color, and you can freeze water to make ice cubes. Your body can transform cheeseburgers and burritos into muscle, fat, and bone. Chemists seek to understand changes such as these.

**Chemistry** is the study of what substances are made of, how they behave, and how they can be transformed. It is the study of matter and how matter changes. In this first unit, you will investigate matter and how it can be changed. You will learn to describe and explain what happens when matter is changed and you will begin to understand what changes are possible.

## LESSON SUMMARY

### What is chemistry?

Chemistry is the study of the substances in the world around you. It is the study of matter and how matter can be changed. The modern study of chemistry emerged from the experimentation and effort of the alchemists. The alchemists invented useful tools and discovered many valuable laboratory techniques in their efforts to create gold out of ordinary substances.

**KEY TERMS**

hypothesis

property

chemistry

## Exercises

### Reading Questions

1. How did the alchemists contribute to the modern study of chemistry?
2. What is chemistry?

### Reason and Apply

3. **PROJECT** www Use the library or the Internet to research the development of alchemy in one of these regions: China, India, the Middle East, Greece, Spain, England, or Egypt. Write a two-paragraph essay on the history of alchemy for your chosen region. Be sure to list your sources.

4. **PROJECT** www Use the library or the Internet to research common uses for sodium hydroxide, which is also called *lye*.

5. Write down at least ten changes that you observe in the world around you. Which changes involve chemistry? Explain your reasoning.

# Defining Matter

## THINK ABOUT IT

People tend to value gold over other substances. You don't often see someone wearing aluminum jewelry or putting coal in a high-security bank vault. What is it about gold that makes it unique? Is it possible to create gold from another substance?

This lesson begins to explore the nature of matter as the first step toward proving whether you can or cannot create gold. After all, chemistry is the study of matter and its properties.

### What is matter?

*To answer this question, you will explore*

➲ Defining Matter

➲ Is It Matter?

➲ Measuring Matter

## EXPLORING THE TOPIC

### ➲ Defining Matter

**Matter** is the word chemists use to refer to all the materials and objects in the world. Your desk, this book, and the paper and ink in the book all are matter. These are all things you can see or feel. However, your senses alone are not always enough to tell you if something is matter. For instance, you cannot see the virus that gave you a cold, but it is matter. Conversely, you can see shadows on the ground cast by the light from the sun, but they are not matter.

A gold ring, the ink in this book, and a virus each have *substance*, which means they are made out of material, or "stuff." The amount of substance, or material, in an object is called **mass.** Mass is a property of matter that can be measured. So, although the virus has very little substance, it still has mass.

Another property that a gold ring, the ink in this book, and a virus have in common is that they take up space, which means they have dimensions. The amount of space something takes up is called **volume** and is also a property of matter that can be measured.

So, matter is anything that has mass and volume. You can also say matter is anything that has substance and takes up space. This explains why a virus is matter but a shadow is not.

### ➲ Is It Matter?

It is easy to see that solids and liquids have mass and volume. When water is poured into a container, you can see how much space it takes up. When filled with water, the container has more mass. You can see this difference if you use a balance. On a two-pan balance, a cup with water will be lower than an identical cup with no water because it has more mass. Thus, water is matter.

When identifying matter, gases can be misleading. For example, most of the time you do not see or feel the air around you and it may seem as if nothing is there. However, when you fill a balloon with carbon dioxide gas, you can see that the carbon dioxide gas inside occupies space. On a two-pan balance, the pan with a balloon filled with carbon dioxide gas will be lower than the pan with an empty balloon because the balloon filled with carbon dioxide gas has more mass. Therefore, gases do have mass and volume even though gases might be harder to detect than solids and liquids.

This side is lighter.    This side is heavier.    This side is lighter.    This side is heavier.

What about other things, like heat and sound? Are they matter? When you heat soup, it does not gain mass, and sound may "fill the room," but it doesn't have mass or volume. Sound is the movement of air against your eardrums. Without matter, sound can't exist, but sound itself is not matter. Similarly, heat can't exist without some form of matter, but heat itself is not matter. Both sound and heat are referred to as types of energy, and energy by itself is not matter. Likewise, feelings and thoughts are not matter, although you could argue that they require an interaction of matter and energy.

**Example**

### Is It Matter?

Classify wind, music, and clouds. Are they matter? Explain your answer.

#### Solution

Wind is the movement of air. Air is matter, but the movement of air is not matter because movement has no mass or volume. Therefore, wind is not matter.

Music is sound, which is the movement of air against your eardrums. Music does not have mass or volume. Therefore, music is not matter.

Clouds are made of water droplets, which have mass and volume. Therefore, clouds are matter.

This electronic balance will measure mass to the nearest thousandth of a gram.

# ⬀ Measuring Matter

### MEASURING MASS

To find the mass of something, you weigh it. In everyday life, things are usually weighed in pounds, lb, or kilograms, kg. For example, you've probably weighed yourself and know approximately how many pounds you weigh. And when you go to the grocery store, you might buy a pound of butter or flour. However, chemists measure mass in kilograms, kg, and grams, g (1 kg = 1000 g). [For review of this math topic, see **MATH Spotlight**: SI Units of Measure on page A-0.]

In the chemistry classroom, an electronic balance is used to measure mass. The balance is precise and shows more digits (places) than you may need. You have to decide how many places to pay attention to when using this balance. For example, when you weigh yourself, you measure to the nearest kilogram. You would state your mass as 70 kg, not 69.903 kg. But if you are measuring the mass of a penny, you may want to be more exact. The mass of the copper penny to the nearest hundredth of a gram is 3.11 g. The mass to the nearest gram is 3 g.

This beaker measures to the nearest tenth of a liter.

This graduated cylinder measures to the nearest milliliter.

Read the measurement at the bottom of the meniscus. Make sure your eye is at the water level.

### MEASURING VOLUME

You can measure the volume of a regularly shaped object by measuring its dimensions, such as length, width, and height, and using a mathematical formula. Volume measured this way is reported in cubic units, for example, cubic meters, $m^3$, or cubic centimeters, $cm^3$.

Chemists often measure the volume of gases and liquids in liters, L, and milliliters, mL (1 L = 1000 mL). The volume of a large soda bottle is 2 L, or 2000 mL. This is the volume of the bottle, and if the bottle is full, it is also the volume of liquid inside. Chemists use a variety of special containers for measuring volume. For approximate measurements, they use beakers. For more precise measurements, they use graduated cylinders. [For review of this math topic, see **MATH Spotlight**: Accuracy, Precision, and Significant Digits on page A-1.]

You can measure the volume of a sample of water by pouring it into a graduated cylinder and reading the number of milliliters, mL, on the side. However, the surface of the water is curved because of the way it adheres, or clings, to the sides of the cylinder. The curvature of the top of a liquid in a container is called a **meniscus.** It is most accurate to read the water level at its lowest point, the bottom of the meniscus, because most of the liquid is at this level. To get an accurate measurement, you must read the graduated cylinder at eye level.

## LESSON SUMMARY

### What is matter?

**KEY TERMS**

matter

mass

volume

meniscus

Chemists seek to understand matter and its properties. Matter can be defined as anything that has mass and volume. Mass is the amount of substance or "stuff" in a material or object. Solids, liquids, and gases all have mass and volume and are classified as matter. To measure the mass of something, you use a balance. Volume is the amount of space taken up by matter. To find the volume of something, you can use a container such as a beaker or graduated cylinder, or you can calculate the volume using dimensions, such as length, height, and width.

## Exercises

### Reading Questions

1. Explain the difference between mass and volume.

2. Describe how you could prove that a bicycle is matter.

### Reason and Apply

3. Someone might claim that air is not matter because you can't see it. Write a paragraph showing how you would prove to this person that air is matter.

4. The Sun is considered matter, but sunlight is not considered matter. Explain why this is so.

5. Sit somewhere and observe your environment. From your observations, make a list of ten things that are matter and a list of ten things that are not matter. Explain your reasoning.

6. Give examples of five words that describe the movement of matter.

# Mass and Volume

## THINK ABOUT IT

Suppose you have two samples of gold, a gold ring and a gold nugget. Is there more gold in the ring or in the nugget? They feel similar in weight, and they look fairly similar in volume. While your senses can give you valuable information, they can't tell you exactly how much gold you have in each sample.

**How do you determine the masses and volumes of different substances?**

*To answer this question, you will explore*

- Measuring Volume
- Comparing Mass and Volume

## EXPLORING THE TOPIC

## Measuring Volume

As you learned in Lesson 3, volume is a measure of size, or how much space each sample takes up.

There are two common ways of measuring the volume of solids: (1) by measuring their dimensions and using a geometric formula and (2) by water displacement. The first method is convenient if the object has a regular shape. The second method is more convenient for irregularly shaped objects.

### USING GEOMETRIC FORMULAS TO DETERMINE VOLUME

If a solid is rectangular, you can find its volume by measuring its three dimensions—length, width, and height—and multiplying these three values. Volume measured in this way is reported in *cubic* units such as cubic centimeters, $cm^3$; cubic meters, $m^3$; or cubic inches, $in^3$. The formula for volume is

$$V = lwh$$

Consider these two solid blocks. You can use the formula to figure out their volumes.

$$V = lwh$$
$$= 1.0 \text{ cm} \cdot 1.0 \text{ cm} \cdot 1.0 \text{ cm}$$
$$= 1.0 \text{ cm}^3$$

$$V = lwh$$
$$= 1.0 \text{ cm} \cdot 0.5 \text{ cm} \cdot 2.0 \text{ cm}$$
$$= 1.0 \text{ cm}^3$$

1.0 cm

1.0 cm

2.0 cm

0.5 cm

Notice that the two solids have different dimensions, but they have the same volume, 1.0 cubic centimeter.

*The water level rises by 5.0 mL because the cube has a volume of 5.0 cm³.*

## USING WATER DISPLACEMENT TO DETERMINE VOLUME

The second method of measuring the volume of a solid object is called **water displacement.** First, pour water into a graduated cylinder. Next, add the object whose volume you are measuring. An object takes up space, so it displaces some of the water when it is placed in the graduated cylinder. This causes the water level to rise by an amount equal to the volume of the object. If you read the volume of water in the graduated cylinder before you submerge the object and again after submerging it, the volume of the object is the difference between these two volumes.

15.0 mL      17.0 mL      19.0 mL

**Important to Know**  One milliliter is exactly equal in volume to one cubic centimeter: 1 mL = 1 cm³.

Imagine you place a gold ring and a gold nugget in graduated cylinders partly filled with water. The graduated cylinder is marked in milliliters, mL. You observe that the water level rises by 2.0 mL in one container and by 4.0 mL in the other. The ring has a volume of 2.0 mL or 2.0 cm³ and the nugget 4.0 mL or 4.0 cm³.

Of course, the water displacement method works only for solids that do not dissolve in water! The solid object also needs to sink so that it is completely submerged. If the object floats, the volume reading will not be accurate.

## ⟳ Comparing Mass and Volume

Imagine that you have two objects with the same volume that are *not* made from the same material. You can compare their masses using a balance. The masses of 1.0 cm³ of gold and 1.0 cm³ of plastic are given in the table below along with the masses of the same volumes of wood, glass, and copper.

### CONSUMER CONNECTION

There are many different types of plastic. Soft-drink bottles are made of polyethylene terephthalate, also called PETE. Plastic pipes and outdoor furniture are made of polyvinyl chloride, also called PVC. Styrofoam cups are made of polystyrene.

Jeffrey Hamilton/Getty Images

1.0 cm³ of gold has much more mass than 1.0 cm³ of plastic.

| Material | Volume | Mass |
|----------|--------|------|
| gold | 1.0 cm³ | 19.3 g |
| plastic | 1.0 cm³ | 0.9−1.5 g |
| glass | 1.0 cm³ | 2.2−3.1 g |
| copper | 1.0 cm³ | 9.0 g |
| wood | 1.0 cm³ | 0.2−1.4 g |

The mass of plastic, glass, and wood varies depending on the type.

Notice that 1.0 cm³ of gold is heavier than the same volume of any of the other materials. Thus, two objects can have exactly the same volume but different masses.

The mass of the plastic depends on the type of plastic being considered. The mass of a 1.0 cm³ sample of plastic varies between 0.9 g and 1.5 g. Wood and glass also show variations. For instance, if you compare a 1.0 cm³ sample of solid wood from an oak tree with a 1.0 cm³ sample of solid wood from a pine tree, they will have different masses. The oak sample will be considerably heavier than the pine. A 1.0 cm³ sample of pure gold will always have a mass of 19.3 g because there is only one type of pure gold.

## Example 1

Try to work out the answers yourself before reading the solutions.

### Cubes of the Same Volume

Suppose you have two cubes. One cube is made of solid gold and the other cube is made of solid plastic. The sides of the cubes each measure 2.0 cm in length.

   **a.** What is the volume of each cube?

   **b.** How much water will each cube displace?

### Solution

   **a.** The volume of a cube is equal to length times width times height.

$$V = lwh$$
$$= 2.0 \text{ cm} \cdot 2.0 \text{ cm} \cdot 2.0 \text{ cm}$$
$$= 8.0 \text{ cm}^3$$

   **b.** Because 1 cm³ = 1 mL, you can reason that 8.0 cm³ = 8.0 mL. So each cube will displace the same amount of water, or 8.0 mL.

## Example 2

### Cubes of the Same Mass

One cube is made of solid copper and another cube is made of solid glass. They have exactly the same mass. How do their volumes compare?

### Solution

If you look at the table of masses on page 14, you will see that 1.0 cm³ of copper has much more mass than 1.0 cm³ of glass. Thus, for the two cubes to have the same mass, the copper cube must be much smaller than the glass cube.

Larger cube of glass

Smaller cube of copper

*The small cube of copper exactly balances the mass of the larger cube of glass.*

## LESSON SUMMARY

### How do you determine the masses and volumes of different substances?

The volume of a solid object is determined either by water displacement or by measuring its dimensions and using a geometric formula for volume. The mass of any solid sample is found by weighing it on a balance or scale. The relationship between mass and volume depends on the type of material being considered. Two objects may have the same volume but different masses, or two objects may have the same mass but different volumes.

**KEY TERM**

water displacement

## Exercises

### Reading Questions

1. Describe two ways to determine the volume of a solid object.

2. If you have two objects with equal volume, do they have the same mass? Explain your thinking.

### Reason and Apply

3. Can you predict the volume of an object just by looking at it? Explain.

4. Can you predict the mass of an object just by looking at it? Explain.

5. If you stretch a rubber band, does it still have the same mass? Explain.

6. You submerge a piece of clay in water and measure the total volume. You change the shape of the clay and put it back into the same amount of water.
   a. The total volume does not change. Explain why.
   b. Does the total mass change? Explain.

7. Describe how you might find the volume of these items:
   a. pancake mix      d. a penny
   b. hair gel         e. lemonade
   c. a shoe box

8. Draw two things with the same mass but different volumes.

9. Draw two things with the same volume but different masses.

10. Use the illustrations to answer these questions.
    a. What is the volume of the liquid inside the container?
    b. What is the volume of the rock in mL? In cm³?

11. Suppose that you have two cubes of exactly the same volume. You weigh them and find a mass of 8.91 g for one cube and 8.88 g for the other cube even though they are made of the same material. How is this possible?

# Density

## THINK ABOUT IT

Gold and lead are metals that are soft and bendable. They are easy to tell apart at a glance because lead is dull gray. But suppose someone tried to trick you by coating a block of lead with a thin layer of gold. How could you prove that this block was a fake?

**How can you use mass and volume to determine the identity of a substance?**

*Gold is a soft, shiny, bendable, metal.*

*To answer this question, you will explore*

- The Definition of Density
- Calculating Density
- Identifying Matter Using Density

## EXPLORING THE TOPIC

## The Definition of Density

*This copper has been shaped to form hollow metal pipes.*

Which is heavier, gold or copper? It depends on the amount of each that you have. If you compare a tiny gold earring and a large copper pipe, the copper will be heavier. If you compare the same volume of each, gold is *always* heavier than copper. That is because there is more matter in 1.0 cm³ of gold than in 1.0 cm³ of copper. You could also say that gold is *denser* than copper. The mass of a substance per unit of volume is called its **density.**

Imagine comparing 1.0 cm³ cubes of wood, water, and copper from larger samples of each material. All three cubes have the same volume, but each cube has a different mass and a different density.

**Example 1**

## Differences in Density

Consider the objects on balances shown in these illustrations. How can differences in density account for what you observe in these pictures?

*The balance is uneven. The two solid cubes have the same volume but different masses.*

*The balance is level. The two solid cubes have the same mass but different volumes.*

### Solution

Because the cubes on the first balance are exactly the same size, for the balance to be uneven one cube must be denser than the other. So the two cubes are made of different materials.

On the second balance, the larger cube must have a lower density than the smaller cube. Again, the two cubes are made of different materials.

## ⟳ Calculating Density

You can determine the density of a substance without actually cutting out a cubic centimeter sample. Suppose a large chunk of gold has a mass of 309 g and a volume of 16.0 cm³. Divide the mass of the gold by its volume to find its mass *per* cubic centimeter. This is equal to the density.

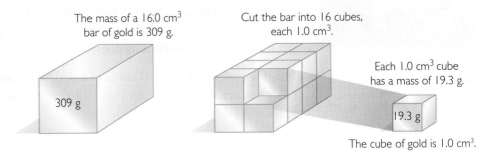

The mass of a 16.0 cm³ bar of gold is 309 g.

Cut the bar into 16 cubes, each 1.0 cm³.

Each 1.0 cm³ cube has a mass of 19.3 g.

309 g

19.3 g

The cube of gold is 1.0 cm³.

$$309 \text{ g} \div 16.0 \text{ cm}^3 = 19.3 \text{ g/cm}^3$$

The density of gold is 19.3 grams per cubic centimeter, or 19.3 g/cm³. This answer is rounded to three significant digits. [For review of this math topic, see **MATH Spotlight**: Accuracy, Precision, and Significant Digits on page A-1.]

### Density

The mathematical formula for density is

$$D = \frac{m}{V}$$

where *D* is the density, *m* is the mass, and *V* is the volume.

**Example 2**

### Density of a Gold Ring

Suppose you have a gold ring that weighs 7.50 g and has a volume of 0.388 mL. Does it have the same density as a big piece of gold?

*Solution*

Use the formula to find the density of the ring.

Start with the formula. $D = \dfrac{m}{V}$

Substitute the values of $m$ and $V$. $D = \dfrac{7.50 \text{ g}}{0.388 \text{ mL}}$

Solve for $D$. $= 19.3 \text{ g/mL} = 19.3 \text{ g/cm}^3$

The ring has the same density as any other sample of gold, big or small. The density of a material does not depend on the shape or size of the sample.

The Bridgeman Art Library/Getty Images

## Identifying Matter Using Density

**Densities of Some Metals**

| Metal | Density |
|---|---|
| copper | $9.0 \text{ g/cm}^3$ |
| zinc | $7.1 \text{ g/cm}^3$ |
| gold | $19.3 \text{ g/cm}^3$ |
| lead | $11.4 \text{ g/cm}^3$ |
| aluminum | $2.7 \text{ g/cm}^3$ |
| brass | $8.4 \text{ g/cm}^3$ |

**Important to Know** The density of a substance does not change with its size or its shape. A solid gold bracelet will have the same density as a solid gold brick, and the density of one penny is the same as the density of two pennies.

Every substance has certain properties, such as color, hardness, melting point temperature, and density, that depend on the type of matter, not on the amount or size of the sample. These properties are called **intensive properties.** Intensive properties do not change if the quantity of the substance changes. Therefore, they can be used to help identify that substance.

On the other hand, **extensive properties,** such as mass or volume, do change depending on the amount of matter. Extensive properties alone can't be used to help identify a type of matter. For example, knowing that a metal earring has a mass of 3 g doesn't help you identify what the earring is made of.

Because density is an intensive property, it can be used to help identify the type of matter that an object or sample is made of. First, determine the density of the object and then compare it with known density values in a reference table like this one to help identify the type of matter.

**Big Idea** Each specific type of substance has a particular density. You can identify a substance by its intensive properties, including density.

### THE FAKE BAR OF GOLD

If someone tried to trick you by coating a block of lead with a thin layer of gold, how could you prove the bar is a fake? One approach would be to scratch the bar to reveal that the inside isn't gold. Another method would be to use density to prove that the bar isn't gold.

Fake gold?

Density = ?
Volume = 647.5 cm³
Mass = 7500 g

A block of pure gold will have a density of $19.3 \text{ g/cm}^3$.

## How can you use mass and volume to determine the identity of a substance?

**KEY TERMS**

density

intensive property

extensive property

To identify substances, you examine their intensive properties, qualities that do not depend on size or amount. Intensive properties include color, hardness, and density. Density is the mass of a substance per unit of volume. If two substances have different densities, then they are probably made of different types of matter.

## Exercises

### Reading Questions

1. In your own words, define density.

2. Explain how density can be used to determine if the golden penny is made of solid gold.

### Reason and Apply

3. How does the density of aluminum compare with the density of gold? What does this tell you about the amount of matter within each?

4. If two objects have the same mass, what must be true? Choose the correct answer(s).
   (A) They have the same volume.
   (B) They are made of the same material.
   (C) They contain the same amount of matter.
   (D) They have the same density.

5. Two objects each have a mass of 5.0 g. One has a density of 2.7 g/cm$^3$ and the other has a density of 8.4 g/cm$^3$. Which object has a larger volume? Explain your thinking.

6. A piece of metal has a volume of 30.0 cm$^3$ and a mass of 252 g. What is its density? What metal do you think this is?

7. A glass marble has a mass of 18.5 g and a volume of 6.45 cm$^3$.
   a. Determine the density of the marble.
   b. What is the mass of six of these marbles? What is the volume? What is the density?
   c. How does the density of one marble compare with the density of six of the marbles?

## CHAPTER 1

# Defining Matter

# SUMMARY

### KEY TERMS

**hypothesis**

**property**

**chemistry**

**matter**

**mass**

**volume**

**meniscus**

**water displacement**

**density**

**intensive property**

**extensive property**

# Alchemy Update

Can a copper penny be turned into gold through chemical processes?

The "golden" penny looks like gold and has almost exactly the same volume as a true gold penny. However, the golden penny has a much lower mass than a true gold penny. This means the density of the "golden" penny is less than the density of gold. It cannot be a solid gold penny.

The concept of density has provided the evidence needed to confirm that the gold-colored penny is not actually made of gold. But if it is not gold, then what is it? Many questions remain to be answered. Is it still possible to make gold some other way? And what is it about gold that makes it gold?

## REVIEW EXERCISES

1. Explain how you would determine the volume of a powdered solid, a liquid, and a rock.

2. Use your own words to define *matter*.

3. Will an object with a higher density displace more water than an object with a lower density? Explain why or why not.

4. How does the density of one penny compare with the density of two pennies?

1 Penny
Mass = 2.6 g
Volume = 0.36 cm³

2 Pennies
Mass = 5.2 g
Volume = 0.72 cm³

5. A small pebble breaks off of a huge boulder. The pebble has the same density as the boulder. In your own words, explain how this can be true.

6. Archeologists discover a silver crown in an ancient tomb. When they place the crown in a tub of water, it displaces 238.1 cm³ of water. The density of silver is 10.5 g/cm³. If the crown is really silver, what will its mass be?

# CHAPTER 2

# Basic Building Materials

*In this chapter, you will study*

- how chemists use chemical formulas to track changes in matter

- how matter is conserved in chemical processes

- the organization of the elements into the periodic table

*Chemists separate substances into their simplest components in order to understand the nature of matter. Shown here is brightly colored, almost pure sulfur found near hot springs.*

Matter is composed of components called elements. There are a limited number of elements in existence. Everything in our world, natural and synthetic, is made of individual elements or some combination of them. Chemical formulas indicate what element or elements a substance is made of.

*A New Language*

---

*A New Language*

# Chemical Names and Symbols

# ⊃ Names and Symbols

**Elements** are the building materials of all matter. In total, there are about 115 known elements. A few are shown here.

Metals

Mercury, Hg     Copper, Cu     Gold, Au     Sodium, Na     Iron, Fe

Nonmetals

Carbon, C     Chlorine, Cl     Iodine, I     Phosphorus, P

From top left: Harry Taylor/Getty Images, Dmitry Gool/iStockphoto/Thinkstock, Dirk Wiersma/Science Source, Martyn F. Chillmaid/Science Source, Sawayasu Tsuji/iStockphoto, Kae Deezign/Shutterstock, Charles D. Winters/Science Source, Andrew Lambert Photography/Science Source, Charles D. Winters/Science Source

Here are some things you might notice. The **chemical symbols** for the elements consist of one or two letters. The first letter is always capitalized. If there is a second letter, it is lowercase. An element can be a solid, liquid, or gas. Some elements are metals, some are not. Sometimes, the symbol for an element is an abbreviation for its name. Other times, the symbol comes from another source, such as a word in another language. For example, the symbol for iron is Fe from the Latin word *ferrum,* and the symbol for gold is Au from the Latin word *aurum.*

Elements combine in specific ratios to form **compounds.** A compound is represented by a **chemical formula.** For example, sodium chloride, or table salt, is NaCl, which tells you that salt is made of the elements sodium, Na, and chlorine, Cl, in a 1:1 ratio. The chemical formula for carbon dioxide is $CO_2$. The subscript number "2" in the formula indicates that the elements carbon and oxygen are combined in a 1:2 ratio. (If the subscript is 1, you normally don't write it.)

Symbol for oxygen

Chemical formula for carbon dioxide

Symbol for carbon     Subscript number

Compounds can differ greatly in appearance and behavior from the elements that they are composed of. For example, sodium is a shiny metal and chlorine is a gas. They are very different from sodium chloride, table salt.

**Big Idea**   **All matter is made up of elements or compounds, or mixtures of these.**

These minerals contain copper, Cu. Notice that not one of them is copper colored.

Copper oxide,
$Cu_2O(s)$

Copper sulfate,
$CuSO_4(s)$

Copper sulfide,
$CuS(s)$

Copper carbonate hydroxide,
$Cu_2(CO_3)(OH)_2(s)$

From left: Andrew Lambert Photography/Science Photo Library, Ewa Brozek/iStockphoto, A. RIZZI/De Agostini Picture Library/Getty Images, Millard H. Sharp/Science Source

*Sino Images/Getty Images*

*Water is unusual, because we can encounter it in three phases of matter all at once: liquid, solid ice, and gas in the atmosphere.*

## ⮑ Physical Form

Elements and compounds can exist as solids, liquids, or gases. These forms are called the **phases** of matter, and are represented by using $(s)$, $(l)$, or $(g)$ after the chemical formula. For example, water in the gas phase is written as $H_2O(g)$, water in the liquid phase is written as $H_2O(l)$, and water in the solid phase as ice is written as $H_2O(s)$.

There is another symbol for physical form, the symbol $(aq)$ for **aqueous.** A substance is aqueous when it dissolves and forms a clear mixture with water. Many familiar liquids are actually aqueous solutions, such as grape juice, vinegar, and ocean water.

> **Important to Know** When solid sugar, $C_{12}H_{22}O_{11}(s)$, appears to "melt" in your mouth, it does not become liquid sugar, $C_{12}H_{22}O_{11}(l)$. It dissolves and becomes aqueous sugar, $C_{12}H_{22}O_{11}(aq)$.

*Christopher Ewing/iStockphoto*

*An orange is a tasty mixture of compounds.*

Most substances are a mixture of compounds. For example, an orange is a mixture of water, $H_2O$, fructose, $C_6H_{12}O_6$, citric acid, $C_6H_8O_7$, limonene, $C_{10}H_{16}$, and many other compounds. Notice that in mixtures, the components are *not* combined in a specific ratio. Some oranges are sweeter or more sour or more juicy because they have more fructose, citric acid, or water than other oranges. But the ratios of the elements in compounds never change. Water is always water, $H_2O$.

## LESSON SUMMARY

### What do chemical names and symbols tell you about matter?

**KEY TERMS**

element

chemical symbol

compound

chemical formula

phase

aqueous

All matter is composed of elements. Each element has a unique symbol. Elements combine with one another to form compounds. A chemical formula specifies which elements are present in a compound. The letters in chemical formulas are symbols for the elements, and the subscript numbers indicate the amount of each element in that substance. A chemical formula may also include the lowercase letters $(s)$, $(g)$, $(l)$, and $(aq)$, which stand for solid, gas, liquid, and aqueous.

### Reading Questions

1. Describe the difference between an element and a compound.
2. What is meant by physical form?

### Reason and Apply

3. How many elements are included in the chemical formula for sodium nitrate, $NaNO_3$? Name them.
4. What is the difference between $NaOH(s)$ and $NaOH(aq)$?
5. You see a ring with a stone that looks like a diamond but wonder why it's so cheap. The jeweler says the stone is a type of diamond called cubic zirconia. How can chemical symbols show that cubic zirconia is not a diamond?
6. You find two containers on a chemical shelf, one labeled $Cu_2O(s)$ and a second labeled $CuO(s)$. Are these substances the same or different? Explain.

# The Copper Cycle

## THINK ABOUT IT

At this point, you are probably convinced that a copper penny cannot be transformed into real gold in a high-school chemistry classroom. However, the question remains, what happened to the copper penny?

### What happens to matter when it is changed?

*To answer this question, you will explore*

➲ The Copper Cycle

➲ Evidence of Chemical Change

➲ Interpreting Observations

## EXPLORING THE TOPIC

➲ ## The Copper Cycle

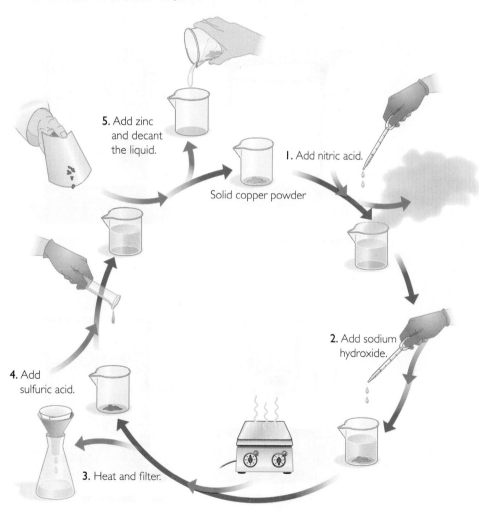

**5.** Add zinc and decant the liquid.

Solid copper powder

**1.** Add nitric acid.

**2.** Add sodium hydroxide.

**4.** Add sulfuric acid.

**3.** Heat and filter.

To make sense of the copper penny experiment, it is important to learn more about how copper can be transformed. Copper itself is an element, represented by the chemical symbol $Cu(s)$. As a solid powder, copper has a distinctive orange-brown appearance.

In class, you transformed copper powder through a series of chemical reactions. If all worked well, copper should have reappeared at the end of the experiment. But what happened between the beginning and the end?

The diagram on page 27 shows the steps of the copper cycle. Start at the top and follow the arrows.

## ⊃ Evidence of Chemical Change

The five steps of the copper cycle are shown in these photos. In each step, a substance is changed into a different substance by mixing it with a liquid or by heating it. These changes are called **chemical changes** or **chemical reactions.** You can usually tell when a chemical reaction occurs because there is evidence of new substances forming. For instance, in class you observed some color changes when two of the liquids turned a bright blue color. You also observed the release of a brown gas and the formation of a dark-colored solid. These are all possible signs that new substances are forming. After a chemical reaction, you have new substances with properties that are different from the starting substances.

*Copper powder*

*Step 1: After adding nitric acid*

*Step 2: After adding aqueous sodium hydroxide*

*Step 3: After heating*

*Step 4: After adding sulfuric acid*

*Step 5: After adding zinc*

## ⊙ Interpreting Observations

Look again at the first step in the copper-cycle diagram. When nitric acid was added to the copper, a blue liquid formed and a toxic brown gas left the beaker. But where did the copper powder go? You may have wondered if the copper left the beaker with the brown gas. However, copper reappeared at the end of the experiment. This is evidence that the copper doesn't leave the beaker at any point in the cycle.

Was copper put back in again? If copper was not added in again and it did not leave with the brown gas, the only other explanation is that it never left the beaker at all. This would mean that the copper is present in copper-containing compounds at each step of the cycle. Copper must be present somewhere in the blue solutions and the blue solid that were observed. So, the copper was not created or destroyed as a result of its chemical journey.

Over many centuries, chemists have carried out procedures like the copper cycle. In doing so, they have learned about the properties of substances and even discovered new substances. They have also learned how to predict change. Chemists know that pouring nitric acid on copper produces a clear blue liquid and a brown gas. They have even determined the identity of the blue liquid and the brown gas. Perhaps further investigation will allow you to track the changes to the copper penny as well.

**Example**

### Rusting Nail

An iron nail that stays in contact with water and air starts to form a reddish-brown coating called rust.

   **a.** Is this a chemical change?

   **b.** Is rust an element, a compound, or a mixture? Explain your thinking.

   **c.** How could you gather more evidence to support your answer to part b?

*Solution*

Rust forms on the surface of the nail, so you can deduce that it is probably caused by exposure to water and air.

Lukas Hejtman/Shutterstock

   **a.** There is evidence of a new substance being formed on the surface of the nail. There is a color change and the rust is softer than iron. Therefore, a chemical change has taken place.

   **b.** Rust is probably a compound, a combination of the iron from the nail plus one or more elements from the air or water.

   **c.** To gather more evidence for chemical change, you could complete several experiments. For example, you could place a nail in pure water or in pure oxygen to test what the iron is reacting with. (Scientists have done this and found that iron rusts only in the presence of both oxygen and water. They have determined that rust is actually iron oxide hydroxide.)

## What happens to matter when it is changed?

When one substance is transformed, or changed, into another substance, a chemical change has occurred. A chemical change is also called a chemical reaction. When chemical reactions occur, new substances with new properties are produced. Evidence of new substances, such as the production of a gas, a color change, or the formation of a solid in solution, is a sign that a chemical change has taken place. When substances undergo chemical changes, the elements themselves are not created or destroyed.

**KEY TERMS**

chemical change

chemical reaction

## Exercises

### Reading Questions

1. Explain what a chemical reaction is. What are some possible signs that a chemical reaction is taking place?

2. In your own words, explain how it might be possible to start with copper and end up with copper after a series of chemical reactions.

### Reason and Apply

3. Look again at the copper-cycle diagram. After adding zinc, the blue solution turns colorless, and copper appears as a solid. What do you think happened to the zinc? Explain your thinking.

4. Suppose you use sugar, butter, eggs, and flour to make some cookie dough. You bake the dough in the oven until the cookies are done. Do the ingredients undergo a chemical change? Give evidence to support your answer.

*Are cookies an example of a chemical change?*

5. Baking soda is a white powder used for baking or cleaning. When you mix baking soda, $NaHCO_3$, with vinegar, $C_2H_4O_2$, you get a clear colorless liquid and bubbles of $CO_2$.

   a. Specify the phase of each of the compounds.
   b. Is this a chemical change? Give evidence.
   c. Where is the sodium, Na, before the change? After the change?

6. **PROJECT**  Research how copper is extracted from a compound found in nature. Write a paragraph describing the process.

# Conservation of Matter

## THINK ABOUT IT

The element copper can be mixed with other substances to make a colorful assortment of compounds. What are these compounds? How can you demonstrate that they all contain copper?

### What happens to elements in a chemical change?

*To answer this question, you will explore*

- ⤵ Translating the Copper Cycle
- ⤵ Tracking an Element
- ⤵ Conservation of Matter

## EXPLORING THE TOPIC

### ⤵ Translating the Copper Cycle

Chemical names and chemical formulas are powerful tools you can use to keep track of matter. In fact, you can use them to figure out what was made at various steps in the copper cycle.

There are several approaches you can take to figure out what was made at each step of the cycle. First, you can compare the appearance of the compounds that were obtained in each step with compounds that you have seen before. For example, in an earlier class you examined samples of copper compounds that looked like those shown in the two photos here.

Visual observation of compounds can give you some clues, but it is not enough to make a definite identification. For example, it is hard to tell the difference between copper sulfate and copper nitrate through observation because they are both blue. In fact, several copper compounds are blue.

Another approach is to examine the chemical names and formulas of the substances that were mixed together. The new substances are formed from parts of the starting materials. For example, when sodium *hydroxide*, NaOH, is added to the beaker in the second step of the cycle, a compound called copper *hydroxide*, $Cu(OH)_2$, is produced. Because a blue solid is formed and hydroxide, $OH^-$, is one of the starting ingredients, you can deduce that copper hydroxide is the product.

### ADDING NITRIC ACID TO COPPER

Start at the beginning of the copper cycle. See if you can figure out what was created after the first step by translating each step into chemical symbols and formulas.

*Copper nitrate,
$Cu(NO_3)_2(s)$*

*Copper sulfate, $CuSO_4(s)$*

*Copper hydroxide,
$Cu(OH)_2(s)$*

Clive Streeter/Getty Images

Ewa Brozek/iStockphoto

© Dorling Kindersley,
Courtesy of the Science
Museum, London

**Step 1:** Nitric acid is added to copper powder. A clear blue solution and a brown gas are formed.

Nitric acid    Copper

We know the chemical names that go along with the first half of this step. Nitric acid *is added to* copper powder *resulting in* a blue solution and a brown gas. Using chemical formulas, you can write this as:

**HNO₃(aq)** is added to **Cu(s)** resulting in ____?____ and ____?____ .
            (blue solution)    (brown gas)

So far in class, you've seen two blue liquids. One was copper sulfate and the other was copper nitrate.

Nitric acid, $HNO_3(aq)$, was added to the copper powder, so you can deduce that the blue solution that formed was copper nitrate, $Cu(NO_3)_2(aq)$, and not copper sulfate, $CuSO_4(aq)$.

**HNO₃(aq)** is added to **Cu(s)** resulting in **Cu(NO₃)₂(aq)** and ____?____ .

How about the brown gas? The brown gas must contain some combination of H, N, O, or Cu because these are the only starting ingredients. Copper, Cu, does not form gaseous compounds, but the other three elements do combine to form several different gases. The chemical formulas and colors of these gases can be found in reference books and are listed in the table below.

Brown gas

Aqueous copper nitrate,
$Cu(NO_3)_2(aq)$

### Gases Containing H, N, or O

| Gas | Color |
| --- | --- |
| $H_2(g)$, hydrogen gas | colorless |
| $N_2(g)$, nitrogen gas | colorless |
| $O_2(g)$, oxygen gas | colorless |
| $H_2O(g)$, water vapor | colorless |
| $NH_3(g)$, ammonia gas | colorless |
| $NO(g)$, nitrogen oxide gas | colorless |
| **$NO_2(g)$, nitrogen dioxide gas** | **brown** |

**ENVIRONMENTAL CONNECTION**

Nitrogen dioxide, $NO_2(g)$, is a part of smog. It causes the red-brown color in the skies above cities with large amounts of air pollution. Nitrogen dioxide irritates the eyes, nose, throat, and respiratory tract. Continued exposure can cause bronchitis.

The only *brown* gas in this table is nitrogen dioxide, $NO_2(g)$. The completed chemical sentence for Step 1 is:

**$HNO_3(aq)$** is added to **$Cu(s)$** resulting in **$Cu(NO_3)_2(aq)$** and **$NO_2(g)$**.

This reaction also produces one more compound: water, $H_2O$. You would not have noticed this because it is clear and colorless. So the final chemical sentence for Step 1 is:

**$HNO_3(aq)$** is added to **$Cu(s)$** resulting in **$Cu(NO_3)_2(aq)$** and **$NO_2(g)$** and **$H_2O(l)$**.

Notice that all the elements in the starting ingredients also appear in the products. No elements are created or destroyed.

You can deduce the products for the other steps in the copper cycle in a similar way.

## ⮕ Tracking an Element

Once you have figured out the products of Steps 2, 3, and 4 of the copper cycle, you can track the journey of copper through the cycle. The illustration below shows the copper compounds that form at each step of the cycle.

5. Add zinc, $Zn(s)$, and decant the liquid.

Solid copper powder $Cu(s)$

1. Add nitric acid, $HNO_3(aq)$.

Nitrogen dioxide $NO_2(g)$

Copper sulfate $CuSO_4(aq)$

Copper nitrate $Cu(NO_3)_2(aq)$

4. Add sulfuric acid, $H_2SO_4(aq)$.

Copper oxide $CuO(s)$

2. Add sodium hydroxide, $NaOH(aq)$.

Copper hydroxide $Cu(OH)_2(s)$

3. Heat and filter.

**HISTORY**
**CONNECTION**

Pennies were last made of solid copper in 1836. Pennies made from 1962 to 1982 are 95% copper and 5% zinc. These pennies have a density of 8.6 g/cm$^3$, which is just slightly less than the density of copper, 9.0 g/cm$^3$. Since 1982, pennies have been made mostly of zinc with a copper coating. These pennies have a density of 7.2 g/cm$^3$, which is very close to the density of pure zinc, 7.1 g/cm$^3$.

The symbol Cu is found at each stage of the cycle. Thus, the element copper is somehow combined in each of these compounds. And of course, the solid that forms at the end of the experiment is elemental copper, Cu($s$).

What you have observed with copper is true of other elements as well. For example, nickel can be taken through a similar cycle, where various substances are added to nickel powder, Ni($s$). Just as with the copper cycle, you can recover elemental nickel, Ni(s), from compounds that contain nickel.

## ⮕ Conservation of Matter

The copper-cycle experiment brought you full circle, back to where you started. You took a sample of the element copper and added substances to it. After several steps, you ended up with copper powder once again. No matter what was done to the copper, the copper was always there in some form. In other words, it was not created or destroyed by the chemical transformations. Over many centuries, scientists have gathered evidence that matter can never be destroyed or created through chemical transformation. There is so much evidence that this is considered a scientific law.

The **law of conservation of mass** states that mass cannot be gained or lost in a chemical reaction. In other words, matter cannot be created or destroyed.

It is possible to prove that no copper was gained or lost during the copper-cycle experiment by measuring the mass of the copper powder at the beginning and again at the end. If you did the experiment perfectly, you would end up with exactly the same amount of copper powder that you started with. However, in real life, the mass of the copper at the end of this experiment is a bit less than the mass of the copper at the beginning due to several factors. Little amounts of copper are lost along the way, because of spills, measurement errors, and sticking to the filter paper or beaker. These small errors are hard to avoid. In addition, some copper compounds remain in the discarded solutions. Nevertheless, mass is still conserved; even if the copper isn't visible in the beaker at the end, it still exists somewhere.

Law of Conservation of Mass
Matter cannot be created or destroyed.

## LESSON SUMMARY

### What happens to elements in a chemical change?

Chemical names and formulas are used to keep track of elements and compounds as they undergo chemical or physical changes. When elemental copper is tracked through a series of reactions, the symbol Cu shows up in the formulas of the new compounds that are made. Elemental copper can be combined with other substances to form new compounds, but it is not destroyed by the chemical transformations. This concept is known as the law of conservation of mass, which states that matter cannot be created or destroyed.

**KEY TERM**

law of conservation of mass

**Reading Questions**

1. How can chemical names and symbols help you figure out what copper compound was made in each step of the copper cycle? Give an example.

2. Explain the law conservation of mass in your own words.

**Reason and Apply**

3. **Lab Report** Write a lab report for the Lab: The Copper Cycle. In your report, include the title of the experiment, purpose, procedure, observations, and conclusions.

> (Title)
>
> **Purpose:** (Explain what you were trying to find out.)
>
> **Procedure:** (List the steps you followed.)
>
> **Observations:** (Describe what you observed during the experiment.)
>
> **Conclusions:** (What can you conclude about what you were trying to find out? Provide evidence for your conclusions.)
>
> **Analysis:** (Explain what happpened to the copper during the experiment.)

4. Explain how the copper-cycle experiment supports the claim that copper is an element—a basic building block of matter.

5. Nickel sulfate, $NiSO_4(aq)$, is a green solution. Nickel chloride, $NiCl_2(aq)$, is a yellow solution. And hydrochloric acid, $HCl(aq)$, is a clear, colorless solution. If you add nickel, $Ni(s)$, to hydrochloric acid, $HCl(aq)$, what color solution do you expect to form? Explain your reasoning.

6. In the final step of the copper cycle, zinc, $Zn(s)$, is added to copper sulfate, $CuSO_4(aq)$. Elemental copper appears as a solid. Explain what you think happens to the elemental zinc.

7. Matter cannot be created or destroyed. List at least two long-term impacts that this concept has for us on this planet.

Germanium

Germanium

Germanium

# Create a Table

## Purpose

To create your own periodic table of the elements from data given on element cards.

## Materials

- Create a Table card deck

## Instructions

1. Work in groups of four with one set of cards.

2. Find Be, Mg, Ca, and Sr in the deck of cards, and arrange them in a column as shown to the right. These cards are all yellow. Look for similarities and differences in these cards. Find at least one pattern or trend, and describe it to your group.

| Be |
|----|
| Mg |
| Ca |
| Sr |

3. With your group, decide how to organize the rest of the cards into a table. Try to organize them in a way that produces as many patterns as possible.

## Questions

1. What characteristics did you use to decide how to sort the cards?

2. What patterns appear in your arrangement? List at least four.

3. Where did you put H and He? What was your reasoning for their placement?

4. Did you notice any cards that didn't quite fit or that seemed out of order? Explain.

5. **Making Sense** To the left are possible cards for the element germanium.
   a. Where does germanium belong in the table?
   b. Which card seems most accurate to you? What is your reasoning?
   c. What would you add to the three empty corners to complete the card?

*Create a Table*

# Properties of the Elements

## THINK ABOUT IT

In the late 1860s, a Russian chemist and teacher named Dmitri Mendeleyev was looking for a way to organize the elements known at the time. He wanted to make it easier to remember and understand their chemical behavior. He started by organizing the elements into groups based on similarities in their properties. You can understand Mendeleyev's organization of the elements by examining the information he used to sort them.

### How is the periodic table organized?

*To answer this question, you will explore*

- Properties of the Elements
- A Table of Elements

## EXPLORING THE TOPIC

## Properties of the Elements

Mendeleyev was intrigued by patterns in the properties of the elements. For example, tin, Sn, sodium, Na, and magnesium, Mg, are all shiny, silvery solids at room temperature.

*Tin, Sn*

*Sodium, Na*

*Magnesium, Mg*

However, by itself, visual appearance is not a reliable characteristic for sorting the elements into groups. For instance, many elements are shiny, silvery solids at room temperature. And while both oxygen and neon are colorless gases, they are not similar enough in their other properties to be grouped together. Mendeleyev focused on three properties besides appearance to sort the elements: reactivity, the formulas of chemical compounds that form when the element combines chemically, and atomic mass.

### REACTIVITY

Mendeleyev focused on the reactivity of elements to sort them. **Reactivity** is a property that describes how easily an element will combine with other substances to form new compounds. An element that is highly reactive combines rapidly

## HISTORY CONNECTION

It took thousands of years for human beings to discover the elements that are the building blocks of matter. Ancient civilizations discovered and used seven metals: gold (Au), copper (Cu), silver (Ag), lead (Pb), tin (Sn), iron (Fe), and mercury (Hg). These seven elements are called the *metals of antiquity*. Many of the early elements were discovered because they are less reactive than other elements and therefore more likely to be found in their pure forms.

with other substances. An explosion, smoke, or a flash of light is a sign that a reaction is proceeding quickly. For instance, when the metal sodium, Na, comes into the slightest contact with water, it reacts vigorously. Sodium even reacts with water vapor in the air. Magnesium, Mg, is another metal that reacts with water. Sodium and magnesium are both shiny, silvery metals that react with water. Based on their properties, these two metals might belong in the same group.

## FORMULAS OF COMPOUNDS

Mendeleyev also paid attention to which elements combine with which, and he noted the ratios in which their atoms combine. For example, magnesium, Mg, can combine with chlorine, Cl. When it does, it forms the compound magnesium chloride, $MgCl_2$. This chemical formula indicates that atoms of magnesium and chlorine combine in a 1:2 ratio.

Sodium, Na, also reacts with chlorine, but it combines in a 1:1 ratio, forming sodium chloride, NaCl. Perhaps magnesium and sodium do not belong in the same group after all.

Examine the table, which shows some of the elements that react with chlorine and the compounds that are formed.

You can sort the elements into three groups according to the formulas of compounds with chlorine. In the Activity: Create a Table, you sorted these groups into separate columns, as Mendeleyev did.

### Some Elements That React with Chlorine

| Element | Symbol | Compound |
|---------|--------|----------|
| magnesium | Mg | $MgCl_2$ |
| sodium | Na | NaCl |
| aluminum | Al | $AlCl_3$ |
| hydrogen | H | HCl |
| calcium | Ca | $CaCl_2$ |
| indium | In | $InCl_3$ |
| gallium | Ga | $GaCl_3$ |

## ATOMIC MASS

Mendeleyev used another property, called atomic mass, to sort the elements. Each element is made of a different kind of atom. (You will study atoms in Lesson 11: Atomic Pudding.) All atoms of the same element have approximately the same mass. The mass of an atom is called its **atomic mass** and is measured in **atomic mass units,** or **amu.**

Atomic mass will be explained in more detail later in the unit. For now, simply keep in mind that each element has an average atomic mass that is expressed as a decimal number.

You can place the elements in order of their atomic masses. However, sorting the cards just by atomic mass doesn't tell you which elements are similar in their properties.

# ➲ A Table of Elements

Mendeleyev combined all of these sorting tactics to create his table. He put the elements with similar reactivity and chemical formulas of compounds into columns. Mendeleyev also sorted the elements in order of their atomic masses. He placed the lighter elements at the top of the columns and the heavier elements at the bottom. When he placed the columns next to each other, the atomic masses increased from left to right as well as from top to bottom.

### Dmitri Mendeleyev's Periodic Table of the Elements

|  | Group I | Group II | Group III | Group IV | Group V | Group VI | Group VII | Group VIII | |
|---|---|---|---|---|---|---|---|---|---|
| 1 | H = 1 | | | | | | | | |
| 2 | Li = 7 | Be = 9 | B = 11 | C = 12 | N = 14 | O = 16 | F = 19 | | |
| 3 | Na = 23 | Mg = 24 | Al = 27 | Si = 28 | P = 31 | S = 32 | Cl = 35 | | |
| 4 | K = 39 | Ca = 40 | __ = 44 | Ti = 48 | V = 51 | Cr = 52 | Mn = 55 | Fe = 56  Co = 59 | |
|  | | | | | | | | Ni = 59  Cu = 63 | |
| 5 | Cu = 63 | Zn = 65 | __ = 68 | __ = 72 | As = 75 | Se = 78 | Br = 80 | | |
| 6 | Rb = 85 | Sr = 87 | Yt = 88 | Zr = 90 | Nb = 94 | Mo = 96 | __ = 100 | Ru = 104  Rh = 106 | |
|  | | | | | | | | Pd = 106 | |
| 7 | Ag = 108 | Cd = 112 | In = 113 | Sn = 118 | Sb = 122 | Te = 125 | I = 127 | | |
| 8 | Cs = 133 | Ba = 137 | Di = 138 | Ce = 140 | __ | __ | | | |
| 9 | __ | | __ | __ | __ | __ | | __ | |
| 10 | __ | __ | Er = 178 | La = 180 | Ta = 182 | W = 184 | __ | Os = 195  Ir = 197 | |
|  | | | | | | | | Pt = 198 | |
| 11 | Au = 199 | Hg = 200 | Tl = 204 | Pb = 207 | Bi = 208 | | __ | __ | |
| 12 | | | | Th = 231 | __ | U = 240 | __ | __ | |

Note: Mendeleyev's symbol for iodine, "J," has been changed to "I" to match modern symbols.

Mendeleyev organized the 63 elements that were known at the time into a table. In his table, the average atomic masses of the elements increase as you proceed across each row and down each column. The elements in each column have similar physical properties and reactivity, and they tend to form compounds with other elements in the same ratios. The table Mendeleyev created, with elements organized into rows and columns, became known as the **periodic table of the elements.**

The periodic table is an extremely useful organization of the elements. Mendeleyev was even able to predict the existence and properties of as-yet undiscovered elements based on gaps he located in his table.

| **Big Idea** | Elements are arranged on the periodic table based on similarities in their chemical and physical properties. |
|---|---|

## LESSON SUMMARY

### How is the periodic table organized?

**KEY TERMS**

reactivity

average atomic mass

atomic mass unit, amu

periodic table of the elements

Dmitri Mendeleyev created one of the first organized tables of the elements. He sorted the elements based on their properties, specifically reactivity, the formulas of compounds created when elements chemically combine, and atomic mass. He organized the elements into a table so that their atomic masses increased in order across a row from left to right and down a column. The elements in each column of the table have similar properties. The table Mendeleyev created came to be called the periodic table of the elements.

### Reading Questions

1. List three properties of the elements that are useful in sorting the elements.

2. Do you expect carbon, C, to be more similar to nitrogen, N, oxygen, O, or silicon, Si? Why?

### Reason and Apply

3. **PROJECT** www Use a reference book or the Internet to look up the average atomic masses and properties of silicon, Si, germanium, Ge, tin, Sn, phosphorus, P, antimony, Sb, sulfur, S, and selenium, Se.
   a. Organize these elements into rows and columns.
   b. List two properties that the elements in each column have in common.

4. **PROJECT** www Use a reference book or the Internet to look up some of the properties of iron, barium, and phosphorus. Explain why nails are made of iron but they are never made of barium or phosphorus.

5. Suppose you have equal amounts of calcium, Ca, in two beakers. In one beaker, you react the calcium with oxygen, O. In the other beaker, you react the calcium with sulfur, S. The reaction with oxygen forms the compound calcium oxide, CaO.
   a. What do you predict is the chemical formula of the compound formed from the reaction between calcium and sulfur?
   b. Which compound has more mass, the compound containing calcium and oxygen, or the compound containing calcium and sulfur? Explain your thinking.

6. **PROJECT** www Use the Internet to research Dmitri Mendeleyev, the Russian chemist credited with the discovery of the periodic table of the elements. Write a brief paragraph describing Mendeleyev's life and work. Include how he became a chemistry professor and how he came up with the idea for the periodic table.

7. Find at least two different versions of the periodic table and bring a copy of each to class.
   a. Write down what you think makes these two versions similar to each other.
   b. Write down what you think makes these two versions different from each other.

# The Periodic Table

## THINK ABOUT IT

The elements copper, Cu, and gold, Au, share many similarities. Both are relatively unreactive elements. They are soft, so it is easy to bend and shape them. They are called *coinage metals* because they have been made into coins by many cultures. Copper and gold have high values as jewelry because they remain shiny for many years. Is the similarity in their properties related to their locations on the periodic table?

**What information does the periodic table reveal about the elements?**

*To answer this question, you will explore*

- The Modern Periodic Table
- Trends in Properties

## EXPLORING THE TOPIC

## The Modern Periodic Table

Scientists have detected around 115 different elements on the planet. Each is unique. Yet groups of elements have similar properties. Recall from Lesson 9: Create a Table that Dmitri Mendeleyev constructed a table based on patterns in the properties of the elements. His table has been replaced over the decades with many updated versions, such as the one shown on pages 42 and 43. The modern periodic table is a storehouse of valuable information about the elements. Over time, you will learn how to make use of the information that is contained there.

### ELEMENT SQUARES

Each element has a square on the periodic table. Within each square is information about that element including its name and symbol. The whole number in each square is called the **atomic number.** Hydrogen is the first element in the table and has the atomic number 1. Helium is the second element and has the atomic number 2. Each atomic number corresponds to a different element.

The decimal number in each square on the periodic table is the average atomic mass in amu.

Here is a square from the periodic table.

# Periodic Table of the Elements

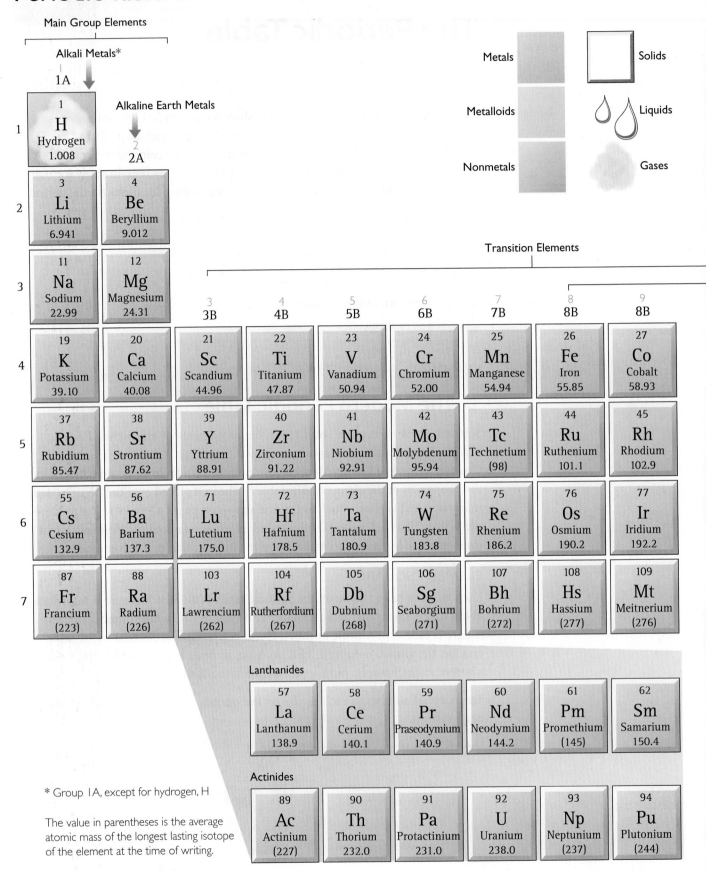

Main Group Elements

Halogens    Noble Gases

| | Atomic number |
|---|---|
| 1 | |
| H | Symbol |
| Hydrogen | Name |
| 1.008 | Average atomic mass |

| 18 8A |
|---|

| 13 3A | 14 4A | 15 5A | 16 6A | 17 7A | |
|---|---|---|---|---|---|
| | | | | | 2 He Helium 4.003 |
| 5 B Boron 10.81 | 6 C Carbon 12.01 | 7 N Nitrogen 14.01 | 8 O Oxygen 16.00 | 9 F Fluorine 19.00 | 10 Ne Neon 20.18 |
| 13 Al Aluminum 26.98 | 14 Si Silicon 28.09 | 15 P Phosphorus 30.97 | 16 S Sulfur 32.07 | 17 Cl Chlorine 35.45 | 18 Ar Argon 39.95 |

| 10 8B | 11 1B | 12 2B | | | | | | |
|---|---|---|---|---|---|---|---|---|
| 28 Ni Nickel 58.69 | 29 Cu Copper 63.55 | 30 Zn Zinc 65.39 | 31 Ga Gallium 69.72 | 32 Ge Germanium 72.64 | 33 As Arsenic 74.92 | 34 Se Selenium 78.96 | 35 Br Bromine 79.90 | 36 Kr Krypton 83.80 |
| 46 Pd Palladium 106.4 | 47 Ag Silver 107.9 | 48 Cd Cadmium 112.4 | 49 In Indium 114.8 | 50 Sn Tin 118.7 | 51 Sb Antimony 121.8 | 52 Te Tellurium 127.6 | 53 I Iodine 126.9 | 54 Xe Xenon 131.3 |
| 78 Pt Platinum 195.1 | 79 Au Gold 197.0 | 80 Hg Mercury 200.6 | 81 Tl Thallium 204.4 | 82 Pb Lead 207.2 | 83 Bi Bismuth 209.0 | 84 Po Polonium (209) | 85 At Astatine (210) | 86 Rn Radon (222) |
| 110 Ds Darmstadtium (281) | 111 Rg Roentgenium (280) | 112 Cn Copernicium (285) | 113 Nh Nihonium (286) | 114 Fl Flerovium (289) | 115 Mc Moscovium (289) | 116 Lv Livermorium (292) | 117 Ts Tennessine (293) | 118 Og Oganesson (294) |

| 63 Eu Europium 152.0 | 64 Gd Gadolinium 157.3 | 65 Tb Terbium 158.9 | 66 Dy Dysprosium 162.5 | 67 Ho Holmium 164.9 | 68 Er Erbium 167.3 | 69 Tm Thulium 168.9 | 70 Yb Ytterbium 173.0 |
|---|---|---|---|---|---|---|---|
| 95 Am Americium (243) | 96 Cm Curium (247) | 97 Bk Berkelium (247) | 98 Cf Californium (251) | 99 Es Einsteinium (252) | 100 Fm Fermium (257) | 101 Md Mendelevium (258) | 102 No Nobelium (259) |

Stockbyte/Getty Images

## PARTS OF THE PERIODIC TABLE

Most modern periodic tables have 18 vertical columns and 7 horizontal rows. The vertical columns are called **groups,** or families. Hydrogen, H, is in Group 1A, along with lithium, Li, and five other elements in that column. Some of the groups have specific names as shown below.

The horizontal rows of the table are called **periods** because patterns repeat periodically, or over and over again, in each row. There are only two elements in Period 1, hydrogen and helium. However, there are eight elements in Periods 2 and 3, and 18 elements in Period 4.

Chemists also have names for sections of the periodic table. Between Group 2A and Group 3A, for example, is where the transition elements fit in.

In addition, two rows of elements are usually shown at the bottom of the table. These elements are called the lanthanides and actinides. If you examine the atomic numbers of these elements, you'll see that they belong in the sixth and seventh rows. If they were included where they belong, the table would look like this.

Most periodic tables show these two rows at the bottom so that everything will fit onto one page.

## Trends in Properties

Once the elements are arranged according to their general properties, many other patterns or trends can be found. These three drawings illustrate some of the trends contained within the periodic table.

### SOLIDS, LIQUIDS, AND GASES

Most of the elements are solids at room temperature. Several elements are gases at room temperature, but only a few are liquids at or near room temperature.

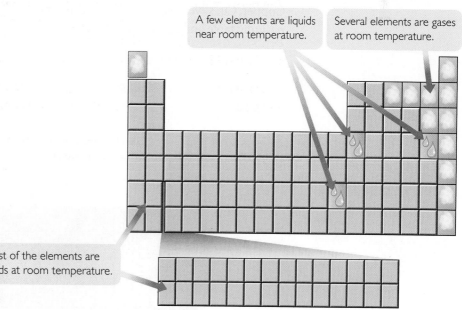

A few elements are liquids near room temperature.

Several elements are gases at room temperature.

Most of the elements are solids at room temperature.

### METALS, METALLOIDS, AND NONMETALS

The majority of the elements are **metals.** On most periodic tables, there is a stair-step line that divides the table. Metals are found to the left and **nonmetals** are found to the right of the stair-step line. The elements found along the stair-step line are called **metalloids.** Metalloids have properties like those of both metals and nonmetals.

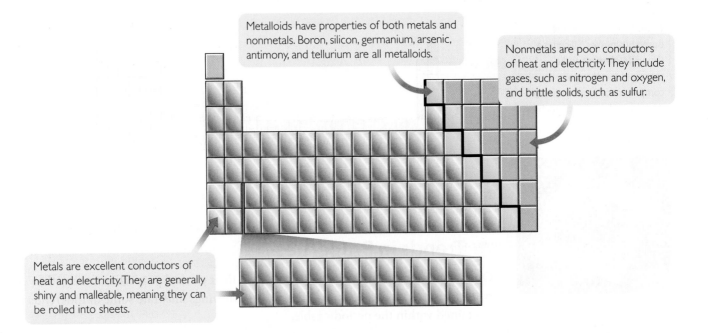

Metalloids have properties of both metals and nonmetals. Boron, silicon, germanium, arsenic, antimony, and tellurium are all metalloids.

Nonmetals are poor conductors of heat and electricity. They include gases, such as nitrogen and oxygen, and brittle solids, such as sulfur.

Metals are excellent conductors of heat and electricity. They are generally shiny and malleable, meaning they can be rolled into sheets.

## REACTIVITY

Elements in the lower left and upper right of the periodic table are the most reactive, with the exception of the noble gases in Group 8A, which are very unreactive. Copper, Cu, silver, Ag, and gold, Au, are metals that are in the middle of the periodic table and are not very reactive.

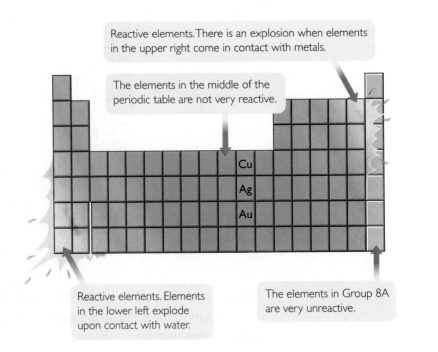

Reactive elements. There is an explosion when elements in the upper right come in contact with metals.

The elements in the middle of the periodic table are not very reactive.

Reactive elements. Elements in the lower left explode upon contact with water.

The elements in Group 8A are very unreactive.

Example 1

## Iodine, I

Find iodine, I, on the periodic table.

**a.** Find iodine's atomic number, average atomic mass, period, and group.

**b.** Would you expect iodine to be a solid, liquid, or gas at room temperature?

**c.** Is iodine a metal, metalloid, or nonmetal? How can you tell?

**d.** Do you expect iodine to be reactive? Explain.

### Solution

Iodine is in the lower-right area of the main group elements.

**a.** The atomic number is 53. Average atomic mass is 126.9 amu. Iodine is in Period 5 and Group 7A, halogens.

**b.** Iodine is a solid at room temperature.

**c.** Iodine is a nonmetal, because it is to the right of the stair-step line.

**d.** Yes, you can expect it to be reactive, though not as reactive as elements above it in Group 7A.

Example 2

## Coinage Metals

Which element would make the best coin: phosphorus, P, silver, Ag, potassium, K, or xenon, Xe? Explain your thinking.

### Solution

Xenon, Xe, is a gas, so it is definitely not a candidate for making a coin. Phosphorus, P, is a nonmetal that is dull and brittle. It would be difficult to shape into a coin. Silver, Ag, is a shiny, malleable metal, so it would make the best coin. Potassium, K, is also a metal, but it is too soft and reactive, so it would not make a good coin. A good coin should not react with other substances.

**KEY TERMS**

atomic number

group

alkali metal

alkaline earth metal

halogen

noble gas

period

main group elements

transition elements

lanthanides

actinides

metal

nonmetal

metalloid

## LESSON SUMMARY

### What information does the periodic table reveal about the elements?

The periodic table is an organized chart of the elements. Each element square contains valuable information, including the element name, symbol, atomic number, and average atomic mass. These elements are arranged in vertical columns called groups, or families, and horizontal rows called periods. Most elements are solids and metals, except for those in the upper right of the table. The most reactive elements are located in the lower left and upper right of the table, excluding the noble gases, which are unreactive and located in the last column on the right.

### Reading Questions

1. Describe how reactivity changes as you go down Group 1A.

2. Choose two different properties and describe how they vary across a period.

### Reason and Apply

3. You will need a handout of the periodic table.
   a. On your periodic table, clearly label the alkali metals, the alkaline earth metals, the halogens, and the noble gases. (If you wish, you may color them and provide a color key at the top.)
   b. Label the main group elements, the transition elements, and the lanthanides and actinides.

4. Name two elements that have properties similar to those of beryllium, Be, and have average atomic masses higher than 130.

5. Which of these elements are solids at room temperature?
   a. fluorine, F       b. titanium, Ti       c. lead, Pb
   d. oxygen, O         e. potassium, K       f. silicon, Si

6. Which of these elements are nonmetals?
   a. bromine, Br       b. carbon, C          c. boron, B
   d. thallium, Tl      e. phosphorus, P      f. aluminum, Al

7. Which two of these elements are the least reactive? Explain your thinking.
   a. chlorine, Cl      b. barium, Ba         c. copper, Cu
   d. rubidium, Rb      e. potassium, K       f. mercury, Hg

8. Can you make jewelry out of each of the elements listed below? Explain your thinking.
   a. copper, Cu        b. neon, Ne
   c. sodium, Na        d. platinum, Pt

## CHAPTER 2

# Basic Building Materials

# SUMMARY

### KEY TERMS

element

chemical symbol

compound

chemical formula

phase

aqueous

chemical change (chemical reaction)

law of conservation of mass

reactivity

atomic mass

atomic mass units, amu

periodic table of the elements

atomic number

group

alkali metal

alkaline earth metal

halogen

noble gas

period

main group elements

transition elements

lanthanides

actinides

metal

nonmetal

metalloid

average atomic mass

# Alchemy Update

Can an element, such as copper, be transformed into gold through chemical processes? Copper and gold are in the same group on the periodic

table and have similar properties. However, they are distinct elements with many differences, such as their appearances. While elements can react to form new compounds, from what you have learned so far it does not appear possible to change one element into another element.

## REVIEW EXERCISES

1.  Make a list of all the information you can extract from the periodic table for the element gold.

2.  Explain the law of conservation of mass and how it relates to the copper cycle.

3.  What is a chemical formula, and what does it tell you?

4.  A filament for a light bulb must conduct electricity. Which of the elements listed below might be useful as a light bulb filament? Explain your thinking.
    **A.** tungsten, W     **B.** sulfur, S     **C.** bromine, Br

## Element Profile

PROJECT     Research an element. Write a report including

- Your element's name, symbol, and description.

- A list of your element's uses.

- A description of how your element is mined or obtained.

CERN/Science Source

# CHAPTER **3**

# A World of Particles

## *In this chapter, you will study*

- models of the atom

- how atoms differ from one another

- nuclear reactions

- how elements are created

*Matter is made up of atoms, and atoms are made up of even smaller, subatomic particles. This image shows the interactions of subatomic particles in a bubble chamber of liquid hydrogen.*

All matter is composed of tiny particles called atoms. Based on their observations, chemists agree that atoms themselves are composed of even tinier structures: a nucleus with electrons orbiting around it. So what exactly are atoms, and what does the structure of atoms tell you about matter?

# Models of the Atom

## THINK ABOUT IT

The drawing depicts a very tiny sample of gold taken from a gold ring.

0.0000000000041 meter

Joel Sartore/National Geographic/Getty Images

The spheres in the cube of gold are so small that they cannot be seen. What are the spheres, and what does this drawing tell you about the element gold?

### How are the smallest bits of matter described?

*To answer this question, you will explore*

- Atoms: Small Bits of Matter
- Models of the Atom
- Simple Atomic Model

## EXPLORING THE TOPIC

## Atoms: Small Bits of Matter

Titanium, Ti

Sulfur, S

Titanium sulfide, $TiS_2$

Imagine you break a piece of matter in half, and then break it in half again and again. How many times can you do this? Can you keep going, getting ever smaller? Around 460 B.C.E., the Greek philosopher Democritus wondered the same thing. He thought that if he could just keep breaking matter in half he would eventually end up with the smallest bit of matter possible.

Democritus proposed that all matter was composed of tiny particles that could not be divided further. Today, we use the word **atom** to describe these bits of matter. Of course, atoms are too small to be seen. Democritus' idea was disregarded for the next two thousand years, in part because Democritus did not have evidence to support it.

In 1803, the British scientist John Dalton suggested that the idea of atoms could help explain why elements come together in specific ratios when they form compounds. He imagined atoms of different elements combining to form compounds in the ratios specified by the chemical formulas of the compounds. For example, to form the compound titanium sulfide, $TiS_2$, titanium and sulfur atoms combine in a 1:2 ratio.

Dalton had more than an idea about atoms. He conducted experiments and made observations to back up his idea. His observations provided strong evidence to support his explanation of how matter behaves.

In science, the word *theory* indicates that an explanation is supported by overwhelming evidence. The word *theory* allows room for doubt and revision, but indicates a greater degree of certainty than the word does in everyday use. The **atomic theory** states that all matter is made up of atoms. The atomic theory helps us make accurate predictions about the behavior of matter.

## Models of the Atom

Since Dalton's time, scientists have created many **models** to describe atoms and their parts. Models are simplified representations of something you want to explain. For example, a model airplane is a small representation of a larger aircraft. Models take many forms. They can be a plan, a physical structure, a

# The Atomic Model through Time

**EXPERIMENT 1**

In 1803, John Dalton studied how elements combine chemically to form compounds. He observed that elements combine in whole-number ratios to form compounds and that matter is not created or destroyed in chemical reactions. Dalton reasoned that elements are made of tiny, indivisible spherical particles called atoms.

**EXPERIMENT 2**

In 1897, J. J. Thomson, a British scientist, zapped atoms with electricity. He observed that negatively charged particles were removed. Thomson reasoned that atoms contain negatively charged particles, which he called electrons.

**EXPERIMENT 3**

In 1911, Ernest Rutherford, a New Zealand–born scientist, shot tiny positively charged particles, called alpha particles, at thin gold foil. He observed that most of the alpha particles went through the foil, but a few bounced back. Rutherford reasoned that there must be something small, massive, and positively charged in an atom, which he called the nucleus.

**SOLID SPHERE MODEL**

The atom is a solid sphere that cannot be divided up into smaller particles or pieces.

**PLUM PUDDING MODEL**

— Positive fluid

— Negative electron

The atom can be divided into a fluid (the "pudding") and electrons. Most of the atom is made of fluid. The fluid spreads out in the atom and is positively charged. The electrons are very tiny and negatively charged.

**NUCLEAR MODEL**

— Negative electron

— Positive nucleus

The atom can be divided into a nucleus and electrons. The nucleus occupies a small amount of space in the center of the atom. The nucleus is dense and positively charged. The electrons circle around the nucleus. The electrons are tiny and negatively charged. Most of the atom is empty space.

1803    1897    1911

drawing, a mathematical equation, or even a mental image. A model that represents the structure of an atom is called an *atomic model*.

Dalton pictured the atom as a hard, solid sphere. Over the next two hundred years, scientists gathered evidence to support and expand on Dalton's model of the atom. It became clear that the atom was more than just a solid sphere.

But how did scientists gather evidence about something too small to be seen? Scientists found they could learn more about atoms and their structure by shooting small pieces of matter at them or by heating them in a flame. Observations from these experiments provided evidence that helped scientists make changes and refine the model of the atom.

**EXPERIMENT 6**

In 1927, Werner Heisenberg, a German scientist, proposed a cloud model of the atom. Heisenberg suggested that the location of an electron could not be specified precisely. Instead, it is only possible to talk about the probability of where an electron might be. This led to a cloud model of the atom. The electron cloud indicates where you will most likely find a single electron.

**EXPERIMENT 4**

In 1913, Niels Bohr, a Danish scientist, developed a model of the atom that explained the light given off when elements are exposed to flame or electric fields. He observed that only certain colors of light are given off. For example, hydrogen atoms give off red, blue-green, and blue light. Bohr reasoned that the electrons orbit around the nucleus at different distances like planets orbiting the Sun. The electrons in these orbits have different energies. When an electron falls from an outer to an inner orbit, the color of the light given off depends on the energies of the two orbits.

**EXPERIMENT 5**

In 1918, Rutherford made a further contribution. He found he could use alpha particles as bullets to knock off small positively charged particles, which he called protons. He reasoned that the nucleus must be a collection of protons.

**EXPERIMENT 7**

In 1932, a British physicist, James Chadwick, found that the nucleus also included uncharged, or neutral, particles, which he called neutrons. He reasoned that the neutrons were important in holding the positively charged protons together.

**SOLAR SYSTEM MODEL**

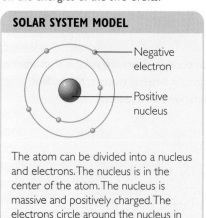

Negative electron

Positive nucleus

The atom can be divided into a nucleus and electrons. The nucleus is in the center of the atom. The nucleus is massive and positively charged. The electrons circle around the nucleus in specified orbits. The electrons are tiny and negatively charged. Different electrons are in orbits at different distances from the nucleus.

**PROTON MODEL**

The nucleus contains protons. The protons are tiny and positively charged. The electrons circle around the nucleus. The electrons are tiny and negatively charged. Most of the atom is empty space.

**ELECTRON CLOUD MODEL**

An electron cloud surrounds the nucleus. The cloud is made up of fast-moving electrons:
The nucleus is made up of protons: and neutrons:

1913          1918          1927     1932

The model of the atom was refined and changed as scientists gathered new evidence. This is what science is all about—a continual process of gathering new knowledge to improve our understanding of the world.

## ⟳ Simple Atomic Model

The six models shown on the timeline all have something valuable to offer in terms of visualizing matter at an atomic level. At right is a simple atomic model of an atom. In the very center of the atom is the **nucleus.** The nucleus consists of positively charged **protons,** and **neutrons,** which have no charge. The **electrons** are even tinier than the protons and neutrons, and they orbit the nucleus. In this particular atom, the electrons are located at two different distances from the nucleus.

Electrons

Nucleus

Each electron has a charge of −1. The neutrons are neutral and so have no charge. Each proton has a charge of +1. A neutral atom has no overall charge. It has equal numbers of positive protons and negative electrons.

**Big Idea**   An atom has a nucleus made of protons and neutrons, with electrons orbiting the nucleus.

© Marvin E. Newman/Getty Images

*The nucleus occupies a very tiny volume. If an atom were the size of a baseball stadium, the nucleus would be smaller than a baseball.*

## LESSON SUMMARY

### How are the smallest bits of matter described?

Long ago, some philosophers imagined that matter was made up of tiny particles called atoms. Over time, scientists gathered evidence from experimental observations to create models of the atom. Today, we know the atom is made up of protons, neutrons, and electrons. The protons and neutrons are in the center of the atom, in the nucleus. Electrons are outside the nucleus. They are much smaller than the protons and neutrons. In a neutral atom, the positive charges on the protons are equal to the negative charges on the electrons.

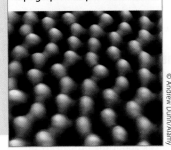
**KEY TERMS**

atom

atomic theory

model

nucleus

proton

neutron

electron

### Reading Questions

1. What evidence caused Thomson to change Dalton's solid sphere model into the plum pudding model?

2. What evidence caused Rutherford to change Thomson's plum pudding model into the nuclear model?

3. What evidence caused Bohr to change Rutherford's nuclear model into the solar system model?

### Reason and Apply

4. Positive and negative charges are attracted to one another. Which of the following are attracted to a negative charge: an electron, a proton, a neutron, a nucleus, an atom? Explain your thinking.

5. Hydrogen and helium are different elements. How can you use the plum pudding model to show how atoms of the two elements might be different from one another?

6. Suppose you discovered protons shortly after Thomson discovered electrons. How would you revise the plum pudding model to include protons? Draw a picture of your revised model of the atom.

7. Draw a solar system model showing one electron, one proton, and one neutron.

8. **PROJECT** Use the Internet or another resource to find out how the size of an atom compares with the size of its nucleus. Is the diameter of an atom 10 times, 1,000 times, or 100,000 times the diameter of the nucleus?

When charged particles are placed near each other, they move toward or away from one another. Similar charges repel, or move away from, one another. Opposite charges attract, or move toward, one another. This photo shows iron filings that have oriented themselves around a magnet's positive and negative ends.

*Arthur S. Aubry/Photodisc/Getty Images*

9. The nuclear model and the solar system model both show atoms with electrons circling around the nucleus.
   a. How do these two models differ?
   b. How are these two models similar?
   c. How can you refine the solar system model so that the atoms do not look flat?

10. **PROJECT** The ancient Greeks discarded the atomic theory because there was no evidence to support it. Try to provide evidence that atoms do indeed exist. Use the Internet to help you.

11. The ancient Greeks claimed that atoms were the smallest pieces of matter. Were they correct? Explain your thinking.

12. Give an example that shows how science is a process of gathering evidence and refining models.

*Atoms by Numbers*

# Atomic Number and Atomic Mass

## THINK ABOUT IT

The element copper is made up of copper atoms. Likewise, gold is made up of gold atoms. We know that copper and gold are different elements. But what makes a copper atom different from a gold atom?

**How are the atoms of one element different from those of another element?**

*To answer this question, you will explore*

- Atomic Number
- Atomic Mass
- The Periodic Table and Atomic Models

## EXPLORING THE TOPIC

### Atomic Number

In Lesson 9: Create a Table, you learned how Mendeleyev arranged the elements in order of increasing atomic mass. Around 1913, Henry Moseley, a British scientist, discovered an amazingly simple and important property of the elements as well. He determined that the atoms of each element differ by one proton from the atoms of the element before it on the periodic table. The first element, hydrogen, H, has one proton, and the second element, helium, He, has two protons. The third element, lithium, Li, has three protons, beryllium, Be, has four protons, and so on.

The **atomic number** is equal to the number of protons in the nucleus of an element. The elements on the periodic table are arranged in order by their atomic numbers. So the element iron, Fe, with atomic number 26, has 26 protons in its nucleus.

**Big Idea** If you know the number of protons in an atom, you know its atomic number and what element it is.

For a neutral atom, the atomic number is also equal to the number of electrons. This is because the overall charge of a neutral atom is 0. Protons have a $+1$ charge and electrons have a $-1$ charge. If the overall charge on an atom is 0, then the number of protons must be equal to the number of electrons.

### Atomic Mass

The atomic mass is the mass of a single atom. The protons and neutrons account for almost all of the mass of an atom. Electrons have a much tinier mass by

© skodonnell/iStockPhoto

comparison. The atomic mass is approximately equal to the total mass of the neutrons and protons because electrons have so little mass.

Every proton in every atom, whether it is an atom of gold or an atom of oxygen, has the same mass. Scientists assign a value of one atomic mass unit, 1 amu, to the mass of a single proton. Neutrons have almost exactly the same mass as protons, so each neutron also has a mass of 1 amu. To determine the mass of a single atom, you add the number of protons and neutrons.

Lithium, Li

Atomic number: 3
Average atomic mass: 6.941 amu

3 protons = 3 amu
+ 4 neutrons = 4 amu
atomic mass = 7 amu

Carbon, C

Atomic number: 6
Average atomic mass: 12.01 amu

6 protons = 6 amu
+ 6 neutrons = 6 amu
atomic mass = 12 amu

## HOW MANY NEUTRONS?

If you look at the atomic models, you will notice that the numbers of protons and electrons are exactly the same as the atomic number, but the number of neutrons is sometimes different from the number of protons. So, how can the periodic table tell you how many neutrons are in an atom? If you know the mass of an atom *and* you know how many protons it has, you can find out how many neutrons it has by subtracting the number of protons from the atomic mass.

## AVERAGE ATOMIC MASS

It turns out that not every atom of an element is identical. So, the decimal number in each element square of the periodic table is the average atomic mass of that element in atomic mass units, amu. This number can also be used to estimate the number of neutrons in a nucleus. Simply round the average atomic mass to the nearest whole number and subtract the atomic number (number of protons). For example, lithium has an average atomic mass of 6.941 amu, so a typical lithium atom probably has a mass of 7 amu. The atomic number of lithium is 3, so there are 3 protons. This accounts for 3 amu of the mass. The other 4 amu must be due to 4 neutrons.

In Lesson 13: Subatomic Heavyweights, you will investigate how the average atomic mass is arrived at and how atoms of an element may differ from one another.

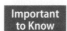 **Important to Know** When considering atomic mass, it is necessary to know whether you are focusing on the mass of one particular atom or the average mass of a group of atoms.

## The Periodic Table and Atomic Models

To draw an atomic model of a specific element, you must know the numbers of protons, neutrons, and electrons. You find this information on the periodic table. This illustration shows you how to get information about atomic structure from each square of the periodic table to build a basic atomic model of an element.

**Build an Atomic Model in Three Steps**

**Start with the protons**

The number of protons in the nucleus is equal to the atomic number. This number is always a whole number.

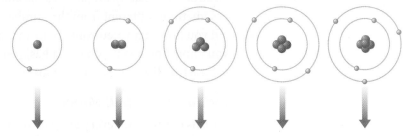

**Add electrons**

In neutral atoms, the number of electrons is equal to the number of protons. The electrons are placed in circles outside of the nucleus.

**Add neutrons**

You can estimate the number of neutrons by subtracting the number of protons from the average atomic mass, rounded to the nearest whole number.

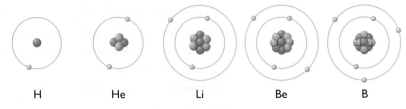

| H | He | Li | Be | B |

**HISTORY CONNECTION**

Ernest Rutherford, a New Zealand–born scientist, is generally credited with the discovery of the proton. He was awarded the Nobel Prize in Chemistry in 1908, and his image appears on the New Zealand 100 dollar note.

## Copper and Gold Atoms

How is an atom of gold, Au, different from an atom of copper, Cu?

*Solution*

You can find the information you need on the periodic table. The atomic number of copper is 29. So neutral copper atoms have 29 protons and 29 electrons.

To estimate the number of neutrons in the atom, round the atomic mass to 64 and subtract the atomic number.

The atomic number is the number of protons.

29
Cu
Copper
63.55

Round the average atomic mass to 64.

$64 - 29 = 35$

Number of neutrons

Number of protons = 29
Number of electrons = 29
Number of neutrons ≈ 35

You can follow the same steps for gold atoms.

The atomic number is the number of protons.

79
Au
Gold
197.0

Round the average atomic mass to 197.

$197 - 79 = 118$

Number of neutrons

Number of protons = 79
Number of electrons = 79
Number of neutrons ≈ 118

So a gold atom has 50 more protons, 50 more electrons, and about 83 more neutrons than a copper atom.

## How are the atoms of one element different from those of another element?

The periodic table reveals information about atomic structure. The atomic number of an element is equal to the number of protons in each of its atoms. The atomic number is also equal to the number of electrons in a neutral atom of an element. You can identify an element by the number of protons in the nucleus of an atom of the element. The protons and neutrons account for almost all the mass of an atom. Therefore, the mass of an atom in amu is approximately equal to the number of protons plus the number of neutrons. You can estimate the number of neutrons in an atom by subtracting the number of protons from the average atomic mass.

**KEY TERM**

atomic number

## Exercises

### Reading Questions

1. What does the atomic number tell you?

2. What does the atomic mass tell you?

### Reason and Apply

3. If you have a sample of atoms and each atom has 12 protons in its nucleus, which element do you have?

4. If you want to identify an element, what one piece of information would you ask for? Explain your thinking.

5. Why does carbon, C, have a larger atomic mass than boron, B, even though they each have six neutrons?

6. Make a table like the one below. Use a periodic table to fill in the missing information.

| Element | Chemical symbol | Atomic number | Number of protons | Number of electrons | Number of neutrons | Average atomic mass |
|---------|----------------|---------------|-------------------|---------------------|--------------------|--------------------|
| nickel | | | | | | |
| | Ne | | | | | |
| | | | | | | 24.31 |
| | | 15 | | | | |
| | | | | | 30 | |

7. Draw a simple atomic model for an atom of neon, Ne.

8. Place the following elements in order from lowest number of protons to highest number of protons: S, Mg, N, Na, Se, Sr. Then give the following information about a neutral atom of each: name, atomic number, number of protons, number of electrons, group number.

# Isotopes

## THINK ABOUT IT

It might surprise you that all atoms of the element copper are not exact replicas of one another. Although all copper atoms have many similarities, their masses can differ by 1 or 2 amu. What is the same and what is different about these atoms?

**How can atoms of the same element be different?**

*To answer this question, you will explore*

→ Isotopes

→ Average Atomic Mass

## EXPLORING THE TOPIC

### → Isotopes

#### VARIATIONS IN THE NUCLEUS

Carbon is the sixth element on the periodic table. Its average atomic mass is listed as 12.01 amu. Why is the atomic mass listed as 12.01 rather than simply as 12?

Most of the carbon atoms that are found in nature do have a mass of 12 amu. However, for every 100 carbon atoms, it is typical to find one carbon atom with a mass of 13 amu and even more rarely a carbon atom with a mass of 14 amu. This makes the *average* atomic mass 12.01 amu. Consider the structures of these three varieties of carbon atom. A model of each is shown here.

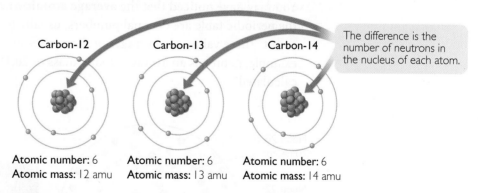

Carbon-12    Carbon-13    Carbon-14

The difference is the number of neutrons in the nucleus of each atom.

Atomic number: 6    Atomic number: 6    Atomic number: 6
Atomic mass: 12 amu    Atomic mass: 13 amu    Atomic mass: 14 amu

Compare the nucleus of each atom. Each has 6 protons. That is what makes them all carbon atoms. Recall that each atom of an element has the same number of protons as every other atom of that element.

However, one carbon atom has 6 neutrons, another carbon atom has 7 neutrons, and the third has 8 neutrons. The mass of an atom is the sum of the number of protons and neutrons, so one carbon atom has a mass of 12 amu and the others have masses of 13 and 14 amu. These atoms are referred to as carbon-12, carbon-13, and carbon-14 to show that they are all atoms of carbon but have different masses.

**TECHNOLOGY
CONNECTION**

A mass spectrometer is used to determine the isotopic composition of an element. It makes an extremely accurate measurement of the mass of an individual atom, using the principle that a heavier particle will travel a straighter path through a magnetic field than a lighter particle.

Atoms of an element that have different numbers of neutrons are called **isotopes.** Carbon has three isotopes. In nature, almost all the elements have at least two isotopes. A few elements have as many as ten isotopes.

### ISOTOPE SYMBOLS

Symbols for the three isotopes of carbon are shown below. Notice that the mass of each isotope is shown as a superscript number (on top). This number is a whole number. It is the sum of the numbers of protons and neutrons in the atom and is sometimes called the **mass number.** The subscript number (on the bottom) is the atomic number of the element, in this case, 6. The number of neutrons in each atom is equal to the top number minus the bottom number.

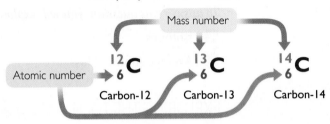

Isotope Symbols for Carbon

These three isotopes are nearly identical in their properties. For example, they all form the same compounds.

**Big Idea** **Isotopes are atoms of the same element but with different numbers of neutrons.**

## Average Atomic Mass

You may have noticed that the **average atomic mass** values in each square of the periodic table are decimal numbers, usually to the nearest hundredth of a unit. These values are averages of the masses of the isotopes in a sample. For example, neon has an average atomic mass of 20.18 amu. How is this average calculated?

The circle contains a sample of 100 atoms of neon.

Noel Hendrickson/Getty Images

About 90% of all neon atoms have an atomic mass of 20 amu, 9% have an atomic mass of 22 amu, and 1% have an atomic mass of 21 amu. By considering a random sample of 100 neon atoms, you can calculate their average atomic mass like this:

$$\text{average atomic mass} = \frac{\text{total mass}}{\text{number of atoms}}$$

$$= \frac{(90)(20\,\text{amu}) + (9)(22\,\text{amu}) + (1)(21\,\text{amu})}{100}$$

$$= 20.19\,\text{amu}$$

As you can see, this number is nearly identical to 20.18, the average atomic mass listed for neon on the periodic table.

The percentage of each isotope of an element that occurs in nature is called the *natural abundance* of the isotope. For example, the natural abundance of neon-20 is about 90.48%.

> **Important to Know** The mass of an isotope refers to the mass of a single specific atom of an element. The average atomic mass given on the periodic table is the average of the masses of all the isotopes in a large sample of that element.

## Example

### Isotopes of Copper

There are two different isotopes of copper. The isotope names and symbols are given here.

$$^{63}_{29}\text{Cu} \qquad ^{65}_{29}\text{Cu}$$

copper-63     copper-65

a.  Explain why both symbols have 29 as the bottom number.

b.  Explain how the two isotopes are different from each other.

c.  Scientists have found the natural abundances of each isotope: 69% copper-63 and 31% copper-65. Explain why the average atomic mass listed on the periodic table for copper is 63.55.

*Solution*

a.  Copper's atomic number is 29, so both isotope symbols have a subscript 29 indicating 29 protons.

b.  The two isotopes have different atomic masses. Both isotopes have 29 protons, so copper-63 has 34 neutrons and copper-65 has 36 neutrons.

c.  You could consider a sample of 100 atoms of copper and calculate their average mass. The average mass of the 100 atoms is determined by adding the masses of the 100 atoms and dividing this total by 100.

$$\frac{69(63\,\text{amu}) + 31(65\,\text{amu})}{100} = \frac{6362\,\text{amu}}{100} = 63.6\,\text{amu}$$

This is very close to the value found on the periodic table.

*(continued)*

Otherwise, you can convert each isotope's percent natural abundance to a decimal number, then multiply this by the isotope's mass number. Do this for each isotope, then add the products:

$$69\% = \frac{69}{100} = 0.69 \qquad 31\% = \frac{31}{100} = 0.31$$

$$(0.69)(63 \text{ amu}) + (0.31)(65 \text{ amu}) = 63.6 \text{ amu}$$

## LESSON SUMMARY

### How can atoms of the same element be different?

KEY TERMS

isotope

mass number

average atomic mass

Elements are composed of nearly identical atoms, each with the same number of protons. However, not every atom of an element has the same number of neutrons in its nucleus. Atoms of an element with different numbers of neutrons are called isotopes. Because neutrons account for part of the mass of an atom, isotopes have different masses. The average atomic mass is an average of the masses of all the different isotopes, taking natural abundance into account.

## Exercises

### Reading Questions

1. Explain the differences between atomic number and atomic mass.

2. Explain the difference between the average atomic mass given on the periodic table and the mass of an atom.

### Reason and Apply

3. How are potassium-39, potassium-40, and potassium-41 different from each other? Write the isotope symbols for the three isotopes of potassium.

4. How many protons, neutrons, and electrons are in each?
   **a.** fluorine-23    **b.** $^{59}_{27}\text{Co}$    **c.** molybdenum-96

5. An isotope of iron, Fe, has 26 protons and 32 neutrons.
   **a.** What is the approximate mass of this isotope?
   **b.** How would you write the symbol for this isotope?

6. Find the element phosphorus, P, on the periodic table.
   **a.** What is the average atomic mass of phosphorus?
   **b.** What is its atomic number?
   **c.** Predict which isotope you would find in greatest abundance for phosphorus.

7. Chlorine, Cl, is 76% chlorine-35 and 24% chlorine-37. Determine the average atomic mass of chlorine.

8. Lithium, Li, is 7.6% lithium-6 and 92.4% lithium-7. Determine the average atomic mass of lithium.

9. Which isotope of nitrogen is found in nature? Explain your reasoning.
   **a.** $^{7}_{14}\text{N}$    **b.** $^{14}_{7}\text{N}$    **c.** $^{15}_{6}\text{N}$

# Stable and Radioactive Isotopes

## THINK ABOUT IT

Some isotopes are found in nature and others exist only because they were created in the laboratory or under unusual conditions. If you went digging in a copper mine and analyzed the samples, you would find copper atoms with either 34 or 36 neutrons. But do any other isotopes of copper exist?

**What types of isotopes do the various elements have?**

*To answer this question, you will explore*

➲ Naturally Occurring Isotopes

➲ Stable and Radioactive Isotopes

## EXPLORING THE TOPIC

## ➲ Naturally Occurring Isotopes

When an atom or element can be found somewhere on Earth, it is called *naturally occurring*. Most chemists agree that there are about 92 naturally occurring elements on the periodic table. The rest of the elements on the table have existed only as a result of human activity or unusual circumstances.

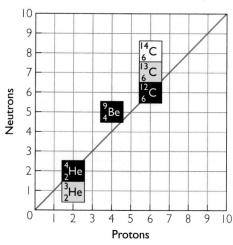

The element beryllium, Be, has only one naturally occurring isotope. This means all beryllium atoms have four protons and five neutrons. The element helium, He, has two naturally occurring isotopes. Its atoms have two protons and either one or two neutrons. Helium and beryllium are shown plotted on a graph of neutrons versus protons. Notice that helium-4 lies on the diagonal line, because its neutrons and protons are equal in number.

Now consider an element like carbon, C, with more than one isotope. Its isotopes, carbon-12, carbon-13, and carbon-14, are also on the graph. Notice that carbon's isotopes line up vertically. This is because they all have 6 protons. One of carbon's isotopes, with 6 protons and 6 neutrons, lies on the diagonal. The other two isotopes have more neutrons than protons in their nuclei.

### ISOTOPES OF THE FIRST 95 ELEMENTS

The graph on page 66 has been expanded to include *all* of the isotopes of the first 95 elements. As before, each square plotted on the graph represents a different isotope. Take a few minutes to study the graph.

Isotopes of the Elements

Here are some things you might notice:

- Some elements have only one isotope. However, *most* elements have more than one isotope.

- The squares that lie on the diagonal line represent isotopes with equal numbers of protons and neutrons.

- Except in hydrogen-1 and helium-3 atoms, the number of neutrons in an atom's nucleus is always equal to or greater than the number of protons. So most of the points lie on or above the diagonal line.

- Radioactive isotopes are not common. (You will learn more about radioactivity later in this lesson.)

- No element has more than 10 isotopes. Most elements have between 1 and 6 isotopes.

- The number of neutrons is roughly equal to the number of protons for atoms up to atomic number 20.

- The collection of plotted points curves up, away from the diagonal line. Beyond atomic number 20, elements have considerably more neutrons than protons.

- The majority of radioactive isotopes are elements with atomic numbers above 80.

- The elements with even numbers of protons have more isotopes than elements with odd numbers of protons. Even numbers of neutrons are also more common than odd numbers.

## Stable and Radioactive Isotopes

When atoms are not stable, they emit small particles. This means that small bits of the nucleus come flying out of the atom. These less stable isotopes are called **radioactive isotopes.** They decay over time as particles are spontaneously emitted from the nucleus in a process called *radioactive decay.* You will learn more about radioactive decay in upcoming lessons.

As you saw in the graph, hydrogen, H, argon, Ar, and cerium, Ce, are examples of elements that have radioactive isotopes shown with white squares. There are only a few white squares scattered throughout the graph until you reach polonium, Po, element number 84. However, technetium, Tc, promethium, Pm, and every element from polonium and beyond, have *only* radioactive isotopes.

### STABILITY STARTS IN THE NUCLEUS

A stable isotope has a stable nucleus. Stability is related to the balance between the number of protons and the number of neutrons in the nucleus. An atom that doesn't have enough neutrons will disintegrate. So will an atom with too many neutrons. The larger the atom, the more neutrons it takes to make a stable nucleus. For example, isotopes of element number 74, tungsten, W, require at least 110 neutrons to be stable.

© Roger Ressmeyer/CORBIS

**Example**

### Hafnium-144

Do you expect the isotope hafnium-144 to exist in nature? Explain your thinking.

*Solution*

The task here is to find out if there is a square on the isotope graph that corresponds to hafnium-144. The way to do this is to figure out how many protons and neutrons are in hafnium-144. Use the periodic table to find the atomic number of hafnium, Hf. The atomic number is 72, so a hafnium atom has 72 protons. An isotope of hafnium with a mass of 144 has $144 - 72 = 72$ neutrons. There is no square on the graph corresponding to 72 protons and 72 neutrons. So a hafnium atom with only 72 neutrons does not exist in nature.

**Example**

(continued)

### Naturally Occurring Isotopes

Sn   Cs   Nd   Tb   Yb   Re   Hg   At   Th

*(graph: Neutrons on the y-axis from 70 to 125, Protons on the x-axis from 50 to 90)*

These are the naturally occurring isotopes of hafnium.

There is no isotope of hafnium that has 72 protons and 72 neutrons.

- ■ Most abundant isotope
- ▫ Other isotopes found in nature
- ▫ Radioactive isotope

## LESSON SUMMARY

### What types of isotopes do the various elements have?

**KEY TERM**

radioactive isotope

Most elements have more than one naturally occurring isotope. Most of these isotopes are stable. Stable isotopes have stable nuclei, with just the right balance of protons and neutrons. In atoms with atomic numbers up to 20, the number of neutrons is roughly equal to the number of protons. Atoms with atomic numbers between 20 and 84 require progressively more neutrons in the nucleus to attain stability. Unstable isotopes are called radioactive isotopes. The nucleus of an unstable isotope will decay and emit radioactive particles. All elements beyond atomic number 83 are unstable, and therefore, all their isotopes are radioactive.

**Exercises**

### Reading Questions

1. Explain the relationship between the words *atom* and *element*.

2. Explain the relationship between the words *atom* and *isotope*.

### Reason and Apply

3. Find these elements on the isotope graph on page 66. How many stable isotopes does each element have?

oxygen, O          neodymium, Nd          copper, Cu          tin, Sn

*(continued)*

4. Use the isotope graph to determine which of these isotopes would be found in nature.

$$^{24}_{12}\text{Mg} \qquad ^{81}_{35}\text{Br} \qquad ^{152}_{60}\text{Nd} \qquad ^{195}_{78}\text{Pt} \qquad ^{238}_{92}\text{U}$$

5. What does the diagonal line on the graph represent?

6. Use the isotope graph to find the isotopes described below. In each case, give the isotope name and the isotope symbol.
   a. Find three isotopes with equal numbers of neutrons and protons.
   b. Find two isotopes with the same number of protons.
   c. Find two isotopes with the same number of neutrons.
   d. Find two isotopes with the same atomic mass units.

7. Is an atom with a nucleus of 31 protons and 31 neutrons a stable isotope? Why or why not?

8. Draw basic atomic models for all the stable isotopes of oxygen, O.

9. Name five elements on the periodic table that have only radioactive isotopes. Determine the neutron-to-proton ratio for each of their isotopes, in decimal form.

10. How many protons could a stable atom with 90 neutrons have? Which elements would these be?

11. Would an atom with 60 protons and a mass number of 155 be stable?

12. Which of the following isotopes are likely to be found in nature? Identify the element by name if it is an isotope that might be found in nature.

$$^{162}_{63}\,?\!\!-\!\!- \qquad ^{75}_{33}\,?\!\!-\!\!- \qquad ^{112}_{56}\,?\!\!-\!\!- \qquad ^{260}_{88}\,?\!\!-\!\!-$$

13. Choose the best answer. The diagonal line on the isotope graph on page 66 represents nuclei with
   (A) the same number of protons.
   (B) the same number of neutrons.
   (C) the same number of protons as neutrons.
   (D) the same number of protons plus neutrons.

14. Choose the best answer. In general, elements with an even atomic number
   (A) have only one isotope.
   (B) have more isotopes than those with an odd atomic number.
   (C) have the same number of protons as neutrons.
   (D) are halogens.

Sort cards into two piles.

# Nuclear Quest

## Purpose

To explore nuclear chemistry.

## Instructions

Play Nuclear Quest. You will need a game board, a pair of dice, Nuclear Quest cards, Gamma Radiation cards, and a game piece for each player.

## Goal

Each player rolls the dice to move along the periodic table, then draws a card from the Nuclear Quest pile and does what the card says. The goal of the game is to discover element 117, give it your own name, and find out if scientists have given it an "official" name. This is accomplished by using nuclear chemistry to proceed through the periodic table.

## Making Sense

Write a paragraph describing what you learned about nuclear chemistry from this game.

# Nuclear Reactions

## THINK ABOUT IT

When the nucleus of an atom is changed in some way, a nuclear reaction is taking place. Nuclear reactions take place in the Sun and sometimes on Earth. Scientists can also make nuclear reactions happen in laboratories and in nuclear power plants.

### What are nuclear reactions?

*To answer this question, you will explore*

⮑ Nuclear Changes

⮑ Radioactive Decay

⮑ Fission and Fusion

## EXPLORING THE TOPIC

### ⮑ Nuclear Changes

Changes in the nucleus are called **nuclear reactions.** Nuclear reactions differ greatly from chemical reactions. Nuclear reactions involve changes to the nucleus and can change one element into another element. The nucleus of an atom is not so easy to change.

Some nuclear reactions don't require any help to get started. But changing the nucleus of an atom *on purpose* can require massive amounts of energy, so you would not be able to do this in your chemistry lab. There are several ways the nucleus of an atom can change. The nucleus can lose particles, it can be split into smaller nuclei, or it can combine with another nucleus or particle. Nuclear changes can happen only under certain circumstances. They happen without intervention in radioactive elements, scientists make them happen, with great difficulty, in a nuclear reactor, and they happen all the time in the cores of stars.

*Nuclear chemistry occurs in the fiery cores of stars.*

### ⮑ Radioactive Decay

Recall that radioactive isotopes are unstable. By ejecting a particle, the nucleus of a radioactive atom can become stable. The process of ejecting or emitting pieces from the nucleus of an atom is called **radioactive decay.**

Scientists have identified the various *subatomic particles,* particles smaller than the atoms themselves, that are emitted from or shot out of the nucleus during radioactive events. These include alpha particles and beta particles.

## ALPHA DECAY

**Alpha decay** involves the ejection of an alpha particle from the nucleus of an atom. An **alpha particle** consists of two protons and two neutrons and can be represented by the Greek letter α, alpha. Because an alpha particle has two protons, it is the same as the nucleus of a helium atom.

Alpha particle

$^{238}_{92}$U nucleus (unstable)

$^{234}_{90}$Th nucleus (stable)

Gamma ray

An alpha particle carries two protons away from the nucleus of an atom. This changes the identity of the element. So, when an atom of uranium, U, atomic number 92, undergoes alpha decay, an atom of thorium, Th, atomic number 90, is formed. This is why alpha decay causes you to move two spaces backward on the Nuclear Quest game board. Notice that a gamma ray is often released as a result of alpha decay.

© Hulton-Deutsch Collection/CORBIS

Periodic Table

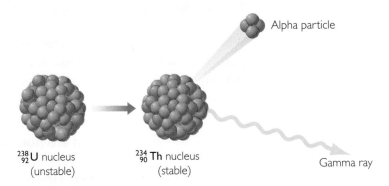

**Big Idea** The elemental identity of an atom is determined by the number of protons it has.

## BETA DECAY

**Beta decay** involves the ejection of an electron from the nucleus of an atom. The ejected electron is called a **beta particle.** Beta particles can be represented by the Greek letter β, beta. You may wonder how an electron got into the nucleus of an atom. In fact, electrons *do not* exist by themselves within the nucleus. However, in beta decay a neutron can split into two parts, becoming an electron and a proton. The electron is ejected and the proton stays behind in the nucleus.

$^{14}_{6}$C nucleus
(unstable)

$^{14}_{7}$N nucleus
(stable)

Beta particle

Gamma ray

*A neutron splits into a proton and an electron.*

Removal of an electron from the nucleus of an atom through this unique process leaves that atom with one less neutron, one more proton, and a new identity. This is why beta decay causes you to move one space forward in the Nuclear Quest game. Beta decay can also result in the release of gamma rays.

## Example

### Plutonium-241

The radioactive isotope plutonium-241, $^{241}_{94}$Pu, decays by emitting a beta particle. What isotope is formed?

### Solution

The atomic number of plutonium is 94, so it has 94 protons. In beta decay, a neutron becomes a proton. This increases the number of protons in the nucleus by one, but decreases the number of neutrons by one. The atomic mass stays the same because protons and neutrons have the same mass. The atomic number of the new atom is 95, and the new mass is still 241 amu. Element 95 is americium, Am.

$$^{241}_{94}\text{Pu} \rightarrow \beta + {^{241}_{95}}\text{Am}$$

The new isotope is americium-241.

### HALF-LIFE

Carbon-14 has a **half-life** of 5730 years. This means that it will take 5730 years for half of a carbon-14 sample to decay to nitrogen-14. It will take another 5730 years for half of the remaining carbon-14 to decay, and so on. By measuring the remaining amount of carbon-14 compared to the amount of stable carbon, the age of the sample can be determined.

Rod Beverley/Shutterstock

## RADIATION

The particles and rays that are emitted during nuclear reactions are forms of **radiation.** So alpha and beta particles are both forms of radiation. Gamma rays are often emitted during radioactive events. **Gamma rays** are a kind of radiation similar to light, microwaves, and x-rays except they are much higher in energy, so they can be very dangerous. Gamma rays are represented by the Greek letter γ, gamma. When gamma rays are emitted, the identity of the emitting atom does not change.

*A sheet of paper, or human skin, stops alpha particles, and a sheet of aluminum foil stops beta particles. Gamma rays can penetrate several inches of lead before stopping.*

Naturally occurring radiation does exist, with low levels coming from sources in the Earth, as well as from the Sun and beyond. Humans have learned how to use certain types of radiation for a variety of scientific purposes. However, radiation from nuclear reactions can be harmful to your health.

Even though radiation can be dangerous, it can also be used to help people. For example, some cancer patients receive radiation treatments to kill cancerous cells so that it kills those cells but doesn't damage the rest of the body. Radiation can also be used to kill bacteria and viruses on food, particularly meat and vegetables. This helps reduce the risk of food poisoning.

## ⟳ Fission and Fusion

The splitting apart of a nucleus is called **fission.** The result is two atoms with smaller nuclei. Fission can occur spontaneously when the nucleus of an atom is unstable. Scientists can also make nuclear fission happen by shooting neutrons at a nucleus. Fission reactions are accompanied by radiation.

© moodboard/Corbis

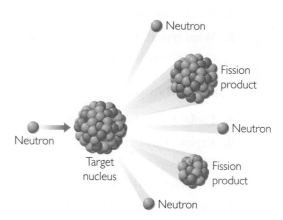

Another type of nuclear reaction is **fusion.** During fusion, nuclei join to form a larger nucleus, resulting in an atom of a different element. Temperatures of around 100,000,000 °C are required before nuclei can successfully fuse together. So the natural conditions on Earth do not support fusion.

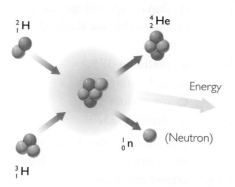

*Two isotopes of hydrogen fuse to become helium.*

**KEY TERMS**

nuclear reaction

radioactive decay

alpha decay

alpha particle

beta decay

beta particle

half-life

radiation

gamma ray

fission

fusion

## LESSON SUMMARY

### What are nuclear reactions?

Radioactive decay, fission, and fusion are all forms of nuclear reactions. When radioactive decay is taking place, particles are emitted from the nucleus. Fission is the process by which the nucleus breaks into smaller nuclei. Fusion is the process by which nuclei combine to make a larger nucleus. Nuclear reactions can involve changes in atomic number, which mean changes in atomic identity. Radiation, such as the emission of alpha particles, beta particles, and gamma rays, accompanies decay and fission. All forms of radiation can cause harm to humans, but some forms are more damaging than others.

### Reading Questions

1. What is a nuclear reaction?

2. What is the difference between alpha decay and beta decay?

3. What type of radiation is most harmful to living things: alpha radiation, beta radiation, or gamma radiation? Why?

### Reason and Apply

4. Explain why the mass of an atom changes when an alpha particle is emitted.

5. Explain why the mass number of an atom does not change when a beta particle is emitted.

6. An alpha particle is not a neutral atom. It has a charge of $+2$. Why is this the case?

7. Suppose each of these isotopes emits a beta particle. Give the isotope name and symbol for the isotope that is produced. Place a checkmark next to the symbol for isotopes that are stable. (Consult the isotope graph on page 66 of Lesson 14: Isotopia.)
   a. potassium-42
   b. iodine-131
   c. iron-52
   d. sodium-24

8. Suppose each of these isotopes emits an alpha particle. Give the isotope name and symbol for the isotope that is produced. Place a checkmark next to the isotopes produced that are stable. (Consult the isotope graph on page 66 of Lesson 14: Isotopia.)
   a. platinum-175
   b. gadolinium-149
   c. americium-241
   d. thorium-232

9. The following items were found at an ancient campsite: an ax with a stone blade and a wooden handle, a clay pot, ashes of a campfire, a jawbone of a deer, and an arrowhead. Which objects could be used to determine the age of the site? Explain.

10. Scientists determine that the wooden beams of a sunken ship have 67% of the concentration of carbon-14 that is found in the leaves of a tree alive today. Find the age of the ship and explain how you determined it. List three possible sources of error in your determination.

11. What fraction of the original $^{14}_{6}C$ would be in a wooden ax handle that was 17,190 years old?

    (A) $\dfrac{1}{4}$     (B) $\dfrac{3}{4}$     (C) $\dfrac{1}{8}$     (D) $\dfrac{1}{2}$

12. Suppose $^{197}_{119}Pt$ undergoes beta decay. The isotope that forms then also undergoes beta decay. What is the final product?

    (A) $^{195}_{120}Ir$     (B) $^{197}_{118}Au$     (C) $^{195}_{119}Hg$     (D) $^{197}_{121}Tl$

# Formation of Elements

## THINK ABOUT IT

The element gold is essentially a collection of identical atoms. Gold has only one stable isotope. So, every gold atom has 79 protons, 79 electrons, and 118 neutrons. Perhaps it is possible to make gold atoms by using nuclear processes to add or subtract protons, neutrons, and electrons. To understand if this is possible, it is useful to consider what processes lead to the creation of new elements.

### How are new elements formed?

*To answer this question, you will explore*

➲ Making New Elements

➲ Writing Nuclear Equations

➲ Nuclear Chain Reactions

## EXPLORING THE TOPIC

## ➲ Making New Elements

Some nuclear reactions occur without intervention on Earth when radioactive isotopes decay. It is much more difficult to change the nucleus of a stable atom. However, there are places with the right conditions and enough available energy to change these nuclei. Nuclear changes take place continuously in the Sun and other stars. And on Earth, scientists can carry out certain nuclear reactions in specially designed facilities like nuclear reactors or particle accelerators.

New elements can be created through nuclear fission or nuclear fusion. Whether a nucleus is split apart or joined with another nucleus, both processes result in the formation of different elements. And both processes require an energetic push to get them started.

**Big Idea** The only way to change one element into another is to change the number of protons in the nucleus.

### MAKING ELEMENTS IN THE STARS

The creation of new elements through nuclear chemistry is called *nucleosynthesis*. Most nucleosynthesis takes place far from Earth, inside stars.

The Sun is a giant ball of hydrogen, H, and helium, He. The amazing light and heat energy that radiates from the Sun is the result of continuous fusion reactions. However, the Sun is not hot enough to produce elements beyond helium on the periodic table. Higher temperatures are needed for the formation of larger atoms.

**ASTRONOMY CONNECTION**

The Sun turns 600 million tons of hydrogen into helium every second! The energy released as a result of this nuclear fusion provides the brilliant sunlight we see every day and the intense heat that warms our planet, even from 93 million miles away. There is enough hydrogen in the Sun to keep the Sun burning for another 5 billion years.

Stars do not exist forever. Some stars collapse in on themselves after millions of years. Other stars explode and become what astronomers call *supernovas.* Where does gold come from? Small stars like our own can produce only helium. Larger, hotter stars can produce heavier elements up to iron. Only a supernova explosion can produce elements heavier than iron, including gold.

## Writing Nuclear Equations

Nuclear reactions can be expressed as **nuclear equations.** You can use nuclear equations to track the number of neutrons and protons in the atoms, along with the identity of the products. Nuclear equations use isotope symbols to represent each particle. The alpha decay of an isotope of barium-140 to form xenon-136 can be written as a nuclear equation using $\alpha$ for the alpha particle:

$$^{140}_{56}\text{Ba} \rightarrow \alpha + ^{136}_{54}\text{Xe}$$

An alpha particle is the same as a helium nucleus, so it can also be written this way:

$$^{140}_{56}\text{Ba} \rightarrow ^{4}_{2}\text{He} + ^{136}_{54}\text{Xe}$$

In the second equation, you can track the protons and the mass that is removed from the barium nucleus. Notice that the equation is balanced numerically. The mass numbers on both sides are equal: $140 = 4 + 136$. The numbers of protons are also equal on both sides: $56 = 2 + 54$. In radioactive decay, the starting isotope is called the **parent isotope** and the resulting isotope is called the **daughter isotope.** In this equation, $^{140}_{56}\text{Ba}$ is the parent isotope and $^{136}_{54}\text{Xe}$ is the daughter isotope.

Beta decay can be shown in two ways as well, because a beta particle is the same as an electron and is sometimes shown as $e^-$, or in isotope form as $^{0}_{-1}e$.

$$^{140}_{56}\text{Ba} \rightarrow \beta + ^{140}_{57}\text{La} \qquad \text{or} \qquad ^{140}_{56}\text{Ba} \rightarrow ^{0}_{-1}e + ^{140}_{57}\text{La}$$

Recall that the beta particle, $\beta$, comes from a neutron in the nucleus splitting into a proton and an electron. The new element $^{140}_{57}\text{La}$ has a higher atomic number because it has one more proton in its nucleus than barium.

Fusion equations have two starting isotopes coming together to form a new product. When a carbon-12 isotope fuses with a nitrogen-14 isotope to form an isotope of aluminum, the equation looks like this:

$$^{12}_{6}\text{C} + ^{14}_{7}\text{N} \rightarrow ^{26}_{13}\text{Al}$$

Notice once again that the equation is balanced numerically on both sides.

**Example**

### Radium-222

Write the equation representing the alpha decay of radium-222. What is the daughter isotope?

*Solution*

Find radium, Ra, on the periodic table. Its atomic number is 88. Now you can write the isotope symbol for radium, $^{222}_{88}\text{Ra}$.

*(continued)*

An alpha particle is the same as a helium nucleus. Because the alpha particle is emitted, it goes on the right side of the equation. Balancing the equation gives the resulting nucleus, which has atomic number 86, and which is radon, Rn.

$$_{88}^{222}\text{Ra} \rightarrow _{2}^{4}\text{He} + _{86}^{218}\text{Rn}$$

The daughter isotope is radon-218.

## Nuclear Chain Reactions

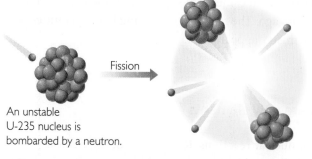

Fission

An unstable U-235 nucleus is bombarded by a neutron.

A barium nucleus, a krypton nucleus, three neutrons, and energy are produced.

Nuclear fission can be provoked by striking nuclei with fast-moving particles, such as neutrons. And, sometimes, one nuclear reaction can lead to others. A nuclear **chain reaction** can take place when the neutrons emitted strike surrounding nuclei, causing them to split apart as well.

For example, natural uranium consists of three isotopes: uranium-238, uranium-235, and uranium-234. All of these uranium isotopes are mildly radioactive but are still fairly stable. Uranium isotopes can be provoked into undergoing fission reactions. Scientists do this by firing a neutron at a uranium atom. This results in the formation of two smaller isotopes and the emission of three other neutrons, which in turn cause other uranium isotopes to undergo fission. The result is a chain reaction releasing enormous amounts of energy.

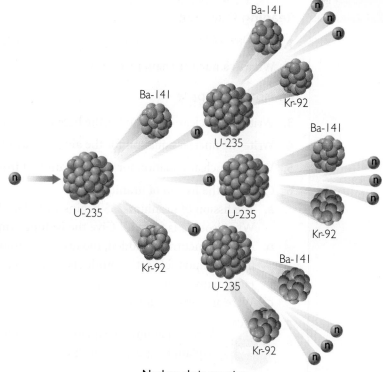

Nuclear chain reaction

Scientists have learned how to control and harness nuclear fission here on Earth. Nuclear fission is used to produce electricity, power submarines, and propel a variety of large ships. Uranium-235 is the isotope most commonly used for nuclear power.

If a nuclear chain reaction is controlled and the energy is released slowly, it can be used to generate electricity. If the energy is released all at once, the result is a nuclear explosion.

## MAKING NEW GOLD

Most of the elements found on Earth are the products of fusion reactions in stars that exploded billions of years ago. The elements have been here for a very, very long time. There has been little change in the amounts of these elements. Small changes to the amounts of each element occur over time due to radioactive decay and fission. The gold we have on Earth came from supernova explosions that may have taken place billions of years ago. The prospect of making large quantities of new gold through nuclear chemistry appears slim.

## LESSON SUMMARY

### How are new elements formed?

**KEY TERMS**

nuclear equation

parent isotope

daughter isotope

chain reaction

Elements are formed through nuclear reactions. Radioactive decay, nuclear fission, and nuclear fusion are all possible sources of new elements. The elements we have on Earth came from fusion reactions in stars and supernova explosions long ago. It is difficult to cause nuclear fusion to occur on Earth because large amounts of energy are required to fuse nuclei. Nuclear fission, on the other hand, has been harnessed as a source of power. The power that is generated by nuclear plants is a result of a controlled fission chain reaction.

## Exercises

### Reading Questions

1. Describe four processes that result in the formation of new elements.

2. What is a nuclear chain reaction?

### Reason and Apply

3. Write the nuclear equation for the beta decay of cerium-141.

4. Write the nuclear equation for the alpha decay of platinum-191.

5. Write a nuclear equation for the formation of iron-54 through fusion.

6. Consider the fission of uranium-235.
   a. The fission of uranium-235 begins with the addition of a neutron to the nucleus. What isotope is formed? Give the isotope symbol.
   b. After the neutron is added, the uranium atom is more unstable and undergoes fission. A possible set of products is krypton-94, barium-139, and 3 neutrons. How many protons are in these products?
   c. Were any protons lost?

7. **PROJECT**
   **www**
   Write a paragraph explaining how a nuclear reactor works. Be sure to explain the purpose of the control rods.

CERN/Science Source

# CHAPTER 3

## A World of Particles

# SUMMARY

### KEY TERMS

atom
atomic theory
model
nucleus
proton
neutron
electron
atomic number
isotope
mass number
average atomic mass
radioactive isotope
nuclear reaction
radioactive decay
alpha decay
alpha particle
beta decay
beta particle
half-life
radiation
gamma ray
fission
fusion
nuclear equation
parent isotope
daughter isotope
chain reaction

# Alchemy Update

Can a copper atom be transformed into a gold atom through nuclear processes?

$^{197}_{79}$Au

Elements are collections of similar atoms. The number of protons in the nucleus determines the identity of an atom. For instance, all atoms of gold have 79 protons. However, atoms of an element may have different numbers of neutrons. The number of neutrons is related to the stability of the atom. Stable gold atoms have 118 neutrons.

Nuclear fission, nuclear fusion, and radioactive decay can result in new elements. But nuclear fusion and nuclear fission are difficult processes to control and often involve large amounts of energy. Nuclear reactions do take place in the stars where atoms are created. So far it is not yet possible or practical to create gold atoms through nuclear reactions here on Earth.

## REVIEW EXERCISES

1. Describe the different processes that might result in a change in atomic identity.

2. Give an example of evidence used by scientists to revise the atomic model.

3. Explain how to use the periodic table to deduce the number of protons, neutrons, and electrons of an atom of a specified element.

4. Write a nuclear equation for the hypothetical fusion of a copper atom with another nucleus to make gold. Can this happen? Why or why not?

## Nuclear Power

**PROJECT**

**WWW**

Research nuclear power. Find out how a nuclear power plant works. Write a report including

- an explanation (with a simple drawing) of how electric power is produced from a nuclear power plant.

- the major benefits and risks of nuclear power.

# CHAPTER **4**

# Moving Electrons

*In this chapter, you will study*

- the systematic arrangement of electrons in an atom

- how ionic compounds are formed

- valence electrons and ionic bonding

- how the periodic table reflects electron arrangement

*When heated, different metal atoms produce flames of different colors. That is what causes the many brilliant colors of a fireworks display.*

While the nucleus of an atom can be difficult to change, the electrons are a different story. Electrons can be moved around within an atom or transferred between atoms. When atoms transfer electrons, they become ions with positive and negative charges, and form ionic compounds. Knowing the arrangement of electrons within an atom can help you to predict which ionic compounds will form.

# Flame Tests

## Purpose

To identify metal atoms in a variety of compounds by using a flame test and to provide evidence for the presence of certain atoms within compounds.

## Materials

- Bunsen burner
- set of tongs
- copper wire
- penny
- set of 11 solutions: sodium carbonate, $Na_2CO_3$; potassium nitrate, $KNO_3$; copper (II) nitrate, $Cu(NO_3)_2$; strontium nitrate, $Sr(NO_3)_2$; potassium chloride, KCl; sodium chloride, NaCl; copper (II) sulfate, $CuSO_4$; strontium chloride, $SrCl_2$; sodium nitrate, $NaNO_3$; copper (II) chloride, $CuCl_2$; potassium sulfate, $K_2SO_4$
- 11 pieces of nichrome wire, each with a loop at one end

## Procedure and Observations

1. For each solution, follow these steps and record your observations.
   - Place the loop end of the nichrome wire into the solution.
   - Place the loop end with the solution on it into the flame.
   - Record the color of the flame in your table.
   - Place the nichrome wire back in the correct solution.

2. Using tongs, place the penny and the copper wire into the flame one at a time and observe the results.

## Analysis

1. Group the substances based on the color of flame produced. What patterns do you notice in the groupings?

2. **Making Sense** Can a flame test be used to identify a metal atom in a compound? Why or why not? What about a nonmetal atom?

3. **If You Finish Early** Copper oxide, CuO, is a black solid. It doesn't look at all like the element copper. What color flame would it produce? Draw a model of copper oxide to explain the color of the flame that it would produce.

*Technicolor Atoms*

# Flame Tests

## THINK ABOUT IT

On the Fourth of July, you can watch colorful fireworks without considering the chemistry behind them. But each color that bursts forth in the sky is associated with a particular chemical compound. For example, any green sparkles you might see are probably due to a compound such as barium sulfate. But what's responsible for the color? Is it the entire barium sulfate compound, or is it one of the atoms in the compound?

**What evidence is there that certain atoms are present in a compound?**

*To answer this question, you will explore*

- Flame Tests
- Evidence for Atoms in Compounds
- Excited Electrons

## EXPLORING THE TOPIC

### Flame Tests

Fireworks originated in China about 2000 years ago. The legend surrounding the discovery suggests that fireworks were discovered by a Chinese alchemist who mixed charcoal, sulfur, and potassium nitrate and accidentally produced a colorful gunpowder. Since then, the noise and the bright colors associated with fireworks have been used in celebrations all over the world.

In a chemistry lab, it is relatively easy to obtain the colorful flames associated with fireworks. You can do so by heating certain compounds in a flame, such as the flames of a Bunsen burner. The flame colors produced by heating four different compounds are shown here.

*Different compounds produce different colors in a flame test. The metal in the compound may determine color.*

## Compound Colors

| Red | Orange | Yellow-orange | Green | Blue-green | Pink-lilac |
|---|---|---|---|---|---|
| lithium nitrate $LiNO_3$ | calcium nitrate $Ca(NO_3)_2$ | sodium nitrate $NaNO_3$ | barium nitrate $Ba(NO_3)_2$ | copper nitrate $Cu(NO_3)_2$ | potassium nitrate $KNO_3$ |
| lithium chloride $LiCl$ | calcium chloride $CaCl_2$ | sodium chloride $NaCl$ | barium chloride $BaCl_2$ | copper chloride $CuCl_2$ | potassium chloride $KCl$ |
| lithium sulfate $Li_2SO_4$ | calcium sulfate $CaSO_4$ | sodium sulfate $Na_2SO_4$ | barium sulfate $BaSO_4$ | copper sulfate $CuSO_4$ | potassium sulfate $K_2SO_4$ |
| lithium $Li$ | calcium $Ca$ | sodium $Na$ | barium $Ba$ | copper wire $Cu$ | potassium $K$ |

The colors for several more compounds are provided in the table. Take a moment to examine the data. What patterns do you notice?

Notice that each metal atom, Li, Ca, Na, and so on, is associated with a specific flame color. Lithium compounds all make a red flame, while the barium compounds all make a green flame. The nonmetal atoms in these compounds do not seem to affect the color of the compound. So $CaSO_4$ and $CaCl_2$ and $Ca(NO_3)_2$ all produce the same color flame, but $CaCl_2$ and $NaCl$ and $CuCl_2$ do not. The metal atom must somehow be responsible for the color of the flame.

## Evidence for Atoms in Compounds

Chemists have found these flame color patterns to be quite helpful. A **flame test** can be used to quickly confirm the presence of certain metal atoms in an unknown sample. So, a potassium compound can quickly be distinguished from a calcium compound by heating samples of the compounds.

Also, the color patterns associated with the flame tests clearly indicate whether certain metal atoms are present within the compounds. For example, copper metal turns the flame blue-green and so does copper chloride. This is evidence that there are copper atoms in the copper chloride.

## Excited Electrons

Heating metal compounds in a Bunsen burner does not destroy the atoms in them or create new elements. This is because the temperature of a flame is not high enough to change the nuclei of metal atoms. However, the heat does have an impact on the electrons of these metal atoms.

ASTRONOMY
CONNECTION

Each element has a light "signature." Astronomers can deduce the composition of a star from the colors of light that it emits. This photo shows the light emitted by the star Arcturus, one of the brightest stars in the night sky. When the light is separated by a prism in this way, you can see that a particular shade of blue is missing because it has been absorbed by iron in the star's corona.

Dr Juerg Alean/Science Source

Sodium atom, Na

1. "Excited" electron moves away from nucleus.

Light energy

2. When electron moves back, it releases light energy.

Electrons are located at different average distances from the nucleus of an atom. When heated, some electrons of metal atoms get "excited" and move to distances farther from the nucleus. This move is only temporary. When the electrons move back to their original distance from the nucleus, they release energy in the form of colored light. This is what you see during a flame test.

The fact that it is relatively easy to affect the electrons in atoms opens up new possibilities for exploration. Perhaps the solution to creating substances with the properties of gold rests in altering the electrons in some way. The nuclei of atoms cannot be changed easily, so it is time to explore ways in which the electrons in atoms can be changed. The rest of this unit explores the role of the electrons in the chemistry of atoms and compounds.

**Example**

### Flame Colors

Which of these compounds will give similar flame colors when heated?

$NaCl$ $\quad$ $CaCl_2$ $\quad$ $SrCl_2$ $\quad$ $Sr(NO_3)_2$ $\quad$ $Cu(NO_3)_2$

*Solution*
The color of the flame depends on the metal. The only two compounds with the same metal are strontium chloride, $SrCl_2$, and strontium nitrate, $Sr(NO_3)_2$.

**KEY TERM**

flame test

LESSON SUMMARY

#### What evidence is there that certain atoms are present in a compound?

Many metal atoms produce a characteristic colored flame when compounds containing those atoms are heated. The colors are a result of light energy produced when excited electrons return to their original distance from the nucleus in the metal atoms. Flame tests provide evidence that certain atoms are present in compounds and that they are not destroyed during chemical changes. The flame test also demonstrates that it is easier to add or remove electrons in atoms, which can be done with a small amount of energy, than to alter the protons and neutrons in the nucleus, which requires a great deal of energy.

**Reading Questions**

1. How did the flame test provide evidence that specific atoms are present in compounds?

2. Explain what is responsible for the colors during a flame test.

**Reason and Apply**

3. PROJECT
    Find out why fireworks are so colorful. What substances are used to produce the colors?

4. Predict the color of the flame for the compound sodium hydroxide, NaOH. Explain your reasoning.

5. Imagine that you were in charge of creating a red and purple fireworks display. Name two combinations of compounds you could use.

6. What evidence do you have from flame tests that copper is responsible for producing a flame with a blue-green color?

7. Does nitrate produce a colored flame? Explain your thinking.

8. Would it matter whether you did a flame test with sodium chloride, NaCl, in solid form or sodium chloride as an aqueous solution? Explain.

9. What flame colors would be produced by these compounds? Explain your choices.
   a. $Na_2CO_3$  b. $Ba(OH)_2$  c. $KOH$  d. $K_2CO_3$  e. $BaO$

10. If two chemical samples both produce an orange flame upon testing, which statement is true?
    (A) The two samples contain identical compounds.
    (B) The samples both contain chlorine atoms.
    (C) The samples both contain calcium atoms.
    (D) The samples both contain potassium atoms.
    (E) The samples contain different compounds.

11. What evidence supports the claim that chloride, $Cl^-$, does not cause the flame to have a color?
    (A) Lithium chloride, LiCl, and sodium chloride, NaCl, have different colors.
    (B) Ammonium chloride, $NH_4Cl$, does not cause the flame to have a color.
    (C) Sodium chloride, NaCl, and sodium nitrate, $NaNO_3$, both produce flames with a yellow-orange color.
    (D) All of the above.

*Life on the Edge*

# Valence and Core Electrons

## THINK ABOUT IT

Lithium, Li, and sodium, Na, are both located in Group 1A on the periodic table because they have very similar properties. Both are soft, silvery metals that react with water. Both form compounds with chlorine in a 1:1 ratio—lithium chloride, LiCl, and sodium chloride, NaCl. What do lithium atoms have in common with sodium atoms that make them behave similarly?

---

**Why do elements in the same group in the periodic table have similar properties?**

---

*To answer this question, you will explore*

⮕ Electron Shells

⮕ Patterns in Atomic Structure

⮕ Valence and Core Electrons

## EXPLORING THE TOPIC

### ⮕ Electron Shells

Recall that elements that are grouped together in columns on the periodic table share similar properties.

Consider lithium, Li, and sodium, Na, from Group 1A on the periodic table. You can get the following information from these two element squares on the periodic table:

<table>
<tr><td align="center">3<br>**Li**<br>Lithium<br>6.941</td><td align="center">11<br>**Na**<br>Sodium<br>22.99</td></tr>
</table>

3 protons, 3 electrons<br>Average atomic mass = 6.941 amu

11 protons, 11 electrons<br>Average atomic mass = 22.99 amu

This basic information doesn't provide any evidence of similarities between atoms of lithium and atoms of sodium. However, examining the structures of these atoms in more detail reveals some patterns.

Lithium, Li

3p

Sodium, Na

11p

Although electron configurations are easier to draw in two dimensions, keep in mind that atoms are not flat. Electron shells are really surfaces of spheres. Each model is useful in understanding different aspects of the atom.

In these models, the electrons are shown orbiting in concentric circles around the nucleus. Chemists call these circles *electron shells,* and these atomic models are called *shell models.*

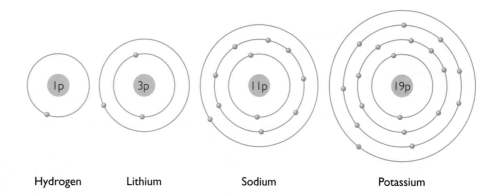

Hydrogen   Lithium   Sodium   Potassium

Lithium has a total of three electrons in two shells. Sodium, on the other hand, has a total of 11 electrons in three shells. But the atoms of both elements do have one thing in common: a single electron in their outer shells. If this feature is responsible for the similar properties of lithium and sodium, you would expect the atoms of the other elements in the same group to also have a single electron in their outer shells.

The atoms of two more Group 1A elements are shown in the illustration. Compare the models of hydrogen and potassium with those of lithium and sodium. The pattern holds: Each of the Group 1A elements has one electron in the outer shell of its atoms. This pattern also extends to other Group 1A elements not shown here, such as rubidium and cesium.

## ⮌ Patterns in Atomic Structure

Examining the shell models of elements beyond lithium on the periodic table reveals further patterns in the outermost electrons. Shell models for the first 14 elements are shown in this illustration. A repeating pattern emerges from the diagram.

H   He   Li   Be   B   C   N   O   F   Ne   Na   Mg   Al   Si

1 electron in outer shell
2 electrons in outer shell
3 electrons in outer shell

*Comparing electron arrangements reveals patterns in atomic structure.*

Starting with lithium, notice how the number of electrons in the outer shell repeats in a regular pattern, going from one to eight and then starting over again.

When these electron shell models are organized according to the periodic table, even more is revealed. Examine this table. The atoms in the first row, or Period 1, have only one shell. The atoms in Period 2 have two shells. The atoms in each new period of the periodic table have an additional electron shell.

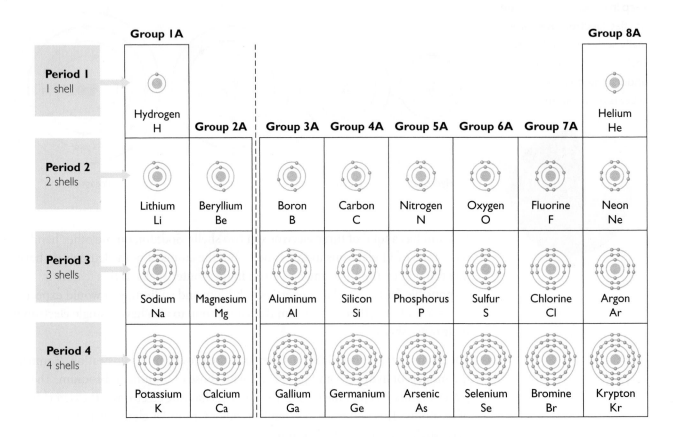

It turns out that for main group elements, Groups 1A through 8A, everything you need to know about the placement of an atom's electrons can be decoded from the periodic table.

**Big Idea** | The arrangement of elements on the periodic table reflects the arrangement of electrons in the atoms of those elements.

In the first column of the table, every element has one electron in its outermost electron shell. In the second column, each has two electrons in the outermost shell, and so on. In the eighth column, each element has eight electrons in its outermost shell. This feature turns out to be the key to many of the properties of the atoms.

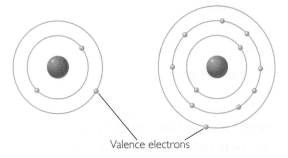

Valence electrons

## Valence and Core Electrons

The outermost shell that electrons occupy is very important in chemistry and is referred to as the **valence shell**. The electrons in that shell are called the **valence electrons**. The electrons that are located in all of the inner shells are known as **core electrons**.

For main group elements, the number of electrons in the valence shell corresponds to the group number on the periodic table. So, every Group 4A element has four electrons in its outermost shell, while every Group 8A element has eight electrons in its outermost shell.

Elements with the same number of electrons in their outermost shells have similar properties. In other words, elements with the same number of valence electrons have similar properties.

## Example

### Sulfur Atoms

Find the element sulfur, S, on the periodic table.

- **a.** What group is sulfur in?
- **b.** How many valence electrons does a sulfur atom have?
- **c.** Explain why selenium, Se, has properties similar to those of sulfur.

### Solution

Sulfur is not a metal, so it is located toward the right side of the periodic table.

- **a.** Sulfur is in Group 6A of the periodic table.
- **b.** The number of valence electrons is the same as the group number, so sulfur has six valence electrons.
- **c.** Selenium, Se, is also in Group 6A. Even though it has many more total electrons than sulfur has, selenium also has six valence electrons. This is why sulfur and selenium have similar properties.

## LESSON SUMMARY

### Why do elements in the same group in the periodic table have similar properties?

Elements that are grouped together in columns on the periodic table share similar properties. For main group elements, atoms in the same group have the same number of electrons in their outermost shells. The outermost shell is called the valence shell, and the electrons in that shell are called valence electrons. Electrons in the inner shells are called core electrons. The number of valence electrons in an atom of a main group element corresponds to the element's group number. So, the periodic table is a direct reflection of the atomic structures of the atoms.

**KEY TERMS**

valence shell

valence electron

core electron

## Exercises

### Reading Questions

1. How can you determine the arrangement of an element's electrons from its position on the periodic table?

2. If you know what group an element is in, what can you predict about its electron arrangement?

*(continued)*

**Reason and Apply**

3. What do beryllium, Be, magnesium, Mg, and calcium, Ca, all have in common?

4. Draw a shell model for boron, B. Identify the core and valence electrons.

5. For main group elements, how does the number of core electrons vary across a period?

6. Consider the elements carbon, C, and silicon, Si.
   a. How many electrons does an atom of each of these elements have?
   b. Draw shell models for atoms of each of these elements.
   c. How many valence electrons do atoms of each of these elements have?
   d. How many core electrons do atoms of each of these elements have?
   e. Why are the properties of carbon and silicon similar?

7. Provide the following information for element number 17. For each answer, explain how you know.
   a. the element's name, symbol, and group number
   b. the number of protons in the nucleus
   c. the number of neutrons in the nucleus
   d. the number of electrons in a neutral atom of the element
   e. the number of electrons that are valence electrons in a neutral atom
   f. the number of electrons that are core electrons in a neutral atom
   g. the names of three other elements with the same number of valence electrons

8. Answer Exercise 7, parts a through g, for element number 50.

9. If an element is located in Group 4A of the periodic table, what can you conclude about atoms of this element?
   (A) It has four valence electrons.
   (B) It has four electron shells.
   (C) It has four core electrons.
   (D) It has properties similar to carbon.
   (E) Both A and D.

10. What do the elements in Period 3 have in common?
    (A) Their atoms have the same number of valence electrons.
    (B) Their atoms have the same number of electron shells.
    (C) Their atoms have the same properties.
    (D) None of the above.

# Ions

## THINK ABOUT IT

Some atoms are more chemically stable than others. In other words, they don't readily combine with other atoms to form new compounds. The atoms of the noble gas elements are considered the most chemically stable atoms on the periodic table. The most reactive elements are located just before and just after the noble gases, in Groups 1A and 7A. Perhaps the electron arrangements of these atoms are related to the atoms' reactivity.

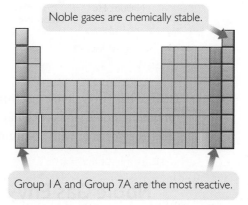

Noble gases are chemically stable.

Group 1A and Group 7A are the most reactive.

**How is chemical stability related to the arrangements of electrons in atoms?**

*To answer this question, you will explore*

⮑ Noble Gases

⮑ Noble Gas Envy

⮑ Patterns in Ion Charges

## EXPLORING THE TOPIC

### ⮑ Noble Gases

Some elements are chemically stable, or rarely react with other elements. Others are reactive and combine readily with other elements to form compounds. What makes some elements chemically stable and others reactive?

---

**INDUSTRY CONNECTION**

Noble gases are so stable that they rarely form compounds. At one time it was thought that they *never* formed compounds with other atoms. But in 1962 a British scientist, Neil Bartlett, created xenon hexafluoroplatinate, a yellow solid, by accident. Chemists have not yet been able to create any compounds with either helium or neon. Shown here is the "light signature" of xenon.

Noble gases are the most chemically stable elements. Shell models for the noble gas elements helium, He, neon, Ne, argon, Ar, and krypton, Kr, are shown below. Take a moment to compare the valence electrons of each.

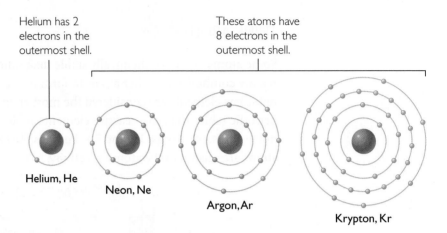

Helium has 2 electrons in the outermost shell.

These atoms have 8 electrons in the outermost shell.

Helium, He

Neon, Ne

Argon, Ar

Krypton, Kr

Helium has two valence electrons, which is the maximum for the first shell. The remaining noble gases each have eight valence electrons. The stability of the noble gases is associated with the number of valence electrons they have.

## ⬭ Noble Gas Envy

Consider two highly reactive elements, sodium, Na, and fluorine, F. The outer shell of a fluorine atom, F, has seven electrons. This is just one short of the eight electrons that neon, Ne, has in its outer shell. The outer shell of a sodium atom, Na, has one electron. This is just one more electron than neon has in total.

11p
12n

9p
10n

Sodium, Na
1 valence electron

Fluorine, F
7 valence electrons

Now examine what happens to these two atoms when they combine to form a compound. Sodium, Na, gives one electron to fluorine, F.

11p
12n

9p
10n

Sodium, Na
1 valence electron

Fluorine, F
7 valence electrons

Now both atoms have an electron arrangement like that of neon, Ne.

Na⁺       F⁻

This sodium atom has a +1 charge because it has lost an electron.

This fluorine atom has a −1 charge because it gains one electron.

Sodium fluoride, NaF

Both ions now have an electron arrangement like the noble gas neon. They form an ionic compound, NaF.

The movement of an electron from one atom to another atom alters the balance of charges on both atoms. Fluorine now has more electrons than protons. Sodium has more protons than electrons. The atoms are now called **ions** because they possess a charge. The sodium ion has a charge of +1 and its symbol is written as Na⁺. The fluorine ion has a charge of −1 and its symbol is written as F⁻. Because Na⁺ and F⁻ have opposite charges, they attract one another. So the movement of an electron from one atom to the other forms a new compound, sodium fluoride, NaF. As a result of this electron transfer, sodium atoms are now bonded to fluorine atoms.

## Example

### Calcium Oxide, CaO

Consider the compound calcium oxide, CaO.

   **a.** Draw shell models for calcium, Ca, and oxygen, O.

   **b.** With arrows, show how electrons can be transferred so that each atom has the same electron arrangement as that of a noble gas.

   **c.** What are the charges on the calcium, Ca, and oxygen, O, ions after electrons have been transferred?

### Solution

   **a.** You can draw a shell model for a neutral atom, an atom that has no charge, using the element number and the element's position on the periodic table. Calcium is in Period 4 and Group 2A, so it has four shells and two valence electrons. Oxygen is in Period 2 and Group 6A, so it has two electron shells and six valence electrons.

Calcium, Ca            Oxygen, O

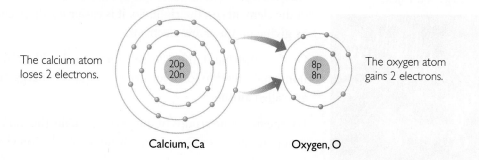

The calcium atom loses 2 electrons.

The oxygen atom gains 2 electrons.

Calcium, Ca         Oxygen, O

<br/>

**Example**  (continued)

b. If calcium transfers two electrons to oxygen, they will both have noble gas electron configurations. The electron arrangement of oxygen now resembles that of neon, Ne. The electron arrangement of calcium now resembles that of argon, Ar.

c. Because calcium has lost two electrons, it now has more protons than electrons, giving it a $+2$ charge. The ion can be written as $Ca^{2+}$. Oxygen has gained 2 electrons, so its charge is $-2$. Its ion can be written as $O^{2-}$.

In the calcium and oxygen example, each ion has an electron arrangement that is identical to that of a noble gas. It is reasonable to assume that there is some advantage to having this sort of electron arrangement.

Noble gases are very stable the way they are, without reacting or exchanging any electrons with other elements. Apparently, other atoms can achieve some of the stability of the noble gases by exchanging electrons and becoming ions.

## ⮕ Patterns in Ion Charges

When atoms lose or gain electrons, they become ions, with negative or positive charges. Below is a portion of the periodic table with the various charges on the ions filled in. In every case, the ion that forms has an electron arrangement identical to the noble gas it is nearest to on the table.

Take a moment to examine the table. What patterns do you notice?

**Ion Charges on Main Group Elements**

| 1A | 2A | 3A | 4A | 5A | 6A | 7A | 8A |
|----|----|----|----|----|----|----|----|
| 1+ | | | | | | | |
| 1 $H^+$ | 2+ | 3+ | 4+ (or 4−) | 3− | 2− | 1− | 2 He |
| 3 $Li^+$ | 4 $Be^{2+}$ | 5 $B^{3+}$ | 6 $C^{4+}$ | 7 $N^{3-}$ | 8 $O^{2-}$ | 9 $F^-$ | 10 Ne |
| 11 $Na^+$ | 12 $Mg^{2+}$ | 13 $Al^{3+}$ | 14 $Si^{4+}$ | 15 $P^{3-}$ | 16 $S^{2-}$ | 17 $Cl^-$ | 18 Ar |
| 19 $K^+$ | 20 $Ca^{2+}$ | 31 $Ga^{3+}$ | 32 $Ge^{4+}$ | 33 $As^{3-}$ | 34 $Se^{2-}$ | 35 $Br^-$ | 36 Kr |

Notice that the elements on the left side of the table, which are mostly metals, tend to form ions with a positive charge. Ions with a positive charge are called **cations.** The valence shells of these atoms have fewer electrons in them than those of the elements on the right. So, it is easier for these elements to form compounds by giving up these few electrons.

On the other hand, the elements on the right side of the table, which are mostly nonmetals, tend to form ions with a negative charge. Ions with a negative charge are called **anions.**

For the first four groups of the periodic table, the ion charge is the same as the group number. For Groups 5A through 7A, the ion charge is negative and goes from $-3$ to $-1$.

**CONSUMER CONNECTION**

Neon signs contain neon gas. When electricity is run through the gas, the electrons get excited. When they come back down to their normal energy level, they emit red light. Real neon signs glow red-orange. Signs that glow other colors are usually made of colored glass and filled with argon.

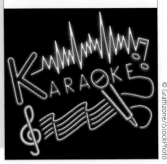

© Grafizone/iStockPhoto

## How is chemical stability related to the arrangements of electrons in atoms?

Noble gases have particularly stable atoms. This is attributed to their electron arrangements. Helium has two electrons, which is the maximum for the first shell. The remaining noble gases all have eight electrons in their outermost shells. Other atoms on the periodic table are more reactive than the noble gases. However, these reactive atoms reach greater chemical stability by gaining or losing valence electrons. Atoms tend to lose or gain electrons to attain an electron arrangement similar to that of a noble gas. Atoms do this by combining with other atoms to form new compounds. When an atom loses or gains electrons, it becomes an ion, with a charge. Ions with a positive charge are called cations. Ions with a negative charge are called anions.

**KEY TERMS**

ion

cation

anion

## Exercises

### Reading Questions

1. Explain the difference between an anion and a cation.
2. Explain what is meant by noble gas envy.

### Reason and Apply

3. How many electrons, protons, and neutrons does $Li^+$ have?
4. Give two similarities and two differences between Cl and $Cl^-$.
5. Give two similarities and two differences between Be and $Be^{2+}$.
6. Which noble gas is closest to magnesium, Mg, on the periodic table? What must happen to a magnesium atom for it to have an electron arrangement similar to that of a noble gas?
7. Which noble gas is closest to sulfur, S, on the periodic table? What must happen to a sulfur atom for it to have an electron arrangement similar to that of a noble gas?
8. List four ions that have the same number of electrons as neon, Ne.
9. List four ions that have the same number of electrons as argon, Ar.
10. What charge would an arsenic, As, ion have?
11. What is the symbol of an ion with 22 protons, 24 neutrons, and 18 electrons?
12. When chlorine gains an electron to become a chloride ion with a −1 charge, it ends up with the electron arrangement of argon. Why doesn't it become an argon atom?
13. Explain why the elements on the right side of the periodic table gain electrons instead of losing them.
14. What periodic patterns do you notice for the charges on the ions?
15. Which of these ions have the correct charge? Choose all that apply.
    **a.** $Na^{2+}$  **b.** $Li^+$  **c.** $Al^{4+}$  **d.** $Ca^{2+}$  **e.** $Ga^{3+}$
16. Which of these ions have the same number of electrons as $S^{2-}$? Choose all that apply.
    **a.** $Cl^-$  **b.** $Ca^{2+}$  **c.** $Na^+$  **d.** $O^{2-}$  **e.** $P^{3-}$

*Getting Connected*

# Ionic Compounds

## THINK ABOUT IT

Sodium and chlorine atoms combine to form sodium chloride, NaCl, when sodium transfers a valence electron to chlorine. There are no other compounds that form between sodium and chlorine. For example, $NaCl_4$ and $Na_3Cl$ are not possible compounds. Why is NaCl the compound that forms, rather than $NaCl_4$ or $Na_3Cl$?

**How can valence electrons be used to predict chemical formulas?**

*To answer this question, you will explore*

➲ Ionic Compounds

➲ The Rule of Zero Charge

➲ More Complex Ionic Compounds

## EXPLORING THE TOPIC

## ➲ Ionic Compounds

There are millions of different compounds on the planet. Each compound is a result of a combination of atoms and has a chemical formula. However, some combinations of atoms are simply not possible. Valence electrons are the key to figuring out chemical formulas and to determining which compounds are possible.

### COMPOUNDS BETWEEN METALS AND NONMETALS

Atoms combine with other atoms to achieve the stability of the noble gases. One way that atoms accomplish this is to transfer valence electrons to other atoms. When atoms gain or lose electrons, the resulting atoms are no longer neutral.

Noble gases

Nonmetals

Metals

They become ions with charges. The compounds that form in this way are called **ionic compounds.** Ionic compounds, like the noble gases, are very stable.

Ionic compounds form between metal atoms and nonmetal atoms. The metal atoms, on the left side of the periodic table, tend to give up electrons and become cations, which have positive charges. The nonmetal atoms, on the right side of the periodic table, tend to accept electrons and become anions, which have negative charges. Because the cation and anion have opposite charges, they are strongly attracted to one another.

**Big Idea** Ionic compounds form when valence electrons are transferred between atoms.

## FORMULAS OF SIMPLE IONIC COMPOUNDS

The table below contains data on three simple ionic compounds. These compounds are all composed of a metal atom and a nonmetal atom in a 1:1 ratio. Pay particular attention to the total charge and the number of valence electrons on each of the atoms that combine.

### Simple Ionic Compounds

| Compound | Metal | Cation | Number of valence electrons | Nonmetal | Anion | Number of valence electrons |
|----------|-------|--------|-----------------------------|----------|-------|-----------------------------|
| NaCl | Na | $Na^+$ | 1 | Cl | $Cl^-$ | 7 |
| CaO | Ca | $Ca^{2+}$ | 2 | O | $O^{2-}$ | 6 |
| GaAs | Ga | $Ga^{3+}$ | 3 | As | $As^{3-}$ | 5 |

Note that in each case, the charge on the cation is equal to and opposite the charge on the anion.

## ⟳ The Rule of Zero Charge

You can see a distinct pattern to the ions that form in ionic compounds. Every time a metal atom and a nonmetal atom bond, they form a compound with an overall zero charge. This is known as the **rule of zero charge.**

Sodium chloride  Calcium oxide  Gallium arsenide

Notice how the ions in each compound add up to zero.

The charges on the metal cations and nonmetal anions add up to zero. The rule of zero charge is useful in predicting the chemical formulas for ionic compounds, which form between a metal and a nonmetal.

## PREDICTING IONIC BONDING USING THE PERIODIC TABLE

You may have noticed that the atoms that combine to form ionic compounds come from very predictable places on the periodic table. Sodium, with one valence electron, can combine with chlorine, fluorine, or bromine, each with seven valence electrons. In fact, any Group 1A element can combine with any Group 7A element to form an ionic compound.

**CONSUMER CONNECTION**

All ionic compounds form crystals. If you look at grains of table salt with a magnifying glass, you will see that the pieces are shaped like tiny boxes.

© cezars/iStockPhoto

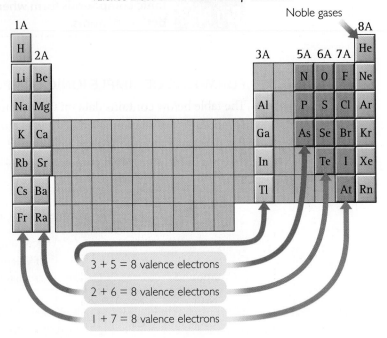

Valence Electrons in Main Group Elements

3 + 5 = 8 valence electrons

2 + 6 = 8 valence electrons

1 + 7 = 8 valence electrons

The drawing illustrates how you can use the periodic table to decide which atoms will combine in a 1:1 ratio to form an ionic compound. There are dozens of possible combinations. For example, an atom in Group 3A can form a compound with an atom in Group 5A.

## ➲ More Complex Ionic Compounds

All of the compounds discussed in this lesson so far have a metal-to-nonmetal ratio of 1:1. However, it is also possible to have compounds with different ratios. For example, one magnesium atom can combine with two chlorine atoms to form magnesium chloride, $MgCl_2$. Before bonding, each magnesium atom has two valence electrons. These two electrons are transferred to two different chlorine atoms, each of which needs one more electron to have the same number of electrons as the noble gas argon, Ar.

Tobias Titz/Getty Images

Magnesium gives up two electrons to form $Mg^{2+}$.

Each chlorine atom accepts one electron to form $Cl^-$.

$Mg^{2+} + Cl^- + Cl^-$ produces $MgCl_2$ with zero charge.

Notice in this table that the atoms combine in such a way that the resulting compound has a zero charge.

**More Complex Ionic Compounds**

| Compound | Metal | Cations | Nonmetal | Anions |
|----------|-------|---------|----------|--------|
| $MgCl_2$ | Mg | $Mg^{2+}$ | Cl | $Cl^-$ $Cl^-$ |
| $Na_2O$ | Na | $Na^+$ $Na^+$ | O | $O^{2-}$ |
| $Al_2O_3$ | Al | $Al^{3+}$ $Al^{3+}$ | O | $O^{2-}$ $O^{2-}$ $O^{2-}$ |

Note that the charges on cations are equal to and opposite the charges on anions. For example, for aluminum oxide, $+3 + 3 - 2 - 2 - 2 = 0$. The total charge on the five atoms in the compound adds up to zero.

You can see that sometimes more than one atom of a particular element must combine to result in a compound with a zero charge. So, ionic compounds in ratios of 1:2, 2:1, or even 2:3 may result.

**Example**

### Sodium Sulfide

The chemical formula for sodium sulfide is $Na_2S$.

   **a.** What is the charge on each sodium ion, Na? On each sulfur ion, S?

   **b.** Show that the charges on the ions add up to zero.

   **c.** Is $Na_2Cl$ a possible ionic compound? Why or why not?

*Solution*

   **a.** Sodium, Na, is in Group 1A and has one valence electron. Sodium atoms give up one electron to have a charge of $+1$. Sulfur, S, is in Group 6A and has six valence electrons. Sulfur atoms gain two electrons to have a charge of $-2$.

   **b.** Check the charge on the compound:

Two sodium cations, each with a $+1$ charge

One sulfur anion with a $-2$ charge

$$2(+1) \quad + \quad (-2)$$

$$(+2) + (-2) = 0$$

The charges on the ions add up to zero, as they should.

   **c.** Each sodium atom has one valence electron and each chlorine atom has seven valence electrons. If two sodium atoms gave up electrons to one chlorine atom, the chlorine atom would have nine valence electrons, which is not a stable number resembling a noble gas. So $Na_2Cl$ is not a compound that is likely to form. Also, such a compound would not have a neutral charge.

## How can valence electrons be used to predict chemical formulas?

Metal atoms and nonmetal atoms combine to form ionic compounds through electron transfer. The metal atoms give up electrons. They form ions with a positive charge, called cations. The nonmetal atoms accept electrons and form ions with a negative charge, called anions. You can use the positions of the elements on the periodic table to predict the ionic compounds they can form. When an ionic compound forms, the total charge on the atoms adds up to zero. This is known as the rule of zero charge.

**KEY TERMS**

ionic compound

rule of zero charge

## Exercises

### Reading Questions

1. How can you use valence electrons to predict which ionic compounds will form?
2. How does the rule of zero charge help you predict the formula of an ionic compound?

### Reason and Apply

3. Lithium nitride has the formula $Li_3N$.
   a. What is the charge on the lithium ion?
   b. What is the charge on the nitrogen ion?
   c. Show that the charges on the ions add up to zero.
   d. What is the total number of valence electrons in all the atoms in $Li_3N$?

4. Aluminum arsenide has the formula AlAs.
   a. What is the charge on the aluminum ion?
   b. What is the charge on the arsenic ion?
   c. Show that the charges on the ions add up to zero.
   d. What is the total number of valence electrons in all the atoms in AlAs?

5. For each of these compounds, show that the charges on the ions add up to zero.
   a. KBr      b. CaO      c. $Li_2O$      d. $CaCl_2$      e. $AlCl_3$

6. What are the total numbers of valence electrons in all the atoms in each of these compounds?
   a. KBr      b. CaO      c. $Li_2O$      d. $CaCl_2$      e. $AlCl_3$

7. Explain why the following compounds do not form.
   a. $NaCl_2$      b. CaCl      c. AlO

# Salty Eights

## Purpose

To make and name as many ionic compounds as possible.

  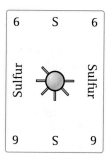

## Materials

- Salty Eights cards

## Instructions

1. Deal eight cards to each player and place the rest in a draw pile. On your turn, you must put down an ionic compound or a noble gas. If you cannot put one down, draw cards until you can. Play until one player has used all of his or her cards.

2. Keep score with a data table like this one. Point values are on the Rules card. List your compounds in the table. Write the metal first and the nonmetal second. If you use more than one atom of an element, don't forget to put the right number in the formula (for example, write $Li_2S$ for lithium sulfide).

| Cation | Anion | Compound formula | Compound name | Point value |
|--------|-------|------------------|---------------|-------------|
|        |       |                  |               |             |
|        |       |                  |               |             |

|  |  |  |  | Total: |
|--|--|--|--|--------|

**Bonus Question** Determine the compound with the highest point value you could make with the cards. Name that compound.

*Salty Eights*

# Formulas for Ionic Compounds

## THINK ABOUT IT

### HISTORY
### CONNECTION

Salt has been a valuable trade item since ancient times. For thousands of years—before refrigeration—humans depended largely on salt as a preservative for foods, especially meat and fish. The wealth of a city, country, or individual could often be linked to the amount of salt possessed or stored. People who are "worth their salt" are considered worthwhile and valuable. Salzburg, Austria, shown here, is named for its many salt mines, which contributed to its great wealth.

There are at least 50 common metal elements. These combine with over 15 nonmetal elements to make a wide variety of ionic compounds. The task of predicting compounds might seem complex given the number of possible ways to combine a metal and a nonmetal. However, if you know how many valence electrons each atom has, you can reliably predict which ionic compounds can be made.

**How can you predict chemical formulas and name ionic compounds?**

*To answer this question, you will explore*

➡ Predicting Chemical Formulas

➡ Naming Ionic Compounds

## EXPLORING THE TOPIC

## ➡ Predicting Chemical Formulas

An ionic compound that is made up of only one kind of metal atom and one kind of nonmetal atom is known as a *salt*. Table salt, NaCl, is the most familiar example. To predict formulas for salt compounds, follow these guidelines:

• Make sure you are combining two different elements—a metal and a nonmetal.

• No matter how many atoms you use, the total number of valence electrons must add up to 8, 16, 24, or another multiple of 8.

• Make sure that the charges on the metal cations and nonmetal anions in your ionic compound add up to zero (rule of zero charge).

This process applies only to the main group elements. Ionic compounds using transition elements are covered in Lesson 23: Alchemy of Paint.

**Example 1**

### Practice Writing Chemical Formulas

Each of these cards represents an atom.

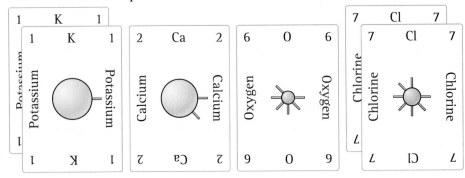

(continued)

**a.** Which atoms are metals and which are nonmetals? How can you tell?

**b.** What ionic compounds can you make from two of these cards?

**c.** What ionic compounds could you make from three cards but only two different elements?

*Solution*

**a.** The atoms from the left side of the periodic table, with fewer valence electrons, are the metals. So, potassium, K, and calcium, Ca, are metal atoms. The atoms from the right side of the periodic table, which have closer to eight valence electrons, are the nonmetals. These include oxygen, O, and chlorine, Cl.

**b.** Look at the number of spokes on each card. The number of spokes represents the number of valence electrons. To create ionic compounds, combine a metal with a nonmetal. The number of spokes must add up to 8. Two ionic compounds are possible with these cards. These are shown here.

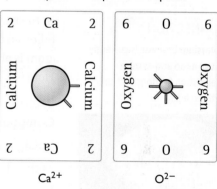

**c.** If two atoms with one valence electron combine with one atom with six valence electrons, the result is eight valence electrons. One possible compound is shown.

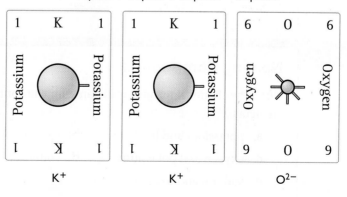

*(continued)*

You can also make $CaCl_2$. Notice that the valence electrons add up to 16, which is a multiple of 8. The charges add up to zero.

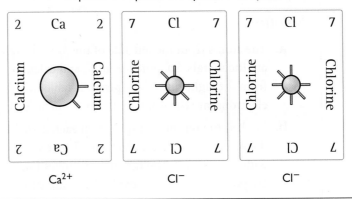

2 spokes + 7 spokes + 7 spokes = 16 spokes

$Ca^{2+}$     $Cl^-$     $Cl^-$

# Naming Ionic Compounds

In the names of ionic compounds, the name of the metal atom comes first, followed by the name of the nonmetal atom. However, the name of the nonmetal atom is altered slightly by replacing the last part of the word with the letters "-ide." Here is a list of ionic compounds with their names. Take a moment to examine them.

| Compound | Name | Compound | Name |
|---|---|---|---|
| NaCl | sodium chloride | $Al_2O_3$ | aluminum oxide |
| $MgF_2$ | magnesium fluoride | GaP | gallium phosphide |
| $Li_2S$ | lithium sulfide | | |

So in ionic compounds, sulfur becomes sulfide, nitrogen becomes nitride, and bromine becomes bromide.

Notice that the chemical name for ionic compounds does not have anything to do with the subscript numbers in the chemical formula. For example, $MgCl_2$ is simply magnesium chloride, *not* magnesium dichloride.

**Example 2**

## Naming Compounds

Write the chemical formula and name for the compound created from each pair of elements.

**a.** potassium and bromine
**b.** oxygen and calcium
**c.** oxygen and potassium
**d.** sodium and chlorine
**e.** sodium and oxygen

*(continued)*

*Solution*

To write the formula for the compound, make sure the atoms are in the correct ratio so that the compound follows the rule of zero charge. To name the compound, write the metal atom first, followed by the nonmetal atom. Change the ending of the nonmetal atom's name to "-ide."

**a.** KBr, potassium bromide

**b.** CaO, calcium oxide

**c.** $K_2O$, potassium oxide

**d.** NaCl, sodium chloride

**e.** $Na_2O$, sodium oxide

## LESSON SUMMARY

### How can you predict chemical formulas and name ionic compounds?

There are several important guidelines to follow in creating ionic compounds. Metal atoms are combined with nonmetal atoms. Next, the total number of valence electrons adds up to eight or a multiple of eight. Finally, the charges on the metal cations and nonmetal anions in ionic compounds add up to zero. When naming ionic compounds made from two different types of elements, the name of the metal atom comes first, followed by the name of the nonmetal atom. In addition, the ending of the name of the nonmetal atom is changed to "-ide."

**Exercises**

### Reading Questions

1. Explain how to use the periodic table to determine the charges on ions.

2. Explain how to use the periodic table to determine the correct formulas for ionic compounds.

### Reason and Apply

3. Which chemical formulas are consistent with the guidelines for creating ionic compounds? Explain your thinking.
   **a.** LiCl  **b.** $LiCl_2$  **c.** $MgCl$  **d.** $MgCl_2$  **e.** $AlCl_3$

4. Give examples of six ionic compounds with a metal-to-nonmetal ratio of 1:1. Specify the total number of valence electrons for each compound. Name each compound.

5. Give examples of three ionic compounds with a metal-to-nonmetal ratio of 2:1. Specify the total number of valence electrons for each compound. Name each compound.

6. Give examples of three ionic compounds with a metal-to-nonmetal ratio of 1:2. Specify the total number of valence electrons for each compound. Name each compound.

*(continued)*

7. Predict the formulas for the ionic compounds that are formed when these metal and nonmetal elements are combined. Name each compound.
   - **a.** Al and Br
   - **b.** Al and S
   - **c.** Al and As
   - **d.** Na and S
   - **e.** Ca and S
   - **f.** Ga and S

8. For each compound, write the cation and anion with the appropriate charge. Then write the chemical formula for each compound.

   Example: sodium fluoride, $Na^+$, $F^-$, $NaF$
   - **a.** magnesium oxide
   - **b.** rubidium bromide
   - **c.** strontium iodide
   - **d.** beryllium fluoride
   - **e.** aluminum chloride
   - **f.** lead sulfide

# Polyatomic Ions

## THINK ABOUT IT

So far, we have considered ionic compounds made of only two elements, a metal and a nonmetal. However, there are some ionic compounds that consist of more than two elements. For example, sodium hydroxide, NaOH, which is commonly found in drain cleaner, consists of one metal element and *two* nonmetal elements. You might wonder about the O and the H. Are they both ions?

### What is a polyatomic ion?

*To answer this question, you will explore*

⮂ Polyatomic Ions

⮂ Predicting Chemical Formulas for Polyatomic Ions

## EXPLORING THE TOPIC

### ⮂ Polyatomic Ions

| Polyatomic ion | Name |
|---|---|
| $OH^-$ | hydroxide |
| $NO_3^-$ | nitrate |
| $CO_3^{2-}$ | carbonate |
| $SO_4^{2-}$ | sulfate |
| $PO_4^{3-}$ | phosphate |
| $BrO_3^-$ | bromate |
| $NH_4^+$ | ammonium |

Some ionic compounds contain ions that consist of two or more elements. These ions are called **polyatomic ions.** In contrast, ions that have only one element are called **monatomic ions.** *Mono* means "one" and *poly* means "many." Calcium sulfate, $CaSO_4$, is an example of a compound made up of a monatomic ion and a polyatomic ion. The calcium ion is monatomic and the sulfate ion is polyatomic.

Monatomic ion

Calcium cation

Polyatomic ion

Sulfate anion

$CaSO_4$
Calcium sulfate

It is important to keep in mind that polyatomic ions are a *group* of atoms that stay together. In calcium sulfate, the entire group of atoms in the sulfate ion has a negative charge.

### NAMES OF POLYATOMIC IONS

Ionic compounds with polyatomic ions follow specific naming rules. Each polyatomic ion has its own name, as shown in the table. When naming a compound, you simply insert the polyatomic ion name at either the beginning or ending of the chemical name. The cation is first and the anion is second. For example, a

**CONSUMER**
**CONNECTION**

Sodium hydroxide, or lye, has many uses. It is used to straighten or curl hair, but if left in too long, it can damage hair and skin.

compound that ends with $NO_3$ is a nitrate. $NaNO_3$ is called sodium *nitrate*, and $Ca(NO_3)_2$ is called calcium *nitrate*.

Sodium cation

Nitrate anion

NaNO₃

Sodium nitrate

| Compound | Chemical formula |
|----------|------------------|
| calcium sulfate | $CaSO_4$ |
| magnesium hydroxide | $Mg(OH)_2$ |
| sodium nitrate | $NaNO_3$ |
| calcium nitrate | $Ca(NO_3)_2$ |
| ammonium carbonate | $(NH_4)_2CO_3$ |

**Important to Know** A polyatomic ion is treated as a unit. For example, the formula for calcium nitrate is written as $Ca(NO_3)_2$ with parentheses around $NO_3$. Notice that it is not written as $CaN_2O_6$.

## Predicting Chemical Formulas for Polyatomic Ions

Several chemical formulas with polyatomic ions are listed in the table. Notice that in some cases there are parentheses around the polyatomic ion, with a subscript after the second parenthesis. Why are the formulas written this way? How do you determine the subscript?

Recall that you can use the rule of zero charge to determine the chemical formulas associated with ionic compounds. For example, a single calcium ion combines with a single sulfate ion because the charges add up to zero:

$$Ca^{2+} + SO_4^{2-} \text{ becomes } CaSO_4, \text{ calcium sulfate}$$

However, like simple ionic compounds, compounds with polyatomic ions do not always combine in a 1:1 ratio. When there is more than one of the same polyatomic ion in a formula, the ion is enclosed in parentheses and a subscript number indicates how many ions are in the compound. For example, a single calcium ion combines with *two* nitrate ions:

$$Ca^{2+} + NO_3^- + NO_3^- \text{ becomes } Ca(NO_3)_2, \text{ calcium nitrate}$$

Below are two more examples of ionic compounds containing polyatomic ions. Each of these compounds consists of three ions. One compound has monatomic and polyatomic ions, while the other has only polyatomic ions. The rule of zero charge applies to both.

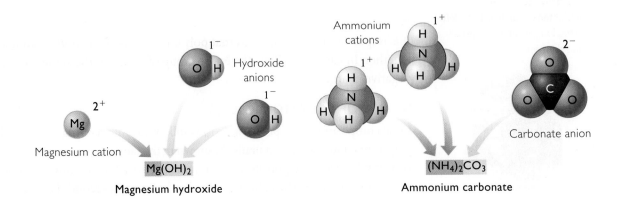

Hydroxide anions

Ammonium cations

Carbonate anion

Magnesium cation

Mg(OH)₂

Magnesium hydroxide

(NH₄)₂CO₃

Ammonium carbonate

## LESSON SUMMARY

### What is a polyatomic ion?

Some ions are composed of more than one atom. These are called polyatomic ions. The entire cluster of atoms shares the charge on a polyatomic ion. If there are two or more of the same polyatomic ion in a chemical formula, parentheses are placed around that ion and a subscript number indicates how many ions are present. To determine the formula of a compound with polyatomic ions, you can use the rule of zero charge. Each polyatomic ion has a unique name that is used when naming the compound it is in.

**KEY TERMS**

polyatomic ion

monatomic ion

## Exercises

### Reading Questions

1. What is a polyatomic ion?
2. How can you tell from a chemical formula if there is a polyatomic ion in a compound?

### Reason and Apply

3. Write the name for each ionic compound listed here.
   a. $NH_4Cl$
   b. $K_2SO_4$
   c. $Al(OH)_3$
   d. $MgCO_3$

4. Write the chemical formula for each compound listed here.
   a. lithium sulfate
   b. potassium hydroxide
   c. magnesium nitrate
   d. ammonium sulfate

5. Sodium cyanide, NaCN, contains a cyanide ion. What is the charge on the cyanide ion?

6. Calcium phosphate, $Ca_3(PO_4)_2$, contains phosphate ions. What is the charge on a phosphate ion?

7. Which chemical formula does not represent a possible compound with sulfate, $SO_4^{2-}$? Explain your answer.
   a. $Na_2SO_4$     b. $KSO_4$     c. $Al_2(SO_4)_3$     d. $CaSO_4$

*Alchemy of Paint*

# Transition Metal Chemistry

## THINK ABOUT IT

For thousands of years, human beings have expressed themselves through painting. Over time, people have discovered pigments with a wide variety of brilliant colors. Almost all paint pigments contain a transition metal cation.

**What types of compounds are made from transition metals?**

*To answer this question, you will explore*

⊃ Transition Metal Compounds

⊃ Charges on Transition Metal Cations

## EXPLORING THE TOPIC

### ⊃ Transition Metal Compounds

The transition metals are named for their location in the middle of the periodic table. Many of the metals that we use in our daily life, such as copper, iron, nickel, silver, and gold, are located in this part of the periodic table.

Transition metals

Transition metal compounds tend to be brightly colored. Hence, they are commonly found as *pigments* used to color paints. For example, red ochre is a pigment made from iron (III) oxide, $Fe_2O_3$. It is thought to be the first pigment ever used by human beings for creating art.

The ancient alchemists worked with the transition metals a great deal, mostly because these metals were closer in their properties to gold than to other metals. As a result, alchemists occasionally discovered paint pigments while trying to create gold.

The table on the next page lists various paint pigments that you can buy at an art store. Examine the data. Pay particular attention to the different charges on the transition metal cations.

Notice that the chemical names are a bit unusual. In the middle of each chemical name for a transition metal compound is a Roman numeral: I, II, III, or IV (meaning 1, 2, 3, or 4). This Roman numeral indicates what the charge is on the transition metal cation. Thus, cobalt (II) oxide has a +2 charge on the cobalt ion, and manganese (IV) carbonate has a +4 charge on the manganese ion.

**HISTORY CONNECTION**

Ionic compounds containing transition metals were used to create ancient cave paintings. This cave painting is from Lascaux, France, and is approximately 17,000 years old.

© The Gallery Collection/Corbis

**Some Pigments Containing Ionic Compounds with Transition Metals**

| Color | Pigment name | Chemical name | Chemical formula | Cation | Anion |
|---|---|---|---|---|---|
| blue | cobalt blue | cobalt (II) oxide | $CoO$ | $Co^{2+}$ | $O^{2-}$ |
| earth tone | red ochre | iron (III) oxide | $Fe_2O_3$ | $Fe^{3+}$ | $O^{2-}$ |
| dark green | viridian | chromium (III) oxide | $Cr_2O_3$ | $Cr^{3+}$ | $O^{2-}$ |
| brown | umber | manganese (IV) dioxide | $MnO_2$ | $Mn^{4+}$ | $O^{2-}$ |
| blue-green | malachite | copper (II) carbonate | $CuCO_3$ | $Cu^{2+}$ | $CO_3^{2-}$ |
| white | titanium white | titanium (IV) dioxide | $TiO_2$ | $Ti^{4+}$ | $O^{2-}$ |
| red | cuprite | copper (I) oxide | $Cu_2O$ | $Cu^+$ | $O^{2-}$ |

**FINE ART CONNECTION**

Some paint pigments have more complicated chemical compositions than the compounds shown in the table. For example, the pigment Egyptian blue, shown below, is calcium copper silicate, with the formula $CaCuSi_4O_{10}$. There are three different cations and one type of anion.

Note that copper, Cu, has a charge of $+1$ in copper (I) oxide, $Cu_2O$, and a charge of $+2$ in copper (II) carbonate, $CuCO_3$.

There is an older naming system for the transition metal ionic compounds. In this system, $Fe_2O_3$ and $FeO$ were called ferric oxide and ferrous oxide. The "-ic" ending indicated the higher ion charge and the "-ous" ending referred to the lower ion charge. So $Co^{2+}$ was cobaltous and $Co^{3+}$ was cobaltic.

## ⮞ Charges on Transition Metal Cations

Recall that you can determine the charges on the main group metals and nonmetals from their positions on the periodic table. For example, all of the alkali metals, Group 1A, form cations with $+1$ charges. All of the halogens, Group 7A, form anions with $-1$ charges. However, you cannot simply determine the charge of a transition metal ion from its location on the periodic table. Instead, it is necessary to use chemical formulas to determine the charge of the ion.

To determine the charges on transition metal cations, you use the charges on ions that you do know and apply the rule of zero charge. For example, iron combines with oxygen to form both $Fe_2O_3$ and $FeO$. Oxygen is a main group anion, so you can use the periodic table to determine that oxygen atoms form ions with a $-2$ charge. The rule of zero charge states that the charges on the ions in the compound should add up to zero. Working backward, you can determine the charge on the iron cations in each compound.

The charges on the three oxygen anions add up to $-6$. Each iron cation must have a charge of $+3$.

The charge on the single oxygen anion is $-2$. The iron cation must have a charge of $+2$.

**Important to Know** Unlike the main group metals, most transition metals can form several ions with different charges.

These two different iron compounds are formally named iron (III) oxide and iron (II) oxide.

## Transition Metal Compounds

Determine the charge on the transition metal cation in each of the compounds given. Then name the compound.

**a.** $Ag_2S$      **b.** $Fe(NO_3)_3$

*Solution*

You can determine the charge on each transition metal cation from the charges on the anions using the rule of zero charge.

**a.** Sulfur anions have a charge of $-2$. So each silver cation must have a charge of $+1$, $Ag^+$. The compound name is silver (I) sulfide.

**b.** The polyatomic ion is nitrate with a $-1$ charge. There are three nitrate ions, so the iron cation must have a charge of $+3$, $Fe^{3+}$. The compound name is iron (III) nitrate.

### CONSUMER CONNECTION

Transition metal ions are responsible for colors in different-colored gem stones. For example, chromium ions, $Cr^{3+}$, are present in red ruby, green emerald, and pink topaz. Iron ions, $Fe^{3+}$, are present in citrine and yellow sapphire, while $Fe^{2+}$ ions make sapphires blue. The transition metals in crystals absorb certain colors of light while allowing other colors to pass through and be seen.

© buckaroohh/iStockPhoto

## LESSON SUMMARY

### What types of compounds are made from transition metals?

Transition metals bond with nonmetal atoms to form ionic compounds. Unlike the main group atoms, most transition metals can have more than one ion charge. The best way to determine the charge on a transition metal cation is to work backward from the anion, whose charge is known. When naming ionic compounds, a Roman numeral is used to indicate the charge on the transition metal cation. Colorful paint pigments are frequently composed of transition metal compounds.

## Reading Questions

1. What does the Roman numeral in a chemical name indicate?
2. Explain how you determine the charge on a transition metal cation from the chemical formula.

## Reason and Apply

3. Determine the charge on the transition metal cation in each of the compounds listed. Then name each compound.
   - **a.** $HgS$
   - **b.** $CuCO_3$
   - **c.** $NiCl_2$
   - **d.** $Co(NO_3)_3$
   - **e.** $Cu(OH)_2$
   - **f.** $FeSO_4$

4. Write the cation and anion in each compound, then determine the correct chemical formula.
   - **a.** copper (II) sulfide
   - **b.** nickel (II) nitrate
   - **c.** iron (II) carbonate
   - **d.** cobalt (II) sulfate
   - **e.** iron (III) carbonate
   - **f.** chromium (VI) oxide

5. Cobalt violet is a paint pigment discovered in 1859. If the cation for this compound is $Co^{2+}$ and the anion is $PO_4^{3-}$, what is the chemical formula?

*Shell Game*

# Electron Configurations

## THINK ABOUT IT

Recall that the chemistry of the elements is closely related to the number of valence electrons in their atoms. The valence electrons are found in the outermost electron shell of an atom.

---

**What does the periodic table indicate about the arrangements of electrons?**

---

*To answer this question, you will explore*

➲ Subshells in Atoms

➲ Electron Configurations

➲ Connecting the Periodic Table to Electron Arrangements

➲ Noble Gas Shorthand

## EXPLORING THE TOPIC

### ➲ Subshells in Atoms

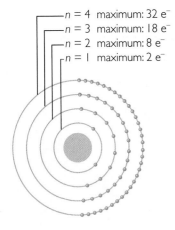

$n = 4$  maximum: 32 e⁻
$n = 3$  maximum: 18 e⁻
$n = 2$  maximum: 8 e⁻
$n = 1$  maximum: 2 e⁻

*Each electron shell can have a maximum number of electrons.*

Electrons are arranged into shells numbered $n = 1, 2, 3$, and so on. The number of electron shells in an atom is the same as the number of the period where the element is located on the periodic table. Each shell has a maximum number of electrons. For instance, the $n = 2$ shell cannot have more than 8 electrons.

Scientific evidence has led chemists to propose that electron shells are further divided into electron subshells. Imagine magnifying the basic atomic model and finding that each shell is composed of subshells. Notice that the number of subshells that a shell has is equal to $n$.

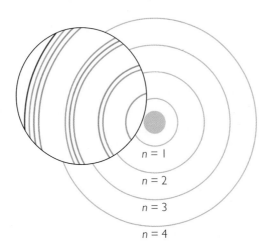

$n = 1$
$n = 2$
$n = 3$
$n = 4$

*Electron shells are further divided into subshells.*

## THE s, p, d, AND f SUBSHELLS

The subshells have special names. They are called the s, p, d, and f subshells. Just like the basic shells, each subshell has a maximum capacity of electrons. The s subshells can have a maximum of 2 electrons, p subshells can have a maximum of 6 electrons, and d subshells can have a maximum of 10 electrons. Finally, f subshells can have a maximum of 14 electrons.

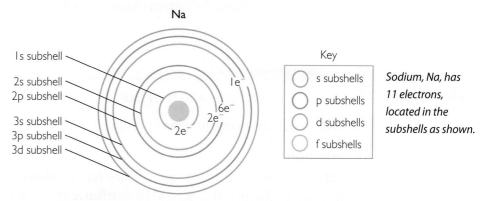

Notice that the name of each subshell is labeled using both the basic shell number and the subshell letter (1s, 2s, 2p, and so on).

**Example 1**

### Electron Arrangements

Use the illustration of the subshells in a sodium atom above to help you answer these questions:

**a.** How many total electrons are there in a sodium, Na, atom? Which shells are they in?

**b.** How many valence electrons does sodium have? Which subshell are they in?

**c.** How many electrons are there in the 3s subshell of sodium? In the 3p subshell?

#### Solution

The atomic number of sodium is 11.

**a.** There are a total of 11 electrons in a neutral sodium atom. The electrons are in shells, $n = 1$, $n = 2$, $n = 3$.

**b.** Sodium has one valence electron, located in the 3s subshell.

**c.** There is one electron in the 3s subshell of sodium, and none in the 3p subshell.

## ➲ Electron Configurations

It can be time-consuming to draw subshell models of the atoms to show the arrangements of the electrons, especially for atoms with large atomic numbers. Chemists have developed a shorthand notation called an **electron configuration** to keep track of the electrons in an atom. The electron configurations for the first ten elements are shown here.

H $1s^1$          C $1s^2 2s^2 2p^2$

He $1s^2$         N $1s^2 2s^2 2p^3$     ← Nitrogen has a total of 7 electrons.

Li $1s^2 2s^1$    O $1s^2 2s^2 2p^4$

Be $1s^2 2s^2$    F $1s^2 2s^2 2p^5$

B $1s^2 2s^2 2p^1$    Ne $1s^2 2s^2 2p^6$

Electrons in each subshell.

Li:  1s$^2$  2s$^1$

Subshells

Each subshell is written using the shell number and the subshell letter. In addition, the number of electrons in each subshell is indicated with a superscript number.

Notice that the superscript numbers add up to the total number of electrons for that atom.

The sequence in which electrons fill up the subshells is 1s, 2s, 2p, 3s, 3p. After the element argon, the pattern changes slightly.

**Example 2**

### Electron Configuration of Sulfur

Write the electron configuration for sulfur, S.

*Solution*

Sulfur is located in the third row in Group 6A. The atomic number of sulfur is 16, so there are 16 electrons that need to be distributed in subshells, beginning with the 1s subshell.

The electron configuration of sulfur is $1s^2 2s^2 2p^6 3s^2 3p^4$.

## ⮕ Connecting the Periodic Table to Electron Arrangements

An outline of the periodic table appears here with color-coding to show the subshell for the outermost electron of each element. For example, any element located in the green area will have its outermost electron(s) in a p subshell.

Periodic Table—subshells

Key
- s-block
- p-block
- d-block
- f-block

As you proceed across the periodic table from one element to the next, one additional proton and one additional electron are added, along with one or more neutrons. Each additional electron goes into a specific subshell. If an atom is located in the orange areas of the table, the last electron is placed into an s subshell. If an atom is located in the blue area of the table, the last electron is placed into a d subshell. And so on.

The elements in each block have related properties. The elements in the s-block are reactive metals. The elements in the d-block tend to form colorful compounds that are used as pigments. The elements in the p-block tend to form colorless compounds.

### DECODING THE TABLE

To write out an electron configuration for a specific element, you can simply "read" from the periodic table, moving across from left to right and then down to the next row. For example, the sequence of subshells for argon, Ar, is 1s, 2s, 2p, 3s, 3p. The electron configuration for argon is $1s^2 2s^2 2p^6 3s^2 3p^6$.

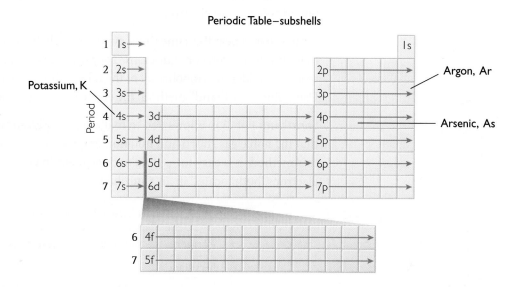

Periodic Table—subshells

Everything runs smoothly until you reach the fourth row of the periodic table. After argon, you might expect the next electron to be in the 3d subshell. However, this does not happen. The next element is potassium, K. Like the other elements in Group 1A, potassium has one electron in the s subshell. So, you place an electron in 4s before 3d. The electron configuration for potassium is $1s^2 2s^2 2p^6 3s^2 3p^6 4s^1$. The electron configuration for arsenic is $1s^2 2s^2 2p^6 3s^2 3p^6 4s^2 3d^{10} 4p^3$.

You may have noticed that you only have to look at the ending of each electron configuration to figure out the identity of the element associated with it. The ending provides you with the exact spot on the periodic table where you can find the element.

**Important to Know** The s subshells fill with electrons before the d subshells from the previous shell. For example, the 4s subshell fills before the 3d subshell, the 5s subshell fills before the 4d subshell, and so on.

---

**Example 3**

### Electron Configuration of Cobalt

Write the electron configuration for cobalt, Co.

#### Solution

Locate cobalt on the periodic table. It is element number 27 and is located in the fourth period of the periodic table.

Simply trace your finger across the periodic table of subshells, writing the subshells as you go. When you get to cobalt, stop writing. Every subshell up to the 4s subshell is completely filled. In addition, cobalt has seven electrons in the 3d subshell. The answer is $1s^2 2s^2 2p^6 3s^2 3p^6 4s^2 3d^7$.

You can check that you have the correct electron configuration by adding the superscript numbers to make sure there are 27 electrons.

# ➲ Noble Gas Shorthand

Depending on the element, the electron configuration can be lengthy to write. Plus, each element just repeats the electron configuration of the previous element but adds one more electron.

Rather than repeat the same thing every time, chemists have devised a quicker way to write out electron configurations. They use the noble gas at the end of each period as a placeholder to symbolize all of the filled subshells before that place on the table. Using this "shorthand" method, the electron configuration of cobalt is $[Ar]4s^2 3d^7$.

Shorthand notation allows you to make some interesting comparisons. Notice that the noble gas shorthand notation emphasizes the valence electrons. Using this method, it is easy to see that each element in Group 2A has two valence electrons, both located in an s subshell.

## Group 2A Elements

| Element | Symbol | Electron configuration |
|---------|--------|------------------------|
| beryllium | Be | $[He]2s^2$ |
| magnesium | Mg | $[Ne]3s^2$ |
| calcium | Ca | $[Ar]4s^2$ |
| strontium | Sr | $[Kr]5s^2$ |
| barium | Ba | $[Xe]6s^2$ |
| radium | Ra | $[Rn]7s^2$ |

**Example 4**

## Electron Configuration of Selenium

Find the element Selenium, Se, element number 34, on the periodic table.

  a. What is the electron configuration of selenium?
  b. Write the electron configuration using noble gas shorthand.
  c. In what subshells are selenium's valence electrons?

### Solution

  a. Selenium is in the p-block, in Period 4. The electron configuration of selenium is $1s^2 2s^2 2p^6 3s^2 3p^6 4s^2 3d^{10} 4p^4$.

  b. The noble gas that comes before selenium is argon, Ar. So, the noble gas shorthand for this configuration is $[Ar]4s^2 3d^{10} 4p^4$.

  c. Selenium's valence electrons are in subshells 4s and 4p.

## LESSON SUMMARY

### What does the periodic table indicate about the arrangements of electrons?

Electrons in atoms are arranged into basic shells labeled $n = 1, 2, 3$, and so on. These shells are divided into subshells. The number of subshells in each shell is equal to *n*. The subshells are referred to as s, p, d, and f subshells. The s, p, d, and f subshells can have a maximum of 2, 6, 10, and 14 electrons, respectively. Chemists use electron configurations to specify the arrangements of electrons in subshells. The periodic table provides the information needed to write electron configurations.

## Exercises

### Reading Questions

1. What are electron subshells?

2. What is an electron configuration?

3. How is the arrangement of electrons in an atom related to the location of the atom on the periodic table?

### Reason and Apply

4. How many subshells are in each shell: $n = 1$, $n = 2$, $n = 3$, $n = 4$?

5. What is the total number of subshells for elements in Period 5 of the periodic table?

6. Draw a subshell model for each of these elements, putting the electrons in their appropriate places.
   **a.** sodium, Na       **b.** neon, Ne       **c.** carbon, C       **d.** vanadium, V

7. What is the outermost subshell for bromine, Br?

8. Name an element with electrons in the f subshell.

9. Consider the element with the atomic number 13.
   **a.** What is the electron configuration for the element with atomic number 13?
   **b.** How many valence electrons does element number 13 have? How do you know?
   **c.** How many core electrons does element number 13 have? How did you figure that out?

10. Explain why the chemical properties of argon, krypton, and xenon are similar, even though there are 18 elements between argon and krypton and 32 elements between krypton and xenon.

11. Write the electron configuration for each of these atoms. Then write it using the noble gas shorthand method.
   **a.** oxygen       **b.** chlorine       **c.** iron       **d.** calcium
   **e.** magnesium   **f.** silver         **g.** silicon     **h.** mercury

12. You should be able to figure out the identity of an atom from its electron configuration alone. Describe at least two ways you could do this.

13. Which elements are described by these electron configurations?
   **a.** $1s^2 2s^2 2p^6 3s^2 3p^6 4s^2 3d^4$
   **b.** $1s^2 2s^2 2p^6 3s^2 3p^2$
   **c.** $1s^2 2s^2 2p^3$
   **d.** $1s^2 2s^2 2p^6 3s^2 3p^6 4s^2 3d^{10} 4p^6 5s^2 4d^{10} 5p^6 6s^1$
   **e.** $1s^2 2s^2 2p^6 3s^2 3p^6 4s^2 3d^{10} 4p^6 5s^2 4d^{10} 5p^6 6s^2 4f^{14} 5d^{10} 6p^2$
   **f.** $[Kr] 5s^2 4d^9$

# CHAPTER **4**

## Moving Electrons

# SUMMARY

### KEY TERMS

**flame test**

**valence shell**

**valence electron**

**core electron**

**ion**

**cation**

**anion**

**ionic compound**

**rule of zero charge**

**polyatomic ion**

**monatomic ion**

**electron configuration**

## Alchemy Update

What do the electrons of an atom have to do with its properties?

The outermost electron shell contains an atom's valence electrons. The number of valence electrons in main group elements is the same as their group number. Valence electrons are responsible for many of an element's properties, so elements in the same group have similar properties. While you cannot change copper atoms into gold atoms, you can combine elements into compounds with new properties.

The electrons in an atom are fairly easy to change, unlike the parts of a nucleus. Atoms become more stable when they transfer electrons to other atoms, forming ionic compounds.

### REVIEW EXERCISES

1. Explain the importance of valence electrons.

2. What are ionic compounds, and how do they form?

3. How is the arrangement of electrons within atoms connected to the periodic table?

4. Write the electron configuration for aluminum. Then write it using noble gas shorthand.

5. For each compound listed, identify the anion and cation and the charge on each.
   a. magnesium chloride, $MgCl_2$
   b. calcium nitrite, $Ca(NO_2)_2$

6. Three polyatomic ions are listed here.

   silicate, $SiO_3^{2-}$  chlorite, $ClO_2^-$  bicarbonate, $HCO_3^-$

   a. Write the chemical formulas for sodium silicate, sodium chlorite, and sodium bicarbonate.
   b. Write the chemical formulas for calcium silicate, calcium chlorite, and calcium bicarbonate.

### CONSUMER
### CONNECTION

Sodium bicarbonate, $NaHCO_3$, is commonly known as baking soda. When it is used in baking, it reacts with other ingredients to release carbon dioxide. For example, baking soda is what causes the batter to rise and bubble when you make pancakes.

# CHAPTER 5

# Building with Matter

*Sugar dissolves in water, but copper does not. This property gives clues about how the atoms in these substances are bonded to each other.*

### In this chapter, you will study

- the basic types of bonds between atoms

- the role of electrons in bonding

- the properties and types of elements associated with each type of bond

- the formation of new substances through bonding

When atoms form bonds, new substances with new properties are formed. There are four basic types of bonds found in the world around us. Each type of bond is a result of a different distribution of electrons within the substance. Further, each type of bond is associated with certain properties. To determine the type of bond in a substance, you can use the chemical formula of that substance.

Ken Karp Photography

## Materials

- bulb with wires
- 9-volt battery with snap connector
- wire with stripped ends
- 100 mL beakers
- paper clips
- salt, NaCl
- sand, $SiO_2$
- paraffin wax, $C_{20}H_{42}$
- calcium chloride, $CaCl_2$
- copper, Cu
- copper (II) sulfate, $CuSO_4$
- aluminum foil, Al
- sucrose, $C_{12}H_{22}O_{11}$
- distilled water, $H_2O$

# Classifying Substances

### Purpose

To investigate the properties of substances.

### Predictions and Data

Predict whether each substance will conduct electricity and whether it will dissolve in water. Make a table for your predictions and data.

### Procedure

1. Assemble your conductivity tester as shown.

2. Take your conductivity tester to each station. Test the pure substance for conductivity.

3. Next, observe the substance in water. Did the substance dissolve?

4. Finally, test the substance mixed with the distilled water in the second beaker for conductivity. Record your results in a data table.

Touch these ends to a solid or liquid.

### Analysis

5. Group the substances according to their properties. How do your results compare with your predictions? What do the substances in each group have in common? Summarize your findings in a paragraph.

# Classifying Substances

## THINK ABOUT IT

If you accidentally drop a piece of jewelry into a tub of water, you should have no trouble getting it back again. However, if you drop a cube of sugar into the water, it will dissolve. Some substances dissolve in water and others do not. So, dissolving is one property of matter that you can use to sort substances into general groups.

### How can substances be sorted into general categories?

*To answer this question, you will explore*

➲ Dissolving and Conductivity

➲ Testing and Sorting Substances

## EXPLORING THE TOPIC

### ➲ Dissolving and Conductivity

There are millions of different substances in our world. Even chemists do not know the identities of all the possible compounds. However, many of the known substances can be sorted into general categories by looking at two properties: dissolving and conductivity.

#### DISSOLVING SUBSTANCES IN WATER

We live on a watery planet. Most substances come into contact with water at some point. Some substances, like sugar, **dissolve** in water to make an aqueous solution. Other substances, like gold, do not dissolve in water and will remain unchanged when placed in water. A substance that dissolves in water is **soluble** in water. A substance that does not dissolve is **insoluble** in water.

Many substances can be sorted into one of two categories based on whether they dissolve in water.

#### CONDUCTIVITY

By examining a second property, it is possible to sort matter even further. The property of **conductivity** has to do with how well electricity flows through a substance. Electrical conductivity requires the movement of charge as either ions or

© cfdesignz/iStockphoto

electrons. Copper wire is a great conductor of electricity. Your entire home is wired with hundreds of feet of copper or aluminum wiring that carries electricity from the power lines outside to the electrical outlets in your home. The human body is another good conductor of electricity. However, in the body, electricity moves as ions, not as electrons, as is the case in wiring.

Some substances do not conduct electricity. The electrical wires in your house are covered with a coating that does not conduct electricity. That way, you can plug in a stereo or a lamp and not get an electrical shock.

*The light bulb doesn't light up because there's a break in the wire.*

*Even though there is a break in the wire, the bulb lights up because the liquid in the beaker conducts electricity.*

Electrical conductivity can be tested by setting up a simple electrical circuit. Wires connect the terminals of a battery to a light bulb. When the circuit is complete, and an electrical current is flowing, the bulb will light up. If the flow of current is interrupted by the presence of a substance that does not conduct electricity, the light bulb will not light up. In the figure above, the red arrows show the direction of flow of positive charge. The electrons flow in the opposite direction.

## Testing and Sorting Substances

You can use a two-step test to sort substances by both properties. First, drop a substance into water to find out if it dissolves. Next, test it for conductivity. This allows you to sort all matter into four basic categories as shown in the illustration below.

Oleg Dolenko/Shutterstock

Examples of substances in each of the four categories are given in the table. Take a moment to look for patterns in the chemical formulas of the substances in each category. Note whether substances are made from metal atoms, nonmetal atoms, or some combination.

### Types of Substances

| Properties | dissolves: yes conducts: yes | dissolves: yes conducts: no | dissolves: no conducts: yes | dissolves: no conducts: no |
|---|---|---|---|---|
| Examples | salt, NaCl; calcium chloride, $CaCl_2$; copper sulfate, $CuSO_4$ | water, $H_2O$; sugar, $C_{12}H_{22}O_{11}$; ethanol, $C_2H_6O$ | gold, Au; copper, Cu; aluminum, Al | sand, $SiO_2$; paraffin, $C_{20}H_{42}$ |
| Types of atoms | metal and nonmetal atoms | nonmetal atoms only | metal atoms only | nonmetal atoms only |

Here are some patterns you might notice:

- The substances that are *soluble* in water and *conduct* electricity are made up of a metal element and one or more nonmetal elements. Remember, these are called ionic compounds.
- The substances that are *soluble* in water but *do not conduct* electricity are often made up of carbon, hydrogen, and oxygen atoms joined together. These are all nonmetal elements.
- The substances that are *insoluble* in water but *do conduct* electricity are made of metallic elements.
- The substances that are *insoluble* in water and *do not conduct* electricity are made up of nonmetal atoms.

Notice that only substances that contain metal atoms will conduct electricity. Also, substances that are made entirely of metal atoms will not dissolve in water. Many ionic compounds dissolve in water.

In Lesson 26: Electron Glue, you will discover that these two properties, solubility and conductivity, are directly related to the manner in which the individual atoms in these substances are linked.

**Important to Know** Ionic compounds conduct electricity only when they are dissolved in water. Dry ionic solids do not conduct electricity.

## Example

### Predicting Properties

Predict whether the following substances dissolve in water and whether they conduct electricity.

**a.** lead, Pb

**b.** potassium bromide, KBr

*(continued)*

*Solution*

   **a.** Lead is a metal element. Because it is made only of metal atoms, it does not dissolve in water, but it does conduct electricity.

   **b.** Potassium bromide consists of a metal and a nonmetal. It is an ionic compound. Therefore, it does dissolve in water, and the mixture does conduct electricity. The dry compound does not conduct electricity.

## LESSON SUMMARY

### How can substances be sorted into general categories?

**KEY TERMS**

dissolve

soluble

insoluble

conductivity

Most substances on the planet can be sorted into four categories based on two properties: whether they are soluble in water and whether they conduct electricity. As you will discover in coming lessons, the properties of conductivity and solubility are directly related to the way in which atoms are connected to each other.

**Reading Questions**

**1.** What does insoluble mean?

**2.** Describe a way to determine whether a substance conducts electricity.

**Reason and Apply**

**3.** What generalization can you make about a substance that is soluble in water and conducts electricity once it has dissolved? Explain your thinking.

**4.** Do all solid substances containing metals conduct electricity? Explain your reasoning.

**5.** If a substance does not conduct electricity as a solid, does this mean that it will not conduct electricity if it dissolves? Explain your reasoning.

**6.** Predict whether each substance listed will conduct electricity, dissolve in water, and/ or conduct electricity once it has dissolved. Explain your thinking in each case.

   **a.** $C_3H_6O(l)$       acetone

   **b.** $Ti(s)$           titanium

   **c.** $LiNO_3(s)$      lithium nitrate

   **d.** $CuZn(s)$       bronze

# Bonding

## THINK ABOUT IT

**HISTORY
CONNECTION**

Stephanie Kwolek started working for DuPont in 1946. In 1965, she succeeded in creating synthetic fibers of exceptional strength due to their extremely strong bonds. This led to Kevlar, the material used in bulletproof vests.

All of the objects in our everyday world, including ourselves, are made up of individual atoms. But what holds those atoms together? Why don't objects just crumble into piles of individual atoms? Something must be holding the atoms together. And why is the desk in your classroom solid, while water simply runs through your fingers? Something about the way atoms are connected must give substances the properties we observe.

### How are atoms connected to one another?

*To answer this question, you will explore*

⮕ Bonds: The "Glue" between Atoms

⮕ Models of Bonding

⮕ Relating Bonds and Properties

## EXPLORING THE TOPIC

### ⮕ Bonds: The "Glue" between Atoms

Chemists call the attraction that holds atoms together a **chemical bond.** As you will discover, several different types of chemical bonds exist. All bonds involve the electrons in some way. A chemical bond is an attraction between the positive charges on the nucleus of one atom and the negative charges on the electrons of another atom. This attraction is so great that it keeps the atoms connected to one another.

### ⮕ Types of Bonding

Recall from Lesson 25 that most substances can be divided into four categories, based on their physical properties. These four categories can be explained by different models of bonding.

A bond is a force of attraction, so it is not possible to see the actual bonds between atoms. However, a model can help to explain how atoms are bonded in substances. Models can also help us to understand how bonding accounts for certain properties of substances that we observe.

. . . . . . . . . . . . . . . . . . . . . . . . . . . . . . . . . . . . . . . . . . . . .

**Big Idea**   There are four main models of chemical bonds between atoms.

. . . . . . . . . . . . . . . . . . . . . . . . . . . . . . . . . . . . . . . . . . . . .

The four models of bonding are called *ionic, molecular covalent, metallic,* and *network covalent.* Take a moment to locate the valence electrons in each model on the next page. The red spheres represent the nuclei of atoms and the core electrons, while the blue areas suggest where the valence electrons are located.

# Four Models of Bonding

## MODEL 1: IONIC

**Properties of ionic substances:**

Dissolve in water

Conduct electricity when dissolved

Tend to be brittle solids

Made of metal and nonmetal atoms combined

In **ionic bonding,** the valence electrons are transferred *from one atom to another. Metal atoms transfer their valence electrons to nonmetal atoms.*

## MODEL 2: MOLECULAR COVALENT

**Properties of molecular covalent substances:**

Some dissolve in water, some do not

Do not conduct electricity

Some are liquids or gases

Made entirely of nonmetal atoms

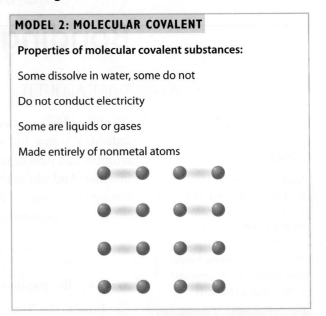

In **molecular covalent bonding,** *the valence electrons are shared between pairs or groups of atoms. This creates small stable units, called molecules, within the substance.*

## MODEL 3: METALLIC

**Properties of metallic substances:**

Do not dissolve in water

Conduct electricity

Bendable, malleable solids

Made entirely of metal atoms

In **metallic bonding,** *the valence electrons are free to move about the substance.*

## MODEL 4: NETWORK COVALENT

**Properties of network covalent substances:**

Do not dissolve in water

Do not conduct electricity

Extremely hard solids

Made entirely of nonmetal atoms

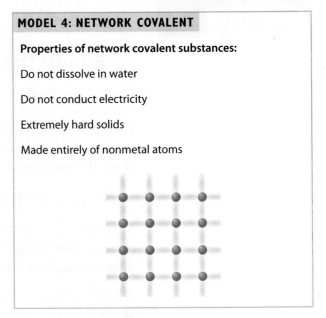

**Network covalent bonding** *is similar to molecular covalent bonding, but the valence electrons are shared throughout the entire substance.*

You have already been introduced to ionic compounds. These all have ionic bonding in which metal atoms transfer valence electrons to nonmetal atoms. The resulting oppositely charged ions are strongly attracted to each other. This attraction is what holds the ions together.

In **covalent bonding,** the nucleus of one atom is attracted to the valence electrons of another atom. Unlike in ionic bonding, one atom does not transfer an electron to the other. Instead, both atoms *share* the valence electrons between them.

Methane, CH₄

Covalent bonding can happen in two different ways. In molecular covalent bonding, the atoms bond to form individual clusters called **molecules,** such as the methane molecule shown here.

In network covalent bonding, the valence electrons are shared between atoms but form a highly regular extended network, creating a very durable structure. Diamond consists of carbon atoms that are covalently bonded in a network.

In a metal, the valence electrons are distributed throughout the substance in what is sometimes called a "sea" of electrons. The valence electrons are free to move throughout the substance. The atoms are bonded by the attraction between the positively charged atoms and the negatively charged "sea" of electrons.

Diamond consists of carbon atoms that are covalently bonded in a network.

In a metal, positively charged atoms and a negatively charged "sea" of atoms create bonds.

## Relating Bonds and Properties

Some properties of substances, such as solubility and conductivity, are directly related to the type of bonds the atoms in the substances have. Therefore, it is possible to match the bond type to the physical properties observed in different types of substances. Examine what happens to each type of substance when it is struck by a hypothetical hammer.

| Ionic substances | Network covalent substances | Metallic substances | Molecular covalent substances |
|---|---|---|---|

Hard but brittle, tend to fracture along planes of atoms

Durable, rigid, difficult to break

Bendable, malleable

Often gases or liquids, or soft solids

Notice that the hardest substance is a solid with network covalent bonding. This is because bonding in these substances is in an organized network.

Bonding can also help to explain the properties of dissolving and conductivity. Examine the illustration below representing dissolving. Water is represented by the lighter blue areas. Ionic solids and molecular covalent substances dissolve in water. Metallic solids and network covalent solids do not.

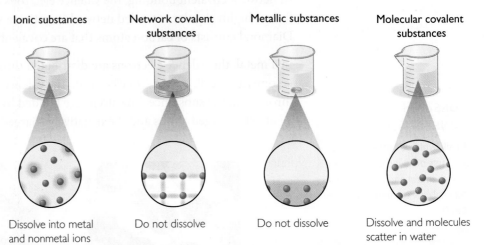

Ionic substances — Dissolve into metal and nonmetal ions

Network covalent substances — Do not dissolve

Metallic substances — Do not dissolve

Molecular covalent substances — Dissolve and molecules scatter in water

Conduction requires the movement of a charged particle, either an ion or an electron. Metals conduct electricity because the valence electrons are free to move throughout the solid. Ionic compounds that have been dissolved in water conduct electricity because the cations and anions are free to move in the solution. Network covalent solids and molecular covalent substances do not conduct electricity. The charge cannot move in these substances because the electrons are "stuck" between the atoms and are not available to move.

The periodic table is a valuable tool in figuring out bonding. You can use the table to determine if the elements in a compound are metals, nonmetals, or both.

- Ionic compounds, such as salts, are made from metal and nonmetal elements.
- Metallic compounds, such as brass, are made only of metal atoms.
- Network covalent compounds, such as diamonds, and molecular covalent compounds, such as methane, are made from nonmetals.

**INDUSTRY CONNECTION**

Metals can be combined into alloys. For example, CuZn is brass, a copper-zinc alloy. Chrome-moly is an alloy of steel, chrome, and molybdenum, used to make bicycle frames that are stronger and lighter than those made of steel. Steel itself is an alloy of iron and carbon.

Masakazu Watanabe/Aflo/Getty Images

**Example**

### Identifying the Type of Bonding

Determine the bonding in each of the following substances. What general physical properties can you expect of each substance?

**a.** magnesium chloride, $MgCl_2$

**b.** rubbing alcohol, $C_3H_8O$

*Solution*

**a.** Magnesium chloride, $MgCl_2$, is an ionic compound with ionic bonding because it is made of a metal and a nonmetal element. It is probably brittle, dissolves in water, and conducts electricity when dissolved.

**b.** Rubbing alcohol, $C_3H_8O$, is a molecular covalent substance with molecular covalent bonding because it is made entirely of nonmetal atoms and is a liquid.

KEY TERMS

chemical bond

ionic bonding

molecular covalent bonding

metallic bonding

network covalent bonding

covalent bonding

molecule

## LESSON SUMMARY

### How are atoms connected to one another?

Atoms in substances are held together by chemical bonds. Chemists have identified four main models of bonding within substances: ionic, network covalent, molecular covalent, and metallic. Many properties of a substance correspond to the type of bonding between the atoms of that substance.

## Exercises

### Reading Questions

1. Explain why substances do not simply crumble into piles of atoms.

2. Name the four models of bonding and explain them in your own words. Be specific about the location of the valence electrons.

### Reason and Apply

3. Determine the type of bonding in each substance.
   **a.** zinc, $Zn(s)$
   **b.** propane, $C_3H_8(l)$
   **c.** calcium carbonate, $CaCO_3(s)$

4. Based on physical properties, which of these substances is an ionic compound? Explain your reasoning.
   **A.** hair gel    **B.** silver bracelet    **C.** motor oil    **D.** baking soda

5. In the copper cycle from Lesson 7, you saw that nitrogen dioxide, $NO_2(g)$, formed when copper, $Cu(s)$, was dissolved in nitric acid. How would you classify the bonding in $NO_2(g)$? Explain.

6. Which statement is true?
   **(A)** Aqueous solutions of calcium chloride, $CaCl_2$, conduct electricity.
   **(B)** Glass, made of silicon dioxide, $SiO_2$, does not dissolve in water.
   **(C)** Ethanol, $C_2H_6O$, dissolves in water but does not conduct electricity.
   **(D)** Brass, also called copper zinc, $CuZn$, conducts electricity.
   **(E)** All of the above are true.

7. Suppose you have a mixture of sodium chloride, $NaCl$, and carbon, $C$. Explain how you can use water to separate the two substances.

8. Explain why copper is used as wire, but copper chloride is not.

9. Explain why carbon is a solid and not a gas.

10. Will each of these substances dissolve in water? Explain your thinking.
    **a.** $Ca$    **b.** $NaNO_3$    **c.** $Si$    **d.** $CH_4$    **e.** $CuSO_4$

### SAFETY Instructions

Wear safety goggles at all times.

The solution contains acid. Handle carefully. Rinse the nickel strips after they have been in the copper solution.

## Materials

- copper sulfate plating solution
- 250 mL beaker
- 2 nickel strips (cut into about 1-by-3-in. strips)
- 1.5-volt D-cell battery with holder
- 2 insulated wires with alligator clips

# Electroplating Metals

## Purpose

To use electrochemistry to extract metals from ionic compounds in a solution.

## Procedure

1.  Set up the electroplating apparatus as shown. Observe what happens.

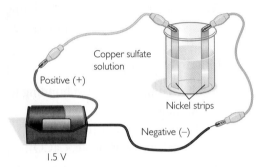

2.  Switch the sides of the battery to which the two alligator clips are attached. Wait at least one minute or until you notice a change.

3.  Reverse the wiring back to its original position.

## Observations

1.  What did you observe when you hooked up the nickel strips to the battery?

2.  What happened when you reversed the flow of electricity?

3.  Where does the copper come from that ends up on the nickel strip?

4.  What is in the copper sulfate solution?

5.  Write a short paragraph explaining your observations.

6.  **Making Sense** Are copper atoms and copper ions the same element? Explain your thinking.

7.  **If You Finish Early** Consider a sample of gold chloride, $AuCl_3$. Explain what procedure you might follow in order to extract solid gold from the compound.

# Electroplating Metals

## THINK ABOUT IT

Metals are important in our lives. They are used for everything from electrical wiring, jewelry, and soda cans, to cars, airplanes, and bridges. Most metals are dug out of the ground as ionic compounds. In ionic compounds, the metal atoms are bonded to nonmetal atoms. Through the ages, people have struggled to extract the pure metals from these compounds by figuring out ways to separate the metal atoms from the nonmetal atoms.

### How can you extract an element from a compound?

*To answer this question, you will explore*

➲ The Quest for Precious Metals

➲ Electroplating Elemental Metals

## EXPLORING THE TOPIC

### ➲ The Quest for Precious Metals

Sometime around 6000 B.C.E., metals came into widespread use. Before that, most tools and implements were made from earth, bone, wood, and stone. The first metals to be discovered were gold, copper, and silver. Later, lead, tin, iron, and mercury were added to the list. Most of these metals are not readily found in nature in pure form.

The first metals that were used were the ones people stumbled upon in their pure form in the earth. Then, people discovered that, using heat, they could extract metals from other compounds dug out of the earth. For example, if iron (III) oxide, $Fe_2O_3$, is heated in the presence of charcoal, the oxygen in the compound is removed as $CO_2$ gas. This leaves behind solid iron, Fe. This process requires very high temperatures.

*Molten iron being poured out to solidify.*

For nearly 7000 years, only a handful of metals was in widespread use. Other metals, such as aluminum, Al, could not be extracted easily by heating and remained unavailable.

### ➲ Electroplating Elemental Metals

#### USING ELECTRICITY TO EXTRACT METALS

Over time, the construction of better furnaces has resulted in the extraction of a variety of metals. More recently, it was discovered that electricity could be used to extract metals from compounds. Recall that electrons are transferred from metal atoms to nonmetal atoms in ionic compounds. Scientists discovered they could "give" electrons back to the metal ions, re-forming the neutral metal atoms once again. For example, copper metal can be extracted from a solution of copper sulfate by running an electrical current through the solution. A simple setup is shown here.

*In this setup the charge moves through the wires as shown.*

Once the battery is hooked up, one nickel strip has a positive charge and the other has a negative charge. In this electrical circuit, the electrons move from the negative terminal of the battery to the negatively charged nickel strip. Electrons also move from the other nickel strip toward the positive terminal of the battery. Ions moving in solution complete the circuit.

The positive copper ions, $Cu^{2+}$, are attracted to the nickel strip with the negative charge. Electrons are added to the copper cations in solution. Each copper ion that gains enough electrons becomes elemental copper. When the copper cations become elemental copper atoms, they come out of solution and coat the negatively charged nickel strip. This process is called **electroplating.**

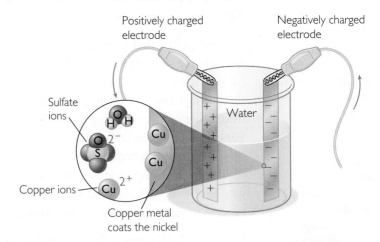

When the experiment shown is completed, the negatively charged nickel becomes coated with the pure copper metal.

You can reverse the process and produce metal cations from metal atoms. The battery has a positive and a negative terminal. By switching the wires in the test setup, you can change the direction the electrical charge moves. This causes the reaction to reverse. The copper coating comes off of the first nickel strip and reenters the solution. Copper from the solution plates onto the other nickel strip.

## ELECTROPLATING GOLD

Is it possible to design an apparatus similar to the one in the Lab: Electroplating Metals to extract gold atoms from ionic compounds containing gold ions? Yes, by all means. However, gold is one of those few elements that *is* commonly found in nature in its elemental form. Gold *compounds* are seldom found in nature. This is because gold is fairly stable as it is. So harvesting gold atoms from gold ions is not the best way to make a pile of gold bars.

However, electroplating techniques can be used to make a little bit of gold go a long way. Using electroplating or other plating techniques, less expensive metals can be covered with thin layers of gold atoms.

Today, electroplating is an extremely important industrial process. For example, aluminum metal has been in widespread use only recently because chemists finally found a way to extract it from ionic compounds using electrochemical methods. Electroplating is also used to coat electronic parts in computers.

*Copper bowl plated with 24 karat gold*

**Example**

## Plating Gold from $AuCl_3$

Suppose that you have a sample of $AuCl_3(s)$.

**a.** What is the charge on the Au ions in $AuCl_3$?

**b.** What is the chemical name of $AuCl_3$?

**c.** Do you expect $AuCl_3$ to conduct electricity as a solid? Explain your thinking.

**d.** Do you expect $AuCl_3$ to dissolve in water? Explain your thinking.

**e.** Describe how you can extract gold from $AuCl_3$.

*Solution*

**a.** You need to find the charge on the main group element first. Cl is in Group 7B in the periodic table, which means that it has seven valence electrons. So, Cl needs to gain an electron, which will make it $Cl^-$. The Au becomes $Au^{3+}$ because the charges in an ionic compound must add up to zero.

**b.** The gold has a $+3$ charge. So, the name is gold (III) chloride.

**c.** $AuCl_3$ is an ionic solid. Ionic solids do not conduct electricity. However, if you dissolve $AuCl_3$ in water, the solution will conduct electricity.

**d.** Because $AuCl_3$ is ionic, you can assume it will dissolve in water, forming $Au^{3+}$ and $Cl^-$ ions.

**e.** Gold can be plated onto a piece of nickel by using a battery to move electrons through a solution containing $AuCl_3$.

## How can you extract an element from a compound?

KEY TERM

electroplating

Most metal atoms are found in nature combined with other atoms, in the form of ionic compounds. Some metals can be extracted from ionic compounds through heating. Elements can also be extracted from ionic compounds by using electricity to move electrons between atoms. Using electroplating, you can create a variety of substances that could be considered "as good as gold."

## Exercises

### Reading Questions

1. Describe three ways in which people have managed to obtain pure metals to use for making tools and other objects.

2. Explain how to plate copper metal onto a nickel strip. Write a simple procedure and draw a labeled sketch showing how to carry it out.

3. Explain how you can *remove* a metal coating from an object.

### Reason and Apply

4. **Lab Report** Complete a lab report for the copper plating experiment. In your report, give the title of the experiment, purpose, procedure, observations, and conclusions.

> (Title)
>
> **Purpose:** (Explain what you were trying to find out.)
>
> **Procedure:** (List the steps you followed.)
>
> **Observations:** (Describe what you observed during the experiment.)
>
> **Conclusions:** (What can you conclude about what you were trying to find out? Provide evidence for your conclusions.)

5. Explain how the copper plating experiment supports the claim that $Cu^{2+}$ is just a Cu atom that is missing two electrons.

6. If you measure the mass of the nickel strip before and after plating copper, do you expect the mass to change? Explain your thinking.

7. Suppose that a classmate claimed that the nickel in the electroplating lab changed into copper. Provide an argument to prove that this is not true.

# CHAPTER 5

# Building with Matter

# SUMMARY

## KEY TERMS

**dissolve**

**soluble**

**insoluble**

**conductivity**

**chemical bond**

**ionic bonding**

**molecular covalent bonding**

**metallic bonding**

**network covalent bonding**

**covalent bonding**

**molecule**

**electroplating**

# Alchemy Update

How can substances with new properties be made?

Sodium, Na          Fluorine, F

A bond is an attraction between the nucleus of one atom and the electrons of another atom. When atoms bond together, new substances with new properties are made. Conversely, elements can be extracted from compounds by breaking the bond.

There are four models of bonding, depending on the location of electrons within the atoms in the substance: ionic, molecular covalent, metallic, and network covalent. Certain properties are associated with each type of bonding.

Atoms form bonds with other atoms, creating a variety of substances with new properties. It is possible to make compounds that look like or behave like gold, such as iron pyrite, $FeS_2$, also called fool's gold. Also, you can use chemical bonding to make substances that are even more valuable than gold.

## REVIEW EXERCISES

1. What does the phrase "as good as gold" mean as it applies to this class?

2. Describe how you could experimentally determine the type of bonding present in a substance.

3. Predict whether isopropanol, $C_3H_8O(l)$, will conduct electricity. State your reasoning.

4. Make a table like this one. Fill it with the model of bonding that fits the category.

|  | Dissolves | Does not dissolve |
|---|---|---|
| **Conducts** |  |  |
| **Does not conduct** |  |  |

# UNIT
# 1

# Alchemy | REVIEW

**M**atter is anything that has substance and takes up space. Mass is a measure of the substance of matter. Volume is a measure of the space that matter takes up. The density of a substance is the mass per unit of volume.

Everything on the planet is composed of elements or compounds made from combinations of these elements. The elements are organized according to their properties and atomic structure into a chart called the periodic table. The periodic table holds a wealth of information about the elements and their atoms. Elements that are located in the same column of the periodic table tend to have similar behavior and reactivity. The noble gases are very stable elements with extremely low reactivity.

79

**Au**

Gold

197.0

Matter is composed of individual building blocks called atoms. Atoms are much too small to be seen, so experimental evidence has led to various models of the atom. In the center of each atom is a dense nucleus consisting of protons and neutrons. Most of the mass of an atom is located in its nucleus. Most atoms have more neutrons than protons in their nuclei, which keeps the nucleus more stable.

— Negative electron

— Uncharged neutron
— Positive proton

Electrons are located at specific distances from the nucleus, called shells. The electrons in the outermost shell of an atom are called its valence electrons and are key to its behavior. Atoms will sometimes lose or gain valence electrons, forming ions with positive and negative charges, and achieving the electron arrangements of the noble gases.

A compound is a substance that contains atoms of more than one type of element, bonded together. There are millions of compounds in existence and many yet to be discovered.

A bond is an attractive force that keeps atoms connected to one another. The attraction occurs between the positively charged nuclei and the negatively charged electrons in different atoms. Usually the valence electrons are involved in bonding.

There are four main models for bonding: ionic, molecular covalent, metallic, and network covalent. These models can be used to explain difference in solubility and conductivity.

## REVIEW EXERCISES

### General Review

*Write a brief and clear answer to each question. Be sure to show your work.*

1. What is the difference between an element and a compound?

2. Consider two objects that each weigh 20.0 g. One is made of lead and has a density of 11.4 g/mL. The second object is made of aluminum and has a density of 2.7 g/mL. Which object takes up more space? What volume does each occupy?

3. Use the periodic table to help you find the atomic symbol, atomic number, group number, number of protons, and number of electrons for these elements:

   **a.** lithium  **b.** bromine  **c.** zinc
   **d.** sulfur   **e.** barium   **f.** carbon

4. Describe nuclear fission and nuclear fusion.

5. What is an isotope? How can you figure out the most common isotope of an element?

6. Describe how you would use the periodic table to help you predict the type of bond between two atoms.

7. What are cations and anions?

8. Sort the following atoms into metals and nonmetals. Then name three ionic compounds that can be formed from some combination of these atoms.

   Na  S  Cl  Sr  Mg  Se  I  Cu

9. Predict the type of bonding in each of the following substances. Provide their chemical names and predict which ones will conduct electricity.

   **a.** $MgCl_2$  **b.** $O_2$
   **c.** AgOH    **d.** Pt

10. Explain how you can determine the charge on a transition metal ion.

11. A substance does not dissolve in water and does not conduct electricity. What model of bonding accounts for these observations?

12. What is required to change one element into another element? Is it possible to do this in a laboratory?

## STANDARDIZED TEST PREPARATION

### Multiple Choice

*Choose the best answer.*

1. The density of zinc is 7.1 g/cm$^3$ and the density of copper is 9.0 g/cm$^3$. What statement correctly describes the density of a zinc block coated with copper?

   **(A)** The block has a density of 7.1 g/cm$^3$.
   **(B)** The block has a density less than 7.1 g/cm$^3$.
   **(C)** The block has a density between 7.1 g/cm$^3$ and 9.0 g/cm$^3$.
   **(D)** The block has a density of 9.0 g/cm$^3$.

**2.** According to an ancient story, Archimedes had his "crowning moment" when he discovered the trickery of a goldsmith by using density. The goldsmith had received an order for a pure gold crown but crafted one from a mixture of metals. The volume of this crafted crown would have differed from the volume of a pure gold crown. If the mass of a pure gold crown was 1.2 kg, and the density of gold is 19.3 g/cm³, what would be the volume of a pure gold crown?

(A) 0.062 cm³      (B) 62.2 cm³

(C) 23.16 cm³      (D) 16.08 cm³

**3.** The shell model below has 3 protons, 3 neutrons, and 3 electrons. What statement correctly describes this atom?

(A) The atomic number of this element is 6 because there are 3 protons and 3 neutrons; and the mass number is 3 because there are 3 protons.

(B) The atomic number of this element is 3 because there are 3 protons, 3 neutrons, and 3 electrons; and the mass number is 9 because there are 3 of each subatomic particle.

(C) The atomic number of this element is 3 because there are 3 protons; and the mass number is 6 because there are 3 protons and 3 neutrons.

(D) The atomic number is 6 because there are 3 protons and 3 electrons; and the mass number is 6 because there are 3 protons and 3 electrons.

**4.** A shell model for a carbon atom is shown. Which statement correctly describes the electrons in carbon?

(A) Carbon has 6 core electrons and no valence electrons because there are a total of 6 electrons in the neutral atom.

(B) Carbon has 12 electrons because the mass number is 12.

(C) Carbon has a core similar to the helium atom, but carbon has 4 valence electrons.

(D) Carbon has a core similar to the neon atom, but carbon has 4 valence electrons.

**5.** The element in the shell model below is a poor conductor of heat and electricity and is very reactive. Which element would have similar properties to this one?

(A) Na, sodium

(B) Br, bromine

(C) Ar, argon

(D) S, sulfur

*Use the periodic table shown to answer Exercises 6 and 7.*

**6.** Two groups of the periodic table are highlighted in the figure. What are the names given to these groups?

(A) Alkali metals and noble gases

(B) Alkali metals and halogens

(C) Alkaline earth metals and noble gases

(D) Alkaline earth metals and halogens

**7.** What type of bonding would you expect to find in substances made when an element from one highlighted group is combined with an element from the other highlighted group?

(A) Ionic only

(B) Molecular covalent only

(C) Network covalent only

(D) Ionic and network covalent

8. If an alkaline earth metal formed a compound with a polyatomic phosphate ion, $PO_4^{3-}$, what would be the subscript given to the metal in the chemical formula for the compound?

(A) 3

(B) 2

(C) 1

(D) There would be no subscript written in the chemical formula.

9. What is present in the nucleus of an atom of $^{226}_{88}Ra$?

(A) 88 neutrons and 138 protons

(B) 88 electrons and 226 protons

(C) 88 protons and 226 neutrons

(D) 88 protons and 138 neutrons

10. Thallium's atomic number is 81 and the average atomic mass is 204.4 amu. Thallium has two isotopes, thallium-203 and thallium-205. Which statement correctly describes thallium?

(A) A naturally occurring sample of thallium contains both isotopes.

(B) Thallium-203 isotope has 122 protons, and thallium-205 has 124 protons.

(C) The most common atom of thallium has a mass of 204.48 amu.

(D) Thallium-203 and thallium-205 both have 81 neutrons.

11. An element has three naturally occurring isotopes. One of the isotopes of the element has a mass number of 24 and is 79% abundant in nature. Another of the isotopes has a mass number of 25 and is 10% abundant in nature. The last isotope of the element has a mass number of 26 and is 11% abundant in nature. Based on this information, what is the most likely identity of this element?

(A) Beryllium, Be

(B) Sodium, Na

(C) Magnesium, Mg

(D) Aluminum, Al

*Use the graph below of the radioactive decay of carbon-14 to answer Exercises 12 and 13.*

12. What is the approximate half-life of carbon-14?

(A) 50 years

(B) 5,730 years

(C) 11,460 years

(D) 17,190 years

13. If the original sample of carbon-14 were 10.0 g, how much would remain after two half-lives?

(A) 234.0 g

(B) 1.25 g

(C) 5.00 g

(D) 2.50 g

14. A sheet of aluminum would protect a person from what type of radiation exposure?

(A) A sheet of aluminum would protect from alpha, beta, and gamma radiation.

(B) A sheet of aluminum would protect from alpha and beta, but not gamma radiation.

(C) A sheet of aluminum would protect from alpha and gamma, but not beta radiation.

(D) A sheet of aluminum would not protect a person from any form of radiation.

Use the diagram of the copper cycle below to answer Exercises 15–19.

15. Which of these statements correctly describes the law of conservation of mass?

   (A) The law of conservation of mass would be supported if the amount of copper at the beginning and end of a copper-cycle experiment were accurately measured.

   (B) The law of conservation of mass states that good laboratory techniques will help prevent the accidental loss of lab materials.

   (C) The law of conservation of mass states that small amounts of mass are lost in the course of an experiment because of spills, measurement errors, and other factors.

   (D) The law of conservation of mass does not apply to gases or aqueous solutions.

16. Which of the following are solids produced in the copper cycle? (Choose all that apply.)

   (A) Nitrogen dioxide
   (B) Copper hydroxide
   (C) Sulfuric acid
   (D) Copper oxide

17. What is true about the copper ions present in the copper cycle?

   (A) Only copper (I) is present in the ionic compounds.
   (B) Only copper (II) is present in the ionic compounds.
   (C) Both copper (I) and copper (II) are present in the ionic compounds.
   (D) Only elemental copper is present in the copper cycle.

18. What is the charge of the copper atoms in the copper solid, $Cu(s)$?

   (A) $+2$    (B) $+1$
   (C) $0$     (D) $-1$

19. Which substances, as written below, would most likely conduct electricity? (Choose all that apply.)

   (A) $NO_2(g)$
   (B) $Cu(NO_3)_2(aq)$
   (C) $NaOH(aq)$
   (D) $Cu(OH)_2(s)$

20. The acetate ion $C_2H_3O_2^-$ has a charge of $-1$. What is the correct formula for calcium acetate?

   (A) $CaC_2H_3O_2$
   (B) $Ca_2C_2H_3O_2$
   (C) $Ca(C_2H_3O_2)_2$
   (D) $C_2H_3O_2Ca$

21. Which of the following compounds will generally *not* form?

   (A) $KBr$
   (B) $NeBr_2$
   (C) $AlBr_3$
   (D) All of these compounds will form.

22. Four models of chemical bonding are illustrated in the diagram. The red spheres represent the nuclei and core electrons of atoms and blue areas represent where valence electrons are located.

Model 1        Model 2        Model 3        Model 4

Which statement below correctly identifies the type of bonding for the compound listed?

   (A) $MgCl_2(s)$ bonds according to Model 1 because valence electrons are shared between the atoms.

   (B) $CO_2(g)$ bonds according to Model 1 because valence electrons are shared between the atoms.

   (C) $Li(s)$ atoms are bonded according to Model 4 because valence electrons are free to move about the substance.

   (D) $N_2(g)$ bonds according to Model 4 because valence electrons are shared throughout the entire substance.

# Water Filtration

## Work Groups of 4

### Purpose

To use an engineering design process to design, build, and refine a simple water filtration system.

### Problem

Your teacher will provide a sample of water that contains different substances. Some of the substances will be visible like sand, dirt, leaves, metals, or compounds. Some substances will be dissolved and not visible. Your job is to design a procedure to filter the water to make it cleaner and colorless. Your goal in the end is to have one container with clear, colorless water and other containers that hold the substances you separate out.

### Safety Notes

- Be careful to treat the water and all the substances observing all the safety rules you would use in any lab.
- Not all substances or organisms that are in the water can be seen or separated out with filtration—so even after filtration, your water sample will not be safe for drinking or other human use such as washing.
- Remember never to taste or touch chemicals or mixtures in a laboratory setting.
- As a part of this project you will need to make a list of all safety measures you should use when creating your filtration system.

### Review

Before you begin, review the handout, *Engineering Design* (following this project description), and reflect on what you have learned so far in your class. For this project, you will focus on identifying the problem, brainstorming for solutions, developing a procedure, and if possible, testing and refining that plan.

Here are some chemistry topics and skills that were covered in Unit 1: Alchemy that you might reflect on before working on this project:

| Laboratory safety | Properties of matter | Classification of matter |
|---|---|---|
| Use of scientific tools | Development of a hypothesis | Mixtures |
| Observation | Phases of matter | Properties of elements |
| Prediction | Physical and chemical properties | |

# Engineering Project

**Part 1:** Identify the Problem and Brainstorm Solutions

1. Identify the Problem: What is the problem you are trying to solve? Write up a description of what you need to accomplish.

2. Brainstorm: In your Group, brainstorm some ideas to solve this problem. Here are some questions to consider:
   - What physical properties of matter might be important to consider when creating a water filter?
   - What are some examples of filters you know about from your own life? How do you think they work? How could you find out more information?
   - What materials do you think you would need to filter your water?
   - What are some substances or organisms that you might not be able to remove from water by filtration only?

3. Research: In the library and online, research different ways that you could create a simple water filtration system. Create a list of all of your sources to turn in with your project. Record your research notes so they can be used later when creating your plan. Here is a website you might try as a start. If you find others, make sure to record the source and the website URL on your list:

   **PBSkids.org**   http://pbskids.org/zoom/activities/sci/waterfilter.html

**Part 2:** Create a Prototype

1. Based on your research and brainstorming, write a lab procedure for how to separate your mixture.

2. Write a list of materials needed for your procedure.

3. Write a list of all safety considerations needed for your procedure.

4. Submit your procedure to your teacher for review and approval. Include your list of resources and research materials.

5. Once approved, conduct your separation procedure in your school laboratory with teacher supervision.

**Part 3:** Refine and Retest

1. Ask your classmates to help you think of ways to improve your procedure. What changes might you make based on their feedback?

2. Make changes to your procedure based on feedback and the results of your testing.

3. Submit your revised procedure to your teacher for review and approval. Include your list of resources and research material.

4. Once approved, use your revised procedure to separate your mixture.

5. Once your procedure and design are refined, your teacher may ask you to lead a small group of students in demonstrating the process you designed.

## Possible Presentation Options:
- Oral presentation describing your procedure
- Poster describing your procedure in detail with diagrams
- A protocol pamphlet describing your procedure with diagrams
- A website that outlines your project including your procedure and diagrams

# Engineering Design

Engineers apply scientific and mathematical principles to create or improve such things as materials, technology, medicine, and processes that help society. Engineers work in a variety of disciplines and design products that affect many areas of our lives. For example, engineers might design a structure like a building or a bridge. They might create new designs for planes, trains, and cars or design improvements for computers, televisions, or cell phones. Engineers can also work to create new methods of using and producing materials used in all areas of our lives including agriculture, environmental protection, medicine, clothing, and personal items.

Engineers, like scientists, use systematic methods and processes to solve problems as well as to improve and create new technologies. Many of the skills and practices used by scientists are also used by engineers. One way to think about the difference between science and engineering is that science allows us to understand the world better and engineering allows us to develop better tools and processes for the world. Engineers use science to create and develop their designs. Both scientists and engineers make use of similar process and skills. However, in engineering design, there are some differences. Here are some general engineering design practices. Some might overlap with scientific practices you've learned about already.

## Engineering Design Process

Engineering is a field that makes practical use of science and mathematics. As in scientific methods, there is no exact engineering method, just steps engineers use when designing solutions to problems. Below we list several of these steps. Engineers will make use of science and math to:

- Identify or determine a real world problem that needs a solution.
- Brainstorm different ways to solve the problem.
  Once they have many ideas, engineers will determine which ideas to explore more deeply.
- Create a plan or prototype.
  A prototype is a preliminary idea or model that can be tested as a solution to the problem being studied. In science, models are also developed to explain the natural world.
- Complete the plan and/or develop the prototype.
- Test the prototype and obtain feedback on what works and what needs improvement.
- Implement changes based on feedback collected. Refine the prototype to work better than the previous version.
- Test the plan or prototype again and again, incorporating additional feedback until a successful product is made.

This process goes through many cycles. Rarely are prototypes perfect at first. Just like scientists, engineers discover throughout the entire engineering design process and adjust their work based on these discoveries.

# Unit 2 | Smells

Alex Cao/Getty Images

*Each flower puts out a scent that attracts specific insects. Not all flowers smell pleasant. The corpse flower, which blooms once every three years, smells like decaying flesh.*

## In this unit, you will learn

- **how atoms form molecules**
- **to predict the smell of a compound**
- **to interpret molecular models**
- **how the nose detects different molecules**
- **about amino acids and proteins**

## Why Smells?

The sense of smell is a familiar and important part of our lives. It helps us detect pleasant smells as well as unpleasant smells that alert us to possible dangers. Sometimes, our sense of smell helps us to detect things that we can't see. Sometimes, we detect something all the way across the room or even in another room. But how does it work? The way atoms are connected in molecules, and the structures of the molecules, have a great deal to do with the properties of those molecules. When you understand the chemistry of molecules, you will begin to answer the question, "What is the chemistry of smell?"

# CHAPTER 6

# Speaking of Molecules

### In this chapter, you will study

- different representations for molecules

- bonding patterns in molecular covalent compounds

- how molecular structure relates to smell

*Our sense of taste is closely related to our sense of smell. For this reason, many chefs have very sophisticated senses of smell.*

The nose is able to identify hundreds of different smells. Most substances that smell are made of molecules. So, how does the nose differentiate one molecule from another? When you analyze the molecular composition and structure of compounds that smell, a variety of patterns emerges. Knowing the number and types of atoms in molecules, and the way atoms are connected in molecules, is vital to understanding the chemistry of smell.

# Molecular Formulas

## THINK ABOUT IT

The topic of smell is both friendly and familiar. Most of us have had many experiences with smells: smells that attract, smells that disgust, and smells that evoke strong memories of vacations, holidays, or other experiences. It turns out that our noses are fantastic chemical detectors. Learning more about your nose and how it works will help you learn more about chemistry.

### What does chemistry have to do with smell?

*To answer this question, you will explore*

➥ The Sense of Smell

➥ Classifying Smells

➥ Molecular Formula, Name, and Smell

## EXPLORING THE TOPIC

### ➥ The Sense of Smell

Someone opens a perfume bottle across the room. After a few seconds, you notice the sweet smell of the perfume. How did the smell get to your nose and how did your nose detect it?

The smells that come from a vial of perfume and end up in your nose are made of molecules. Molecules are groups of nonmetal atoms that are covalently bonded together. In this unit, you will examine mechanisms that your nose uses to detect and distinguish various smells.

To be consistent, you will always work with the consensus smell. In other words, you will work with the smell reported by the majority of your classmates.

### ➥ Classifying Smells

To compare smells, we need a common language for describing them. How would you describe the smells of the items in these pictures?

Scientists classify smells by placing similar-smelling substances in the same category. For example, apples, bananas, strawberries, and pineapples might all fit into a category called "fruity" smelling. Roses, geraniums, and lilies might all fit into a category called "floral" smelling. Further, fruity- and floral-smelling substances might be grouped together in a larger category called "sweet" smelling.

Not all scientists agree on the specific names for smell categories or on the number of different categories that should exist. As you begin studying smells, you will start with three categories: sweet, minty, and fishy.

## Molecular Formula, Name, and Smell

A closer investigation reveals connections among a substance's chemical name, its chemical formula, and its smell. For a molecular covalent compound, the chemical formula is also called the **molecular formula,** because it describes the makeup of each molecule.

Examine the drawings of the five vials. Each vial contains a cotton ball soaked in a different molecular compound. See if you can find any connections between the smells and their chemical names and molecular formulas.

| Minty | Sweet | Sweet | Minty | Fishy |
|---|---|---|---|---|
| 1 | 2 | 3 | 4 | 5 |
| $C_{10}H_{14}O$ | $C_7H_{14}O_2$ | $C_8H_{16}O_2$ | $C_{10}H_{16}O$ | $C_8H_{19}N$ |
| L-carvone | Ethyl pentanoate | Hexyl acetate | D-pulegone | Diisobutylamine |

Here are some patterns that you might notice.

### ALL THE MOLECULES

- contain carbon and hydrogen atoms
- have more hydrogen atoms than carbon atoms
- are composed of three different types of atoms

### SWEET-SMELLING MOLECULES

- contain two oxygen atoms
- have twice as many hydrogen atoms as they do carbon atoms
- have two-word names ending in "-ate"

### MINTY-SMELLING MOLECULES

- contain one oxygen atom
- contain ten carbon atoms

- have even numbers of carbon and hydrogen atoms
- have names ending in "-one"

**FISHY-SMELLING MOLECULES**

- contain a nitrogen atom
- do not contain oxygen atoms

Perhaps *every* molecule with two oxygen atoms smells sweet. Perhaps *every* molecule that smells fishy contains a nitrogen atom. These are reasonable hypotheses based on very little data. You can test these hypotheses by gathering more data.

**Example**

### Predicting Smell

How would you expect a molecule with the molecular formula $C_9H_{18}O_2$ to smell? Explain.

*Solution*

The molecule has two oxygen atoms. Based on data given so far, this suggests that the molecule smells sweet. It would be useful to know the name to confirm this. So far, we have seen that the names of sweet-smelling molecules end in "-ate."

## LESSON SUMMARY

### What does chemistry have to do with smell?

The smells that your nose detects are made of molecules. The molecular formulas and chemical names of molecular substances are directly related to the way these substances smell. You may be able to predict the smell of a substance by examining its chemical formula, its chemical name, or both.

**KEY TERM**

molecular formula

**Exercises**

### Reading Questions

1. How do scientists classify smells? What purpose might this serve?
2. From the evidence so far, how are molecular formulas related to smell? How are chemical names related to smell?

### Reason and Apply

3. Name five substances from home that do not fit into the smell categories of sweet, minty, or fishy. What new categories would you put them in?
4. Make a list of five things from home that have no smell.
5. In a paragraph or two, write about *one* of the following. You may use a cartoon format instead, if you wish.
   a. Describe a memory that you have that relates to smell. Describe the smell itself and what it reminds you of. (Avoid using adjectives that are not informative, such as "weird.")

*(continued)*

    **b.** If you were a smell superhero, how would you use your smell superpowers or your enhanced sense of smell to assist the world?

    **c.** What would your life be like if you couldn't smell?

**6.** Try tasting something while having your nose plugged and your eyes closed. You may wish to work with a partner who can provide you with a variety of different-flavored jelly beans or other food items to sample.

    **a.** Describe your experience. How difficult is it to taste something when you can't smell it and you don't know what it is?

    **b.** How do you think taste is related to smell?

**7.** Predict the smells of these molecules. Explain your reasoning.

    **a.** methyl octenoate, $C_9H_{18}O_2$

    **b.** monoethylamine, $C_2H_7N$

    **c.** ethyl acetate, $C_4H_8O_2$

**8.** What do you think 1,5-pentanedithiol ($C_5H_{12}S_2$) would smell like? How confident of your answer are you? Explain why.

**9.** What makes something smell the way it does? Discuss your own hypothesis about why different things smell different. Include questions you still have.

# Structural Formulas

## THINK ABOUT IT

Would you rather smell sweaty gym socks or fresh pineapple? The choice is pretty clear. But what if you were asked if you would rather smell ethyl butyrate or hexanoic acid? How would you know which compound is safe to sniff? Both molecules have the same molecular formula: $C_6H_{12}O_2$.

Alberto Pérez Veiga/Shutterstock

© RedHelga/iStockphoto

**How can molecules with the same molecular formula be different?**

*To answer this question, you will explore*

- Predicting Smells
- Structural Formula
- Orientations of Structural Formulas

## EXPLORING THE TOPIC

### Predicting Smells

In Lesson 28: Sniffing Around, you arrived at some generalizations that relate smell to molecular formula and chemical name. What happens if you use these generalizations to make predictions about some new compounds? Consider the five new compounds listed in the table.

| Vial | Chemical name | Molecular formula | Predicted smell | |
|------|---------------|-------------------|-----------------|---|
| 1 | pentyl acetate | $C_7H_{14}O_2$ | sweet | All three have two oxygen atoms. Two end in "-ate" and one in "acid": sweet smelling? |
| 2 | pentanoic acid | $C_5H_{10}O_2$ | sweet | |
| 3 | propyl acetate | $C_5H_{10}O_2$ | sweet | |
| 4 | ethylthiol | $C_2H_6S$ | a new smell category | Contains a sulfur atom: new smell category? |
| 5 | hexylamine | $C_6H_{15}N$ | fishy | Contains a nitrogen atom: fishy? |

The predictions based on molecular formula and chemical name alone are reasonably accurate. Four out of five are correct. Ethylthiol has a skunk smell, so it belongs in a new smell category. The only prediction that is incorrect is the pentanoic acid in vial 2. It has a molecular formula *identical* to the molecular formula of a sweet-smelling molecule, but it has a very putrid smell. Apparently, two compounds can have identical formulas, different names, and very different smells. Perhaps this is why its name does not end in "-ate" like the other sweet-smelling compound.

As you continue to study molecules, some of the patterns discovered here will hold up, while others will be revised or abandoned. You can continue to refine your ideas about smell so that you can better understand the chemistry of smell.

**Big Idea** Science involves coming up with ideas based on observations and then refining these ideas based on further observation.

## Structural Formula

As it turns out, two compounds can have identical molecular formulas but smell much different. This is because having the same molecular formula does not guarantee that the molecules themselves are identical.

To show how the individual atoms in a molecule are bonded to each other, chemists use what is called a structural formula. A **structural formula** is a diagram or drawing that shows how the atoms in a molecule are arranged and where they are connected. The lines between the atoms represent covalent bonds holding the atoms together. Take a moment to examine the structural formulas for ethyl butyrate and hexanoic acid below. What differences do you see?

These compounds have the same molecular formula, but different smells.

$C_6H_{12}O_2$

**Ethyl butyrate**, pineapple smell          **Hexanoic acid**, sweaty gym sock smell

Notice that the oxygen atoms are located in different places within the two molecules. In one molecule, the oxygen atoms are in the middle of the molecule, whereas in the other molecule, the oxygen atoms are at one end.

### ISOMERS

In some compounds, the atoms can be connected only a certain way. For instance, the formula $C_2H_6$ has only one structural formula.

$C_2H_6$

As the number and types of atoms in a molecule increase, there are more possible ways to connect the atoms. Molecules with the same molecular formula and different structural formulas are called **isomers.** The formula $C_2H_6O$ has two isomers, shown here.

$C_2H_6O$

Ethyl butyrate and hexanoic acid are considered isomers of each other. The formula $C_6H_{12}O_2$ has many more isomers. In fact, there are more than 25 different molecules with the molecular formula $C_6H_{12}O_2$, each with unique properties.

Drawings of structural formulas are flat and two-dimensional. However, molecules are not flat. They take up three dimensions in space.

Ethanol

## ⮑ Orientations of Structural Formulas

There are several ways to draw the *same* structural formula. Consider these two drawings.

These two molecules have the same molecular formula and each atom has the same connections. The structural formula has just been turned 90° in space. These molecules are *not* isomers; they have the same structural formula.

Here are four structural formulas. How many different compounds are represented?

Molecule 1          Molecule 2          Molecule 3          Molecule 4

If you build three-dimensional models of these molecules, you will find that there are only three different compounds here. The first and the third structural formulas are identical. All the atoms in the first and third structures are connected in the same way.

The illustration on the next page shows how molecule 1 can be turned and rotated to match molecule 3 without taking the model apart.

*Start with molecule 1.*     *Turn it around.*     *Rotate the carbon atom on the left.*
*This is the same as molecule 3.*

Structural formulas show only the sequence in which one atom is connected to the next. The way atoms are connected in a molecule is like a charm bracelet. The middle charm is always between the same two charms regardless of how you hold the bracelet. It does not matter if the oxygen atom is drawn on the left or right side of the molecule. It also does not matter if the O—H is drawn horizontally or vertically. It is only the number and type of connections that matter.

**Example**

## Isomers of $C_3H_9N$

How many different isomers are represented below?

Molecule 1   Molecule 2   Molecule 3   Molecule 4

*Solution*

There are three isomers. Molecules 2 and 3 are identical. So there are three different compounds represented here.

LESSON SUMMARY

### How can molecules with the same molecular formula be different?

Compounds with identical molecular formulas may have different structures. Chemists use drawings called structural formulas to show how the atoms in molecules are connected to each other. Each line in a structural formula represents a covalent bond between atoms. When molecules have the same molecular formula but different structural formulas, they are called isomers. Isomers represent different compounds with different properties. Because smell is a property, compounds with the same molecular formula may smell different from each other because they are different substances.

**KEY TERMS**

structural formula

isomer

**Reading Questions**

1. What information does a structural formula provide?

2. What are isomers?

**Reason and Apply**

3. If you are given a structural formula of a molecule, do you also know its molecular formula? Explain your thinking.

4. The words *isotope* and *isosceles* also have the prefix "iso-." How are their meanings similar to that of the word *isomer*?

5. Structural formulas for six molecules are shown below. Write the molecular formula for each of the molecules.

a.
```
        O   H   H
        ||  |   |
H—O—C—C—C—H
            |   |
            H   H
```

b.
```
    H   O   H   H   H
    |   ||  |   |   |
H—C—C—C—C—C—H
    |       |   |   |
    H       H   H   H
```

c.
```
    H   O   H   H
    |   ||  |   |
H—C—C—C—C—H
    |       |   |
    H     H—C—H H
            |
            H
```

d.
```
    H   H   O       H
    |   |   ||      |
H—C—C—C—O—C—H
    |   |           |
    H   H           H
```

e.
```
    H   H   H   H   H
    |   |   |   |   |
H—C—C—C—N—C—H
    |   |   |   |   |
    H   H   H       H
```

f.
```
            H
            |
H—C—O—C—H
    ||      |
    O       H
```

6. Two molecules have the same molecular formula, yet one smells sweet and the other smells putrid. Explain how you think this might be possible.

7. For each of the molecules in Exercise 5, draw an isomer of the compound.

*HONC If You Like Molecules*

# Bonding Tendencies

## THINK ABOUT IT

To understand the chemistry of smell, it is necessary to understand how molecules are put together. If you examine the structural formulas of molecules, you can see patterns in the way the atoms are connected. For example, hydrogen atoms are always arranged around the outside of the molecule. Plus, hydrogen atoms always have only one line connecting them to other atoms, while carbon atoms always have more than one line connecting them to other atoms.

**What are the rules for drawing structural formulas?**

*To answer this question, you will explore*

➲ The HONC 1234 Rule

➲ Drawing Structural Formulas

## EXPLORING THE TOPIC

### ➲ The HONC 1234 Rule

The structural formulas for three molecules are shown here. Take a moment to count the number of times each type of atom—hydrogen, oxygen, nitrogen, and carbon—is connected to other atoms.

$C_6H_{12}O_2$

$C_{10}H_{18}O$

$C_4H_{11}N$

- Every **Hydrogen** atom has **one** line connecting it to other atoms.
- Every **Oxygen** atom has **two** lines connecting it to other atoms.
- Every **Nitrogen** atom has **three** lines connecting it to other atoms.
- Every **Carbon** atom has **four** lines connecting it to other atoms.

This information is sometimes referred to as the **HONC 1234 rule.** Within most molecules, hydrogen makes one bond, oxygen makes two bonds, nitrogen makes three bonds, and carbon makes four bonds. The bonds in a structural formula are represented by lines. One line connecting two atoms is called a *single bond*. A pair of lines connecting the same two atoms, as in C=O, is called a *double bond*.

The HONC 1234 rule tells you how four of the most common nonmetal atoms will bond. You will learn about the bonding of other nonmetal atoms like sulfur, S, and chlorine, Cl, in later lessons.

Example 1

## HONC 1234

Are the following molecules correct according to the HONC 1234 rule? If not, what is wrong with them?

$$
\begin{array}{ccc}
 & \text{H} \quad \text{H} \quad \text{H} & \quad\quad \text{H} \quad\quad \text{H} \\
 & | \quad\; | \quad\; | & \quad\quad | \quad\quad\; | \\
\text{H}-\text{C}-\text{C}-\text{O}-\text{H} & \quad \text{H}-\text{C}-\text{H}-\text{C}-\text{H} \\
 & | \quad\; | \quad\; | & \quad\quad | \quad\quad\; | \\
 & \text{H} \quad \text{H} \quad \text{H} & \quad\quad \text{H} \quad\quad\; \text{H}
\end{array}
$$

Molecule 1          Molecule 2

### Solution

Both molecules are incorrect according to the HONC 1234 rule. In molecule 1, there is an oxygen atom with four bonds. It should have only two bonds. In molecule 2, there is a hydrogen atom with two bonds. It should have only one. The hydrogen atom cannot go in the middle of the molecule, between two carbon atoms.

## ⮕ Drawing Structural Formulas

The HONC 1234 rule is all you need to draw the structural formulas for thousands of different molecules correctly. Some examples are given here and on top of the next page.

### STRUCTURAL FORMULA FOR $C_2H_6$

Consider a molecule with two carbon atoms and six hydrogen atoms. Its molecular formula is $C_2H_6$. What structural formula would this molecule have? Start by connecting the two carbon atoms with a single bond. The "carbon backbone" is generally a good place to start when drawing molecules.

**Step 1:** Connect the carbon atoms:

C—C

**Step 2:** Add hydrogen atoms. Make sure each hydrogen atom has just one bond and each carbon atom has a total of four bonds:

$$
\begin{array}{cc}
\text{H} & \text{H} \\
| & | \\
\text{H}-\text{C}-\text{C}-\text{H} \\
| & | \\
\text{H} & \text{H}
\end{array}
$$

The structural formula shown above is consistent with the HONC 1234 rule. This is the only structure you can make out of this molecular formula. It represents a substance called *ethane*.

### STRUCTURAL FORMULA FOR $C_2H_6O$

Consider a molecule with two carbon atoms, six hydrogen atoms, and one oxygen atom. Its molecular formula is $C_2H_6O$. What structural formula could this molecule have?

### ASTRONOMY CONNECTION

At room temperature, ethane, $C_2H_6$, is a flammable gas. On Earth, ethane is one of the components of natural gas. It has also been detected in the atmospheres of Jupiter, Saturn, Uranus, and Neptune.

NASA/GSFC

Start by connecting the carbon atoms. Then add the oxygen atom.

**Step 1:** Connect the carbon atoms:

C—C

**Step 2:** Add the oxygen atom to the carbon chain:

C—C—O

**Step 3:** Add hydrogen atoms last:

$$\begin{array}{ccc} \text{H} & \text{H} & \\ | & | & \\ \text{H}-\text{C}-\text{C}-\text{O}-\text{H} \\ | & | & \\ \text{H} & \text{H} & \end{array}$$

Make sure your structural formula follows the HONC 1234 rule. Notice that hydrogen atoms are generally the last atoms you add when creating a structural formula.

This molecule is called *ethanol* or *ethyl alcohol*.

### A SECOND STRUCTURE—AN ISOMER

Ethanol is not the only possible structure for this particular molecular formula. The atoms in $C_2H_6O$ can also be arranged a different way, to form a completely different molecule. This new structure is created by placing the oxygen atom between the two carbon atoms.

**Step 1:** Connect the carbon and oxygen atoms:

C—O—C

**Step 2:** Add hydrogen atoms:

$$\begin{array}{ccc} \text{H} & & \text{H} \\ | & & | \\ \text{H}-\text{C}-\text{O}-\text{C}-\text{H} \\ | & & | \\ \text{H} & & \text{H} \end{array}$$

This structure *also* follows the HONC 1234 rule and is a correct structural formula for $C_2H_6O$. This molecule is called *dimethyl ether*. Ethanol and dimethyl ether are isomers. They have the same molecular formula but different structural formulas. Dimethyl ether has different properties from ethanol, including smell.

Not all structural formulas are simple to figure out. As the number and type of atoms increase, there are more possible ways to connect the atoms.

**Example 2**

### Structural Formula for $C_4H_{11}N$

Draw a structural formula for the molecular formula $C_4H_{11}N$.

*Solution*

Start by making a chain of the carbon atoms. Place the nitrogen atom anywhere in the chain. Finish by adding hydrogen atoms so that every carbon atom has four bonds and the nitrogen atom has three bonds. Two possible solutions are shown here, but many more are possible.

## What are the rules for drawing structural formulas?

The individual atoms in molecules are not connected randomly. Each atom in a molecule has a tendency to bond a specific number of times. The HONC 1234 rule describes the bonding patterns of hydrogen, oxygen, nitrogen, and carbon atoms. The HONC 1234 rule can be a valuable tool for creating structural formulas from molecular formulas.

**KEY TERM**

HONC 1234 rule

## Exercises

### Reading Questions

1. What is the HONC 1234 rule?

2. Explain why one molecular formula can represent more than one structural formula.

### Reason and Apply

3. Use the HONC 1234 rule to create possible structural formulas for molecules with these molecular formulas. Remember that it is easiest to start with the carbon atoms.
   **a.** $C_3H_8O_2$     **b.** $C_4H_{11}N$     **c.** $C_4H_{10}$     **d.** $C_5H_{12}O$

4. Is each structural formula correct according to the HONC 1234 rule? For any molecules that don't follow the HONC 1234 rule, repair the incorrect structural formula. (*Note:* There may be more than one correct way to repair a formula.)

5. Think about how molecules might interact with your nose. Why do you think molecules with different structural formulas have different smells?

# Featured ACTIVITY

## Materials

- Lewis dot puzzle pieces

*Three hydrogen atoms combine with one nitrogen atom to form*

*...a molecule of NH₃.*

H:N:H
H

*Lewis dot structure for NH₃*

# Connect the Dots

## Purpose

To investigate the role of electrons in covalent bonding.

## Instructions

The puzzle pieces you have been given represent Lewis dot symbols. The puzzle pieces allow you to pair up electrons and create molecules.

1.  Use the puzzle pieces to construct the molecules given below. Then draw the Lewis dot structure for each molecule.

    $PH_3$ $\qquad$ HOCl $\qquad$ $F_2$ $\qquad$ $CH_3Cl$

2.  Use the puzzle pieces to create more molecules, following the directions given below. For each molecule, draw the Lewis dot structure and write the molecular formula.

    a.  Use one S atom and as many H atoms as you need.

    b.  Use one Si atom and as many F atoms as you need.

    c.  Use two O atoms and as many H atoms as you need.

3.  Use the puzzle pieces to construct a molecule with the molecular formula $C_2H_6$. Draw its Lewis dot structure and its structural formula.

4.  Use the puzzle pieces to construct all the possible isomers of $C_3H_8O$. Draw Lewis dot structures for each isomer. Do the molecules follow the HONC 1234 rule?

5.  Use the puzzle pieces to design your own molecule with at least five carbon atoms. Draw its Lewis dot structure. What is the molecular formula of your designer molecule? Does it obey the HONC 1234 rule?

6.  Find a puzzle piece for each type of atom. Put hydrogen and helium aside. Sort the rest of the puzzle pieces according to the periodic table. Record your sort by copying it into a grid. Include the symbols and the dots.

7.  List two patterns that you notice.

8.  **Making Sense** Using what you've learned, explain why the HONC 1234 rule works.

9.  **If You Finish Early** Draw the Lewis dot structures of two different molecules with the molecular formula $C_2H_7N$.

# Lewis Dot Symbols

## THINK ABOUT IT

The HONC 1234 rule is a great trick to help you figure out the structures of molecules. But why does it work? Why does carbon connect with four atoms, while hydrogen can connect with only one atom? To answer these questions, let's take a closer look at bonds.

### How does one atom bond to another in a molecule?

*To answer this question, you will explore*

⮞ The Covalent Bond

⮞ Lewis Dot Symbols and Structures

⮞ Bonded Pairs and Lone Pairs

## EXPLORING THE TOPIC

## ⮞ The Covalent Bond

All of the smelly compounds we have studied so far have been molecules. In fact, most of the substances that smell are made up of molecules, not ionic, metallic, or network covalent compounds.

The bonds that are found in molecules are called covalent bonds. Covalent bonds are formed between the atoms of nonmetallic elements. There are only about 15 nonmetallic elements in the periodic table. That is not many compared to the number of metallic elements.

The atoms involved in a covalent bond *share* a pair of valence electrons between them. The drawing below shows the difference between covalent bonds and ionic bonds.

Covalent bonding: electron sharing

Ionic bonding: electron transfer

In covalent bonds, the nonmetal atoms are *sharing* electrons. As a result, the nonmetal atoms are tightly bound together, but the atoms do not become ions with charges.

**Big Idea** A covalent bond is one in which nonmetal atoms *share* one or more pairs of electrons with one another. An ionic bond is one in which a metal atom *gives up* electrons to a nonmetal atom.

# Lewis Dot Symbols and Structures

Structural formulas are one way to show the structure of a molecule on paper. Each line in a structural formula can be replaced with a pair of dots to represent the electrons that are being shared.

Each line represents a covalent bond, or a pair of shared electrons.

Structural formula

Lewis dot structure

A drawing of a molecule that uses dots to represent the valence electrons is called a **Lewis dot structure.** Above is a molecule of dimethyl ether, $C_2H_6O$. Both the structural formula and the Lewis dot structure represent the same molecule.

Lewis dot structures keep track of every valence electron in every atom of a molecule. If you break the molecule apart into its individual atoms, you can see where each valence electron came from.

A **Lewis dot symbol** is the symbol of an element with dots to show the number of valence electrons that a single atom of that element has. Examine these Lewis dot symbols for hydrogen, oxygen, nitrogen, and carbon. Notice that some electrons are not paired up. These electrons are referred to as *unpaired electrons* and are available for bonding. The unpaired electrons form covalent bonds with other atoms. Lewis dot symbols can help to explain the HONC 1234 rule and the bonding of nonmetal atoms.

| I valence electron | 6 valence electrons | 5 valence electrons | 4 valence electrons |
|---|---|---|---|
| H· | ·Ö: | ·N̈· | ·C̈· |
| I electron available for sharing | 2 electrons available for sharing | 3 electrons available for sharing | 4 electrons available for sharing |

Recall that you can determine the number of valence electrons in an atom by locating its element on the periodic table. Hydrogen is in Group 1A and has one valence electron. Carbon is in Group 4A and has four valence electrons. Once you know the number of valence electrons an atom has, it is easy to draw a Lewis dot symbol for it.

The Lewis dot symbol is named after Gilbert Newton Lewis, a chemist who was a professor at the University of California at Berkeley. Lewis first introduced the idea of shared electrons and covalent bonds in 1916. He was the first scientist to use a system of dots to represent valence electrons in atoms.

# Bonded Pairs and Lone Pairs

Lewis dot symbols can be used to draw Lewis dot structures and structural formulas for molecules. Imagine bringing together the Lewis dot symbols for hydrogen and carbon to form methane, $CH_4$.

Because carbon bonds four times, it will bond with four hydrogen atoms. Once they are bonded, every valence electron in all of the atoms is paired up. The pairs of electrons between atoms are called **bonded pairs.** There are four bonded pairs in the methane molecule shown at the bottom of the previous page.

A pair of electrons in a molecule that are not shared between atoms is called a **lone pair** of electrons. In a molecule of ammonia, the nitrogen atom has one lone pair of electrons.

$$H:\overset{\displaystyle ..}{N}:H$$
$$H$$

**Example**

### Lewis Dot Structure of $PCl_3$

Examine the Lewis dot structure of $PCl_3$, phosphorus trichloride.

$$:\overset{..}{\underset{..}{Cl}}:\overset{..}{P}:\overset{..}{\underset{..}{Cl}}:$$
$$:\overset{..}{\underset{..}{Cl}}:$$

**a.** Draw the Lewis dot symbols for phosphorus and chlorine.

**b.** How many bonds does a phosphorus atom make? How many does each chlorine atom make?

**c.** How many bonded pairs does this molecule have? How many lone pairs?

**d.** How many covalent bonds does this molecule have?

*Solution*

**a.** The periodic table can help you figure out how many valence electrons phosphorus and chlorine have. Phosphorus is in Group 5A and chlorine is in Group 7A, so they have five and seven valence electrons, respectively.

$$\cdot\overset{..}{P}\cdot \qquad \cdot\overset{..}{\underset{..}{Cl}}:$$

**b.** Phosphorus has three unpaired electrons, so it makes three bonds. Chlorine has only one unpaired electron, so it makes one bond.

*The three bonded pairs of electrons are circled.*
*All of the other pairs of electrons are lone pairs.*

**c.** The molecule has three bonded pairs and ten lone pairs.

**d.** Each bonded pair represents one covalent bond. There are three covalent bonds in the molecule.

LESSON SUMMARY

### How does one atom bond to another in a molecule?

Molecules are made of nonmetal atoms that are covalently bonded. A covalent bond is a bond in which a pair of electrons is shared by two atoms. Lewis dot symbols keep track of the number of valence electrons in each atom. They can help you to predict covalent bonding between atoms. A Lewis dot structure shows how the atoms in an entire molecule are bonded together. The valence electrons that are involved in bonding are called a bonded pair. The valence electrons that are paired up in a molecule, but do not take part in bonding, are called lone pairs.

## Exercises

### Reading Questions

1. How are an ionic bond and a covalent bond different? Similar?

2. What is a Lewis dot structure? Explain how you would create one.

### Reason and Apply

3. Draw the Lewis dot symbols for these elements:

   Te    I    K    Bi    In    Pb
   a. Arrange them in order of their group numbers on the periodic table.
   b. Determine how many covalent bonds each element would make.

4. Germanium, antimony, selenium, and bromine each bond to a different number of hydrogen atoms.

   $GeH_4$    $SbH_3$    $H_2Se$    HBr
   a. Draw Lewis dot symbols for Ge, Sb, Se, and Br.
   b. Draw a Lewis dot structure for each molecule.
   c. Explain the pattern in the number of hydrogen atoms.

5. Draw Lewis dot structures for the molecules listed here.
   a. $TeCl_2$    b. HI    c. $AsBr_3$    d. $SiF_4$    e. $F_2$

6. How many lone pairs does each of the molecules in Exercise 5 have?

7. In your own words, explain why the HONC 1234 rule works.

8. Draw Lewis dot puzzle pieces for Si, P, S, and Cl. What rule would you make for Si, P, S, and Cl? What would be the name of this bonding rule?

# Octet Rule

## THINK ABOUT IT

When atoms bond covalently to form molecules, they share electrons to obtain an electron arrangement similar to that of a noble gas atom. Lewis dot structures can help you to discover how atoms share electrons to form molecules.

### How do atoms bond to form molecules?

*To answer this question, you will explore*

➤ The Octet Rule

➤ Double and Triple Bonds

## EXPLORING THE TOPIC

### ➤ The Octet Rule

When carbon, nitrogen, oxygen, and fluorine combine with hydrogen, they form these molecules:

$$\begin{array}{cccc} & \text{H} & & \\ & | & & \\ \text{H} - \text{C} - \text{H} & \text{H} - \text{N} - \text{H} & \text{H} - \text{O} - \text{H} & \text{H} - \text{F} \\ & | & | & \\ & \text{H} & \text{H} & \end{array}$$

All of the above compounds are extraordinarily different. However, their Lewis dot structures reveal a striking similarity. Once carbon, nitrogen, oxygen, and fluorine are bonded, they are each surrounded by eight valence electrons.

Eight valence electrons

Methane     Ammonia     Water     Hydrogen fluoride

This tendency to bond until eight valence electrons surround an atom is called the **octet rule.** (The word *octet* comes from *octo,* which is Latin for "eight.") After these atoms are bonded, they all resemble atoms of a noble gas in their electron arrangements. Recall that the electron arrangements of the noble gases are very stable.

Carbon atom shares eight electrons

Hydrogen atom shares two electrons

Note that hydrogen is an exception to the octet rule. Each hydrogen atom shares two electrons, not eight. A covalently bonded hydrogen atom resembles the noble gas helium, He, which has only two valence electrons.

$F_2$

$SCl_2$

$CH_4O$

### PERIODIC TRENDS

To predict covalent bonding in molecules, you can use the periodic table. In the illustration, Lewis dot symbols are shown for nonmetal elements in the second, third, and fourth rows. Notice that the elements in the same group have similar Lewis dot symbols and bond in a similar way. For example, selenium, Se, which is below sulfur, S, and oxygen, O, has two unpaired electrons and forms two bonds, just like oxygen.

*Elements in the same group have the same number of valence electrons and therefore have similar Lewis dot symbols. Helium is an exception.*

## Double and Triple Bonds

There is more than one way to satisfy the octet rule through bonding. Quite a few of the structural formulas you have examined so far have a bond with two lines. This type of bond is called a **double bond.** In a Lewis dot structure, a double bond is represented by four dots instead of the usual two. A double bond contains four electrons that are shared between atoms. Methyl methanoate, $C_2H_4O_2$, is an example of a molecule that contains a double bond. Its structural formula and Lewis dot structure are shown here.

A double bond means four shared electrons.

The carbon and oxygen atoms with double bonds are surrounded by a total of eight valence electrons each, just like the atoms with single bonds.

---

### Carbon Dioxide

Draw the Lewis dot structure and structural formula for carbon dioxide, $CO_2$.

*Solution*

**Step 1:** Start with the Lewis dot symbols. Bring the atoms together.

Incorrect

:Ö· ·Ċ· ·Ö:  ⟶  :Ö:Ċ:Ö:

**Step 2:** Check to see if the octet rule is satisfied. This Lewis dot structure isn't correct. Move the remaining unpaired electrons to create double bonds.

Incorrect                                    Correct

:Ö:Ċ:Ö:  ⟶  :Ö::C::Ö:

**Step 3:** Check your molecule again to see that each atom in it satisfies the octet rule and still has the correct number of valence electrons.

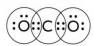

Each atom is now surrounded by a total of eight valence electrons.

**Step 4:** To make the structural formula, replace each pair of dots with a line.

O=C=O

---

If two atoms can share four electrons, can they share six electrons, or even more? The answer is yes.

N≡N    O=O    H—H

:N⋮⋮⋮N:    :Ö::Ö:    H:H

Triple bond  Double bond  Single bond

Nitrogen gas, $N_2$, has a **triple bond** between the two nitrogen atoms, so each nitrogen atom has six shared valence electrons. Notice that there are also two lone pairs in this molecule. Oxygen gas, $O_2$, has a double bond between the oxygen atoms. The four lone pairs in the oxygen molecule have been adjusted slightly in this Lewis dot structure to space the electrons more evenly around the atoms. The placement of the dots in a Lewis dot structure can vary as long as the number of pairs of electrons is correct. Quadruple bonds are also possible, but quite rare.

Some of the most common substances in the world around you are molecules with double and triple bonds. The air you breathe is composed mainly of nitrogen gas and oxygen gas. In fact, the air is 78% nitrogen gas.

## LESSON SUMMARY

### How do atoms bond to form molecules?

**KEY TERMS**

octet rule

double bond

triple bond

When nonmetal atoms bond, they share electrons to obtain the same electron arrangement as a noble gas atom. Nonmetal atoms will share electrons with other atoms so that both atoms share a total of eight valence electrons each. This is called the octet rule. Hydrogen still fits the sharing pattern, but it ends up with only two valence electrons like the noble gas helium. Atoms can also satisfy the octet rule by forming double and triple bonds in which they share four or six valence electrons.

## Exercises

### Reading Questions

1. Explain why nitrogen bonds with hydrogen to form $NH_3$, but not $NH_2$ or $NH_4$. Use Lewis dot structures to support your argument.

2. What is the octet rule, and how can you use it to create a molecular structure?

### Reason and Apply

3. List three nonmetal elements that combine with only one fluorine atom to satisfy the octet rule.

4. List two nonmetal elements that combine with three hydrogen atoms to satisfy the octet rule.

5. Draw Lewis dot structures for these molecules. Notice that in part d and part f, the formulas are written in a way that emphasizes the structure of the molecule.
   a. $CF_4$
   b. $CH_3Cl$
   c. $SiCl_2H_2$
   d. $CH_3OH$
   e. $HOCl$
   f. $CH_3NH_2$

6. Use the octet rule to draw Lewis dot structures for all the stable molecules with the molecular formula $C_3H_8O$. (*Hint:* There are three total molecules.)

7. Consider the molecules $C_2H_2$, $N_2H_2$, and $H_2O_2$.
   a. Draw a Lewis dot structure for each one. What pattern do you notice?
   b. What can you do to check that your Lewis dot structures are correct? Name at least two ways.

8. Consider the molecules $C_2H_4$ and $N_2H_4$. Draw a Lewis dot structure for each of the molecules.

9. Which is more likely to exist in nature, a molecule of $CH_3$ or a molecule of $CH_4$? Explain your reasoning.

# Functional Groups

## THINK ABOUT IT

It makes sense that the structure of a molecule would affect its properties. The atoms in ethyl butyrate (pineapple smell) and hexanoic acid (dirty sock smell) are connected differently, so they behave differently when they enter your nose. But what about molecules that smell similar? What is it about these structures that causes them to have similar properties?

**What does the structure of a molecule have to do with smell?**

*To answer this question, you will explore*
↪ Relating Smell to Molecular Structure
↪ Functional Groups
↪ Classifying Molecules

## EXPLORING THE TOPIC

## ↪ Relating Smell to Molecular Structure

### PUTRID-SMELLING MOLECULES

Consider two molecules in the same smell category. Both hexanoic acid and butyric acid smell putrid. The first smells like stinky feet while the second smells like a carton of spoiled milk. Take a moment to compare the structural formulas of hexanoic acid and butyric acid.

Notice that both molecules have two oxygen atoms bonded on the end in an identical fashion. They both contain the structural feature highlighted.

Perhaps all putrid-smelling molecules have this same structural feature. Two more compounds that smell putrid are shown here.

You can see that this same structural feature is present in isopentanoic acid and isobutyric acid as well.

### SWEET-SMELLING MOLECULES

Now look at the structural formulas for two sweet-smelling molecules.

Ethyl butyrate, $C_6H_{12}O_2$

Ethyl acetate, $C_4H_8O_2$

Just like the putrid-smelling molecules, these sweet-smelling molecules have an identical structural feature, which is highlighted in yellow. This feature is similar to the one found in putrid-smelling molecules, but it is slightly different.

## ⟳ Functional Groups

The structural features that groups of molecules have in common are called **functional groups.** A functional group often stands out as an unusual or unique portion of a molecule. The functional group found in the putrid-smelling molecules is called a *carboxyl* functional group. The functional group found in the sweet-smelling molecules is called an *ester* functional group.

Two other functional groups are associated with fishy smells and minty smells.

Amine group, fishy

Ketone group, minty

---

**Example 1**

### Functional Group and Smell

The structural formulas of two molecules are shown here. Predict how the molecules will smell. What is your reasoning?

*Solution*

This molecule contains an ester functional group. It is probably a sweet-smelling molecule.

This molecule contains a carboxyl functional group. It is probably a putrid-smelling molecule.

## OTHER FUNCTIONAL GROUPS

There are other functional groups besides the four discussed so far. For instance, a molecule may have a hydroxyl group (−OH). Molecules containing this feature are called *alcohols*. A molecule containing an aldehyde group is similar to a ketone, except that the carbon that is double bonded to the oxygen is between a carbon and a hydrogen.

$$-\overset{|}{\underset{|}{C}}-O-H \qquad \qquad -\overset{\overset{O}{\|}}{C}-H$$

Hydroxyl group         Aldehyde group

Another type of molecule is one containing a single oxygen located between two carbon atoms. Molecules containing this feature are called *ethers*.

$$-\overset{|}{\underset{|}{C}}-O-\overset{|}{\underset{|}{C}}-$$

Ether group

Some molecules have no functional groups. For example, *alkanes* consist of only carbon and hydrogen atoms connected with single bonds.

$$-\overset{|}{\underset{|}{C}}-$$

Alkane

## ⤳ Classifying Molecules

The easiest way to classify molecules is by functional group. All the molecules that have ester functional groups are referred to as *esters*. All the sweet-smelling compounds you have encountered so far contain an ester functional group and have two-word names that end in "-yl" and "-ate." Eth**yl** butyr**ate**, eth**yl** acet**ate**, and hex**yl** acet**ate** all smell sweet.

The putrid-smelling compounds have a carboxyl group and names that end with "-ic acid." Butyr**ic acid** and hexano**ic acid** both smell putrid.

You may be able to identify the smell of a molecule by paying attention to its chemical name. Examine the compounds in the table. Each of these molecules has a unique smell even though they all have three carbon atoms and at least six hydrogen atoms. Their names are also very different.

| Name | Compound type | Functional group | Smell | Common use or source |
|---|---|---|---|---|
| propane $C_3H_8$ | alkane | no functional group <br> H—C—C—C—H (with H atoms) | no smell or gasoline-like | fuel in camp stoves |
| propionic acid $C_3H_6O_2$ | carboxylic acid | carboxyl <br> H—C—C—C—O—H (with H atoms and O) | putrid | in sweat |

| Name | Compound type | Functional group | Smell | Common use or source |
|---|---|---|---|---|
| ethyl formate $C_3H_6O_2$ | ester | ester | sweet | in raspberries |
| acetone $C_3H_6O$ | ketone | ketone | minty | nail polish remover |
| isopropanol $C_3H_8O$ | alcohol | hydroxyl | medicinal | rubbing alcohol |
| timethylamine $C_3H_9N$ | amine | amine | fishy | in bad breath |

Compounds that have the same functional group tend to have similar properties. For example, esters smell sweet, dissolve in water, and change phase from liquid to gas fairly easily. There are exceptions. For example, larger alkanes tend to have a gasoline smell, while smaller alkanes like propane in the table have almost no smell.

## LESSON SUMMARY

### What does the structure of a molecule have to do with smell?

Based on what the lessons have covered so far, compounds that have similar smells also appear to have similar structural features. These features are called functional groups. Functional groups have names. Compounds are frequently named according to the functional groups they contain. If you identify the functional group in a compound, you may also be able to predict how that compound will smell.

**KEY TERM**

functional group

## Reading Questions

1. What is a functional group?

2. What information would you want to have to predict the smell of a compound? Explain why.

## Reason and Apply

3. Explain why $C_2H_4$ has fewer hydrogen atoms than $C_2H_6$.

4. Create a four-carbon molecule that is
   a. an alkane
   b. a carboxylic acid
   c. an alcohol
   d. an ester

5. Consider a compound called hexanol. Chemists can tell from its name that it has six carbons (hex-) and it is an alcohol (-ol).
   a. What is the molecular formula of hexanol?
   b. Draw a possible structural formula for hexanol.
   c. Is the molecular formula or the structural formula more useful in determining the smell of hexanol? Explain.

6. **Research** Examine the ingredients lists on some household products, such as shampoo, lotion, or cleanser. Write down the names of compounds and functional groups that you were able to identify.

7. If you were a chemist and you wanted to invent a new smell, what would you think about doing or creating?

8. If you were a chemist and you wanted to change the smell of a molecule, what might you try to do to that molecule?

9. Explain what you think is going on in this cartoon.

Pentyl propionate          Octanoic acid

*(continued)*

10. Structural formulas for six molecules are shown here. Identify (by name) the functional group in each molecule.

11. Write the molecular formula for each of the molecules in Exercise 10.

# Ester Synthesis

## THINK ABOUT IT

Body odor, stinky shoes, moldy carpets, dog odors, musty basements, bad breath—these are just a few of the smells that we consider unpleasant. But the structures of these molecules are not so different from the structures of sweet-smelling substances. Molecules that have a carboxyl functional group are very similar to molecules that have an ester functional group. Perhaps there is a way to change one type of molecule into the other.

**How can a molecule be changed into a different molecule by using chemistry?**

*To answer this question, you will explore*

⮕ Transforming Smells

⮕ Chemical Reactions

⮕ Chemical Synthesis

## EXPLORING THE TOPIC

## ⮕ Transforming Smells

Every year, consumers spend millions of dollars to purchase products that will deal with unwanted odors. Often, these products do nothing more than cover up a foul smell with a pleasing smell. But some products use chemistry to change the molecular structure of the smelly molecules.

Butyric acid, $C_4H_8O_2$, is a carboxylic acid with a putrid smell. However, the difference between this molecule and a sweet-smelling molecule is minimal. Butyric acid can be chemically changed into a sweet-smelling ester.

*How can this . . .*          *become this?*

Butyric acid          Methyl butyrate

You can see that to accomplish this task, you would have to somehow remove the hydrogen atom that is attached to an oxygen atom in the butyric acid molecule and attach in its place a carbon atom with three hydrogen atoms. This procedure is possible in a chemistry lab.

---

**CONSUMER CONNECTION**

The putrid-smelling compound butyric acid is formed when butter goes bad. In fact, *butyric* means "from butter." Butyric acid is also found in Parmesan cheese and vomit.

© Corbis

---

# Chemical Reactions

In the ester synthesis lab, you mixed some substances together with a catalyst and heated them, following a specific set of instructions. The result was a molecule that no longer smelled bad.

**CONSUMER CONNECTION**

An entire industry is devoted to the creation of fragrances and flavors for use in our foods and commercial products. Some companies even manufacture room odorizers to inject a pleasant aroma into public environments or businesses. These aromas include the smell of freshly brewed coffee, baking bread, or an evergreen forest. The idea is to create environments that evoke pleasant memories or experiences for the user—and to stimulate the appetite or the pocketbook.

Butyric acid + Methanol $\xrightarrow[\text{(Heat)}]{H_2SO_4}$ Sweet-smelling compound

When butyric acid and methanol are mixed and heated, no change is visible to the eye. However, your nose tells you that a dramatic change in properties has occurred. The foul smell of sweaty socks disappears and the sweet smell of apples fills the room.

The change in smell is clear evidence that a chemical reaction has occurred. A **chemical reaction** is the process by which matter is changed so that new substances are formed. For new substances to form, the atoms must rearrange. Some bonds between atoms must be broken and new bonds must be formed. The result of a chemical reaction is a new compound with entirely different properties. In this case, one of the new properties is a different smell.

The detection of a sweet-smelling compound is evidence that you have made an ester. In fact, methyl butyrate smells like apples. In Lesson 35: Making Scents, you will examine this transformation in greater detail.

# Chemical Synthesis

Chemists often work in the laboratory with the goal of producing a specific compound. This process is called synthesis. **Synthesis** is the process of producing a chemical compound, usually by combining two simpler compounds. Many of the products that we have come to depend on are products that chemists have synthesized, like plastics, fabrics, cosmetics, deodorants, cleaning products, vitamins, and medications.

Sometimes chemists work to synthesize a molecule that is very rare or hard to find. For instance, many lifesaving, anti-cancer compounds have been discovered in plant and animal life in the rainforests of South America. It may be difficult to harvest these compounds or find enough of a substance to treat the many patients who are in need. Chemists study the structure of a compound like this and work to create it in the laboratory. Most of the medicines in use today are synthesized in laboratories.

*The U.S. National Cancer Institute has identified 3000 plants containing compounds with anti-cancer properties. Most are found in the rainforest. Here, a guide points out a medicinal plant.*

Sometimes the point of synthesis is to create a new molecule that is *not* found in nature. All of the plastics that we have in our lives are synthesized from small molecules that have been strung together through chemical reactions. These compounds are valued for their amazing properties. Many plastics are durable, flexible, and easily molded into a variety of useful shapes—everything from shoes to computer keyboards.

## LESSON SUMMARY

### How can a molecule be changed into a different molecule by using chemistry?

To transform a putrid-smelling molecule into a sweet-smelling molecule, it is necessary to transform a carboxylic acid molecule into an ester molecule. This transformation requires a chemical reaction in which bonds between certain atoms are broken and new bonds are made. The result of any chemical reaction is a different compound or compounds than the compound you started with. And when the functional group changes, the smell changes. Chemists control chemical reactions in the laboratory to synthesize specific molecules with valuable properties.

**KEY TERMS**

chemical reaction

synthesis

## Exercises

### Reading Questions

1. What would you do to cause an alcohol and an acid to react?

2. What is chemical synthesis?

### Reason and Apply

3. In class, why do you think the foul smell of butyric acid disappeared when you mixed it with methanol and heated the mixture in the Lab: Ester Synthesis?

*(continued)*

4. The structural formulas for the molecules used in the Lab: Ester Synthesis are given in the table. Identify the functional groups.

| Test tube | Organic acid | Structural formula | Alcohol | Structural formula |
|---|---|---|---|---|
| 1 | acetic acid | | isopentanol | |
| 2 | acetic acid | | butanol | |
| 3 | butyric acid | | ethanol | |

5. Explain how you can convert acetic acid into an ester. Be specific about how the molecule needs to change.

6. **Lab Report** Write a lab report for the Lab: Ester Synthesis. In your report, include the title of the experiment, purpose, procedure, observations, and conclusions.

**Lab: Ester Synthesis**

**Purpose:** (Explain what you were trying to find out.)

**Procedure:** (List the steps you followed.)

**Observations:** (Explain what you observed during the experiment.)

**Conclusions:** (What can you conclude about what you were trying to find out? Provide evidence for your conclusions.)

*Making Scents*

# Analyzing Ester Synthesis

## THINK ABOUT IT

When you heat a mixture of a carboxylic acid and an alcohol, a chemical reaction takes place. You can smell the outcome. A really putrid-smelling substance is transformed into one that smells sweet. An ester has been produced. But what has happened to the original substances that were mixed together?

**What happened to the molecules during the creation of a new smell?**

*To answer this question, you will explore*

➡ Chemical Equations

➡ The Role of the Catalyst

➡ Ester Synthesis

## EXPLORING THE TOPIC

### ➡ Chemical Equations

Chemists use **chemical equations** to keep track of changes in matter. These can be certain physical changes or changes due to chemical reactions. A chemical equation usually uses chemical formulas to describe what happens when substances are mixed together and new substances with new properties are formed. In class, you used structural formulas to describe what happened to the molecules that were combined.

Consider what happens if you add butyric acid to ethanol and create a pineapple smell. A description of the reaction is given here. Each of these substances consists of a collection of molecules that are structurally different. The structural formulas for the molecules in each of the substances are also shown along with their chemical formulas.

Adalberto Rios Szalay/
Sexto Sol/Getty Images

*Butyric acid, $C_4H_8O_2$, is added to ethanol, $C_2H_6O$, to produce water, $H_2O$, and ethyl butyrate, $C_6H_{12}O_2$.*

In a chemical reaction, the substances that are mixed together are called the **reactants.** The new substances that are produced are called the **products.**

The reactants in this chemical reaction are butyric acid and ethanol. One product that is easily detected by smell is ethyl butyrate. However, this is not the only product. Although it is difficult to observe with the five senses, water is also a product of this chemical reaction. The chemical equation helps to make this clear. The chemical equation shows that in a chemical reaction, no matter is created or destroyed.

**Big Idea** Matter cannot be created or destroyed. Matter is conserved.

The next illustration highlights the areas of the reactant molecules that change during the chemical reaction. You can see that a hydroxyl, $-OH$ group, breaks off from the butyric acid molecule. In addition, the $H-O$ bond in the ethanol molecule must also break. A new bond forms as the larger molecular pieces come together to form ethyl butyrate.

Finally, you can see that the $H-$ and $-OH$ pieces that are left over combine to form $H_2O$, water.

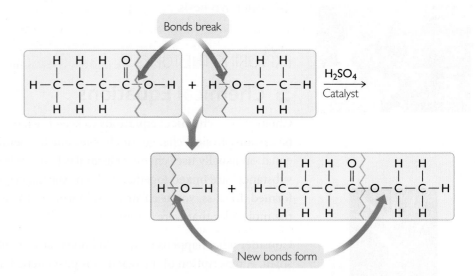

**CONSUMER CONNECTION**

Esters are found in a number of household products, from cleansers and air fresheners to hand lotions and shampoos. Their main role is to make a product smell a certain way, although some are detergents as well.

For every butyric acid molecule that reacts with an ethanol molecule, one molecule of water is produced along with one molecule of ethyl butyrate.

The complete chemical equation for this chemical reaction is

$$C_4H_8O_2 + C_2H_6O \longrightarrow H_2O + C_6H_{12}O_2$$

There are the same number of each type of atom on both sides of the equation. All of the matter is accounted for.

## ⮕ The Role of the Catalyst

There's one more ingredient in the chemical reaction that we have not yet discussed. When you try to carry out this reaction in a lab, you will find that simply mixing butyric acid and ethanol does not produce ethyl butyrate. In fact,

Clive Streeter/Getty Images

nothing happens. To make the chemical reaction happen, it is necessary to add sulfuric acid, $H_2SO_4$, and heat the mixture.

$$C_4H_8O_2 + C_2H_6O \xrightarrow{H_2SO_4} H_2O + C_6H_{12}O_2$$

The sulfuric acid does not get used up during the reaction. It is still present at the end. In this reaction, the sulfuric acid is simply helping the reaction along. Chemists refer to a substance that assists a chemical reaction as a **catalyst.** A catalyst speeds up a reaction, but the catalyst itself is not consumed by the reaction. In chemical equations, a catalyst is written above the arrow.

## Ester Synthesis

Many different acids and alcohols can be brought together to form an ester and water. The general description of this reaction is

$$acid + alcohol \longrightarrow water + ester$$

Take another look at the illustration on page 180 showing the bonds breaking. It is not necessary for every single bond in the acid and alcohol molecules to break apart to form the ester. All of the C−H bonds remain connected during this reaction.

During chemical reactions, the major changes in the molecules often occur where the functional groups are located. The part of the molecule that is easiest to change is the functional group. This makes sense with what you already know: The functional group is directly related to the properties of a molecule.

## LESSON SUMMARY

### What happened to the molecules during the creation of a new smell?

When chemical reactions occur, new compounds with different properties and structures are produced. The substances that are combined are called the reactants. The substances that are produced are called the products. When chemical reactions occur, bonds break and new bonds form. Chemical equations use chemical formulas to track the changes that occur during chemical reactions. Because matter is conserved, all atoms are accounted for in a chemical equation. Some chemical reactions require a catalyst to help them get started or proceed more rapidly. A catalyst is not consumed by the chemical reaction it is assisting.

**KEY TERMS**

chemical equation

reactant

product

catalyst

## Exercises

### Reading Questions

1. Explain why converting a carboxylic acid to an ester might be useful.

2. Describe what happens during a chemical reaction.

3. What is a catalyst?

*(continued)*

**Reason and Apply**

4. Write a paragraph answering the question, "How can scientists use chemistry to create compounds with specific smells?" Include evidence to support your answer.

5. Below are the structural formulas for four esters. Write the correct molecular formula for each one.

a.
```
     H   H   H   H   H       O   H   H
     |   |   |   |   |       ‖   |   |
 H—C—C—C—C—C—O—C—C—C—H
     |   |   |   |   |           |   |
     H   H   H   H   H           H   H
```

b.
```
                         H
                         |
                     H—C—H
     H   O       H   H   |   H
     |   ‖       |   |   |   |
 H—C—C—O—C—C—C—C—H
     |           |   |   |   |
     H           H   H   H   H
```

c.
```
     O       H   H   H   H   H   H   H   H
     ‖       |   |   |   |   |   |   |   |
 H—C—O—C—C—C—C—C—C—C—C—H
             |   |   |   |   |   |   |   |
             H   H   H   H   H   H   H   H
```

d.
```
     H   O       H   H   H   H
     |   ‖       |   |   |   |
 H—C—C—O—C—C—C—C—H
     |           |   |   |   |
     H           H   H   H   H
```

6. Ester molecules are named for the alcohol and acid that form them using the convention

(alcohol name)yl (acid name)ate

For example, methanol + ethanoic acid combine to make methyl ethanoate. Name the esters formed from

a. isopropanol + methanoic acid
b. caproic acid + butanol
c. salicylic acid + ethanol

# CHAPTER 6

## Speaking of Molecules

# SUMMARY

### KEY TERMS

molecular formula

structural formula

isomer

HONC 1234 rule

Lewis dot structure

Lewis dot symbol

bonded pair

lone pair

octet rule

double bond

triple bond

functional group

chemical reaction

synthesis

chemical equation

reactant

product

catalyst

# Smells Update

The investigation into the chemistry of smell has challenged you to explore how molecules are put together. So far you know that the smell of a substance is related to its molecular formula, chemical name, and any functional groups present.

This is based on the data investigated so far. Further investigation may lead you in new directions.

## REVIEW EXERCISES

1. What functional groups are present in each of these molecules?

   a.
   $$\begin{array}{c} \text{H}\ \text{H}\ \ \ \ \ \ \ \text{O}\ \ \text{H}\ \text{H}\ \text{H}\ \text{H}\ \text{H} \\ \vert\ \ \vert\ \ \ \ \ \ \ \vert\vert\ \ \vert\ \ \vert\ \ \vert\ \ \vert\ \ \vert \\ \text{H}-\text{C}-\text{C}-\text{O}-\text{C}-\text{C}-\text{C}-\text{C}-\text{C}-\text{C}-\text{H} \\ \vert\ \ \vert\ \ \ \ \ \ \ \ \ \ \ \ \vert\ \ \vert\ \ \vert\ \ \vert\ \ \vert \\ \text{H}\ \text{H}\ \ \ \ \ \ \ \ \ \ \ \text{H}\ \text{H}\ \text{H}\ \text{H}\ \text{H} \end{array}$$

   b.
   $$\begin{array}{c} \text{H}\ \ \text{H}\ \ \text{O}\ \ \text{H}\ \ \text{H} \\ \vert\ \ \vert\ \ \vert\vert\ \ \vert\ \ \vert \\ \text{H}-\text{C}-\text{C}-\text{C}-\text{C}-\text{C}-\text{H} \\ \vert\ \ \vert\ \ \ \ \ \ \vert\ \ \vert \\ \text{H}\ \ \text{H}\ \ \ \ \ \ \text{H}\ \ \text{H} \end{array}$$

   c.
   $$\begin{array}{c} \text{H}\ \ \text{H}\ \ \text{H}\ \ \text{H}\ \ \text{O} \\ \vert\ \ \vert\ \ \vert\ \ \vert\ \ \vert\vert \\ \text{H}-\text{C}-\text{C}-\text{C}-\text{C}-\text{C}-\text{O}-\text{H} \\ \vert\ \ \vert\ \ \vert\ \ \vert \\ \text{H}\ \ \text{H}\ \ \text{H}\ \ \text{H} \end{array}$$

   d.
   $$\begin{array}{c} \text{H}\ \ \text{H}\ \ \text{H} \\ \vert\ \ \vert\ \ \vert \\ \text{H}-\text{C}-\text{C}-\text{C}-\text{N}-\text{H} \\ \vert\ \ \vert\ \ \vert\ \ \vert \\ \text{H}\ \ \text{H}\ \ \text{H}\ \ \text{H} \end{array}$$

2. Draw the Lewis dot structure and the structural formula for each of these molecules.
   a. $SiF_4$  b. $CO_2$  c. $CH_4$  d. $SF_2$
   e. $C_2H_4$  f. $C_2H_2$  g. $C_2H_6$

3. How many lone pairs are in each of these molecules?
   a. $CO_2$  b. $SiF_4$  c. $CH_4$

4. Draw at least two structural formulas for the molecular formula $C_3H_6O$.

5. From what you've learned so far, how is molecular structure related to smell?

---

## Functional Groups

**PROJECT**

Research a functional group. Choose a functional group and find out as much as you can about molecules that contain it. Create a poster that has these details:

- A large drawing of the functional group along with its name
- The structural formulas for at least five molecules that possess your functional group, along with their chemical names
- A brief description of the properties associated with this group of compounds

---

# CHAPTER 7

# Building Molecules

### In this chapter, you will study

- three-dimensional molecular models

- the role of valence electrons in determining molecular shape

- the receptor site theory and have a chance to develop your own model of how a nose works

*Molecular models help chemists to picture something too tiny to see.*

Two compounds with the same functional group may have different smells. In addition to molecular structure, the *shape* of a molecule seems to be related to its smell. Molecular shape is determined by the bonds between atoms and the locations of electrons. Because we experience different molecular shapes as different smells, our experience tells us that the nose can somehow detect these differences.

# Ball-and-Stick Models

## THINK ABOUT IT

The structural formula is a valuable source of information about a molecule. You can use the structural formula to identify functional groups in a molecule. Knowing the functional group also helps you to predict molecular properties, including smell. However, sometimes only knowing the functional group is not enough to predict the smell of a compound.

**What three-dimensional features of a molecule are important in predicting smell?**

*To answer this question, you will explore*

🡒 New Smell Molecules

🡒 Three-Dimensional Models

## EXPLORING THE TOPIC

### 🡒 New Smell Molecules

Below are three compounds, each with a distinctive smell. In spite of their wide range of smells, each of the three compounds contains the same functional group.

Geraniol, $C_{10}H_{18}O$

Menthol, $C_{10}H_{20}O$

Fenchol, $C_{10}H_{18}O$

Take a moment to locate their functional groups. Each of these molecules has a hydroxyl functional group, and each name ends in "-ol." All three molecules are alcohols. So, why are the three smells so different? Something besides just functional group affects the smell of a compound. The answer lies in the three-dimensional shape of the molecules of each compound.

© loooby/iStockphoto

# ⮞ Three-Dimensional Models

It is difficult to show a three-dimensional drawing of a molecule on a flat piece of paper. With a molecular model set, you can build a three-dimensional structure.

Ken Karp Photography

carbon atom

oxygen atom

nitrogen atom

hydrogen atom

bond

lone pair paddle

Notice that the bonds in a model kit look like little sticks. And the atoms are small spheres. These models are called **ball-and-stick models.**

A picture of a ball-and-stick model for ethyl acetate, $C_4H_8O_2$, is shown here. The carbon atoms are shown in black, the hydrogen atoms are white, and the oxygen atoms are red.

The carbon atoms are not in a straight line.

Some atoms are pointing away from you.

Some atoms are pointing toward you.

Illustrations of ball-and-stick models have some drawbacks, in that some atoms may be partially visible or entirely hidden. However, the illustrations convey more information about molecular shape than structural formulas do. Using a real model is the best way to examine the three-dimensional shape of a molecule. The two representations for a molecule of citral, $C_{10}H_{16}O$, are shown for comparison. Look for similarities and differences between them.

Structural formula

Ball-and-stick model

Citral, $C_{10}H_{16}O$

Notice that the structural formula shows the citral molecule as flat with all of the carbon atoms in a line. The ball-and-stick representation, on the other hand, shows that the carbon atoms are not arranged in a line, but are connected in a zigzag fashion. Both representations contain much of the same information, but the ball-and-stick model adds information about the way the atoms are arranged in space.

**Example**

## Ball-and-Stick Model

Examine the drawing of this ball-and-stick model of isopentylacetate, $C_7H_{14}O_2$.

a. Draw the structural formula of this compound.

b. What functional group is in the compound?

c. Predict a smell for this compound.

*Solution*

a. The structural formula of this compound is

b. This compound has an ester functional group.

c. Because it has an ester group and its name ends in "-ate," it probably smells sweet. (In fact, this molecule smells like bananas.)

Ball-and-stick models for the three compounds introduced at the beginning of this lesson are below.

Geraniol, $C_{10}H_{18}O$

Menthol, $C_{10}H_{20}O$

Fenchol, $C_{10}H_{18}O$

The overall shapes of the molecules differ greatly. The carbon atoms in geraniol are connected in a long chain, while the menthol and fenchol molecules both contain ring structures. Perhaps these overall shapes are related to the different smells of these three alcohols.

## LESSON SUMMARY

### What three-dimensional features of a molecule are important in predicting smell?

**KEY TERM**

ball-and-stick model

Molecules are three-dimensional. A ball-and-stick model kit is a tool that you can use to construct models of molecules. Ball-and-stick models allow you to see how the various atoms in molecules are arranged in space, as well as the overall shape of each molecule. Molecular shape may have something to do with smell.

**Exercises**

### Reading Questions

1. What are the differences between a structural formula and a ball-and-stick model?

2. What is your hypothesis about why the three alcohols smell different even though they have the same functional group?

*(continued)*

**Reason and Apply**

**3.** What model pieces do you need to build a ball-and-stick model of geraniol?

**4.** Consider this model. Its molecular formula is $C_7H_{14}O_2$.

   **a.** Draw the Lewis dot structure.
   **b.** Draw the structural formula.
   **c.** What is the functional group in the molecule?
   **d.** What can you predict about the name and smell of this compound?

**5.** Consider this model. Its molecular formula is $C_3H_6O_2$.

   **a.** Draw the Lewis dot structure.
   **b.** Draw the structural formula.
   **c.** What is the functional group in the molecule?
   **d.** What can you predict about the name and smell of this compound?

**6.** Consider this model. Its molecular formula is $C_2H_7N$.

   **a.** Draw the Lewis dot structure.
   **b.** Draw the structural formula.
   **c.** What is the functional group in the molecule?
   **d.** What can you predict about the name and smell of this compound?

**7.** What evidence do you have that the structural formula may not always be useful in predicting smell?

# Two's Company

## Purpose
To use three-dimensional models to visualize small molecules.

## Materials
- gumdrops, marshmallows, and toothpicks
- ruler
- ball-and-stick molecular model set

## Part 1: Gumdrop Molecules

1. Create a methane molecule using gumdrops, marshmallows, and toothpicks.

2. Make sure every pair of electrons in the molecule is as far away as possible from every other pair of electrons. Use a ruler to check the distances.

3. Draw Lewis dot structures for the following molecules:
   **a.** $CH_4$     **b.** $NH_3$     **c.** $H_2O$

4. How many pairs of electrons are located around the central atom of each molecule?

5. Besides the identity of the central atom, what is different about these three molecules?

6. Using gumdrops and toothpicks, create ball-and-stick models of $NH_3$ and $H_2O$.

7. Did you remember to include lone pairs? Fix your models if you have to so that lone pairs are represented. Do the lone pairs affect the shape of the molecule?

8. Compare your three gumdrop models. Describe any similarities.

## Part 2: Ball-and-Stick Models

1. Use the molecular model sets to create models of $CH_4$, $NH_3$, $H_2O$, and HF. Use black for carbon, white for hydrogen, red for oxygen, and blue for nitrogen.

2. Add the appropriate lone pair paddles to your models.

3. How many lone pair paddles would you need for an atom of neon? Explain your answer.

4. Draw sketches of your three-dimensional models. What is the shape of each molecule if you ignore the lone pair paddles?

5. **Making Sense** Explain how the lone pairs affect the shapes of these molecules.

Ken Karp Photography

Ken Karp Photography

# Electron Domains

## THINK ABOUT IT

You may have noticed that the atoms in the molecular model kits have a certain number of holes and that the sticks only fit in certain places. So, when you connect the atoms, the molecules automatically end up with the correct three-dimensional shape. But what is it that determines the shape of a molecule?

**How do electrons affect the shape of a molecule?**

*To answer this question, you will explore*

⮕  Shapes of Molecules

⮕  Electron Domains

## EXPLORING THE TOPIC

### ⮕ **Shapes of Molecules**

These illustrations show models of ethanol, $C_2H_6O$.

Ethanol

*Ball-and-stick model*                    *Structural formula*

What accounts for the three-dimensional structure of ethanol? The answer lies in the valence electrons. They determine the ultimate shape of a molecule.

To understand molecular shape, it is useful to begin by examining a simple molecule such as methane, $CH_4$. While the structural formula is flat and cross-shaped, the ball-and-stick model shows that methane has a three-dimensional structure. Both models show that there are four bonds.

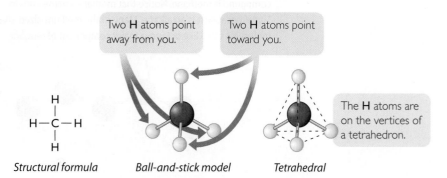

Two **H** atoms point away from you.

Two **H** atoms point toward you.

The **H** atoms are on the vertices of a tetrahedron.

*Structural formula*          *Ball-and-stick model*          *Tetrahedral*

The word used to describe the shape of the methane molecule is **tetrahedral.** A tetrahedral molecule has a symmetrical shape, with one atom exactly in the center. The distance between any other two atoms bonded to the central atom is the same.

You can prove to yourself that a ball-and-stick model of methane is symmetrical in every direction by spinning the molecule as shown in the illustrations. The molecule looks exactly the same no matter what direction it is spun.

*One **H** atom sticks straight up. Three **H** atoms rest on the table.*

*You can spin the molecule around and it looks identical.*

*You can grab one of the **H** atoms on the table and spin the molecule around and it still looks identical.*

## Electron Domains

### METHANE, CH₄

The illustration here shows a methane molecule with the shared valence electrons superimposed, or laid, over the ball-and-stick model. The bonded electron pairs occupy space between the carbon atom and each of the four hydrogen atoms. The space occupied by the electrons is called an **electron domain.** An electron domain can consist of a bonded pair of electrons, a lone pair of electrons, or multiple bonded pairs of electrons.

Bonded pairs of electrons take up space. This space is called an electron domain.

*A methane molecule has four electron domains.*

### CONSUMER CONNECTION

Molecules containing *only* carbon and hydrogen atoms are generally referred to as hydrocarbons. The simplest hydrocarbon compound is methane. Notice that methane's name ends in "-ane" because it is an alkane. Many of the medium-sized alkane molecules smell like octane, a major component of gasoline.

Octane

Each electron domain in the tetrahedral methane molecule is the same distance from the other three electron domains. They are as far apart from one another as possible. Because of this arrangement, the hydrogen atoms are as far apart from one another as possible and the distance between any two hydrogen atoms is the same. Additionally, in the tetrahedral molecule, if you measure the angle formed between any two hydrogen atoms and the carbon atom, you'll find that all these bond angles have equal measures. This is not accurately shown in the structural formula.

This H−C−H angle appears to be 90°.

This H−C−H angle appears to be 180°.

*Each **H—C—H** bond angle is actually 109°.*

This tendency for electron pairs to be as far apart from one another as possible is called **electron domain theory.** This theory is also referred to as *valence shell electron pair repulsion theory,* or *VSEPR theory.*

## AMMONIA, NH₃

The structural formula and ball-and-stick model for ammonia, NH₃, are shown below. The ball-and-stick model shows that the hydrogen atoms are all located on one side of the nitrogen atom.

*Structural formula*   *Ball-and-stick model*   *Pyramidal*

The word used to describe the shape of the ammonia molecule is **pyramidal.**

The nitrogen atom has a lone pair of electrons. Therefore, while there are only three bonds, there are four electron domains. The four electron domains are arranged in a tetrahedral shape to get as far apart as possible, similar to what happens in methane.

*An ammonia molecule has four electron domains, just like methane.*

This electron domain is associated with a lone pair of electrons.

These electron domains are associated with bonded pairs of electrons.

## WATER, H₂O

Take a look at one more molecule—water, $H_2O$. The word used to describe the shape of a water molecule is **bent.**

Structural formula         Ball-and-stick model

A water molecule has four electron domains, two bonding pairs and two lone pairs. The four electron domains of water are arranged in a tetrahedral shape, as in methane and ammonia. This results in a bent shape for the water molecule. In all three molecules, the bond angles are 109° or close to this.

*A molecule of water has four electron domains, just like methane.*

Bent

## BALL-AND-STICK MODELS WITH LONE PAIR PADDLES

In a molecular model kit, the lone pairs are sometimes represented as plastic paddles. This helps you to visualize where the final shape of a molecule comes from.

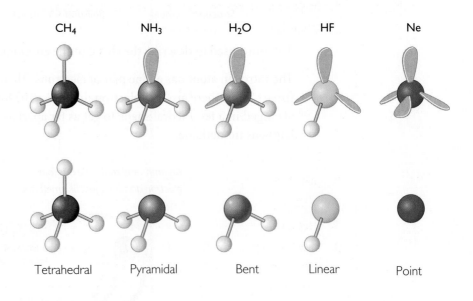

CH₄      NH₃      H₂O      HF      Ne

Tetrahedral      Pyramidal      Bent      Linear      Point

Notice that you name the shape of the molecule based only on the arrangement of atoms. The lone pairs are not considered.

## Phosphine, PH$_3$

What is the shape of phosphine, PH$_3$? Explain your thinking.

### Solution

Begin by drawing the Lewis dot structure and a structural formula of phosphine. A phosphorus atom has five valence electrons, and each hydrogen atom contributes one electron, for a total of eight electrons. There are three bonding pairs and one lone pair. The four atoms in phosphine are arranged in a pyramidal shape.

$$H:\overset{..}{\underset{..}{P}}:H \qquad H-\overset{..}{P}-H$$
$$\underset{H}{} \qquad \underset{H}{|}$$

## LESSON SUMMARY

### How do electrons affect the shape of a molecule?

**KEY TERMS**

tetrahedral shape

electron domain

electron domain theory

pyramidal shape

bent shape

The three-dimensional shape of a molecule is determined by the various electron pairs in the molecule. *Both* bonded pairs of electrons *and* lone pairs of electrons affect the shape of a molecule. Each electron pair takes up space, called an electron domain. The final shape of a molecule is determined by the fact that electron domains in a molecule are located as far apart from one another as possible.

**Reading Questions**

1. In your own words, describe a tetrahedral shape.

2. What is meant by electron domain theory?

**Reason and Apply**

3. If you were going to predict the three-dimensional structure of a small molecule, what would you want to know?

4. Predict the three-dimensional structure of H$_2$S. Explain your thinking.

5. List three molecules that have a tetrahedral shape.

6. List three molecules that have a bent shape.

7. What is the shape of arsine, AsH$_3$? Explain your thinking.

8. Predict the three-dimensional shape of HOCl. Explain your thinking.

9. Draw a methane, CH$_4$, molecule and show how it fits inside a tetrahedron. Do the same for ammonia, NH$_3$, and water, H$_2$O.

*Let's Build It*

# Molecular Shape

## THINK ABOUT IT

So far you have considered the shapes of small molecules containing single bonds. All of the molecules you've studied have had four electron domains. However, small molecules that have only three or even two electron domains also exist. These molecules have double and triple bonds. So, there are more molecular shapes to consider.

### How can you predict the shape of a molecule?

*To answer this question, you will explore*

➲  More Shapes

➲  Larger Molecules

## EXPLORING THE TOPIC

### ➲ More Shapes

#### DOUBLE BONDS

Formaldehyde, $CH_2O$, is a very simple molecule that contains one double bond. The double bond counts as only one electron domain even though it contains two pairs of bonded electrons. This means that the central atom in the formaldehyde molecule (the carbon atom) has only three electron domains surrounding it, spread out into a flat triangle. The phrase used to describe the underlying shape of the formaldehyde molecule is **trigonal planar.**

The double bond is one electron domain.

Structural formula

There are three electron domains around the central carbon atom.

Lewis dot structure

*The three electron domains spread out into the shape of a flat triangle.*

Trigonal planar

Both ammonia, $NH_3$, and formaldehyde, $CH_2O$, have four atoms but have very different shapes. The difference in shape is due to the number of electron domains around the central atom.

*The three hydrogen atoms are below the nitrogen atom to make room for the lone pair on nitrogen.*

Ammonia

Formaldehyde

*There are no lone pairs, so the three electron domains spread out into a triangle.*

Pyramidal            Trigonal planar

## TWO DOUBLE BONDS

In a molecule with a double bond, the double bond counts as one electron domain. So a molecule like carbon dioxide, with two double bonds, has only two electron domains around the central atom that have to be as far apart as possible. This shape is described as **linear.**

Each double bond is one electron domain.

There are two electron domains around the central carbon atom.

The electron domains are as far apart as possible.

Linear

**Important to Know** The four electrons in a double bond remain between two atoms, so they are considered one electron domain. An electron domain can have two, four, or six electrons.

Both water, H$_2$O, and carbon dioxide, CO$_2$, have three atoms. However, the shape of water is bent while carbon dioxide is linear. The difference in shape is due to the number of electron domains around the central atom.

Water

Bent

Carbon dioxide

Linear

## Example 1

### Ethene, C$_2$H$_4$

Determine the molecular shape of ethene, C$_2$H$_4$.

### Solution

Begin by drawing the Lewis dot structure for ethene to determine the number of electron domains around each carbon atom. Next, translate that into a three-dimensional representation of the molecule. In this case, there are no lone pairs. The molecule is flat. Each half of ethene is trigonal planar.

*There are three electron domains around each carbon atom.*

*This molecule is flat when turned 90°.*

### INDUSTRY
### CONNECTION

Ethene, C$_2$H$_4$, is also called ethylene. It is a colorless gas with a slightly sweet odor. Ethene is used in the manufacture of many common products like polyethylene plastic and polystyrene. Ethene also can be used as a plant hormone to control the ripening of fruit or the opening of flowers.

Photodisc/Getty Images

## TRIPLE BONDS

Another example of a linear molecule is hydrogen cyanide, HCN. It also has two electron domains around the central atom, like carbon dioxide. One electron domain is a single bond between carbon and hydrogen, and the other is a triple bond between carbon and nitrogen.

$$H-C\equiv N$$

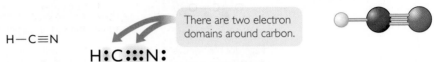

There are two electron domains around carbon.

*Hydrogen cyanide has one single bond and one triple bond. This results in two electron domains around the central atom.*

**Example 2**

### Ethyne, $C_2H_2$

Determine the molecular shape of ethyne, $C_2H_2$.

*Solution*

Begin by drawing the Lewis dot structure of ethyne to determine the number of electron domains around the carbon atoms. To satisfy the octet rule and the HONC 1234 rule, the carbon atoms must have a triple bond between them. Each carbon atom has two electron domains surrounding it. Ethyne is a linear molecule.

There are two electron domains around each carbon atom.

## Larger Molecules

The shapes of all of these small molecules help to explain why chains of carbon atoms are crooked rather than straight. The illustration shows a ball-and-stick model for heptanoic acid, $C_7H_{14}O_2$, with seven carbon atoms in a chain. The arrangement of atoms around six of the carbon atoms is tetrahedral. Most of the molecule is a series of overlapping tetrahedral shapes.

### INDUSTRY
### CONNECTION

Ethyne, $C_2H_2$, is also called acetylene. It is the highly flammable gas used by welders in oxyacetylene torches. Miners' helmets used to have carbide lamps attached to them. A carbide lamp is based on the reaction between calcium carbide, $CaC_2$, and water, $H_2O$, to produce acetylene, $C_2H_2$. The acetylene gas burns to produce a bright light in the dark cave.

This piece of the molecule is a tetrahedral shape.

This piece of the molecule is a trigonal planar shape.

This piece of the molecule is a bent shape.

Heptanoic acid, $C_7H_{14}O_2$

The various electron domains in this molecule cause its overall shape to be crooked, or zigzag.

## How can you predict the shape of a molecule?

The shape of a molecule is affected by the location and number of its electron domains. Lewis dot structures help you to determine the number of electron domains in a molecule. An atom involved in a double bond or triple bond will have fewer electron domains surrounding it. This results in molecules that have trigonal planar and linear shapes. Large molecules can have various geometric shapes within them. Areas around double bonds are flat, and those around triple bonds are linear.

**KEY TERMS**

trigonal planar shape

linear shape

## Exercises

### Reading Questions

1. What shapes are possible if a molecule has three electron domains?

2. What shapes are possible for a molecule with three atoms? Explain your thinking.

### Reason and Apply

3. Describe the shape of each of these molecules. (Use Lewis dot structures, the periodic table, and the HONC 1234 rule to assist you.)

    $Cl_2$        $CO_2$        $H_2O$

4. Which of the molecules in Exercise 3 have the same shape?

5. What shape or shapes do you predict for a molecule with two atoms? A molecule with three atoms? Four atoms? Five atoms? You can draw Lewis dot structures or electron domains to assist you in answering the questions.

6. Predict the shapes of these molecules:

    $CF_4$        $NF_3$        $H_2Se$        $H_2CS$

7. Consider the butane, $C_4H_{10}$, molecule.
    a. Draw the Lewis dot structure for butane.
    b. How many electron domains does the molecule have?
    c. What shape would you predict for $C_4H_{10}$?
    d. Explain why the carbon atom chain is not straight.

# Space-Filling Models

## THINK ABOUT IT

Methyl octenoate, $C_9H_{16}O_2$, is a compound that smells like violets. A ball-and-stick model of methyl octenoate shows that it is a series of overlapping tetrahedral shapes stuck together. There is a trigonal planar segment in the area of the double bonds. But how would you describe the shape of the *whole* molecule? And does the shape of the whole molecule have anything to do with its smell?

**How is the shape of a molecular compound related to its smell?**

*To answer this question, you will examine*

➲ Space-Filling Models

➲ Relating Shape and Smell

## EXPLORING THE TOPIC

## ➲ Space-Filling Models

Most of the smell molecules you have encountered are considerably larger than three, four, or five atoms. The best way to look at the overall shape of these molecules is with a different type of model, called a **space-filling model.**

Take a moment to compare the ball-and-stick model of methyl octenoate with the space-filling model. In a space-filling model, the sticks between atoms have been eliminated. There is no space between atoms. Instead, bonded atoms are shown slightly overlapping.

Methyl octenoate, $C_9H_{16}O_2$

Ball-and-stick model

Space-filling model

No space between atoms

Double bonds are harder to see.

In some ways, a space-filling model could be considered more accurate than a ball-and-stick model. In reality, a stick has no resemblance to a bond. In a molecule, the atoms *share* electrons with neighboring atoms. This would suggest that the atoms are located very close to, or overlapping, one another. However, in an illustration of a space-filling model, you can't see multiple bonds, and some atoms may be hidden behind others.

## ➲ Relating Shape and Smell

Space-filling models for molecules that smell sweet, minty, and like camphor (piney) are shown here. Take a moment to look for patterns that may indicate a connection between the shapes of these molecules and their smells.

### SWEET-SMELLING MOLECULES

Methyl octenoate, $C_9H_{16}O_2$        Citronellol, $C_{10}H_{20}O$

Geraniol, $C_{10}H_{18}O$

### MINTY-SMELLING MOLECULES

Carvone, $C_{10}H_{14}O$        Pulegone, $C_{10}H_{16}O$

Menthone, $C_{10}H_{18}O$

Fenchol, $C_{10}H_{18}O$          Camphor, $C_{10}H_{16}O$

The sweet-smelling molecules are all long and stringy. The minty-smelling molecules all have a six-carbon ring structure. They have a shape that resembles a frying pan. The camphor-smelling molecules are a tight cluster of atoms in the shape of a ball. So far, there appear to be three smell categories that are directly related to the molecular shape.

Long, stringy

Frying pan–shaped

Sweet smell          Minty smell

Ball-shaped

Camphor smell

## LESSON SUMMARY

### How is the shape of a molecular compound related to its smell?

A space-filling model is a three-dimensional model that shows how the atoms in a molecule are arranged in space and how they fill this space. The shape of a molecule appears to be related to its smell. For example, the sweet-smelling molecules explored here are all long and stringy in shape. The minty-smelling molecules are frying pan–shaped, and the camphor-smelling molecules are ball-shaped. More data will certainly be helpful to confirm a connection between a molecule's shape and its smell.

**KEY TERM**

space-filling model

**Reading Questions**

1. How is a space-filling model useful?

2. How is a space-filling model different from a ball-and-stick model?

**Reason and Apply**

3. Draw a possible structural formula for a molecule that smells sweet and has the molecular formula $C_9H_{20}O$. What general shape does this molecule have?

4. If someone told you a molecule had a six-carbon ring, what smell would you predict for that compound?

5. Draw a possible structural formula for a molecule that smells minty and has the molecular formula $C_{10}H_{16}O$. What general shape does this molecule have?

6. If someone told you a molecule is ball-shaped, what else would you want to know to predict the smell of the molecule?

7. Which do you think has more influence on smell: the functional group that is present or the shape of a molecule? Explain why you think so.

# Shape and Smell

## THINK ABOUT IT

So far the investigation into the chemistry of smell has turned up evidence that molecular formula, functional group, chemical name, and the overall shape of molecules can help predict smell. Which pieces of information are most useful? What should you consider first?

**What chemical information is most useful in predicting the smell of a compound?**

*To answer this question, you will explore*

➲ A Summary of the Smell Investigation So Far

➲ General Rules for Predicting Smell

## EXPLORING THE TOPIC

## ➲ A Summary of the Smell Investigation So Far

Clearly, the relationship between molecules and smell is quite complex. Often, when you smell and examine new molecules, you find out new information and refine your hypotheses. Here is what you might conclude about smell so far.

### MOLECULAR FORMULA

Molecular formulas can give some indication of the way compounds smell.

| 2 O atoms | sweet or putrid |
| --- | --- |
| 1 N atom | fishy |
| Only C and H | gasoline or no smell |
| 1 O atom | sweet, minty, or camphor |

Molecular formula is most useful in predicting fishy smells. It does not work so well for sweet, putrid, minty, or camphor smells.

### FUNCTIONAL GROUP

Functional groups are even better than molecular formulas in narrowing down a smell. For example, compounds with a carboxyl functional group smell putrid. Compounds with an ester functional group smell sweet.

Carboxylic acids, putrid

Esters, sweet

Todd Gipstein/National Geographic/Getty Images

CONSUMER
CONNECTION

Citronella oil has been used for decades as an insect and animal repellent. It is found in candles, lotions, gels, and other products designed to repel mosquitoes and other pests.

Because alcohols and ketones might smell sweet, minty, or like camphor, functional group is not always correct in predicting smells.

## CHEMICAL NAME

Molecules are named according to their functional groups. A carboxylic acid molecule, for instance, will have a chemical name ending in "-ic acid."

### Molecular Names and Smells

| Compound type | Name ending | Examples | Smell |
|---|---|---|---|
| carboxylic acid | -ic acid | butyric acid, $C_4H_8O_2$ <br> pentanoic acid, $C_5H_{10}O_2$ | putrid |
| ester | -yl -ate | ethyl acetate, $C_4H_8O_2$ <br> propyl acetate, $C_5H_{10}O_2$ | sweet |
| amine | -ine | hexylamine, $C_6H_{15}N$ <br> diisobutylamine, $C_8H_{19}N$ | fishy |
| alkane | -ane | methane, $CH_4$ <br> pentane, $C_5H_{12}$ <br> hexane, $C_6H_{14}$ | no smell or gasoline |
| alcohol | -ol | citronellol, $C_{10}H_{20}O$ <br> menthol, $C_{10}H_{20}O$ | sweet, minty, camphor |
| ketone | -one | menthone, $C_{10}H_{18}O$ <br> pulegone, $C_{10}H_{16}O$ | sweet, minty, camphor |

Because the functional group is not useful in predicting smells for alcohols and ketones, the chemical name is also not very useful.

## MOLECULAR SHAPE

The overall shape of a molecule can be an important link to smell. Camphor-smelling compounds are made of ball-shaped molecules. Minty-smelling compounds are flat in overall shape. And compounds made of long and stringy molecules often smell sweet.

Long, stringy — **Citronellol,** sweet smell

Flat/frying pan–shaped — **Carvone,** minty smell

Ball-shaped — **Fenchol,** camphor smell

## ⬤ General Rules for Predicting Smell

Each piece of chemical information appears to be valuable at different times, depending on the smell category. It is helpful to look for general guidelines you can use to predict smell.

This table summarizes the chemical information of each of the five smell categories you have encountered. In each category, the most useful information for predicting that smell has been highlighted.

### Summary of Chemical Information

| Smell classification | Molecular formula info | Shape(s) | Functional group(s) |
|---|---|---|---|
| **sweet** | 2 O atoms in some, 1 O atom in others | stringy | hydroxyl and ester |
| **minty** | rings, 1 O atom in all | frying pan–shaped | ketone and hydroxyl |
| **camphor** | double rings, 1 O atom in all | ball-shaped | ketone and hydroxyl |
| **putrid** | 2 O atoms in all | stringy | carboxyl |
| **fishy** | 1 N atom in all | stringy | amine |

The only category that causes a little trouble is the sweet-smelling category. Not every compound made of stringy molecules is sweet smelling. But, if the compound has stringy molecules and is not a carboxylic acid or an amine, then it is sweet smelling.

**Example**

## Predict a Smell

Examine the following molecular compound and predict its smell. What information is most useful in predicting the smell? What type of compound is this?

### Solution

Here is what you can deduce:

- Molecular shape: Because the molecule is stringy, it probably does not have a minty or camphor smell. Its smell may be sweet, putrid, or fishy.

- Molecular formula: There is no N atom, so its smell can't be fishy. It has two O atoms, so it might have a sweet or putrid smell.

- Functional group: The carboxyl functional group indicates that the molecule smells putrid.

- The putrid smell and carboxyl group indicate that this is a carboxylic acid.

The functional group was most useful in predicting the smell of this compound.

## What chemical information is most useful in predicting the smell of a compound?

By looking for patterns in the chemical information of molecules, it is possible to come up with general guidelines for certain smell categories. Some chemical information is more useful than other information depending on the smell category. Molecular shape is an accurate predictor for minty and camphor compounds. Functional group can be used to predict the smells of amines and carboxylic acids. Each piece of chemical information, including chemical formula and name, allows you at least to narrow down the possible smells of a compound.

## Exercises

### Reading Questions

1. Can the molecular formula of a compound help you to predict its smell? Explain your reasoning.

2. What one piece of chemical information would you want in order to predict the smell of a molecule? Explain your choice.

### Reason and Apply

3. What is the minimum amount of information you need to know to determine if a compound smells sweet?

4. If someone tells you a compound smells sweet, what can you assume about its molecules? What can't you assume about its molecules?

5. If someone told you a compound was made of molecules that had a stringy three-dimensional shape, what else would you want to know to determine that compound's smell?

# Receptor Site Theory

## THINK ABOUT IT

Imagine you are sipping a cool glass of lemonade on a hot summer day. The lemonade contains some fresh mint leaves for extra flavor. You detect both the lemon smell and the minty smell. The molecules associated with these two smells are quite different. What is happening inside your nose that allows you to detect both smells at the same time and to tell them apart? How does the nose know these molecules are different?

### How does the nose detect and identify different smells?

*To answer this question, you will explore*

→ Receptor Site Theory

→ Phase Change and Molecular Stability

## EXPLORING THE TOPIC

### → Receptor Site Theory

The chemistry of smell is a fairly new science, and much remains to be discovered about how the sense of smell works. A number of theories exist to explain what happens in the nose when a smell is detected. While scientists are still uncertain about all of the details about how smell works, one theory seems to fit the evidence better than the others. This theory is called the **receptor site theory.** It is a widely accepted theory of how the interior of the nose detects different smells.

The receptor site theory uses the "lock and key" model. In this model, molecules are like "keys" that fit only into certain "locks." In other words, molecules have specific shapes that will fit only certain receptor sites, and the lining of the nose is covered with receptor sites.

### HEALTH CONNECTION

Many medications work by using a lock and key mechanism. The molecules found in a medication may be designed to *stimulate* a receptor site, resulting in the desired physical response. Or a medication may have been designed to *block* receptor sites, preventing a certain unwanted physical response.

*Molecules have specific shapes that fit into specific receptor sites in the nose, the way a key fits into a specific keyhole of a lock.*

The lemony-smelling and minty-smelling molecules in the compounds of the lemonade might fit into different receptor sites as shown in the illustration below.

**Citral,** sweet-smelling

**Carvone,** minty-smelling

**Fenchol,** camphor-smelling

In this model, each type of molecule fits only into its own receptor sites, not into the others. According to receptor site theory, this is how the nose distinguishes between two molecules.

Scientists are unsure how many receptor sites are in the nose or how many different shapes the sites accommodate. However, they do know that receptor sites are made of large, very intricate protein molecules. These protein molecules consist of hundreds of atoms bonded together. They are much more elaborate and complex than the simple drawings of receptor sites shown in the illustration.

Scientists think that once a smell molecule has "docked" in a receptor site, nerves are stimulated and send a message to the brain. If minty receptor sites are stimulated, then the brain registers a minty smell.

It is possible that smell is much more complex than just described. A specific smell might be the result of a combination of different molecules stimulating a combination of receptor sites.

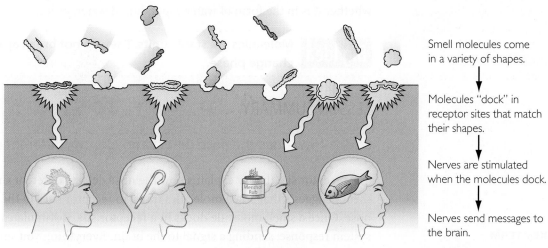

Smell molecules come in a variety of shapes.

↓

Molecules "dock" in receptor sites that match their shapes.

↓

Nerves are stimulated when the molecules dock.

↓

Nerves send messages to the brain.

*The brain interprets the messages as smells.*

# Phase Change and Molecular Stability

## SMELLS ARE GASES

When you bring an onion home from the supermarket, it doesn't give off much of a smell. But when you slice or chop an onion, the odor can become quite overpowering. What's happening? When you cut up the onion, your knife is cutting through some of the cell walls of the onion and releasing a lot of liquid.

One of the molecules that is in this liquid and is responsible for the distinctive smell of onions is allyl propyl disulfide. Its structural formula is shown here. Its molecular formula is $C_6H_{12}S_2$. This is the molecule your nose detects when you smell onions.

Allyl propyl disulfide, $C_6H_{12}S_2$

But something else has to happen for you to smell the onion. Molecules of allyl propyl disulfide have to get from the onion into your nose. The only way that can happen is if some of the molecules travel through the air to your nose. To do this, the compound changes phase to become a gas or vapor that floats through the air and enters your nostrils. There is no other possible explanation for your ability to smell an onion several feet away.

You cannot smell something unless some portion of it is in the form of a gas. This means that everything you are able to smell is a substance that has become a gas.

## MOLECULES ARE STABLE UNITS

Because your nose is detecting the shape and functional group of a molecule, this means that the molecule must enter your nose in one piece. When molecules change phase, they do not break apart into pieces or into individual atoms. Molecules go from the solid phase, to the liquid phase, to the gas phase without breaking apart into individual atoms. This is because molecules are stable units, with strong covalent bonds between the atoms. For example, water is still water whether it is in the form of water vapor, liquid water, or ice.

**Big Idea** Molecules are stable units. They do not break apart when they change phase.

## LESSON SUMMARY

### How does the nose detect and identify different smells?

The receptor site theory explains how the nose detects different smells. This theory suggests that smell molecules travel to receptor sites in the lining of the nose. Once a smell molecule docks in a receptor site, it stimulates a nervous-system response, sending a signal to the brain. Everything you smell is in the form of molecules in the gas phase.

**Important to Know** A substance does not have to boil in order for some of its molecules to become a gas. At the surface of molecular liquids and solids, some molecules escape and become a gas.

**KEY TERM**

receptor site theory

**Reading Questions**

1. Explain the receptor site theory and how it applies to smell detection.

2. What does phase change have to do with smell?

**Reason and Apply**

3. Here is a structural formula for an active ingredient in muscle ointment. How does this compound smell? How does your nose detect the smell?

4. A chemist creates a new molecule that has a completely different three-dimensional shape from other molecules humans have ever encountered. Would you be able smell it? Select the best answer.
   (A) Probably not, because synthetic compounds do not have a smell.
   (B) Definitely, because all esters smell sweet.
   (C) Probably not, because our noses would not have developed receptor sites to detect it.
   (D) Definitely, because parts of the molecule can break off to fit into receptor sites in our noses.

5. Some smells are similar, such as popcorn and freshly baked bread. Explain how you might account for similar smells using the receptor site theory.

6. How might the receptor site theory explain why a dog has a better sense of smell than a person?

7. Why do you think you might "get used to" a smell and hardly detect it? Use the receptor site model in your explanation.

8. Are some smells "faster" than others? Explain.

# CHAPTER 7

# Building Molecules

# SUMMARY

## KEY TERMS

ball-and-stick model

tetrahedral shape

electron domain

electron domain theory

pyramidal shape

bent shape

trigonal planar shape

linear shape

space-filling model

receptor site theory

# Smells Update

The smell of a substance sometimes has more to do with molecular shape than with the functional groups that are present. In this chapter, you learned about three-dimensional molecular structure and related it to smell. Ball-and-stick and space-filling models show how atoms are arranged within a molecule. The actual shape of a molecule is determined by the locations of electron domains.

No matter what shape a molecule has, it must fit into an appropriate receptor site in order for the nose to detect its smell.

## REVIEW EXERCISES

1. Examine the ball-and-stick model for propyl acetate.
   a. What is the molecular formula for this molecule?
   b. Draw the Lewis dot structure and the structural formula.
   c. What functional group is in the molecule?
   d. Predict the smell of the compound.

2. Predict the shape of each molecule. Support your answer.
   a. ammonia, $NH_3$
   b. silicon chloride, $SiCl_4$
   c. hydrogen sulfide, $H_2S$
   d. hydrogen cyanide, HCN
   e. formaldehyde, $CH_2O$

3. Explain what determines a molecule's shape.

4. Draw a possible Lewis dot structure for $C_4H_{10}O$.

5. From what you've learned so far, how is molecular shape related to smell?

# Other Smell Classifications

 **PROJECT** Research a smell classification not discussed in class, such as musk, woody, spicy, nutty, leather, or tobacco. Create a poster that includes

- A drawing describing the smell classification and sources of the smell

- The structural formulas and ball-and-stick models of at least five molecules in the smell classification that you investigated, along with their chemical names

- A brief description of the properties associated with this group of molecules

# CHAPTER 8

# Molecules in Action

## In this chapter, you will study

- polarity

- intermolecular interactions

- why some substances do not smell

- how phase, molecular size, polarity, shape, and bonding patterns relate to smell

*The water droplets on a leaf are the result of interactions between molecules in the water and the surface of the leaf.*

Molecules interact with each other in a variety of ways. The process of smelling involves interactions between the molecules that enter the nose and the molecules that make up the lining of the nose. One key to molecular interactions is the fact that atoms in a molecule often do not share electrons equally. This unequal sharing creates partial positive and negative charges on different atoms in the molecule. These partial charges influence how molecules interact with each other and help to determine properties like phase, smell, and whether the substance will dissolve in another substance. Molecular size, shape, and bonding patterns all contribute to smell.

*Attractive Molecules*

# Attractions between Molecules

## THINK ABOUT IT

Not everything has a smell. In fact, there are more substances on the planet that we *don't* smell than substances that we *do* smell. If everything around us had a smell, our noses would probably be overwhelmed most of the time.

**Why do some molecules smell while others do not?**

*To answer this question, you will explore*

➲ Compounds That Do Not Have a Smell

➲ Polar Molecules

➲ Intermolecular Force

## EXPLORING THE TOPIC

## ➲ Compounds That Do Not Have a Smell

Take a big whiff of air. Perhaps you are able to detect some odors—a freshly mowed lawn, somebody's food cooking. But the air itself does not smell. Small molecules, such as nitrogen, oxygen, carbon dioxide, and water vapor, obviously fit inside the receptor sites. So, why can't we smell them?

Perhaps these molecules are too small to stay docked in a receptor site. But some small molecules do have a smell. Hydrogen sulfide, $H_2S$, smells like rotting eggs and ammonia, $NH_3$, smells pungent.

*Some small molecules do not have a smell.*

*But $H_2S$ and $NH_3$ both have strong smells.*

So the molecular size of a compound does not explain whether we can smell it. It turns out that a molecular property called *polarity* is involved.

Buck Forester/Flickr/Getty Images

## Polar Molecules

When two different materials are rubbed together, some electrons can transfer from one material to the other. The result is an imbalance of positive and negative charges. One material has an excess of positive charge and the other an excess of negative charge. This is known as static electricity.

Suppose that you rub a plastic wand on a piece of cloth. The plastic wand will end up with a charge. In class you tested a number of liquids by holding a charged wand near them and observed that some of the streams bend toward the wand.

In this table, all of the liquids in the first column are attracted to a charged wand. They are made up of **polar molecules.** The liquids in the second column are not attracted to a charged wand. They consist of **nonpolar molecules.**

These molecules are all **polar.** They all have a smell.

| Polar molecules | Nonpolar molecules |
| --- | --- |
| water | mineral oil |
| vinegar | hexane |
| nail polish remover | motor oil |
| rubbing alcohol | |
| antifreeze | |

These molecules are all **nonpolar.** The two oils do not have a smell.

Water is a polar molecule. As this illustration shows, there are partial charges at different locations on the water molecule. These **partial charges** are much smaller than the charge on an individual electron or proton. However, the charges are large enough to cause a stream of water to be attracted to a charged wand.

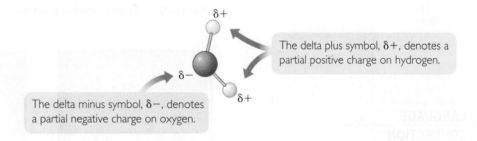

The delta plus symbol, δ+, denotes a partial positive charge on hydrogen.

The delta minus symbol, δ−, denotes a partial negative charge on oxygen.

The oxygen atom has a partial negative charge, and the hydrogen atoms each have a partial positive charge.

## Intermolecular Force

### EXPLAINING THE CHARGED WAND

When a wand with a negative charge is placed next to a stream of water, the water molecules orient themselves in space so that their positive ends are lined up in the direction of the wand. As a result, the whole stream of water is attracted to, or pulled toward, the charged wand. The illustration on the next page shows a possible way to depict what happens to the molecules.

Martyn F. Chillmaid/Science Photo Library/Science Source

Water molecules        Charged wand

The water molecules have oriented their positive sides in the direction of the wand.

Water molecules next to a charged wand

Water molecules interacting

## LANGUAGE CONNECTION

The saying "Oil and water don't mix" is one way to remember that a nonpolar substance and a polar substance tend not to dissolve in each other.

## INTERMOLECULAR BEHAVIOR

The partial charges on polar molecules cause more than an attraction to a charged wand. They also cause attractions between molecules. As the molecules in a polar liquid tumble around, they tend to align with each other because partial negative charges are attracted to partial positive charges. This attraction between individual molecules is an **intermolecular force.** The prefix *inter-* means "between" and the force is a force of attraction.

Intermolecular attractions can be used to explain many observable properties.

*Observation:* Water beads up on waxed paper. Oil spreads out.

*Explanation:* The water molecules are polar, while the oil molecules are nonpolar. The individual water molecules are attracted to each other and "cling together." Individual oil molecules are not attracted to one another to the same extent, and they spread out.

*Observation:* Water is a liquid at room temperature, but methane is a gas at room temperature. They have roughly the same molecular mass.

*Explanation:* Water is polar and methane is nonpolar. The individual water molecules are attracted to each other and stay together as a liquid. The attractions between methane molecules are much weaker. The methane molecules spread throughout the room as a gas.

*Observation:* Methanol dissolves easily in water. Oil floats on top of water but does not dissolve.

*Explanation:* The polar methanol molecules are attracted to the polar water molecules and go into solution. The nonpolar oil molecules are not attracted to the polar water molecules.

## LESSON SUMMARY

### Why do some molecules smell while others do not?

**KEY TERMS**

polar molecule

nonpolar molecule

partial charge

intermolecular force

Molecules can be divided into two classes: polar and nonpolar. Polar molecules are attracted to a charged wand because they have partial charges distributed within the molecule. Nonpolar molecules are not attracted to a charged wand. Polarity is responsible for intermolecular attractions that affect many properties of molecules, possibly including smell properties.

## Exercises

### Reading Questions

1. Explain in your own words what a polar molecule is.

2. What are intermolecular attractions?

### Reason and Apply

3. **Lab Report** Write a lab report for the Lab: Attractions Between Molecules. In your conclusion, explain why some liquids were attracted to the wand and others were not, and why some liquids beaded up on waxed paper while others did not.

(Title)

**Purpose:** (Explain what you were trying to find out.)

**Procedure:** (List the steps you followed.)

**Results:** (Explain what you observed during the experiment.)

**Conclusions:** (What can you conclude about what you were trying to find out? Provide evidence for your conclusions.)

*(continued)*

4. Methanol, $CH_3OH$, is attracted to a charged wand. The hydrogen atoms have a partial positive charge and the oxygen atom has a partial negative charge.

   a. Draw a picture showing how you predict the methanol molecules are oriented when they are attracted to a wand with negative charges.

   b. Do you expect methanol to bead up or spread out on waxed paper? Explain your thinking.

   c. Do you expect methanol to dissolve in water? Explain your thinking.

5. Hexane, $C_6H_{14}$, is not attracted to a charged wand and it spreads out on waxed paper. Do you expect hexane to dissolve in water? Explain your thinking.

6. Explain why no liquids are repelled from a charged wand.

# The BARE ESSENTIALS of POLARITY

You don't have to go to the ends of the earth to find polar molecules. They're all over the place. A polar molecule is just a molecule with a difference in electrical charge between two ends.

Polarity in molecules is caused by differences in electronegativity between atoms. Electronegativity describes the ability of an atom to attract bonding electrons toward itself.

Bonded pair of electrons

HEY!

Chlorine is more electronegative than hydrogen. So the bonded pair of electrons in HCl spends more time near chlorine.

Electronegativity values tend to increase as you move "northeast" on the periodic table, and decrease as you move "southwest."

KEEP OUT

The noble gases are often not assigned electronegativity values. They rarely bond to other atoms.

When two atoms with different electronegativity values bond, the bonding electrons spend more time around the more electronegative atom, creating a PARTIAL NEGATIVE CHARGE on that atom. The other atom then has a PARTIAL POSITIVE CHARGE, and the bond is polar.

When atoms with equal electronegativity values bond, they form nonpolar bonds. The electron-attracting strength of each atom is the same.

However, if the electronegativities of two bonded atoms are different, then their bond will be polarized—maybe a little...

...maybe a lot.

Because the elements have such varying electronegativities and can bond in many different combinations, there is really a continuum of polarity in bonding. We can break the continuum down into three categories.

**NONPOLAR COVALENT**

$O_2$

$H_2$

The clearest examples of nonpolar covalent bonds are those between identical atoms, such as in $H_2$, $N_2$, $O_2$, or $Cl_2$. Bonds between atoms with nearly the same electronegativity value, such as carbon and hydrogen, can also be considered nonpolar.

**POLAR COVALENT**

Partial negative charge

HF

Partial positive charge

In a polar covalent bond, two atoms share bonded pairs of electrons somewhat unequally. The electrons are more attracted to one atom than the other. Examples include bonds between carbon and oxygen atoms, or between hydrogen and fluorine atoms.

**IONIC**

Negative charge

Come back!

NaCl

Positive charge

A large difference in electronegativity results in the winner-take-all situation of ionic bonding. The more electronegative atom takes the bonding electrons and becomes a negative ion, while the other atom becomes a positive ion. The opposite charges on the ions attract each other.

Polar bonds between atoms create dipoles. The word dipole can refer to (1) the polarity of an individual polar bond between atoms, (2) the net polarity of an individual polar molecule that may have several polar covalent bonds within it, and (3) the polar molecule itself.

## Confusing? Here are some examples:

An $N_2$ molecule isn't a dipole and it doesn't have any dipoles.

Bond: Nonpolar
Molecule: Nonpolar

HCl has a dipole and it is a dipole.

Bond: Polar
Molecule: Polar

$CO_2$ has two dipoles but the $CO_2$ molecule itself is not a dipole. Its polar bonds balance each other out and make the molecule nonpolar overall.

Bonds: Polar
Molecule: Nonpolar

$H_2O$ has two dipoles. Because of its bent shape, it also has a dipole in the sense of an overall polarity.

Bonds: Polar

Molecule: Polar

The polarity of molecules can affect many of their other properties, such as their solubility, their boiling and melting points, and their odor.

Mmmmm... you smell PENGUINY.

Why are we in this comic strip? Penguins and polar bears don't even live at the same poles!

Suits me!

# Electronegativity and Polarity

## THINK ABOUT IT

Hydrogen chloride, HCl, is a colorless but very toxic gas. Its smell is described as a suffocating, acrid odor. Like most other small molecules that smell, HCl molecules are polar. But what makes an HCl molecule polar? Where do the partial charges come from on the atoms in a polar molecule?

**What makes a molecule polar?**

*To answer this question, you will examine*

➲ Electronegativity

➲ Nonpolar Molecules

➲ Electronegativity and Bonding

## EXPLORING THE TOPIC

### ➲ Electronegativity

The hydrogen atom and the chlorine atom in hydrogen chloride, HCl, form a covalent bond by sharing a pair of electrons. This cartoon represents HCl as a penguin and a polar bear. The bonded pair of electrons is represented as two scoops of ice cream. Although the penguin and the polar bear are sharing the ice cream, they are not sharing it equally.

*Differences in atoms' electronegativity lead to differences in sharing electrons.*

Similarly, the hydrogen atom and the chlorine atom in a hydrogen chloride molecule do not share the bonded pair of electrons equally. The chlorine atom attracts the shared electrons much more strongly than the hydrogen atom does. As a result, the shared electrons spend more time near the chlorine atom than they do near the hydrogen atom. Because of this displacement of the electrons, the hydrogen atom has a partial positive charge and the chlorine atom has a partial negative charge.

The tendency of an atom to attract shared electrons is called **electronegativity.** An atom that has a large electronegativity value strongly attracts shared electrons. In this case, the chlorine atom, like the polar bear, is stronger in attracting electrons, so it is more electronegative than the hydrogen atom. The atoms that are more electronegative are the ones that end up with a partial negative charge. The atoms that are less electronegative end up with a partial positive charge. The result is a polar bond.

A polar molecule is called a **dipole,** because it has two poles: a positive end and a negative end. A dipole can be shown with an arrow starting at the positive end and pointing to the negative end of the molecule. The polar bond itself is also called a dipole.

## ● Nonpolar Molecules

When two atoms with identical electronegativities bond together, the attraction of the shared electrons is identical. As a result, the molecule is nonpolar. For example, $H_2$ and $Cl_2$ are both nonpolar molecules. They have no partial charges.

*A contest between two polar bears of equal strength or two penguins of equal strength would result in a tie.*

There is another way to end up with a nonpolar molecule. Examine the next illustration. Why is carbon dioxide, $CO_2$, a nonpolar molecule?

*$CO_2$ is a nonpolar molecule.*

The two dipoles in $CO_2$ balance each other, and there is no partial positive end to the molecule. So the overall molecule is nonpolar.

# ➔ Electronegativity and Bonding

There are two types of covalent bonds: polar covalent bonds and nonpolar covalent bonds. When two atoms with different electronegativities bond together, the result is a polar covalent bond.

*Different electronegativities result in polar bonds.*

When the electronegativities of the atoms that bond are identical, the result is a nonpolar covalent bond.

*Identical electronegativities result in nonpolar bonds.*

If the electronegativities of the two atoms differ greatly, it is possible for electrons to be pulled entirely toward one of the atoms in a bond. The result is an ionic bond.

*Extremely different electronegativities result in ionic bonds.*

Ionic compounds represent the extreme of polar bonds, in which electrons are transferred to the more electronegative atom in the pair.

## LESSON SUMMARY

### What makes a molecule polar?

Electronegativity is a measure of the ability of an atom to attract the electrons that are involved in a bond. Different atoms have different electronegativities. When two different kinds of atoms bond together, they do not attract the bonding electrons equally. The electrons are attracted to the atom with greater electronegativity. The result is a polar covalent bond. When atoms with identical electronegativities bond together, the electrons are shared equally, and the result is a nonpolar covalent bond. If the two atoms involved in a bond differ greatly in electronegativity, the result is electron transfer and an ionic bond.

**KEY TERMS**

electronegativity

dipole

## Exercises

### Reading Questions

1. What is the difference between a polar covalent bond and a nonpolar covalent bond?

2. What is the difference between a polar covalent bond and an ionic bond?

3. Explain why carbon dioxide is a nonpolar molecule even though its bonds are polar.

### Reason and Apply

4. Are these molecules polar covalent or nonpolar covalent?
   **a.** $N_2$    **b.** HF    **c.** $F_2$    **d.** NO    **e.** FCl

5. Using polar bears and penguins, create an illustration showing an ammonia, $NH_3$, molecule. (*Hint:* You may wish to start with a Lewis dot structure.)

6. Is the molecule HOCl polar or nonpolar? Use a Lewis dot structure to explain your thinking.

7. Use electronegativity to explain why some molecules are attracted to a charged wand.

# Electronegativity Scale

## THINK ABOUT IT

Some atoms are "greedier" than other atoms when it comes to sharing the bonding electrons between them. A bond between two atoms with very different electronegativities is more polar than a bond between atoms with similar electronegativities. How can the polarity of different bonds be compared?

### How can electronegativity be used to compare bonds?

*To answer this question, you will explore*

➲ Electronegativity Scale

➲ Diatomic Molecules

## EXPLORING THE TOPIC

## ➲ Electronegativity Scale

Chemists have assigned each atom a number, called an *electronegativity value.* This number corresponds to the tendency of an atom to attract bonding electrons. By using these numbers, it is possible to compare the polarity of different bonds. The table here shows the electronegativity value for an atom of each element in the periodic table.

Electronegativity scale

\* Electronegativity values for the lanthanides and actinides range from about 1.10 to 1.50.

Notice that the values generally increase from left to right and bottom to top. Noble gases are often not assigned values because they generally do not form compounds.

A scale for electronegativity was first proposed by Linus Pauling in 1932. The scale chemists use today ranges in value from 4.0 down to 0. The electronegativity scale allows you to compare individual atoms. For example, fluorine, with an electronegativity value of 3.98, attracts shared electrons more strongly than hydrogen, with an electronegativity value of 2.10.

# Diatomic Molecules

The electronegativity scale also allows you to compare the polarity of bonds. In a polar covalent bond, the electrons tend to spend more time around the more electronegative atom. Several substances with only two atoms are shown below. Molecules with two atoms are called **diatomic molecules.** By looking at the numerical difference between the electronegativities of the two atoms, it is possible to compare the polarity of one bond to another. Bonds that have a greater difference in electronegativity between the two atoms are more polar.

Differences in electronegativity

When the difference in electronegativity between the two atoms is very large, the bond is no longer considered covalent. In that case, the electrons are transferred from the less electronegative atom to the more electronegative atom. A cation and an anion form that are attracted to one another in an ionic bond. As shown on this electronegativity scale, when the difference between the electronegativities of two bonded atoms is greater than about 2.1, the bond is considered ionic:

### Example

### Potassium and Chlorine

Predict the type of bond you would find between potassium and chlorine.

### Solution

Use the electronegativity scale to find the electronegativities of potassium, K, and chlorine, Cl. It shows that the values are 0.82 for K and 3.16 for Cl. The difference is 2.34, so the bond is ionic.

## How can electronegativity be used to compare bonds?

Electronegativity values can help you to compare and classify different bonds. If there is no difference in electronegativity between the two atoms, the bond is considered nonpolar covalent. The greater the electronegativity difference, the more polar the bond. The electrons in a polar bond tend to spend more time around the more electronegative atom. When the difference between the electronegativities of two bonded atoms is greater than about 2.1, the bond is considered ionic and one atom gives up an electron to the other atom, forming two ions.

**KEY TERM**

diatomic molecule

## Exercises

### Reading Questions

1. Explain how electronegativity values help you to determine the polarity of a bond between two atoms.

2. How can you determine which atom in a covalent bond is partially positive?

### Reason and Apply

3. Consider the following pairs of atoms. Place each set in order of increasing bond polarity. Describe the trend.
   **a.** Li−F      Na−F      K−F      Rb−F      Cs−F
   **b.** Mg−O      P−S      N−F      K−Cl      Al−N

4. Place a partial positive or a partial negative charge on each atom in the following pairs of atoms. Describe the trend.
   **a.** H−B    **b.** H−C    **c.** H−N    **d.** H−O    **e.** H−F

5. Is hydrogen always partially positive when bonded to another atom? Explain.

6. Name three pairs of atoms with ionic bonds. For each pair, show the difference in electronegativity between the two atoms.

7. Name three pairs of atoms with polar covalent bonds. For each pair, show the difference in electronegativity between the two atoms.

8. Describe or draw what happens to the electrons in a polar covalent bond, a nonpolar bond, and an ionic bond.

9. What do we mean when we say that bonding is on a continuum?

## *I Can Relate*

# Polar Molecules and Smell

### THINK ABOUT IT

Polar molecules have a potent smell. The smell of ammonia, $NH_3$, in some window cleaners is quite strong. In contrast, the methane gas, $CH_4$ does not have a smell even though its $C-H$ bonds are all polar. So, why can't you smell $CH_4$ gas?

**What does polarity have to do with smell?**

*To answer this question, you will examine*

➲ Polarity of Molecules

➲ Nonpolar Molecules

➲ Polarity and Smell

### EXPLORING THE TOPIC

## ➲ Polarity of Molecules

If the *bonds* between atoms of a molecule are polar, what about the molecule as a whole unit? One way to tell whether a molecule is polar is to examine its overall shape. Notice that each of the three molecules shown here has an irregular shape, or some sort of asymmetry. Asymmetric molecules are polar.

Ammonia, $NH_3$  Water, $H_2O$  Hydrogen fluoride, HF

Direction of dipole

*Formaldehyde, $CH_2O$*

Ammonia, water, and hydrogen fluoride are all polar. In these molecules, there is a partial negative charge on the nitrogen, oxygen, and fluorine atoms because these atoms are more electronegative than hydrogen atoms. The hydrogen atoms have partial positive charges. To figure out which end of a molecule has a partial negative charge and which end has a partial positive charge, check the electronegativities of the atoms and the overall shape of the molecule.

Consider formaldehyde, $CH_2O$, shown here. The numbers next to the atoms are their electronegativities.

The oxygen end of the formaldehyde molecule has a partial negative charge because the oxygen atom is the most electronegative atom in the formaldehyde molecule. Similarly, the hydrogen atoms have partial positive charges because they are the least electronegative atoms in the molecule.

**Example 1**

### Phosphine, $PH_3$

Is phosphine, $PH_3$, polar? What is the direction of the dipole?

*Solution*

A Lewis dot structure of $PH_3$ shows four pairs of electrons, three bonding pairs and one lone pair. These four pairs of electrons are arranged around the phosphorus, P, atom in a tetrahedral shape.

Because the molecule is asymmetrical, it is polar. You can find the electronegativity values for phosphorus and hydrogen in the electronegativity scale in Lesson 44: Thinking (Electro)Negatively. The electronegativity value for P is 2.19 and for H is 2.10. Because the electronegativity of the phosphorus atom is greater, the P atom attracts the electrons more strongly and has a partial negative charge. The H atoms have partial positive charges. The overall dipole is shown.

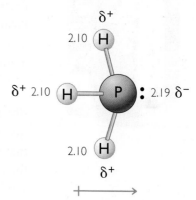

## ➲ Nonpolar Molecules

Diatomic molecules with two identical atoms are nonpolar. The electrons between the two identical atoms are shared equally. However, these are not the only kind of nonpolar molecules. Some molecules are symmetrical. The symmetry in these molecules can make them nonpolar even though they have polar bonds within them. The polarities of the individual polar bonds balance each other out due to the shape of the molecule.

$$\overset{\delta^-}{\ddot{O}} = \overset{\delta^+}{C} = \overset{\delta^-}{\ddot{O}}$$

$$CO_2 \qquad CH_4 \qquad CF_4$$

Nonpolar molecules

# Polarity and Smell

Small molecules that are polar have a smell. Small molecules that are nonpolar do not have a smell. What does polarity have to do with the smell of small molecules?

| Polar: Have a smell | Nonpolar: Do not have a smell | Exceptions |
|---|---|---|
| HF, HCl, $H_2S$, $CH_3F$, $CH_2F_2$, $CHF_3$, $NH_3$, $PH_3$, HOCl | $N_2$, $O_2$, $CO_2$, $CH_4$, $CF_4$ | $F_2$, $Cl_2$, $Br_2$, and $CCl_4$ are nonpolar but have a smell. $H_2O$ is polar but does not have a smell. |

Polar molecules have properties that distinguish them from nonpolar molecules. Polar molecules dissolve in water and are attracted to other polar molecules. One well-accepted theory is that polar molecules *dissolve* in the mucous membrane of the nose and then attach to receptor sites. The mucus is composed largely of water.

## RECEPTOR SITES IN THE NOSE

Receptor sites in the nose contain large polar protein molecules. Scientists believe that small polar molecules are attracted to the polar receptor sites after they enter the nose. You could think of polarity as working somewhat like a magnet, with small polar molecules "sticking" to the polar part of a receptor site. Nonpolar molecules are not attracted to the same extent as polar molecules, so they may not be detected by the nose.

**ENVIRONMENTAL CONNECTION**

Chlorotrifluoromethane, also known as Freon 13, was developed early in the 20th century and used widely as a refrigerant. However, use of Freon 13 and other chlorofluorocarbons (CFCs) was significantly reduced in the late 1980s due to their effects on the ozone layer of the atmosphere.

*Ammonia, $NH_3$, dissolves in water because ammonia and water are both polar liquids.*

---

**Example 2**

## Chlorotrifluoromethane, $CCIF_3$, and Tetrafluoromethane, $CF_4$

Compare chlorotrifluoromethane with tetrafluoromethane. Explain why one of these molecules has a smell and the other does not.

### Solution

Both molecules have a tetrahedral shape. In tetrafluoromethane, each $C-F$ bond is polar. However, the four $C-F$ bonds are distributed symmetrically so that the molecule as a whole is nonpolar. Therefore, it does not have a smell.

**Example 2**

*(continued)*

Chlorotrifluoromethane is not as symmetrical. This is because one of the F atoms is replaced by a Cl atom. The molecule is polar because F and Cl have different electronegativities. Therefore, it has a smell.

*Tetrafluoromethane is symmetrical and nonpolar.*

*Chlorotrifluoromethane is asymmetrical and polar.*

## NATURE CONNECTION

Many animals, such as elephants, seem to be able to smell water at great distances.

## LESSON SUMMARY

### What does polarity have to do with smell?

Small asymmetrical molecules with polar bonds are polar. Molecules that are symmetrical in every way are nonpolar. Theories suggest that the polarity of a molecule may help it to "stick" in a receptor site in the nose. Further, polar substances are more soluable in water. This may allow polar substances to dissolve in the mucous membrane of the nose, whereas nonpolar substances are simply exhaled undissolved and undetected.

## Exercises

### Reading Questions

1. How can you determine if a molecule is polar?
2. Describe one theory of why small nonpolar molecules do not have a smell.

### Reason and Apply

3. For each of the molecules listed, draw a Lewis dot structure and indicate the shape of each molecule. Decide whether these substances smell. Explain your reasoning for each.
   **a.** $H_2Se$    **b.** $H_2$    **c.** Ar    **d.** HOF    **e.** $CHClF_2$    **f.** $CH_2O$
4. For each of the polar molecules in Exercise 3, draw the dipole.
5. Water is an exception to our rule about small molecules. It is a polar molecule, yet humans don't smell it. What do you think is going on?
6. Do you think it might be useful if you could smell the air? Explain your thinking.
7. Methane, $CH_4$, gas leaks can be very dangerous and can be explosive. Why do you think dimethyl sulfide, $C_2H_6S$, is added to natural gas that is used in homes and buildings?

*Sniffing It Out*

# Phase, Size, Polarity, and Smell

## THINK ABOUT IT

You can smell chicken frying or someone's perfume from across a room, but you can't smell a piece of paper or a binder on your desk. And cookies right out of the oven are much easier to smell than cold ones. How can you use what you have learned about molecules to understand these observations?

**What generalizations can you make about smell and molecules?**

*To answer this question, you will explore*

➲ Bonding and Smell

➲ Molecular Size and Smell

## EXPLORING THE TOPIC

## ➲ Bonding and Smell

In Unit 1: Alchemy, you sorted matter into four classes based on whether the substance dissolved in water and whether the substance conducted electricity. This sorting led to four models of bonding: ionic, molecular covalent, metallic, and network covalent. How does the property of smell relate to bonding?

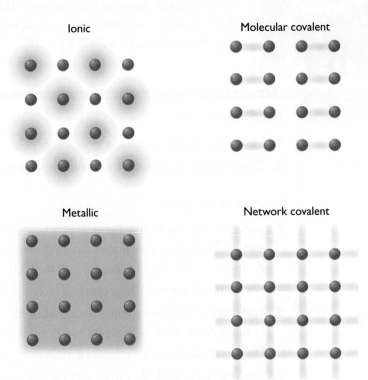

Ionic                    Molecular covalent

Metallic                 Network covalent

*Red spheres represent nuclei and core electrons. Blue areas represent bonding electrons.*

Recall that all the substances you smelled in this unit were molecules with nonmetal atoms. This evidence suggests that substances that have a smell fit into the molecular covalent category.

**Chemical Bonding and Smell**

| Substance type and bonding | Smell? | Phase | Examples | |
|---|---|---|---|---|
| | | | Name | Formula |
| **ionic** <br> metals bonded to nonmetals | **no** | solid | sodium chloride (table salt) <br> calcium oxide (lime) <br> calcium carbonate (chalk) | NaCl <br> CaO <br> $CaCO_3$ |
| **molecular covalent** <br> nonmetals bonded in molecules | **yes,** with some exceptions | gas, liquid, or solid | nitrogen <br> ammonia <br> menthol | $N_2$ <br> $NH_3$ <br> $C_{10}H_{18}O$ |
| **metallic** <br> elemental metals | **no** | solid | gold <br> copper <br> aluminum | Au <br> Cu <br> Al |
| **network covalent** <br> nonmetals bonded in a network | **no** | solid | carbon (diamond) <br> silicon dioxide | C <br> $SiO_2$ |

Molecular covalent substances exist as gases, liquids, or solids at room temperature. Substances in the other three bonding categories are strictly solids (with a few exceptions, such as liquid mercury). It is reasonable to propose that substances that are ionic, metallic, or network covalent do not have a smell because they do not form gases easily at room temperature. This means that they cannot get into the receptor sites in our noses.

**Big Idea** Only molecular covalent substances form gases at room temperature.

**Example 1**

### Bonding and Smell

Predict whether you would be able to smell the following substances. Explain your reasoning.

    **a.** propanol, $C_3H_8O$     **b.** iron, Fe     **c.** copper sulfate, $CuSO_4$

*Solution*

    **a.** Propanol consists of nonmetal atoms, so it is a molecular covalent substance. It probably has a smell.

    **b.** Iron is a metallic solid. It does not have a smell.

    **c.** Copper sulfate consists of both metal and nonmetal atoms, so it is an ionic solid. It does not have a smell.

# Molecular Size and Smell

In the previous lesson, you learned that small molecules have a smell only if they are polar. What about medium-sized and large molecules? Take a moment to examine the data in this table.

## Molecular Size, Polarity, and Smell

| Molecular size and polarity | Smell? | Phase | Examples | |
|---|---|---|---|---|
| | | | Name | Formula |
| small nonpolar molecules | **no** | gas | nitrogen | $N_2$ |
| | | | carbon dioxide | $CO_2$ |
| | | | methane | $CH_4$ |
| **small polar molecules** | **yes** | gas | hydrogen sulfide | $H_2S$ |
| | | | ammonia | $NH_3$ |
| | | | fluoromethane | $CH_3F$ |
| **medium-sized polar and nonpolar molecules** | **yes** | liquid | octane | $C_8H_{18}$ |
| | | | geraniol | $C_{10}H_{18}O$ |
| | | | carvone | $C_{10}H_{14}O$ |
| | | | pentylpropionate | $C_8H_{16}O_2$ |
| large polar and nonpolar molecules | **no** | solid | 1-triacontyl palmitate (beeswax) | $C_{46}H_{92}O_2$ |
| | | | polystyrene | $C_{8000}H_{8000}$ |
| | | | cellulose | $C_6H_{10}O_5$ |

Many pieces of information can be gathered from the table. First, small molecules are gases at room temperature, medium-sized molecules are liquids, and large molecules are solids. One reason for this sorting by size is that large molecules tend to have more intermolecular attractions. This means that larger molecules are less likely to be gaseous and are less likely to get to your nose.

Note that a molecule is considered medium-sized if it has about 5 to 19 carbon atoms and large if it has 20 or more carbon atoms. Carbon is a unique element because it bonds easily to other carbon atoms and forms four bonds. This makes it ideal for forming long and complex molecules.

In general, all medium-sized molecules have a smell, whether they are polar or nonpolar. The medium-sized molecules form a liquid because they are attracted to one another, but they are volatile, meaning they can go easily into the gas phase. Once these molecules get inside the nose, the intermolecular attractions between the molecules and the receptor sites allow you to detect a smell.

## HEATING AND COOLING

What happens when you warm or cool molecular substances? Small molecules are already gases at room temperature. But larger molecules are liquids or solids. These molecules must undergo a phase change in order to be smelled. Warming

**Important to Know** Some substances smell only because they are contaminated with small molecules. For example, plastic itself does not have a smell, but it is often contaminated with small molecules that are used to make the plastic.

## BIOLOGY CONNECTION

The grassy odor that is emitted when the lawn is mowed is cis-3-hexenal. This molecule is unstable and tends to rearrange to form trans-2-hexenal. Scientific evidence connects trans-2-hexenal with healing from stress.

© Jim Craigmyle/Corbis

a molecular liquid or solid will generally give it a more intense aroma. This is because heating a substance increases the rate at which a substance becomes a gas. Cooling a substance has the opposite effect.

When there are more molecules in the air, you are more likely to detect them. So the temperature of a substance will usually affect the strength of its smell. For example, you are more likely to smell warm food than cold leftovers from the fridge.

**Example 2**

### Predict the Smell

Below is the structural formula of a substance called 2-acetylthiazole. Predict whether this substance will have a smell at room temperature. Give evidence to support your answer.

$$H$$
$$|$$
$$C-N \qquad H$$
$$\qquad \qquad | $$
$$C \qquad C-C-C-H$$
$$H-C \qquad S \qquad || \quad |$$
$$\qquad \qquad O \quad H$$

2-acetylthiazole

### Solution

This molecule's formula is $C_5H_5NSO$. It has five carbon atoms. This is a medium-sized molecule, so it probably has some sort of smell. (In fact, it does. It is described as having a hazelnut or popcorn smell.)

## LESSON SUMMARY

### What generalizations can you make about smell and molecules?

Humans can smell some molecular substances. We cannot smell ionic, metallic, or network covalent substances. This is because the first requirement for a substance to be smelled is that it be in a gaseous form. The size of the molecule seems to be another important factor in predicting smell. Most medium-sized molecules will smell. Most large-sized molecules will not smell. Much of this is due to the effect of size on phase change. For small molecules, polarity also plays a role. Small nonpolar molecules do not have a smell. Small polar molecules do have a smell.

### BIOLOGY
### CONNECTION

There is scientific evidence that some fish are better at smelling and tasting underwater than dogs are, out in the air. This explains why fish will often turn away from or avoid a fisherman's lure. It is said that a salmon's sense of smell can detect when a person puts their hand in the water 100 ft upstream.

Darryl Leniuk/Digital Vision/Getty Images

**Reading Questions**

1. Explain what types of substances you would expect to have a smell and why.

2. Explain what types of substances you would expect *not* to have a smell and why.

**Reason and Apply**

3. Predict whether you would be able to smell these substances. Explain your reasoning.

   decanol, $C_{10}H_{22}O$      lead, Pb      iron oxide, $Fe_2O_3$      potassium chloride, KCl

4. Predict whether you would be able to smell these substances. Explain your reasoning.

   neon, Ne      ethane, $C_2H_6$      decane, $C_{10}H_{22}$      methylamine, $CH_5N$

5. If a substance is capable of becoming a gas under normal conditions, should you be able to smell it? Explain.

6. Molecules that have a structure containing more than about 20 carbon atoms rarely smell. Explain why this is so.

7. Your cotton T-shirt smells when you take it out of the clothes dryer. Explain why.

8. Occasionally you might go into a restaurant or a room that smells and after a while you stop noticing the odor. Explain what you think is going on.

9. From a biological point of view, why do you think it is useful for humans to be able to smell certain substances and not others?

## CHAPTER 8

# Molecules in Action

# SUMMARY

## KEY TERMS

**polar molecule**

**nonpolar molecule**

**partial charge**

**intermolecular force**

**electronegativity**

**dipole**

**diatomic molecule**

# Smells Update

Some substances, like hot apple cider, have a smell. Other substances, like a granite tabletop, do not. The process of smelling is an interaction between molecules. To be smelled, a substance must become airborne and enter the nose. Once it is in the nose, it must dissolve and be detected. Receptor sites are composed of large, folded protein molecules with a polar area that attracts polar molecules.

## Summary of Investigation into Smell

|  | Has a smell | Does not have a smell |
|---|---|---|
| **Bonding in compound** | molecular covalent | ionic, metallic, network covalent |
| **Size of compound** | medium-sized molecules and small polar molecules (see polarity) | large molecules |
| **Polarity of compound** | small polar molecules | small nonpolar molecules |
| **Phase of compound** | gas phase | no molecules in the gas phase |

## REVIEW EXERCISES

1. Which element is more electronegative?
   **a.** Zn or Br      **b.** Li or Cs      **c.** Au or Al

2. Place the following pairs of atoms in order of increasing bond polarity. Explain your reasoning.

   H−I         H−Cl         H−F

3. Draw the structural formula and place a partial positive and a partial negative charge on each atom in these molecules. Which molecules are polar? Explain your reasoning.
   **a.** HCl      **b.** $CH_4$      **c.** $H_2O$

4. At cold temperatures, hydrogen bromide, HBr, is a liquid.
   **a.** Draw the Lewis dot structure for HBr. Label the dipole.
   **b.** Would HBr be attracted to a charged wand? Would it form a round drop on waxed paper? Explain your reasoning.
   **c.** In HBr, what type of bond exists between H and Br?
   **d.** Do you expect HBr to have a smell? Explain.

5. Use what you know about molecules to explain why you can't smell perfume while the bottle is closed but can smell it once the bottle is opened.

6. Which can you smell better, cold bread or warm bread right out of the oven? Explain why.

## Sense of Smell Study

PROJECT Plan and conduct an experiment to compare how the sense of smell differs among different groups of people.

- **Create a goal statement.** Write a sentence or two stating the goal of your study. What do you hope to discover or explore?

- **Choose subjects to study.** Pick two categories of people to study and compare. These categories should be clear and easy to determine (for example, children under the age of 12 versus adults, vegetarians versus meat-eaters, women versus men).

- **Write a proposal.** Write several sentences stating how you propose to accomplish your goal, including how you will conduct your study, how you will randomly choose your sample of participants, and how you will set up the control variables. Clear your proposal with your teacher before conducting your study.

- **Conduct your study.** Keep your data organized in a table or chart. Keep careful notes of everything that you do and what you observe.

- **Write up your results and conclusions.** Write up the results of your study and any conclusions you have come to based on your collected data. Include your data table. If possible, create a graph using your data.

L-Carvone

D-Carvone

Archie McPhee and Company

# CHAPTER 9

# Molecules in the Body

### In this chapter, you will study

- mirror-image isomers
- amino acid molecules
- protein molecules

*These molecules are mirror images of each other and have different properties. D-Carvone produces the scent of dill, the other the fragrance of mint. The two unite in this unusual treat.*

Compounds that you smell are sensed by receptor sites in your nose. Receptor sites are made of large molecules called proteins. The protein molecules consist of amino acid molecules that are linked together to form long chains. There are a great variety of protein molecules, made of various combinations of different amino acids. Protein molecules are essential to nearly all body functions, including the sense of smell.

# Mirror-Image Isomers

## THINK ABOUT IT

Limonene is a molecule that comes in two forms. The structural formulas for both forms are identical. Yet one smells like pine needles, while the other has the lemon-orange fragrance of citrus fruits. How can two molecules with the same molecular formula and the same structural formula have such distinct smells?

**What are mirror-image isomers?**

*To answer this question, you will explore*

➲ Mirror Images of Molecules

➲ Properties of Mirror-Image Isomers

## EXPLORING THE TOPIC

### ➲ Mirror Images of Molecules

Many things in nature are mirror images of each other: your ears, for example, or your eyebrows. While your ears may look physically identical, your left ear would not work well on the right side of your head. Your hands and feet are other good examples. Consider your left and right hands. They are made of the same parts and have the same overall structures. But if you try to place your left hand on your right hand with your palms down, the thumbs will point in opposite directions.

Your hands are mirror images of each other. That is, when you observe your left hand in a mirror, it will look identical to your right hand and vice versa. However, you cannot *superimpose* them on each other. No matter how you position your left hand, it is never identical to your right hand.

#### MOLECULES WITH A HANDEDNESS

Just as your right hand cannot be superimposed on your left hand, some molecules have a "handedness."

The two limonene molecules described earlier have identical structures that are mirror images. The mirror images are not superimposable on one another. They are actually different molecules. You could say that one molecule is left-handed and the other is right-handed. They are isomers of one another.

Determining if two molecules are mirror images or the same molecule can be visually challenging. A three-dimensional model of the molecule and a mirror can help.

CCl₄

Consider the mirror image of a simple tetrahedral molecule, tetrachloromethane, CCl₄. It is fairly easy to see that the mirror image can be superimposed on the original molecule. (In other words, you can imagine picking it up and turning or sliding it so that it matches the original.) So, tetrachloromethane does not have a mirror-image isomer. The molecule and its mirror image are identical.

Now consider dichloromethane, CCl₂H₂, and chlorofluoromethanol, CHClFOH. Take a moment to decide if either of the mirror images can be superimposed on its original molecule.

*A can is identical to its mirror image, but a screw is not. The screw's thread spirals in the opposite direction in the mirror image.*

CCl₂H₂ — CHClFOH

The mirror image of dichloromethane, CCl₂H₂, can be superimposed on the original molecule. However, when there are four different groups attached to a carbon atom as in the second molecule, the mirror image cannot be superimposed on the original molecule.

The two molecules of chlorofluoromethanol are **mirror-image isomers.** They have different properties.

---

**Example**

## Objects in a Mirror

Which of the objects listed here look identical in a mirror? Which would look different? Explain any differences.

**a.** a can

**b.** a shoe

**c.** fluoromethane, CH₃F

**d.** 2-chlorobutane, C₄H₉Cl

*(continued)*

### Solution

A and C look identical in a mirror. B and D do not.

In the 2-chlorobutane molecule, the second carbon atom from the left has four different groups attached to it: $CH_3$, H, Cl, and $CH_2CH_3$. Therefore, this molecule has a mirror-image isomer. When the H and Cl atoms of the two mirror-image isomers are in the same location, the $CH_3$ and $CH_2CH_3$ groups are in opposite locations.

## ➲ Properties of Mirror-Image Isomers

One way to figure out if a molecule has a mirror-image isomer is to check to see if any of the carbon atoms are attached to four different groups.

A structural formula for limonene, $C_{10}H_{16}$, is shown below (left). The illustration below (right) shows that nine of the carbon atoms in a limonene molecule are not attached to four *different* groups.

The remaining carbon atom, circled in the diagram, is attached to a hydrogen atom and three carbon "groups." The two sides of the carbon ring are different from one another and considered different groups. You can conclude that limonene has a mirror-image isomer because it contains a carbon atom attached to four different groups of atoms. The two isomers are referred to as D-limonene and L-limonene.

The mirror-image isomers of limonene are shown on top of the next page. The main difference between them has to do with the position of the $CCH_2CH_3$ group highlighted in yellow. If you flip the L-limonene molecule around and try to superimpose it onto the D-limonene molecule, the $CCH_2CH_3$ group will point toward you in one isomer and away from you in the other isomer. It is as if you have aligned your thumbs on your right and left hands, only to find one palm down and the other palm up.

$$\underset{CH_3}{\overset{H}{\diagdown}}C=C\underset{CH_3}{\overset{H}{\diagup}}$$

*Cis* isomer

$$\underset{CH_3}{\overset{H}{\diagdown}}C=C\underset{H}{\overset{CH_3}{\diagup}}$$

*Trans* isomer

D-limonene             L-limonene

## DIFFERENT SMELLS

Just as your left foot does not fit into your right footprint, D-limonene does not fit into the smell receptor site for L-limonene.

Because the mirror-image isomers do not fit into the same receptor site, you would expect them to have distinct smells. Indeed, this is the case: D-limonene smells like citrus while L-limonene smells like pine. The two molecules have the same molecular formula, the same structural formula, and the same shape, but they are mirror images of one another.

## LESSON SUMMARY

### What are mirror-image isomers?

**KEY TERM**

mirror-image isomer

Nature is full of mirror images that are not identical. Some objects and some molecules can be superimposed on their mirror images. For other objects, the mirror image is not superimposable. The mirror image cannot be positioned such that it is identical to the original object or molecule. Molecules with mirror images that cannot be superimposed are called mirror-image isomers. These molecules fit into different smell receptor sites and therefore have different smells.

## Exercises

### Reading Questions

1. What do you have to do to the mirror image of the letter "D" to superimpose the mirror image onto the original letter?

2. What types of molecules have mirror-image isomers?

*(continued)*

**Reason and Apply**

3. List five objects that look different in a mirror and five that do not.

4. Draw mirror images of the capital letters in the English alphabet. Explain why symmetrical objects do not have mirror-image isomers.

5. Draw mirror images of the lowercase letters in the English alphabet. Which lowercase letters are mirror images of one another?

6. Draw the structural formulas for $CH_2O$ and $ClHCO$. Explain why these two molecules do not have mirror-image isomers.

7. Draw structural formulas for the molecules listed below. Draw the mirror image of each. Circle the molecules with mirror-image isomers.
   a. $CH_4$
   b. $CFH_3$
   c. $CClFH_2$
   d. $CBrClFH$

8. Examine the molecules here. Which have mirror-image isomers? Explain your reasoning.
   a. butyric acid          b. ethyl formate          c. 2-butanol

9. Two structural formulas are shown.

Citronellol

Geraniol

   a. Write the molecular formulas for citronellol and geraniol.
   b. Molecules with the structural formula for citronellol can smell either like insect repellant or like flowers. Explain why there are two distinct smells.
   c. All molecules with the structural formula for geraniol smell like roses. Explain why there is only one smell for this molecule.

# Amino Acids and Proteins

## THINK ABOUT IT

Smelly molecules are small or medium-sized molecules that travel as a gas into your nose. Once in your nose, these molecules interact with receptor sites, which send a signal to your brain. Receptor sites are also built of atoms bonded together to create a "pocket" into which a specific smell molecule can fit.

### What is a receptor site made of?

*To answer this question, you will explore*

- Amino Acids
- Proteins

## EXPLORING THE TOPIC

### Amino Acids

**Amino acids** are the building materials for many of the structures in the body, especially those related to the functioning of cells. They are important molecules in all living organisms. As the name implies, amino acids have two functional groups: an amine group and a carboxyl group. The general formula for an amino acid is $C_2H_4NO_2R$, where R is simply a placeholder in the formula for a variety of possible groups of atoms. The groups consist of carbon and hydrogen, and sometimes nitrogen and/or oxygen.

*The structure of an amino acid molecule*

Here are structural formulas for three amino acids with the R groups highlighted. Notice that everything else about them is identical.

| Glycine | Alanine | Valine |

> ### BIOLOGY
> ### CONNECTION
>
> The human body uses 20 amino acids. Of those 20 amino acids, about half can be produced by the body and aren't necessary in our diets. The others cannot be produced by the body, so we need to get them from food. Sources of these amino acids—in the form of protein—include meat, eggs, grains, legumes, and dairy products.
>
>
>
> Foodcollection RF/Getty Images

You might notice that the carbon atom in the center of the amino acid molecule has four different groups attached to it. Therefore, amino acids have mirror-image isomers and can be left-handed or right-handed molecules. There's a trick to figuring out the handedness of an amino acid. The illustration shows you how.

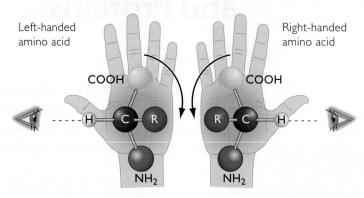

Left-handed amino acid

Right-handed amino acid

*Left-handed amino acids: When you point the H atom toward your eye, the COOH group, the R group, and the NH₂ group are arranged clockwise. This is sometimes referred to as the CORN rule for **COOH**, **R**, and **NH₂**.*

The human body uses only the left-handed amino acids, but both mirror images are available. How did life begin selecting only one of the two mirror images? This is a question of heated debate. Currently, no one has a good explanation.

There are 20 amino acids, with different R groups, that your body needs. Your body produces about half of these amino acids. You get the others as part of your diet. Six of the amino acids are listed in this illustration. These amino acids are placed into two categories: either hydrophobic ("water fearing") or hydrophilic ("water loving"). Take a moment to determine the difference.

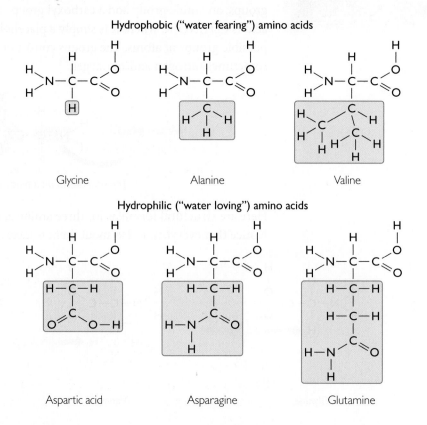

Hydrophobic ("water fearing") amino acids

Glycine          Alanine          Valine

Hydrophilic ("water loving") amino acids

Aspartic acid          Asparagine          Glutamine

**BIOLOGY**
**CONNECTION**

The average human can distinguish among approximately 10,000 different odors. The number of receptor sites in the nose is estimated to be around 1000 by some researchers. This means that some smells must be stimulated by a combination of different receptors, much as a musical chord on a piano has a distinct sound.

The R groups for the amino acids classified as hydrophobic contain only carbon and hydrogen atoms. These R groups are nonpolar. They are not attracted to water molecules. The R groups for the amino acids classified as hydrophilic contain atoms such as oxygen and nitrogen. These R groups are polar and are attracted to water molecules. As you will see, the nature of the R group is important when amino acids are linked to form proteins.

## Proteins

As shown in the next illustration, the carboxyl group of one amino acid can link to the amine group on another amino acid. The bond between two amino acids is called an **amide bond,** or a **peptide bond.** In the process, two hydrogen atoms and an oxygen atom break off from the amino acids and form a water molecule. In this way, it is possible to form long chains of amino acids called **proteins.** Because they are very large molecules, proteins are often called *macromolecules*.

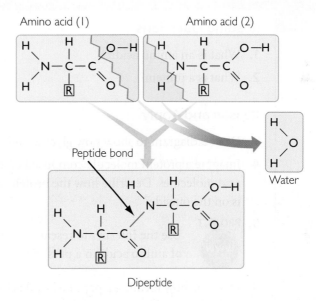

Amino acid (1)    Amino acid (2)    Water

Peptide bond

Dipeptide

Proteins are essential in all living organisms. Protein molecules send messages throughout your body, they help the immune system fight off infections, and they are building blocks of many parts of your body, including organs, hair, and muscles. Proteins have a wide variety of structures and functions because many different sequences of amino acids are possible.

Ken Karp Photography

As the chain of amino acids forms, it twists and bends. In the photograph, each bead represents an amino acid. The entire chain of beads represents a protein molecule. Notice that amino acids can repeat.

The R groups on each amino acid play a major role in determining the resulting structure of the protein. The hydrophobic R groups tend to cluster in the center of the molecule. The hydrophilic R groups tend to be on the outside of the molecule, near the watery environment.

Smell receptor sites are protein molecules that have a pocket that fits a particular smell molecule. When the smell molecule attaches to the pocket, it changes the structure of the protein just a bit. This small change is enough to affect a neighboring protein, which triggers a signal to the brain.

## What is a receptor site made of?

There are 20 different amino acid molecules that are essential building blocks of the human body. Amino acids have both a carboxyl functional group and an amine functional group. Amino acids all have mirror-image isomers, but only left-handed amino acid molecules are used by the human body. The carboxyl group on one amino acid can link with the amine group on another amino acid to form a peptide bond. In this way, a long chain of amino acids called a protein forms. A large variety of protein molecules exists because the amino acids can be linked in a variety of sequences. Some amino acids in a protein chain are polar and others are nonpolar. This affects the folding of a chain of amino acids in the protein. Smell receptor sites are composed of proteins.

**KEY TERMS**

amino acid

peptide bond (amide bond)

protein

## Exercises

### Reading Questions

1. What is an amino acid?

2. What is a protein?

### Reason and Apply

3. Draw a diagram to show how glycine and alanine combine to form a peptide bond.

4. Imagine a protein molecule composed of 200 glycine molecules and 200 aspartic acid molecules. Describe how the protein is folded. What is on the inside and what is on the outside?

5. **PROJECT**

   **www** Use the Internet to research amino acids. Describe three different sources of amino acids in a person's diet.

6. Describe how forming a peptide bond is like forming an ester from an acid and an alcohol.

# Molecules in the Body

# SUMMARY

## KEY TERMS

**mirror-image isomer**

**amino acid**

**peptide bond (amide bond)**

**protein**

# Smells Update

Receptor sites exist throughout living systems. They serve as the mechanism for triggering many biological responses and processes, including smell. Receptor sites can be very specific, allowing only certain molecules to dock. So, the molecular structure of the receptor site is important. Some molecules have mirror-image isomers that have different properties, including smell. Amino acid molecules exist as mirror-image isomers. Amino acids combine to form a vast array of proteins, including those that make up receptor sites.

## REVIEW EXERCISES

1. How do you determine if two molecular models represent mirror-image isomers?

2. Draw the structural formula for these molecules. Which have mirror-image isomers? Explain your reasoning.
   **a.** CBrClFH      **b.** $CH_4$      **c.** $CH_2Cl_2$

3. Explain how you could use your feet to show that the "handedness" of a molecule can determine if it fits into a receptor site.

4. What are amino acids and what purpose do they serve in your body?

5. Using the receptor site model, explain how a person can smell L-carvone but not its mirror-image isomer D-carvone.

---

## Modeling a Receptor Site

PROJECT    Build a space-filling model of a smell molecule from a compound in one of the five smell classifications you have studied. Use clay, plaster, sand, or other materials to build a receptor site for this smell molecule. Describe other molecules that might fit into the receptor site you built. Then create a space-filling model of a molecule that will not fit into the receptor site you built.

Your project should include

- the receptor site model that you built, labeled with the smell that it detects and a description of molecules that fit into it

- a space-filling model of a molecule that fits into the receptor site and one that does not (label which is which)

---

# Smells | REVIEW

S mell chemistry is an active area of research. Scientists are still refining theories about how smell works. However, one thing is clear: Smell is all about molecules. To understand smell, it is important to understand what molecules are and how they are put together.

The smell of a compound is greatly dependent on its composition and structure. This is true of other molecular properties besides smell. You do not smell some molecules because they are too large or heavy to become airborne, or because they are nonpolar and do not stick in receptor sites. Other molecules have a type of smell that depends on what atoms or functional groups are present. The smell of a molecule may also depend on the shape of the entire molecule.

By exploring the topic of smell, you have studied covalent bonding and the general rules governing how atoms come together to form the amazing structures called molecules.

## REVIEW EXERCISES

### General Review

*Write a brief and clear answer to each question. Be sure to show your work.*

1. What are isomers?

2. Draw two isomers with the molecular formula $C_2H_4O$.

3. A molecule can be described using a molecular formula, a structural formula, a ball-and-stick model, or a space-filling model. What information does each of these provide?

4. Consider the structural formula for methyl salicylate. Name two functional groups that are in this molecule.

5. Here is a ball-and-stick model of methyl salicylate.

a. Does it follow the HONC 1234 rule?

b. Write the molecular formula for methyl salicylate.

c. Describe the overall shape of this molecule.

d. Find the oxygen atom that is bonded to two carbon atoms. Explain why this bond is bent.

6. A polar molecule has the molecular formula $C_3H_8O$.

   a. Draw the structural formula.

   b. Draw the Lewis dot structure and label the shared electrons and the lone pairs.

   c. Identify the dipoles.

   d. Name the geometric shape around each carbon atom.

   e. Discuss what the name of this compound might be. Explain your reasoning.

7. What is a dipole? Explain why methane, $CH_4$, has four dipoles but has no overall dipole.

8. Table salt, NaCl, has no smell because

   A. It is not made up of molecules.

   B. It is an ionic compound.

   C. No part of it enters the gas phase at room temperature.

   D. All of the above.

   E. None of the above.

## STANDARDIZED TEST PREPARATION
## Multiple Choice

*Choose the best answer.*

1. Based on its name and chemical formula, predict the smell of formic acid, $CH_2O_2$.

   (A) Sweet        (B) Minty

   (C) Fishy        (D) Putrid

2. Which of the following compounds are covalent molecules? (Choose all that apply.)

   (A) $NH_4C_2H_3O_2$     (B) $C(CH_3)_4$

   (C) $CH_2BrF$        (D) $Na_2CO_3$

3. Which statement best describes what happens to the valence electrons in a covalent bond?

   (A) Valence electrons are shared between two metal atoms.

   (B) Valence electrons are shared between two nonmetal atoms.

   (C) Valence electrons are transferred from a metal atom to a nonmetal atom.

   (D) Valence electrons are transferred from a non-metal atom to a metal atom.

4. Which of the following best describes how two different compounds made up of molecules that are isomers of each other might smell?

   (A) The two compounds will most likely smell different from each other because isomers have the same molecular formula but different structural formulas.

   (B) The two compounds will most likely smell the same because isomers have the same molecular formula but different structural formulas.

   (C) The two compounds will most likely smell the same because isomers have the same molecular formula and the same structural formulas.

   (D) The two compounds will most likely smell different from each other because isomers have different molecular formulas and the same structural formulas.

**5.** Which of the following statements best describes the molecule below?

(A) The molecular formula for the molecule is $C_9H_{15}O$.

(B) The functional group is an ester.

(C) The shape around all of the carbon atoms is linear.

(D) The name of the compound will most likely end with "-one."

**6.** Which statement about the Lewis dot symbol of an element is true?

(A) The total number of dots surrounding the symbol is the number of valence electrons for that element.

(B) The number of dots surrounding the symbol will be the same for all elements of the same group on the periodic table.

(C) For elements in the same period on the periodic table, the number of dots surrounding the element symbol will increase as you move from left to right on the periodic table.

(D) All of the above statements are true.

**7.** Which statement correctly describes the Lewis dot structure of $CF_4$?

(A) The Lewis structure for $CF_4$ has a total of 40 dots in the structure because there are 5 atoms.

(B) In the Lewis structure for $CF_4$, carbon is the central atom with four bonded pairs of electrons, and each fluorine atom has a total of three bonded pairs of electrons.

(C) In the Lewis structure for $CF_4$, carbon is the central atom with four bonded pairs of electrons, and each fluorine atom has three lone pairs of electrons.

(D) In the Lewis structure for $CF_4$, the fluorine atom shares seven electrons with each carbon atom.

**8.** Which statement best describes water, $H_2O$?

(A) Water is a bent molecule that has two lone pairs around the central oxygen atom.

(B) Water is a linear molecule that has two lone pairs around the central oxygen atom.

(C) Water is a bent molecule with no lone pairs around the central oxygen atom.

(D) Water is a bent molecule with no lone pairs around the central hydrogen atom.

**9.** Which of the molecules below have a double bond in their structural formula?

$$H_2, N_2, O_2, Cl_2, C_2H_2, C_2H_4, C_2H_6, H_2CO$$

(A) $H_2, N_2, O_2, Cl_2$

(B) $C_2H_2, C_2H_4, C_2H_6, H_2CO$

(C) $O_2, C_2H_4, H_2CO$

(D) $H_2, N_2, Cl_2, C_2H_6$

**10.** How many electrons are shared between the carbon and nitrogen in hydrogen cyanide, HCN?

(A) Six electrons

(B) Five electrons

(C) Four electrons

(D) Three electrons

**11.** Based on the functional group, what is the predicted smell for the compound below?

(A) The molecule has a carboxyl group, so it likely smells putrid.

(B) The molecule has a carboxyl group, so it likely smells minty.

(C) The molecule has an ester group, so it likely smells sweet.

(D) The molecule has a ketone group, so it likely smells minty.

12. Which of the following correctly describes electron domains and molecular shape?

   (A) Only bonded pairs of electrons affect the shape of a molecule.
   (B) Only lone pairs of electrons affect the shape of a molecule.
   (C) Electron pairs have a tendency to repel one another due to their negative charge.
   (D) Electron pairs have a tendency to be as close to one another as possible.

13. What molecular shapes are represented in this picture of ethanol, $C_2H_5OH$?

   (A) Pyramidal and bent
   (B) Tetrahedral and pyramidal
   (C) Tetrahedral and bent
   (D) Pyramidal and linear

14. Aspects of the receptor site theory are given below. What can you conclude about smell molecules from the receptor site theory?

   I: Smell molecules come in a variety of molecular shapes.
   II: Smell molecules change phase to become a gas or vapor that floats through the air to enter your nose and be detected.
   III: Smell molecules "dock" in receptor sites that match their molecular shapes.
   IV: Nerves are stimulated when the molecules dock and send messages to the brain, which are interpreted as a particular smell.

   (A) Smell molecules must break apart into individual atoms when they become a gas.
   (B) Smell molecules must lose their functional groups when they become a gas.
   (C) Smell molecules must change their shape to fit the "dock" in the receptor sites.
   (D) Smell molecules must keep their same structural formula when they become a gas.

15. What is the *bond* polarity of the C−H bonds in methane, $CH_4$? Refer to the electronegativity scale at the beginning of Lesson 44 to answer.

   (A) The C−H bonds are nonpolar covalent bonds because the electronegativity difference between carbon and hydrogen is large.
   (B) The C−H bonds are polar covalent bonds because the electronegativity difference between carbon and hydrogen is large.
   (C) The C−H bonds are ionic bonds because the electronegativity difference between carbon and hydrogen is large.
   (D) The polarity of C−H bonds is small because the electronegativity difference between carbon and hydrogen is small.

16. Which of these molecules are polar? (Choose all that apply.)

   (A) $NH_3$
   (B) $CCl_4$
   (C) $C_2H_6$
   (D) $CH_3OH$

17. Which of the choices below illustrates the concept "like dissolves like?"

   (A) Iodine, $I_2$, is more soluble in water than in oil.
   (B) Iodine, $I_2$, is more soluble in oil than in water.
   (C) Vinegar, $CH_3COOH$, is more soluble in oil than in water.
   (D) Sodium chloride, NaCl, is more soluble in oil than in water.

**18.** Which of the following cannot be explained by describing the intermolecular forces between molecules?

(A) Water molecules form drops rather than spreading out on wax paper.

(B) Water is a liquid at room temperature while methane is a gas at room temperature, even though both water and methane have similar molecular mass.

(C) The O−H bond in water is polar because of differences in electronegativity of the atoms.

(D) Oil tends to float on water while ethanol tends to dissolve in water.

**19.** Which group of elements has the highest electronegativity values?

(A) C, H, O

(B) O, Cl, F

(C) Ne, H, C

(D) Cl, Br, I

**20.** Which of the following molecules has a mirror-image isomer?

(A) $CH(CH_3)FBr$

(B) $CHF_3$

(C) $CH_2F_2$

(D) $CF_4$

### HISTORY

#### CONNECTION

In 2004, Richard Axel and Linda Buck shared a Nobel prize for their advances in understanding the mechanism of smell. Axel and Buck found that mammals have approximately 1000 different genes for olfactory receptors, which led to a better understanding of smell.

# Scent-Detection Device

## Work Groups of 4

### Purpose

To use the process of engineering design to plan and design an imaginary scent-detection device.

### Problem

Your engineering design team is charged with creating a plan for a scent-detection device to be used to detect certain smell molecules. You may have seen technicians from a gas company use a device to detect the presence of leaking gas. Some of those devices detect the odor-causing molecules in natural gas. Using the knowledge you have gained so far in the course, your job is to brainstorm and develop a plan or protocol for a scent-detection device. You are not responsible for creating a device; you are just designing a plan for its development. You will determine the types of odors your device will detect and who would most likely benefit from the creation of this detector.

### Review

Review the handout, *Engineering Design* (found after the Unit 1 Engineering Project), and reflect on the material that has been presented thus far in your chemistry class. For this project, you will focus on brainstorming solutions, developing a plan, and refining that plan based on feedback.

In Unit 2: Smells, you learned about molecular structure and properties. Throughout the unit, these chemistry concepts were related to the concept of smell. In Chapter 7, Lesson 41, you were introduced to the receptor site theory of how the interior of the nose works to detect different smells. You should review the different smell categories and the receptor site theory as background. The following list gives some of the many topics and skills covered in Unit 2: Smells. You may find these helpful as you work to engineer a solution to the problem:

- Smell classification
- Molecular structure and smell
- Molecular geometry and smell
- Polarity and smell
- Phase and smell
- Molecular size and smell
- Structural formulas

- Shape vs smell
- Intermolecular forces
- Functional groups and smells
- Isomers
- "Lock and key" mechanisms
- Absence of smell

**Part 1:** Identify the Problem and Brainstorm Solutions

1. What is the problem you are trying to solve? Write up a description of what you need to accomplish. What type of smell or smells do you want your device to detect?

2. In your Group, brainstorm some ideas to solve this problem. Here are some questions to consider:
   - What properties of a substance are most important when determining the smell of the substance?
   - What properties of the individual molecule are most important when determining smell?

- How can you use both the macroscopic properties and molecular properties to determine the smell of a substance?
- What types of smells will your imaginary device detect?
- Based on the smell you are interested in detecting, do your answers to the previous three questions need to change? If so, how?
- Why are the structural formula and the shapes of the molecules important to consider?
- What type of materials could you use to make your device?
- What other questions do you have?

3. Research: In the library and online, research smell detection. Create a list of all of your sources to turn in with your project. You might want to focus your research on current ways that scientists detect chemicals for protection in areas like public places or airports.

## Part 2: Create a Plan or Prototype

For this section, you will be creating a prototype of your device.

1. Based on your research and brainstorming, create a diagram of a device that might be able to detect smell. For this step, you could also consider creating a 3D-model.

2. Write up a description of how your device would be able to detect smell. What would your device measure? How would your device collect and report the data?

3. Write a list of materials and possible safety considerations needed for the creation of your device.

## Part 3: Refine and Retest

1. Have your classmates provide feedback for your plan for your device. Ask them to help you think of ways to improve your plan. What changes might you make based on their feedback?

2. Make changes to your plan, diagram and description based on feedback and further research.

3. How would you test your procedure to see if the changes have made an improvement? What will you observe to determine if your scent device is improved?

## Presentation

Choose one of these options to present your scent device plan, or get permission from your teacher to present in a different way:

- Oral presentation describing your procedure
- Poster describing your procedure in detail with your diagrams and model
- A website that outlines your project including your procedure and diagrams

# Unit 3 | Weather

Stockbyte/Getty Images

Water changes phase and is transported around the planet. It may then show up as snow or rain.

## In this unit, you will learn

- about proportional relationships

- about temperature scales and how thermometers work

- the effects of changing temperature, pressure, and volume on matter

- about the behavior of gases

- how to read weather maps and make weather predictions

Wes Walker/Lonely Planet Images/Getty Images

## Why Weather?

Thunderstorms dump great quantities of rain, fog seeps into a bay at nightfall, warm temperatures entice us to the beach, and hurricanes devastate coastal communities. For better or worse, the weather is a part of our everyday lives. Physical change is at the core of weather. Weather occurs when matter undergoes changes in location, density, phase, temperature, volume, and pressure. When you understand the relationships between these changes in matter, you will begin to answer the question, "What is the chemistry of weather?"

# CHAPTER **10**

# Physically Changing Matter

## In this chapter, you will study

- how to read basic weather maps
- proportional relationships
- density and phase changes in matter
- temperature scales and thermometers
- the relationship between volume and temperature of gases

*A view of Earth's atmosphere from space shows weather in action.*

Weather is the result of interactions between Earth, the atmosphere, water, and the Sun. It is about the movement of matter around the planet. Water, for example, changes phase from liquid to gas and then back to liquid as it moves from oceans to the atmosphere, to clouds, and down to Earth as rain. The atmosphere is composed of different gases, so the study of the behavior of gases is important in understanding weather. The gas laws are mathematical equations that describe the relationship between changing temperature, volume, and pressure of gases. Understanding the behavior of matter in relation to temperature allows weather experts to make predictions about the weather.

# Weather Science

## THINK ABOUT IT

You have probably seen a weather forecast, either on television or in the newspapers. You may have wondered—what are highs and lows? What are fronts? What does the jet stream have to do with the weather?

### What causes the weather?

*To begin to answer this question, you will explore*

➲ Earth, Air, Water, and Sun

➲ Weather Maps

## EXPLORING THE TOPIC

### ➲ Earth, Air, Water, and Sun

From outer space, you can see clouds above the surface of the planet. A satellite picture of Earth also clearly shows that our planet's surface is mostly water.

This lesson launches an investigation into the connection between chemistry and weather. **Weather** refers to clouds, winds, temperature, and precipitation in a region at any given time. It differs from *climate,* which describes the overall weather trends of a region over long periods of time. Weather is caused by the interaction of Earth, the atmosphere, water, and the Sun. It is about the movement of matter.

A planet must have an *atmosphere* to have weather. An atmosphere is a layer of gases surrounding the planet. In contrast, the Moon has no weather because it barely has any atmosphere. The atmosphere on Earth is often referred to as *air.* Air is composed of a mixture of gases. Earth's atmosphere consists of nitrogen (78%), oxygen (21%), argon (1%), carbon dioxide (0.04%), and traces of other gases. There is some water vapor, $H_2O(g)$, in the atmosphere, but the amount varies from place to place, from day to day, and even from hour to hour. On average, it makes up about 1% of the atmosphere.

The weather we experience is caused by physical changes in the atmosphere. **Physical changes** alter the form a substance takes but do not change the identity of the substance. Examples of physical changes are changes in volume, temperature, shape, size, and pressure. These are different from chemical changes, which produce new substances. The heating of Earth by the Sun is the root cause of most of the physical changes in Earth's atmosphere.

Heating and reshaping a substance are both physical changes. Mixing chemicals to form a new compound is a chemical change.

The **phase changes** of water between solid, liquid, and gas are types of physical changes that play an important role in weather. For example, rain is evidence that a phase change from gas to liquid has taken place.

**Big Idea** Weather is the result of physical changes to matter.

## ⤷ Weather Maps

Meteorologists use maps to keep track of atmospheric conditions and to predict the weather. Weather information for a specific day is shown on the maps here. What patterns do you observe as you compare the six maps?

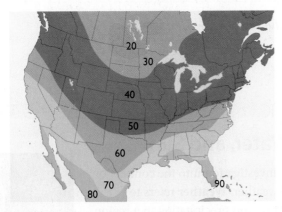

*Temperature highs in degrees Fahrenheit*

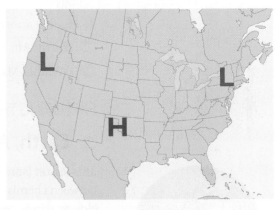

*Air pressure highs and lows*

*Precipitation: rain and snow*

*Cloud cover*

*Fronts*

*Jet stream*

In addition to familiar features such as temperature and precipitation, weather maps often show a curve called the *jet stream*. The jet stream shows the location of high-altitude winds. These winds are at least 57 miles per hour (up to 190 mi/h) and are in the upper atmosphere above 20,000 feet in altitude (that's 4 miles up). These winds have a great effect on what happens in the air below them and generally "steer" storms.

Here are some features to take note of:

- In North America, temperatures are colder in the north and warmer in the south.
- The shape of the jet stream coincides with the contour lines on the temperature map.
- If it is raining or snowing, then clouds must be present, but clouds do not necessarily mean it will rain or snow.
- Fronts are located in areas of low pressure.
- Clouds are more likely to form near fronts, so it is more likely to rain and snow there.
- Warm fronts and cold fronts are associated with low-pressure systems.
- There is no precipitation for regions with high pressure. High pressure is associated with sunny days. A region with high pressure can be either hot or cold.

## Example

### Weather Maps

Describe the weather in these areas using the six weather maps in this lesson. Include the temperature, air pressure, cloud cover, precipitation, and storm fronts.

   **a.** West Coast     **b.** middle of the United States

*Solution*

   **a.** West Coast: The temperature is about 60 °F, slightly colder to the north. The air pressure is low. There are storm fronts and it is cloudy, with rain to the south and clearing to the north.

   **b.** Middle of the United States: It is about 40 °F and sunny. The air pressure is high, there are no clouds, and there is no precipitation.

## LESSON SUMMARY

### What causes the weather?

**KEY TERMS**

weather

physical change

phase change

Weather is about the movement of air and water. The gases in our atmosphere and the water on Earth's surface undergo physical changes and move around the planet. Water exists in all three phases (solid, liquid, and gas) on Earth. Water shows up as snow, ice, rain, clouds, fog, and water vapor. Meteorologists use detailed maps to monitor the physical changes taking place in our atmosphere and to keep track of the movement of moisture and air around the planet. Information on weather maps includes temperature, air pressure, cloud cover, precipitation, warm and cold fronts, and the location of the jet stream.

### Reading Questions

1. What substances are necessary for a planet to have weather?

2. Describe three variables that meteorologists study to make weather predictions.

3. What is a physical change?

### Reason and Apply

4. Name three types of maps that meteorologists use. Describe the maps and the information that they contain.

5. Consult a newspaper, the Internet, or a weather report on television.
   a. Name three states that are currently experiencing a cold front (or warm front) moving through.
   b. Which states are experiencing high-pressure systems? What is the weather forecast in those states?
   c. If you were traveling to New York tomorrow, would you pack a raincoat? Explain.

6. Nitrogen, $N_2(g)$, oxygen, $O_2(g)$, carbon dioxide, $CO_2(g)$, and water vapor, $H_2O(g)$, are all gases found in the atmosphere.
   a. Draw the structural formula for each and include lone pair electrons.
   b. Draw the Lewis dot structure for water.
   c. Which molecules are polar and which molecules are nonpolar?
   d. Which one of the four substances is naturally found as a liquid, a solid, and a gas on Earth?

**ENVIRONMENTAL CONNECTION**

On other planets, the compounds that are involved in the weather may be quite different from those found on Earth. The atmosphere of Mars is mostly carbon dioxide. The polar ice caps on Mars are made of frozen carbon dioxide.

Science Source

# Measuring Liquids

## THINK ABOUT IT

After a rainstorm, rain in the gutter or in a puddle can be several inches deep, yet on the sidewalk and grass, it may not be that deep. The water runs off the sidewalks, runs down hills, collects in depressions, and soaks into grass. So, how do you determine how much rain fell?

### How do meteorologists keep track of the amount of rainfall?

*To answer this question, you will explore*

⟳ Rain Gauges

⟳ Volume Versus Height

⟳ Proportional Relationships

## EXPLORING THE TOPIC

## ⟳ Rain Gauges

Suppose you have three rain gauges that are different sizes. You place them next to one another out in the rain. After a few hours, these containers might look like this:

A          B          C

Sample data for these three containers are given in the table below.

| Container | Area of the base (cm²) | Height of rain (cm) | Volume of rain (cm³) |
|-----------|-----------------------|---------------------|----------------------|
| A | 10.0 cm² | 4.0 cm | 40.0 cm³ |
| B | 2.5 cm² | 4.0 cm | 10.0 cm³ |
| C | 2.0 cm² | 4.0 cm | 8.0 cm³ |

You can see that the volume of rain in each container is quite different. However, the *height* of rain is the same. This is why meteorologists report height of rainfall, in centimeters or inches, not volume of rainfall. The height of rainfall is the same for these containers, but the volume is not.

# Volume Versus Height

Imagine that you place a rain gauge with a square base of 2.0 cm² in a steady rain for several hours. Each hour, you go outside to measure how much rain has collected in the rain gauge.

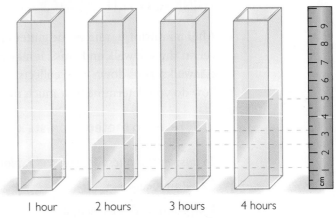

*Rain gauge with a 2.0 cm² base*

The graph shows how volume changes as height changes, or volume versus height. *Versus* means *compared with* and is abbreviated *vs.* Both the volume and the height of the rain in the gauge increase as more rain is collected. Take a moment to examine the data.

The graph shows that volume increases in a steady and predictable way in relationship to height. And once some data points are known, others can be predicted.

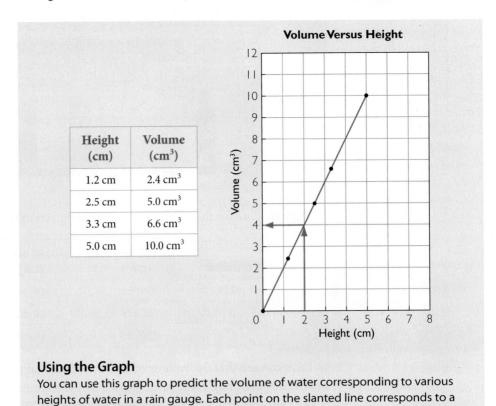

| Height (cm) | Volume (cm³) |
|---|---|
| 1.2 cm | 2.4 cm³ |
| 2.5 cm | 5.0 cm³ |
| 3.3 cm | 6.6 cm³ |
| 5.0 cm | 10.0 cm³ |

## Using the Graph
You can use this graph to predict the volume of water corresponding to various heights of water in a rain gauge. Each point on the slanted line corresponds to a different volume and height.

Suppose you want to predict the volume of water when the height is 2.0 cm. You can either use the graph to find the answer or use math to calculate it. [For review of this math topic, see **Math Spotlight**: Graphing on page A-8.]

**Method 1:** First, find the height, 2.0 cm, on the *x*-axis. Move vertically from this spot until you reach the slanted line. Now move horizontally until you reach the *y*-axis. This point on the line corresponds to a volume of 4.0 cm³.

**Method 2:** Notice that for every data point in the table, the volume is always 2 times the height. The volume for a height of 2.0 cm is 4.0 cm³. Similarly, the volume for a height of 6 cm would be $6 \cdot 2 = 12$ cm³.

## ➲ Proportional Relationships

The relationship between a container's volume and its height is a **proportional** one. This means that volume and height are related to one another by a single number called the **proportionality constant.** When the height is multiplied by this number, you get the volume. In the example we just used, the height and volume of rain in the rain gauge were related by the number 2.0. In other words, for this container, the volume is always 2.0 times the height, and 2.0 is the proportionality constant.

Volume of rain = (proportionality constant) · (height of rain)

A proportionality constant can be represented by the letter *k* in a math formula.

Volume = $k$ · height

In the case of the rain gauge, the proportionality constant is also equal to the area of the base.

Volume = 2.0 cm² · height

Proportionality constant, *k*

The graph provides another way to look at proportional relationships. Whenever two variables are proportional to each other, the graph of the data points is *always* a straight line passing through the origin (0, 0). [For review of this math topic, see **MATH Spotlight**: Ratios and Proportions on page A-11.]

Volume and height are proportional for containers with a uniform shape and parallel walls.

*Volume* is *proportional to height.*     *Volume* is not *proportional to height.*

Amount of rainfall is not proportional to time. This is because rainfall is not necessarily steady. It can rain a lot and then a little.

## How do meteorologists keep track of the amount of rainfall?

Everywhere around the world, the amount of rainfall is measured by its height. If volume were used to report rainfall, each meteorologist would report a different number, depending on the area of the base of the rain gauge used. Height and volume for a specific rain gauge are related by a proportionality constant, $k$.

**KEY TERMS**

proportional

proportionality constant

## Exercises

### Reading Questions

1. Suppose you have a cylindrical rain gauge with a base area of 2.0 cm². Explain two different ways you can determine the volume when the height of water is 3.0 cm.

2. Explain in your own words why meteorologists prefer to measure rain in inches or centimeters, not in milliliters or cubic centimeters.

### Reason and Apply

3. If the amount of rainfall increases, do both the volume and height of water in the rain gauge keep track of this increase? Explain your thinking.

4. If you use a beaker for a rain gauge and the weather station uses a graduated cylinder, will both instruments give the same volume? The same height?

5. If a large washtub, a dog's water dish, and a graduated cylinder were left outside during a rainstorm, would the three containers all have the same volume of water in them after the storm? Explain why or why not.

6. Inches are used in the United States, but centimeters are used by scientists and in the rest of the world. Look at a ruler that is marked in both inches and centimeters.
   a. Make a graph of centimeters versus inches so that the *y*-axis is centimeters and the *x*-axis is inches.

   b. Convert 12 in. to centimeters.
   c. How many inches is 10 cm?
   d. How many inches is 1 cm?

7. A student placed the same empty rain gauge outside before five different rainstorms. She measured the height and volume of the water in the gauge after each storm. Her results are in the table.

| Storm number | Height (cm) | Volume (cm³) |
|:---:|:---:|:---:|
| 1 | 1.0 cm | 2.5 cm³ |
| 2 | 2.5 cm | 6.3 cm³ |
| 3 | 0.5 cm | 1.1 cm³ |
| 4 | 8.0 cm | 20.0 cm³ |
| 5 | 5.0 cm | 12.5 cm³ |

   a. Which rainstorm dropped the most rain? Did you use height or volume to answer this question? Why?
   b. What pattern do you notice?
   c. Draw a graph of the data to show that the volume and height are proportional.
   d. Explain why the data points do not all lie exactly on a straight line.
   e. Predict the volume for a height of 6.0 cm. Show your work.

# Density of Liquids and Solids

## THINK ABOUT IT

Much of the country depends on snowfall for the year's water supply. The mountains get many feet of snow every winter, and this snow eventually melts and travels from creek to stream to river to reservoir. Scientists measure the snowpack, or the total amount of snow that has accumulated on the ground, to predict the amount of water that will be available for consumption the rest of the year. But is the amount of water in a foot of snow in the mountains equal to the amount in a foot of rainfall?

### How much water is present in equal volumes of snow and rain?

*To answer this question, you will explore*

➲ Density and Phase

➲ Converting Snowfall to Rainfall

➲ The Density of Ice

## EXPLORING THE TOPIC

## ➲ Density and Phase

Just as rainfall is measured in inches, snowfall can also be measured in inches. All you need is a ruler. However, scientists who study water distribution are more interested in how much liquid water the snow represents.

When snow melts, its volume decreases. However, its mass is still the same. This is because the same water molecules are present in the frozen snow and in the melted snow.

If the volume of the sample changes but its mass stays the same, then mathematically, the density of the sample has to change. The density of snow must be lower than the density of water.

Recall that density is the mass per unit volume.

$$D = \frac{m}{V}$$

Unlike water, snow has a wide range of densities. Snow can be fluffy or packed. So, the volume of liquid water contained in any snow sample depends on the density of the snow being considered.

Snow          Water

*Mass is not lost, but the volume decreases.*

AP Photo/Public Opinion, Noelle Haro-Gomez

## MASS AND VOLUME ARE PROPORTIONAL

The equation for density relates mass and volume.

As long as the density of the substance doesn't change, its mass and volume are proportional. The proportionality constant here is the density, $D$.

$$m = DV$$

Because mass and volume are proportional to each other for a specific substance, the data points lie on a straight line that goes through the origin. Examine the graphs of mass versus volume for rain, snow, and ice. How does the slope of the line relate to the density?

| Form | Density |
|------|---------|
| rain | 1.0 g/mL |
| snow | 0.1−0.5 g/mL |
| ice | 0.92 g/mL |

**Mass Versus Volume of Water**

Rain

Ice

Snow with a density of 0.5 g/mL

### Using the Graph

The line for rain is steepest because liquid water has the greatest density. For every 1.0 mL increase in rain volume, the mass of the water increases by 1.0 g. The line for snow is the least steep because the density of snow is less than that of both rain and ice. For snow, a 1 mL increase in volume corresponds to only a 0.5 g increase in mass.

**Big Idea**   Matter changes density when it changes phase.

Florence Low/California Department of Water Resources

Philippe Psaila/Science Photo Library

*Scientists called hydrologists measure snowpack by weighing an aluminum tube filled with snow.*

## ➔ Converting Snowfall to Rainfall

### CALCULATING THE DENSITY OF SNOW

One way to determine the volume of water in a snowpack is to melt the snow. But melting a large amount of snow is inconvenient.

Another way to study the water volume of the snowpack is to determine the density of the snow and calculate the volume of liquid water it represents. When new snow has fallen in the mountains, hydrologists take samples of the snow with a large aluminum tube. They push the empty tube all the way down into the snow. Then they weigh the tube with the snow still inside. The density of the snow is the mass divided by the volume.

## Volume of Water from Snow Melt

Suppose you have a volume of 20 mL of snow with a density of 0.25 g/mL. What volume of liquid water do you get when you melt this amount of snow?

### Solution

When the snow melts, its mass will not change. Use the density equation to convert between mass and volume.

Use the volume and density of snow to find its mass.

$$m_{snow} = D_{snow} \, V_{snow}$$
$$= (0.25 \text{ g/mL}) \cdot (20 \text{ mL}) = 5 \text{ g}$$

The mass of snow is equal to the mass of water.

$$m_{snow} = m_{water} = 5 \text{ g}$$

Rearrange the density equation to find the volume of water.

$$V_{water} = \frac{m_{water}}{D_{water}} = \frac{5 \text{ g}}{1.0 \text{ g/mL}} = 5 \text{ mL}$$

When 20 mL of snow with a density of 0.25 g/mL melts, you get 5 mL of water.

You can also solve this problem using a graph. The graph here shows mass versus volume for liquid water, as well as snow with a density of 0.25 g/mL.

**Converting Snow to Liquid Water**

Mass (g) vs Volume (mL)

Liquid water

Snow

### Using the Graph

The arrows on the graph show how to convert from snow volume to liquid water volume. You begin with 20 mL of snow, which is equivalent to 5 g of snow. When snow melts, the mass remains the same. 5 g of liquid water is equivalent to 5 mL of snow.

### SPORTS
### CONNECTION

Skiers and snowboarders enjoy fresh "powder." This is the term for untouched, freshly fallen snow with a particularly low density. Powder is perfect for landing on because it forms a natural cushion. This means that it does not hurt as much when you land on it compared with snow that is compacted.

© Ralph A. Clevenger/CORBIS

# ⤵ The Density of Ice

The density of ice is 0.92 g/mL. It is less than the density of liquid water, which is 1.0 g/mL. Because water becomes less dense when it freezes, its volume increases. This unique property of water is due to a special kind of intermolecular force called *hydrogen bonding*. In both liquid and solid water, the hydrogen atoms in one water molecule attract the highly electronegative oxygen atoms in another. When water freezes, the molecular motion slows down and the molecules lock into a hexagonal structure due to hydrogen bonding.

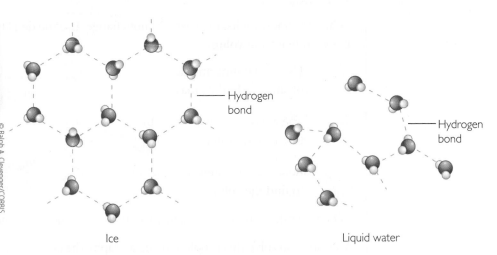

Ice                Liquid water

## LESSON SUMMARY

### How much water is present in equal volumes of snow and rain?

The density of snow is less than the density of liquid water. You can use mathematical equations or graphs to convert the volume of snow to the volume of liquid water. To use graphs, you plot two lines of mass versus volume, one for snow and another for liquid water. The slope of each line is the density of that substance. To use mathematical equations, you have to know the density of the snow and the density of liquid water.

## Exercises

### Reading Questions

1. Explain how to use mathematical equations to convert the volume of snow in a snowpack to the volume of water in the snowpack.

2. Explain how to use graphs to convert the volume of snow in a snowpack to the volume of water in the snowpack.

### Reason and Apply

3. How are snow and ice different?

4. When ice melts, will the liquid water occupy more or less volume than the ice? Explain your thinking.

5. Suppose that you melt 24 mL of ice. What is the volume of liquid water that results?

*(continued)*

**6.** Which has more water for equal volumes of snow: snow with a density of 0.5 g/mL or snow with a density of 0.25 g/mL? Explain your thinking.

**7.** Suppose you have a box with a volume of 17.5 mL.
   **a.** If you fill this box with ice, what mass of ice do you have? (The density of ice is 0.92 g/mL.)
   **b.** If you fill this box with liquid water, what mass of liquid water do you have? (The density of liquid water is 1.0 g/mL.)

**8.** Suppose that you have a box that is full and contains 500 grams of a substance.
   **a.** What is the volume of the box if the substance inside is corn oil? (The density of corn oil is 0.92 g/mL.)
   **b.** What is the volume of the box if the substance inside is lead? (The density of lead is 11.35 g/mL.)

**9.** Lead, Pb, is more dense than iron, Fe.
   **a.** Which occupies a larger volume: 4.3 g of Pb or 4.3 g of Fe? Explain your thinking.
   **b.** Which has a larger mass: 2.6 mL of Pb or 2.6 mL of Fe? Explain your thinking.

# Thermometers

## Purpose

To examine how the volume of a liquid and a gas change in response to temperature.

## Part 1: Liquid Thermometer

### Procedure

Follow the handout instructions to build a thermometer like the one shown here. Then mark the liquid level in the straw for these conditions.

- room temperature
- vial warmed by your hand
- ice water
- ice water with 1 tablespoon of salt per 200 mL water
- boiling water (do not allow the thermometer to touch the bottom of the beaker)

### Observations and Analysis

1. What did you observe? What is happening to the liquid in the vial to make it move up and down in the straw?

2. Create a scale for the thermometer. Assign numbers for the places you marked on the straw for boiling water and ice water. What numbers did you choose and why?

3. Based on your newly created temperature scale, estimate the temperature in the room. How did you arrive at your answer?

## Part 2: Gas Thermometer

### Materials

- 250 mL beakers (3)
- ice
- 10 mL graduated cylinder
- food coloring
- test-tube holder—wire
- hot plate

### Procedure

1. Set up three beakers: one with ice water, one with room-temperature water, and one with hot water (cooled enough so that it is no longer steaming). Add 1–2 drops of food coloring to each beaker.

2. Hold the 10 mL graduated cylinder with a test-tube holder. Invert the graduated cylinder and immerse it in each beaker for one minute.

### Making Sense

Describe how a thermometer works.

# Thermometers

## THINK ABOUT IT

On a single day, a thermometer in Anchorage, Alaska, might read −15 degrees, while a thermometer in Sydney, Australia, reads 35 degrees, and a thermometer in your classroom reads 68 degrees. It is interesting to consider how a small glass tube with some red liquid inside can indicate how hot or cold it is.

### How is temperature measured?

*To answer this question, you will explore*

➲ Changes in Volume

➲ Creating Temperature Scales

➲ Fahrenheit Versus Celsius

## EXPLORING THE TOPIC

## ➲ Changes in Volume

*This thermometer shows the two most widely used temperature scales, Fahrenheit and Celsius.*

Matter responds to increasing temperature in a variety of ways. A wad of bubble gum gets softer, a "mood" ring changes color, and the air in a balloon expands. By systematically tracking any one of these changes, you could make a thermometer.

However, to make a *good* thermometer, it is better to track a physical change that is reproducible and easily measurable.

*The liquid crystals in a mood ring change color in response to changes in temperature.*

One way to make a thermometer is to fill a thin glass or plastic tube with a liquid. As the liquid in the thermometer is heated, its volume and height increase. As the liquid is cooled, its volume and height decrease. Because volume and height are proportional, you can detect the increase in volume as an increase in height.

In boiling water

In ice water

At room temperature

## GAS THERMOMETERS

Gases also expand and contract upon heating and cooling. A gas thermometer requires a container with a flexible volume. Suppose you have a sample of air in hot water trapped in an inverted graduated cylinder. If you heat the trapped air sample, some of the air will bubble out of the cylinder. This is evidence that the air is expanding. If you cool the trapped air sample in ice water, water will rise into the cylinder. This is evidence that the air sample contracts as it cools, leaving room for some water to move up into the cylinder. The expansion and contraction of air is important in understanding the weather.

*In hot water*    *In room-temperature water*    *In ice water*

**Big Idea** Most matter expands in volume as it is heated and contracts as it is cooled. In solids, the change is usually very small. In gases, the change is dramatic.

## ⮕ Creating Temperature Scales

To make a thermometer, you could put a colored liquid in a glass or plastic tube. Then you need to construct a scale, or number line, on the tube to assign numbers to different temperatures. You can construct a scale by measuring two temperatures, such as the boiling point and freezing point of water. It is typical to measure water because water is easily accessible and always boils and freezes at the same temperatures at sea level.

The temperature at which a substance melts or freezes is called the **melting point,** or the **melting temperature.** The temperature at which it boils or condenses to a liquid is called the **boiling point,** or the **boiling temperature.** Each time you measure the height of the liquid at the melting point and boiling point of water, you will get the same two heights on your thermometer.

Once you have two marks, you can divide the length between the marks into a round number of intervals. Each interval is one unit of temperature. The intervals

*These two thermometers are different sizes but have the same temperature scale. Both show the same temperature.*

provide numerical values for other temperatures besides the melting and boiling points of water. For instance, you can use your thermometer to measure the room's temperature. Someone else who makes a thermometer using the same points and intervals that you used will be using the same scale, and you can accurately compare temperatures.

## Fahrenheit Versus Celsius

One commonly used temperature scale is the Celsius scale. This scale sets the melting temperature of ice at 0° and the boiling temperature of water at 100° as measured near sea level as opposed to in the mountains (water boils at a lower temperature when you are at higher altitudes). This scale uses degrees Celsius, °C, as units. The Swedish astronomer Anders Celsius created this temperature scale in 1747. Most of the world and most scientists use the Celsius temperature scale.

In contrast, a weather report in the United States gives temperature in degrees Fahrenheit, °F. German physicist Daniel Gabriel Fahrenheit had proposed his temperature scale in 1724. On the Fahrenheit scale, the melting temperature of ice is 32 °F and the boiling temperature of water is 212 °F.

There are 180 Fahrenheit degrees between 0 °C and 100 °C. This means that a Fahrenheit degree unit is a smaller change in temperature than a Celsius degree unit.

In this course, you will use the Celsius scale. It is useful to know how to convert between the two scales. The formula given here allows you to convert from degrees Celsius, C, to degrees Fahrenheit, F.

$$F = \frac{9}{5}C + 32$$

[For review of this math topic, see **MATH Spotlight**: SI Units of Measure on page A-0.]

## Weather Forecast

The weather forecast in Tokyo, Japan, calls for a 60% chance of precipitation with temperatures reaching 30 °C, while in Washington, D.C., the weather forecast calls for a 70% chance of precipitation with temperatures reaching 50 °F.

    **a.** Which city will be warmer? Justify your answer by converting the temperatures to the same scale.

    **b.** Assuming that there is precipitation in each of these cities, will rain or snow be seen in each place? Explain your thinking.

*Solution*

To compare temperatures, you convert them to the same scale.

    **a.** You can convert 30 °C to °F:

$$F = \frac{9}{5}C + 32$$

Substitute the known value. $\quad F = \frac{9}{5}(30) + 32$

Solve for the unknown value. $\quad F = 86$

So, 30 °C = 86 °F. You could also convert 50 °F to °C:

Tokyo is warmer because 86 °F (30 °C) is higher than 50 °F (10 °C).

    **b.** Because ice and snow melt at 32 °F (0 °C), you would expect rain rather than snow in both cities.

## LESSON SUMMARY

### How is temperature measured?

Matter changes volume in response to changes in temperature. Matter generally expands when heated and contracts when cooled. This property can be used to measure temperature: The volume of liquid in a thermometer changes as the temperature changes. To set a temperature scale, you need to start with at least two points. Once these two temperature points are noted on a scale, it is possible to figure out where any other temperature would be on that scale.

### Reading Questions

**1.** Explain how the height of a liquid can be used to measure temperature.

**2.** Describe in your own words how to construct a temperature scale.

### Reason and Apply

**3.** What are the advantages of the Celsius temperature scale over the Fahrenheit temperature scale?

*(continued)*

4. When the temperature is 0 °C, is it also 0 °F? Explain.

5. Convert −40 °C to °F. Show your work.

6. Which is larger: one Celsius degree or one Fahrenheit degree? Explain.

7. The doctor tells you that your body temperature is 40 °C. Are you sick? Show your work and explain your answer.

8. You will be traveling to Hawaii where the forecast is for a temperature of 31 °C during the day, dropping to 28 °C overnight. Your friend recommends that you bring clothing for warm weather. Is this a good recommendation? Why or why not?

9. Create a graph comparing the Celsius and Fahrenheit scales.
   a. Plot the freezing point and the boiling point of water on your graph. Draw a straight line connecting the two points.
   b. Use the graph to determine the temperature in °F when it is 10 °C.
   c. If the weather forecast says it is 55 °F in Washington, D.C., what is the temperature in °C?

*Absolute Zero*

# Kelvin Scale

## THINK ABOUT IT

The coldest temperature recorded on Earth was in Antarctica: a chilling −89 °C. The temperature is even lower on other planets that are farther away from the Sun. Researchers have recorded temperatures as low as −235 °C on the surface of Triton, a moon of Neptune!

### How cold can substances become?

*To answer this question, you will explore*

⮕ Absolute Zero

⮕ Kelvin Scale

⮕ Molecules in Motion

## EXPLORING THE TOPIC

### ⮕ Absolute Zero

The volumes of most substances decrease as the temperature decreases. But there is a limit to how much a substance can shrink. It does not make sense that matter would have a negative volume. Hypothetically, the lowest temperature possible would correspond to a volume of *zero*.

Consider an example. The volume of a gas inside a flexible container is measured as the gas is cooled. Several data points are then plotted on a graph.

**Volume Versus Temperature**

−273 °C

Volume (mL) vs. Temperature (°C)

**Using the Graph**
This graph shows that as the temperature decreases, the volume decreases in a predictable way. The data points lie more or less on a straight line. If you extend this line to the *x*-axis, you can determine the theoretical temperature of the gas at zero volume.

**HISTORY**
**CONNECTION**

The French inventor and physicist Guillaume Amontons first proposed the existence of absolute zero in 1702. Although he became deaf in his early teens, he did not allow this to stop him from having a productive scientific career. In addition to his work on temperature and pressure, Amontons developed the gas thermometer.

Sheila Terry/Science Photo Library

More precise measurements reveal −273.15 °C as the temperature at zero volume. Scientists hypothesize that this value corresponds to the lowest temperature possible, and call it **absolute zero.** In actuality, as the temperature is lowered, a gas would condense to a liquid and then to a solid well above this temperature, so zero volume is a property that is still a hypothesis.

## Kelvin Scale

On the Celsius scale, the temperature at which the volume of gas is theoretically zero is −273 °C. If you want to create a scale where the temperature is 0 when the volume is 0, you must add 273 to each temperature. This new temperature scale is called the **Kelvin scale.** One kelvin is the same size as one Celsius degree.

$$K = C + 273$$

$$C = K - 273$$

For example, to convert 20 °C to kelvins, add 273. This temperature corresponds to 293 K.

Note that the word *degree,* or the symbol for degree, °, is not used with the Kelvin scale. The unit of temperature on the Kelvin scale is the kelvin, K.

**Volume Versus Temperature**

273 K = 0 °C

### Using the Graph
The Kelvin scale sets the zero point at absolute zero. As a result, all temperatures are positive when they are expressed in kelvins. Notice that this graph is the same as the previous graph, but the *x*-axis has been shifted by 273. The value for 0 °C is 273 K.

Room temperature is often considered to be 20 °C, or 68 °F, so 293 K is a temperature you will often find in problems. If you subtract 273 from a temperature in kelvins, you will be back at the Celsius temperature.

## Molecules in Motion

Gases are quite different from liquids and solids. The molecules and atoms in solids and liquids are held close together by forces between molecules and atoms. Liquids lie at the bottom of any container they are in, and solids generally maintain their own shape without a container. In contrast, the molecules in gases disperse to fill whatever space they are in. This indicates that there are no forces between the molecules in a gas.

So, why is it that gas molecules are found throughout their container and not just at the bottom of it? And why is it that the space a gas occupies expands so dramatically as temperature increases? To explain many of the properties of gases, scientists rely on a model called the **kinetic theory of gases.** This model proposes that gas particles—atoms or molecules—are in constant motion. The word *kinetic* comes from a Greek word, *kinetikos,* meaning "moving."

**Big Idea**  On a particle level, all matter is in constant motion.

These illustrations show the motion of three gas particles at three different times, based on the kinetic theory of gases.

Three gas particles are moving in a container.

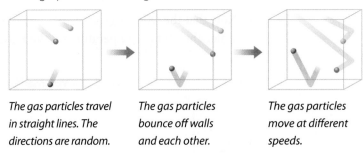

*The gas particles travel in straight lines. The directions are random.*

*The gas particles bounce off walls and each other.*

*The gas particles move at different speeds.*

One characteristic of gases described by the kinetic theory is that not all gas particles are moving at the same speed. For example, the gas particle shown in blue moves more from frame to frame than the one shown in gray. However, the kinetic theory focuses on the average speed for all the particles at a given temperature. Indeed, the **temperature** of a gas can be defined as a measure of the *average* kinetic energy of the gas particles. When the temperature increases, the average speed of the gas particles increases. Scientists hypothesize that if you could cool matter to absolute zero, the atoms in the substance would stop moving.

**Big Idea**  Temperature is a measure of the average speed of the atoms or molecules in a sample.

## LESSON SUMMARY

### How cold can substances become?

The Kelvin scale assigns a value of 0 K to the hypothetical temperature of a gas with zero volume. This point is called absolute zero and is at −273 °C. Scientists consider this to be the lowest hypothetical temperature that matter can reach. They hypothesize that all motion stops at absolute zero. The kinetic theory of gases describes the motions of the gas particles. The atoms or molecules in a gas are constantly moving at an average speed that increases with increasing temperature.

## Reading Questions

1. What is absolute zero? Why is it considered a hypothetical temperature?

2. What advantages does the Kelvin scale have over the Celsius scale?

3. How does the kinetic theory of gases explain temperature?

## Reason and Apply

4. What are the freezing and boiling temperatures of water in degrees Celsius, kelvins, and degrees Fahrenheit?

5. Which unit is the smallest: one Celsius degree, one kelvin, or one Fahrenheit degree? Explain your thinking.

6. Would you describe each of these temperatures as warm, hot, or cold?
   **a.** 100 K     **b.** 60 °C     **c.** 250 K     **d.** 25 °C     **e.** 300 K
   **f.** −100 °C     **g.** 400 K

7. Convert each of the Kelvin temperatures in Exercise 6 to degrees Celsius, and vice versa.

8. What do you think is the highest temperature that can be reached by a substance? What is your reasoning?

9. Here are a few common temperatures on the Fahrenheit scale. Convert each of these to the Kelvin scale.
   **a.** 95 °F (a hot day)     **b.** 350 °F (oven temperature)     **c.** 5 °F (freezer)

10. The temperature on the surface of Venus is 736 K. Convert this temperature into degrees Fahrenheit and degrees Celsius. Describe what the atmosphere of Venus might be like.

11. Choose the best answer. According to the kinetic theory of gases, particles in a sample of gas
    **(A)** Move randomly, in curved paths, at different speeds, and bounce off each other and the walls of their container.
    **(B)** Move randomly, in straight-line paths, at different speeds, and bounce off each other and the walls of their container.
    **(C)** Move randomly, in straight-line paths, at the same speed, and bounce off each other and the walls of their container.
    **(D)** Move randomly, in curved paths, at the same speed, and bounce off each other and the walls of their container.

12. Use the kinetic theory of gases to explain why gases expand upon heating and shrink upon cooling.

*Sorry, Charlie*

# Charles's Law

## THINK ABOUT IT

The vent for a heater in your house is typically placed near the floor. This is because hot air rises and cooler air descends. If the vent were near the ceiling, the room would still be cold near the floor because the hot air would remain near the ceiling. So hot air must be less dense than cold air. For a similar mass, it occupies a larger volume than cooler air.

### How can you predict the volume of a gas sample?

*To answer this question, you will explore*

➾ Predicting Gas Volume

➾ Charles's Law

## EXPLORING THE TOPIC

## ➾ Predicting Gas Volume

A *piston* is a movable part that traps a sample of gas within a cylinder. It can move up and down if the volume of the gas changes. The illustration shows how the volume of nitrogen gas changes as the temperature changes. At first, the volume of the gas is 600 mL at a temperature of 300 K. When the cylinder is cooled, the gas contracts and the piston moves down. Then when the cylinder is heated, the gas expands and the piston moves up.

**HISTORY CONNECTION**

Hydrogen gas is less dense than air. Jacques Charles took advantage of this when he invented the hydrogen balloon. In August 1783, terrified villagers destroyed the balloon when it landed outside of Paris. Charles built another hydrogen balloon, and he and a friend made the first piloted flight that December.

$V_1 = 600$ mL
$T_1 = 300$ K

$V_2 = 400$ mL
$T_2 = 200$ K

$V_3 = 800$ mL
$T_3 = 400$ K

Subscripts are used to indicate corresponding volumes and temperatures for the same gas under different conditions. For example, $V_1$ is the volume at the first temperature, $T_1$. Notice that if you determine the ratio of the volume of the gas to the temperature of the gas, the ratio is the same at all three of these temperatures.

$$\frac{V_1}{T_1} = 2.0 \text{ mL/K} \qquad \frac{V_2}{T_2} = 2.0 \text{ mL/K} \qquad \frac{V_3}{T_3} = 2.0 \text{ mL/K}$$

In fact, if nothing else changes, the ratio of volume to temperature for this gas will be 2.0 mL/K at any temperature. The relationship between the volume and

## HISTORY
## CONNECTION

The French Montgolfier brothers invented the first hot air balloon capable of carrying passengers. They demonstrated their creation in June 1783. The brothers thought that they had discovered a new gas that was lighter than air. In fact, the gas in their balloon was just air. It was less dense than the surrounding air because it was heated.

© Gianni Dagli Orti/Corbis

**Important to Know** Each gas sample has a unique value for the proportionality constant, $k = V/T$, depending on the amount of gas in the sample.

temperature for this sample of gas is proportional. So the volume and temperature for this gas are related by the proportionality constant, $k = V/T$, or 2.0 mL/K. With this proportionality constant, you can determine the volume of this gas at any temperature. [For review of this math topic, see **MATH Spotlight**: Ratios and Proportions on page A-11.]

## CHANGING THE AMOUNT OF GAS

Suppose you start with a *different* amount of gas in the same cylinder. For this new sample of gas, the proportionality constant, $k = V/T$, is 1.5 mL/K.

$V_1 = 450$ mL

$T_1 = 300$ K

$\dfrac{V_1}{T_1} = 1.5$ mL/K

$V_2 = 300$ mL

$T_2 = 200$ K

$\dfrac{V_2}{T_2} = 1.5$ mL/K

$V_3 = 600$ mL

$T_3 = 400$ K

$\dfrac{V_3}{T_3} = 1.5$ mL/K

Again, you can use $k = V/T$ for this gas sample to calculate the volume at any temperature.

If you graph volume versus temperature for the two gas samples described previously, the result will be two different lines with different slopes.

**Volume Versus Temperature**

600 mL of gas at 300 K

$k = 2.0 \dfrac{mL}{K}$

450 mL of gas at 300 K

$k = 1.5 \dfrac{mL}{K}$

Volume (mL)

Temperature (K)

### Using the Graph

Because the relationship between temperature and volume is proportional, the lines pass through the origin. The slope of each line is equal to the proportionality constant for that sample. You can use the graph to predict the volume of either gas sample at any temperature.

# ↪ Charles's Law

Jacques Charles was a French inventor, scientist, mathematician, and very active balloonist. Because of his interest in hot air balloons, he gave a lot of thought to the relationship between the volume and temperature of a gas. In 1787, he described the proportionality of gas volume and temperature mathematically. This description is now known as **Charles's law.**

---

**Charles's Law**

If pressure and the number of particles of a gas stay the same, then volume is proportional to the Kelvin temperature.

$$V = kT \text{ or } k = \frac{V}{T}$$

---

## Example 1

### Calculate Gas Volume

Imagine that a cylinder with a piston contains 10 mL of air at 295 K. What will the volume of gas in the cylinder be at 550 K?

*Solution*

Calculate the volume by using Charles's law or by drawing a graph. You can predict that the volume will increase when the temperature increases. So the answer should be greater than 10 mL.

**Using the Formula**

Find the value of $k$.

$$k = \frac{V}{T} = \frac{10 \text{ mL}}{295 \text{ K}} = 0.034 \text{ mL/K}$$

Solve for the volume at 550 K.

$$V = kT = 0.034 \text{ mL/K} \cdot 550 \text{ K} = 18.5 \text{ mL}$$

**Graphical Analysis**

**Volume Versus Temperature**

First, plot the data point for the initial conditions. Draw a line through this point and the origin, since at 0 K the $y$-value is theoretically 0 mL. Use the line to find the volume when the $x$-value is 550 K. Both methods give the same answer. At 550 K, the volume is 18.5 mL.

## HISTORY
### CONNECTION

On May 6, 1937, the airship *Hindenburg* caught fire as it landed at an airfield in New Jersey. The *Hindenburg* was kept afloat with massive chambers filled with hydrogen, which is less dense than air but is also highly flammable. Thirty-six people died when a fire on board spread rapidly.

**Example 2**

## Calculate Gas Temperature

A balloon contains 500 mL of helium at 20 °C. You want to increase the volume to 550 mL. What temperature in °C will give you a volume of 550 mL for this sample of gas?

### Solution

You can predict that the temperature will be higher for a larger volume.

Convert the temperature to kelvins.

$$K = C + 273$$
$$= 20 + 273$$
$$= 293 \text{ K}$$

Determine the proportionality constant, $k$.

$$k = \frac{V}{T} = \frac{500 \text{ mL}}{293 \text{ K}} = 1.7 \text{ mL/K}$$

Rearrange the equation in terms of $T$ and solve for temperature at 550 mL.

$$T = \frac{V}{k} = \frac{550 \text{ mL}}{1.7 \text{ mL/K}} = 324 \text{ K}$$

Convert back to degrees Celsius.

$$C = K - 273 = 324 - 273 = 51 \text{ °C}$$

A temperature of 51 °C is needed to give a volume of 550 mL. The temperature is higher for a larger volume, as predicted.

---

**Important to Know** To solve problems using Charles's law, temperature values must first be converted to the Kelvin scale.

[For review of this topic, see **Math Spotlight**: Solving Equations on page •••.]

## LESSON SUMMARY

### How can you predict the volume of a gas sample?

The volume and temperature of a sample of gas are proportional to each other if nothing else changes (such as pressure or the amount of gas). This relationship between the volume and temperature of a gas in kelvins is described by the equation $V = kT$ and is known as Charles's law. The proportionality constant, $k$, is different for each different amount of gas.

**KEY TERM**

Charles's law

### Reading Questions

1. Explain how to determine the proportionality constant, $k$, for a sample of gas.
2. Explain how to determine the volume of a gas at a certain temperature using the proportionality constant, $k$.

### Reason and Apply

3. A gas sample in a cylinder has a volume of 620 mL at 293 K. If you allow the piston to move while you heat the gas to 325 K, what will the volume of the gas be? Check your answer by drawing a graph.

4. A gas sample in a cylinder has a volume of 980 mL at a temperature of 27 °C. If you allow the piston to move while you heat the gas to 325 K, what will the volume of the gas be at 325 K?

5. A gas sample in a cylinder with a piston has a volume of 330 mL at 280 K. What temperature will you need to heat it to in order to change the volume to 220 mL?

6. A 2.0 L gas sample at 20 °C must be cooled to what temperature for the volume to change to 1.0 L? Show at least two different ways to solve this problem.

7. Imagine that you have a huge helium balloon for a parade. Around noon, it is 27 °C when you fill the balloon with helium gas to a volume of 25,000 L. Later in the day, the temperature drops to 15 °C.
   a. What is the proportionality constant, $k = V/T$, at the beginning of the day?
   b. Calculate the volume of the balloon when the temperature has dropped to 22 °C.
   c. What will the proportionality constant, $k = V/T$, be at the end of the day when the temperature is 15 °C? Explain your answer.

8. Would this cup make a good rain gauge? Explain your thinking.

Petr Jilek/Shutterstock

# Density, Temperature, and Fronts

### Purpose

To investigate the movement of air masses and their role in determining the weather.

### Part 1: Weather Maps

Reexamine the weather maps from Lesson 49: Weather or Not to answer these questions.

1.  Examine the first map at the left. What relationships do you see between fronts, clouds, and precipitation?

2.  Examine the second map. What relationship do you see between fronts and lows? Between fronts and highs?

3.  Describe any connections you find between fronts and the jet stream.

4.  Where would you expect to see warm and cold air masses on each map?

### Part 2: Warm and Cold Fronts

1.  Why is a cold air mass denser than a warm air mass?

2.  Explain why clouds might form when a warm air mass collides with a cold air mass.

3.  **Making Sense** What does air density have to do with weather fronts?

4.  **If You Finish Early** Eighty percent of the air in our atmosphere is made up of nitrogen gas, $N_2$, while only 1% of the air is made up of water vapor. Why doesn't it rain liquid nitrogen instead of rainwater?

# Density, Temperature, and Fronts

## THINK ABOUT IT

Meteorologists have identified large masses of air that have a consistent temperature and water content. These **air masses** form in particular locations on Earth's surface and cover thousands of square miles. They have a great influence on the weather. Weather fronts occur at the boundaries of these air masses, where warm and cold air meet.

### How do weather fronts affect the weather?

*To answer this question, you will explore*

↻ Air Density

↻ Weather Fronts

↻ Relationships between Weather Variables

## EXPLORING THE TOPIC

## ↻ Air Density

### SMALL AIR MASSES

To understand a large mass of air on Earth's surface, it is helpful to examine the behavior of a small sample of gas. When a gas sample gets warmer, it expands to fill a larger volume. Notice that as the temperature increases, there are fewer molecules per milliliter of volume in the cylinders. The number of molecules does not change, but the spacing between those molecules does change. The density, $D = m/V$, decreases because the same mass, $m$, is divided by a larger volume, $V$, resulting in a smaller number.

200 K          300 K          400 K

*Air expands and becomes less dense as it is heated.*

### LARGE AIR MASSES

In Unit 1: Alchemy, you found that every substance has a specific density. For example, the density of gold is 19.3 g/mL, and the density of water is 1.0 g/mL.

This is true only at a specified temperature. The density of a substance changes as the temperature changes. For liquids and solids, this change in density is very small. For gases, the change is much larger.

When air gets warmer, it expands as well as rises, because warm air is less dense than colder air. This is what happens when a warm air mass meets a cold air mass. This fact is key to understanding weather fronts and storms.

## ⮑ Weather Fronts

### WARM AND COLD AIR MEET

When a cold air mass meets a warm air mass, two different weather situations can arise. These two situations are shown in the illustrations and are described here.

*Cold front. Cold air overtakes warm air.*

*Warm front. Warm air overtakes cold air.*

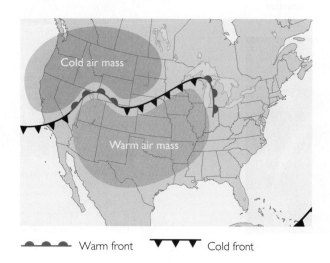

Warm front ▼▼▼ Cold front

*Clouds at the leading edge of a cold front*

## COLD FRONTS

A cold front occurs when cold air overtakes warm air. The warm air is suddenly pushed up as the advancing cold air forces its way underneath it. The warm air cools at higher altitudes, which causes water vapor in the air to condense into tiny liquid cloud droplets. The clouds associated with cold fronts can be seen directly in the area of the advancing front. They form quickly and are often thick, puffy clouds, like the ones seen before thunderstorms. Heavy rains are often associated with cold fronts.

## WARM FRONTS

A warm front occurs when warm air overtakes cold air. The warm air gradually flows up and over the cold air because the warm air is less dense. The warm air cools at higher altitudes and water vapor condenses to form clouds. The clouds associated with warm fronts usually form ahead of the place where the air masses meet. You will see high, wispy clouds when a warm front is still hundreds of miles away. As the warm front gets closer, the clouds get thicker and lower, and precipitation falls. Steady, light rain over a large area is associated with warm fronts.

*Clouds at the leading edge of a warm front*

## ⮕ Relationships between Weather Variables

The clouds, high- and low-pressure systems, precipitation, and the jet stream can be added to a weather map that contains some fronts. Take a moment to draw some conclusions about how these weather variables relate to one another.

### NATURE
**CONNECTION**

Cold fronts are responsible for squalls, tornadoes, strong winds, and other destructive weather.

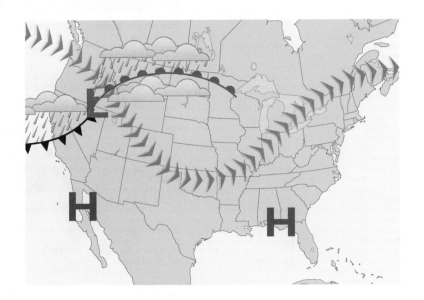

- Fronts occur at the boundaries between warm and cold air masses.
- Clouds and steady, light rain form ahead of a warm front. Clouds and heavy showers form at and behind a cold front.
- On weather maps, Ls are closely associated with fronts while Hs appear to be away from the fronts. Hs represent areas of high pressure and are associated with clear skies. Ls represent areas of low pressure and are associated with storms and cloudy skies.

When air masses form over Earth, they get their properties from the area beneath them. In other words, if an air mass forms over water, it will tend to be full of moisture; if it forms over land, it will tend to be dry.

## LESSON SUMMARY

### How do weather fronts affect the weather?

When a cold air mass and a warm air mass meet, the warm air rises and layers on top of the cold air. This is because warm air is less dense than cold air. A cold front occurs when a cold air mass overtakes a warm air mass, and a warm front occurs when a warm air mass overtakes a cold air mass. Clouds and precipitation occur with both warm and cold fronts. Warm fronts produce steady, light rain, and cold fronts produce sudden, heavy showers.

**KEY TERM**

air mass

---

**Exercises**

### Reading Questions

1. Explain why hot air rises.
2. Explain the difference between a warm front and a cold front.

### Reason and Apply

3. Suppose you have two gas samples in flexible containers with the same outside pressure and the *same amount of gas* in each. Sample A is at a temperature of 25 °C and Sample B is at 5 °C. Which of these statements is true?
   **(A)** Sample A occupies a larger volume and has lower density.
   **(B)** Sample A has higher density and a smaller volume.
   **(C)** Sample B has molecules moving at a greater average speed.

4. A cold front is approaching your hometown and is due to arrive tomorrow. What kind of weather would you expect to observe?

5. A warm front is approaching your hometown and is due to arrive tomorrow. What kind of weather would you expect to observe?

6. The continental polar air mass overtakes the maritime tropical air mass.
   **a.** What kind of front develops?
   **b.** What happens to the air masses when they meet?
   **c.** What sort of weather would you expect and where?

*(continued)*

7. **PROJECT**  Look in the newspaper or on the Internet. Find a recent weather map with at least one warm front and one cold front.

   a. Find a warm front on the map. What weather is predicted for tomorrow in the region of the warm front?

   b. What weather is predicted in the direction in which the warm front is moving?

   c. Find a cold front on the map. Describe the weather forecast given for the region near the cold front.

   d. What weather is predicted for tomorrow in the region close to the cold front?

   e. What weather is predicted in the direction in which the cold front is moving?

# CHAPTER 10

## Physically Changing Matter

# SUMMARY

### KEY TERMS

**weather**

**physical change**

**phase change**

**proportional**

**proportionality constant**

**melting point**
 **(melting temperature)**

**boiling point**
 **(boiling temperature)**

**absolute zero**

**Kelvin scale**

**kinetic theory of gases**

**temperature**

**Charles's law**

**air mass**

# Weather Update

The water and gases that make up Earth's atmosphere change density and temperature as they move about the planet. Physical changes of matter cause the weather we experience. Most matter expands when heated and contracts when cooled. When heated enough, matter changes phase, from solid to liquid to gas. The density of gases is dramatically lower than the density of either liquids or solids. Changes in temperature and density play a role in the weather.

## REVIEW EXERCISES

1.  What is 50 °F in kelvins?

2.  Use the kinetic theory of gases to explain temperature.

3.  A gas sample in a flexible container has a volume of 650 mL at a temperature of 27 °C. If the pressure stays the same when the sample is heated to 80 °C, what will be the new volume?

4.  You fill a balloon with helium to a volume of 3 L in the morning, when the temperature is 23 °C. At the end of the day, the temperature drops to 10 °C.
    a.  What is the proportionality constant, $k = V/T$, at the start of the day?
    b.  What is the proportionality constant at the end of the day? Explain your thinking.
    c.  What will the volume of the balloon be when the temperature is 10 °C?

5.  a.  What is the volume of 5.2 g of solid $CO_2$ if the density of solid $CO_2$ is 1.56 g?
    b.  What will the volume of the 5.2 g(s) be if all of the solid changes to a gas? The density of $CO_2(g)$ is 0.0019 g/mL. Explain your reasoning.

## Different Thermometers

 PROJECT ✓ Different thermometers use different properties of matter to measure temperature. Research how one of these thermometers works. Write a description and include a diagram.

Liquid crystal thermometer

Infrared ear thermometer

Digital resistance thermometer

Cooking thermometer

Galileo density thermometer

# CHAPTER **11**

# Pressing Matter

## In this chapter, you will study

- the density of gases

- the behavior of gas particles

- the relationships between gas pressure, temperature, and volume

- high- and low-pressure weather systems

*The hot air inside a balloon has extremely low density and it causes the balloon to rise. Density differences cause warm air masses to rise in the atmosphere in a similar way.*

The atoms and molecules in solids, liquids, and gases are always in motion. However, gas molecules are free to move to fill the entire space of their container. Gas molecules collide with one another and with anything else they come in contact with, causing pressure. When samples of gases are placed in containers, it is possible to measure the relationships between gas pressure, temperature, and volume. The air pressure of the atmosphere has a great deal of influence on the weather.

# Gas Density

## THINK ABOUT IT

Earth is unique in that it is the only planet in the solar system with so much liquid water on its surface. The weather and the life on Earth depend on the movement of moisture around the planet. The phase changes of water are mostly responsible for this movement. When water changes phase from liquid to gas, the airborne water molecules spread out in the atmosphere and the density changes dramatically.

**How do the densities of a solid and a gas compare?**

*To answer this question, you will explore*

→ Density of Gases

→ The Molecular View

→ Comparing Density and Phase of Water

## EXPLORING THE TOPIC

## → Density of Gases

An airplane glides easily through the air at around 500 miles per hour. The density of air must be very low for objects to pass so easily through it. Determining the density of a gas is not as straightforward as determining the densities of liquids and solids. This is because it is not easy to measure the mass of a gas. It is hard to put a gas on a balance, and many gases, like helium, rise up in the air around them.

One approach is to measure the mass of a substance while it is a liquid or a solid, and then turn it into a gas. Provided you don't let any escape, the substance will have the same mass as a gas as it did as a liquid or a solid. You can then measure the volume of the gas to determine its density.

### SUBLIMATION OF DRY ICE

One substance that you can use to study phase changes is carbon dioxide, $CO_2$. Carbon dioxide turns *directly* from a solid to a gas at the extremely low temperature of $-78\,°C$ ($-108\,°F$). This means the temperature in your classroom is warm enough to turn solid carbon dioxide into a gas.

When a solid changes directly into a gas without forming a liquid, the phase change is called **sublimation.** In class, you gathered data on the density of $CO_2(g)$ by finding the mass of a chunk of $CO_2(s)$ and then placing it into a plastic bag and allowing it to sublime. The volume of the gas is huge compared to the volume of the same mass of solid.

*Charles D. Winters/Science Source*

*$CO_2(s)$ is commonly referred to as "dry ice" because it does not go through a liquid phase when it changes to a gas.*

*One gram of solid $CO_2(s)$ occupies about 0.6 mL.*

*One gram of gaseous $CO_2(g)$ occupies about 500 mL.*

The density of the carbon dioxide changes dramatically as the phase changes. $CO_2(s)$ has a density of 1.56 g/mL. In contrast, $CO_2(g)$ has a very low density of 0.0019 g/mL. The volume of the gas is nearly 800 times as large as the volume of the solid!

> **Big Idea** The densities of gases are extremely low compared to the densities of liquids and solids.

## ⮕ The Molecular View

Why did the bag puff up so much as the $CO_2(s)$ sublimed? Because you cannot see gases, you must depend on hypotheses about how the molecules in a gas behave. Consider four possible hypotheses and the corresponding models for the sublimation of solid carbon dioxide.

CO₂(s)          Hypothesis A          Hypothesis B          Hypothesis C          Hypothesis D

Hypothesis A

**Hypothesis A:** The $CO_2$ molecules become larger in the gas phase compared with the solid phase.

**Evidence against Hypothesis A:** You can move freely through a gas, so the molecules must not take up much space. The mass stays the same, so it is difficult to imagine how the molecules would get bigger. The evidence doesn't support Hypothesis A.

Hypothesis B

**Hypothesis B:** The $CO_2$ molecules rise to the top of the container.

**Evidence against Hypothesis B:** The bag filled with $CO_2(g)$ was pushed out in all directions. This indicates that the molecules are spread throughout, not just at the top. For example, the air in a room is everywhere, not just near the ceiling. The evidence doesn't support Hypothesis B.

Hypothesis C

**Hypothesis C:** The molecules go to the outer edges of the bag, pushing the sides out.

**Evidence against Hypothesis C:** If gases went to the outer edges of their containers, you would walk into a room and find all the air on the walls and none in the middle. Since you can breathe everywhere in the room, this indicates that gas molecules are spread throughout. The evidence doesn't support Hypothesis C.

Hypothesis D

**Hypothesis D:** The $CO_2$ molecules spread apart from one another and bounce around inside the bag.

**Evidence supporting Hypothesis D:** This model is consistent with the large volume and the low density of the gas. This model is also consistent with the bag staying puffed out in all directions and offering resistance when you push on it. This evidence supports Hypothesis D.

Model D can still be improved upon. It shows ten molecules in a small cubic space. In actuality, the gas molecules are much tinier, much farther apart, much more numerous, and moving very fast. When a gas is trapped in a bag, the moving molecules bang into the bag and push it out. If you were to remove the bag, the molecules would quickly spread throughout the room. This is what happens to water when it **evaporates,** or changes into a vapor, and spreads out in the atmosphere.

| **Big Idea** | A sample of gas expands to fill whatever space it is in. |

## Comparing Density and Phase of Water

The table shows the density for some different forms of water. Water has three phases: solid, liquid, and gas. But these phases can take a few different forms. For example, snow and ice both represent solid forms of water.

**Different Forms of Water**

| Substance | Chemical formula | Density (g/mL) | Volume of 1000 g (mL) |
|---|---|---|---|
| rain | $H_2O(l)$ | 1.0 g/mL* | 1,000 mL |
| water vapor | $H_2O(g)$ | 0.0008 g/mL* | 1,250,000 mL |
| ice | $H_2O(s)$ | 0.92 g/mL† | 1,087 mL |
| snow | $H_2O(s)$ | 0.1–0.5 g/mL | 2,000–10,000 mL |
| clouds | $H_2O(l)$ | 0.001–0.002 g/mL | 500,000–1,000,000 mL |

*The density is for 25 °C. †The density is for 0 °C.

Notice what happens to the *volume* of 1000 grams of water as it goes from liquid to water vapor. The volume increases more than a thousand-fold when evaporation occurs.

The density landscape illustration shows the relationship between the location of matter and its density.

*Clouds could be considered a mixture of water vapor and liquid water. It takes 15 million cloud droplets to form one drop of rain. The extremely low density of clouds explains why airplanes can move easily through them.*

### ASTRONOMY CONNECTION

The lowest temperature ever recorded on Earth was −129 °F in 1983 in Antarctica. At this temperature, $CO_2(g)$ condenses to form solid dry ice. The southern polar ice cap on Mars contains dry ice because the temperature there is below the sublimation temperature of $CO_2$.

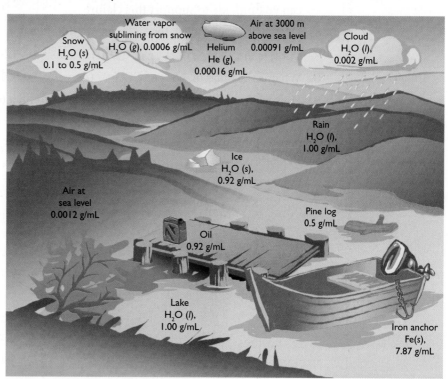

Substances that are less dense float in substances that are denser. For example, ice floats in liquid water. When the temperature of a mass of air increases, the air expands, resulting in a lower density. This change in density causes the warmer air to rise above cooler air. These changes are related to the movement of matter on the planet and are essential to causing different weather.

## LESSON SUMMARY

### How do the densities of a solid and a gas compare?

Gases form when solids sublime and liquids evaporate. During both sublimation and evaporation, molecules spread far apart from each other. This increases the volume of the substance by about 1000 times. It also decreases the density of the substance dramatically. Solids are generally denser than liquids. Gases are much less dense than both solids and liquids. Gases expand to fill whatever container they are in.

**KEY TERMS**

sublimation

evaporation

## Exercises

### Reading Questions

1. How does the density of a gas compare with the density of a solid?

2. Draw a molecular view of carbon dioxide gas.

### Reason and Apply

3. In your own words, define sublimation and evaporation.

4. Describe three ways to show that gases exist.

5. When water freezes, the water molecules move apart very slightly. What evidence can you provide to support this claim?

6. Draw a molecular view for water vapor, liquid water, and ice.

7. What is the volume of 6.4 g of $CO_2(s)$? (density = 1.56 g/mL)

8. What is the mass of 3.5 L of $CO_2(g)$? (density = 0.0019 g/mL)

9. How many grams of $CO_2(s)$ would you need to fill a 6.5 L bag with $CO_2(g)$?

10. A person has 15 g of dry ice and wants to completely fill a bag that has a volume of 8 L with carbon dioxide gas. Is this enough dry ice to fill the bag with $CO_2(g)$? Explain.

# Air Pressure

## THINK ABOUT IT

Earth is surrounded by the *atmosphere,* a layer of air nearly 80 miles thick. The atmosphere is made up predominantly of colorless, odorless gases, which you can move through. The air molecules are tiny, and they are spaced far apart, so most of the time it feels as if nothing is there. However, the weight of all that air above you exerts a great deal of pressure on you.

**What evidence do we have that gases exert pressure?**

*To answer this question, you will explore*

⟳ Evidence of Air Pressure

⟳ Explanation for Air Pressure

⟳ The Atmosphere

## EXPLORING THE TOPIC

## ⟳ Evidence of Air Pressure

When you fill a car's tires with air, the rubber is stretched quite tightly. The pressure exerted by the air in the tires is high enough to push the entire car up off the ground. Indeed, the air pressure is holding up more than 2000 pounds. How can air be so powerful?

The ability of car tires to remain inflated while holding up so much weight provides evidence that air is pushing on the inside of the tire in all directions. This "pushing" property of a gas is called **pressure.** Pressure is defined as a force over a specific surface area. A gas exerts pressure on all surfaces it comes in contact with.

There are many ways to demonstrate that gases exert pressure on the things around them. Here are two examples.

### SUBMERGED PAPER

When a plastic cup is turned upside down and submerged in water, the paper inside the cup stays dry. There is air trapped in the cup that pushes out in all directions. The air pushes on the water with enough pressure to keep the water out of the cup.

### BALLOON IN A BOTTLE

Suppose you try to blow up a balloon inside a bottle. Even with a great deal of effort, the balloon will inflate only a tiny bit. The balloon does not inflate because there is air inside the bottle. Even though it looks like

### HISTORY
### CONNECTION

Atmospheric pressure was demonstrated in 1654 when Otto von Guericke, of Magdeburg, Germany, used a vacuum pump to remove almost all of the air from the space between two copper hemispheres that were 1.6 ft in diameter. The air pressure holding the two halves together was so strong that two teams of horses could not pull them apart. However, when air was let back in, the hemispheres fell apart easily.

"nothing is there," the bottle is full of air, and when you try to inflate the balloon, you are pushing on this air, which takes a lot of work!

# ➲ Explanation for Air Pressure

Collisions of gas molecules with surrounding objects are what we experience as air pressure. There are huge numbers of molecules in an automobile tire, many more than are shown in the illustration. The tiny push from each one adds up to a lot, enough to hold a car up.

The pressure of the air inside your tires is expressed in pounds per square inch, or $lb/in^2$. This unit is sometimes written as *psi,* for pounds per square inch. That is, each square inch of surface area on the tire experiences a certain number of pounds of force from the air inside. A tire on a racing bicycle might be inflated to a pressure of $100\ lb/in^2$.

*A tire filled with air is firm because the air molecules are colliding with the inside walls of the tire, thereby pushing the walls out.*

## Example

### Air Pressure in Car Tires

Suppose a car weighs 2000 pounds. Each of its four tires touches the road over an area that is about a 4-by-4-inch square. What air pressure do you need in each tire to hold up the car?

### Solution

Each of the four tires needs to push up ¼ of the 2000 pounds, or 500 pounds. A 4-by-4-inch square has an area of $16\ in^2$, so each tire is in contact with a $16\ in^2$ area of the ground. Pressure is force per unit area.

$$Pressure = force/area$$
$$= 500\ lb/16\ in^2$$
$$= 31\ lb/in^2$$

The pressure in each tire needs to be at least $31\ lb/in^2$.

# The Atmosphere

Earth's atmosphere is a mixture of gases, mostly nitrogen, with some oxygen, carbon dioxide, water vapor, and argon. This mixture of gases is what we call air. The density of Earth's atmosphere changes as you travel up in altitude—the air becomes thinner at higher elevations. This concept will be covered more extensively in later lessons.

. . . . . . . . . . . . . . . . . . . . . . . . . . . . . . . . . . . . . . . . . . . . . . . . . . . . . . . .

**Big Idea**  The atmosphere is a mixture of gases, including gaseous water.

. . . . . . . . . . . . . . . . . . . . . . . . . . . . . . . . . . . . . . . . . . . . . . . . . . . . . . . .

Although we barely notice it, the air around us exerts pressure on us all the time. **Atmospheric pressure** is air pressure that is always present on Earth as a result of air molecules colliding with objects on the planet. The pressure due to the atmosphere is the equivalent of a 14.7-pound weight pushing on every square inch of surface. We say that the pressure due to the atmosphere is 14.7 pounds per square inch, or 14.7 lb/in$^2$. This pressure is measured at sea level.

Scientists also use a unit called an **atmosphere,** or **atm,** to measure air pressure.

---

**Atmospheric Pressure**  At sea level and 25 °C, there is 1 atm of pressure.

$$1 \text{ atm} = 14.7 \text{ lb/in}^2$$

---

Atmospheric pressure played a role in all of the demonstrations you completed in class. The air pressure mat is particularly good at demonstrating how much pressure the air around us exerts on objects. If the mat has a surface area of 100 square inches and standard air pressure from the atmosphere is 14.7 lb/in$^2$, the total pressure on the mat from the air is 1470 pounds! No wonder no one can lift it.

## LESSON SUMMARY

### What evidence do we have that gases exert pressure?

Earth is surrounded by an atmosphere of gases. We call this mixture of gases air. Although we cannot see it, the air around us exerts pressure. Gas molecules are constantly moving and colliding with anything they come in contact with. Air pressure is defined as the force over a surface area. This force is caused by the collisions of the gas molecules. Gas pressure is measured in pounds per square inch, $lb/in^2$, or in atmospheres, atm. At sea level and 25 °C, the atmospheric pressure is 14.7 $lb/in^2$, or 1 atm.

## Exercises

### Reading Questions

1. Give three pieces of evidence that air exerts pressure.

2. Explain what causes air pressure.

### Reason and Apply

3. High up in space, there are no molecules to collide with the outside walls of a balloon. Therefore, the air pressure inside the balloon is much greater than the air pressure outside of the balloon. Describe what would happen to an inflated balloon if it were suddenly put in space.

4. Calculate the weight in pounds of the atmosphere pushing down on the back of your hand.

5. When you fly in an airplane, it is common to feel painful pressure in your ears. Explain what might be happening in terms of air pressure.

6. What is the minimum pressure you need in bike tires to hold up a person who weighs 150 lb? Assume that the tires touch the ground in a square that is 1.5 in by 1.5 in.

7. The air pressure from the atmosphere measures 0.5 atm at an altitude of 18,000 ft. How much pressure is this in pounds per square inch? Calculate the weight in pounds of the atmosphere pushing down on the back of your hand at this altitude.

8. PROJECT

    Go to a library or do a Web search to find out how gases, such as helium or oxygen, are transported for industrial use.

9. If a car runs over the tip of your foot, it doesn't break your toes. Explain why.

⚠

**SAFETY Instructions**

The cap on the tip of the syringe should always be pointed down, away from eyes. Wear safety goggles.

# Boyle's Law

### Purpose

To observe and quantify the relationship between gas pressure and gas volume.

### Materials

- 50 mL plastic syringe with cap screwed on tight
- bathroom scale

### Procedure

1. Start with the syringe at 50 mL. Make sure the cap is on tight.

2. Hold the syringe so that the tip is on top of a bathroom scale.

3. Repeat these steps for at least five different volumes.
   - One person depresses the plunger by a few milliliters.
   - A second person reads the exact volume.
   - A third person reads the number of pounds that is exerted on the bathroom scale.
   - Everyone records the volume and weight data in a table like this one.

### Data

| Trial | Volume (mL) | Weight that you apply (lb) | Pressure that you apply (lb/in$^2$) | Atmospheric pressure (lb/in$^2$) | Total pressure (lb/in$^2$) |
|---|---|---|---|---|---|
| 1 | | | | 14.7 lb/in$^2$ | |
| 2 | | | | 14.7 lb/in$^2$ | |

### Analysis

1. Estimate the cross-sectional area inside of the syringe in square inches. Determine the pressure applied in pounds per square inch. Calculate the rest of the values needed to complete your table.

2. Graph the total pressure versus volume. What happens to the pressure of the gas as the volume of the gas decreases?

3. **Making Sense** Using today's observations, explain how the pressure and volume of a gas change in relation to each other.

*Feeling under Pressure*

# Boyle's Law

## THINK ABOUT IT

A flat basketball does not bounce very well. However, you can push air into the ball with a pump. The air you trap inside pushes on the skin of the ball with a lot of pressure. The amount of air you push into the ball determines how firm and bouncy the ball is. What would happen if you pushed the same amount of air into a smaller ball?

**How does gas volume affect gas pressure?**

*To answer this question, you will explore*

➲ The Syringe and Scale

➲ Graphing Pressure–Volume Data

➲ Boyle's Law

## EXPLORING THE TOPIC

### ➲ The Syringe and Scale

In class, you examined how the pressure of a gas trapped inside a container is related to the volume the gas occupies. A syringe makes a good container for such an investigation, because it can be sealed off and the volume of air trapped inside the syringe can be measured.

If you push down on the plunger of the syringe, you can feel the pressure of the gas inside the syringe. Air molecules take up space and as you squeeze them into a smaller space, they push back more and more. This is because the same number of molecules occupies a smaller volume, resulting in more collisions between the molecules and the container. More collisions mean greater gas pressure.

**GATHERING DATA**

You can learn more about the relationship between the volume of a gas and its pressure by using a capped syringe and a bathroom scale. As you push down on the plunger of the syringe, the gas pressure inside the syringe increases, and the

The basketball used by professional players is inflated to a pressure sufficient to make it rebound to a height of 5.4 feet when it is dropped from a height of 6 feet. This test is conducted in a basketball factory, and then that air pressure is stamped on the outside of the ball.

weight on the scale increases. You can record this quantitative, or numerical, data. To calculate the pressure of the gas, divide the measured weight on the scale by the surface area of the plunger in pounds per square inch, or $lb/in^2$. Weight is a measure of force.

Pressure applied = weight that you apply divided by area, plus atmospheric pressure.

Area of the plunger in contact with the gas is the cross-sectional area inside the plunger.

Gas pressure = pressure applied

**Important to Know** Because the atmosphere is also pushing on the syringe, you need to add 14.7 $lb/in^2$ to the pressure that you apply.

Note that the pressure you apply plus atmospheric pressure is equal to the gas pressure inside the syringe. When you push down, the gas pushes back with the same pressure that is being applied to it.

Data for this experiment are given in the table. Notice that as the volume decreases, the pressure of the gas inside the syringe increases.

**Pressure and Volume Data**

| Trial | Volume (mL) | Total pressure ($lb/in^2$) |
|-------|-------------|----------------------------|
| 1 | 100 mL | 15 $lb/in^2$ |
| 2 | 80 mL | 19 $lb/in^2$ |
| 3 | 60 mL | 25 $lb/in^2$ |
| 4 | 40 mL | 38 $lb/in^2$ |
| 5 | 30 mL | 50 $lb/in^2$ |
| 6 | 20 mL | 75 $lb/in^2$ |

**Big Idea** If you squeeze a sample of gas into a smaller container, its pressure will increase.

## ⮞ Graphing Pressure–Volume Data

Many of the relationships you have explored so far, like mass versus volume and volume versus temperature of a gas, are proportional relationships. The relationship between pressure and volume is not directly proportional. A graph of the data for this experiment results in a curved line that does not go through the origin.

**Gas in a Syringe**

Gas pressure and gas volume are **inversely proportional** to one another. In other words, when one variable increases, the other decreases, and vice versa.

When the volume of the gas in the syringe is very small, the gas pressure is quite high. When the volume of the gas is large, the gas pressure is quite low. However, as long as there are gas molecules in a container, neither the gas pressure nor the gas volume can ever reach zero.

## ⊃ Boyle's Law

In 1662, a British scientist named Robert Boyle discovered that gas pressure and volume are inversely proportional to each other. This relationship is known as **Boyle's law.**

---

**Boyle's Law**

The pressure $P$ of a given amount of gas is inversely proportional to its volume $V$, if the temperature and amount of gas are not changed. The relationship is

$PV = k$ or $P = k\left(\dfrac{1}{V}\right)$, where $k$ is the proportionality constant.

---

Examine the Pressure and Volume Data table on page 305. If you multiply pressure by volume, the product will always be around 1500. (It will vary slightly due to rounding and slight errors in measurement.) Boyle's law expresses this relationship as $PV = k$. In this experiment, the proportionality constant, $k$, is 1500 mL · lb/in². [For a review of this math topic, see **MATH Spotlight**: Ratio and Proportions on page A-11.]

Boyle's law can be used to solve problems involving gas pressure and gas volume. Once you have one set of pressure and volume measurements for a gas, you can determine the proportionality constant. With the proportionality constant, you can determine the pressure of the gas at any volume.

## Gas in a Syringe

Determine what the pressure of the gas inside this syringe would be if the volume were reduced to 10 mL.

### Solution

Start with Boyle's law.

$$P = k\left(\frac{1}{V}\right)$$

Substitute values for $k$ and $V$.

$$= 1500 \text{ mL} \cdot \text{lb/in}^2\left(\frac{1}{10 \text{ mL}}\right)$$

Solve.

$$= 150 \text{ lb/in}^2$$

You can check this answer: $PV = 150 \text{ lb/in}^2 \cdot 10 \text{ mL} = 1500 \text{ mL} \cdot \text{lb/in}^2$.

## LESSON SUMMARY

### How does gas volume affect gas pressure?

**KEY TERMS**

inversely proportional

Boyle's law

If you squeeze a gas into a smaller volume, it will exert more pressure on its container because a smaller space means more collisions between the molecules and the container. If you let a gas spread out into a huge volume, the gas pressure will get very low because there will be fewer collisions between the molecules and the container. Gas pressure and gas volume are inversely proportional. This means that as one variable increases, the other decreases, and vice versa. This relationship is described by Boyle's law: $P = k(1/V)$.

### Reading Questions

1. Explain how you can use a scale to measure the pressure of the gas in a syringe.

2. Describe how pressure varies as the volume of a trapped gas changes.

### Reason and Apply

3. **Lab Report** Write a lab report for the Lab: Boyle's Law. In the results section, provide two graphs of the data. Plot pressure versus volume and plot pressure versus the inverse of the volume, $1/V$. Also discuss how accurate your results are and what factors might contribute to experimental error in this lab. In the

Title

Purpose: (Explain what you were trying to find out.)

Procedure: (List the steps you followed.)

Results: (Explain what you observed during the experiment.)

Conclusions: (What can you conclude about what you were trying to find out? Provide evidence for your conclusions.)

*(continued)*

conclusions section, explain the relationship between pressure and volume. Explain how your data relate to Boyle's law.

4. You are using a bicycle pump to fill a bicycle tire with air. It gets harder and harder to push the plunger on the pump the more air is in the tire. Explain what is going on.

5. How is the relationship between gas pressure and gas volume different from the relationship between gas volume and gas temperature?

6. Imagine you blow up a balloon before going scuba diving. You put on your gear and descend to 30 ft with the balloon. The total pressure from the ocean at that depth is 2 atm. Assume temperature is constant.
   a. What happens to the volume of the balloon?
   b. What happens to the pressure of the air on the inside of the balloon?

7. This table shows experimental data for gas volume and pressure.

| Trial | Volume (mL) | Pressure applied (lb/in$^2$) |
|---|---|---|
| 1 | 60 mL | 10 lb/in$^2$ |
| 2 | 40 mL | 15 lb/in$^2$ |
| 3 | 30 mL | 20 lb/in$^2$ |
| 4 | 15 mL | 40 lb/in$^2$ |
| 5 | 10 mL | 60 lb/in$^2$ |

   a. Show that $P = k(1/V)$ for each trial. Remember to add 14.7 lb/in$^2$ to the values in the last column to account for atmospheric pressure.
   b. Show that $PV = k$ for each trial.
   c. Create a graph of $P$ versus $V$. What does the graph tell you about the relationship between gas pressure and volume?
   d. Create a graph of $P$ versus $1/V$. What does the graph tell you about the relationship between gas pressure and inverse volume?
   e. Use the graphs from parts c and d to estimate the volume when the pressure is 30 lb/in$^2$.
   f. Use the graphs from parts c and d to estimate the pressure when the volume is 50 mL.

# Gay-Lussac's Law

## THINK ABOUT IT

Gases are trapped in balloons, soda cans, pickle jars, bicycle tires, and in your lungs if you hold your breath. Some of these containers are flexible and can vary in size, while others are rigid. A gas sample can be described by pressure, volume, and temperature. The gas might be heated, cooled, compressed, taken underwater, left outside overnight, or taken up in an airplane.

**How does gas pressure change in flexible and rigid containers?**

*To answer this question, you will explore*

⮞ Gay-Lussac's Law

⮞ Gas Law Problems

## EXPLORING THE TOPIC

### ⮞ Gay-Lussac's Law

Gases are sometimes stored in rigid gas containers. For example, underwater divers wear a metal gas tank strapped to their backs. To supply air to the diver for a sufficient amount of time, a lot of air is pushed or compressed into the tank. The result is that the gas inside the tank is at a high pressure.

With rigid containers the pressure of the air inside the container changes if the temperature changes. Indeed, there is a risk of an explosion if you heat the gas inside a pressurized tank.

© Robert Yin/Corbis

*A scuba diver carries a tank of compressed gas.*

### PRESSURE VERSUS TEMPERATURE

Consider a tank containing helium. If this tank is placed in the sun and the temperature increases, the pressure of the gas inside the tank also increases.

**Helium Tank Data**

| Temperature (K) | Volume (L) | Pressure inside tank (atm) | P/T (atm/K) |
|---|---|---|---|
| 293 K | 55 L | 130 atm | 0.44 |
| 313 K | 55 L | 138 atm | 0.44 |
| 283 K | 55 L | 125 atm | 0.44 |
| 240 K | 55 L | 106 atm | 0.44 |

## COOKING CONNECTION

A pressure cooker can cook food in a fraction of the time it takes with a regular pot. Because the lid fits tightly, a higher pressure is achieved, resulting in a higher temperature.

**Pressure Versus Temperature: Helium Gas in a Tank**

### Using the Graph

Because the pressure and the Kelvin temperature are proportional, the graph is a straight line through the origin. If you used the Celsius temperature, the graph would still be a straight line, but it would not go through the origin.

**Big Idea** If you heat a sample of gas in a closed rigid container, its pressure will increase.

If you graph the helium tank data, the points lie on a straight line. The straight line through the data points and the point (0, 0) indicate that there is a proportional relationship between pressure and temperature. Joseph Louis Gay-Lussac, a French scientist, described this proportional relationship in the early 1800s.

**Gay-Lussac's Law** The pressure $P$ and the Kelvin temperature $T$ of a gas are proportional when the volume and the amount of gas do not change. The relationship is $P = kT$. The value of the proportionality constant, $k$, depends on the specific gas sample.

### Example 1

#### Propane Gas Tank

A 5.0 L propane gas tank on a camping stove contains propane gas at a pressure of 2.0 atm when the temperature is 14 °C (about 57 °F) in the morning. During the day, the tank warms up to a temperature of 34 °C (about 93 °F). What is the pressure of the propane gas inside the tank when the gas temperature is 34 °C?

#### Solution

The volume of the tank does not change. The temperature of the gas increases, causing the pressure of the gas to increase.

**Example 1**

*(continued)*

You can use the subscript 1 for the first set of conditions, and subscript 2 for the second set of conditions. These are sometimes called initial conditions and final conditions. You need to find $P_2$, the final pressure.

**Initial Conditions**

$P_1 = 2.0$ atm

$T_1 = 14\,^\circ C$

$V_1 = 5.0$ L

**Final Conditions**

$P_2 = ?$

$T_2 = 34\,^\circ C$

$V_2 = 5.0$ L

The variables that are changing are $P$ and $T$. Use Gay-Lussac's law: $P = kT$. Recall that the temperature must be in kelvins.

Convert the temperature to the Kelvin scale.

$$K = 273 + C$$
$$T_1 = 273 + 14 = 287 \text{ K (in the morning)}$$
$$T_2 = 273 + 34 = 307 \text{ K (during the day)}$$

Determine the value of $k$ using $P_1$ and $T_1$.

$$k = \frac{P_1}{T_1} = \frac{2.0 \text{ atm}}{287 \text{ K}} = 0.0070 \text{ atm/K}$$

Use $k$ to solve for $P_2$.

$$P_2 = k\,T_2$$
$$= 0.0070 \text{ atm/K} \cdot 307 \text{ K}$$
$$= 2.15 \text{ atm}$$

At $34\,^\circ C$, the pressure of the propane gas in the tank has increased to 2.15 atm.

## ➲ Gas Law Problems

Gas pressure, gas volume, and gas temperature can change depending on the conditions. But no matter what, the gases must obey the gas laws!

There are two general types of containers: rigid and flexible. A rigid container has a volume that doesn't change. A flexible container can vary in size.

Rigid containers

*A can and a glass bottle are rigid containers.*

Flexible containers

*A balloon and a syringe are flexible containers.*

The type of container has an effect on what happens to the gas inside when the conditions change. For example, Boyle's law and Charles's law apply when a gas sample is in a flexible container, which can change volume. Gay-Lussac's law applies for gas samples in rigid containers.

So far you've been introduced to three of the gas laws. These gas laws apply when any two of the variables, $P$, $V$, or $T$, are changed and everything else stays the same.

| | | | |
|---|---|---|---|
| Charles's law | $V = kT$ | $P$ and amount of gas do not change. | $k = \dfrac{V}{T}$ |
| Gay-Lussac's law | $P = kT$ | $V$ and amount of gas do not change. | $k = \dfrac{P}{T}$ |
| Boyle's law | $P = k\left(\dfrac{1}{V}\right)$ | $T$ and amount of gas do not change. | $k = PV$ |

Note that the proportionality constant, $k$, is a generic symbol and is different for each gas sample and for each gas law.

## Example 2

### Balloon with Air

Imagine that you fill a balloon with air to a volume of 1.5 L. The air is at a temperature of 20 °C. You place the balloon in a refrigerator for half an hour until the air in the balloon is at a temperature of 10 °C. What is the new volume of the balloon?

### Solution

First figure out which gas law to use. The external pressure on the balloon remains the same because the air pressure in the refrigerator is the same as the air pressure outside. Use Charles's law. You can predict that $V$ will decrease because $T$ decreases.

| **Initial Conditions** | **Final Conditions** |
|---|---|
| $P_1 = 1.0$ atm | $P_2 = 1.0$ atm |
| $T_1 = 20\,°C$ | $T_2 = 10\,°C$ |
| $V_1 = 1.5$ L | $V_2 = ?$ |

Convert temperature to the Kelvin scale.

$$K = 273 + C$$
$$T_1 = 273 + 20$$
$$= 293\text{ K}$$

$$T_2 = 273 + 10$$
$$= 283\text{ K}$$

Determine the value of $k$.

$$k = \frac{V_1}{T_1} = \frac{1.5\text{ L}}{293\text{ K}} = 0.0050\text{ L/K}$$

Use $k$ to determine $V_2$.

$$V_2 = k\,T_2$$
$$= 0.0050\text{ L/K} \cdot 283\text{ K}$$
$$= 1.4\text{ L}$$

The balloon has shrunk to a volume of 1.4 L.

## How does gas pressure change in flexible and rigid containers?

Gases can be trapped in an assortment of containers and subjected to various conditions of pressure, temperature, and volume. The gas laws allow you to calculate new values for gas temperature, gas pressure, and gas volume when two of these three variables change while the other remains the same. Charles's law and Gay-Lussac's law are proportional relationships: $V = kT$ and $P = kT$. Boyle's law is different because $P$ and $V$ are inversely proportional: $P = k(1/V)$. The proportionality constant, $k$, is the key to solving gas law problems.

**KEY TERM**

Gay-Lussac's law

## Exercises

### Reading Questions

1. Describe three ways in which you can change a gas sample.
2. Explain the difference between flexible and rigid containers.

### Reason and Apply

3. A scuba-diving tank holds 18 L of air at a pressure of 40 atm. If the temperature does not change, what volume would this same air occupy if it were allowed to expand until it reached a pressure of 1.0 atm?

4. Imagine you fill a balloon with air to a volume of 240 mL. Initially, the air temperature is 25 °C and the air pressure is 1.0 atm. You carry the balloon with you up a mountain where the air pressure is 0.75 atm and the temperature is 25 °C.
   a. When the balloon is carried up the mountain, what changes? What stays the same?
   b. The air pressure on the outside of the balloon has decreased. Can the air pressure on the inside decrease so that the pressures are equal? Why or why not?
   c. What happens to the volume occupied by the air inside the balloon? Explain your thinking.
   d. Solve for the new volume of the balloon.

5. The air inside a 180 mL glass bottle is at 1.0 atm and 25 °C when you close it. You carry the glass bottle with you up a mountain where the air pressure is 0.75 atm and the temperature is 5 °C.
   a. The air pressure on the outside of the glass bottle has decreased. What happens to the volume of air inside of the bottle? Explain your thinking.
   b. Do you expect the temperature of the air inside the bottle to cool to 5 °C when you're at the top of the mountain? Explain your thinking.
   c. What happens to the pressure inside the glass bottle? Explain your thinking.
   d. Solve for the new pressure of the gas.

*Be the Molecule*

# Molecular View of Pressure

## THINK ABOUT IT

Imagine you have three balloons: one filled with sand, a second filled with water, and a third filled with air. The air balloon has the least mass but it has the largest volume. It is also the most uniform and spherical, indicating that the gas molecules are pushing out in all directions.

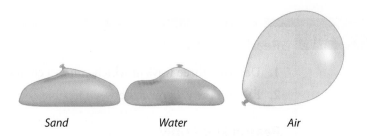

| Sand | Water | Air |

**How do molecules cause gas pressure?**

*To answer this question, you will explore*

- Kinetic View of Pressure
- Changing Gas Pressure

## EXPLORING THE TOPIC

### Kinetic View of Pressure

Inside the sand balloon, the atoms that make up sand are tightly packed together. Inside the water balloon, the forces between the water molecules keep them close to each other but not quite as close as the molecules in solids. Inside the air balloon, however, there is a lot of movement and activity. The gas molecules are moving rapidly and bouncing off each other and the walls of the balloon.

**Big Idea**   Gas pressure is caused by the collisions of molecules or atoms with all surfaces such as the walls of a container or your skin.

### COLLISIONS CAUSE PRESSURE

According to the **kinetic theory of gases,** gas molecules are in constant motion. The gas molecules in a sample move at many hundreds of miles per hour in straight line paths. Even though they are very tiny, the molecules inside the container hit each other and the walls of their container. There are so many molecules hitting the walls that the collisions add up to a measurable pressure.

Note that gas pressure is caused by molecules hitting the walls of the container only. When gas molecules collide with each other, this may change their speed or direction, but it does not affect the overall gas pressure.

BSIP/Science Source

The illustration shows a gas trapped in a container with a piston that can move up and down. The temperature and pressure of the gas inside the container are indicated on a thermometer and pressure gauge. Notice that the gas inside the container exerts enough pressure on the piston to hold it up. The piston does not fall downward, even though the gas molecules are very far apart.

## Changing Gas Pressure

Look again at the illustration. Suppose you keep the temperature the same but decrease the volume of the cylinder. What happens to the gas molecules inside the container? There is less space for the same number of molecules. The pressure increases. If you keep the volume the same but heat the gas inside the container, the gas molecules move faster. The pressure increases again. The drawings here illustrate these two changes.

---

**CHANGING VOLUME**

**Action:** The piston is pushed downward to decrease the volume. The temperature is held constant by allowing heat to exchange with the surroundings.

**Observation:** The gas pressure increases.

**Explanation:** When the gas molecules are forced into a smaller space, they hit the walls of the container more often. The gas pressure is higher. This change is associated with Boyle's law.

---

**CHANGING TEMPERATURE**

**Action:** The gas is heated. The volume is held constant by clamping the piston so that it cannot move.

**Observation:** The gas pressure increases.

**Explanation:** At higher temperatures, the gas molecules move faster. They hit the walls of the container harder and more often, which increases the gas pressure. This change is associated with Gay-Lussac's law.

## How do molecules cause gas pressure?

The collisions of gas molecules with the walls of a container cause gas pressure. If the volume decreases while the temperature stays the same, the molecules collide with the walls more frequently. This causes an increase in gas pressure. If the volume stays the same and the temperature increases, the molecules move faster. They collide with the walls more frequently and they hit the walls harder. This causes an increase in gas pressure.

**KEY TERM**

kinetic theory of gases

## Exercises

### Reading Questions

1. Use the kinetic theory of gases to explain why increasing the gas volume decreases the gas pressure.

2. Use the kinetic theory of gases to explain why decreasing the gas temperature decreases the gas pressure.

### Reason and Apply

3. Suppose that the pressure of a gas in a cylinder has increased. What might have changed to cause this? Explain your thinking.

4. Why is it dangerous to heat a gas in a sealed container?

5. Can you decrease the volume of a gas to zero? Why or why not?

6. When you fly in a commercial airplane, you often feel the change in air pressure in your ears. It feels painful. Use the kinetic theory of gases to explain what you think is happening.

7. At sea level, the pressure of trapped air inside your body is 1 atm. It is equal to the pressure of air outside your body at sea level. Imagine that you do a deep sea dive. You descend slowly to a depth where the pressure outside your body is 3.5 atm.
   a. How does the volume of your lungs compare at 3.5 atm with the volume at sea level?
   b. Why is it dangerous to hold your breath and ascend quickly?

8. As a helium balloon floats up into the sky, the pressure of the atmosphere around it decreases as the balloon's altitude increases. If the gas temperature does not change, what do you expect will happen to the volume of the balloon? Explain your thinking.

9. A gas is trapped in a cylinder with a piston as shown. The gas pressure is 1.0 atm, the volume is 500 mL, and the temperature is 300 K.
   a. Draw the cylinder after the volume has been changed to 1000 mL. The gas temperature is 300 K.
   b. Does the pressure inside the cylinder increase or decrease? Explain your thinking.
   c. Which gas law applies?

10. **PROJECT**

Look up information on steam engines in a book or on the Web. Explain how a steam engine works.

# Combined Gas Law

## THINK ABOUT IT

In some circumstances, more than two gas variables change at once. For example, if you released a balloon at sea level and it rose into the atmosphere, you would be dealing with three changing gas variables: pressure, temperature, and volume. What effect does this have on the balloon?

**What are the relationships among pressure, volume, and temperature for a sample of gas?**

*To answer this question, you will explore*

➡ Changing Pressure, Volume, and Temperature

➡ Combined Gas Law

## EXPLORING THE TOPIC

➡ ## Changing Pressure, Volume, and Temperature

### VARIATIONS IN PRESSURE AND TEMPERATURE WITH ALTITUDE

Both the temperature and the pressure of the atmosphere decrease with altitude. These two conditions (temperature and pressure) will naturally vary as a balloon rises and will affect the overall volume of the balloon. The sample of gas inside the balloon responds to the conditions outside the balloon.

The illustration indicates the approximate average pressure and temperature of the atmosphere at various altitudes. At an altitude of 40,000 ft, the air pressure is only 0.2 atm and the air temperature is around a chilly −57 °C (−70 °F).

### WEATHER BALLOONS

Weather balloons are large balloons filled with helium or hydrogen gas that carry instruments aloft to study the atmosphere at high altitudes. As the weather balloon rises, the instruments measure atmospheric conditions like the temperature, relative humidity, and pressure. This information is relayed back to a

| Altitude (ft) | Temp | Pressure |
|---|---|---|
| 45,000 | | 0.15 atm |
| 40,000 | −57 °C | 0.20 atm |
| 35,000 | | 0.25 atm |
| 30,000 | −45 °C | 0.30 atm |
| 25,000 | −35 °C | 0.40 atm |
| 20,000 | | 0.50 atm |
| 15,000 | | 0.60 atm |
| 10,000 | −5 °C | 0.70 atm |
| 5,000 | 5 °C | 0.85 atm |
| Sea level 0 | 17 °C | 1.00 atm |

## METEOROLOGY
### CONNECTION

A weather balloon has three parts: the elastic balloon, the instrument package, called a *radiosonde,* and the parachute. Weather balloons are released twice a day from sites all around the world. There are about 900 release sites worldwide; 95 of these sites are in the United States. Data from weather balloons are fed into National Weather Service supercomputers across the country. This information is used to help predict weather around the country.

weather station by a transmitter. Wind speed and wind direction can be calculated from the collected data.

To follow the changes to the gas inside a weather balloon, you must first know the starting conditions of the balloon. Imagine a clear cool morning, where the air temperature is 17 °C and the atmospheric pressure is 1 atm. A team of meteorologists inflates a weather balloon with helium to a volume of 8000 L. If released, this weather balloon will rise no matter what the outside conditions. This is because the balloon is full of helium and helium is less dense than air.

## CONFLICTING OUTCOMES

Consider the effect of changing conditions on the volume of the rising weather balloon. Charles's law indicates that the volume of the balloon will shrink with decreasing temperature. However, Boyle's law indicates that as the air pressure outside the balloon decreases, the balloon will expand in volume.

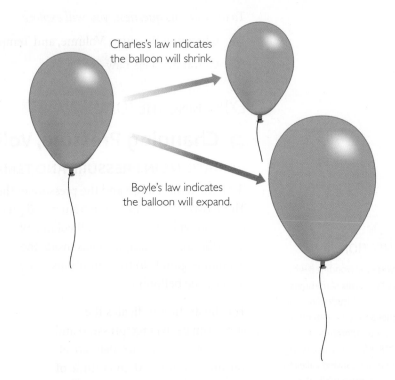

Charles's law indicates the balloon will shrink.

Boyle's law indicates the balloon will expand.

There are two conflicting outcomes. Will the balloon expand or shrink?

## Combined Gas Law

For a sample of gas, the relationship among gas pressure, temperature, and volume can be expressed in a mathematical equation called the **combined gas law.**

**Combined Gas Law**
If you know the temperature, $T$, pressure, $P$, and volume, $V$, of a gas, you can determine $k$. Then you can use $k$ to determine the volume of the gas for other pressures and temperatures.

$$k = \frac{PV}{T}$$

## Weather Balloon at 25,000 Ft

A weather balloon is filled at sea level with 8000 L of helium. The pressure at sea level is 1 atm and the temperature is 17 °C. Calculate the volume of the balloon at 25,000 ft when the atmospheric pressure has decreased to 0.4 atm.

### Solution

Because pressure, volume, and temperature are all changing, you need to use the combined gas law. The first step in using the combined gas law is to make note of all the values for $P$, $V$, and $T$. Notice that we already know five out of the six values. Remember to convert °C to K. Then it is a matter of solving for $V_2$.

| $P_1 = 1.0$ atm | $P_2 = 0.4$ atm |
|---|---|
| $V_1 = 8000$ L | $V_2 = ?$ |
| $T_1 = 290$ K | $T_2 = 238$ K |

Use the combined gas law.

$$k = \frac{P_1 V_1}{T_1} = \frac{P_2 V_2}{T_2}$$

Determine the proportionality constant, $k$

$$k = \frac{P_1 V_1}{T_1}$$

$$= \frac{1.0 \text{ atm} \cdot 8000 \text{ L}}{290 \text{ K}}$$

$$= 27.6 \frac{\text{atm} \cdot \text{L}}{\text{K}}$$

Use $k$ to solve for $V_2$.

$$V_2 = \frac{k T_2}{P_2}$$

$$V_2 = \frac{\left(27.6 \frac{\text{atm} \cdot \text{L}}{\text{K}}\right)(238 \text{ K})}{0.4 \text{ atm}}$$

$$V_2 = 16{,}422 \text{ L}$$

The balloon has expanded to about 16,000 L at this altitude.

[For a review of this math topic, see **MATH Spotlight**: Accuracy, Precision, and Significant Digits on page A-1.]

### PRESSURE VERSUS TEMPERATURE

The example shows that decreasing air pressure has a greater effect on a weather balloon than decreasing temperature does. The volume of the balloon continues to expand as the balloon goes up in altitude, in spite of the decreasing temperature. The balloon eventually ruptures. There is a parachute on every weather balloon so that the weather instruments can be carried safely to the ground.

### What are the relationships among pressure, volume, and temperature for a sample of gas?

If there is a situation involving a gas in which volume, temperature, and pressure are all changing, then you can use the combined gas law to determine the effects of changing two variables on the third variable. The value of the proportionality constant, $k$, is equal to $PV/T$. Therefore, $P_1V_1/T_1 = P_2V_2/T_2$. Remember that the combined gas law works only in situations where the amount of gas involved remains the same.

**KEY TERM**

combined gas law

## Exercises

### Reading Questions

1. What is the combined gas law? When do you apply it?

2. Does the temperature of the atmosphere have any effect on the volume of a weather balloon as it rises? Explain.

### Reason and Apply

3. A gas collected in a flexible container at a pressure of 0.97 atm has a volume of 0.500 L. The pressure is changed to 1.0 atm. The amount of gas and the temperature of the gas do not change.
   a. Will the volume of the gas increase or decrease? Explain.
   b. Which equation should you use to calculate the new volume of the gas?
   c. Calculate the volume of the gas at a pressure of 1.0 atm.

4. The gas pressure in an automobile tire was 1.0 atm at 21 °C. After an hour of driving, the tire heated up to 55 °C. Assume that the tire remained at constant volume and no gas escaped from the tire.
   a. Did the gas pressure inside the tire increase or decrease? Explain.
   b. Which equation should you use to calculate the new gas pressure of the tire?
   c. Calculate the gas pressure of the tire at 55 °C.

5. The helium inside a balloon has a volume of 1.5 L, a pressure of 1.0 atm, and a temperature of 25 °C. The balloon floats up into the sky where the air pressure is 0.95 atm and the temperature is 20 °C.
   a. Will the volume of the balloon increase or decrease? Explain.
   b. Which equation should you use to calculate the new volume of the balloon?
   c. Calculate the volume of the balloon at 20 °C and 0.95 atm.

6. Someone leaves a steel tank of nitrogen gas in the sun. The gas pressure inside the tank was 150 atm at 27 °C to begin with. After several hours, the internal temperature rises to 55 °C. What is the gas pressure in the tank now?

7. A car airbag is an example of a trapped gas. The airbag is designed to inflate to 65 L if you have an accident in a location where the temperature is 25 °C and the air pressure is 1.0 atm.
   a. Determine the size of the airbag if you have an accident in the mountains where the temperature is −5 °C and the air pressure is 0.8 atm.
   b. Suppose that you have an accident in a location in which the airbag inflates to 60 L. What has to be true about the temperature and air pressure in this location?

Ken Karp Photography

**SAFETY Instructions**

Safety goggles must be worn at all times.

# High and Low Air Pressure

### Purpose

To find the connection between air pressure and the weather forecast.

### Materials

- 2 L plastic bottle, or vacuum chamber and vacuum pump
- warm tap water
- long safety matches

### Procedure

1. Put a small amount (about 15 mL) of warm water into the plastic bottle.

2. Light a match. Blow it out and then hold it inside the bottle to collect some smoke. Quickly remove the match and put the cap tightly on the bottle. Shake the chamber to add moisture to the air inside.

3. Squeeze the bottle and release. Pump a few times while observing the air inside the bottle.

4. Repeat the experiment, but this time use 10 mL of cold water. Next, repeat the experiment with a dry bottle. Do not add water. Just create smoke, close the bottle, and squeeze and release.

### Observations

1. What did you observe inside the bottle when you squeezed and released the bottle?

2. What happens to the pressure, volume, and temperature of the air inside the bottle when you squeeze it? When you release it?

3. What gas law best explains how the properties of the gas inside the bottle change during this experiment? Explain your choice.

4. How does this gas law explain what's going on in the bottle?

5. What did you observe when you used a dry bottle?

*Cloud in a Bottle*

# High and Low Air Pressure

### THINK ABOUT IT

Weather maps typically contain information about high air pressure and low air pressure. You often hear meteorologists refer to high- and low-pressure systems. But how does this information help you figure out whether it will be rainy or sunny?

**How are areas of high and low air pressure related to the weather?**

*To answer this question, you will explore*

- High and Low Air Pressure
- Cloud Formation

### EXPLORING THE TOPIC

## High and Low Air Pressure

Meteorologists measure air pressure at different locations on Earth's surface. These readings of air pressure are summarized on an air-pressure map like this one.

| H | High-pressure system | ▼▼▼ Cold front | 938 Isobar |
| L | Low-pressure system | ▲▲▲ Warm front | ))))))) Jet stream |

The lines labeled with numbers on the map are called *isobars.* They mark areas with the same air pressure. The numbers correspond to air pressure measurements in millibars. One atmosphere is equal to 1013.25 millibars. Notice that each new line represents an air pressure difference of 4 millibars.

Low-pressure areas are associated with lower numbers, high-pressure areas with higher numbers. However, high and low air pressure are measured relative to the surrounding air. In other words, a high-pressure air system is not associated with a certain number but simply shows an area that stands out as having higher air pressure than the areas around it. If you examine the map, you can also see that the isobars encircle areas with high- and low-pressure systems.

The map also shows the jet stream as a series of blue arrows traveling from west to east across the continent. Warm and cold fronts are also included. Take a moment to look for patterns.

- The jet stream weaves between the Hs and Ls instead of going through the middle of them.
- The jet stream appears to follow the general curvature of the isobar lines. For this particular map, what this means is that the jet stream will guide western storms up into Canada.
- The location of the weather fronts is also associated with air pressure. Notice that the weather fronts appear to be centered on the low-pressure areas.
- There are no weather fronts near the high-pressure areas, so you wouldn't expect to see any storms in these areas.

### WEATHER ASSOCIATED WITH HIGH AND LOW PRESSURE

In general, low-pressure areas are associated with fronts, clouds, and precipitation. This is because when warm and cold air masses meet, the warm air rises, leaving behind an area of low pressure. Areas of lower air pressure coincide with areas where fronts are located.

The centers of all storms are areas of low air pressure. When forecasters say a low-pressure area is moving toward your region, cloudy weather and precipitation usually are on the way.

In contrast, high-pressure areas are places where skies are clear and pleasant. When forecasters say that a high-pressure system has moved into your area, you can expect clear skies and sunny days. High-pressure centers mean that denser air must be sinking, which inhibits precipitation and cloud formation.

# Cloud Formation

When you look up at the sky and see big, puffy, white clouds, it's hard to believe that they are just a bunch of condensed water droplets suspended in the air. Because they are a mixture of water droplets and air, our eyes see them as white or sometimes as gray. But what causes clouds to form in the first place?

Because there is no volume or container associated with Earth's atmosphere, it makes sense to explain weather by focusing on pressure and temperature changes in air masses.

Warm air rises because it is less dense than cold air. As it rises, its temperature and pressure decrease. As the water vapor cools, it condenses into droplets, forming clouds.

*A cloud is made up of millions of suspended water droplets that condense from atmospheric water vapor.*

Enough water vapor must be present in the air for clouds to form. This is why more clouds form over rainforests than over deserts. On warm days, large puffy clouds are observed at lower altitudes because there are many water molecules in the air. On cold days, if there are clouds at all, they tend to be wispy and high in the sky because there are so few water molecules in the air.

## EARTH SCIENCE CONNECTION

North of the equator, winds around high-pressure areas move in a clockwise direction. The winds around low-pressure areas move in a counterclockwise direction. This is one reason why the jet stream loops up and over highs and down and under lows as it travels across the United States. South of the equator, winds around high- and low-pressure systems move in the opposite directions.

## LESSON SUMMARY

### How are areas of high and low air pressure related to the weather?

Air pressure varies across the surface of Earth. On air pressure maps, areas of high and low air pressure are indicated with Hs and Ls. Areas of high pressure indicate clear skies and sunny days. Areas of low pressure are associated with storms. For clouds to form, there must be moisture in the air. If this moisture is in a warm air mass, it travels upward, cools off, and undergoes a phase change, forming clouds composed of tiny water droplets. Clouds form under conditions of low temperature and low pressure.

## Exercises

### Reading Questions

1. What weather is associated with high air pressure? Low air pressure?

2. Describe how a cloud forms. Start with water in a lake.

### Reason and Apply

3. One atmosphere of pressure, 1 atm, is equivalent to 1013.25 millibars. Convert the pressure from millibars to atmospheres for the two highs and the two lows on the air pressure map on page 322. What do you notice about these values?

4. **PROJECT** Find out about different types of clouds and the conditions under which they form. Try to explain how the type of cloud relates to the conditions.

www

# CHAPTER **11**

## Pressing Matter

# SUMMARY

### KEY TERMS

**sublimation**

**evaporation**

**pressure**

**atmospheric pressure**

**atmosphere (atm)**

**inversely proportional**

**Boyle's law**

**Gay-Lussac's law**

**kinetic theory of gases**

**combined gas law**

# Weather Update

Air pressure at Earth's surface is constantly changing because air molecules are moving from place to place. Areas of high pressure are associated with clear skies and warm temperatures. The gas molecules in regions of high pressure are moving rapidly and expand into areas of lower pressure. Areas of low pressure are associated with clouds and storms. The gas molecules in regions of low pressure are moving more slowly and water molecules are condensing. Air pressure keeps changing as water evaporates and condenses, which changes the number of air molecules.

Studying gas behavior in small containers helps us to explain the behavior of gases in the atmosphere. Three gas laws relate gas pressure to temperature and volume—Charles's law, Gay-Lussac's law, and Boyle's law. The combined gas law relates pressure, volume, and temperature.

## REVIEW EXERCISES

1.  Use the kinetic theory of gases to explain how temperature affects the pressure of a gas.

2.  Explain why a weather balloon pops when it reaches a sufficiently high altitude.

3.  A cylinder with a movable piston contains 26.5 L of air at a pressure of 1.5 atm. If the temperature is kept the same but the volume of the container is expanded to 50.0 L, what is the new pressure of the gas?

4.  A gas sample in a rigid container has a pressure of 3.0 atm at 350 K. If the temperature is raised to 450 K, what will the new gas pressure be?

5.  A sample of gas in a flexible container has a volume of 7.5 L, a pressure of 2.5 atm, and a temperature of 293 K. What will the new pressure be if the volume expands to 10.0 L and the temperature is raised to 315 K?

## High and Low Pressure

PROJECT

In a newspaper or on the Web, find a weather map of the United States.

1.  Find the areas of high and low pressure.

2.  Explain where the air is moving up and where the air is moving down.

3.  Where do you expect to find stormy weather? Explain your thinking.

# CHAPTER 12

# Concentrating Matter

## In this chapter, you will study

- variations in the atmosphere with changing altitude

- how to determine the numbers of gas molecules in a sample

- the relationships between pressure, temperature, volume, and number of molecules in a gas

- humidity

- conditions that lead to hurricane formation

*Lightning is an extreme weather phenomenon that typically occurs during thunderstorms.*

**G**as molecules are too small and numerous to count. However, the mathematical relationships between gas temperature, pressure, volume, and number of molecules allow scientists to determine the exact number of molecules in a gas sample. Knowing the number of water molecules in a particular volume of air helps scientists to predict when precipitation will occur. Air pressure and air temperature greatly affect the amount of water vapor in the air. When moist air cools, the water vapor in the air changes from a gas to a liquid. When extreme conditions of temperature and pressure occur, hurricanes can form.

# Pressure and Number Density

## THINK ABOUT IT

Airplanes fly at about 35,000 ft (~6.6 mi) above Earth. At this altitude, the air pressure is less than 0.25 atm and the air temperature is less than $-45\ °C$. It is very cold, and the air is very thin up there! You might wonder how many oxygen molecules are available at this air pressure to breathe.

**How is the number of gas molecules in a sample related to pressure?**

*To begin to answer this question, you will explore*

⟡ Pressure and Number Density

⟡ Measuring Air Pressure

## EXPLORING THE TOPIC

## ⟡ Pressure and Number Density

The illustration provides information on the *troposphere*, or lower atmosphere, where clouds form and weather takes place. The containers with spheres show how the density of air molecules decreases with altitude.

Chemists use the variable $n$ to refer to the number of molecules, atoms, or particles they are considering. The number of air molecules per unit of volume is referred to as the **number density,** $n/V$.

The number density of air molecules is relatively low at high altitudes. This is why it gets more difficult to breathe at higher altitudes. There are fewer air molecules per volume of air.

If there are fewer gas molecules in a certain volume, then the air pressure will be lower. Air pressure is proportional to number density. This relationship can be expressed by the equation

$$P = k\left(\frac{n}{V}\right)$$

The decrease in air pressure with increasing altitude is due mainly to a decrease in the number density of gas molecules.

**Big Idea** | **The more gas molecules you have in a space, the greater the gas pressure will be.**

**Example**

### Bicycle Tire

Use the kinetic theory of gases to explain why the air pressure increases as you pump up a bicycle tire.

### Solution

As you pump up a tire, the number of gas molecules per unit of volume becomes greater and greater. This increases the number of collisions between molecules and the walls of the tire. The pressure inside the tire increases until the tire is almost rigid.

## → Measuring Air Pressure

There are a number of different ways to measure air pressure. Meteorologists and chemists commonly use either a manometer or a barometer. Both instruments measure the height of a liquid in a tube.

### MANOMETER

A *manometer* consists of a liquid in a U-tube as shown in the illustration. If the U-tube is partially filled with water, the water levels on both sides of the U will be at equal heights. This is because the air pressure is the same on both sides.

Imagine that you seal off one side of the U-tube with a stopper so that the number of molecules in the sealed-off area cannot change. The heights of the water will be the same, at least at first. However, if the outside air pressure increases, it will push on the water in the open end causing the water to move up on the side with the stopper, compressing the air on that side. The air pressure of the trapped gas increases

## HISTORY
## CONNECTION

Evangelista Torricelli invented the barometer. Pressure is still sometimes reported in torrs. A *torr* is equivalent to a millimeter of mercury, or mm Hg. 1 atm = 760 mm Hg = 760 torr

because $n/V$ increases. If the outside air pressure decreases, the water in the open end will rise. The air pressure of the trapped gas also decreases because $n/V$ is smaller.

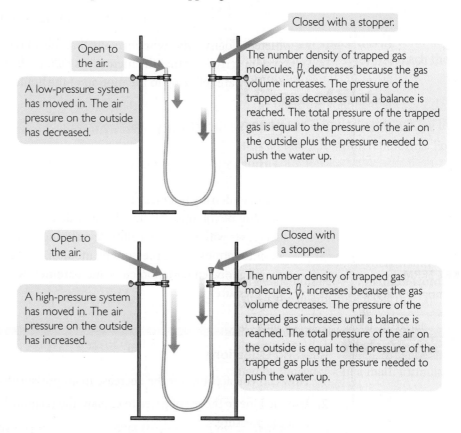

Open to the air.

A low-pressure system has moved in. The air pressure on the outside has decreased.

Closed with a stopper.

The number density of trapped gas molecules, $\frac{n}{V}$, decreases because the gas volume increases. The pressure of the trapped gas decreases until a balance is reached. The total pressure of the trapped gas is equal to the pressure of the air on the outside plus the pressure needed to push the water up.

Open to the air.

A high-pressure system has moved in. The air pressure on the outside has increased.

Closed with a stopper.

The number density of trapped gas molecules, $\frac{n}{V}$, increases because the gas volume decreases. The pressure of the trapped gas increases until a balance is reached. The total pressure of the air on the outside is equal to the pressure of the trapped gas plus the pressure needed to push the water up.

In a U-tube manometer, the water level on the closed side *rises* when the air pressure in the atmosphere increases. The water level on the closed side *falls* when the air pressure in the atmosphere decreases.

## BAROMETER

A *barometer* consists of a tube filled with liquid, which is inverted in a dish of the same liquid. The surface of the pool of liquid is exposed to the atmosphere. There is almost no gas pressure inside the tube.

The barometer shown in the illustration measures the difference in pressure on two surfaces of a liquid. The pressure on the liquid inside the tube is close to zero, so it is the mass of liquid in the tube that balances the air pressure on the outside.

**Important to Know** The liquid does not spill out of the inverted tube because the force of the air pushing on the surface of the liquid is greater than the weight of the liquid pushing down in the inverted tube.

Vacuum

The height of mercury in the tube is measured.

When atmospheric pressure *increases*, the level of mercury in the tube *rises*. When atmospheric pressure decreases, the level of mercury *falls*.

Atmospheric pressure

Pool of mercury

Mercury is often the liquid of choice for barometers due to its high density. Air pressure is sometimes expressed in inches of mercury or millimeters of mercury.

$$1 \text{ atm} = 29.9 \text{ in Hg} = 760 \text{ mmHg}$$

Meteorologists will sometimes say that the mercury is *rising* when a high-pressure system arrives, or that the mercury is *falling* when a low-pressure system moves in. They are referring to the motion of the liquid mercury in the inverted tube.

## LESSON SUMMARY

### How is the number of gas molecules in a sample related to pressure?

**KEY TERM**

number density

Number density is a measure of the number of molecules, atoms, or particles per unit of volume, $n/V$. The number density of air molecules in the atmosphere decreases with increasing altitude resulting in lower gas pressure. It is possible to monitor atmospheric pressure by monitoring the level of a liquid in a tube that is sealed at one end. Manometers and barometers are two instruments that measure gas pressure.

## Exercises

**Reading Questions**

1. How can you account for the decrease in air pressure with increasing altitude?
2. Use the kinetic theory of gases to explain the relationship between number density and gas pressure.
3. Explain the difference between a manometer and a barometer.

**Reason and Apply**

4. What effect does changing the number of gas particles in a container have on the gas pressure in the container? Use the kinetic theory of gases in your explanation.
5. Name three ways that you can increase the pressure of a tire.
6. The boxes in the illustrations show tiny samples of air. Assume they are at the same temperature.

A          B          C

   a. List the samples in order of increasing gas pressure. Explain your reasoning.
   b. Sketch a volume of air that has a pressure in between the pressures in boxes A and B.
7. Explain how the mercury level in a barometer changes when a high-pressure system moves into a region.
8. Explain why the liquid in the barometer does not spill out.

# The Mole and Avogadro's Law

## THINK ABOUT IT

Suppose you have two balloons, one filled with helium and the other with carbon dioxide. The pressure, temperature, and volume of the two gases are identical. However, the masses of the balloons are different. While the helium balloon floats, the carbon dioxide balloon sinks. How can you figure out the number of gas particles in each balloon?

**How do chemists keep track of the number of gas particles?**

*To begin to answer this question, you will explore*

➡ Counting Gas Particles

➡ Avogadro's Law

## EXPLORING THE TOPIC

## ➡ Counting Gas Particles

If you take two air samples of identical volume in the same room, you might predict that both samples will contain the same number of air molecules. A smaller air sample from this room will contain fewer air molecules than a larger air sample.

Because the air molecules are spread uniformly over the space the gas occupies, the number of air molecules, *n*, is proportional to the volume, *V*, the gas occupies.

### A MOLE OF GAS PARTICLES

Gas particles are far too numerous to count individually. So chemists have defined a counting unit for small particles called a **mole.** The abbreviation for mole is mol. A mole represents a very large number of gas particles. A mole is equal to 602,000,000,000,000,000,000,000 particles. This is 602 sextillion! This number is also called **Avogadro's number,** after Amedeo Avogadro, a 19th century Italian scientist.

The number of gas particles in a sample is referred to in terms of moles of gas. Keep in mind that when referring to 1 mole, 2 moles, or half a mole of gas particles, these quantities all refer to very large numbers. For example, 2 moles of gas particles is 1,204,000,000,000,000,000,000,000 particles (1,204 sextillion). Half a mole of gas particles is 301,000,000,000,000,000,000,000 (301 sextillion).

### STANDARD TEMPERATURE AND PRESSURE, STP

Chemists have figured out how to calculate the number of gas particles in a sample of gas if they know the temperature, volume, and pressure of the sample. However, when comparing gases, it is convenient to define just one set of conditions. Chemists have chosen a gas pressure of 1 atm and a gas temperature of 273 K as the conditions for comparing gases. These conditions are called **standard temperature and pressure,** or **STP.** At STP, 1 mol of gas particles occupies a volume of 22.4 L.

**Example 1**

### Two Moles of Air Molecules

Suppose you want to collect 2.0 mol of air molecules at STP. What size box do you need?

### Solution

Because 1 mol of air molecules has a volume of 22.4 L at STP, 2 mol has a volume of 44.8 L. So, you need a box with a volume of 44.8 L.

**Big Idea** The mole is a counting unit. One mole of gas particles at standard temperature and pressure occupies a volume of 22.4 liters.

## ⊃ Avogadro's Law

Several cylinders each contain different gas samples at STP. The data for each cylinder is shown in the table on the next page. Notice that the volumes of the samples vary.

A few molecules (or atoms, in the case of helium and neon) are drawn in the table to help you visualize the gas. Obviously, only a few gas particles are drawn because it is impossible to draw 602 sextillion. However, the numbers of molecules and atoms are in correct proportion to each other so that you can make comparisons. Take a moment to look for patterns.

Notice that the mass of the gas sample depends on the identity of the gas. A hydrogen molecule has a mass of 2.0 amu, while each neon atom has a mass of 20 amu. So 0.25 mol of neon gas has ten times the mass of 0.25 mol of hydrogen gas.

**Various Gases at STP**

| Sample | 1 | 2 | 3 | 4 | 5 |
|---|---|---|---|---|---|
| Gas | $H_2$ | Ne | $CO_2$ | $O_2$ | He |
| | | | | | |
| *n* (moles) | 0.25 | 0.25 | 0.50 | 0.50 | 1.0 |
| Volume | 5.6 L | 5.6 L | 11.2 L | 11.2 L | 22.4 L |
| Mass | 0.50 g | 5.0 g | 22.0 g | 16.0 g | 4.0 g |

The volume of each cylinder is determined by the number of gas particles. For example, a 22.4-liter sample at STP always has twice as many gas particles as an 11.2-liter sample, regardless of the identity of the gas.

If two gases have the same pressure, volume, and temperature, then they have the same number of gas particles *independent of the identity of the gases*. This generalization is known as **Avogadro's law.** Avogadro's law is extremely useful. Amedeo Avogadro first proposed this hypothesis in 1811. The reverse is also true: If two gas samples with the same pressure and temperature have the same number of particles, then they occupy the same volume.

**Big Idea**  If two gas samples have the same pressure, temperature, and volume, they contain the same number of particles, even if you are comparing two different types of gases.

---

**Example 2**

### Helium and Carbon Dioxide Balloons

Suppose you have two balloons, one filled with helium and the other with carbon dioxide. The pressure, temperature, and volume of the two gases are identical.

   **a.** Why is the mass of the carbon dioxide balloon greater?

   **b.** What do you know about the number of atoms in the balloons?

*Solution*

   **a.** Even though the number of $CO_2$ molecules is identical to the number of He atoms, the mass of the carbon dioxide balloon is greater because individual molecules of carbon dioxide, $CO_2$, have a greater mass than atoms of helium, He.

*(continued)*

**b.** There are more atoms in the $CO_2$ balloon. The number of molecules of $CO_2$ is equal to the number of atoms of He according to Avogadro's law. Because each molecule of $CO_2$ consists of three atoms, there are three times as many atoms in the $CO_2$ balloon than in the He balloon.

## LESSON SUMMARY

### How do chemists keep track of the number of gas particles?

**KEY TERMS**

mole

Avogadro's number

standard temperature and pressure (STP)

Avogadro's law

Gas particles move randomly and are distributed uniformly in the space they occupy. At standard temperature and pressure, STP, the gas pressure is 1.0 atm and the temperature is 273 K. At STP in a volume of 22.4 L, there is 1 mol of gas particles. One mole represents 602,000,000,000,000,000,000,000 particles. This is 602 sextillion, also called Avogadro's number. Avogadro's law states that equal volumes of any gas at the same pressure and temperature have the same number of gas particles, independent of the identity of the gas.

### Reading Questions

1. Explain why chemists invented the unit called a mole.
2. Explain Avogadro's law.

### Reason and Apply

3. Suppose you have 22.4 L of the following gases at STP: neon, Ne, argon, Ar, and xenon, Xe.
   **a.** How many atoms are there in each gas sample?
   **b.** What is the number density of atoms, $n/V$, for each sample?
   **c.** Which sample has the largest mass? Explain your reasoning.
   **d.** Which sample has the largest mass density, $m/V$?

4. Suppose you have 22.4 L of the following gases at STP: hydrogen, $H_2$, nitrogen, $N_2$, and carbon dioxide, $CO_2$.
   **a.** How many molecules are there in each gas sample?
   **b.** What is the number density of molecules, $n/V$, for each sample?
   **c.** Which sample has the largest mass? Explain your reasoning.
   **d.** Which sample has the largest number of atoms? Explain your reasoning.
   **e.** Which sample has the largest mass density, $m/V$?

5. Which has more atoms: 8.0 g of helium, He, or 40.0 g of argon, Ar? Explain your reasoning.

6. Which has more particles: a balloon filled with 10 L of oxygen, $O_2$, gas, or a balloon filled with 15 L of hydrogen, $H_2$, gas? Explain your reasoning. Assume STP.

7. At 25 °C, which balloon has the greater volume: an oxygen, $O_2$, balloon at 1.2 atm with a mass of 16.0 g, or a helium, He, balloon at 1.2 atm with a mass of 2.0 g? Explain your reasoning.

# Ideal Gas Law

## THINK ABOUT IT

Climbing Mount Everest is extremely difficult without bringing along a tank of oxygen to help you breathe. As you climb higher and higher, the air pressure gets lower and lower. At 29,000 ft, on top of Mount Everest, the air pressure is only 0.33 atm. You might wonder how many oxygen molecules are available at this pressure.

**How can you calculate the number of moles of a gas if you know *P*, *V*, and *T*?**

*To begin to answer this question, you will explore*

⮕ Pressure, Volume, Temperature, and Number

⮕ The Ideal Gas Law

## EXPLORING THE TOPIC

### ⮕ Pressure, Volume, Temperature, and Number

Four variables describe gas samples: pressure, volume, temperature, and number of moles of atoms or molecules. Any two variables can be related to each other mathematically if the other two are kept the same. You might begin to wonder if you can relate all four variables in a single equation. The combined gas law relates three of the variables: pressure, volume, and temperature.

So far you've learned that at STP, there is 1.0 mol of gas molecules in a volume of 22.4 liters for *any* gas sample.

### INCREASING THE PRESSURE

Suppose you want to adjust this sample of gas so that it has a pressure of 2.0 atm. In what ways can you accomplish this?

You can change the pressure to 2.0 atm in different ways.

$P = 1.0$ atm
$V = 22.4$ L
$T = 273$ K
$n = 1.0$ mol

*Original gas sample*

$P = 2.0$ atm
$V = 11.2$ L
$T = 273$ K
$n = 1.0$ mol

***Decrease V.*** *Squeeze the gas at 273 K into a volume of 11.2 L.*

$P = 2.0$ atm
$V = 22.4$ L
$T = 546$ K
$n = 1.0$ mol

***Increase T without changing V.*** *Heat the gas in 22.4 L to a temperature of 546 K.*

$P = 2.0$ atm
$V = 22.4$ L
$T = 273$ K
$n = 2.0$ mol

***Increase n without changing V.*** *Add 1.0 mol of gas molecules to the container at 273 K at a volume of 22.4 L.*

So while the pressure is the same in each case, you can see how the other variables have changed.

| Target pressure | Volume | Temperature | Moles |
|---|---|---|---|
| 2.0 atm | 11.2 L | 273 K | 1.0 mol |
| 2.0 atm | 22.4 L | 546 K | 1.0 mol |
| 2.0 atm | 22.4 L | 273 K | 2.0 mol |

## INCREASING THE VOLUME

Suppose you want to adjust the original sample of gas so that it has a volume of 44.8 L. How can you accomplish this?

You can produce a sample of gas with a volume of 44.8 L in several ways.

$P = 1.0$ atm
$V = 22.4$ L
$T = 273$ K
$n = 1.0$ mol

Original gas sample.

$P = 0.5$ atm
$V = 44.8$ L
$T = 273$ K
$n = 1.0$ mol

**Increase V.** *Expand 1.0 mol of gas molecules at 273 K until the pressure is 0.5 atm.*

$P = 1.0$ atm
$V = 44.8$ L
$T = 546$ K
$n = 1.0$ mol

**Increase T.** *Heat 1.0 mol of gas molecules at 1.0 atm to a temperature of 546 K.*

$P = 1.0$ atm
$V = 44.8$ L
$T = 273$ K
$n = 2.0$ mol

**Increase n.** *Add 1.0 mol of gas molecules to the container at STP, to arrive at 2.0 total moles.*

| Pressure | Target volume | Temperature | Moles |
|---|---|---|---|
| 0.5 atm | 44.8 L | 273 K | 1.0 mol |
| 1.0 atm | 44.8 L | 546 K | 1.0 mol |
| 1.0 atm | 44.8 L | 273 K | 2.0 mol |

It is apparent that these variables are mathematically related. In fact, if you combine the equation for the combined gas law with the relationship described by Avogadro's law, the result is a mathematical relationship known as the ideal gas law.

## ⮑ The Ideal Gas Law

The **ideal gas law** relates the pressure, volume, temperature, and number of moles for a gas sample.

---

Ideal Gas Law

The product of the pressure, $P$, and volume, $V$, of a gas is proportional to the product of the number of moles, $n$, and the temperature, $T$.

$$PV = nRT$$

The two products are related by the constant, $R$.

$$R = 0.082 \frac{\text{L} \cdot \text{atm}}{\text{mol} \cdot \text{K}}$$

---

The **universal gas constant,** $R$, is the same for all gases.

Because $R$ has units of L · atm/mol · K, you can use this value of $R$ only if volume is measured in *liters,* pressure in *atmospheres,* temperature in *kelvins,* and number of gas molecules in *moles.* When these units are used, $R$ is always equal to 0.082 L · atm/mol · K.

If you know three of the four variables for any gas sample, you can use the ideal gas law to calculate the fourth variable. This is most useful in determining the moles of gas molecules in a gas sample because the pressure, volume, and temperature can be easily measured.

**Example**

### Moles of Air on Mount Everest

How many moles of air are in a 0.5 L breath on top of Mount Everest? The pressure is 0.33 atm and the temperature is 254 K.

*Solution*

The ideal gas law relates all these quantities. Insert the values for $P$, $V$, $T$, and $R$ into the equation and solve for $n$.

$$PV = nRT$$

$$(0.33 \text{ atm})(0.5 \text{ L}) = n\left(0.082 \frac{\text{L} \cdot \text{atm}}{\text{mol} \cdot \text{K}}\right)(254 \text{ K})$$

$$n = \frac{(0.33 \text{ atm})(0.5 \text{ L})}{\left(0.082 \dfrac{\text{L} \cdot \text{atm}}{\text{mol} \cdot \text{K}}\right)(254 \text{ K})}$$

$$n = 0.008 \text{ mol of air}$$

So there is only 0.008 mol of air molecules in a 0.5 L breath of air atop Mount Everest.

### How can you calculate the number of moles of a gas if you know *P*, *V*, and *T*?

Any sample of gas can be described by four variables: pressure, volume, temperature, and moles. The ideal gas law relates these four variables to each other mathematically: $PV = nRT$, where $R$ is the universal gas constant. The value of $R$ is 0.082 L · atm/mol · K for all gas samples if units of atmospheres, liters, moles, and kelvins are used. If you know three of the four variables for any gas sample, you can use the ideal gas law to calculate the fourth variable.

**KEY TERMS**

ideal gas law

universal gas constant, $R$

## Exercises

### Reading Questions

1. What is the ideal gas law?
2. Describe when you might want to use the ideal gas law.

### Reason and Apply

3. How many moles of hydrogen, $H_2$, gas are contained in a volume of 2 L at 280 K and 1.5 atm?

4. What volume would 1.5 mol of nitrogen, $N_2$, gas occupy at standard temperature and pressure?

5. Find the pressure of 3.40 mol of gas if the gas temperature is 40.0 °C and the gas volume is 22.4 L.

6. How many moles of helium, He, gas are contained in a 10,000 L weather balloon at 1 atm and 10 °C?

7. Suppose you have 1.0 mol of gas molecules in 22.4 L at STP. Describe three ways you can get a gas pressure of 0.50 atm.

8. Will the pressure of helium, He, gas be the same as the pressure of oxygen, $O_2$, if you have 1 mol of each gas, each at a volume of 22.4 L and each at 273 K? Explain your thinking.

# Humidity, Condensation

## THINK ABOUT IT

Exercising can make you work up a sweat. Sweating helps to regulate your body temperature. When a breeze blows across your sweaty forehead, you may notice that your skin feels cooler. However, on a humid day, with a lot of moisture in the air, it seems as if no amount of sweating helps you to cool down.

### What is humidity and how is it measured?

*To answer this question, you will explore*

➔ Evaporation and Condensation

➔ Humidity

➔ Relative Humidity

## EXPLORING THE TOPIC

## ➔ Evaporation and Condensation

After a summer rainstorm, puddles of water on the ground often disappear quickly. The rainwater is evaporating. Recall that evaporation is a phase change from a liquid to a gas.

Evaporation is the reverse process of condensation, when water vapor becomes a liquid. These two processes, evaporation and condensation, are both occurring wherever water is present. In fact, there is a competition between the two processes that can result in net evaporation or net condensation.

The rates of both evaporation and condensation depend mainly on temperature and the amount of water vapor already in the air.

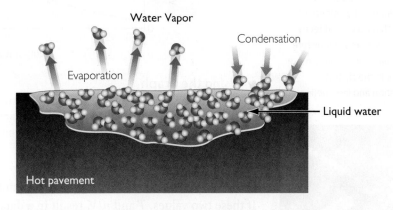

Cloud formation, rain and snowfall, fog, frost, and the appearance of dew are all events associated with the condensation of water vapor out of the atmosphere.

### BIOLOGY CONNECTION

Only mammals sweat. Humans and other primates have sweat glands all over their bodies. Dogs, cats, and mice have sweat glands only on their feet. Rabbits have sweat glands around their mouths. Dogs pant when they are hot, and the evaporation cools them down. Cats lick their paws and chests. The water evaporates from their fur, cooling them.

Juniors/SuperStock

**Important to Know** The water in rain puddles does not have to boil to go into the gas phase. Evaporation takes place on the surface of a liquid all the time. Try leaving a glass of water on a table. The water will gradually disappear, leaving the glass empty.

You might have noticed a tendency for fog to form on cool nights after a warm, humid day. This is because a lot of water has evaporated during the day. When the temperature decreases, the amount of water vapor in the air exceeds the maximum allowable for the cooler temperature and fog forms.

# Humidity

Water vapor is present in the air around us all the time. There is more or less water vapor in the air depending on the weather conditions. Warmer air masses can have more water vapor than colder air masses.

**Humidity** is the number density of water vapor in the air in moles per unit of volume, or $n/V$. Sometimes, it is expressed as grams per unit of volume (such as $cm^3$ or liters). For a given temperature, there is a limit to the number density of water molecules that can be in the air. At a certain point, the air becomes saturated, and no further net evaporation takes place. The graph shows the maximum number density for air temperatures between $-1\,°C$ and $40\,°C$. You can see that the maximum humidity depends on temperature.

On a cool day, around $10\,°C$ ($50\,°F$), for instance, the maximum number density of water molecules in the air is approximately 0.5 mol per 1000 liters of air. This amount represents the maximum humidity for this temperature. The actual humidity may be at or below this value.

**Water Vapor Density Versus Temperature**

Points on the curve correspond to 100% relative humidity.

Water vapor density (mol/1000 L)

Temperature (°C)

## Using the Graph
This graph represents an inequality. The actual value for humidity can be any point on or below the curve. The graph shows that the total amount of water vapor that can be in the air increases with increasing temperature.

If the temperature of the air drops or the humidity rises, precipitation can occur. If these two values, $T$ and $n/V$, result in a data point above the curve, there will be precipitation or condensation of some sort. The condensation of water on the

*When air comes in contact with a cold surface, its temperature drops, causing some of the water vapor in the air to condense on the surface.*

outside of a glass of ice water is a sign that the air next to the glass has cooled enough for the water vapor to change phase to a liquid.

Like all gases, water vapor exerts a pressure. This pressure is part of the total pressure exerted by all the gases in the atmosphere. Each gas exerts a **partial pressure,** and these partial pressures add up to the atmospheric pressure.

## ⮌ Relative Humidity

When meteorologists consider humidity, they focus on the relative humidity. **Relative humidity** is the percent of the maximum humidity for a specified temperature. When the air contains the maximum amount of water vapor for the temperature of the air, it is at 100% relative humidity.

The terms *humidity* and *relative humidity* are often used interchangeably in weather reports. Meteorologists might say the humidity today is at 35% when they are really talking about the relative humidity.

**Example**

### Relative Humidity

Imagine that on a certain day the water vapor density is 1.6 mol/1000 L and the temperature is 30 °C.

a. What is the relative humidity?

b. At night, the temperature drops to 20 °C. Do you expect there will be precipitation?

### Solution

a. You can read the maximum water vapor density at 30 °C on the graph Water Vapor Density Versus Temperature. It is 1.7 mol/1000 L. This corresponds to 100% humidity. The relative humidity is the measured water vapor density divided by the maximum possible water vapor density.

$$\frac{1.6}{1.7} = 0.94 = 94\%$$

b. The point (20 °C, 1.6 mol/1000 L) on the graph is above the 100% humidity curve, so you can expect precipitation.

Matter that is in contact with an evaporating liquid will cool off. Sweating helps cool you because the evaporation of water on your skin transfers heat away from the body. When there is more water vapor in the air, it becomes harder for sweat to evaporate. This is why you often feel much hotter in humid air than in dry air.

## What is humidity and how is it measured?

Humidity refers to the amount of water vapor in the air. It is a measure of the number density of water molecules. Air temperature affects how much water vapor can be in the air. Warmer air can contain more water vapor than colder air. There is an upper limit to the amount of water vapor that can be in the air at any given temperature. This upper limit is called maximum humidity, or 100% relative humidity. At 100% relative humidity, no more water can evaporate into because the air is saturated. You might expect rain or fog when the relative humidity is close to 100%.

## Exercises

### Reading Questions

1. What does humidity measure?
2. Explain what is meant by relative humidity.

### Reason and Apply

3. What is the maximum vapor density possible at 30 °C?

4. Is it possible for the water vapor density in the air to reach 10 mol per 1000 L at a temperature of 40 °C? Explain why or why not.

5. Use the graph Water Vapor Density Versus Temperature to predict the water vapor density at 100% humidity for 35 °C.

6. What is the relative humidity if there is 0.5 mol of water vapor per 1000 L of air at 40 °C?

7. When the humidity is 25% and the temperature is 30 °C, what is the water vapor density?

8. On one day, 100% humidity corresponds to 1.9 mol of water vapor per 1000 L of air. On another day, 100% humidity corresponds to only 1.0 mol of water vapor per 1000 L of air. How can two different water vapor densities both be at 100% humidity?

9. Explain why it is easy to get dehydrated when you are exercising at a temperature of 0 °C and a relative humidity of 20%.

10. Do you feel cooler at 100% humidity at 30 °C, or at 50% humidity at 30 °C? Explain your thinking.

11. Suppose the humidity is 65% during the day when the temperature is 30 °C.
    a. If the temperature drops to 25 °C at night, do you expect fog? Explain.
    b. If the temperature drops to 20 °C at night, do you expect fog? Explain.

12. On a cold winter day the relative humidity is 50% outdoors but only 5% indoors. Which answer best explains what is going on?
    (A) It must be raining outside.
    (B) It must be snowing outside.
    (C) The heater is on inside, so the same humidity is a lower relative humidity.
    (D) The heater is on inside, and it is evaporating some of the humidity in the air.

# Extreme Physical Change

## THINK ABOUT IT

Water and air provide our planet with its weather and support plant and animal life. Weather itself is a result of a dynamic interplay of physical change between water, Earth's atmosphere, and energy from the Sun. There are times when these physical changes create conditions that can be catastrophic. A hurricane is one example of an extreme weather phenomenon that can be quite destructive.

### What are hurricanes and what causes them?

*To answer this question, you will explore*

⊃  The Properties of Hurricanes

⊃  Hurricane Formation

⊃  Global Warming

## EXPLORING THE TOPIC

### ⊃ The Properties of Hurricanes

A *hurricane* is an enormous tropical rainstorm with powerful winds that forms over the ocean. It has a distinctive shape to it, resembling a giant rotating pinwheel. The wind speeds in a hurricane are over 75 mi/h and may even exceed 150 mi/h. The storm may bring torrential rains, destructive winds, and flooding to areas of land that it passes over. The sheer size of these storms is shown in this photo of a hurricane in the Gulf of Mexico.

NASA/Jeff Schmaltz, MODIS Land Rapid Response Team

These giant storms may be from 125 to 1000 mi across and 15 mi high. Further, they are characterized by very low air pressure at their centers. In general, the lower the air pressure, the larger the storm. Consider the data in the table for three hurricanes in the same year.

| Hurricane name | Year | Wind speed (mi/h) | Pressure (atm) | Category |
|---|---|---|---|---|
| Hurricane Rita | 2005 | 173 | 0.885 | 5 |
| Hurricane Ike | 2008 | 145 | 0.923 | 4 |
| Hurricane Sandy | 2012 | 115 | 0.928 | 3 |

Hurricanes are categorized according to their strength, as indicated in the table. They are accompanied by a surge in coastal waters called a *storm surge.* A storm surge can reach 20 ft or more in height, causing major coastal flooding.

## Hurricane Formation

Hurricanes form only over very warm ocean waters of at least 80 °F. In addition, they require a great deal of moisture in the air. For these two reasons, hurricanes originate southeast of the United States, in the tropical ocean waters near the equator.

You have probably heard people refer to "hurricane season." Most hurricanes form during summer and fall when the waters are warmest in tropical zones. For the United States, hurricane season stretches from June through November.

A storm must grow through several stages before it is considered a hurricane. A future hurricane starts out as a tropical depression, which is a clearly defined low-pressure system with winds below 38 mi/h. Some tropical depressions continue to build and become tropical storms. In tropical storms, the air circulates around a low-pressure system with winds between 38 and 74 mi/h. Finally, a tropical storm that builds to wind speeds beyond 75 mi/h is considered a hurricane.

It can take several days for a tropical storm to develop into a hurricane. It begins with moist warm air from the ocean surface rising rapidly. The water vapor in this warm air condenses and forms storm clouds. The heat released by this condensation warms the cooler air above it, causing the air to rise even more. This process continues, with more and more warm air moving up into the

*These satellite images show the stages of hurricane formation.*

developing storm. A spiral wind pattern begins to develop, and the hurricane takes on its characteristic shape. In the very center of the storm, an "eye" develops and cool air descends, creating a calm storm-free area. The eye of a hurricane may

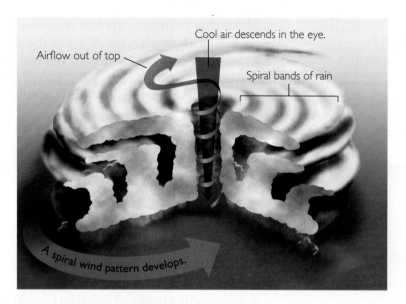

Airflow out of top

Cool air descends in the eye.

Spiral bands of rain

A spiral wind pattern develops.

be between 12 and 60 mi in diameter. When the eye of the storm passes over an area, people will experience calm, clear conditions, as if there is no hurricane at all. The eye is surrounded by the eyewall, which is the most violent part of the storm. Near the ocean surface, spiral bands of rain stretch out for miles.

## ➲ Global Warming

Earth is currently in a warming phase. Several recent hurricane seasons in the United States were unusually destructive. The number and severity of storms caused scientists to speculate that there is a connection between the increasing number and severity of hurricanes and the warming of the entire planet. In other words, Earth may be undergoing long-term climate change.

Experts think that the warming of the planet is worsened by our practice of using petroleum products for our energy needs. Burning gasoline, coal, and natural gas releases enormous amounts of carbon dioxide into our atmosphere. Increases in carbon dioxide have jumped up rapidly in recent years. This increase causes a greater amount of heat to be trapped in the Earth's atmosphere and is linked to an increase in average air temperature.

*The effects of global warming can be seen in the loss of habitat for polar animals.*

*Dramatic effects of climate change are shown in these two photos of Glacier Bay National Park, taken 60 years apart.*

Field, W. O. 1941. Muir Glacier. From the Glacier Photograph Collection. Boulder, Colorado USA: National Snow and Ice Data Center/World Data Center for Glaciology. Digital media. (Left)

Weiss and Overpeck, University of Arizona

Meteorologists hypothesize that increases in ocean temperatures of only a few degrees may have dramatic effects on the weather of our planet, including the possibility of increased frequency and intensity of hurricanes. One degree may not sound like much, but it is sufficient to raise ocean levels by a few inches and to cause more rainfall and stronger storms.

Studies show a steady increase in global temperature over the past 120 years. Scientists hope energy conservation will help to reverse or slow down this warming trend. Automobile makers are actively involved in creating vehicles that run on alternative energy sources.

## LESSON SUMMARY

### What are hurricanes and what causes them?

A hurricane is a large and powerful tropical storm with intense spiraling winds. Hurricanes form over the warm tropical waters near the equator. At the very center of a hurricane is an area of extremely low air pressure. The heat released when moist warm air rises and condenses into storm clouds powers hurricane formation. Also, the low pressure caused by condensation draws more air into the storm from areas of relatively higher pressure. This cycle of evaporation and condensation of very large amounts of water is a key feature of hurricanes. Experts studying climate changes due to global warming are monitoring data about increases in the frequency and intensity of hurricanes.

## Exercises

### Reading Questions

1. What conditions are necessary for hurricane formation?
2. Write a creative paragraph describing what life on the planet may be like if global warming continues at its present pace.

### Reason and Apply

3. **PROJECT**

    Find data showing that ocean temperature is related to the wind speed of a hurricane.

4. **PROJECT**

    Discuss three pieces of evidence that global warming is occurring.

# CHAPTER 12

# Concentrating Matter

# SUMMARY

### KEY TERMS

**number density**

**mole**

**Avogadro's number**

**standard temperature and pressure (STP)**

**Avogadro's law**

**ideal gas law**

**universal gas constant, $R$**

**humidity**

**partial pressure**

**relative humidity**

# Weather Update

Air pressure is directly related to the number density, or number of particles per unit of volume, of the gas molecules in a sample of air. Avogadro's law states that two gas samples have the same number of molecules if temperature volume and pressure are the same. Avogadro's law and the combined gas law together form the ideal gas law. The ideal gas law relates pressure, volume, temperature, and the number of moles by a proportionality constant. This constant, $R$, is called the universal gas constant.

*Extreme weather can occur if conditions support rapid evaporation of water molecules followed by rapid condensation of these molecules.*

The amount of water vapor in the air plays a big role in determining the weather. Water vapor density, or humidity, is dependent on air temperature and air pressure. When air is at 100% humidity, chances are it is raining, snowing, or densely foggy because the air is saturated with water molecules.

## REVIEW EXERCISES

1. Explain how you can determine the number of molecules in a breath of air.

2. What type of weather do you predict if the relative humidity is 100%?

3. Suppose that you have 22.4 L of helium, He, gas and 22.4 L of neon, Ne, gas at STP. Which of these statements is *false*?
   **(A)** Both samples contain the same number of atoms.
   **(B)** Both samples have the same number density, $n/V$.
   **(C)** The two samples have different masses.
   **(D)** The two samples have the same mass density, $D$.

4. How many moles of nitrogen gas, $N_2(g)$, are contained in 2 liters at 350 K and 1.5 atm?

5. Suppose the water vapor density is 1.5 mol/1000 L at a temperature of 35 °C, and the maximum water vapor density at 35 °C is 2.2 mol/1000 L. What is the relative humidity?

---

## Global Climate Change

PROJECT

Earth's overall climate is currently in a warming phase. Research some of the causes and effects of global warming on Earth.

- What evidence do scientists have that average global temperatures are increasing?

- What are some possible causes of global warming?

- How might global warming affect the severity and frequency of storms in a region?

- What are some possible effects of global warming on different living things and their environments?

---

StockbyteGetty Images

# Weather | REVIEW

## UNIT 3

© Dennis MacDonald/Alamy

The atmosphere surrounding our planet is a mixture of gases, including water vapor. The weather that we experience every day is almost entirely caused by physical changes to these gases. Energy from the Sun causes water molecules to change phase and evaporate into the air. This water vapor moves around, sometimes changing phase again and forming clouds, rain, ice, snow, dew, fog, and so on.

By studying small samples of gas in containers, we can learn more about how gases in the atmosphere behave. The kinetic theory of gases explains the behavior of individual gas particles under varying conditions. For example, when a gas in a rigid container is heated, its pressure increases because the molecules move faster, increasing the force and number of collisions with whatever they contact. When a sample of gas is squeezed into a smaller volume, it also exerts more pressure. This is because the same number of rapidly moving molecules must occupy a smaller space. The pressure, volume, and temperature of a gas sample are related by the combined gas law.

Meteorologists track the variables that affect the gases in the atmosphere. For example, they record water vapor density, or humidity, and areas of differing air pressure. They study air masses and fronts where warm and cold air masses collide. They combine all the information and use it to make predictions about the weather.

## REVIEW EXERCISES

### General Review

*Write a brief and clear answer to each question. Be sure to show your work.*

1. Suppose that you have a 500 mL sample of liquid water and a 500 mL sample of ice. The ice has a density of 0.92 g/mL. Which has a greater mass? Explain your reasoning.

2. Convert 25 °F to kelvins. Show your work.

3. Use the kinetic theory of gases to explain why a gas expands when heated and contracts when cooled.

4. Suppose that you fill a balloon with 600 mL of air when the temperature is 27 °C. Later in the day, the temperature drops to 21 °C. Determine the new volume of the balloon.

5. A cylinder with a movable piston contains 2.5 L of gas at a pressure of 2 atm. If the volume is

decreased to 2.1 L and the temperature stays the same, what is the new pressure of the gas?

6. Which graph represents Boyle's law? Explain your thinking.

**A.**

**B.**

**C.**

7. A propane gas tank contains propane gas at a pressure of 2.00 atm and a temperature of 30 °C. If the temperature is increased to 37 °C, what is the new pressure of the gas?

8. Suppose that you fill a balloon with 75 L of air at 28 °C and a pressure of 1 atm. You then take the balloon to the top of a mountain where the pressure is 0.8 atm and the temperature is 18 °C. What is the new volume of the balloon?

9. Suppose that you have 22.4 L of nitrogen gas, $N_2(g)$, at STP.

a. What are the temperature and pressure of this sample?

b. How many molecules of $N_2$ are in the sample?

c. How many nitrogen atoms are in the sample?

d. What is the number density, $n/V$, of nitrogen molecules in the sample?

10. Suppose that you have a 750 mL rigid container of hydrogen gas, $H_2(g)$, at 250 K and a 750 mL rigid container of nitrogen gas, $N_2(g)$, at 250 K. A pressure gauge on each container shows that they both have the same internal pressure.

a. What do you know about the number of individual molecules of gas in each container? Explain your reasoning.

b. Which container would have the greater mass? Explain your reasoning.

11. Suppose that you have 77.0 L of hydrogen, $H_2$, gas at a pressure of 3.5 atm and a temperature of 2 °C. How many moles of hydrogen are in this sample?

12. Would you feel cooler if the temperature were 28 °C at 100% relative humidity, or 28 °C at 45% relative humidity? Explain your reasoning.

13. Name the gas laws. Write the formula associated with each one, and list the factors that must stay constant for each law to apply.

## STANDARDIZED TEST PREPARATION

### Multiple Choice

*Choose the best answer.*

1. Which option has only units of pressure?

(A) atm, lb/in², in Hg

(B) lb/in², in Hg, mL/K

(C) in Hg, mL/K, atm

(D) mL/K, atm, lb/in²

2. A cylindrical rain gauge contains a volume of 6 mL for a height of 2 cm of rain. What is the volume of rain for a height of 9 cm in this rain gauge?

(A) 9 mL
(B) 18 mL
(C) 27 mL
(D) 54 mL

*Use the graph below for Exercises 3 and 4.*

**Mass Versus Volume of Water**

3. For equal volume samples of snow, ice, and rain, which statement is correct in describing the mass of the samples?

(A) The masses of each will be the same because in each case, the substance is water, $H_2O$.
(B) The mass of snow will be smallest because the snow has the highest density.
(C) The mass of the ice will be smallest because the ice has the lowest density.
(D) The mass of the rain will be largest because the rain has the highest density.

4. What is the volume of 15.0 g of ice?

(A) 7.5 mL
(B) 13.8 mL
(C) 15.0 mL
(D) 16.3 mL

5. The density of solid nitrogen is 1.026 g/cm$^3$ and the density of nitrogen gas at STP is 0.00125 g/cm$^3$.

What is the volume of gas produced when 50 mL of solid nitrogen sublimes?

(A) 39 L
(B) 41 L
(C) 61 L
(D) 64 L

6. The average temperature at the top of Mt. Denali in Alaska can drop below $-50\,°F$. What is this temperature in Kelvins?

(A) 125 K
(B) 263 K
(C) 227 K
(D) 283 K

7. Which of the following correctly describes the movement of gas particles according to the kinetic theory of gases?

(A) All gas particles move with the same speed at a given temperature.
(B) All gas particles move in curved lines in random directions.
(C) The average speed of gas particles increases with decreasing temperature.
(D) Gas particles are in constant motion at temperatures above 0 K.

8. Imagine that a gas sample is in a cylinder with a piston. Assuming that the pressure and number of molecules stays the same, what will happen to the volume of the gas if the temperature increases?

(A) The volume decreases because the particles are closer together.
(B) The volume decreases because the particles have collided less.
(C) The volume increases because the particles are moving faster.
(D) The volume increases because the particles are less attracted to each other.

9. A balloon is filled with air to a volume of 1.35 L at a temperature of 22 °C. The balloon is taken to a birthday picnic on a sunny day. After several hours in the Sun, the air inside of the balloon was warmed to 32 °C. What is the new volume of the balloon?

(A) 2.0 L
(B) 1.3 L
(C) 1.4 L
(D) 0.93 L

10. What phase change occurs when a cloud forms from water vapor?

(A) Evaporation
(B) Condensation
(C) Melting
(D) Freezing

11. A weather balloon expands as it travels to a higher altitude. Which of the following correctly describes the gas particles **inside** of the balloon?

   (A) There is more space between the gas particles inside the balloon because the atmospheric pressure is less at higher altitudes.

   (B) There is more space between the gas particles inside the balloon because the atmospheric pressure is greater at higher altitudes.

   (C) There are more collisions between the gas particles inside of the balloon due to the lower atmospheric pressure.

   (D) The average speed of the gas particles inside the balloon increases due to the decrease in temperature at higher altitudes.

12. A 0.75 L balloon has a pressure of 1.0 atm. The balloon is placed inside a vacuum chamber, in which air outside the balloon can be removed. After air has been removed from the chamber, the pressure inside the balloon is reduced to 0.50 atm. The temperature remains unchanged. What is the new volume of the balloon?

   (A) 0.38 L        (B) 0.50 L
   (C) 0.67 L        (D) 1.5 L

13. Which hypothesis best illustrates the particle view of what occurs when solid carbon dioxide sublimes?

Hypothesis A        Hypothesis B

$CO_2(s)$

Hypothesis C        Hypothesis D

   (A) Hypothesis A. The gas molecules individually expand when carbon dioxide sublimes.

   (B) Hypothesis B. The gas molecules float to the top of the container when carbon dioxide sublimes.

   (C) Hypothesis C. The gas molecules move to the surfaces of the container when carbon dioxide sublimes.

   (D) Hypothesis D. The gas molecules move relatively far apart from one another when carbon dioxide sublimes.

14. An 8.5 L scuba tank is filled with air to a pressure of 150 atm at 26 °C. The scuba tank is cooled to a temperature of 15 °C. What is the pressure inside of the tank?

   (A) 17 atm
   (B) 86 atm
   (C) 140 atm
   (D) 160 atm

15. Which of the following statements correctly describes the relationship between gas measurements?

   (A) When temperature and the amount of gas do not change, the pressure and volume of the gas are inversely proportional.

   (B) When pressure and the amount of gas do not change, the volume and temperature are inversely proportional.

   (C) When the amount of gas and volume do not change, the pressure and temperature are inversely proportional.

   (D) When the temperature and pressure do not change, the number of particles and volume are inversely proportional.

16. A weather balloon is filled with helium gas to a volume of 7500 L at sea level where the pressure is 1 atm and the temperature is 24 °C. The balloon travels to the top of the stratosphere where the temperature is −15 °C and the pressure is 0.001 atm. What is the volume of the balloon while at this altitude?

   (A) $4.7 \times 10^6$ L
   (B) $6.5 \times 10^6$ L
   (C) $8.6 \times 10^6$ L
   (D) $1.2 \times 10^7$ L

**17.** Suppose you have two balloons, one filled with helium, He, and one filled with argon, Ar. Each balloon has a volume of 22.4 L at STP. Which of the following correctly describes these samples?

(A) The helium balloon contains fewer gas molecules because the atomic number of helium is less than that of argon.

(B) The helium balloon contains fewer gas molecules because the mass number of helium is less than that of argon.

(C) Each balloon contains the same number of gas particles because the balloons are the same volume, at the same temperature and pressure.

(D) Each balloon contains a different number of gas particles because each balloon is filled with a different gas.

**18.** Complete the data set for a sample of gas.

| Pressure | Volume | Temperature | Moles |
|----------|--------|-------------|-------|
| 0.75 atm | 11.2 L | 300 K | ? |

(A) 0.34 mol      (B) 0.50 mol
(C) 2.93 mol      (D) 244 mol

**19.** How many moles of $N_2$ gas are there in a container that holds 11.2 L at 0 °C and 1 atm?

(A) 0.25 mol      (B) 0.50 mol
(C) 1.0 mol      (D) 2.0 mol

**20.** Find the pressure of 2.5 mol of gas if the gas temperature is 32.0 °C and the gas volume is 67.2 L.

(A) 10.2 atm      (B) 1.07 atm
(C) 0.930 atm      (D) 0.0976 atm

# Hair Dryer Hot Air Balloon

## Work Groups of 4

### Purpose

To use the process of engineering design to plan and design a hot air balloon using a hair dryer.

### Problem

Your engineering design team is charged with creating a hair dryer hot air balloon. Your job will be to research possible ways to create a functional hot air balloon with everyday items and to test its effectiveness.

### Safety Notes

- Be careful handling materials that may be hot. It is important to follow safety rules for electricity use.
- Review fire safety rules and have appropriate fire safety materials on hand.
- As a part of this project, you will need to make a list of all safety measures you should use when creating your hot air balloon.

### Review

Before you begin, review the handout, *Engineering Design* (found after the Unit 1 Engineering Project), and reflect on what you have learned so far in your class. For this project, you will focus on identifying the problem, brainstorming for solutions, developing a procedure and a design, and if possible, testing and refining the procedure and design.

In Unit 3: Weather, you studied properties of gases. You might want to review these concepts when considering your hot air balloon design. The following table shows some of the many topics and skills that were covered in Unit 3: Weather. You may find these helpful as you work to engineer a solution to the problem:

| Weather science | Combined gas law | Boyle's law | Gas density |
|---|---|---|---|
| Thermometers | Physical change | High and low air pressure | Gay-Lussac's law |
| Air pressure | Charles's law | Density | Ideal gas law |

**Part 1:** Identify the Problem and Brainstorm Solutions

1. **Identify the Problem:** What is the problem you are trying to solve? Write up a description of what you need to accomplish.

2. **Brainstorm:** In your Group, brainstorm some ideas to solve this problem. Here are some questions to consider:
   - How does a large hot air balloon work?
   - What properties of gases are most important for hot air balloons to work?
   - How might you use a hair dryer to make your hot air balloon?
   - Which gas laws should you consider when creating your design?
   - How will you control the temperature and the air flow for the balloon?

- How does air/gas density play a role in the function and design of the balloon?
- What are the best materials to create you balloon?
- How might the density of the materials used affect the balloon?

3. Research: In the library and online, research different ways that you could create a hot air balloon using a hair dryer. Create a list of all of your sources to turn in with your project. Record your research notes so they can be used later when creating your plan. Here is a website you might try as a start. If you find others, make sure to record the source and the website URL on your list:

**PBS Kids**   http://pbskids.org/zoom/activities/sci/hotairballoon.html

## Part 2: Create a Prototype

1. Based on your research and brainstorming, write a lab procedure for how to create a hot air balloon using a hair dryer. Include a diagram of your design with explanations of how it works.

2. Write list of materials needed to make your hot air balloon.

3. Write a list of all safety considerations needed for your procedure and design.

4. Submit your procedure and design to your teacher for review and approval. Include your list of resources and research material.

5. Once approved, use your procedure and design to create a hot air balloon.

6. Test your balloon to see how well it works.

## Part 3: Refine and Retest

1. Have your classmates provide feedback to you about your hot air balloon. Ask them to help you think of ways to improve your design. What changes might you make based on their feedback?

2. Make changes to your procedure and design based on feedback and the results of your testing.

3. Submit your revised procedure and design to your teacher for review and approval. Include your list of resources and research material.

4. Once approved, use your procedure and design to create a hot air balloon.

5. Test your balloon to see how well it works.

6. Once your procedure and design is refined, your teacher may ask you to lead a small group of students in demonstrating your hot air balloon.

## Possible Presentation Options

- Oral presentation describing your balloon design and procedure with diagrams and a demonstration of the working balloon
- Poster describing your balloon design and procedure with diagrams and a demonstration of the working balloon
- A protocol pamphlet balloon design and procedure with diagrams and a demonstration of the working balloon
- A website that outlines balloon design and procedure with diagrams and a video demonstration of the working balloon on the website

# Unit 4 | Toxins

Some animals produce toxins for self-defense, or to paralyze their prey. However, many toxins also have medicinal uses. For example, certain compounds in the venom of this African saw-scaled viper can help to treat a person having a heart attack.

© Tony Phelps/naturepl.com

Charles D. Winters/Science Source

## In this unit, you will learn:

- how toxins are defined

- how chemists determine toxicity

- the mechanisms by which toxic substances act in our bodies and what this has to do with chemical reactions

## Why Toxins?

Chemical reactions help our bodies to process food and create new tissues. However, some chemical reactions have toxic and harmful outcomes. The toxicity of a substance is highly dependent on the dose size of that substance. Sometimes, a small amount of a compound, such as a vitamin, can be therapeutic, but a large amount can damage your health. This unit investigates chemical changes by exploring how toxic substances are measured and tracked through their transformations.

# CHAPTER 13

# Toxic Changes

*Biologists define toxins as harmful substances that come from living organisms, like these fly agaric mushrooms.*

### In this chapter, you will study

- how to write, balance, and interpret chemical equations

- definitions of chemical and physical change

- how to classify chemical reactions

- what happens to mass when a chemical change occurs

Chemical changes are necessary for life. However, some substances interact with the body in ways that have unhealthy effects. To monitor and understand toxic interactions, it is necessary to use chemical equations to track changes in matter.

# Chemical Equations

## THINK ABOUT IT

In the world around you, matter undergoes many changes. Some of these changes have very little impact on living things on the planet. Other changes may be vital and necessary for life. Still other changes may threaten the health and well-being of plant and animal life. Chemists use chemical equations to keep track of all types of changes in matter, including those that are beneficial and those that are unsafe.

**How do chemists keep track of changes in matter?**

*To answer this question, you will explore*

➥ Chemical Equations

➥ Toxic Substances and Their Effects

## EXPLORING THE TOPIC

### ➥ Chemical Equations

A chemical equation is a chemical "sentence" that describes change, using numbers, symbols, and chemical formulas. Chemical equations describe what happens when a single substance is changed, or when two or more substances are combined and a change occurs. Once you understand how to decode chemical equations, you will be able to use them to predict what you might observe when substances are mixed.

#### INTERPRETING A CHEMICAL EQUATION

In some reactions that take place in your body, the element chromium is safe and even necessary. In other reactions, it is toxic. Consider what happens if you ingest chromium metal and it reacts with the hydrochloric acid in your stomach. A chemical equation can help you decode this reaction.

The substances you start with are called **reactants.**

The substances you end up with are called **products.**

$$2HCl(aq) + Cr(s) \longrightarrow CrCl_2(aq) + H_2(g)$$

Aqueous hydrochloric acid

Solid chromium

Aqueous chromium (II) chloride

Hydrogen gas

This reaction requires twice as many **HCl** molecules as **Cr** atoms.

**Chemical equation:** $2HCl(aq) + Cr(s) \longrightarrow CrCl_2(aq) + H_2(g)$

Lisa Maree Williams/Getty Images

**Interpretation:** Hydrochloric acid reacts with solid chromium to produce a solution of chromium (II) chloride and bubbles of hydrogen gas.

A chemical equation can help you anticipate what you will observe when the reactants are combined. Examine the same equation, but focus on what you expect to observe.

$$2HCl(aq) + Cr(s) \longrightarrow CrCl_2(aq) + H_2(g)$$

According to the equation, when hydrochloric acid and chromium react, the solid chromium will disappear. You would expect to see the formation of a new aqueous solution as well as some evidence that a gas was produced.

Solid chromium disappears. Hydrogen gas bubbles form and escape. A blue solution of chromium (II) chloride forms.

Hydrochloric acid HCl(aq)

Hydrogen H₂(g)

Chromium (II) chloride CrCl₂(aq)

Chromium Cr(s)

Hydrochloric acid is added to solid chromium.

All the hydrogen has escaped. The beaker now contains only aqueous chromium (II) chloride.

Sometimes, the changes that take place when substances are mixed are not visible to the eye. For example, death from poisoning occurs within minutes of swallowing a solution of sodium cyanide, NaCN. One successful treatment for this type of poisoning is injection of an *antidote,* a remedy that counteracts the poison. The antidote for sodium cyanide is a solution containing sodium thiosulfate, $Na_2S_2O_3$. But if you were to observe the reaction between sodium cyanide and sodium thiosulfate in a beaker, you would not be able to tell that a reaction had occurred because both reactants and both products are clear, colorless liquids.

. . . . . . . . . . . . . . . . . . . . . . . . . . . . . . . . . . . . . . . . . . . . . . . . . . . . . . .

**Big Idea**   Chemical equations keep track of changes in matter.

. . . . . . . . . . . . . . . . . . . . . . . . . . . . . . . . . . . . . . . . . . . . . . . . . . . . . . .

## ⮌ Toxic Substances and Their Effects

Toxic substances enter the body in limited ways. The most common methods are through ingestion (eating or swallowing), inhalation (breathing something in), or contact with the skin. Once toxic substances enter the body, they react in a variety of ways.

Toxic substances can react with water in mucous membranes, with oxygen carried through the blood, or with stomach acid. Some toxic substances have an immediate negative effect on the well-being of a person or living thing. Other toxic substances may stay a long time in the body, becoming part of the body's chemistry, perhaps damaging the body many years later.

Toxic substances may be molecular, ionic, or metallic substances. Many small molecular substances react with the body and create acid products, which can damage and irritate tissue or upset the acidity of the blood. Toxic metals often react to form ionic compounds, which move throughout the body and compete with "good" metals that are useful to the body. Some ionic compounds form solids that clog the body's filtering systems. Of course, these are just a few of the possible types of toxic reactions that occur. Chemical equations are the main tool you will use to track these changes.

## LESSON SUMMARY

### How do chemists keep track of changes in matter?

A chemical equation tracks changes in matter. The left side of the equation contains the chemical formulas for the reactants or the substances that are being combined. The right side of the equation contains the chemical formulas for the products or the substances that are produced. The equation also shows the phase of each reactant and product. Decoding an equation allows you to predict what substances may be made when the reactants are combined. Chemical equations often provide more information than what you can observe with your senses.

## Exercises

### Reading Questions

1. What is the difference between a reactant and a product?
2. Are chemicals and chemical reactions important for life? Why or why not?
3. Describe in your own words what a toxic substance is.

### Reason and Apply

4. Both bleach and ammonia are used for cleaning. However, it is very dangerous to mix bleach with ammonia because they react to produce sodium hydroxide and the toxic gas chloramine.

$$NaOCl(aq) + NH_3(aq) \longrightarrow NaOH(aq) + NH_2Cl(g)$$

   a. Write an interpretation of the chemical equation.
   b. What do you expect to observe?

5. Poisoning with mercury chloride can be reversed by chelation therapy. The chelating agent called EDTA, $C_{10}H_{16}N_2O_8$, is injected into the bloodstream. EDTA forms a water-soluble compound with mercury ions, allowing removal from the body through the kidneys.

$$HgCl_2(s) + C_{10}H_{16}N_2O_8(aq) \longrightarrow HgC_{10}H_{12}N_2O_8(aq) + 4HCl(aq)$$

   a. Write an interpretation of the chemical equation.
   b. What do you expect to observe?

6. Describe at least three types of effects that a toxic substance can have on the body.

*Making Predictions*

# Observing Change

## THINK ABOUT IT

Table salt, or sodium chloride, NaCl, can enhance the flavor of soup. Even though the salt dissolves in the soup, you can still taste it. This change is described by a chemical equation:

$$NaCl(s) \longrightarrow NaCl(aq)$$

It is also possible to change the salt in far more dramatic ways. For example, when an electric current is passed through a sodium chloride solution, a toxic, green gas bubbles out of the solution. The change is described by this chemical equation:

$$2NaCl(aq) + 2H_2O(l) \longrightarrow 2NaOH(aq) + Cl_2(g) + H_2(g)$$

**How can you predict what you will observe based on a chemical equation?**

*To answer this question, you will explore*

➔ Predicting Change from Chemical Equations

➔ Information in Chemical Equations

## EXPLORING THE TOPIC

## ➔ Predicting Change from Chemical Equations

While some chemical changes are difficult to see, many changes in matter are accompanied by observable evidence. For example, when a change takes place, you may hear fizzing, a pop, or an explosion. You might see changes in color or physical form. You may smell gases that escape, or you may feel heat. Perhaps there is even a fire.

But what if you don't have a laboratory and a lot of chemicals around? How can you predict what you might observe if you are provided only with a chemical equation?

### SUGAR DISSOLVING

Consider these two chemical equations. What type of change does each describe?

Sugar dissolves in water.

Low heat evaporates the water and leaves solid sugar.

$$C_{12}H_{22}O_{11}(s) \longrightarrow C_{12}H_{22}O_{11}(aq)$$

$$C_{12}H_{22}O_{11}(aq) \longrightarrow C_{12}H_{22}O_{11}(s)$$

The only difference between these two equations is the position of the (*aq*) and (*s*) symbols. Notice that the chemical formula does not change from one side of the equation to the other. The sugar molecule changes form but not identity. The first equation describes dissolving, in which a solid is mixed with water. The second equation describes removing water from an aqueous substance, leaving a solid.

Although the sugar changes appearance when it dissolves, the sugar itself has not changed into a different compound. The chemical equation indicates how the molecules have, or have not, changed.

### SUGAR MELTING

Consider these two equations that involve changes to sugar.

Additional low heat melts sugar.

$$C_{12}H_{22}O_{11}(s) \longrightarrow C_{12}H_{22}O_{11}(l)$$

Melted sugar cools to a hard solid.

$$C_{12}H_{22}O_{11}(l) \longrightarrow C_{12}H_{22}O_{11}(s)$$

Again the formulas are identical on both sides of the arrow, except for the symbols in parentheses, (*l*) and (*s*). These equations describe a phase change from a solid to a liquid and back to a solid. By heating sugar carefully, you can melt it to a clear liquid. When liquid sugar is cooled, it becomes a solid again.

The same compound shows up on both sides of the equation. Even though the tiny grains of sugar have been transformed into a single, large, chunk of solid sugar, the atoms within the sugar molecules have not rearranged into new compounds. Sugar is still sugar.

**Big Idea** When a substance changes phase or dissolves, its chemical formula does not change.

### SUGAR DECOMPOSING

Consider another transformation of sugar.

$$C_{12}H_{22}O_{11}(s) \longrightarrow 12C(s) + 11H_2O(g)$$

Notice that the two sides of the equation look very different. There is no sugar on the right side of the arrow. The sugar has been decomposed. It has been converted into two products: solid carbon and water vapor.

If you heat sugar sufficiently (beyond melting), it will turn black and it will release smoke. It will not taste sweet anymore, and there is no way to turn it back into sugar. These property changes, and the release of smoke, are signs that a more dramatic rearrangement of atoms has occurred.

Higher heat decomposes sugar, turning it into carbon and water vapor.

Adding water will not turn the carbon back into sugar.

$C_{12}H_{22}O_{11}(s) \longrightarrow 12C(s) + 11H_2O(g)$

$12C(s) + 11H_2O(l) \longrightarrow$ no change

This particular rearrangement of atoms is not easy to reverse. In other words, you cannot combine carbon with water and expect to form sugar.

### SUGAR REACTING

Consider one more transformation of sugar, as described by this equation:

$$C_{12}H_{22}O_{11}(s) + 8KClO_3(s) \longrightarrow 12CO_2(g) + 11H_2O(g) + 8KCl(s)$$

Interpreting the chemical formulas tells you that sugar and potassium chlorate are added together. The result is the formation of two gases, carbon dioxide and water vapor, as well as the formation of some solid potassium chloride.

*Sugar and potassium chlorate react explosively.*

The sugar and potassium chlorate are converted to carbon dioxide gas, water vapor, and solid potassium chloride. This change happens very quickly and releases energy in the form of light and heat.

## Changing Salt

Predict what you would observe for the changes described by these equations:

**a.** $NaCl(s) \longrightarrow NaCl(aq)$ **b.** $NaCl(s) \longrightarrow NaCl(l)$

**c.** $2NaCl(l) \longrightarrow 2Na(l) + Cl_2(g)$

*Solution*

**a.** The solid sodium chloride dissolves in water.

**b.** The solid sodium chloride melts to become liquid sodium chloride.

**c.** The liquid sodium chloride decomposes into liquid sodium and chlorine gas.

## ⤷ Information in Chemical Equations

Just as a chemical equation can help you predict what you might observe, it can also provide more information than you can get by observation alone. Consider the reaction described at the beginning of this lesson. The change that happens upon passing an electric current through aqueous sodium chloride is described by this equation:

$$2NaCl(aq) + 2H_2O(l) \longrightarrow 2NaOH(aq) + Cl_2(g) + H_2(g)$$

If you were to observe this reaction without knowing the equation, you would not be able to identify the compounds before and after the reaction. Many solutions are clear and colorless, and you might not realize that two gases are produced. A chemical equation identifies the reactants and products in a reaction. This chemical equation indicates that aqueous sodium chloride combines with water to form aqueous sodium hydroxide, chlorine gas, and hydrogen gas. An equation also identifies the form or phase a substance is in. This equation indicates that a liquid of some sort will be observed before and after the reaction, and gas bubbles will be seen as the reaction takes place.

Some information is not contained in a chemical equation. The equation above, for example, does not indicate that chlorine gas is green and toxic. The reaction of sugar with potassium chlorate, mentioned previously, is quite spectacular. The equation does not indicate that the reaction is explosive or that a purple flame is produced. An equation does not tell you whether a procedure will result in hot or cold sensations. It won't tell you how fast a change will occur or if a color change will be observed. It does not tell you if there will be an explosion, a loud noise, or an overflowing beaker. Only direct observation can provide you with all of the information about what you will see, hear, and feel.

**CONSUMER CONNECTION**

When sugar molecules are allowed to crystallize, they form a hard candy, like lollipops. When lemon juice or fructose is added to the sugar, the molecules do not lock into place as in crystalline candy. Instead, they form soft candy like taffy or caramels, called amorphous candy.

Scott Karcich/iStockphoto

## LESSON SUMMARY

### How can you predict what you will observe based on a chemical equation?

Chemical equations contain valuable information about what is happening during a change in matter. They track the identities of the substances involved in the change, using chemical formulas. They also identify the phase of each substance. By carefully examining a chemical equation, you can predict what you might observe if the procedure were completed. A chemical equation cannot tell you everything that you will observe. It does not inform you about the speed of a reaction, if a color change will occur, or if energy is transferred.

## Exercises

### Reading Questions

1. In words, describe the difference between sugar melting and sugar dissolving in water. The formula for sugar is $C_{12}H_{22}O_{11}$.

2. Use chemical equations to describe the difference between sugar melting and sugar decomposing. The formula for sugar is $C_{12}H_{22}O_{11}$.

### Reason and Apply

3. Describe what you think you would observe for these chemical equations.
   a. $Mg(s) + 2HCl(aq) \longrightarrow H_2(g) + MgCl_2(aq)$
   b. $2H_2O_2(aq) \longrightarrow 2H_2O(l) + O_2(g)$
   c. $2NaCl(aq) + Pb(NO_3)_2(aq) \longrightarrow 2NaNO_3(aq) + PbCl_2(s)$

4. Write a chemical equation for these reaction descriptions.
   a. Solid sodium chloride dissolves in water.
   b. Solid magnesium sulfide is heated to produce solid magnesium and sulfur gas.
   c. Solid titanium is heated in oxygen gas to produce titanium dioxide.

5. It's extremely dangerous to mix bleach and ammonia because highly toxic gas will form. So you should never mix these two chemicals. The two chemical reactions below describe what would happen if bleach and ammonia were mixed. If these reactions were carried out by a professional chemist using proper safety precautions in a controlled environment, what might that chemist observe in each case?

$$NaOCl(aq) + NH_4OH(aq) \longrightarrow NH_2Cl(g) + NaOH(aq) + H_2O(l)$$

$$NaOCl(aq) + 2NH_4OH(aq) \longrightarrow N_2H_4(g) + NaCl(aq) + 3H_2O(l)$$

6. **PROJECT**

   Look up chloramine, $NH_2Cl$, and hydrazine, $N_2H_4$, on the Internet. Describe the toxicity of each compound.

*Spare Change*

# Physical versus Chemical Change

## THINK ABOUT IT

You breathe in oxygen and it dissolves in your blood, where it binds to a molecule called hemoglobin. The hemoglobin transports the oxygen to where it is needed for reactions with carbohydrates. In the end, the oxygen and carbohydrates react to produce carbon dioxide and water. Some of these changes are physical changes and some are chemical changes.

### How are changes in matter classified?

*To answer this question, you will explore*

⮕ Defining Physical and Chemical Change

⮕ Dissolving: A Special Case

## EXPLORING THE TOPIC

### ⮕ Defining Physical and Chemical Change

Examine the chemical equations in the table.

| Physical change | Chemical change |
|---|---|
| $H_2O(l) \longrightarrow H_2O(s)$ | $2Na(s) + Cl_2(g) \longrightarrow 2NaCl(s)$ |
| $Br_2(l) \longrightarrow Br_2(g)$ | $CH_4(g) + 2O_2 \longrightarrow CO_2(g) + 2H_2O(l)$ |
| $I_2(s) \longrightarrow I_2(g)$ | $2KClO_3(s) \longrightarrow 2KCl(s) + 3O_2(g)$ |

### PHYSICAL CHANGE

The changes in the left column are all **physical changes.** In each case, the substance on the left side of the equation changes phase (solid, liquid, or gas), but it does not change its identity. Also, there is only *one* substance involved in each of these equations, and that one substance is present on both sides of the arrow.

$H_2O(l) \longrightarrow H_2O(s)$

*Liquid water freezes to become ice.*

$Br_2(l) \longrightarrow Br_2(g)$

*Liquid bromine evaporates to bromine gas.*

$I_2(s) \longrightarrow I_2(g)$

*Solid iodine sublimes to iodine gas.*

When something changes in a physical way, the chemical formula does not change because you end up with the same substance you started with. Changing pressure or temperature, and grinding a substance into a powder, are other examples of physical change.

### CHEMICAL CHANGE

The three equations in the right column of the table on the previous page are chemical changes. In each case, new substances are created. In the equations, the chemical formula(s) on the left side of each arrow are different from the formula(s) on the right side. New substances are formed during chemical changes, so we can expect the products to have properties significantly different from the reactants.

$2Na(s) + Cl_2(g) \longrightarrow 2NaCl(s)$
*Solid sodium reacts with chlorine gas to produce solid sodium chloride.*

$CH_4(g) + 2O_2(g) \longrightarrow CO_2(g) + 2H_2O(l)$
*Methane gas and oxygen gas burn to produce carbon dioxide gas and water.*

$2KClO_3(s) \longrightarrow 2KCl(s) + 3O_2(g)$
*Solid potassium chlorate is heated to form solid potassium chloride and oxygen gas.*

Heat, light, smoke, smells, bubbles, and changes in color often accompany chemical changes. When chemical change takes place, chemists say that a chemical reaction has occurred. In fact, the terms chemical *change* and chemical *reaction* mean the same thing.

## ⮕ Dissolving: A Special Case

Chemical equations can also represent the process of dissolving. Here are three equations that represent dissolving.

Liquid ethanol dissolves in water: $C_2H_6O(l) \longrightarrow C_2H_6O(aq)$

Gaseous hydrochloric acid dissolves in water: $HCl(g) \longrightarrow HCl(aq)$

Solid copper chloride dissolves in water: $CuCl_2(s) \longrightarrow CuCl_2(aq)$

These equations indicate that the identity of each dissolved substance has not changed. Each equation describes a physical change. The symbol (*aq*), for aqueous, on the product side of the equation means that each substance has dissolved in water.

Here is a different way to write the second and third equations listed above.

$$HCl(g) \longrightarrow H^+(aq) + Cl^-(aq)$$

$$CuCl_2(s) \longrightarrow Cu^{2+}(aq) + 2Cl^-(aq)$$

You cannot always tell by observing a change whether it is a physical change or a chemical change. A chemical equation can provide more information than observation alone.

When hydrogen chloride gas and solid copper chloride dissolve in water, they break apart into ions. The second type of equation shows what happens during this change. These equations showing a substance breaking apart to form ions are necessary for the dissolving of ionic compounds and acids. Note that molecular substances like ethanol, $C_2H_6O$, do not break apart when they dissolve. So there is only one way to write the equation.

Ionic compounds dissolved in water have different properties from ionic solids. For example, ionic solutions conduct electricity, but the solid forms do not. This makes dissolving seem like a chemical change. However, the identity of the substances has not changed. You can recover solid compounds from solutions by evaporating the water they are dissolved in. So you can categorize dissolving as either physical or chemical change.

## Example

### Magnesium Chloride and Water

Write two equations that show what happens when magnesium chloride, $MgCl_2(s)$, dissolves.

### Solution

$$MgCl_2(s) \longrightarrow MgCl_2(aq)$$

$$MgCl_2(s) \longrightarrow Mg^{2+}(aq) + 2Cl^-(aq)$$

## LESSON SUMMARY

### How are changes in matter classified?

In general, changes in matter can be classified as either physical or chemical. Physical changes, including phase changes, involve a change in form without changing the identity of the substance. There is no change in chemical formula. Chemical changes, also called chemical reactions, involve the formation of new substances with new properties. The chemical formulas of these new products are different from the chemical formulas of the reactants. Dissolving can fall into both categories of change, though it is usually considered a physical change.

**KEY TERM**

physical change

## Exercises

### Reading Questions

1. What is the difference between a physical change and a chemical change?

2. Explain why dissolving can be described as either a physical or chemical change.

### Reason and Apply

3. Give five examples of physical changes. Explain why each example is a physical change.

*(continued)*

4. Copy the equations 1–7 onto your paper. Match the chemical equations with their descriptions.

   1. $2CH_4O(l) + O_2(g) \longrightarrow$
      $2CH_2O(l) + 2H_2O(l)$

   2. $NH_2Cl(g) \longrightarrow NH_2Cl(aq)$

   3. $2C_8H_{14}(l) + 23O_2(g) \longrightarrow$
      $16CO_2(g) + 14H_2O(l)$

   4. $H_2SO_4(aq) + CaCO_3(s) \longrightarrow$
      $CaSO_4(aq) + CO_2(g) + H_2O(l)$

   5. $Hg(l) \longrightarrow Hg(g)$

   6. $C_{16}H_{30}O_2(s) + H_2 \longrightarrow$
      $C_{16}H_{32}O_2(s)$

   7. $CaO(s) + H_2O(l) \longrightarrow$
      $Ca(OH)_2(s)$

   A. The sulfuric acid dissolved in raindrops of acid rain reacts with the calcium carbonate in seashells and marble structures. This reaction produces calcium sulfate dissolved in water and carbon dioxide gas.

   B. To balance the effect of acid rain, solid calcium oxide has been added to many lakes. It reacts with water to form solid calcium hydroxide, a strong base that is only slightly soluble in water.

   C. Chloramine is added to our water supply in very small amounts to kill bacteria.

   D. When octane and oxygen gas are burned in our cars, carbon dioxide and water come out in the exhaust.

   E. Methanol, if ingested, reacts with oxygen to form formaldehyde, which is toxic. Water is also formed in this reaction.

   F. Liquid mercury evaporates to produce mercury vapor.

   G. Saturated fatty acids, like palmitic acid, tend to form long solids and clog people's arteries. These saturated fatty acids can be made from unsaturated fatty acid by adding hydrogen gas.

5. For equations 1–7 from Exercise 4, label each one as a physical or chemical change.

6. Classify the following two changes as physical or chemical. Explain your reasoning.

   a. $CaCO_3(s) + H_2SO_4(aq) \longrightarrow CaSO_4(aq) + CO_2(g) + H_2O(l)$

   b. $NaCl(s) \longrightarrow NaCl(l)$

# Conservation of Mass

## THINK ABOUT IT

During chemical and physical changes, the atoms in a substance rearrange. Sometimes, a solid forms or disappears. Sometimes, the substances change appearance or substances with new properties appear. But is matter really "appearing" or "disappearing"?

**How does mass change during a chemical or physical change?**

*To answer this question, you will examine*

➲ Tracking Mass

➲ Conservation of Mass

## EXPLORING THE TOPIC

## ➲ Tracking Mass

In Unit 1: Alchemy, you learned that matter is conserved during chemical reactions. If matter is not created or destroyed during physical and chemical changes, then there should be no measurable difference in mass before and after a change. Three chemical procedures are described here. The mass before and after each change is shown on the balances.

### PHYSICAL CHANGE—DISSOLVING A SOLID

Consider what happens when water is added to solid sodium carbonate.

$$Na_2CO_3(s) \longrightarrow Na_2CO_3(aq)$$

$H_2O(l)$    $Na_2CO_3(s)$      $Na_2CO_3(aq)$

125.00 g    =    125.00 g

*Mass of reactants*      *Mass of products*

The solid seems to disappear into the liquid. However, notice that the total mass doesn't change. It is the same before and after the substances are mixed. The ions in the white solid have spread evenly throughout the water.

This change can also be represented by the chemical equation:

$$Na_2CO_3(s) \longrightarrow 2Na^+(aq) + CO_3^{2-}(aq)$$

Even though this compound breaks apart into ions, the mass is the same before and after the change.

© moodboard/Corbis

**Important to Know** It is difficult to measure the weight of a sample of gas, so it is tempting to conclude that gases do not have mass. However, as you saw in Unit 3: Weather, gases are made of molecules and they do have mass.

## CHEMICAL CHANGE—PRODUCING A SOLID

Consider a chemical change in which a new solid forms. Adding an aqueous solution of sodium chloride to an aqueous solution of silver nitrate produces aqueous sodium nitrate and solid silver chloride.

$$NaCl(aq) + AgNO_3(aq) \longrightarrow NaNO_3(aq) + AgCl(s)$$
$$58\text{ g} \quad + \quad 170\text{ g} \quad = \quad 85\text{ g} \quad + \quad 143\text{ g}$$

Even though a solid is produced, the mass of the products is still equal to the mass of the reactants. No mass was gained or lost.

## CHEMICAL CHANGE—PRODUCING A GAS

Now consider a chemical change that produces a gas as a product. Adding an aqueous solution of hydrochloric acid to solid magnesium metal produces an aqueous solution of magnesium chloride and bubbles of hydrogen gas.

$$2HCl(aq) + Mg(s) \longrightarrow H_2(g) + MgCl_2(aq)$$

*Mass of reactants*      *Mass of products*

According to the balance, some mass appears to have been lost. However, notice that some gas escaped from the beaker. If you were able to trap the gas that escaped and measure its mass, it would account for the difference in masses seen here.

---

**Example**

### Chemical Change—Decomposing Calcium Carbonate

Examine this description of a chemical change along with its chemical equation.

**Verbal description:** Solid calcium carbonate is heated to produce solid calcium oxide and carbon dioxide gas.

**Chemical equation:** $CaCO_3(s) \longrightarrow CaO(s) + CO_2(g)$
$$50\text{ g} \quad = \quad 28\text{ g} \quad + \quad ?$$
mass of reactants = mass of products

What would you expect the mass of the carbon dioxide gas to be? Explain your reasoning.

### Solution

The mass of the products on the right side must equal the mass of the reactants on the left side. So, the mass of the carbon dioxide produced is 22 grams.

## ➲ Conservation of Mass

The fact that the mass of the reactants is equal to the mass of the products follows the law of conservation of mass. Mass is conserved during chemical and physical changes because no atoms are created or destroyed.

### GETTING RID OF GARBAGE

People generate a lot of waste: cups, cans, bottles, candy wrappers, paper, and diapers. Most of this waste just keeps piling up in landfills. Some of the waste is quite toxic. In the United States alone, about 135 million tons of garbage are put into landfills each year. Will it be there forever?

Atoms are not created or destroyed; they are simply rearranged to form new substances. Therefore, everything on the planet is made up of a limited number of atoms. Even your body does not have the same atoms in it that it did several years ago. You are continually generating new cells of every kind and getting rid of the old ones.

The law of conservation of mass is important in considering waste disposal. You cannot get rid of the atoms in the waste that you throw away. Smelly, toxic garbage accumulates in large quantities in waste disposal sites. A small portion of it may biodegrade—that is, get broken down into harmless products by microorganisms—but the process can take decades. The waste will be there unless scientists find ways to convert more of our waste into useful products. Given this fact, it makes sense to reuse and recycle as much as possible.

## LESSON SUMMARY

### How does mass change during a chemical or physical change?

When chemical and physical changes take place, the atoms in substances rearrange. A rearrangement may be a physical change, such as the mixing of two substances, or a chemical change, the formation of new compounds. However, the atoms involved in these rearrangements cannot be created or destroyed. A chemical equation tracks the atoms involved in chemical and physical changes. All of the atoms are accounted for, so the mass of the products is identical to the mass of the reactants. This is known as the law of conservation of mass.

## Exercises

### Reading Questions

1. Explain the law of conservation of mass.
2. Explain how the law of conservation of mass applies to garbage.

*(continued)*

**Reason and Apply**

3. Below is a chemical equation along with a verbal description of the reaction.

   **Verbal description:** Gaseous sulfur trioxide is added to liquid water to produce aqueous sulfuric acid.

   $$\text{Chemical equation: } SO_3(g) + H_2O(l) \longrightarrow H_2SO_4(aq)$$

   a. Was matter lost or gained during this reaction? Explain how you could prove this by taking measurements.
   b. The left side of the equation shows one atom of S, two atoms of H, and four atoms of O. How many atoms of each element are on the right side of the equation? How does this provide evidence for the law of conservation of mass?

4. When an ice cube melts, which of these quantities will change?
   **(A)** The number of atoms it contains
   **(B)** Its mass
   **(C)** Its volume
   **(D)** All of the above
   **(E)** None of the above

5. Explain what happens to the number of atoms, the mass, and the weight of the water in a glass when it evaporates.

6. Write a paragraph to convince a friend that mass is conserved when chemical and physical changes take place. Give evidence.

7. What would you have to do to prove that matter is conserved when a piece of paper is burned?

**LESSON 72**

# Balancing Chemical Equations

## THINK ABOUT IT

The law of conservation of mass states that mass is not lost or gained in a chemical reaction. When you write a chemical equation to describe change, it is important that the equation follows this law. So any equation describing a chemical change must account for every atom involved.

### How do you balance atoms in a chemical equation?

*To answer this question, you will explore*

⮕ Balancing Chemical Equations

⮕ Coefficients Are Counting Units

## EXPLORING THE TOPIC

## ⮕ Balancing Chemical Equations

Imagine you're in business to make ammonia, $NH_3$, for farmers to fertilize their crops. It would be helpful to know how much nitrogen, $N_2(g)$, and hydrogen, $H_2(g)$, to combine so that nothing is wasted. How do you make sure you have the correct amount of nitrogen and hydrogen so that you have enough of each with no extra?

A chemical equation represents the exact ratio in which reactants combine to form products. A chemical equation that accounts for all the atoms involved and shows them combining in the correct ratio is called a *balanced* chemical equation. Learning how to balance chemical equations is a necessary part of working with chemical reactions.

### FORMATION OF AMMONIA

The unbalanced equation below describes how nitrogen gas and hydrogen gas react to make ammonia gas.

To balance this equation, first take an inventory of the atoms on each side.

| Reactants | | $N_2(g)$ | + | $H_2(g)$ | ⟶ | $NH_3(g)$ | Product |

**Inventory of Atoms**

| Reactants | Products | |
|-----------|----------|---|
| 2 N | 1 N | ← Unbalanced |
| 2 H | 3 H | ← Unbalanced |

Next, balance the atoms on each side by adding units of $N_2$, $H_2$, or $NH_3$. The product side needs one more N atom. The only way to accomplish this is to add a whole ammonia molecule, $NH_3$, to the product side.

__ N₂(g)    +    __ H₂(g)    ⟶    **2**NH₃(g)

**Inventory of Atoms**

| Reactants | Products | |
|-----------|----------|---|
| 2 N | 2 N | ← Balanced |
| 2 H | 6 H | ← Unbalanced |

Take a new inventory. The N atoms are now balanced. Now the reactant side needs four more H atoms. The only way to accomplish this is to add two more molecules of H₂ to the reactant side.

__ N₂(g)    +    **3** H₂(g)    ⟶    __ 2NH₃(g)

**Inventory of Atoms**

| Reactants | Products | |
|-----------|----------|---|
| 2 N | 2 N | |
| 6 H | 6 H | ← Balanced |

> **Important to Know** To balance a chemical equation, you can change only the coefficients, the numbers in front of each chemical formula. You cannot change the chemical formulas.

The equation is now balanced. The balanced equation shows that mass is conserved. There are the same numbers of N and H atoms on each side of the arrow.

In the balanced equation above, the 3 in front of the H₂ and the 2 in front of the NH₃ are called **coefficients.** Coefficients indicate how many units of each substance take part in the reaction. If there is no number in front of a chemical formula in an equation, the coefficient is understood to be a 1.

### FORMATION OF RUST

Iron reacts with oxygen in the air to form iron (III) oxide, or rust. Iron (III) oxide is *not* a molecule. It is a highly organized collection of Fe and O atoms that are bonded to each other ionically. It is an ionic solid. There are a huge number of atoms in any piece of iron (III) oxide. But each piece will have two Fe atoms for every three O atoms. So Fe₂O₃ is the **formula unit** of iron oxide.

The unbalanced equation below describes how iron reacts with the oxygen in the air to form iron (III) oxide.

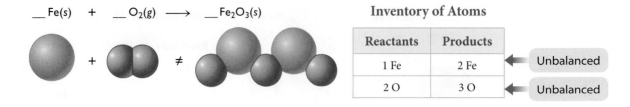

__ Fe(s)    +    __ O₂(g)    ⟶    __ Fe₂O₃(s)

**Inventory of Atoms**

| Reactants | Products | |
|-----------|----------|---|
| 1 Fe | 2 Fe | ← Unbalanced |
| 2 O | 3 O | ← Unbalanced |

Balance the atoms on each side by adding molecules of O₂, atoms of Fe, or formula units of Fe₂O₃. If you add an Fe atom to the reactant side, the Fe atoms are balanced. However, the O atoms are still not equal.

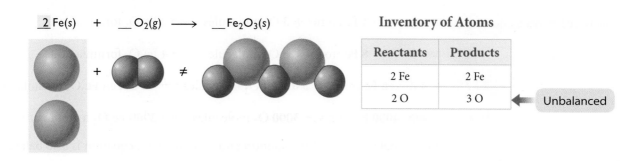

$\underline{2}\,Fe(s) \quad + \quad \underline{\phantom{2}}\,O_2(g) \longrightarrow \underline{\phantom{2}}\,Fe_2O_3(s)$

**Inventory of Atoms**

| Reactants | Products |
|-----------|----------|
| 2 Fe | 2 Fe |
| 2 O | 3 O |

⬅ Unbalanced

If you add two $O_2$ molecules to the reactant side and one $Fe_2O_3$ formula unit to the product side, the oxygen atoms are balanced.

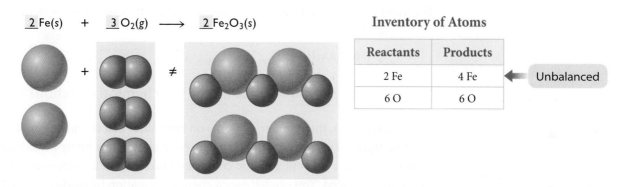

$\underline{2}\,Fe(s) \quad + \quad \underline{3}\,O_2(g) \longrightarrow \underline{2}\,Fe_2O_3(s)$

**Inventory of Atoms**

| Reactants | Products |
|-----------|----------|
| 2 Fe | 4 Fe |
| 6 O | 6 O |

⬅ Unbalanced

But now the iron atoms are unbalanced again. Add two more iron atoms on the reactant side, increasing the total to four. The balanced chemical reaction is shown here.

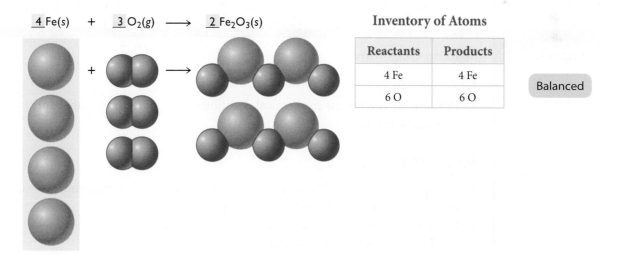

$\underline{4}\,Fe(s) \quad + \quad \underline{3}\,O_2(g) \longrightarrow \underline{2}\,Fe_2O_3(s)$

**Inventory of Atoms**

| Reactants | Products |
|-----------|----------|
| 4 Fe | 4 Fe |
| 6 O | 6 O |

Balanced

It took a few steps, but the equation is finally balanced. Four iron atoms combine with three oxygen molecules to form two formula units of iron oxide.

## ↻ Coefficients Are Counting Units

Once you have a balanced equation, multiplying all the coefficients by any counting unit will also give a balanced equation. You can multiply the coefficients by a dozen, or a thousand, or a million.

Original balanced equation → **4** Fe atoms + **3** $O_2$ molecules ⟶ **2** $Fe_2O_3$ formula units

× 2 → **8** Fe atoms + **6** $O_2$ molecules ⟶ **4** $Fe_2O_3$ formula units

× 12 → **4 dozen** Fe atoms + **3 dozen** $O_2$ molecules ⟶ **2 dozen** $Fe_2O_3$ formula units

× 1000 → **4000** Fe atoms + **3000** $O_2$ molecules ⟶ **2000** $Fe_2O_3$ formula units

Recall that a mole is also a counting unit. So, the next equation is also balanced.

× 602,000,000,000, 000,000,000,000 → **4 mol** Fe atoms + **3 mol** $O_2$ molecules ⟶ **2 mol** $Fe_2O_3$ formula units

The coefficients in chemical equations can be multiples of any counting unit. The counting unit that chemists use most often is the mole. We can use moles to count the number of atoms, molecules, or formula units involved in a reaction.

## LESSON SUMMARY

### How do you balance atoms in a chemical equation?

To conserve mass, the number of atoms on both sides of a chemical equation must be equal. When an equation is balanced, it shows how many molecules, atoms, or formula units of an ionic compound take part in a reaction and how many are produced. You can balance a chemical equation by working with the coefficients in front of the chemical formulas. The coefficients can be any counting unit, including moles. Units of mass or volume *cannot* be used as coefficients in chemical equations.

**KEY TERMS**

coefficient

formula unit

## Exercises

### Reading Questions

1. Why do chemical equations need to be balanced?
2. How are the coefficients in chemical equations different from the subscripts in chemical formulas?

### Reason and Apply

3. Your recipe for banana bread calls for two ripe bananas. However, you have six ripe bananas that you want to use before they go bad.
   a. How can you make banana bread with all six bananas?
   b. How is this related to balancing a chemical equation?

4. Copy and balance these chemical equations.
   a. $K(s) + I_2(s) \longrightarrow KI(s)$
   b. $Mg(s) + Br_2(l) \longrightarrow MgBr_2(s)$
   c. $KBr(aq) + AgNO_3(aq) \longrightarrow KNO_3(aq) + AgBr(s)$
   d. $KClO_3(s) \longrightarrow KCl(s) + O_2(g)$
   e. $C_2H_6(g) + O_2(g) \longrightarrow CO_2(g) + H_2O(l)$
   f. $Al(s) + O_2(g) \longrightarrow Al_2O_3(s)$
   g. $P_4(s) + H_2(g) \longrightarrow PH_3(g)$

# Types of Reactions

## THINK ABOUT IT

Toxins work by reacting with chemicals in the body. These toxic reactions can remove compounds that are important to body function, or they can create new compounds that interfere with normal body processes. Classifying toxic reactions according to the different ways they react can help us to understand how toxic substances work in the body. It can also help us to come up with suitable approaches for dealing with them.

**How do atoms rearrange to form new products?**

*To answer this question, you will explore*

➲ Combination and Decomposition Reactions

➲ Exchange Reactions

## EXPLORING THE TOPIC

## ➲ Combination and Decomposition Reactions

© Stephen Frink/Corbis

Imagine that you are an ocean researcher and you spend your days underwater in a submarine. As you breathe, you use up oxygen gas, $O_2(g)$, and release carbon dioxide gas, $CO_2(g)$. Eventually, the high level of carbon dioxide and low level of oxygen in your submarine will be dangerous to your health. How do you remove carbon dioxide and produce oxygen gas to sustain life in a submarine?

### COMBINATION REACTIONS

Combination reactions are used in air scrubbers on submarines to remove $CO_2(g)$ from the air. In a **combination reaction,** two reactants combine to form a single product. In Reactions 1 and 2, carbon dioxide combines with another compound to form a new product.

**Reaction 1:** Carbon dioxide gas reacts with solid sodium oxide to produce sodium carbonate (washing soda):

$$CO_2(g) + Na_2O(s) \longrightarrow Na_2CO_3(s)$$

**Reaction 2:** Carbon dioxide gas reacts with aqueous sodium hydroxide to produce sodium bicarbonate (baking soda):

$$CO_2(g) + NaOH(aq) \longrightarrow NaHCO_3(aq)$$

A combination reaction is sometimes represented by the general equation

$$A + B \longrightarrow C$$

### DECOMPOSITION REACTIONS

In Reactions 3 and 4, compounds containing oxygen atoms decompose to produce $O_2(g)$ and an ionic solid. These decomposition reactions are two ways to produce oxygen. In a **decomposition reaction,** one reactant decomposes, or breaks down, to produce two or more products.

**Reaction 3:** Solid potassium chlorate decomposes to produce solid potassium chloride and oxygen gas:

$$2KClO_3(s) \longrightarrow 2KCl(s) + 3O_2(g)$$

**Reaction 4:** Solid barium peroxide decomposes to produce solid barium oxide and oxygen gas:

$$2BaO_2(s) \longrightarrow 2BaO(s) + O_2(g)$$

A decomposition reaction is sometimes represented by the general equation

$$A \longrightarrow B + C$$

On the International Space Station, oxygen is produced by a decomposition reaction that uses electrical energy to split water into hydrogen gas and oxygen gas:

$$2H_2O(l) \longrightarrow 2H_2(g) + O_2(g)$$

## ⮕ Exchange Reactions

Calcium is a necessary part of our diet, especially as we age and our bones become more fragile. An example of a calcium supplement is calcium carbonate, $CaCO_3(s)$. However, before the calcium in a vitamin tablet can become bone, it must be made more transportable in the bloodstream. It begins its chemical journey in the stomach.

Examine the reactions involving calcium below. What patterns do you notice in the ways the atoms in the reactants are rearranged to produce the products?

**Reaction 5:** Solid calcium reacts with hydrochloric acid in the stomach to produce aqueous calcium chloride and hydrogen gas:

$$Ca(s) + 2HCl(aq) \longrightarrow CaCl_2(aq) + H_2(g)$$

**Reaction 6:** Solid calcium carbonate reacts with hydrochloric acid (stomach acid) to produce aqueous calcium chloride and carbonic acid:

$$CaCO_3(s) + 2HCl(aq) \longrightarrow CaCl_2(aq) + H_2CO_3(aq)$$

In Reaction 5, calcium atoms replace the hydrogen atoms in the HCl molecules.

$$Ca(s) + 2\ H\ Cl(aq) \longrightarrow CaCl_2(aq) + H_2(g)$$

Ca replaces H.

In Reaction 6, the polyatomic ion, $CO_3^{2-}$, switches places with the chlorine ion, $Cl^-$. The second reaction is a common way in which calcium is made available to your body through chemical changes that take place in your stomach.

$$Ca\ CO_3(s) + 2H\ Cl\ (aq) \longrightarrow CaCl_2(aq) + H_2CO_3(aq)$$

$CO_3^{2-}$ ions change places with $Cl^-$ ions.

In both cases, the two reactants exchange atoms. The first reaction is called a **single exchange reaction** or a single replacement reaction because atoms of an element are exchanged with atoms of another element in a compound. A single exchange reaction is sometimes represented by the general equation

$$A + BC \longrightarrow AC + B$$

The second reaction is called a **double exchange reaction** or a double replacement reaction because two compounds exchange atoms with each other. A double exchange reaction is sometimes represented by the general equation

$$AB + CD \longrightarrow AD + CB$$

**Example**

## Classify Chemical Reactions

Classify each reaction as a combination reaction, a decomposition reaction, a single exchange reaction, or a double exchange reaction.

   **a.**  $H_2(g) + Cl_2(g) \longrightarrow 2HCl(g)$

   **b.**  $CaCO_3(s) \longrightarrow CaO(s) + CO_2(g)$

   **c.**  $Sn(s) + O_2(g) \longrightarrow SnO_2(s)$

   **d.**  $CaI_2(s) + Cl_2(g) \longrightarrow CaCl_2(s) + I_2(g)$

   **e.**  $AgNO_3(aq) + NaOH(aq) \longrightarrow AgOH(s) + NaNO_3(aq)$

## Solution

   **a.**  Two reactants combine to produce one product. This is a combination reaction.

   **b.**  One reactant decomposes to produce two products. This is a decomposition reaction.

   **c.**  Two reactants combine to produce one product. This is a combination reaction.

   **d.**  Chlorine, Cl, replaces iodine, I, in $CaI_2$. This is a single exchange reaction.

   **e.**  Silver, Ag, and sodium, Na, exchange places. This is a double exchange reaction.

Toxins sometimes act in the body in an exchange reaction by replacing an atom or atoms from a compound that exists in the body. This can have the effect of destroying necessary compounds or creating a new compound with harmful properties, or both. One way to remove toxic substances from the environment is by using combination and decomposition reactions.

## LESSON SUMMARY

**KEY TERMS**

combination reaction

decomposition reaction

single exchange reaction
   (single replacement)

double exchange reaction
   (double replacement)

### How do atoms rearrange to form new products?

Chemical reactions can be classified into general categories based on how atoms in the reactants rearrange to form the products. Four general types of chemical change include combination reactions, decomposition reactions, single exchange reactions, and double exchange reactions. Toxins sometimes harm the body through chemical reactions. However, chemical reactions can also be used to remove toxins from the body or the environment.

**Reading Questions**

1. How are combination reactions and decomposition reactions related?
2. What is the difference between a single exchange reaction and a double exchange reaction?

**Reason and Apply**

3. Classify these reactions as combination, decomposition, single exchange, or double exchange. Copy the equations, fill in any missing products, and write a balanced equation for each reaction.
   a. $NaOH(aq) + HNO_3(aq) \longrightarrow NaNO_3(aq) +$ _____ $(l)$
   b. $C_2H_4(g) + Cl_2(g) \longrightarrow$ _____ $(g)$
   c. $Cl_2(g) + MgBr_2(s) \longrightarrow Br_2(s) +$ _____ $(s)$

4. List four molecules and four ionic compounds in the reactions from Exercise 3.

5. Sulfur trioxide gas combines with liquid water to produce aqueous sulfuric acid:

$$\text{_____} (g) + H_2O(l) \longrightarrow H_2SO_4(aq)$$

   Predict the missing reactant, balance the equation, and explain how these toxic substances could enter the body.

6. Solid lithium reacts with aqueous hydrochloric acid to produce hydrogen gas and aqueous lithium chloride:

$$Li(s) + HCl(aq) \longrightarrow H_2(g) + \text{_____}$$

   Predict the missing product, balance the equation, and explain how these toxic substances could enter the body.

7. Aqueous silver nitrate and aqueous sodium hydroxide are mixed to produce aqueous sodium nitrate and solid silver hydroxide:

$$AgNO_3(aq) + NaOH(aq) \longrightarrow NaNO_3(aq) + \text{_____}$$

   Predict the missing product, balance the equation, and explain how the reaction could help remove toxic substances from water.

# CHAPTER 13

## Toxic Changes

## SUMMARY

### KEY TERMS

**physical change**

**coefficient**

**formula unit**

**combination reaction**

**decomposition reaction**

**single exchange reaction**
   **(single replacement)**

**double exchange reaction**
   **(double replacement)**

# Toxins Update

A toxic substance causes an undesirable chemical reaction, producing a harmful or unhealthy change in a living system. Chemical equations keep track of these changes and allow you to predict what you will observe when compounds combine. Becoming familiar with chemical equations is the first step in understanding these chemical changes.

## REVIEW EXERCISES

1.  Why is it difficult to identify a physical or chemical change through observations alone?

2.  Based on the chemical equation for a reaction, can you tell if any of the substances are toxic?

3.  Consider the equation for the formation of a kidney stone.

    $$Na_3PO_4(aq) + 3CaCl_2(aq) \longrightarrow Ca_3(PO_4)_2(s) + 6NaCl(aq)$$

    a.  Is each reactant bonded ionically or covalently? How do you know?
    b.  Is this is a combination reaction, decomposition reaction, single exchange reaction, or double exchange reaction?
    c.  Is this a chemical change or a physical change?
    d.  How does a balanced reaction show that matter is conserved?
    e.  What is the chemical name of the solid that makes up a kidney stone?

4.  Copy and balance these chemical equations.
    a.  $N_2(g) + H_2(g) \longrightarrow N_2H_4(g)$
    b.  $KNO_3(s) + K(s) \longrightarrow K_2O(s) + N_2(g)$
    c.  $H_2SO_4(aq) + NaCN(aq) \longrightarrow HCN(g) + Na_2SO_4(aq)$
    d.  $H_3PO_4(aq) + Ca(OH)_2(aq) \longrightarrow Ca_3(PO_4)_2(aq) + H_2O(l)$
    e.  $C_3H_8(g) + O_2(g) \longrightarrow CO_2(g) + H_2O(l)$
    f.  $H_2S(g) + O_2(g) \longrightarrow SO_2(g) + H_2O(l)$
    g.  $H_2(g) + O_2(g) \longrightarrow H_2O(l)$

---

## Toxins in the Environment

**PROJECT**
**www**
Research a potentially toxic substance (your teacher may assign you one). Find out where in your environment you might find this substance and describe its effects on the body. Prepare a short report.

---

# CHAPTER 14

# Measuring Toxins

### *In this chapter, you will study*

- how lethal doses of toxic substances are measured

- ways to count small objects by weighing them

- converting between counting units (moles) and mass units

- how Avogadro's number is used

*Many compounds, such as the medications shown here, keep us healthy and even cure diseases. But these same substances can be quite harmful if taken in the wrong doses.*

Any substance, even water, can be toxic if too much of it is consumed. Yet many substances that are considered quite harmful are therapeutic when administered in the correct amounts. The dose of a substance is key to determining the safety of that substance. This chapter concentrates on how amounts of matter are measured and counted.

# Toxicity

## THINK ABOUT IT

Poisonous snakebites are medical emergencies. The venom from the bite is very toxic and can be deadly, especially for small children or small animals such as rabbits and mice. However, tiny amounts of snake venom have been used therapeutically as a medicine to control high blood pressure. How much snake venom is toxic and how much is therapeutic?

### How much is too much of a substance?

*To answer this question, you will explore*

➲   How Toxicity Is Measured

➲   Relationship to Body Weight

## EXPLORING THE TOPIC

### ➲  **How Toxicity Is Measured**

You have probably been warned about the dangers of dozens of different substances, from pesticides to poisonous wild mushrooms. You know better than to drink your shampoo or rub gasoline on your skin. But how is toxicity determined in the first place? And how is it that some things that are good for you, like vitamin tablets, can be toxic or even lethal in high doses?

#### MEASURING TOXICITY

For our safety, scientists try to determine the **toxicity** of the products we buy and the substances we come in contact with. Sometimes, they find the toxicity accidentally, when a person is exposed to a harmful substance. Other times, it takes a great deal of evidence before it is determined that a substance is toxic. This was the case with lead-based paints.

It is clearly not a good idea to determine how toxic something is by purposefully exposing humans. Yet it is vital to know how much of a particular substance is dangerous. So, toxicities of most substances are measured by exposing laboratory animals to toxic substances. Typically, rats or mice (sometimes rabbits, dogs, or monkeys) are used. The harm that might be done to the animals must be carefully weighed against potential harm to humans if the toxicities of substances, such as medicines, are unknown.

One way to measure toxicity is to measure the *lethal dose* of a substance, the amount of substance that can cause death. The $LD_{50}$ of a substance is the amount of a substance that causes the death of half, or 50%, of the animals exposed to it.

**BIOLOGY
CONNECTION**

One of the chemicals in the venom of certain species of snakes lowers the blood pressure of their prey to make them unconscious. In tiny doses the same chemical can be used to treat high blood pressure.

Edelcio Muscat/Getty Images

The table provides the $LD_{50}$'s for several common substances in grams, milligrams, or micrograms (mcg) of substance per kilogram of body mass. A microgram is one millionth of a gram.

## Lethal Doses

| Common name | Chemical name and formula | Lethal dose ($LD_{50}$) per kg body mass | Toxic response |
| --- | --- | --- | --- |
| aspirin | acetylsalicylic acid, $C_9H_8O_4$ | 200 mg/kg (rat, oral) | gastric distress, confusion, psychosis, stupor, ringing in ears, drowsiness, hyperventilation |
| table salt | sodium chloride, NaCl | 3 g/kg (rat, oral) 12,357 mg/kg (human, oral) | eye irritation, elevated blood pressure |
| castor beans | ricin—very large protein molecule, molecular mass 63,000 amu | 30 mg/kg (human, oral) 3.0 mcg/kg (human, intravenous) | vomiting, diarrhea, internal bleeding, kidney and liver failure; death within minutes if injected |
| arsenic | arsenic (III) oxide $As_2O_3$ | 15 mg/kg (rat, oral) | *acute:* irritates eyes, skin, respiratory tract; nausea *chronic:* convulsions, tissue lesions, hemorrhage, kidney impairment |
| sugar | glucose, $C_6H_{12}O_6$ | 30 g/kg (rat, oral) | depressed activity, gastric disturbances; if diabetic: heart disease, blindness, nerve damage, kidney damage |
| snake venom | alpha-bungarotoxin, $C_{338}H_{529}N_{97}O_{105}S_{11}$ | 25.0 mcg/kg (rat, intramuscular) | paralysis, suffocation, loss of consciousness, seizures, hemorrhaging into tissues |
| coffee beans | caffeine, $C_8H_{10}N_4O_2$ | 140 mg/kg (dog, oral) 192 mg/kg (rat, oral) | acute kidney failure, nausea, psychosis, hemorrhage, increased pulse, convulsions |

Notice the wide range of doses—from grams, to milligrams (thousandths of a gram), to micrograms (millionths of a gram). A small $LD_{50}$ indicates that it does not take very much of the substance to produce ill effects. Substances with small $LD_{50}$'s should definitely be avoided. Snake venom is one of the most dangerous toxins in the table, with an $LD_{50}$ of 25.0 mcg/kg, while sugar, with an $LD_{50}$ of 30 g/kg is one of the safer substances.

## DOSAGE DETERMINES TOXICITY

You may be surprised to see lethal doses for substances like sugar and aspirin. These substances are usually considered beneficial. However, everything is toxic in a large enough dose, even life-sustaining substances like water and oxygen. And scientists have discovered that many highly toxic substances also have therapeutic value. Botulinum toxin (Botox) can help patients suffering from cerebral palsy. A very large protein isolated from the venom of the blue scorpion has been discovered to inhibit certain cancer tumors.

*Castor oil is made from castor beans, which are the seeds of the castor plant.*

*The blue scorpion is found mainly in Cuba and other areas of the Caribbean.*

## MEDICINE CONNECTION

Many toxins serve as medicines when used in moderation. Atropine, for example, extracted from the deadly nightshade plant, is used to treat cardiac arrest and poisoning by organophosphate insecticides and nerve gases. However, atropine is potentially addictive and an overdose can be fatal.

## ⊃ Relationship to Body Weight

You may be wondering how testing a substance on very small animals, like rats, will determine whether it is toxic to a larger mammal, such as a human. To take size into account, the $LD_{50}$ is reported in milligrams of substance per kilogram of body weight, mg/kg. That way the lethal dose can be estimated for a mammal of any size.

Testing lab animals gives us only an approximate range for safety. Our bodies differ in many ways from the bodies of rats or rabbits. And it may seem particularly unkind to subject animals to testing. However, the alternative would be to risk human life with untested medications. Imagine if the safety of products on store shelves had never been confirmed in any way.

**Example**

### Caffeine

The lethal dose for caffeine is approximately 150–200 mg/kg of body mass. How much caffeine would be lethal for a 120 lb person? How many cups of coffee is this?

*Structural formula for caffeine*

I cup
strong coffee

*Contains ~150 mg caffeine*

### *Solution*

Lethal dose is reported in mg/kg, so first convert pounds to kilograms. There are 2.2 pounds in a kilogram.

| | |
|---|---|
| Convert body weight from pounds to kilograms. | $120 \text{ lb} \cdot \dfrac{1 \text{ kg}}{2.2 \text{ lb}} = 54.4 \text{ kg}$ |
| Multiply the person's weight by the lethal dose. | $54.4 \text{ kg} \cdot \dfrac{150 \text{ mg}}{1 \text{ kg}} = 8160 \text{ mg caffeine}$ |
| Convert number of milligrams to number of cups. | $8160 \text{ mg} \cdot \dfrac{1 \text{ cup}}{150 \text{ mg}} = 54.4 \text{ cups}$ |

So, it would be difficult to ingest a lethal dose of caffeine simply by drinking coffee. (It would take about 54 cups for the 120 lb person!)

[For a review of this math topic, see **MATH Spotlight**: Dimensional Analysis on page A-16.]

## How much is too much of a substance?

Every substance on the planet is a potential toxin. By the same token, many substances that are considered toxic have health benefits when taken in the correct dose. Thus, toxicity is determined by dose. Scientists use a variety of methods to measure toxicity. One measurement is called the $LD_{50}$. This is the amount of a substance that kills 50% of animals exposed to it. It is commonly expressed in milligrams of substance per kilogram of body weight. A low $LD_{50}$ means a more toxic substance.

**KEY TERM**

toxicity

## Exercises

### Reading Questions

1. How can scientists determine the toxicity of a substance?
2. How is toxicity related to body weight?

### Reason and Apply

3. Find at least five products at home with labels that warn of toxicity. Give the name of each product, and answer these questions:
   a. How does each label advise you to avoid harmful exposure to the product?
   b. What does each label tell you to do if a dangerous exposure does occur?
   c. Using all of this information, what can you hypothesize about the chemical properties of each product?
   d. Look at the recommended treatment for dangerous exposure to each product. What chemical and physical processes do you think might be involved in the treatment?

4. Ethanol is grain alcohol. The $LD_{50}$ for ethanol is 7060 mg/kg (rat, oral).
   a. How many milligrams of ethanol would be lethal to a 132 lb adult?
   b. How many glasses containing 13,000 mg of ethanol would be lethal to a 22 lb child?

5. The $LD_{50}$ for vitamin A is 1510 mg/kg (rat, oral).
   a. How many mg of vitamin A would be lethal to a 132 lb adult?
   b. How many vitamin tablets containing 0.40 mg of vitamin A would be lethal to an adult?
   c. **PROJECT**
    Use the Internet or other resource to research vitamin A. What are the benefits of vitamin A?

# Counting by Weighing

## Purpose

To count large numbers of small objects by weighing.

## Procedure

1. Obtain a sandwich bag from your teacher. Your bag may contain items such as rice, beans, or paper clips.

2. Your challenge is to determine the number of objects in your bag without opening it.

3. Brainstorm how you will solve the challenge with the members of your group and decide what tools you might use.

## Questions

1. Describe the method you are going to use to determine the number of objects you have in your bag. Be specific.

2. Find out what substances are in the sandwich bags of the other groups. Make a list of all the substances.

3. Which of the substances will be easiest to count? Explain your reasoning.

4. **Making Sense** What do you think this activity has to do with keeping track of chemical compounds?

5. **If You Finish Early** Are 100 molecules of snake venom equivalent in toxicity to 100 atoms of arsenic? Why or why not?

© STUDIO SATO/amanaimages/Corbis

*Make It Count*

# Counting by Weighing

## THINK ABOUT IT

When a toxin enters your body, it is important for a doctor to know exactly how much toxin there is. One molecule of snake venom has over 100 times as much mass as one atom of arsenic. So, 1 milligram of snake venom contains many fewer particles than 1 milligram of arsenic. To track the toxin, you need an inventory, or a count, of the atoms or molecules. However, atoms are so small that you cannot see them, let alone count them. Luckily, there are other ways of figuring out the number of atoms in a sample besides counting them one by one.

**How can mass be used to count large numbers of small objects?**

*To answer this question, you will explore*

⮕ Using Mass to Count

⮕ Weighing Atoms

## EXPLORING THE TOPIC

## ⮕ Using Mass to Count

Let's explore how to count objects by weighing them. The masses of a number of different small objects are provided in the table.

| Substance | Mass of 1000 pieces | Mass of 10 pieces | Mass of 1 piece |
|---|---|---|---|
| rice grains | 22 g | 0.22 g | 0.022 g |
| lentils | 56 g | 0.56 g | 0.056 g |
| rubber bands | 260.5 g | 2.60 g | 0.260 g |
| paper clips | 500 g | 5.00 g | 0.500 g |
| pennies | 2500 g | 25.00 g | 2.500 g |

It is difficult to accurately measure the mass of an object if the object has a mass that is below the detection limit of the balance. Many electronic balances have 0.001 g as a lower weight limit. So, weighing heavier objects tends to yield more accurate results.

One way around this difficulty is to count out a collection of identical (or nearly identical) objects and weigh the collection. You can then find the mass of a single object by dividing the mass of the collection by the number of objects. This gives you an average mass of one object that is more accurate than if you had weighed the object individually.

Once you know the average mass of a single object, you can use it to determine exactly how many objects you have in a large sample. Consider a sample of pushpins.

| | |
|---|---|
| 10 pushpins have a mass of 4.7 g. | |
| So 1 pushpin has an average mass of 0.47 g. | |
| A large sample of pushpins has a mass of 84.6 g. | |
| Therefore this sample contains | $\dfrac{84.6 \text{ g}}{0.47 \text{ g/pushpin}} = 180$ pushpins |

In addition to helping you overcome the detection limit of the balance, there is a second reason that it is important to find the *average* mass of the objects in a sample: There is a slight variation in the masses of the objects. For example, not all grains of rice are identical in size. Some are slightly larger and others are slightly smaller (like isotopes of atoms). So, an average mass is a more accurate predictor of the mass of a *typical* object.

## Example

### A 5 lb Bag of Rice

How many grains of rice are there in a 5 lb bag of rice? The average mass of a rice grain is 0.0221 g, and the bag contains exactly 5.00 lb. (1 lb = 454 g)

*Solution*

First, determine how many grains of rice are in the bag.

Multiply by the conversion factor.

$$\text{mass of rice in grams} = 5.00 \text{ lb} \cdot \frac{454 \text{ g}}{1 \text{ lb}}$$

$$= 2270 \text{ g rice}$$

This is the mass of all the grains of rice. Next, determine how many grains of rice are in the bag.

$$\text{Total mass of rice} = \text{average mass of 1 grain} \cdot \text{number of grains}$$

Divide the total mass of the rice by the average mass of 1 grain.

$$\text{number of grains} = \frac{\text{total mass of rice}}{\text{average mass of 1 grain}}$$

Calculate the answer.

$$= \frac{2270 \text{ g}}{0.0221 \text{ g}} = 103{,}181 \text{ grains}$$

So, rounded to three significant digits, there are about 103,000 grains.

[For a review of these math topics, see **MATH Spotlight**: Dimensional Analysis on page A-16 and **MATH Spotlight**: Accuracy, Precision, and Significant Digits on page A-1.]

## PERCENT ERROR

Chemists use **percent error** to express how accurate their measurements are. For example, imagine a sandwich bag contains exactly 340 beans. Suppose you calculated the number of beans using mass and came up with 352 beans. To figure out your percent error, you would use this formula:

$$\text{Percent error} = \left| \frac{(\text{observed value} - \text{actual value})}{\text{actual value}} \right| \cdot 100\%$$

In this case, the percent error is

$$\left| \frac{352 - 340}{340} \right| \cdot 100\% = \left| \frac{12}{340} \right| \cdot 100\% = 3.5\%$$

Notice that saying your answer was off by 3.5% is more meaningful than saying you were off by 12 beans, which could be a lot or a little, depending on how many beans you were counting. The smaller the percent error, the more accurate your answer.

## ⟳ Weighing Atoms

You can apply the same method you used to count tiny objects to counting atoms. To determine the number of atoms in a 1.0 g sample, you will need to know the average mass of the atoms in the sample. The average mass of the atoms of each element is given on the periodic table in atomic mass units, or amu.

Atomic mass units are special units used for atoms because the mass of an atom in grams is so tiny. Each atomic mass unit is equal to 0.00000000000000000000000166 g. There are 23 zeros in front of the 166, making this number inconvenient to use.

Consider hydrogen atoms. Hydrogen has an average atomic mass of about 1.0 amu, or 0.00000000000000000000000166 g. How many hydrogen atoms are in a 1.0 g sample?

$$\text{number of H atoms} = \frac{\text{total mass of the sample of H atoms}}{\text{average mass of 1 H atom}}$$

$$= \frac{1.0 \text{ g}}{0.00000000000000000000000166 \text{ g}}$$

$$= 602{,}000{,}000{,}000{,}000{,}000{,}000{,}000 \text{ H atoms}$$

So 1 gram of hydrogen atoms contains 602 sextillion, or 1 mole, of hydrogen atoms.

## LESSON SUMMARY

### How can mass be used to count large numbers of small objects?

It is difficult to count very small objects like atoms and molecules. One method for getting a count of small objects is to weigh a large number of the objects and divide by the average weight of one object. The average atomic mass on the periodic table is equivalent to the mass of one atom of each element. This number is an average because some atoms are isotopes with different masses. The mass of a single atom is expressed in atomic mass units rather than grams. To find the number of atoms in a sample of an element, you can divide the mass of the sample by the average mass of one atom.

### BIOLOGY CONNECTION

Some snakebites can be harmless, others lethal. For example, the bites of garter snakes, found throughout North America, are generally harmless. However, it would take just 1/14,000 ounce of venom from an Australian brown snake, shown here, to kill a person.

William West/AFP/Getty Images

**KEY TERM**

percent error

### Reading Questions

1. Explain how you can use mass to count large numbers of objects.

2. What does the percent error tell you?

### Reason and Apply

3. Recall the method you used to count the objects in the sandwich bag you received in class and the results you obtained.
   a. Explain what you were trying to find out.
   b. Explain the method you used.
   c. Show your calculations and results.
   d. Calculate the percent error. Explain how you might modify your procedure to reduce the percent error.

4. Suppose that you have 50 grams of rice and 50 grams of beans. Which sample has more pieces? Explain your thinking.

5. Suppose that you have 740 tiny plastic beads and 740 marbles. Which sample has more mass? Explain your reasoning.

6. One bean weighs 0.074 g. How many beans are in a 50-pound bag?

7. Suppose you want to fill a 500 g bag with twice as many red jelly beans as yellow. What mass of red jelly beans do you need? A jelly bean weighs 0.65 g.

8. What is the mass in grams of one copper atom?

9. What is the mass in grams of one gold atom?

10. Suppose you have 50 grams of copper and 50 grams of gold. Which sample has more atoms? Explain your thinking.

11. Suppose you use mass to calculate the number of beans. Which of these experimental values has a smaller percent error?
    **(A)** A calculated value of 1342 when the actual value is 1327.
    **(B)** A calculated value of 1327 when the actual value is 1342.
    **(C)** They have the same percent error.
    **(D)** There is not enough information to answer the question.

*Billions and Billions*

# Avogadro's Number

## THINK ABOUT IT

Mercury is a toxic substance that accumulates in the body and damages the central nervous system and other organs. Which do you think would be worse for you, 1 mole of mercury or 10 grams of mercury? To answer this question, it is necessary to understand the relationship between these two measures.

**What is the relationship between mass and moles?**

*To answer this question, you will explore*

⮕ Moles and Grams

⮕ Scientific Notation

## EXPLORING THE TOPIC

### ⮕ Moles and Grams

In the world around you, many different measures are used to specify amounts. For example, at the store you might buy 10 apples or a 5 lb bag of apples. One measure is a counting unit and the other is a unit of weight. To compare the prices, you would probably want to know how many apples are in the 5 lb bag.

Likewise, chemists use both number and mass to specify the amount of a substance. For example, they might know the effect of either 1 mol of mercury atoms or of a 10 g sample of mercury. To compare these two quantities, they would need to know either how many moles of atoms are in each gram of mercury or the mass in grams of one mole of mercury.

The mole, abbreviated mol, is a counting unit used to count a large number of atoms. A mole of atoms is equal to 602 sextillion atoms. This number is also referred to as Avogadro's number.

*One mole of O$_2$, Cu, NaCl, and H$_2$O*

The table compares the average mass of one atom with the mass of 1 mol of atoms for five elements. Examine the table.

Comparing Atoms and Moles

| Element | Average mass of one atom (amu) | Mass of one mole of atoms (g) |
|---|---|---|
| hydrogen, H | 1.00 amu | 1.00 g |
| carbon, C | 12.01 amu | 12.01 g |
| iron, Fe | 55.85 amu | 55.85 g |
| arsenic, As | 74.92 amu | 74.92 g |
| mercury, Hg | 200.59 amu | 200.59 g |

Notice that the numerical value on the periodic table for the average atomic mass of an element is identical to the mass of one mole of atoms of that element. In other words, both values are represented by the same number. What is different about the two measurements is the units. The average mass of *one atom* of arsenic is 74.92 amu, and the mass of *one mole* of arsenic atoms is 74.92 grams. How convenient!

> **Big Idea** The mass of one mole of atoms of an element in grams is numerically the same as the average atomic mass of that element.

### MOLAR MASS

The mass of one mole of atoms of an element is called the **molar mass.** You can find the molar mass of any element by looking up the average atomic mass on the periodic table. For example, the average atomic mass of mercury is 200.6 amu. The mass of 1 mol of mercury atoms is 200.6 g.

---

**Example 1**

### Copper versus Arsenic

Which has more mass, 1 mol of copper atoms or 1 mol of arsenic atoms?

*Solution*

Consult the periodic table to find the molar mass of each substance. The molar mass of copper is 63.55 grams per mole, and the molar mass of arsenic is 74.92 grams per mole. A mole of arsenic atoms has more mass.

---

**Example 2**

### Amount of Mercury, Hg

Which is more toxic, 1 mol or 10 g of mercury, Hg?

*Solution*

By looking on the periodic table, you can determine that the mass in grams of 1 mol of mercury is 200.6 g. So, 1 mol of Hg is more toxic than 10 g of Hg because 1 mol represents a much larger mass.

---

## ⮕ Scientific Notation

Consider the table on the next page. It shows the mass, in grams, of different numbers of atoms of mercury. Notice that the number of atoms and the mass are shown both in longhand and in **scientific notation.** Scientific notation is a shorthand method that uses an exponent to keep track of where the decimal point should be. Small numbers, between 0 and 1, have negative exponents. Numbers greater than 1 have positive exponents. The exponent indicates where the decimal point belongs in the longhand number.

## Number of Atoms and Mass of Mercury Atoms

| Number of atoms | Number in scientific notation | Mass (g) | Mass in scientific notation (g) |
|---|---|---|---|
| 1 | $1 \times 10^0$ | 0.0000000000000000000000033 g | $3.3 \times 10^{-22}$ g |
| 1000 | $1 \times 10^3$ | 0.00000000000000000033 g | $3.3 \times 10^{-19}$ g |
| 1,000,000,000 | $1 \times 10^9$ | 0.00000000000033 g | $3.0 \times 10^{-13}$ g |
| 602,000,000,000,000,000,000,000 | $6.02 \times 10^{23}$ (1 mol) | 200.59 g | $2.0059 \times 10^2$ g |

## ASTRONOMY CONNECTION

Each galaxy in the universe is estimated to have 400 billion stars in it. There are about 130 billion galaxies in the known universe. This is about $5.2 \times 10^{22}$ total stars in the universe, or 0.08 mol.

NASA/JPL-Caltech/ESA/Harvard-Smithsonian CfA

In scientific notation, ten thousand is written $1.0 \times 10^4$. One ten-thousandth is written $1.0 \times 10^{-4}$.

$$10,000 = 1.0 \times 10^4$$

$$1/10,000 = \frac{1}{1.0 \times 10^4} = 1.0 \times 10^{-4} = 0.0001$$

[For a review of this topic, see **MATH Spotlight**: Scientific Notation on page A-14.] The mass of 1000 atoms of arsenic is only 0.0000000000000000000012 g or $1.2 \times 10^{-19}$ g. This many arsenic atoms wouldn't even be visible to the naked eye. On the other hand, 74.92 g of arsenic contains 1 mol, or $6.02 \times 10^{23}$ atoms.

When you are dealing with atoms, numbers tend to be either very tiny or very large. So it is more convenient to use scientific notation.

## LESSON SUMMARY

### What is the relationship between mass and moles?

Chemists keep track of the number of molecules and atoms they are working with by using a unit called the mole. They also use scientific notation to express numbers that are very small or very large. One mole is equal to 602 sextillion. Using scientific notation, this number is written as 602,000,000,000,000,000,000,000, or $6.02 \times 10^{23}$. The mass (in grams) of one mole of atoms of an element can be found on the periodic table. This is called the molar mass and has the same numerical value as the average atomic mass.

**KEY TERMS**

molar mass

scientific notation

## Exercises

### Reading Questions

1. How do you find the average atomic mass of atoms of an element? What unit of measure is this given in?

2. What is meant by the term *molar mass*? What unit of measure is this given in?

3. Why do chemists convert between moles and grams?

(continued)

**Reason and Apply**

4. Give the molar mass for the elements listed.
   a. nitrogen, N
   b. neon, Ne
   c. chlorine, Cl
   d. copper, Cu

5. Which has more mass?
   a. 1 mol of hydrogen, H, or 1 mol of carbon, C
   b. 1 mol of aluminum, Al, or 1 mol of iron, Fe
   c. 1 mol of copper, Cu, or 1 mol of gold, Au
   d. 5 mol of carbon, C, or 1 mol of gold, Au

6. Which contains more atoms?
   a. 12 g of hydrogen, H, or 12 g of carbon, C
   b. 27 g of aluminum, Al, or 27 g of iron, Fe
   c. 40 g of calcium, Ca, or 40 g of sodium, Na
   d. 40 g of calcium, Ca, or 60 g of zinc, Zn
   e. 10 g lithium, Li, or 100 g of lead, Pb

7. Which is more toxic? Explain your reasoning.
   a. 1 mol of beryllium, Be, or 10 g of beryllium
   b. 2 mol of arsenic, As, or 75 g of arsenic
   c. 2 mol of lead, Pb, or 500 g of lead

8. Copy and complete the table.

| Amount | Number of moles | Number of atoms | Number of atoms in scientific notation |
|---|---|---|---|
| 12 g carbon, C | 1 mol | 602,000,000,000,000,000,000,000 atoms | $6.02 \times 10^{23}$ atoms |
| 24 g carbon, C | | | |
| 40 g calcium, Ca | | | |
| 20 g calcium, Ca | | | |

*What's in a Mole?*

# Molar Mass

## THINK ABOUT IT

Lead is a highly toxic substance that can be accidentally ingested by humans and animals. Lead atoms interfere with normal processes in the body, causing disturbances in the nervous system. Suppose you have 100 g of lead (II) carbonate, $PbCO_3$, and 100 g of lead (II) chloride, $PbCl_2$. It is the lead atoms specifically that are toxic. If you are exposed to equal masses of lead carbonate and lead chloride, which substance exposes you to more lead atoms and is potentially more toxic? To find out, you need to determine the mass of 1 mol of each.

---

### How can you convert between mass and moles?

---

*To answer this question, you will explore*

⤸  Molar Mass of Compounds

⤸  Comparing a Mole's Worth

## EXPLORING THE TOPIC

## ⟳ Molar Mass of Compounds

### COUNTING WITH MOLES

Any object can be counted with units such as a dozen or a million or a mole. Consider a dozen cheese sandwiches. Each sandwich has two slices of bread and one slice of cheese. Therefore, a dozen sandwiches would have a total of two dozen slices of bread and one dozen slices of cheese.

In a similar way, you can count atoms of lead in a lead compound. Because a compound consists of bonded atoms, you can simply add the molar masses of each atom in a molecule or formula unit of the compound to obtain the molar mass of the substance. For example, each formula unit of lead (II) chloride, $PbCl_2$, contains one atom of lead and two atoms of chlorine.

Molar mass of lead (II) chloride, $PbCl_2$

1 lead atom

=

1 lead (II) chloride formula unit

2 chlorine atoms

1 mol lead (II) chloride

=

1 mol lead atoms

2 mol chlorine atoms

The molar masses of lead and chlorine are 207.2 g/mol and 35.45 g/mol, respectively.

Adding the molar mass of lead and twice the molar mass of chlorine gives the molar mass of lead chloride, $PbCl_2$, which is 278.1 grams per mole, or 278.1 g/mol.

## Example 1

### The Mass of One Mole

Suppose you have 1 mol of lead, Pb, 1 mol of lead (II) chloride, $PbCl_2$, and 1 mol of lead (II) carbonate, $PbCO_3$. What is the mass of each sample?

*Solution*

You can find the molar mass of lead, Pb, on the periodic table, 207.2 g/mol.

A mole of $PbCl_2$ has 1 mol of lead atoms and 2 mol of chlorine atoms. You find the molar mass of the compound by adding these molar masses together.

$$\text{molar mass of } PbCl_2 = 207.2 + 2(35.45)$$
$$= 278.1 \text{ g/mol}$$

So the mass of 1 mol of lead (II) chloride is 278.1 g. The molar mass of $PbCO_3$ can be found similarly.

$$\text{molar mass of } PbCO_3 = 207.2 + 12.01 + 3(16.00)$$
$$= 267.2 \text{ g/mol}$$

So the mass of 1 mol of lead (II) carbonate is 267.2 g.

[For a review of this math topic, see **MATH Spotlight**: Solving Equations on page A-3.]

## ⊃ Comparing a Mole's Worth

A mole of atoms or molecules is usually an amount that you can hold in your hand if the substance is liquid or solid. A table with a "mole's worth" of a few substances is given below.

**HEALTH CONNECTION**

Lead (II) carbonate is one of the white pigments in lead paint. Lead paint dries quickly and forms a durable shiny coating. It is still used to draw lines on pavement, but it is no longer used to paint houses or furniture due to the high toxicity of lead.

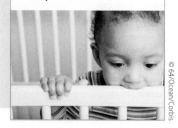

© 64/Ocean/Corbis

**Molar Masses**

| Chemical formula | Molar mass (g/mol) | Moles of what? | Equivalent to |
|---|---|---|---|
| $O_2(g)$ | 32.00 g/mol | oxygen molecules | 22.4 L oxygen gas |
| $Al(s)$ | 26.98 g/mol | Al atoms | 2 aluminum cans |
| $H_2O(l)$ | 18.02 g/mol | $H_2O$ molecules | 18 mL water |
| $He(g)$ | 4.00 g/mol | He atoms | 22.4 L helium gas |
| $NaCl(s)$ | 58.44 g/mol | sodium chloride units | 1/4 cup salt |
| $C_{12}H_{22}O_{11}(s)$ | 342.3 g/mol | sugar molecules | 0.75 lb sugar |

Note that the molar mass of oxygen gas, $O_2(g)$, is double the molar mass of oxygen found on the periodic table because oxygen gas is diatomic. Also note that the volume of a mole of $O_2$ gas is the same as the volume of a mole of any other gas at standard temperature and pressure, or STP.

Example 2

## Toxicity of Lead Compounds

Which is potentially the most toxic: 1 g lead, Pb, 1 g lead (II) chloride, $PbCl_2$, or 1 g lead (II) carbonate, $PbCO_3$?

### Solution

The molar mass of each compound was calculated in Example 1.

$$\text{molar mass of Pb} = 207.2 \text{ g/mol}$$

$$\text{molar mass of } PbCl_2 = 278.1 \text{ g/mol}$$

$$\text{molar mass of } PbCO_3 = 267.2 \text{ g/mol}$$

Because Pb has the lowest molar mass, 1 g lead, Pb, will have the largest number of moles of lead, so it will potentially be the most toxic, followed by lead (II) carbonate, $PbCO_3$.

## LESSON SUMMARY

### How can you convert between mass and moles?

Chemists compare moles of substances rather than masses of substances because moles are a way of counting atoms, molecules, or units in a compound. The molar mass of a substance is the mass, in grams, of one mole of the substance. The molar mass of a compound is the sum of the molar masses of the atoms in the compound.

## Exercises

### Reading Questions

1. Explain how to determine the molar mass of sodium chloride, NaCl.
2. Describe the approximate size of 1 mol of a solid, a liquid, and a gas. Give a specific example of each.

### Reason and Apply

3. Copy this table and use a periodic table to complete the second column

| Chemical formula | Molar mass (g/mol) | Moles of what? |
|---|---|---|
| Ne($g$) | | 1 mol Ne atoms |
| Ca($s$) | | 1 mol Ca atoms |
| $CO_2$($g$) | | 1 mol carbon dioxide molecules |
| $CaCO_3$($s$) | | 1 mol calcium carbonate units |
| $CH_4O$($l$) | | 1 mol methanol molecules |
| $C_2H_6O$($l$) | | 1 mol ethanol molecules |
| $Fe_2O_3$($s$) | | 1 mol iron oxide units |

*(continued)*

4. Which has more moles of molecules, 1.0 g methanol, $CH_4O$, or 1.0 g ethanol, $C_2H_6O$?

5. Which has more moles of metal atoms?
   a. 10.0 g calcium, Ca, or 10.0 g calcium chloride, $CaCl_2$
   b. 5.0 g sodium chloride, NaCl, or 5.0 g sodium fluoride, NaF
   c. 2.0 g iron oxide, FeO, or 2.0 g iron sulfide, FeS

6. How many grams of carbon molecules are in 1 mol of each substance?
   a. methane, $CH_4$
   b. methanol, $CH_4O$
   c. ethanol, $C_2H_6O$

7. What is the mass of 5 mol of iron (III) oxide, $Fe_2O_3$?

8. Which of these has the most chromium?
   (A) 1.0 g chromium (II) chloride, $CrCl_2$
   (B) 1.0 g chromium (III) chloride, $CrCl_3$
   (C) 1.0 g chromium (IV) oxide, $CrO_2$

*Mountains into Molehills*

# Mass-Mole Conversions

## THINK ABOUT IT

When you get a headache, you might reach for one of the many pain relievers on the market. A bottle of aspirin might direct you to take one 325 mg tablet. A bottle of acetaminophen might say to take one 500 mg tablet. Is one of these medications stronger than the other? One way to find out is to compare the number of moles of pain reliever in each dose.

### How are moles and mass related?

*To answer this question, you will explore*

⮞ Relating Mass and Moles

⮞ Converting between Mass and Moles

## EXPLORING THE TOPIC

### ⮞ Relating Mass and Moles

You can find the mass of a substance by using a balance. However, you cannot directly measure the number of moles in a sample. You must convert the mass of a substance to moles to figure out how many atoms or molecules you have. If you want to compare amounts of two substances accurately, it is important to be able to convert between mass and moles.

Typically, you measure the mass of a substance in grams. Medications, like aspirin, are often measured in milligrams. Because there are 1000 milligrams in 1 gram, converting from milligrams to grams is simply a matter of dividing by 1000.

Consider three compounds commonly used in over-the-counter pain relievers.

| Pain reliever | Molecular formula | Molar mass | Adult dose | Moles in a standard dose |
|---|---|---|---|---|
| ibuprofen | $C_{13}H_{18}O_2$ | 206.3 g/mol | 400 mg | 0.0019 mol |
| acetaminophen | $C_8H_9NO_2$ | 151.2 g/mol | 500 mg | 0.0033 mol |
| acetylsalicylic acid (aspirin) | $C_9H_8O_4$ | 180.2 g/mol | 325 mg | 0.0018 mol |

Notice that each pain reliever has a different adult dose. It would appear that acetylsalicylic acid (aspirin), with a smaller, 325 mg dose, is stronger than either acetaminophen, 500 mg, or ibuprofen, 400 mg. But ibuprofen is a heavier molecule than acetylsalicylic acid. One gram of aspirin will not represent the same number of molecules as one gram of ibuprofen. To make

### EARTH SCIENCE CONNECTION

It is estimated that there are $2.0 \times 10^{21}$ grains of sand on the entire Earth. This is only 0.003 mol of sand grains! In contrast, there is approximately 0.0002 mol of water molecules, or $1.2 \times 10^{21}$ molecules, in a single drop of water.

an accurate comparison, you must convert the milligrams of each compound to moles.

Examine the table. Which medication contains the most moles of pain reliever? A 500 mg dose of acetaminophen contains almost twice as many moles of pain reliever as a 325 mg dose of acetylsalicylic acid. So aspirin does appear to be more powerful than acetaminophen. Acetylsalicylic acid has the lowest dose and uses the fewest molecules of pain reliever to get rid of your headache. Ibuprofen is almost identical in potency to acetylsalicylic acid. According to the data, acetaminophen is not as powerful as the other two compounds.

## ⮕ Converting between Mass and Moles

The relationship between the mass in grams of a substance and moles is proportional. The proportionality constant is the molar mass, in g/mol. The molar mass will be different for each substance.

**Mass-Mole Conversions** The molar mass is the mass per mole of a substance. You can use it to convert between the mass, $m$, and the number of moles, $n$.

**Example 1**

### Converting from Moles to Mass

Imagine that a pharmaceutical company has 100 mol of acetaminophen available to be made into 500 mg tablets. The molar mass of acetaminophen is 151.2 g/mol. How many 500 mg tablets can be made?

*Solution*

First, figure out how many grams of acetaminophen there are in 100 mol.

| | |
|---|---|
| Start with the mole to mass formula. | $m = $ molar mass $\cdot\, n$ |
| Substitute values and solve for mass. | $= (151.2 \text{ g/mol}) \, 100 \text{ mol}$ |
| | $= 15{,}120 \text{ g}$ |

Because each 500 mg tablet is 0.5 g, each gram of the compound will make two tablets.

$$(15{,}120 \text{ g}) \, (2 \text{ tablets/g}) = 30{,}240 \text{ tablets}$$

So 30,240 tablets can be made from 100 mol of acetaminophen.

**Example 2**

### Converting from Mass to Moles

Restaurants usually provide small packets of sugar to sweeten coffee or tea. Each sugar packet contains approximately 1.0 g of sucrose, $C_{12}H_{22}O_{11}$. How many moles of sucrose does this represent?

*(continued)*

*Solution*

First, use the periodic table to determine the molar mass of sucrose. The molar mass of sucrose is 342.3 g/mol rounded to one decimal place. To convert from mass to moles, divide the mass in grams by the molar mass.

Rearrange the mole to mass formula to solve for *n*.

$$n = \frac{m}{\text{molar mass}}$$

Substitute values and solve for moles of sucrose.

$$= \frac{1.0 \text{ g}}{342.3 \text{ g/mol}} = 0.003 \text{ mol}$$

So one sugar packet contains 0.003 mol of sucrose.

## LESSON SUMMARY

### How are moles and mass related?

Molar mass values allow you to convert between mass in grams of a substance and moles. To convert mass to moles, divide the mass of a sample in grams by its molar mass. To convert moles to mass, multiply the number of moles of the substance by its molar mass.

## Exercises

### Reading Questions

1. How can you convert between moles of a substance and grams of a substance?
2. Why might a 200 mg tablet of aspirin not have the same effect as a 200 mg tablet of ibuprofen?

### Reason and Apply

3. There are 8.0 mol of H atoms in 2.0 mol of $CH_4O$ molecules. How many moles of H atoms are there in 2.0 mol of $C_2H_6O$ molecules?
4. List these compounds in order of increasing moles of molecules: 2.0 g $CH_4O$, 2.0 g $H_2O$, 2.0 g $C_8H_{18}$. Show your work.
5. Which has more moles of oxygen atoms, 153 g of BaO, or 169 g of $BaO_2$? Show your work.
6. List these compounds in order of increasing mass in grams: 2.0 mol $SiCl_4$, 2.0 mol PbO, 2.0 mol $Fe_2O_3$. Show your calculations.
7. Copper is the third most used metal after iron and aluminum. Copper usage is rapidly expanding as more products are developed that contain electronic components. Suppose you run a company that buys copper compounds and then recycles the copper for resale. Your company wants to get the most pure copper for

*(continued)*

the lowest cost. Three different suppliers want to sell you 1 mol $CuO(s)$, 1 mol $CuCO_3(s)$, and 1 mol $Cu_2O(s)$ for the same price.

a. Which compound has the greatest total mass? Show your work.

b. Which compound has the greatest mass of Cu? Show your work.

c. Assuming it costs the same to extract the copper from each compound, which compound represents the best deal for your company? Explain.

8. Suppose Container A contains 1 mol $C(s)$ and 1 mol $O_2(g)$, and Container B contains 1 mol $CO_2(g)$. Both containers are closed and both are the same size. Compare the containers in each of these ways.

a. number of atoms

b. number of gas molecules

c. mass

d. gas pressure

*How Sweet It Is*

# Comparing Amounts

## THINK ABOUT IT

The safety of artificial sweeteners has been the subject of debate for years. More recently, some schools have banned the sale of both diet and regular soft drinks. Regular soft drinks are sweetened with fructose, while diet soft drinks are usually sweetened with aspartame. If you want to make a healthy decision, you need to consider the toxicities of fructose and aspartame and the amount of each substance in a can of beverage.

**How can you use moles to compare toxicity?**

*To answer this question, you will explore*

➲  Comparing Moles of Sweeteners

➲  Toxicity of Sweeteners

## EXPLORING THE TOPIC

## ➲ Comparing Moles of Sweeteners

It is hard for many of us to imagine a day without sugar. From cereal to flavored drinks and candy, our intake of sugar adds up. So, when the first artificial sweeteners came onto the market in the 1950s, it seemed a good idea to use this low-calorie substitute for a favorite ingredient.

Two sweeteners, fructose, $C_6H_{12}O_6$, and aspartame, $C_{14}H_{18}N_2O_5$, are shown. Notice that they are both medium-sized molecules made up of carbon, hydrogen, and oxygen atoms. The artificial sweetener, aspartame, also has two nitrogen atoms in its molecules.

Fructose
$C_6H_{12}O_6$
*Molar mass: 180.1 g/mol*

Aspartame
$C_{14}H_{18}N_2O_5$
*Molar mass: 294.3 g/mol*

The molar masses of fructose and aspartame represent the mass per mole of molecules. Recall that you can obtain the molar mass of a molecule by adding the molar masses of each atom in the molecule.

40 g
Fructose

0.225 g
Aspartame

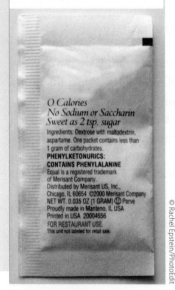

© Rachel Epstein/PhotoEdit

The molar mass of fructose is less than that of aspartame. A 1 g sample of fructose will contain more molecules than a 1 g sample of aspartame because each fructose molecule has a smaller mass than an aspartame molecule.

A can of regular soft drink has about 40.0 g of fructose as a sweetener. A can of diet soft drink has about 0.225 g of aspartame as a sweetener. You can determine the number of molecules of sweetener in each can using the molar mass of each substance.

$$\text{Fructose} = \text{(regular soft drink)}$$
$$n = \frac{40 \text{ g}}{180.1 \text{ g/mol}}$$
$$= 0.22 \text{ mol fructose}$$

$$\text{Aspartame} = \text{(diet soft drink)}$$
$$n = \frac{0.225 \text{ g}}{294.3 \text{ g/mol}}$$
$$= 0.00076 \text{ mol aspartame}$$

A diet soft drink needs only 0.00076 mol of aspartame molecules to replace the sweetness of 0.22 mol of fructose molecules in a regular soft drink. So you need significantly less aspartame to obtain the same sweetness provided by fructose.

## Toxicity of Sweeteners

Artificial sweeteners allow people with diabetes, and others who must avoid sugar, to enjoy foods that would otherwise be dangerous for them to eat. Artificial sweeteners do not cause cavities, and because they are not digested, artificial sweeteners do not provide extra calories to the body. However, the safety of artificial sweeteners continues to be a subject of debate.

To make an informed decision about whether you want to consume artificial sweeteners, it is useful to compare the toxicities of fructose and aspartame. Take a moment to review this information about each sweetener.

| Fructose (sugar) | Aspartame (artificial sweetener) |
|---|---|
| $C_6H_{12}O_6$ | $C_{14}H_{18}N_2O_5$ |
| $LD_{50} = 28.5$ g/kg $= 0.158$ mol/kg | $LD_{50} = 10$ g/kg $= 0.034$ mol/kg |

The $LD_{50}$ of fructose in grams is about three times larger than the $LD_{50}$ for aspartame (30 g per kg versus 10 g per kg). The $LD_{50}$ in moles is almost five times larger for fructose than for aspartame (0.158 mol per kg versus 0.034 mol per kg). So, the sweetner aspartame is more toxic than fructose if you consume similar amounts of each.

**Example**

### How Many Cans?

How many cans of a regular soft drink does a 64 kg (141 lb) person have to drink in one sitting to exceed the lethal dose of fructose? How many cans of diet soft drink does the same person have to drink in a short time to exceed the lethal dose of aspartame?

*Solution*

First, determine how many grams of each sweetener are lethal for a 64 kg person. Then, take this number of grams and divide by the grams of each sweetener in a can of soft drink to get the number of cans of soft drink that would be lethal.

*(continued)*

| | **Regular soft drink** | **Diet soft drink** |
|---|---|---|
| Determine the lethal dose for each sweetener using $LD_{50}$ values. | $64 \text{ kg}\left(25.8\dfrac{\text{g}}{\text{kg}}\right) = 1651 \text{ g}$ | $(64 \text{ kg})\left(10\dfrac{\text{g}}{\text{kg}}\right) = 640 \text{ g}$ |
| Divide by the mass of sweetener in one can to get number of cans. | $(1651 \text{ g})\left(\dfrac{1 \text{ can}}{40 \text{ g}}\right) = 41.3 \text{ cans}$ | $640 \text{ g}\left(\dfrac{1 \text{ can}}{0.225 \text{ g}}\right) = 2884 \text{ cans}$ |

It would take about 41 cans of regular soft drink to reach the lethal dose of fructose and 2884 cans of diet soft drink to reach the lethal dose of aspartame.

**HEALTH**

**CONNECTION**

In 1958, Congress passed the Delaney Clause of the Food, Drug, and Cosmetic Act. The clause prohibited the use of all food additives found to be carcinogenic. As a result, several artificial sweeteners have been banned by the Food and Drug Administration over the years.

It would be difficult to drink enough regular soda or diet soda to reach the lethal dose in a short period of time. So it is highly unlikely that exposure to either fructose or aspartame would be lethal in the *short term*. However, it is important to note that the $LD_{50}$ is not a good measure of health effects from *long-term* exposure. It is probably best to limit intake of both sugar and aspartame.

## LESSON SUMMARY

### How can you use moles to compare toxicity?

Chemists often find it useful to compare moles of substances rather than masses of substances. Moles allow you to compare numbers of molecules or formula units in your sample. Substances with large molar masses contain fewer molecules per gram than substances with smaller molar masses. The health effect of exposure to a toxic substance depends on both the $LD_{50}$ and the amount of the substance.

**Exercises**

### Reading Questions

1. What evidence shows that aspartame is sweeter than fructose?

2. What evidence shows that it would be difficult to exceed the lethal dose of aspartame?

### Reason and Apply

3. There are 25 mg of caffeine, $C_8H_{10}N_4O_2$, in a can of regular soft drink. The $LD_{50}$ for caffeine, $C_8H_{10}N_4O_2$, is 140 mg/kg.
   a. How many cans of regular soft drink can a 65 kg person drink in a short period of time before exceeding the lethal dose?
   b. What is the toxicity of caffeine in moles per kilogram?

4. The $LD_{50}$ for saccharin, $C_7H_5NO_3S$, is 14.2 g/kg. If you have 1 mol of aspartame and 1 mol of saccharin, which would be more toxic? Show your work.

5. **PROJECT** Write an argument for or against the use of artificial sweeteners. Be sure to provide evidence to support your argument. Cite at least two references that you use.

**www**

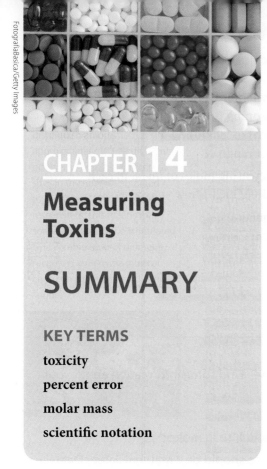

CHAPTER **14**

# Measuring Toxins

# SUMMARY

### KEY TERMS

**toxicity**

**percent error**

**molar mass**

**scientific notation**

## Toxins Update

To keep the public safe, health experts measure the toxicity of substances that people might be exposed to in the course of their daily lives. The most common measurement of toxicity is called lethal dose, or $LD_{50}$, which is expressed in milligrams per kilogram of body weight. The higher the $LD_{50}$ for a substance, the safer the compound is.

Chemical formulas track compounds in counting units called moles. To determine if a certain quantity of matter amounts to a lethal dose, it is necessary to convert between mass and moles using molar mass.

### REVIEW EXERCISES

1. Mass and moles are proportionally related. Explain what this means.

2. Why is it necessary to take a person's weight into account when considering the correct dose for a medication?

3. Why is scientific notation a useful tool for chemists?

4. Consider vitamin $D_3$, $C_{27}H_{44}O$, cholecalciferol. Vitamin $D_3$ helps the body extract calcium and phosphorus from food, plays a role in mental health, and may prevent some cancers. Most of the vitamin D you get is produced in your skin when you are exposed to the Sun.

*Vitamin $D_3$*

Lethal dose: $LD_{50}$ = 42.0 mg/kg (rat, oral)

Recommended Dietary Allowance: RDA = 0.010 mg/day

**a.** Determine the number of moles and molecules of vitamin $D_3$ in the Recommended Dietary Allowance.

**b.** Determine the lethal dose for a 140 lb person.

**c.** Determine the number of moles of vitamin $D_3$ in a lethal dose for a 140 lb person.

d. Determine the number of molecules of vitamin D$_3$ in a lethal dose for a 140 lb person.

e. Would $1.0 \times 10^{20}$ molecules of vitamin D$_3$ represent a lethal dose for a 140 lb person? Explain your reasoning.

f. Would 400 tablets (0.010 mg each) of vitamin D$_3$ be a lethal dose for a 140 lb person? Explain your reasoning.

5. List these compounds in order of increasing moles of metal in each compound: 5.0 g NaCl, 5.0 g AgCl, 5.0 g LiCl. Show your work.

6. Copy and balance these chemical equations.

a. $TiCl_4(s) + Mg(s) \longrightarrow Ti(s) + MgCl_2(s)$

b. $Mg(s) + H_2O(l) \longrightarrow MgO(s) + H_2(g)$

c. $Fe(s) + CuSO_4(s) \longrightarrow FeSO_4(s) + Cu(s)$

d. $Li(s) + H_2O(l) \longrightarrow LiOH(aq) + H_2(g)$

## Lethal Dose of a Toxic Substance

PROJECT

Research a toxic substance (your teacher may assign you one). Find some information about your substance. Your project should include

- the LD$_{50}$ for the substance.

- the lethal dose for a 140 lb person expressed in grams and also in moles.

- two sources for the information you obtained. One source should be a Material Safety Data Sheet (MSDS).

# CHAPTER 15

# Toxins in Solution

*In this chapter, you will study*

- how to track dissolved substances in solutions

- a particle view of solutions

- how to create solutions

*Water is one of the most valuable resources on the planet. It must be carefully monitored to make sure it is safe for public use.*

Many compounds dissolve in water and can become part of our drinking water supply. An aqueous solution is one that contains water, with other ingredients uniformly mixed in. Understanding solutions allows you to monitor and track these substances.

# Solution Concentration

## Purpose
To explore solutions and the concentration of dissolved solids in solution.

## Part 1: Dissolving Solids

### Materials
- sugar, $C_{12}H_{22}O_{11}$, 60 g
- salt, NaCl, 60 g
- water, 200 mL
- 250 mL beakers (2)
- balance
- stirring rod
- 2 spatulas or plastic spoons
- 2 green or red gummy bears

### Procedure

1. Make predictions. How many grams of sugar do you think you can dissolve in 100 mL of water? (1 tsp = 4 g) How many grams of salt? (1 tsp = 6 g)

2. Determine how many grams of sugar will dissolve in 100 mL of water at room temperature. Keep adding sugar and stirring until you notice that the beaker contains solid sugar that will not dissolve.

3. Repeat this process using table salt, NaCl, instead of sugar.

### Calculations

1. Compare the number of *grams* of dissolved solid per liter of solution.

2. Compare the number of *moles* of solid per liter of solution.

3. Explain why the sugar solution has the greater mass of dissolved solid per liter but the salt solution has the greater number of moles of solid per liter.

## Part 2: Gummy Bears

Five gummy bears were placed overnight in aqueous sugar solutions as shown.

A — Water  B — 0.1 mol/L  C — 1.0 mol/L  D — 2.0 mol/L  E — Corn syrup    Dry gummy bear

1. Describe what happens to the bears in the solutions. Which solution had the greatest effect on the gummy bears? Why do you think this is so?

2. What evidence do you have that it is water and not sugar that is moving into and out of the bears?

3. Place a gummy bear in each saturated solution you created. Predict what they will look like after 24 hours. Explain your reasoning.

# Solution Concentration

## THINK ABOUT IT

When a toxic substance is in solid form, keeping track of toxic amounts is fairly simple. But what happens if the toxic substance is dissolved in water? For example, the water that comes out of your tap is probably a solution containing small amounts of dissolved chlorine. Small amounts of chlorine benefit the water supply by killing bacteria, but large amounts would be toxic to people who drink the water. It is important to have methods for tracking substances dissolved in water.

**How can you keep track of compounds when they are in solution?**

*To answer this question, you will explore*

➲  Solutions

➲  Solution Concentration

## EXPLORING THE TOPIC

## ➲ Solutions

A **solution** is a mixture of two or more substances that is uniform throughout. This means it is uniform at the molecular, ionic, or atomic level. The ocean, a soft drink, liquid detergent, window cleaner, and even blood plasma are all examples of solutions. A glass of powdered drink has many dissolved substances in it, but they are mixed so well that the individual substances in the solution are no longer distinguishable.

© Bill Aron/PhotoEdit

*A beverage made from a drink mix has many dissolved substances in it—not just sugar.*

Each solution listed above consists of one or more substances dissolved in water. The water in these solutions is called the **solvent.** There are many other solvents besides water, such as alcohol, turpentine, and acetone. So the solvent is the substance that other substances are dissolved *in.* A solvent does not necessarily have to be a liquid. Solvents can be gases, liquids, or solids. Water is one of the most important solvents on Earth because it is so common in living systems.

A substance that is dissolved in a solvent is called the **solute.** Solutes can be gases, liquids, or solids. In a sugar solution, the sugar is the solute and water is the solvent. In a carbonated soft drink, carbon dioxide gas is one of the solutes dissolved in the soft drink.

*A tossed salad is a heterogeneous mixture.*

A solution is considered a **homogeneous mixture** because of its uniformity. Mixtures that are not uniform throughout are called **heterogeneous.** Salad dressing or mixed nuts are examples of heterogeneous mixtures.

## Solution Concentration

The **concentration** of a solution refers to the amount of a substance dissolved in a specified amount of solution. The concentration of a solution depends on both the quantity of solute and the quantity of solvent. There are several ways of measuring these quantities. A common way to measure concentration is to determine the number of moles of solute per total volume of solution in liters. This is called the **molarity** of a solution and is expressed in moles per liter, mol/L. Notice that solution concentration is the same as number density, $n/V$.

$$\text{Molarity} = \frac{\text{moles of solute}}{\text{liters of solution}}$$

Consider the two solutions in the illustration. Both contain a blue dye. Dyes are generally molecular substances that make a solution appear to be a certain color. The concentration of blue dye is greater in the first solution even though the volume of solution is smaller. You can tell this because the solution on the left is darker than the solution on the right.

Molarity of solution → **1.0 M**
1.0 mol of dye per liter of solution

**0.50 M**
0.50 mol of dye per liter of solution

*Dyes are often molecular solids. This is a sample of blue-colored dye.*

The M is an abbreviation for **molar,** which means moles of solute per liter of solution. The dye solution on the left is 1.0 molar, 1.0 M, while the one on the right is 0.50 molar, or 0.50 M.

If you want to make a solution more concentrated, you need to add solute. If you want to make a solution more dilute, or *less* concentrated, you need to add water.

---

**Example 1**

### Molarity

Determine the molarity of each.

    **a.** 0.10 mol of NaCl in 1.0 L of solution

    **b.** 0.50 mol of NaCl in 1.0 L of solution

    **c.** 0.10 mol of NaCl in 0.50 L of solution

    **d.** 0.10 mol of NaCl in 250 mL of solution

Example 1

(continued)

## Solution

Molarity is the number of moles of solute divided by the liters of solution.

a. $\dfrac{0.10 \text{ mol}}{1.0 \text{ L}} = 0.10 \text{ mol/L} = 0.10 \text{ M}$

b. $\dfrac{0.50 \text{ mol}}{1.0 \text{ L}} = 0.50 \text{ mol/L} = 0.50 \text{ M}$

c. $\dfrac{0.10 \text{ mol}}{0.50 \text{ L}} = 0.020 \text{ mol/L} = 0.020 \text{ M}$

d. $\dfrac{0.10 \text{ mol}}{250 \text{ mL}} = \dfrac{0.10 \text{ mol}}{0.25 \text{ L}} = 0.40 \text{ mol/L} = 0.40 \text{ M}$

## CALCULATING MOLARITY FROM MASS IN GRAMS

In the classroom, you were challenged to create a saturated sugar solution and a saturated salt solution. A **saturated solution** is one that contains the maximum amount of dissolved solute possible in a certain volume of liquid. To make a saturated solution, you must add solid and stir until no more solid dissolves. You can calculate how many grams of each solid you added if you find the mass of the solutions before and after you added the solute. The molar mass of a substance allows you to convert the mass of solute in grams to moles of solute. You can then determine the molarity of each solution.

**Big Idea** Molarity keeps track of the number of dissolved particles in a liquid.

Example 2

## Finding Molarity

What is the concentration in moles per liter of a 250 mL solution that contains 16.0 g of dissolved sugar, $C_{12}H_{22}O_{11}$?

## Solution

First, convert the mass of sugar to moles by dividing by the molar mass of sugar, $C_{12}H_{22}O_{11}$. Next, to determine molarity, divide the number of moles by the volume.

Convert mass to moles.     $\dfrac{16.0 \text{ g}}{342.3 \text{ g/mol}} = 0.047 \text{ mol}$

Calculate the molarity.     $\dfrac{0.047 \text{ mol}}{0.250 \text{ L}} = 0.188 \text{ M}$

solution

solvent

solute

homogeneous mixture

heterogeneous mixture

concentration

molarity

molar

saturated solution

## LESSON SUMMARY

### How can you keep track of compounds when they are in solution?

A solution is a mixture of two or more substances that is uniform throughout. To track substances that are in solution, chemists measure the number of particles per liter of liquid. This is called the concentration or the molarity of the solution. Molarity is a measure of number density and is expressed in units of moles per liter of solution.

## Exercises

### Reading Questions

1. What does it mean that a solution is "uniform throughout"?

2. What is the difference between the concentration of a solution and the volume of a solution?

### Reason and Apply

3. Find at least three solutions at home. Identify the solute and the solvent for each.

4. Find at least three mixtures at home that are not solutions. Identify the substances in each mixture.

5. Place these salt solutions, NaCl($aq$), in order of increasing molarity.
   a. 4.0 mol per 8.0 L
   b. 6.0 mol per 6.0 L
   c. 1.0 mol per 10 L

6. Determine the molarity for each of these salt solutions, NaCl($aq$). Then list the solutions in order of increasing molarity.
   a. 29.2 g per 0.50 L
   b. 5.8 g per 50 mL
   c. 2.9 g in 10.2 mL

7. Determine the molarity for each of these solutions. Then list the solutions in order of increasing molarity.
   a. 25 g $C_6H_{12}O_6$ per 0.25 L
   b. 50 g $C_6H_{12}O_6$ per 0.25 L
   c. 50 g $C_{12}H_{22}O_{11}$ per 0.25 L

8. Which of the substances listed here represent a heterogeneous mixture? Choose all that apply.
   (A) the air in your classroom
   (B) beef stew
   (C) mouthwash
   (D) chocolate pudding

9. How can you increase the molarity of a solution?
   (A) Add solute.
   (B) Add solvent.
   (C) Pour out some of the solution.
   (D) All of the above.

10. When expressing molarity, M stands for
    (A) atoms per liter of water.
    (B) moles per milliliter of solvent.
    (C) moles per liter of solution.
    (D) atoms per milliliter of solute.
    (E) grams per liter of solution.

# Molecular Views

## THINK ABOUT IT

A tiny drop of a concentrated solution of hydrochloric acid, HCl($aq$), will cause a severe burn if it comes in contact with your skin. In contrast, a larger volume of a more dilute HCl($aq$) solution will cause only a minor skin irritation. How can a single drop of a solution be so powerful?

**How can you use molarity to determine the moles of solute?**

*To answer this question, you will explore*

⮕  Particle Views of Solutions

⮕  Molarity Calculations

## EXPLORING THE TOPIC

## ⮕ Particle Views of Solutions

When a toxin is dissolved in water, the amount of toxin you are exposed to depends on both the concentration of the solution and the volume of the solution. To better understand the relationship between solution concentration and solution volume, it is useful to explore solutions with the same volume but different concentrations, as well as solutions with the same concentration but different volumes.

### SAME VOLUME, DIFFERENT CONCENTRATIONS

This illustration shows four solutions of red dye, as well as particle views of each. Each red dot represents a red dye molecule dissolved in water. Take a moment to compare the solutions and molecular views.

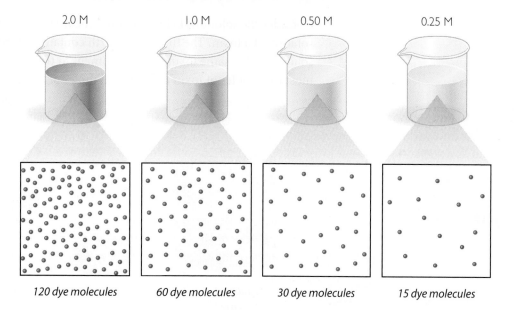

| 2.0 M | 1.0 M | 0.50 M | 0.25 M |

| *120 dye molecules* | *60 dye molecules* | *30 dye molecules* | *15 dye molecules* |

Notice that each time the concentration is halved, there are half as many particles per unit of volume. So, as the solution concentration decreases, equal volumes of solution will contain fewer molecules.

The color of the red dye solution provides evidence to support this conclusion. When there are more dye molecules per unit of volume, the red color of the solution is more intense.

### SAME CONCENTRATION, DIFFERENT VOLUMES

The particle views of the two solutions in the next illustration show what happens when you pour a portion of a solution into another container. Take a moment to compare the solutions and molecular views.

*120 dye molecules*

*30 dye molecules*

There are fewer total dye molecules when you pour a portion of the solution into the smaller container. However, the concentration of the solution has not changed. If you remove 250 mL from a 1 L solution, you have one quarter as many dye molecules. But because the concentration has not changed, the color of the red dye solution does not change.

## ⮑ Molarity Calculations

Consider the solutions below. Suppose the first two contain a concentrated solution of a toxin. The third and fourth contain a more dilute, or watered down, solution of the same toxin. Take a moment to try to put the solutions in order of most total moles of toxin to least total moles of toxin.

| Solution A | Solution B | Solution C | Solution D |
| --- | --- | --- | --- |
| 1000 mL | 100 mL | 100 mL | 200 mL |
| 6.0 M | 6.0 M | 2.0 M | 2.0 M |

The relationship between moles of solute and volume of solution in liters is proportional. Recall that the molarity of a solution in moles/liter is the proportionality constant, $k$, for that solution. So, if you multiply the molarity by the volume in liters, you can determine the number of moles of solute.

moles of solute = $k$ · volume of solution in liters

$$= 6.0 \text{ mol/L} \cdot 1.0 \text{ L} = \mathbf{6 \ mol} \qquad \text{Solution A}$$

$$= 6.0 \text{ mol/L} \cdot 0.10 \text{ L} = \mathbf{0.60 \ mol} \quad \text{Solution B}$$

$$= 2.0 \text{ mol/L} \cdot 0.10 \text{ L} = \mathbf{0.40 \ mol} \quad \text{Solution C}$$

$$= 2.0 \text{ mol/L} \cdot 0.20 \text{ L} = \mathbf{0.20 \ mol} \quad \text{Solution D}$$

You can see that even though the volume of Solution B is smaller than the volume of Solution D, there are more moles of toxin dissolved in Solution B. This is because Solution B is more concentrated than Solution D.

A small drop of highly concentrated toxin may be worse for you than a large quantity of dilute toxin. For example, a 2 mL drop of 12 M HCl solution is much more toxic than 50 mL of 0.10 M HCl solution.

> **Big Idea** The concentration of a solution does not change with the size of the sample.

**Example**

## Comparing Moles of Solute

Two solutions of a toxic substance are shown in the illustration.

50 mL
2.0 M toxin

175 mL
0.50 M toxin

Which beaker contains the larger dose of the toxic substance?

*Solution*

Use the concentration of each solution to determine the moles of solute.

2.0 M solution: 2.0 mol/L · 0.050 L = 0.10 mol

0.50 M solution: 0.50 mol/L · 0.175 L = 0.0875 mol

So, the beaker on the left, the more concentrated solution, contains more moles of toxic substance.

Scientists also measure the concentration of substances in solution using parts per million (ppm) or parts per billion (ppb). When water is tested, the results are often reported in ppm. A level of 1 ppm means that there is 1 mg of a dissolved solute per liter of solution.

## LESSON SUMMARY

### How can you use molarity to determine the moles of solute?

The concentration of a solution does not change with the size of the sample. This is because concentration is a measure of number density, or moles of particles per unit of volume, a property that does not change with sample size. However, a large sample of a solution will contain more total particles than a small sample of the same solution. The relationship between moles of particles in a solution and liters of solution is a proportional one. It is described by the formula, moles of solute = $k$ · volume of solution, where $k$ is the molarity of the solution. This equation can be used to calculate the exact number of moles in a sample of known concentration and volume.

## Exercises

### Reading Questions

1. How can two solutions with different volumes have the same concentration?

2. How can you figure out how many moles of solute you have in a solution with a specific concentration?

### Reason and Apply

3. Draw molecular views for blue dye solutions that are 0.50 M, 0.25 M, and 0.10 M.

4. What portion of 1.0 L of 0.50 M blue dye solution has the same number of moles as 1.0 L of 0.25 M blue dye solution?

5. Glucose and sucrose are two different types of sugar. Consider these aqueous solutions:

1.0 L
1.0 M $C_6H_{12}O_6$
(glucose)

1.0 L
1.0 M $C_{12}H_{22}O_{11}$
(sucrose)

500 mL
1.0 M $C_{12}H_{22}O_{11}$
(sucrose)

a. Which solution has the most molecules? Explain.
b. Which solution has the greatest concentration? Explain.
c. Which solution has the most mass? Explain.

*(continued)*

6. How many moles of sodium cations, $Na^+$, are dissolved in a 1.0 L sample of each solution listed below?
   a. 0.10 M NaCl
   b. 3.0 M $Na_2SO_4$
   c. 0.30 M $Na_3PO_4$

7. Draw particle views for 1.0 M NaCl, 2.0 M NaCl, and 1.0 M $Na_2S$. Use different symbols for each type of ion. Circle the solution(s) with the least total number of ions.

8. Determine the number of moles of solute in each of these aqueous solutions.
   a. 50 L of 0.10 M NaCl
   b. 0.25 L of 3.0 M $C_6H_{12}O_6$
   c. 35 mL of 12.0 M HCl
   d. 300 mL of 0.025 M NaOH

9. How many liters of each solution do you need to get 3.0 mol HCl?
   a. 12.0 M HCl
   b. 2 M HCl
   c. 0.5 M HCl
   d. 0.010 M HCl

*Holey Moley*

# Preparing Solutions

## THINK ABOUT IT

Human blood plasma contains a certain molarity of dissolved salt, NaCl. If you are taken to the hospital in an ambulance and given an intravenous, or IV, solution, it is critical that the salt concentration match that of your blood, about 0.15 M NaCl. Otherwise, the IV solution could be toxic to you.

### How can you create a solution with a specific molarity?

*To answer this question, you will explore*

- Creating Solutions
- Relating Mass, Moles, and Volume

## EXPLORING THE TOPIC

## Creating Solutions

Suppose a lab technician wants to make a 0.15 M sodium chloride solution. How should this solution be prepared?

### PREPARING A 0.15 M SALT SOLUTION

A concentration of 0.15 M means that 0.15 mol of salt is dissolved in 1 L of solution. Therefore, if you want to make 1 L of solution, you will need to measure out 0.15 mol of solid sodium chloride, NaCl.

To know how much salt to weigh out, you must convert moles to mass using the molar mass of the compound. In this case, simply add the molar masses for Na and Cl from the periodic table. The molar mass of NaCl is 58.4 g/mol to one decimal place. Now multiply the moles of NaCl you want by the molar mass to get the number of grams of NaCl you need for your solution.

$$\text{Mass} = \text{molar mass} \cdot n = (58.4 \text{ g/mol})(0.15 \text{ mol}) = 8.76 \text{ g NaCl}$$

If you dissolve 8.76 g NaCl in water so that the total volume of solution is 1 L, you will have a solution that is 0.15 M, or 0.15 mol/L NaCl.

© Tony Freeman/PhotoEdit

**Important to Know** When you dissolve one substance in another, they both take up space. So, to make 1 L of an aqueous solution, you add less than 1 L of water.

Add some water and stir.

Add water to make exactly 1 L of solution.

1 L

*Measure 8.76 g*
**NaCl(s)**

*0.15 M NaCl(aq)*

Example 1

### Dextrose Solution

Sometimes doctors administer an IV solution containing a solution of dextrose to a patient who has low sugar or high sodium in the bloodstream. Dextrose is a type of sugar. The structural formula for dextrose is shown in the illustration.

The IV solution contains 150 g of dextrose in 3 L of solution. What is the molarity of this solution?

Dextrose

*Solution*

You need to convert grams of dextrose to moles of dextrose to determine the molarity. You know there are 150 g of dextrose in 3 L, or 50 g/L.

First, determine the molar mass of dextrose. As you can see from the ball-and-stick model, there are 6 C atoms, 12 H atoms, and 6 O atoms, so the formula is $C_6H_{12}O_6$.

| | |
|---|---|
| Find the molar mass of dextrose, $C_6H_{12}O_6$. | 180.2 g/mol |
| Divide grams of solute by the molar mass to obtain moles. | $n = \dfrac{50 \text{ g}}{180.2 \text{ g/mol}} = 0.28 \text{ mol}$ |

There is 0.28 mol of dextrose in 1 L of solution, so the molarity is 0.28 M.

## ⮕ Relating Mass, Moles, and Volume

Just as it is important to convert mass to moles, it is also important to be able to convert moles to mass. The table below shows data for four samples of sucrose, $C_{12}H_{22}O_{11}$, each with a different volume and molarity. How many grams of sucrose are dissolved in each sample?

To find the moles of solute in each sample, multiply the values in the second and third columns. To find the mass of solute in each sample, multiply the fourth and fifth columns.

Molarity • Volume ⬇                    Moles • Molar mass ⬇

| Sample | Molarity (moles/liter) | Volume (liters) | Number (moles of solute) | Molar mass of solute (grams/mole) | Mass of solute (grams) |
|---|---|---|---|---|---|
| 1 | 1.0 M | 1.0 L | 1.0 mol | 342.3 g/mol | 342.3 g |
| 2 | 0.10 M | 1.0 L | 0.10 mol | 342.3 g/mol | 34.23 g |
| 3 | 1 M | 0.1 L | 0.10 mol | 342.3 g/mol | 34.23 g |
| 4 | 0.10 M | 0.1 L | 0.010 mol | 342.3 g/mol | 03.423 g |

Notice that there is more than one way to make a 0.10 M solution. For example, you can dissolve 0.10 mol to make 1.0 L or you can dissolve 0.010 mol to make 100 mL (0.10 L).

**Example 2**

## Guidelines for Safe Drinking Water

Arsenic, As, is an element that is found naturally in the soil. As a result, tiny levels of arsenic are often found in some water supplies. Like that of all substances, the toxic effect of arsenic is dependent on the dosage you receive. For this reason, the Environmental Protection Agency, or EPA, sets limits on the amounts of dissolved substances that can be present in our drinking water. The EPA has set the upper limit for arsenic in drinking water at 0.00010 g/L.

Is it safe to drink water that has a concentration of 0.000020 M As?

*Solution*

The EPA limit is given in grams/liter, and the concentration of the solution is in moles/liter. So you need to convert moles to grams (or vice versa) to make a comparison.

The proportionality constant between grams of arsenic and moles of arsenic is the molar mass, which you can find on the periodic table. To find the mass of arsenic, multiply the molar mass by the number of moles.

Look up the molar mass of As.  74.9 g/mol

Mass = molar mass · $n$  (74.9 g/mol)(0.000020 mol) = 0.0015 g As

The 0.000020 M arsenic solution has 0.0015 g of As per liter of solution. This amount is 15 times the concentration allowed by the EPA. So, the solution is not safe for drinking.

## LESSON SUMMARY

### How can you create a solution with a specific molarity?

Molar mass values on the periodic table allow you to convert between grams and moles. The amount in grams of solid needed to create solutions of specific molarities can be calculated using the molar mass of the solid. Likewise, the mass of solid dissolved in a solution can be determined if the concentration and volume of the solution are known.

**Exercises**

### Reading Questions

1. Explain how you would prepare a solution of sucrose with a molarity of 0.25.
2. How are mass of solute, moles of solute, and volume of solution related?

### Reason and Apply

3. How many grams of solute do you need to make 1 L of each of the solutions listed?
   a. 0.50 M NaCl
   b. 2.0 M $CaCl_2$
   c. 1.5 M NaOH

*(continued)*

4. Copy this table and complete it for solutions of glucose, $C_6H_{12}O_6$. Remember to convert milliliters to liters.

| Molarity (moles/liter) | Volume (liters) | Number (moles of solute) | Molar mass of solute (grams/mole) | Mass of solute (grams) |
|---|---|---|---|---|
| | 2.0 L | | | 180 g |
| 0.40 M | | 0.10 mol | | |
| 0.10 M | | 0.03 mol | | |
| | 100 mL | | | 0.180 g |

5. What volume of each of these solutions would contain 58.44 g of NaCl?
   a. 0.10 M NaCl
   b. 3.0 M NaCl
   c. 5.5 M NaCl

6. How many grams of fructose, $C_6H_{12}O_6$, are in 1 L of soft drink if the molarity of fructose in the soft drink is 0.75 M?

7. Which is more concentrated: a 1.0 L solution with 20 g of sucrose, $C_{12}H_{22}O_{11}$, or a 1.0 L solution with 20 g of glucose, $C_6H_{12}O_6$?

8. The Environmental Protection Agency has set the upper limit for fluorine, F, in drinking water at 0.0040 g/L of water. Is it safe to drink a glass of water in which the concentration of fluorine is 0.00010 M?

*Is It Toxic?*

# Mystery Solutions

## THINK ABOUT IT

Selling bottled water is a big business. Some companies claim to get their water from mountain streams. Some say that their water is quite pure. Still others claim their water contains beneficial minerals. How could you determine how pure a water sample is?

**Can molarity calculations be used to identify a toxic substance?**

*To answer this question, you will explore*

⮑ Accounting for the Mass of a Solution

⮑ Drinking Water Standards

## EXPLORING THE TOPIC

## ⮑ Accounting for the Mass of a Solution

Suppose that you want to know how pure a sample of water is. Examine the illustration and consider how you might use mass to make a determination of purity.

Three 100 mL samples

What masses do you expect if the samples are all pure water? Recall that mass per volume is equal to density. The density of pure water is 1 g/mL. If a 100 mL sample is pure water, then you expect a mass of 100 g.

The sample in the first beaker may be pure water. The other two samples must contain solutes that add more mass to the same volume, making them denser (and therefore heavier) than pure water.

### THE DENSITY OF A SOLUTION

Suppose you have three samples of salt water with the concentrations shown in the illustration at the top of the next page. Take a moment to consider how the densities of the three solutions compare. Why does the ball float higher in the solution on the right?

The ball floats in each solution.

0.61 M NaCl          0.89 M NaCl          5.10 M NaCl

*It's easy to float in the Dead Sea, which is a salt lake and one of the world's saltiest bodies of water.*

As the density of the solution increases, an object floats higher. That is why when you swim in very salty water, you float with more of your body out of the water. As the concentration increases, there are more units of NaCl per volume of solution. The higher the salt concentration, the denser the solution.

## SAME MASS, DIFFERENT SOLUTES

Imagine that you add 10 g of three different solids to three different beakers and then add enough water to make 100 mL of solution. The solutions all have approximately the same mass—the mass of about 100 mL of water plus 10 g of solute. But do they have the same concentration?

| Compound | Molar mass |
|----------|------------|
| $CuSO_4$ | 249.7 g/mol |
| NaOH | 40.0 g/mol |
| KCl | 74.6 g/mol |

Ions in 100 mL of solution

Copper (II) sulfate          Sodium hydroxide          Potassium chloride
**$CuSO_4$**          **NaOH**          **KCl**
*10 g in 100 mL*          *10 g in 100 mL*          *10 g in 100 mL*

Copper (II) sulfate, $CuSO_4$ has the largest molar mass. That means there are fewer formula units of $CuSO_4$ in 10 grams compared with NaOH. While you've added the same mass to each sample, you have not added the same number of moles. Dissolving 10 grams of the compound with the higher molar mass will result in a lower concentration.

**Example**

### Different Solutions, Same Concentration

Suppose you create 0.10 M solutions of copper (II) sulfate, sodium hydroxide, and potassium chloride. For equal volumes, which solution will have the most mass?

*Solution*

The mass of each solution is the mass of the water plus the mass of solute. The mass of the solute added depends on its molar mass. To determine the mass corresponding to 0.10 mol of each solute, multiply the number of moles by the molar mass of each compound.

$$\text{Mass of } CuSO_4 = (249.7 \text{ g/mol})(0.10 \text{ mol}) = 25.0 \text{ g } CuSO_4$$

$$\text{Mass of } NaOH = (40.0 \text{ g/mol})(0.10 \text{ mol}) = 4.0 \text{ g } NaOH$$

$$\text{Mass of } KCl = (74.6 \text{ g/mol})(0.10 \text{ mol}) = 7.5 \text{ g } KCl$$

So, for equal volumes of solution, the $CuSO_4(aq)$ will have the most mass.

## ⮕ Drinking Water Standards

Environmental Protection Agency standards protect public health by limiting contaminants in drinking water. The maximum contaminant levels for fluorine, iron, and barium in drinking water are given in mg/L in the table.

| Element | Maximum contaminant level |
|---|---|
| fluorine, F | 4.0 mg/L |
| iron, Fe | 0.30 mg/L |
| barium, Ba | 2.0 mg/L |

Based on the numbers in the table, you might conclude that iron, Fe, is the most toxic because the allowable level is the smallest by mass. However, these contaminants interact with your body atom by atom. So you need to compare the contaminants by number of atoms.

Find the concentration of the potential toxins in moles per liter. To do so, convert the maximum contaminant level milligrams to grams, and then divide by the molar mass.

fluorine, F   $\dfrac{0.0040 \text{ g/L}}{19.0 \text{ g/mol}} = 0.00021 \text{ mol/L} = 0.00021 \text{ M} = 2.1 \times 10^{-4} \text{ M}$

iron, Fe   $\dfrac{0.00030 \text{ g/L}}{55.8 \text{ g/mol}} = 0.000054 \text{ mol/L} = 0.000054 \text{ M} = 5.4 \times 10^{-5} \text{ M}$

barium, Ba   $\dfrac{0.0020 \text{ g/L}}{137.3 \text{ g/mol}} = 0.000014 \text{ mol/L} = 0.000014 \text{ M} = 1.4 \times 10^{-5} \text{ M}$

In moles per liter, the maximum contaminant level for barium, Ba, is the smallest numerical value. This means less of it is tolerated in our drinking water, making barium the most toxic of the three elements.

## Can molarity calculations be used to identify a toxic substance?

When a substance is dissolved in water, it adds to the mass of the solution. One mole of one substance may have a very different molar mass than one mole of another substance. So different solutions of the same concentration can be differentiated by weighing them. Comparing contaminants by number of atoms may give a different picture of toxicity than comparing them by mass.

## Exercises

### Reading Questions

1. Explain how you might use mass to determine if a sample of water is contaminated.

2. Explain why the density of a solution increases as the concentration increases.

3. Explain why 0.10 M $CuCl_2$ has a greater density, $m/V$, than 0.10 M KCl.

### Reason and Apply

4. Find the molarity of each solution. Which solution has the highest molarity? The highest density? Give your reasoning.
   a. 1.0 g KCl in 1.0 L water
   b. 5.0 g KCl in 0.5 L water
   c. 10.0 g KCl in 0.3 L water

5. Find the molarity of each solution. Which solution has the highest molarity? The highest density? Give your reasoning.
   a. 10 g $NaNO_3$ in 1.0 L water
   b. 10 g $Cd(NO_3)_2$ in 1.0 L water
   c. 10 g $Pb(NO_3)_2$ in 1.0 L water

6. Find the mass of solute in each solution. Which solution has the highest density? Is it possible to calculate the density of each solution? Why or why not?
   a. 1.0 L of 0.10 M $NaNO_3$
   b. 1.0 L of M 0.10 M $Cd(NO_3)_2$
   c. 1.0 L of 0.10 M $Pb(NO_3)_2$

7. Which would weigh the most? Explain your thinking.
   (A) 1.0 L of 1.0 M NaBr
   (B) 500 mL of 1.0 M KCl
   (C) 1.0 L of 0.5 M NaOH

8. PROJECT    Look up the MSDS or Materials Safety Data Sheet, for sodium chloride, NaCl, and sodium sulfate, $Na_2SO_4$.
   
   www
   a. Describe the toxic effects of each salt.
   b. What are the $LD_{50}$'s (mouse, oral)?
   c. Compare the toxicities based on weight in grams.
   d. Compare the toxicities based on moles.
   e. Compare the toxicities based on volume of 0.10 M solution.

Dori Moreno/Getty Images

# CHAPTER 15

## Toxins in Solution

## SUMMARY

### KEY TERMS

solution

solvent

solute

homogeneous mixture

heterogeneous mixture

concentration

molarity

molar

saturated solution

## Toxins Update

It is easy for toxins to dissolve in and enter our water supply. In fact, it would be challenging to find or create a sample of "pure water"—one that had nothing else in it but $H_2O$ molecules. To track the concentrations of different substances in solutions, scientists use moles of solute per liter of solvent. This is also called the molarity of the solution. If you know the molarity of a solution and its volume, you can determine the exact number of grams, moles, or even molecules or ions of a substance in your water sample. Sometimes, toxic substances in solution are tracked in parts per million, or ppm. A solution with a toxic concentration of 1 ppm would have 1 mL of toxic substance per 1000 L of solution.

### REVIEW EXERCISES

1. Why is it useful to know exactly how many moles or grams of contaminant are in a water sample?

2. Lead is one of the few substances with a zero tolerance level in our drinking water, meaning that there should be no lead at all in our water. Why do you think this is?

3. Consider a 10 mL water sample from a nearby creek that has a 0.002 M lead (II) nitrate, $Pb(NO_3)_2$, concentration.
   a. How many lead ions are in this sample?
   b. How many grams of lead nitrate are in this sample?
   c. How many grams of lead ions are in this sample?

4. Suppose you measure 100 mL each of 1.0 M NaOH, 1.0 M KCl, and 0.5 M $PbCl_2$ solutions.
   a. Which has the greatest concentration?
   b. Which has the most mass?
   c. Which has the most moles?
   d. Which has the most ions?

5. a. Suppose you need to make a 0.5 M solution of sodium hydroxide, NaOH. Explain step by step how you would go about making this solution.
   b. Do you have to use 1 L of water when you make this solution? Explain your reasoning.

6. A mass of 47 g of sulfuric acid, $H_2SO_4$, is dissolved in water to prepare a 0.50 M solution. What is the volume of the solution?

7. Determine the molarity of the solutions created by following each set of directions.
   a. Dissolve 7.30 g NaCl to make 500 mL of solution.
   b. Dissolve 43.84 g NaCl to make 1 L of solution.
   c. Dissolve 25.00 g NaCl to make 1 L of solution.
   d. Dissolve 25.00 g NaCl to make 2 L of solution.

8. How many grams of magnesium sulfate, $MgSO_4$, do you need to make 100 mL of a 0.1 M solution?

# Types of Bonding

**PROJECT**

**www**

Research a toxic substance (your teacher may assign you one). Your report should cover these topics.

- Is the bonding in your substance ionic, covalent, or metallic?

- If your substance is ionic, determine the cation and the anion. Write and label the ions and their charges. Include a particle view drawing of your substance.

- If your substance is covalent, determine if it is polar or nonpolar. Explain your reasoning. Include the structural formula and show the dipole(s).

- If your substance is metallic, find out its density. Include a particle view drawing of your substance.

- Does your substance conduct electricity? Explain your reasoning.

# CHAPTER **16**

# Acidic Toxins

### In this chapter, you will study

- how the acidity of solutions is tracked

- theories of acids and bases

- differences between acids and bases

- the effect of dilution and neutralization on acid-base solutions

- a lab procedure for determining acid concentration

*A vinegaroon is similar to a scorpion, but instead of stinging, it squirts acetic acid, or vinegar, on its prey.*

Some solutions can be classified as either *acids* or *bases*. They have a unique set of properties and are an extremely important part of solution chemistry. Toxic levels of acids and bases can burn the skin and damage living tissue. However, without acids and bases, the body would not function properly. This chapter introduces you to the properties of acids and bases and shows how chemists monitor them.

# Acids and Bases

## THINK ABOUT IT

Solutions have à wide variety of properties. Acetic acid, or vinegar, is used to flavor salad dressing, while citric acid gives a sour taste to lemon juice and orange juice. Ammonium hydroxide is used to clean windows, and sodium hydroxide is used to open clogged drains. All of these solutions are either acids or bases and can be classified into these categories based on their general properties.

**What are the properties of acids and bases?**

*To answer this question, you will explore*

⮑  Acids and Bases

⮑  Indicators

⮑  The pH Scale

## EXPLORING THE TOPIC

### ⮕ Acids and Bases

#### GENERAL PROPERTIES

Acids and bases are special categories of solutions. They are extremely useful to us in our everyday lives precisely because of their unique properties. The term *acid* comes from the Latin word *acidus,* which means sour. Many of the sour tastes in our food come from the acids found in those foods.

For much of history, the term *alkaline* was used instead of *base*. Bases are found in many household cleaners, from soaps to drain openers to oven cleaners. Bases have a bitter taste, as you may have noticed if you have ever accidentally tasted soap. Bases usually feel slippery to the touch. The slipperiness of bases arises from the fact that they are reacting with the fats in your skin and turning them into soap.

In general, acids and bases are toxic, especially large quantities of concentrated solutions. It is important that acids and bases not splash on your skin or in your eyes. Acids and bases are both corrosive and can cause a *chemical burn.* A chemical burn is one in which living tissue is damaged.

### ⮕ Indicators

Many acidic and basic solutions are colorless and odorless, which can make them difficult to detect by their appearance alone. Because these solutions can be toxic, it is useful to be able to monitor them. Luckily, there are molecular substances called **indicators** that change color when they come into contact with acids and bases. If you add a drop or two of an indicator to an unknown solution, you can tell if you have an acid or a base by the color that results.

### CONSUMER
### CONNECTION

Carbonic acid, $H_2CO_3$, and phosphoric acid, $H_3PO_4$, are two acids found in carbonated soft drinks. The phosphoric acid acts as a flavoring and a preservative, while the carbonic acid is a byproduct of carbonation.

© Rob Bouwman/iStockphoto

Cabbage juice is a natural acid-base indicator obtained by boiling or grinding up red cabbage. The cabbage juice changes color as it is added to various solutions, as shown in the illustration. Take a moment to look for patterns.

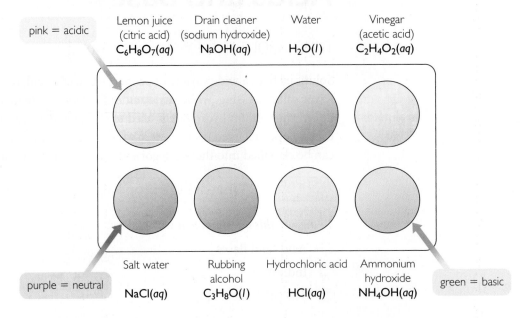

Cabbage juice turns pink when it is in contact with acids, such as citric acid, acetic acid, and hydrochloric acid. It turns green or yellow in contact with bases, such as sodium hydroxide and ammonium hydroxide. And it remains purple or blue in contact with *neutral* substances such as pure water, salt water, and rubbing alcohol. Neutral solutions are neither acidic nor basic. So cabbage juice can be used to indicate if a solution is acidic or basic or neither.

## ⟳ The pH Scale

There are dozens of different types of acid-base indicators, each with a unique color scheme. The colors associated with "universal indicator" are shown in the illustration. Notice that there is a number associated with each color. This number is referred to as the *pH number*, or just as the *pH*.

Another way to measure the pH for a solution is with paper coated with indicator, referred to as pH paper.

One common type of pH paper is called *litmus paper*. Litmus paper turns red in acidic solutions and blue in basic solutions.

The **pH scale** is a number line that assigns number values from 0 to 14 to acids and bases.

Substances with a pH below 7 at 25 °C are **acids.** Substances with a pH above 7 at 25 °C are **bases.** Substances with a pH at or near 7 at 25 °C are considered neutral. Stomach acid is extremely acidic, with a pH around 1. Lemon juice is the

Litmus Paper

Acid          Base

**The pH scale**

| | | | | | | |
|---|---|---|---|---|---|---|
| Stomach acid | Lemon juice | Vinegar | Salt solution, water, rubbing alcohol | Washing soda | Ammonium hydroxide | Drain cleaner |

0     Increasing [$H^+$]     7     Increasing [$OH^-$]     14

Acids     Neutral     Bases

next most acidic substance shown here, with a pH near 2.4. The most basic substance shown here is drain cleaner. The substances on either end of the pH scale are potentially more dangerous and more toxic than substances found in the middle of the scale.

## LESSON SUMMARY

### What are the properties of acids and bases?

**KEY TERMS**

indicator

pH scale

acid

base

Acids and bases are special categories of solutions. The sour tastes in our food come from the acids found in those foods. Bases cause the slippery feel of soaps and detergents. Chemical compounds called indicators help chemists to identify the presence of acids and bases in solution. Indicators change color in response to acids and bases. The pH scale is a number scale that assigns values to acids and bases, between 0 and 14. Substances with a pH below 7 at 25 °C are acids. Substances with a pH above 7 at 25 °C are bases. Substances with a pH at or near 7 at 25 °C are neutral.

## Exercises

### Reading Questions

1. What are some of the observable properties of acids and bases?

2. What is the pH scale?

3. What does it mean to say that a substance has a neutral pH?

### Reason and Apply

4. **Lab Report** Write a lab report for the Lab: Acids and Bases. In your report, give the title of the experiment, purpose, procedure, observations, analysis, and conclusion.

5. Classify each of the following solutions at 25 °C as acidic or basic based on the information provided.
   a. lemon juice tastes sour
   b. a solution of washing soda turns cabbage juice green
   c. a dilute solution of potassium hydroxide feels slippery
   d. a sugar solution has a pH of 7
   e. drain cleaner has a pH of 12

(Title)

**Purpose:** (Explain what you were trying to find out.)

**Procedure:** (List the steps you followed.)

**Observations:** (Describe your observations.)

**Analysis:** (Explain what you observed during the experiment.)

**Conclusion:** (What can you conclude about what you were trying to find out? Provide evidence for your conclusions.)

*(continued)*

6. Examine the list of acids below. List three patterns that you notice in the names and chemical formulas.

   hydrochloric acid, $HCl$      nitric acid, $HNO_3$

   carbonic acid, $H_2CO_3$      acetic acid, $C_2H_4O_2$

   sulfuric acid, $H_2SO_4$      citric acid, $C_6H_8O_7$

7. Examine the list of bases below. List three patterns that you notice in the names and chemical formulas.

   sodium hydroxide, $NaOH$      magnesium hydroxide, $Mg(OH)_2$

   potassium hydroxide, $KOH$      barium hydroxide, $Ba(OH)_2$

   lithium hydroxide, $LiOH$      ammonium hydroxide, $NH_4OH$

8. **PROJECT**   Find three acids and three bases used in your home.

   a. What are these acids and bases used for?

   b. Look up the chemical name and chemical formula of each acid and base.

## Acid and Base Theories

### THINK ABOUT IT

People have known for thousands of years that vinegar, lemon juice, green apples, and many other foods taste sour. However, it is only recently that scientists have presented theories to explain what all these substances have in common. In what ways are these sour-tasting substances similar, and why are they defined as acids?

**How are acids and bases defined?**

*To answer this question, you will explore*

- Chemical Makeup of Acids and Bases
- Acid and Base Theories
- Strong and Weak Acids and Bases

### EXPLORING THE TOPIC

## Chemical Makeup of Acids and Bases

Many substances dissolve in water to produce solutions that are acidic. Because the behavior of all acidic solutions is similar, you might expect some similarity in the chemical makeup of these substances.

Take a moment to look for patterns in the table of common acids.

**BIOLOGY**

**CONNECTION**

The human stomach secretes hydrochloric acid, HCl, which begins the digestion process for large protein molecules and also provides a barrier against bacteria, fungi, and other microorganisms naturally found in food and water. Stomach acid is also necessary for the proper absorption of certain minerals such as iron and zinc as well as some of the B vitamins.

**Common Acids**

| Name | Found in | Chemical formula | Ions in solution |
|------|----------|------------------|------------------|
| hydrochloric acid | stomach acid | $HCl$ | $H^+ \, Cl^-$ |
| nitric acid | acid rain | $HNO_3$ | $H^+ \, NO_3^-$ |
| sulfuric acid | acid rain | $H_2SO_4$ | $H^+ \, H^+ \, SO_4^{2-}$ |
| phosphoric acid | cola | $H_3PO_4$ | $H^+ \, H^+ \, H^+ \, PO_4^{3-}$ |
| acetic acid | vinegar | $CH_3COOH$ | $H^+ \, CH_3COO^-$ |

Here are some patterns you might notice.

- The acids are made of main group nonmetal atoms, such as carbon, oxygen, and chlorine, bonded covalently.
- The only element that they all have in common is hydrogen.
- They all break apart, or **dissociate,** in solution to form at least one hydrogen cation, $H^+$, and an anion.

You might also notice that some of the molecular formulas are written in a way that gives some information about the molecular structure. For example, acetic acid, $C_2H_4O_2$, is written as $CH_3COOH$.

Many other substances dissolve in water to produce solutions that are basic. If acidic solutions have $H^+$ ions in common, what is similar in the chemical makeup of basic solutions?

Take a moment to look for patterns in the names and chemical formulas of the common bases listed in this table.

### Common Bases

| Name | Found in | Chemical formula | Ions in solution |
|------|----------|------------------|------------------|
| sodium hydroxide | drain cleaner | NaOH | $Na^+$ $(OH)^-$ |
| magnesium hydroxide | antacid tablets | $Mg(OH)_2$ | $Mg^{2+}$ $(OH)^-$ $(OH)^-$ |
| ammonia | window cleaner | $NH_3$ | $NH_4^+$ $(OH)^-$ |
| sodium carbonate | washing soda | $Na_2CO_3$ | $Na^+$ $Na^+$ $(HCO_3)^-$ $(OH)^-$ |

Here are some things you might notice.

- Many of the bases contain a metal atom and a hydroxide ion, $OH^-$.
- The metal atoms are in the first two columns of the periodic table.
- Some compounds, such as ammonia and sodium carbonate, produce $OH^-$ when dissolved in water, even though there is no $OH^-$ in the chemical formula.
- Except for the two hydroxides, their names do not reveal that they are bases.

## Acid and Base Theories

### ARRHENIUS THEORY OF ACIDS AND BASES

One of the early theories of acids and bases was first proposed in 1884 by a Swedish chemist named Svante Arrhenius. He defined an acid as a substance that adds hydrogen ions, $H^+$, to an aqueous solution. He defined a base as a substance that adds hydroxide ions, $OH^-$, to an aqueous solution.

### BRØNSTED-LOWRY THEORY OF ACIDS AND BASES

The Arrhenius definitions of acids and bases do not explain how substances without hydroxide ions, $OH^-$, in their formula, like ammonia, $NH_3$, can be bases. In the early 1920s, a Danish chemist, Johannes Brønsted, and an English chemist, Thomas Lowry, proposed a slightly different definition for acids and bases to account for this chemical behavior. They defined an acid as a substance that can donate a proton to another substance. They defined a base as a substance that can accept a proton from another substance. A proton in this case is actually a hydrogen ion, $H^+$. Because a hydrogen atom typically has no neutrons, if you remove the electron from a hydrogen atom to form an ion, all that's left is a proton.

When ammonia, $NH_3$, is added to water, it removes a hydrogen ion, $H^+$, from some of the water molecules, leaving behind some hydroxide ions, $OH^-$. An $NH_3$

molecule accepts a hydrogen ion and becomes $NH_4^+$. This makes $NH_3$ a Brønsted-Lowry base.

$$NH_3(g) + H_2O(l) \longrightarrow NH_4^+(aq) + OH^-(aq)$$

So dissolved ammonia really consists of ammonium and hydroxide ions. For this reason, it is often referred to as ammonium hydroxide solution. Many substances that do not contain $OH^-$ can act as bases.

## ↻ Strong and Weak Acids and Bases

These two illustrations show particle views of a 0.010 M hydrochloric acid, HCl, solution and a 0.010 M formic acid, HCOOH, solution. The water molecules are not shown. Take a moment to examine them.

Particle Views of a Strong and Weak Acid

Hydrochloric acid: strong
0.010 M HCl
pH = 2

Formic acid: weak
0.010 M HCOOH
pH = 2.6

○ H⁺   ● Cl⁻   ◐ HCl          ○ H⁺   ● HCOO⁻   ◐ HCOOH

Notice that there are no molecules of HCl in the solution on the left. The HCl has dissociated completely into $H^+$ and $Cl^-$ ions. However, the solution on the right contains formic acid molecules. Only some of the HCOOH molecules have dissociated into ions. This means that the concentration of $H^+$ ions is smaller in the formic acid solution than in the hydrochloric acid solution, even though the solutions have the same molarities.

Acids that dissociate completely in solution are called **strong acids.** Strong acids include hydrochloric acid, HCl, nitric acid, $HNO_3$, sulfuric acid, $H_2SO_4$, and hydrobromic acid, HBr. Strong acids are good conductors of electricity.

Acids that dissociate only partially in solution are called **weak acids.** These solutions are only moderate conductors of electricity. Some common weak acids are formic acid, HCOOH, acetic acid (vinegar), $CH_3COOH$, citric acid, $C_3H_5O(COOH)_3$, and phosphoric acid, $H_3PO_4$. Weak acids tend to be less corrosive because they do not dissociate completely into ions.

Bases can also be classified as strong or weak. A **strong base** dissociates completely into ions in solution and **weak bases** dissociate only partially. Some examples of strong bases include sodium hydroxide, NaOH, and barium hydroxide, $Ba(OH)_2$. Examples of weak bases include ammonia, $NH_3$, and aniline, $C_6H_5NH_2$.

| Big Idea | Strong acids and strong bases dissociate completely into ions. Weak acids and weak bases dissociate only partially into ions. |

**BIOLOGY**

**CONNECTION**

Formic acid, HCOOH, is the simplest carboxylic acid. It is most famous for its presence in bee stings and ant bites. In fact, the word formic comes from *formica,* the Latin word for ant.

Kim Taylor/Nature Picture Library

## How are acids and bases defined?

**KEY TERMS**

dissociate

strong acid

weak acid

strong base

weak base

The Arrhenius definition of an acid is a compound that dissociates in solution to form a hydrogen ion, $H^+$, and an anion. The Arrhenius definition of a base is a substance that dissociates in solution to form a hydroxide ion, $OH^-$, and a cation. However, not all bases contain $OH^-$. Brønsted and Lowry defined acids as proton donors and bases as proton acceptors. The word "proton" in this case refers to an $H^+$ ion. A strong acid or base dissociates completely into ions in solution. A weak acid or base dissociates only partially into ions in solution.

## Exercises

### Reading Questions

1. According to the Arrhenius theory, what is an acid and what is a base?

2. How is the Brønsted-Lowry theory of acids similar to the Arrhenius theory, and how is it different?

3. What is the difference between a strong acid and a weak acid? Draw a picture showing the particle view of each.

### Reason and Apply

4. Label the substances listed below as acids or bases. In each case, list the ions you would expect to find in solution.
   a. hydroiodic acid, HI
   b. formic acid, HCOOH
   c. rubidium hydroxide, RbOH
   d. hypochlorous acid, HOCl
   e. selenous acid, $H_2SeO_4$
   f. phosphine, $PH_3$
   g. perchloric acid, $HClO_4$
   h. calcium hydroxide, $Ca(OH)_2$

5. Consider a solution of hydrobromic acid, HBr.
   a. If you draw a particle view of this acid with 10 $H^+$ ions, how many $Br^-$ ions would you need? Explain your thinking.
   b. Sketch a particle view of an HBr solution with 10 $H^+$ ions.

6. Consider a solution of magnesium hydroxide, $Mg(OH)_2$.
   a. If you draw a particle view of this base with 10 $Mg^{2+}$ ions, how many $OH^-$ ions would you need? Explain your thinking.
   b. Sketch a particle view of a $Mg(OH)_2$ solution with 10 $Mg^{2+}$ ions.

7. Explain why aqueous washing soda, $Na_2CO_3$, is a basic solution.

8. A solution of 0.10 M hydrochloric acid, HCl, is a better conductor of electricity than 0.10 M acetic acid, $CH_3COOH$. Sketch the ions and molecules in both solutions to explain this observation.

*pHooey!*

# [H$^+$] and pH

## THINK ABOUT IT

During delivery, a baby can sometimes go into distress. So doctors monitor the vital signs to make sure the baby is okay. One test is called a fetal blood test. This test involves taking the pH of the baby's blood. If the baby's blood has a pH of 7.2 or lower, the doctors know that the baby is not getting enough oxygen and is in danger. A pH of around 7.3 indicates that the baby is safe.

### How is pH related to the acid or base concentration of a solution?

*To answer this question, you will explore*

- Acid Concentration
- Relationship between [H$^+$] and pH
- pH of Water and Basic Solutions

## EXPLORING THE TOPIC

### Acid Concentration

If you are testing the acidity of a solution by determining pH, you are determining the hydrogen ion concentration in moles per liter. When a solution has a greater hydrogen ion concentration than water does, it is an acid. Chemists use a standard notation with brackets to symbolize concentration in moles per liter. For example, [H$^+$] symbolizes the concentration of hydrogen ions, H$^+$, in moles per liter.

### PARTICLE VIEWS OF ACIDIC SOLUTIONS

To get an idea of how pH numbers relate to hydrogen ion concentration, it is useful to compare particle views of solutions. The illustration shows the ions in two hydrochloric acid solutions, one that is 0.010 M and the other 0.002 M. How is the hydrogen ion concentration, [H$^+$], related to the pH?

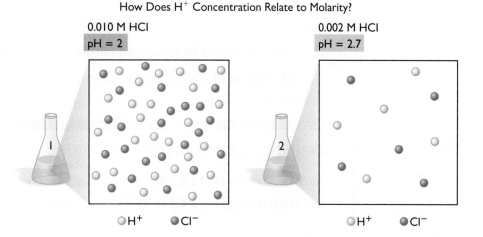

How Does H$^+$ Concentration Relate to Molarity?

0.010 M HCl
pH = 2

0.002 M HCl
pH = 2.7

1

2

○ H$^+$      ● Cl$^-$        ○ H$^+$      ● Cl$^-$

In the illustration, solution 1 has five times as many $H^+$ cations and five times as many $Cl^-$ anions as solution 2. However, the difference in the pH is only 0.7. Small changes in pH signal large differences in hydrogen ion concentrations.

**Big Idea** The greater the concentration of $H^+$ ions in a solution, the lower its pH, and the more acidic it is.

## Relationship between [H⁺] and pH

The hydrogen ion concentration of a solution is directly related to the pH of that solution. The table shows the $H^+$ concentration and the pH number for a few different concentrations of hydrochloric acid.

### The Relationship between [H⁺] and pH Number

| Molarity | [H⁺] | pH |
|---|---|---|
| 1.0 M HCl | $1.0 \times 10^0$ M | 0 |
| 0.1 M HCl | $1.0 \times 10^{-1}$ M | 1 |
| 0.01 M HCl | $1.0 \times 10^{-2}$ M | 2 |
| 0.001 M HCl | $1.0 \times 10^{-3}$ M | 3 |
| 0.0001 M HCl | $1.0 \times 10^{-4}$ M | 4 |
| 0.00001 M HCl | $1.0 \times 10^{-5}$ M | 5 |

Notice that each time the $H^+$ concentration changes by a factor of 10, the pH number changes by 1 unit. You can also see that the pH number is equal to the exponent, without the negative sign. For example, $1.0 \times 10^{-4}$ M HCl has a pH of 4. This is true whenever the $H^+$ concentration is expressed in scientific notation and the coefficient is 1.0.

Each unit of change in pH represents a tenfold difference in $H^+$ concentration. This is called a *logarithmic* relationship. The mathematical formula connecting pH to [H⁺] is

$$pH = -\log [H^+]$$

You can use this formula and a calculator to convert between pH and [H⁺].

[For review of this math topic, see **MATH Spotlight**: Logarithms on page A-17.]

**Example 1**

### Calculating pH for an HCl Solution

**a.** What is the pH of a 0.000001 M HCl solution?

**b.** What is the pH of a $4.2 \times 10^{-3}$ M HCl solution?

**Example 1**

*(continued)*

*Solution*

**a.** Each HCl formula unit dissociates into an $H^+$ and a $Cl^-$ ion. So the concentration of $H^+$ ions is 0.000001 M, which can also be written $1.0 \times 10^{-6}$ M. Because the coefficient is 1.0, the pH is the exponent without the minus sign. The pH is 6.

**b.** Use the relationship pH $= -\log [H^+]$.

$$pH = -\log [4.2 \times 10^{-3}] = 2.4$$

The pH of the $4.2 \times 10^{-3}$ molar HCl solution is between 2 and 3. This makes sense because the $H^+$ concentration is between $1.0 \times 10^{-2}$ M and $1.0 \times 10^{-3}$ M.

## ⤷ pH of Water and Basic Solutions

Basic solutions also have pH numbers associated with them. This is because they also have an $H^+$ concentration and pH value.

The number line below shows the relationship between pH and $H^+$ concentration for acidic, basic, and neutral solutions at 25 °C. The $H^+$ concentration in basic solutions is quite small. For example, a solution with a pH of 14 has a hydrogen ion concentration of $1.0 \times 10^{-14}$ M, or 0.00000000000001 M.

Concentration of $H^+$

### DISSOCIATION OF WATER

It may seem odd that a basic solution has hydrogen ions in it. To understand this, it is necessary to consider water on a molecular level.

Notice that the hydrogen ion concentration in a neutral solution, such as water, is listed as $1.0 \times 10^{-7}$ M $H^+$ at 25 °C. Water molecules  dissociate slightly, forming $H^+$ and $OH^-$ ions. The dissociation is so slight that there are only 2 $H^+$ ions and 2 $OH^-$ ions for every 1 billion water molecules. This translates to an $H^+$ concentration of $1 \times 10^{-7}$ M, or a pH of 7. Water is considered neutral because it contains equal amounts of $H^+$ and $OH^-$.

When a substance dissolves in water and the $H^+$ concentration, $[H^+]$, remains equal to the $OH^-$ concentration, $[OH^-]$, we say that substance is neutral. For example, NaCl dissolves in water to form $Na^+$ cations and $Cl^-$ anions. This does not change the balance of $H^+$ and $OH^-$ ions already present in the water. NaCl(*aq*) is neutral.

When an acid such as HCl dissolves in water, it breaks apart into $H^+$ and $Cl^-$ ions, increasing the ratio of $H^+$ ions to $OH^-$ ions. Likewise, when a substance like NaOH dissolves in water, it breaks apart into $Na^+$ and $OH^-$ ions. This also alters the balance of $H^+$ and $OH^-$ ions in the water. In the case of NaOH($aq$), the hydrogen ion concentration, $[H^+]$, is less than the hydroxide ion concentration, $[OH^-]$, so NaOH is considered a base.

The table shows the concentration of $H^+$ and $OH^-$ ions in a variety of sodium hydroxide solutions.

**The Relationship between $[OH^-]$ and pH Number**

| Molarity | $[OH^-]$ | $[H^+]$ | pH |
|---|---|---|---|
| water | $1.0 \times 10^{-7}$ M | $1.0 \times 10^{-7}$ M | 7 |
| 0.00001 M NaOH | $1.0 \times 10^{-5}$ M | $1.0 \times 10^{-9}$ M | 9 |
| 0.0001 M NaOH | $1.0 \times 10^{-4}$ M | $1.0 \times 10^{-10}$ M | 10 |
| 0.001 M NaOH | $1.0 \times 10^{-3}$ M | $1.0 \times 10^{-11}$ M | 11 |
| 0.01 M NaOH | $1.0 \times 10^{-2}$ M | $1.0 \times 10^{-12}$ M | 12 |
| 0.1 M NaOH | $1.0 \times 10^{-1}$ M | $1.0 \times 10^{-13}$ M | 13 |
| 1.0 M NaOH | $1.0 \times 10^{0}$ M | $1.0 \times 10^{-14}$ M | 14 |

Notice that the $H^+$ concentration decreases as the $OH^-$ concentration increases and vice versa. There is a mathematical relationship between $[H^+]$ and $[OH^-]$:

$$[H^+]\,[OH^-] = 1 \times 10^{-14}$$

In *any* aqueous solution at 25°C, the hydrogen ion concentration multiplied by the hydroxide ion concentration is equal to $1.0 \times 10^{-14}$. For example, when the $OH^-$ concentration is $1 \times 10^{-3}$ M, the $H^+$ concentration is $1 \times 10^{-11}$ M.

$$(1 \times 10^{-11})(1 \times 10^{-3}) = 1 \times 10^{-14}$$

You can also see that adding the values of the exponents always results in a total of −14.

**Example 2**

**Calculating pH for a Solution of Sodium Hydroxide, NaOH**

What is the pH of a 0.000001 M NaOH solution?

*Solution*

A solution with a molarity of 0.000001 M NaOH has an $OH^-$ concentration of $1.0 \times 10^{-6}$ moles per liter. The $H^+$ concentration must be $1.0 \times 10^{-8}$ M because the exponents must sum to −14. The pH is just the exponent of the $H^+$ concentration without the negative sign. So the pH is 8.

## How is pH related to the acid or base concentration of a solution?

The pH number and the hydrogen ion concentration of a solution are related logarithmically: $pH = -\log [H^+]$. This means that when the hydrogen ion concentration changes by a factor of 10, the pH changes by 1. Water and basic solutions also have pH numbers because they contain small concentrations of $H^+$ ions. Water dissociates slightly into $H^+$ and $OH^-$ ions. In pure water, the $H^+$ concentration is $1 \times 10^{-7}$ M and the pH is 7. The hydroxide ion concentration and the hydrogen ion concentration in a solution are related mathematically: $[H^+][OH^-] = 1.0 \times 10^{-14}$.

## Exercises

### Reading Questions

1. What is the relationship between pH and $H^+$ concentration?

2. How is the $H^+$ concentration related to the $OH^-$ concentration in a solution?

### Reason and Apply

3. What pH would you expect for solutions with the concentrations given here?
   a. $[H^+] = 1.0 \times 10^{-4}$ M
   b. $[H^+] = 1.0 \times 10^{-12}$ M
   c. $[OH^-] = 1.0 \times 10^{-8}$ M

4. Determine the pH for solutions with the $H^+$ concentrations given here.
   a. $[H^+] = 0.0014$
   b. $[H^+] = 6.0 \times 10^{-8}$ M
   c. $[H^+] = 4.2 \times 10^{-11}$ M
   d. $[H^+] = 1.5 \times 10^{-1}$ M

5. Which of the solutions in Exercise 4 are acids? Which are bases?

6. What $H^+$ concentration would you expect for the solutions below?
   a. pH = 9
   b. pH = 7.3
   c. pH = 2.9
   d. pH = 10.2

7. What is the pH of a 2.5 M HCl solution?

8. What is the pH of a 0.256 M NaOH solution?

9. PROJECT  The Richter scale is a logarithmic scale for measuring the magnitude of earthquakes. How much more intense is a Richter 7.0 earthquake than a Richter 5.0 earthquake?

*Watered Down*

# Dilution

## THINK ABOUT IT

In Unit 1: Alchemy, a "golden" penny was made with the help of a strong base called sodium hydroxide. In Unit 2: Smells, you added concentrated sulfuric acid to a mixture of chemicals to make a sweet-smelling ester. Whenever you work with very concentrated acids and bases, you are instructed to rinse with large amounts of water if you get any on your skin or clothing. How does this help?

### How does dilution affect acids and bases?

*To answer this question, you will explore*

⮞ Diluting Acidic Solutions

⮞ Approaching pH 7

## EXPLORING THE TOPIC

## ⮞ Diluting Acidic Solutions

If you spill concentrated acid on your skin, it will begin to sting and burn right away. If you spill it on your clothing, you will eventually see holes in the cloth. The best thing you can do for this situation is to put a lot of water on the part of your skin or clothing where the acid spilled.

The process of adding more and more water to a solution is called **dilution.** You could say that the solution is becoming more dilute or "watered down." This changes the number density of the solute particles in solution.

The effect of diluting a concentrated solution can be demonstrated with a glass of grape juice. The color of the grape juice becomes paler as more water is added. And the more dilute the grape juice becomes, the weaker it tastes.

Ken Karp Photography

Acidic solutions are not necessarily colorful like grape juice, but the concept of dilution is the same: The solution becomes less concentrated. When you dilute an acid, you can track what is happening by testing the pH.

The illustration at the top of the next page shows a particle view of what happens when you dilute an acidic solution. A few drops of red dye have been added so that you can visually keep track of the different solutions. The solution is diluted with water in the beaker. The starting solution in the test tube has a pH of 2. The particle-view squares show how the $H^+$ ions and $Cl^-$ ions end up farther apart as more water is added.

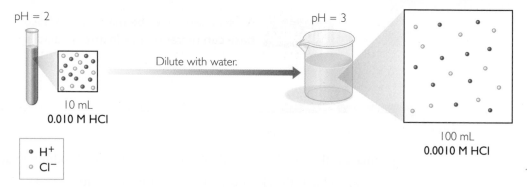

pH = 2

10 mL
**0.010 M HCl**

Dilute with water.

pH = 3

100 mL
**0.0010 M HCl**

• H⁺
∘ Cl⁻

*Mixing the 10 mL of 0.010 M HCl with 90 mL of water dilutes the original solution by a factor of 10.*

Dilution of a strong acid or base can make it safer. This is why you can swim in a swimming pool even though the water contains both acids and bases.

**Example 1**

## Diluting HCl

Imagine you have a 1.0 mL water sample containing 0.10 M hydrochloric acid, HCl. You want to get it to a safe pH of 6. How much water do you need to add?

### Solution

The pH of 0.10 M HCl is 1. The $H^+$ concentration needs to change from $1.0 \times 10^{-1}$ M to $1.0 \times 10^{-6}$, a factor of $10^5$, or 100,000. So you need to add 99,999 mL of water to 1 mL 0.10 M HCl to get it to a safe pH of 6.

**RECREATION CONNECTION**

Swimming pools require a slightly basic pH, between 7.2 and 7.8. If the pH is too low, eye irritation occurs and metal pumps may corrode. If the pH is too high, the water may become cloudy and the chlorine may not work to combat algae. Sodium carbonate is added to pools to keep the pH in the right range. Indicators are used to test the pH.

## ➲ Approaching pH 7

Suppose you continue to dilute the 0.0010 M HCl solution in the large beaker in the previous illustration to a concentration of 0.00010 M HCl and pH of 4. If you continue to dilute, what is the lowest $H^+$ concentration you can obtain? What is the lowest pH you can reach?

If you take 1 mL of 0.00010 M HCl and add 9 mL of water, the new solution will have a concentration of 0.000010 M HCl and a pH of 5. If you take 1 mL of 0.000010 M HCl and add 9 mL of water, the new solution will have a concentration of 0.0000010 M HCl and a pH of 6. You can continue this process, diluting the acid until the pH becomes 7. At this point, the liquid is no longer acidic. It is neutral.

As you dilute an acidic solution, its $H^+$ concentration gets closer and closer to the $H^+$ concentration of water, $1 \times 10^{-7}$ moles per liter. However, the pH does not increase beyond 7, no matter how much you dilute the solution. In other words, the solution does not become basic. You cannot start with an acidic solution and dilute it to make a basic solution.

Likewise, you cannot dilute a basic solution to make an acidic solution. When you dilute a basic solution, the $OH^-$ concentration decreases, the $H^+$ concentration increases, and the pH decreases toward 7.

**Example 2**

### Diluting NaOH

Imagine that you have 10 mL of a 0.010 M solution of sodium hydroxide, NaOH.

  **a.** What is the $H^+$ concentration of this solution?

  **b.** What is the pH?

  **c.** What is the NaOH concentration if you dilute the 0.010 M solution by adding 90 mL of water?

  **d.** Does the pH increase or decrease after diluting the 0.010 M NaOH solution?

  **e.** What is the pH of the solution after adding 90 mL of water?

#### Solution

  **a.** The $OH^-$ concentration is 0.010 mole per liter, which is equivalent to $1 \times 10^{-2}$ M. Because $[H^+][OH^-] = 1 \times 10^{-14}$, the $H^+$ concentration is $1 \times 10^{-12}$ M. The exponents sum to $-14$.

  **b.** $pH = -\log[H^+] = -\log 10^{-12} = 12$

  **c.** If you add 90 mL water to 10 mL of 0.010 M NaOH, then you have 100 mL. The volume has increased tenfold. So the concentration after diluting is 0.0010 M NaOH.

  **d.** As the $OH^-$ concentration decreases, the $H^+$ concentration increases. So the pH decreases when you dilute 0.010 M NaOH to 0.0010 M.

  **e.** The $OH^-$ concentration is now $1.0 \times 10^{-3}$ M. Because $[H^+][OH^-] = 1.0 \times 10^{-14}$, $[H^+] = 1 \times 10^{-11}$ M.

$$pH = -\log[1 \times 10^{-11}] = 11$$

So the pH has decreased, from 12 to 11, as you might expect. Diluting with water makes a solution more neutral.

### LESSON SUMMARY

#### How does dilution affect acids and bases?

When water is added to an acidic solution, the result is a "watering down" of the solution. The $H^+$ concentration of the solution decreases, and the pH increases toward 7. When water is added to a concentrated basic solution, the $OH^-$ concentration decreases. This increases the $H^+$ concentration and the pH decreases toward 7. Acids cannot be made into bases and bases cannot be made into acids by adding water.

**KEY TERM**

dilution

### Reading Questions

1. When you add water to an acidic solution, describe what happens to the pH of that solution.

2. Explain why you cannot turn an acid into a base by diluting it with water.

### Reason and Apply

3. Copy this table and fill in the missing information.

| Solution | [H$^+$] in scientific notation | pH number | [OH$^-$] in scientific notation |
|---|---|---|---|
| 1.0 M HCl | $1 \times 10^0$ | 0 | $1 \times 10^{-14}$ |
| 0.10 M HCl | $1 \times 10^{-1}$ | 1 | $1 \times 10^{-13}$ |
| 0.010 M HCl | | | |
| 0.0010 M HCl | | | |
| 0.00000010 M HCl | | | |
| 0.000000010 M HCl | $1 \times 10^{-7}$ M | | |
| 0.0000000010 M HCl | $1 \times 10^{-7}$ M | | |
| 0.0010 M NaCl | | | |
| 0.00000010 M NaOH | | | $1 \times 10^{-7}$ M |
| 0.0000010 M NaOH | $1 \times 10^{-8}$ M | | $1 \times 10^{-6}$ M |
| 0.000010 M NaOH | | | |

4. How much water do you need to add to 10 mL of a solution of HCl with a pH of 3 to change the pH to 6?

5. Explain why the pH of a $1 \times 10^{-9}$ M HCl solution is 7 even though the exponent is $-9$.

6. Explain why the pH of a NaCl solution does not change if you dilute the solution.

7. Imagine that you have 0.75 L of a 0.10 M HCl solution.
   a. How many moles of H$^+$ are in the 0.75 L?
   b. If you add 0.35 L of water, what is the new concentration of the solution?
   c. What is the pH after adding 0.35 L?

8. Imagine that you have 15 mL of a 0.025 M HCl solution.
   a. How many moles of H$^+$ are in the 15 mL?
   b. If you add 25 mL of water, what is the new concentration of the solution?
   c. What is the final pH after adding 25 mL?

**SAFETY Instructions**

Safety goggles must be worn at all times.

Acids and bases are corrosive. Do not get any on skin or near eyes.

In case of a spill, rinse with large amounts of water.

Ken Karp Photography

**Materials**

- well plate
- set of five labeled dropper bottles: 0.10 M HCl (hydrochloric acid), 0.10 M HNO$_3$ (nitric acid), 0.10 M NaOH (sodium hydroxide), 0.10 M NH$_4$OH (ammonium hydroxide), and bromothymol blue indicator

# Neutralization Reactions

## Purpose

To examine reactions between acids and bases.

## Procedure

1. Add 20 drops of each solution to the well plate as specified in the illustration.

2. Add 1 drop of bromothymol blue indicator to each well. Bromothymol blue is yellow in acid, blue in base, and green in neutral solution.

3. Test with HCl: Try to turn the solutions in wells 1, 2, and 3 green using drops of 0.10 M HCl. Record your observations in a data table.

| Solution in the well plate | Indicator color | Acid, base, or neutral? | Drops of 0.10 M HCl to turn solution green | Indicator color after mixing | Does a reaction occur? |
|---|---|---|---|---|---|
| 20 drops 0.10 M HCl | | | | | |

4. Test with NaOH: Try to turn the solution in wells 4, 5, and 6 green using drops of 0.10 M NaOH.

## Analysis

1. What did you observe when you mixed an acid with a base?

2. Acids react with bases to form an ionic salt and water. Copy this chemical equation and label the acid, base, and ionic salt.

$$HCl(aq) + NaOH(aq) \longrightarrow NaCl(aq) + H_2O(l)$$

3. Complete and balance the equation for the reaction in each well. If no reaction occurred, simply write "no reaction" on the products side of the equation.

4. What combinations of reactants do not yield new products?

5. Name the three salts produced in the reactions.

6. **Making Sense** List three things you learned as a result of performing this lab procedure.

# Neutralization Reactions

## THINK ABOUT IT

There are many advertisements on television about products that are designed to treat "acid indigestion." Some of these substances claim to neutralize extra acid that may have built up in your stomach.

### What happens when acids and bases are mixed?

*To answer this question, you will explore*

⇨ Neutralization Reactions

⇨ Predicting the Products of Neutralization Reactions

## EXPLORING THE TOPIC

### ⇨ Neutralization Reactions

Pure water is neutral because it has equal numbers of $H^+$ and $OH^-$ ions.

Water dissociates into $H^+$ and $OH^-$ ions

$$H_2O(l) \longrightarrow H^+(aq) + OH^-(aq)$$

In pure water at 25 °C, $[H^+]$ and $[OH^-]$ are both $1.0 \times 10^{-7}$ mol/L. But if an acid or a base is added to water, it upsets the balance of ions in solution. Acidic solutions have an excess of $H^+$ and basic solutions have an excess of $OH^-$.

Increasing $[H^+]$

0         7         Increasing $[OH^-]$       14

pH scale

One way to reduce the toxicity of an acidic or basic solution is by dilution. If you add enough water, you can reduce the concentration of the acid or base. Extremely dilute acids and bases have a pH of 7 at 25 °C because this is the pH of pure water.

However, there is another way you can reduce the toxicity of an acidic or a basic solution. To counteract the effects of an acid in solution, you can add a base. Likewise, to counteract the effects of a base in solution, you can add an acid.

The acid and base react in solution and neutralize each other. The chemical equation for the reaction between an acid and base is shown below.

A Neutralization Reaction

| Acid | Base | | Salt | Water |

$$HCl(aq) + NaOH(aq) \longrightarrow NaCl(aq) + H_2O(l)$$

$Cl^-$ changes place with $OH^-$

Notice that this reaction is a double exchange reaction. The $H^+$ and $Na^+$ cations exchange anions. The result is the production of a salt, NaCl, and water. The excess $H^+$ from the acid combines with $OH^-$ from the added base to form $H_2O$. The reaction is referred to as a **neutralization reaction.**

......................................................................................

**Big Idea**   Acids and bases neutralize each other, producing a salt and water.

......................................................................................

So, one way to make an acidic solution safe is by adding a base. Likewise, you can make a basic solution safe by adding an acid.

Not all neutralization reactions produce neutral solutions. For example, if you use an acidic solution that is very concentrated and a basic solution that is not very concentrated, there will be leftover $H^+$ ions after mixing. There will not be enough $OH^-$ ions to neutralize all of the $H^+$ ions. But some neutralization will have taken place. As a result of mixing, the solution will be closer to neutral than either of the starting solutions.

## ⤷ Predicting the Products of Neutralization Reactions

**HEALTH CONNECTION**

Some antacids are made of calcium carbonate. Calcium carbonate is the main component of eggshells and seashells and can also be used as an inexpensive calcium supplement.

Christian Delbert/Shutterstock

Several neutralization reactions are shown. Notice the patterns in the products of the reactions.

$$HCl(aq) + KOH(aq) \longrightarrow KCl(aq) + H_2O(l)$$
$$HNO_3(aq) + NaOH(aq) \longrightarrow NaNO_3(aq) + H_2O(l)$$
$$H_2SO_4(aq) + 2NaOH(aq) \longrightarrow Na_2SO_4(aq) + 2H_2O(l)$$
$$2HCl + Mg(OH)_2(aq) \longrightarrow MgCl_2(aq) + 2H_2O(l)$$
$$H_2SO_4(aq) + Ca(OH)_2(aq) \longrightarrow CaSO_4(aq) + 2H_2O(l)$$

Each reaction results in the production of an ionic compound (a salt) plus water. The salt consists of the cation of the base and the anion of the acid. For example, the cation in potassium hydroxide, KOH, is potassium, $K^+$. The anion in nitric acid, HNO, is nitrate, $NO_3^-$. The $K^+$ and $NO_3^-$ produce the ionic compound, potassium nitrate, $KNO_3$, which is a salt.

Some acids transfer more than one $H^+$ ion. For example, each mole of sulfuric acid, $H_2SO_4$, transfers two moles of $H^+$ ions. This is because there are two hydrogen ions in solution for every one sulfate anion, $SO_4^{2-}$. Likewise, each mole of calcium hydroxide, $Ca(OH)_2$, transfers two moles of $OH^-$ ions. There are two hydroxide ions in each formula unit of calcium hydroxide to balance the charge on the calcium cations, $Ca^{2+}$.

## Producing Potassium Sulfate, $K_2SO_4$

Which of the acids and bases in this table could you mix to make potassium sulfate, $K_2SO_4$? Write a balanced chemical equation for the reaction.

| Acids | Bases |
|-------|-------|
| $HNO_3$ | KOH |
| $H_2SO_4$ | NaOH |
| HBr | $Mg(OH)_2$ |

### Solution

One way to use a neutralization reaction to make potassium sulfate is

$$H_2SO_4(aq) + 2KOH(aq) \longrightarrow K_2SO_4(aq) + 2H_2O(l)$$

Notice that you need 2 mol of KOH to balance the equation properly.

*The appearance of the white solid in suspension gives this remedy the name "milk."*

### STOMACH ACID RELIEF

A common over-the-counter remedy for excess stomach acid uses a neutralization reaction. Milk of magnesia is a white mixture containing magnesium hydroxide, $Mg(OH)_2$. The mixture looks like milk because magnesium hydroxide, a white solid, does not dissolve completely in water. Instead, the white solid is suspended in the liquid.

Magnesium hydroxide is a base. It neutralizes stomach acid, HCl, to produce a salt and water as shown by this reaction.

$$2HCl(aq) + Mg(OH)_2(aq) \longrightarrow MgCl_2(aq) + 2H_2O(l)$$

stomach acid + milk of magnesia $\longrightarrow$ magnesium chloride + water

## LESSON SUMMARY

### What happens when acids and bases are mixed?

A neutralization reaction is a reaction in which an acid and a base react in aqueous solution to produce an ionic compound (a salt) and water. The pH approaches 7 because the $H^+$ from the acid and the $OH^-$ from the base combine to form $H_2O$. It is relatively easy to predict the products of a neutralization reaction. The salt that forms is made from the cation of the base and the anion of the acid.

**KEY TERM**

neutralization reaction

### Reading Questions

1. Describe two ways to make a strong acidic solution safer.
2. What is a neutralization reaction?

David R. Frazier Photolibrary, Inc.

*(continued)*

**Reason and Apply**

3. **Lab Report** Complete a lab report for the Lab: Neutralization Reactions. In your report, give the title of the experiment, purpose, procedure, observations, and conclusion. In your conclusion, explain how you can use chemical equations to predict the products of the reactions you carried out.

> (Title)
>
> **Purpose:** (Explain what you were trying to find out.)
>
> **Procedure:** (List the steps you followed.)
>
> **Observations:** (Describe your observations.)
>
> **Conclusion:** (What can you conclude about what you were trying to find out?)

4. Predict the products for the reactions given below. Be sure to balance each equation.
   a. $HF(aq) + NaOH(aq) \longrightarrow$
   b. $HCl(aq) + Mg(OH)_2(aq) \longrightarrow$
   c. $HF(aq) + NH_4OH(aq) \longrightarrow$
   d. $CH_3COOH(aq) + NaOH(aq) \longrightarrow$
   e. $HNO_3(aq) + Mg(OH)_2(aq) \longrightarrow$
   f. $HNO_3(aq) + NH_4OH(aq) \longrightarrow$
   g. $CH_3COOH(aq) + Mg(OH)_2(aq) \longrightarrow$
   h. $CH_3COOH(aq) + NH_4OH(aq) \longrightarrow$
   i. $H_2SO_4(aq) + HNO_3(aq) \longrightarrow$
   j. $H_2SO_4(aq) + Mg(OH)_2(aq) \longrightarrow$

5. Suppose you mix 1 mol of sulfuric acid, $H_2SO_4$, with 1 mol of sodium hydroxide, NaOH. Why does the pH of the solution remain below 7?

6. Suppose you mix 1 mol of sodium hydroxide, NaOH, with 1 mol of magnesium hydroxide, $Mg(OH)_2$. Why does the pH of the solution remain above 7?

7. Which of these substances might be useful in neutralizing a lake damaged by acid rain?
   **(A)** $H_2SO_4$
   **(B)** $CH_3COOH$
   **(C)** $CaCl_2$
   **(D)** $Ca(OH)_2$

8. Which combination of reactants would result in a neutralization reaction with sodium nitrate, $NaNO_3$, as one of the products?
   **(A)** $Mg(NO_3)_2 + NaOH$
   **(B)** $HNO_3 + NaOH$
   **(C)** $CH_3OH + NaOH$
   **(D)** $HNO_3 + NaCl$

# Titration

## THINK ABOUT IT

When sulfur dioxide, $SO_2$, and nitrogen oxide, NO, two components of air pollution, come in contact with water in the atmosphere, they are converted to sulfuric acid, $H_2SO_4$, and nitric acid, $HNO_3$. The acids then fall as acid rain or snow. Acidic precipitation is highly destructive, sometimes causing the complete elimination of fish and insect species from lakes or streams. Acid rain also makes the soil unsuitable for plant life, killing off whole sections of forest. Water scientists, called hydrologists, regularly study the acidity of lakes and streams to monitor the extent of the acid rain problem.

### How can you use a neutralization reaction to figure out acid or base concentration?

*To answer this question, you will explore*

⮞ Titrations

⮞ Particle Views of Titrations

⮞ Titration Calculations

## EXPLORING THE TOPIC

### ⮞ Titrations

To track the spread of acid rain, scientists take water samples from lakes and test them for $H^+$ concentration. One way to determine the concentration of a strong acid in a water sample is to use a method called a **titration.** A titration is a neutralization reaction that is monitored with an acid-base indicator. For example, a strong base with a known concentration is added to a strong acid sample with an unknown concentration. The base is added until the indicator changes color.

The indicator provides a visual signal that the solution has reached the **equivalence point,** the point at which the moles of base added have neutralized the moles of acid. By keeping track of the exact volume of base that is added to a known volume of acid, you can figure out the unknown concentration of acid in the sample.

The illustration shows a titration setup. The long thin tube is a *burette.* A valve on the burette allows you to regulate how much solution goes into the beaker. During a titration, solution is added from the burette until the indicator changes color. You determine the volume of solution added by noting the change in the height of the solution in the burette.

A Titration Setup

— Base

Acid + indicator

*Phenolphthalein is an indicator that changes from colorless to bright pink in a base.*

## → Particle Views of Titrations

This illustration shows what happens as equal volumes of 1.0 M HCl, hydrochloric acid, and 1.0 M NaOH, sodium hydroxide, are mixed together.

$$HCl(aq) + NaOH(aq) \longrightarrow NaCl(aq) + H_2O(l)$$

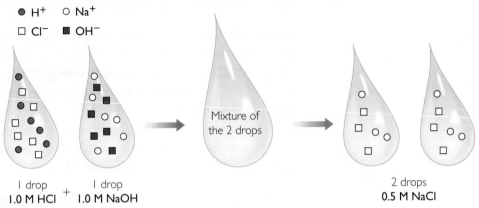

*Neutralization of 1.0 M NaOH by 1.0 M HCl*

When you add one drop of 1.0 M HCl to one drop of 1.0 M NaOH, $H^+$ ions from the acid and $OH^-$ ions from the base combine to form $H_2O$. The $Na^+$ ions and $Cl^-$ ions from the original drops are dissolved in the water. Notice that the concentration of $Na^+$ ions and $Cl^-$ ions has decreased because each ion has spread out over twice the volume, equal to two drops of solution.

The next illustration shows what happens when equal volumes of 1.0 M HCl and 2.0 M NaOH are mixed together.

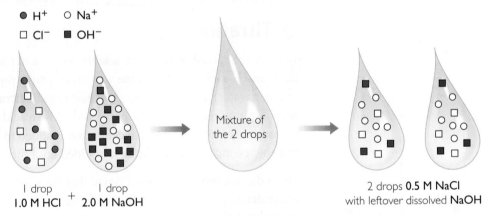

*Neutralization of 1.0 M HCl by 2.0 M NaOH*

In this example, the base is more concentrated than the acid so the two-drop mixture will have leftover dissolved NaOH in addition to dissolved 0.5 M NaCl.

Because the 2.0 M NaOH is twice as concentrated as the 1.0 M HCl, you only need half as much of it to neutralize the HCl.

| Moles of $H^+$ in 1.0 M HCl (10 mL) | Moles of $OH^-$ in 2.0 M NaOH (5 mL) |
|---|---|
| moles $H^+$ = molarity · volume<br>= 1.0 M · 0.010 L<br>= 0.005 mol | moles $OH^-$ = molarity · volume<br>= 2.0 M · 0.005 L<br>= 0.005 mol |

## ⟳ Titration Calculations

Suppose you take a 100 mL water sample from a lake contaminated with sulfuric acid, $H_2SO_4$. You want to determine the concentration of the sulfuric acid in the lake. So you titrate the sample with 2.0 molar sodium hydroxide, 2.0 M NaOH, until the solution is neutral. At the equivalence point, you know that the moles of base you added neutralized moles of acid that were in the unknown solution.

Bromothymol blue indicator is yellow in acidic solution, green in neutral solution, and blue in basic solution.

You need an indicator that changes color when the solution is neutral, so you add a drop of bromothymol blue to the lake water sample. You add NaOH slowly until the indicator turns green. Because you know the volume and molarity of the NaOH added, you can calculate the moles of acid per liter of solution in the lake water sample.

**Example**

### Titration of $H_2SO_4$

A 100 mL sample of aqueous sulfuric acid, $H_2SO_4$, is titrated with 2.0 M NaOH. After adding 50 mL of NaOH, the indicator changes color at pH 7. What was the starting concentration of the $H_2SO_4$?

*Solution*

Begin by writing a balanced chemical equation.

$$H_2SO_4(aq) + 2NaOH(aq) \longrightarrow Na_2SO_4(aq) + 2H_2O(l)$$

sulfuric acid + sodium hydroxide $\longrightarrow$ sodium sulfate + water

Find the total number of moles of NaOH that were used to neutralize the $H_2SO_4$.

moles NaOH = (molarity of NaOH)(volume of NaOH added)

$$= (2.0 \text{ mol/L})(0.050 \text{ L}) = 0.10 \text{ mol NaOH}$$

Since 2 mol of NaOH are needed to neutralize every 1 mol of $H_2SO_4$, the moles of $H_2SO_4$ in the sample must be half the number of moles of the added NaOH.

$$\text{moles of } H_2SO_4 = \frac{1}{2}(\text{moles of NaOH}) = \frac{1}{2}(0.10 \text{ mol}) = 0.05 \text{ mol}$$

Finally, find the molarity of the sulfuric acid solution.

$$\text{molarity of sulfuric acid} = \frac{n}{V}$$

$$= 0.050 \text{ mol}/0.100 \text{ L}$$

$$= 0.50 \text{ M } H_2SO_4$$

## LESSON SUMMARY

### How can you use a neutralization reaction to figure out acid or base concentration?

A titration is a chemical procedure carried out between an acid and a base to determine the concentration of either the acid or the base. A titration is a neutralization reaction that is monitored with an indicator. The volume of acid and the volume of base used in the procedure are carefully recorded. If the molarity of either the acid or the base is known, the molarity of the other can be determined.

## Exercises

### Reading Questions

1. Describe how you might use a titration to figure out the concentration of potassium hydroxide in a water sample.
2. What is the role of an indicator in titration?

### Reason and Apply

3. How many mL of 0.1 M NaOH would be required to neutralize 2.0 L of 0.050 M HCl?
4. The table represents a series of titrations using 0.10 M NaOH to determine the concentrations of five different samples of HCl. Copy the table and fill in the missing values.

| Initial volume of HCl | Volume of 0.10 M NaOH added | Total moles of NaOH | Total moles of HCl | Initial HCl concentration |
|---|---|---|---|---|
| 1.0 L | 1.0 L | 0.10 mol | 0.10 mol | 0.10 M |
| 100 mL | 200 mL | 0.020 mol | 0.020 mol | |
| 50 mL | 200 mL | | | |
| 50 mL | | 0.0025 mol | 0.0025 mol | 0.050 M |
| 100 mL | 73 mL | | | |

5. A student mixes 100 mL of 0.20 M HCl with different volumes of 0.50 M NaOH. Are the final solutions acidic, basic, or neutral? Explain your thinking.
   a. 100 mL of 0.20 M HCl + 20 mL of 0.50 M NaOH
   b. 100 mL of 0.20 M HCl + 40 mL of 0.50 M NaOH
   c. 100 mL of 0.20 M HCl + 60 mL of 0.50 M NaOH

6. Imagine that you use 0.95 M NaOH to titrate several water samples. The volume of base needed to neutralize a specified amount of acid is given. Determine the acid concentration and the pH for each.
   a. 25 mL acid, 46 mL NaOH
   b. 10 mL acid, 17 mL NaOH
   c. 25 mL acid, 12 mL NaOH

# CHAPTER 16

## Acidic Toxins

# SUMMARY

### KEY TERMS

indicator

pH scale

acid

base

dissociate

strong acid

weak acid

strong base

weak base

dilution

neutralization reaction

titration

equivalence point

# Toxins Update

Acids and bases are present in living systems and are a valuable part of the chemistry of life. They are toxic under certain conditions, such as when they are too concentrated or when they upset the pH balance that must be maintained for proper health. The pH number is a measure of the $H^+$ concentration of a solution. The lower the pH, the more acidic a solution is.

There are two main approaches to dealing with toxic acids and bases. One is dilution. This moves the pH toward neutral. Another approach is neutralization. Acids and bases neutralize one another.

## REVIEW EXERCISES

1. What are the main differences between the Arrhenius and Brønsted-Lowry definitions of acids and bases?

2. Name three substances that you might use to neutralize a hydrochloric acid solution. Write the balanced chemical equation for each reaction.

3. Explain how a titration procedure works to help you identify the $H^+$ concentration of a sample solution.

4. Lemon juice has a pH around 2. This is quite acidic and can damage the tissue of the eye.
   a. What is the concentration of hydrogen ions, $[H^+]$, in lemon juice?
   b. What is the concentration of hydroxide ions, $[OH^-]$?

5. Sodium bicarbonate, $NaHCO_3$, is a base.
   a. Explain how sodium bicarbonate can be classified as a base, even though it has no $OH^-$ ion.
   b. Explain why you had sodium bicarbonate on hand during the Lab: The Copper Cycle from Unit 1: Alchemy.

## Dissolving Toxins

PROJECT

www

Choose a toxic substance (your teacher may assign you one). Find some information about your toxin. Your report should cover these topics.

- Does your substance dissolve in water? If it does not dissolve, would it sink or float?

- Make a particle view drawing of your substance in water, even if it does not dissolve.

- Is your substance acidic, basic, or neutral? If your substance cannot be described in this way, explain why.

# CHAPTER 17

# Toxic Cleanup

*Chemical reactions are often used to remove toxic substances from a solution. Here, a solid forms when an ionic solution is added.*

### In this chapter, you will study

- solids in solutions

- soluble and insoluble ionic compounds

- how to determine the product yield of a reaction

- how to use the "mole tunnel" to perform stoichiometric calculations

One way to remove specific ions from a solution is to use chemical reactions to transform these ions into solid products. Solid compounds can be easily filtered out of a solution. The human body uses this same approach with many toxic substances, by forming solids that are then passed out of the body as waste. Mass-mole conversions allow you to determine how to create the maximum amount of product from a chemical reaction.

# Precipitation Reactions

## THINK ABOUT IT

Kidneys

Sometimes, reactions between dissolved substances result in the formation of a solid. In your kidneys, for example, dissolved substances can react to form solid calcium oxalate. If there is enough of this solid, a painful blockage called a kidney stone can form.

### Which substances precipitate from aqueous solutions?

*To answer this question, you will explore*

➲ Precipitation of Ionic Solids

➲ Solubility

➲ Toxicity of Precipitates

## EXPLORING THE TOPIC

### ➲ Precipitation of Ionic Solids

*Kidney stone crystal*

Sometimes certain anions and cations combine and come out of solution as a solid, or a **precipitate**. This kind of reaction is called a **precipitation reaction.**

The formation of kidney stones is an example of a precipitation reaction. If the concentrations of calcium ions, $Ca^{2+}$, and polyatomic oxalate ions, $C_2O_4^{2-}$, in urine become too great, calcium oxalate, $CaC_2O_4$, precipitates as kidney stones.

$$CaCl_2(aq) + Na_2C_2O_4(aq) \longrightarrow 2NaCl(aq) + CaC_2O_4(s)$$

| Dissolved calcium chloride | Dissolved sodium oxalate | Dissolved sodium chloride | Solid calcium oxalate |

The equation for the formation of kidney stones shows that it is a double exchange reaction. Many other similar chemical combinations result in precipitation reactions.

## Example 1

### Precipitation of Lead Iodide

An aqueous solution of lead nitrate, $Pb(NO_3)_2(aq)$, is mixed with an aqueous solution of potassium iodide, $KI(aq)$. The result is a bright yellow solid, lead iodide, $PbI_2(s)$, in a clear solution. Write a balanced chemical equation for this precipitation reaction.

Example 1

(continued)

*Solution*

This is a double exchange reaction. The lead and potassium cations exchange anions with one another. Begin by writing an equation for what you know.

$$Pb(NO_3)_2(aq) + KI(aq) \longrightarrow PbI_2(s) + ?$$

lead        potassium       lead      clear
nitrate       iodide      iodide  solution

The potassium ions, $K^+$, and the nitrate ions, $NO_3^-$, combine in a 1:1 ratio to form $KNO_3$.

$$Pb(NO_3)_2(aq) + KI(aq) \longrightarrow PbI_2(s) + \mathbf{KNO_3(aq)}$$

lead        potassium      lead      potassium
nitrate      iodide     iodide    nitrate

Balance the equation.

$$Pb(NO_3)_2(aq) + 2KI(aq) \longrightarrow PbI_2(s) + 2KNO_3(aq)$$

lead        potassium      lead      potassium
nitrate      iodide     iodide    nitrate

## ⊃ Solubility

Ionic substances vary significantly as to how much they will dissolve in water. For some ionic substances, large quantities dissolve in water. These substances have a high solubility. For other ionic substances, only very small quantities dissolve. These substances have a low solubility. Substances with a low solubility tend to form precipitates in aqueous solutions.

**Big Idea** Some ionic solids are more soluble than others. When a compound reaches the limits of its solubility, undissolved solid is visible.

You can use a solubility table such as the one shown here to determine the solubility of various ionic compounds. To use the table, combine a cation from the rows on the left with an anion from the columns on the right to determine if the compound formed is either very soluble (S), or insoluble/not very soluble (N).

**WEATHER**

**CONNECTION**

Precipitation can refer to either water coming out of the atmosphere as rain or snow or as compounds coming out of an aqueous solution as solids.

## Solubility Trends

| | | Anion | | | | | | |
|---|---|---|---|---|---|---|---|---|
| | | $NO_3^-$ | $Cl^-$ | $OH^-$ | $SO_4^{2-}$ | $CO_3^{2-}$ | $C_2O_4^{2-}$ | $PO_4^{3-}$ |
| Cations | Most alkali metals, such as $Li^+$, $Na^+$, $K^+$, $NH_4^+$ | S | S | S | S | S | S | S |
| | Most alkaline earth metals, such as $Mg^{2+}$, $Ca^{2+}$, $Sr^{2+}$ | S | S | N | S | N | N | N |
| | Some Period 4 transition metals, such as $Fe^{3+}$, $Co^{3+}$, $Ni^{2+}$, $Cu^{2+}$, $Zn^{2+}$ | S | S | N | S | N | N | N |
| | Other transition metals, such as $Ag^+$, $Pb^{2+}$, $Hg^{2+}$ | S | N | N | N | N | N | N |

Recall that in Unit 1: Alchemy, you characterized ionic solids as soluble in water. This is true for many ionic solids. But as you can see from the solubility table, some ionic solids are not very soluble.

**Example 2**

### Predicting Solid Products

Suppose that you combine aqueous calcium nitrate, $Ca(NO_3)_2$, with aqueous sodium phosphate, $Na_3PO_4$, and there is a double exchange reaction. Do you expect a precipitate to form?

*Solution*

First, write a balanced chemical equation for this reaction. You know that the two aqueous cations exchange anions:

$$3Ca(NO_3)_2(aq) + 2Na_3PO_4(aq) \longrightarrow Ca_3(PO_4)_2(?) + 6NaNO_3(?)$$

Next, use the solubility table to determine if there is a precipitate. According to the table, $NaNO_3$ is soluble in water, so it remains dissolved. $Ca_3(PO_4)_2$ is not very soluble, so it forms a precipitate.

$$3Ca(NO_3)_2(aq) + 2Na_3PO_4(aq) \longrightarrow Ca_3(PO_4)_2(s) + 6NaNO_3(aq)$$

### EARTH SCIENCE
### CONNECTION

Stalactites and stalagmites are caused by precipitation of solids from water that drips from cave walls. The word *stalagmite* comes from the Greek word for "drip." These beautiful cave structures are often formed from calcium carbonate.

Philip Schermeister/Getty Images

Reactions in aqueous solutions usually involve ions in solution. To describe what is happening, you can write a **complete ionic equation,** which shows all of the dissolved ions involved in the reaction. Examine the complete ionic equation for the reaction in Example 2.

$$3Ca^{2+}(aq) + 6NO_3^-(aq) + 6Na^+(aq) + 2PO_4^{3-}(aq) \longrightarrow$$
$$Ca_3(PO_4)_2(s) + 6Na^+(aq) + 6NO_3^-(aq)$$

An ion that appears on both sides of a complete ionic equation and does not directly participate in the reaction is called a **spectator ion.** For example, $Na^+$ and $NO_3^-$ are spectator ions in this reaction. To write a more efficient equation for this reaction, you can cancel the spectator ions on each side of the equation:

$$3Ca^{2+}(aq) + \cancel{6NO_3^-(aq)} + \cancel{6Na^+(aq)} + 2PO_4^{3-}(aq) \longrightarrow$$
$$Ca_3(PO_4)_2(s) + \cancel{6Na^+(aq)} + \cancel{6NO_3^-(aq)}$$

The result is a **net ionic equation** that describes the reaction in terms of only the ions that are involved in the reaction:

$$3Ca^{2+}(aq) + 2PO_4^{3-}(aq) \longrightarrow Ca_3(PO_4)_2(s)$$

## ⮕ Toxicity of Precipitates

There are positive and negative aspects to the solubility of toxic substances. On the one hand, if a substance is not very soluble, it might not react with anything, and it might pass through the body relatively unnoticed. For example, some metals that can be ingested go through our bodies without causing harm. (Imagine a child swallowing a nickel.) However, many metal compounds are soluble, and once our bodies absorb them, some can do long-term damage.

### LESSON SUMMARY

#### Which substances precipitate from aqueous solutions?

A precipitate is a solid that forms when a reaction in an aqueous solution produces a compound that is not very soluble. Most precipitation reactions are double exchange reactions involving ionic compounds. The cations exchange anions in solution, and one of the substances precipitates as a solid. Precipitation depends on the solubility of a compound. Some ionic compounds are not very soluble, so they precipitate from aqueous solutions. Precipitates in your body include bones, teeth, and kidney stones.

**KEY TERMS**

precipitate

precipitation reaction

complete ionic equation

spectator ion

net ionic equation

---

## Exercises

### Reading Questions

1. Describe what a precipitate is and how it forms.

2. What does solubility have to do with precipitation from solution?

3. Explain what a spectator ion is.

### Reason and Apply

4. Circle the ionic solids that are soluble in water.

   $LiNO_3$     KCl     $MgCl_2$     $Ca(OH)_2$     RbOH

   $CaCO_3$     $Li_2CO_3$     $PbCl_2$     AgCl

5. Name solutions you can combine that will precipitate each of the compounds listed.
   **a.** $CaCO_3$   **b.** $FePO_4$   **c.** $HgCl_2$   **d.** AgCl

6. Write balanced chemical equations for these reactions. Be sure to indicate which substances are aqueous and which are solid.
   **a.** $NaCl(aq) + Hg(NO_3)_2(aq)$     **b.** $NaOH(aq) + CaCl_2(aq)$
   **c.** $K_2CO_3(aq) + LiNO_3(aq)$     **d.** $Zn(NO_3)_2(aq) + Na_3PO_4(aq)$
   **e.** $NaCl(aq) + Ca(NO_3)_2(aq)$

7. Refer to your answers for Exercise 6. For each equation that represents a precipitation reaction, write both complete and net ionic equations.

Ken Karp Photography

# Mole Ratios

## Purpose

To examine how the ratio of reactants affects the amount of products.

## Materials

- 12 × 75 mm test tubes (10)
- test-tube rack
- marker (to label test tubes)
- 2 dropper bottles with 0.10 M solutions of calcium chloride, $CaCl_2$, and sodium carbonate, $Na_2CO_3$

## Procedure

1.  Number five small test tubes from 1 to 5.

2.  Add drops of 0.10 M $CaCl_2(aq)$ and 0.10 M $Na_2CO_3$ to the test tubes as indicated in the table below.

3.  Swirl each test tube gently to mix the reactants. Allow the solids to settle for about ten minutes while you complete the table and continue the analysis.

| | Tube 1 | Tube 2 | Tube 3 | Tube 4 | Tube 5 |
|---|---|---|---|---|---|
| Drops of 0.10 M $CaCl_2$ | 4 | 6 | 12 | 18 | 20 |
| Drops of 0.10 M $Na_2CO_3$ | 20 | 18 | 12 | 6 | 4 |
| Ratio of $CaCl_2$ to $Na_2CO_3$ | 1:5 | | | | |

## Observations and Analysis

1.  Write the balanced chemical equation for the exchange reaction of aqueous calcium chloride with aqueous sodium carbonate. Predict the ratio of moles of $CaCl_2(aq)$ to $Na_2CO_3(aq)$ that will produce the maximum amount of $CaCO_3(s)$.

2.  After the solids have settled, determine which tube has the most $CaCO_3(s)$. Identify the ratio in the table that resulted in the largest volume of solid.

3.  **Making Sense** How do the balanced chemical equations compare to your observations?

4.  **If You Finish Early** Suppose that you mix calcium chloride, $CaCl_2(aq)$, with sodium oleate, $NaC_{18}H_{33}O_2(aq)$. What products do you expect? What ratio of drops would give you the largest amount of precipitate? (*Note:* One product is calcium oleate, commonly referred to as soap scum.)

*Mole to Mole*

# Mole Ratios

## THINK ABOUT IT

Water treatment plants in our communities process millions of gallons of water a day. These plants remove toxic substances from the water through a series of chemical and physical treatments. For example, impure water often contains dissolved metal cations. These can be removed from water through precipitation of insoluble metal salts. Once the metals are in solid form, they can be safely separated by filtration. Of course, for maximum effectiveness, the treatment plant should remove as much of each metal from the water as possible.

*A worker takes a sample to determine water quality at a treatment plant.*

© Will & Deni McIntyre/CORBIS

**How can you convert all the reactants to products?**

*To answer this question, you will explore*

⮑ Moles of Product

⮑ Excess and Limiting Reactant

## EXPLORING THE TOPIC

### ⮑ Moles of Product

The list of metals that can show up in our water supplies is quite long, and it includes lead, silver, mercury, chromium, copper, zinc, cadmium, and tin. According to the solubility table in the previous lesson, many metals form hydroxides and carbonates that are not soluble. So a good way to remove the more toxic metal cations from our water supplies is to add hydroxides or alkali metal carbonates to precipitate hazardous metal cations.

Imagine a water sample that contains unwanted copper ions. The goal is to figure out how to remove as many of those ions per liter of solution as possible through precipitation. Sodium hydroxide, NaOH, is chosen as a reactant because it is soluble and contains a hydroxide ion. When hydroxide ions enter the solution, solid copper (II) hydroxide, $Cu(OH)_2$, will precipitate. The question remains: How much NaOH should be added, and in what concentration?

The next illustration shows what happens when a solution of 1.0 M $CuSO_4$ is mixed with a solution of 1.0 M NaOH in different ratios. Take a moment to examine the outcomes.

In each case, a total of 24 mL of solution are combined. However, the amounts of 1.0 M $CuSO_4$ and 1.0 M NaOH that are combined vary from test tube to test tube. The combination that produces the maximum amount of solid precipitate is in test tube 3, where 8 mL of 1.0 M $CuSO_4$ combined with 16 mL of 1.0 M NaOH. This results in 0.008 mol of solid copper hydroxide.

After the reaction, all four test tubes contain aqueous sodium sulfate, $Na_2SO_4(aq)$, and a blue solid, copper (II) hydroxide, $Cu(OH)_2(s)$. Therefore, sodium sulfate and copper (II) hydroxide must be products of this reaction. The solutions in three of the test tubes are almost colorless, while the solution in test tube 4 is pale blue. This means there is unreacted $CuSO_4(aq)$ left over in the last test tube.

**Chemical equation:** $CuSO_4(aq) + 2NaOH(aq) \longrightarrow Cu(OH)_2(s) + Na_2SO_4(aq)$

## MOLE RATIO

The coefficients in a chemical equation indicate the proportions in which substances react, in moles or other counting units. The coefficients in this particular equation indicate that for every 1 mol of $CuSO_4$, 2 mol of NaOH are required to make the maximum amount of products. The test tube containing the most amount of solid (test tube 3) corresponds to this 1:2 ratio of reactants. This ratio is often referred to as the **mole ratio.** So, using the ratio given by the balanced chemical equation produces the maximum amount of product.

> **Big Idea**  To get the maximum amount of product from a reaction, reactants must be mixed in the correct proportions.

Example 1

### Precipitation of Silver Hydroxide

An aqueous solution contains 0.020 mol of silver nitrate, $AgNO_3(aq)$. Predict the number of moles of $NaOH(aq)$ you will need to precipitate all of the silver, Ag, as silver hydroxide, AgOH.

*Solution*

First, write a balanced chemical equation for the reaction.

$$AgNO_3(aq) + NaOH(aq) \longrightarrow AgOH(s) + NaNO_3(aq)$$

The mole ratio of $AgNO_3$ to NaOH is 1:1.

The ratio given by the balanced equation is 1:1, and you have 0.020 mol of $AgNO_3$. So you will need 0.020 mol of NaOH to precipitate all of the silver.

## ⟳ Excess and Limiting Reactant

Look again at the four test tubes. Using the molarity and volume data for each reactant, you can determine the number of moles of each reactant. The table shows the moles of each reactant and the ratios in which they are combined.

| Test tube | Moles of $CuSO_4$ | Moles of NaOH | Ratio | Left over | Runs out |
|---|---|---|---|---|---|
| 1 | 0.004 mol | 0.020 mol | 1:5 | NaOH | $CuSO_4$ |
| 2 | 0.006 mol | 0.018 mol | 1:3 | NaOH | $CuSO_4$ |
| 3 | 0.008 mol | 0.016 mol | **1:2** | neither | neither |
| 4 | 0.020 mol | 0.004 mol | 5:1 | $CuSO_4$ | NaOH |

**BIOLOGY CONNECTION**

One of the best ways to test for the presence of heavy metals in a person's body is through hair samples. Human hair is a permanent record of some of the substances that have passed through someone's body. A three-inch strand of hair will give a six-month history of what's going on chemically in the body.

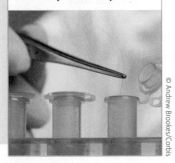

© Andrew Brookes/Corbis

Notice that in test tubes 1, 2, and 4, one of the reactants runs out before the other. Only in test tube 3 have both reactants been used up completely. Test tubes 1 and 2 no longer have copper ions in solution, but the solution is now toxic due to the excess NaOH. Test tube 4 still has copper ions in solution.

If you mix the reactants in a ratio other than the mole ratio, a reaction will still occur, but one of the reactants will run out and the other one will have some left over. The reactant that runs out is called the **limiting reactant,** or **limiting reagent.** This is because it limits how much product you can make.

To remove all of the copper ions from the solution, it is best to mix the reactants in the mole ratio determined by the balanced chemical equation.

If you want to remove copper ions (or any other metal ions) from water in a water treatment plant, it would be best to add sodium hydroxide in the mole ratio specified by the chemical equation. To do so, you will first have to determine the concentration of the copper ions in your water in moles per liter so you can match that concentration with the appropriate concentration of reactant.

**Example 2**

## Precipitation of Calcium Phosphate

Aqueous calcium chloride, $CaCl_2(aq)$, reacts with aqueous sodium phosphate, $Na_3PO_4(aq)$, forming a precipitate of calcium phosphate, $Ca_3(PO_4)_2(s)$.

- **a.** If 12 mol of $CaCl_2(aq)$ react with 8 mol of $Na_3PO_4(aq)$, will there be any reactant left over? If so, which reactant?
- **b.** If 12 mol of $CaCl_2(aq)$ react with 16 mol of $Na_3PO_4(aq)$, will there be any reactant left over? If so, which reactant?
- **c.** If 12 mol of $CaCl_2(aq)$ react with 4 mol of $Na_3PO_4(aq)$, will there be any reactant left over? If so, which reactant?

### Solution

First, write a balanced chemical equation for the reaction. Because atoms are conserved, $NaCl(aq)$ must be the other product.

$$3CaCl_2(aq) + 2Na_3PO_4(aq) \longrightarrow Ca_3(PO_4)_2(s) + 6NaCl(aq)$$

The mole ratio for $CaCl_2$ and $Na_3PO_4$ is 3:2. For any other ratio, there will be leftover reactant.

- **a.** 12 mol $CaCl_2$ to 8 mol $Na_3PO_4$ is a ratio of 3:2. So all the reactants will be converted to products.
- **b.** 12 mol $CaCl_2$ to 16 mol $Na_3PO_4$ is a ratio of 3:4. There will be some $Na_3PO_4$ left over.
- **c.** 12 mol $CaCl_2$ to 4 mol $Na_3PO_4$ is a ratio of 4:1. There will be some $CaCl_2$ left over.

## LESSON SUMMARY

### How can you convert all the reactants to products?

**KEY TERMS**

mole ratio

limiting reactant
(limiting reagent)

Coefficients in chemical equations represent the proportions in which moles of reactants combine and products form. The ratio of two coefficients is also called the mole ratio. To produce the maximum amount of product, moles of reactants should be combined in the exact proportions specified by the coefficients. Mass and volume amounts cannot be substituted for coefficients. If there is not enough of a reactant, resulting in other reactants being left over, that reactant is called a limiting reactant.

**Exercises**

### Reading Questions

1. Why are coefficients important in chemical equations?
2. Explain how to create the maximum amount of product from a reaction.

*(continued)*

**Reason and Apply**

3. For each of the reactions listed, determine the mole ratio of reactants that produces the maximum amount of precipitate. Be sure to balance the equations.

   **a.** $AgNO_3(aq) + NaCl(aq) \longrightarrow AgCl(s) + NaNO_3(aq)$

   **b.** $Cu(NO_3)_2(aq) + K_2CO_3(aq) \longrightarrow CuCO_3(s) + KNO_3(aq)$

   **c.** $ZnCl_2(aq) + NaOH(aq) \longrightarrow Zn(OH)_2(s) + NaCl(aq)$

   **d.** $CaCl_2(aq) + Na_2C_2O_4(aq) \longrightarrow CaC_2O_4(s) + NaCl(aq)$

4. **Lab Report** Write a lab report for the experiment you did in which you found the mole ratio by precipitating solids.

   **(Title)**

   **Purpose:** (Explain what you were trying to find out.)

   **Procedure:** (Describe the steps you followed.)

   **Observations:** (Describe what you observed. Include a data table of your results.)

   **Analysis:** (Write a balanced chemical equation for each reaction.)

   **Conclusion:** (Explain what happened in the different test tubes.)

5. Aqueous silver nitrate reacts with aqueous sodium chloride producing a precipitate of silver chloride.

$$AgNO_3(aq) + NaCl(aq) \longrightarrow AgCl(s) + NaNO_3(aq)$$

   **a.** How many moles of NaCl do you need to react with 0.10 mol of $AgNO_3$?

   **b.** How many grams of NaCl does this represent?

6. Imagine that you have 500 mL of 0.20 M silver nitrate solution. How many milliliters of 0.50 M sodium chloride solution should you add to remove all the silver ions from the solution?

**BIOLOGY**

**CONNECTION**

Calcium phosphate is similar to the solid that forms bones and teeth. The calcium cations and phosphate anions dissolved in your blood deposit as solids to form your skeleton.

# Stoichiometry

## THINK ABOUT IT

Sales of medications bring in billions of dollars per year. Most of these medications are produced through chemical reactions that are performed in a laboratory. In manufacturing the medications, it is important to know exactly how much reactant you need to prepare a specified mass of product. The balanced chemical equation for the reaction serves as a guide. However, the calculation requires conversion from mass to moles and back to mass.

**How do you convert between grams and moles to determine the mass of product?**

*To answer this question, you will explore*

⮕ Gram-Mole Conversions

⮕ Solving Stoichiometry Problems

## EXPLORING THE TOPIC

## ⮕ Gram-Mole Conversions

GVictoria/Shutterstock

One type of medication available is an antacid made of aluminum hydroxide, $Al(OH)_3$. It neutralizes excess acid in the stomach. To make aluminum hydroxide, many manufacturers start with aluminum chloride, which is commercially available in large quantities. When solutions of aluminum chloride are mixed with sodium hydroxide, a precipitate of aluminum hydroxide is formed. The balanced chemical equation for the reaction is

$$AlCl_3(aq) + 3NaOH(aq) \longrightarrow Al(OH)_3(s) + 3NaCl(aq)$$

The coefficients in the equation indicate how many moles to mix together. But in the laboratory, reactants are measured in grams, kilograms, or pounds.

**GOING THROUGH THE MOLE TUNNEL**

A chemical equation is like a recipe for making a product. It indicates how much of the reactants to mix. If you want to make aluminum hydroxide, $Al(OH)_3$, you need one formula unit of aluminum chloride, $AlCl_3$, for every three formula units of sodium hydroxide, NaOH. Before the recipe can be put into practice, it must be converted to mass. This is because in the laboratory, substances are weighed, not counted.

| 1 mol $AlCl_3(aq)$ + | 3 mol $NaOH(aq)$ ⟶ | 1 mol $Al(OH)_3(s)$ + | 3 mol $NaCl(aq)$ |
|---|---|---|---|
| ⬇ | ⬇ | ⬇ | ⬇ |
| ___?___ g $AlCl_3(aq)$ + | ___?___ g $NaOH(aq)$ ⟶ | ___?___ g $Al(OH)_3(s)$ + | ___?___ g $NaCl(aq)$ |

If you know the mass of reactants, you can determine the mass of products that can be made.

This requires a trip through the "mole tunnel." Starting with grams of reactant and the balanced chemical equation, there are three steps you must take to convert from grams of reactants to grams of products.

Reactant A + Reactant B ⟶ Product C + Product D

**Step 1** Convert grams of reactants to moles of reactants.

$$\text{moles} = \frac{\text{grams}}{\text{molar mass}}$$

**Step 3** Convert moles of products to grams of products.

$$\text{grams} = (\text{molar mass})(\text{moles})$$

Mole ratio

Mole tunnel

**Step 2** Relate moles of reactants to moles of products using mole ratios.

To go from one side of the equation to the other, you must go through moles, represented by the tunnel under the equation. The "mole tunnel" in the illustration is one way to remember these steps. These types of calculations, involving quantities of reactants and products, are referred to as gram-mole conversions, or **stoichiometry** calculations. The word *stoichiometry* comes from the Greek words *stoikheion*, meaning "element," and *metrein*, meaning "to measure."

The mole tunnel also works in reverse. If you want a certain mass of a product, you can determine the mass of reactants required.

## ⬭ Solving Stoichiometry Problems

The chemical equation for the formation of aluminum hydroxide, $Al(OH)_3$, shows the reactants combining in a mole ratio of 1:3. The chemical equation also indicates that the mole ratio of the $AlCl_3$ reactant to the $Al(OH)_3$ product is 1:1. So, for every mole of $AlCl_3$ that reacts, one mole of $Al(OH)_3$ is produced. If you want to make 100 mol of $Al(OH)_3$, you will need 100 mol of $AlCl_3$ and 300 mol of NaOH.

The molar masses on the periodic table allow you to convert between grams and moles.

### Molar Masses

Reactants: $AlCl_3$ = 133.5 g/mol    NaOH = 40.0 g/mol

Products: $Al(OH)_3$ = 78.0 g/mol    NaCl = 58.5 g/mol

Next, the mole amounts shown in the chemical equation can be converted to gram amounts. The table summarizes these calculations.

|  | Reactants | | Products | |
|---|---|---|---|---|
|  | $AlCl_3$ | NaOH | $Al(OH)_3$ | NaCl |
| **Moles** | 1 mol | 3 mol | 1 mol | 3 mol |
| **Grams** | 133.5 g | 120.0 g | 78.0 g | 175.5 g |
| **Total mass** | 253.5 g | | 253.5 g | |

**HISTORY CONNECTION**

Early black-and-white photographs, called *calotypes,* were made on paper prepared with silver nitrate. The paper was loaded into a camera under near-total darkness, then exposed to a scene for 10–60 seconds to make a photo negative. Prints could then be produced from the negative. An early photograph of President Abraham Lincoln is shown.

© CORBIS

## From Moles to Grams

Suppose you have 5.00 mol of $AlCl_3$ in solution and an unlimited supply of NaOH. How many grams of $Al(OH)_3$ and NaCl can you make?

$$AlCl_3(aq) + 3NaOH(aq) \longrightarrow Al(OH)_3(s) + 3NaCl(aq)$$

*Solution*

Use mole ratios to determine the moles of product you can make.

$$AlCl_3(aq) + 3NaOH(aq) \longrightarrow Al(OH)_3(s) + 3NaCl(aq)$$

1 : 1

1 : 3

So, with 5.00 mol of $AlCl_3$, you can make 5.00 mol of $Al(OH)_3$ and 15.0 mol of NaCl. Use molar mass to convert these quantities to grams.

Multiply the number of moles by the molar mass.

5.0 mol $Al(OH)_3$ · 75.0 g/mol = 375 g

15.0 mol NaCl · 58.5 g/mol = 878 g

If you start with 5.0 mol $AlCl_3$, you can react it with 15.0 mol NaOH to prepare 375 g $Al(OH)_3$ and 878 g NaCl.

## From Grams to Grams

Suppose you have 500 g of $AlCl_3$ in solution and an unlimited supply of NaOH. How many grams of each product can you make?

*Solution*

You cannot convert directly from grams of reactant to grams of product. You must use the "mole tunnel."

Use molar mass to convert from grams to moles of reactants.

Divide by the molar mass. $\dfrac{500 \text{ g AlCl}_3}{133.5 \text{ g/mol}} = 3.75$ mol $AlCl_3$

Use mole ratios to determine the moles of each product.

So 3.75 mol $AlCl_3$ will make 3.75 mol $Al(OH)_3$ and 11.25 mol NaCl.

Use molar mass to convert between moles and grams.

Multiply the number of moles by the molar mass.

3.75 mol $Al(OH)_3$ · 75.0 g/mol = 281 g $Al(OH)_3$
11.25 mol NaCl · 58.5 g/mol = 658 g NaCl

So, 500 g $AlCl_3$ produces 281 g $Al(OH)_3$ and 658 g NaCl.

## How do you convert between grams and moles to determine the mass of product?

To determine the mass of product produced by a certain mass of reactant (and vice versa), it is necessary to convert mass to moles and then back to mass. This is because the balanced chemical equation is written in moles. Typically, when you're working with chemical reactions, you know the grams of reactant you need to produce a certain number of grams of product. Calculations involving mole ratios and masses of reactants and products are referred to as gram-mole conversions, or stoichiometry calculations.

**KEY TERM**

stoichiometry

## Exercises

### Reading Questions

1. Explain why you need to do gram-mole conversions when carrying out chemical reactions.

### Reason and Apply

2. Consider this balanced chemical equation.

$$Cu(s) + 2AgNO_3(aq) \longrightarrow Cu(NO_3)_2(aq) + 2Ag(s)$$

   copper     silver nitrate     copper (II) nitrate     silver

   a. Find the mole ratio of $AgNO_3$ to $Cu(NO_3)_2$.
   b. Find the mole ratio of $AgNO_3$ to Ag.
   c. Suppose you have 6.0 mol of Cu. How many moles of $AgNO_3$ are needed to react completely with 6.0 mol of Cu? How many mol of $Cu(NO_3)_2$ are produced?
   d. Suppose you want to make 30.0 g Ag. How many moles of Ag is that? How many moles of Cu do you need?

3. Consider this balanced chemical equation.

$$2NaCl(aq) + Pb(NO_3)_2(aq) \longrightarrow 2NaNO_3(aq) + PbCl_2(s)$$

   sodium chloride     lead (II) nitrate     sodium nitrate     lead (II) chloride

   a. Suppose you want to make 30.0 g $PbCl_2$. How many grams of NaCl do you need?
   b. Suppose you have 56 g $Pb(NO_3)_2$. How many grams of NaCl are needed to react completely with 56 g $Pb(NO_3)_2$? How many grams of $PbCl_2$ are produced?

4. An aqueous solution of sodium carbonate reacts with an aqueous solution of calcium sulfate to produce solid calcium carbonate and an aqueous solution of sodium sulfate.

$$2Na_2CO_3(aq) + CaSO_4(aq) \longrightarrow Na_2SO_4(aq) + CaCO_3(s)$$

   sodium carbonate     calcium sulfate     sodium sulfate     calcium carbonate

   a. What type of reaction is this?
   b. How many grams of each reactant are needed to produce 4500 g of $CaCO_3$?

# Limiting Reactant and Percent Yield

## THINK ABOUT IT

Imagine you want to make as many cheese sandwiches as possible from a loaf of bread and a large package of sliced cheese. The package of cheese contains 15 slices. The loaf of bread contains 24 slices. Which ingredient will you run out of first, and how many sandwiches can you make?

**Which reactant determines how much product you can make?**

*To answer this question, you will explore*

⮞ Limiting Reactants

⮞ Solving Limiting Reactant Problems

⮞ Percent Yield

## EXPLORING THE TOPIC

### ⮞ Limiting Reactants

Mixing reactants to form products is a little like making cheese sandwiches from specific quantities of cheese and bread. In the example above, you could make 12 sandwiches before running out of bread. You would have 3 slices of cheese left over. In the world of chemistry, substances rarely come together in the exact mole ratios specified by a chemical equation. One of the reactants usually runs out first. As you learned in Lesson 91: Mole to Mole, the reactant that runs out is called the limiting reactant.

Imagine that you have a beaker with an aqueous solution of sulfuric acid containing a total of 4 mol of $H_2SO_4$. You add 112 g of solid potassium hydroxide, KOH. Will the addition of this amount of KOH($s$) completely neutralize the $H_2SO_4$($aq$)? Will either reactant be left over?

The balanced chemical equation for the reaction you are examining is

$$H_2SO_4(aq) + 2KOH(s) \longrightarrow K_2SO_4(aq) + 2H_2O(aq)$$

You must first find out how many moles are represented by 112 g of KOH. The molar mass of KOH is 56 g/mol, so 112 g KOH is equal to 2 mol of KOH.

The chemical equation shows that you will need 2 mol of KOH to neutralize 1 mol of $H_2SO_4$. There are 4 mol of $H_2SO_4$ in the aqueous solution, and you have only 2 mol of KOH. You will be able to neutralize only 1 mol of $H_2SO_4$. There will be 3 mol of $H_2SO_4$ left over.

The illustration represents what happens in your beaker. Each unit represents 1 mol.

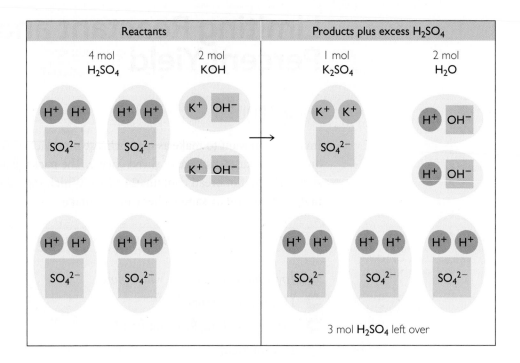

| Reactants | | Products plus excess H₂SO₄ | |
|---|---|---|---|
| 4 mol $H_2SO_4$ | 2 mol KOH | 1 mol $K_2SO_4$ | 2 mol $H_2O$ |

3 mol $H_2SO_4$ left over

There is not enough KOH to react with all of the $H_2SO_4$. KOH is the limiting reactant, and at the end of the reaction, there will be excess $H_2SO_4$.

## ⤴ Solving Limiting Reactant Problems

Knowing the identity of the limiting reactant allows you to figure out the maximum amount of product you can make with the substances you have. Mole ratios from the balanced chemical equation are the key to solving limiting reactant problems.

**Big Idea** The limiting reactant determines how much product you can make from a chemical reaction.

The steps to complete a limiting reactant problem are shown here.

Solving a Limiting Reactant Problem

**Step 1:** Write the balanced chemical equation.

**Step 2:** Determine the molar mass of each compound.

**Step 3:** Determine the number of moles of each reactant that you have.

**Step 4:** Use the mole ratio to identify the limiting reactant.

**Step 5:** Use the limiting reactant to determine the maximum amount of product.

Example 1

## Preparing Magnesium Chloride

Adding hydrochloric acid, HCl, to magnesium hydroxide, $Mg(OH)_2$, produces magnesium chloride, $MgCl_2$, and water. Imagine you start with 10.0 g of $Mg(OH)_2$ and 100 mL of 4.0 M HCl. How much $MgCl_2$ will be produced?

*Solution*

**Step 1:** Write the balanced chemical equation.

$$Mg(OH)_2(s) + 2HCl(aq) \longrightarrow MgCl_2(aq) + 2H_2O(l)$$

**Step 2:** Determine the molar mass of each compound.

Reactants: $Mg(OH)_2$ = 58.3 g/mol     HCl = 36.5 g/mol
Products: $MgCl_2$ = 95.3 g/mol       $H_2O$ = 18.0 g/mol

**Step 3:** Determine the number of moles of each reactant that you have.

$$\text{Moles of } Mg(OH)_2 = \frac{10.0 \text{ g}}{58.3 \text{ g/mol}} = 0.17 \text{ mol}$$

$$\text{Moles of HCl} = (0.100 \text{ L})(4.0 \text{ mol/L}) = 0.40 \text{ mol}$$

**Step 4:** Use the mole ratio to identify the limiting reactant.

The reactants combine in a 1:2 ratio. So you need $0.17 \text{ mol} \cdot 2 = 0.34 \text{ mol}$ of hydrochloric acid, HCl, to react with the 0.17 mol of magnesium hydroxide, $Mg(OH)_2$.

You have 0.40 mol of HCl, so the limiting reactant is $Mg(OH)_2$. When the reaction is complete, there will be 0.06 mol of HCl left over.

**Step 5:** Use the limiting reactant to determine the maximum amount of product.

For each mole of $Mg(OH)_2$, 1 mol of $MgCl_2$ is produced. So, 0.17 mol of $MgCl_2$ is produced. Use the molar mass to determine the mass of $MgCl_2$.

$$(0.17 \text{ mol})(95.3 \text{ g/mol}) = 16.2 \text{ g } MgCl_2$$

So starting with 10.0 g $Mg(OH)_2$ and 100 mL of 4.0 M HCl, you can make 16.2 g $MgCl_2$.

---

**Important to Know** The reactant present in the smallest amount is not necessarily the limiting reactant. The mole ratio specified by the coefficients of the balanced equation must be taken into consideration.

## ⟳ Percent Yield

When you use the limiting reactant to calculate the amount of a product produced in a chemical reaction, you are calculating the **theoretical yield** for that product or what you should be able to produce in theory. For a variety of reasons, including experimental error, reactions rarely produce the predicted theoretical yield when run in the laboratory. The amount of a product produced when the reaction is run is called the **actual yield** for that product and is usually a bit less than the theoretical yield. When you compare the actual yield with the theoretical yield and express it as a percentage, it is called the **percent yield.**

$$\text{Percent yield} = \frac{\text{actual yield}}{\text{theoretical yield}} \cdot 100\%$$

The percent yield tells you how successful or efficient your procedure was in the laboratory. The closer your percent yield is to 100%, the more efficient your procedure was.

Example 2

## Percent Yield of MgCl$_2$

Suppose that you ran the reaction from Example 1 and 15.4 g of MgCl$_2$ are produced. What is the percent yield of your reaction?

### Solution

Calculate the percent yield for the reaction you ran. The theoretical yield was 16.2 g MgCl$_2$.

$$\text{Percent yield} = \frac{\text{actual yield}}{\text{theoretical yield}} \cdot 100\%$$

$$= \frac{15.4 \text{ g}}{16.2 \text{ g}} \cdot 100\%$$

$$= 95.1\%$$

Your percent yield was very close to the theoretical yield.

## LESSON SUMMARY

### Which reactant determines how much product you can make?

**KEY TERMS**

theoretical yield

actual yield

percent yield

In the real world, substances rarely come together in the exact mole ratios specified by a chemical equation. Usually, there is more of one reactant than is required and less of the other. If reactants are not present in the exact mole ratio, one reactant, the limiting reactant, will get used up, therefore limiting the amount of product that can be produced. There will be excess of the other reactant. The percent yield of product is the actual amount of product produced in a chemical reaction as compared with the theoretical yield expressed as a percent. The theoretical yield of a chemical reaction is determined by the limiting reagent and the molar ratios in the chemical equation.

Exercises

### Reading Questions

1. How can you determine how much product you can make from two compounds?
2. What is percent yield?

*(continued)*

**Reason and Apply**

3. Consider the reaction to remove mercury from a water source through precipitation:

$$Hg(NO_3)_2(aq) + 2NaCl(aq) \longrightarrow HgCl_2(s) + 2NaNO_3(aq)$$

   a. On your paper, draw a diagram that shows the correct amount of sodium chloride needed to react with the amount of mercury (II) nitrate represented in the box on the right.
   b. What else is present in the beaker at the end of the reaction, besides the solid $HgCl_2$?
   c. Why aren't there any mercury ions left over at the end?

   The box on the right contains: $NO_3^-$, $NO_3^-$, $NO_3^-$, $Hg^{2+}$, $Hg^{2+}$, $NO_3^-$, $NO_3^-$, $NO_3^-$, $Hg^{2+}$, $NO_3^-$, $Hg^{2+}$, $NO_3^-$

4. Suppose you were trying to remove 50.3 g of mercury (II) nitrate, $Hg(NO_3)_2$, from a water supply using the reaction described in Exercise 3. How many grams of sodium chloride would you need to add?

5. Silver nitrate, $AgNO_3$ reacts with sodium chloride, NaCl, in aqueous solution to form solid silver chloride, $AgCl(s)$, and aqueous sodium nitrate, $NaNO_3(aq)$. Suppose you start with 6.3 g of $AgNO_3$ and 4.5 g of NaCl.
   a. Write a balanced chemical equation for this reaction.
   b. How many moles of each reactant are you starting with?
   c. What is the limiting reactant?
   d. How many grams of each product do you expect to produce?
   e. How many grams of excess reactant do you expect to have when the reaction is complete?

6. In the laboratory, you run a procedure for the reaction described in Exercise 5 and produce 2.9 g of silver chloride. What is the percent yield for your procedure?

**INDUSTRY**

**CONNECTION**

Pharmaceutical companies can calculate percent yield for each chemical reaction involved in creating a medicine. A low percent yield can be an indication of possible problems with the manufacturing process. Correcting these problems helps the manufacturer to make a safer, better product and to make the most of it at the lowest cost of raw materials.

Bernd Auers/Getty Images

# CHAPTER 17

## Toxic Cleanup

# SUMMARY

### KEY TERMS

precipitate

precipitation reaction

complete ionic equation

spectator ion

net ionic equation

mole ratio

limiting reactant (limiting reagent)

stoichiometry

theoretical yield

actual yield

percent yield

# Toxins Update

One way to remove toxic substances from water supplies is to add the appropriate ions to create a solid precipitate that can be filtered out. In other situations, the appropriate ions can be added to dissolve a solid into solution. A solubility table can help identify which ionic compounds can help remove a toxic ion from solution.

How do you know how much reactant you need? Sometimes, it is necessary to translate between moles and mass. The relationship between mass and moles is a proportional one, connected by molar mass. When comparing quantities of reactants and products, it is necessary to consider the mole ratios.

## REVIEW EXERCISES

1.  Write a complete ionic equation and a net ionic equation for three precipitation reactions that involve the formation of solid nickel compounds.

2.  Calcium chloride reacts with sodium phosphate to produce calcium phosphate and sodium chloride. The unbalanced chemical equation is

$$\underline{\quad} \; CaCl_2(aq) + \underline{\quad} \; Na_3PO_4(aq) \longrightarrow$$
$$\underline{\quad} \; Ca_3(PO_4)_2(s) + \underline{\quad} \; NaCl(aq)$$

   a.  Balance the chemical equation for this reaction.
   b.  If you start with 13.5 g of $CaCl_2$ and 9.80 g of $Na_3PO_4$, how many grams of calcium phosphate can you expect to produce?
   c.  How many grams of excess reactant will be left over?
   d.  In the laboratory, you run a procedure for this reaction, and you produce 8.20 g of calcium phosphate. What is the percent yield for your procedure?

3.  Suppose that you mix a slightly soluble compound with water.
   a.  Draw a particle view diagram of the resulting mixture.
   b.  Would the resulting mixture be homogeneous or heterogeneous? Explain.

---

## Removing Toxins

PROJECT   Choose a toxin (your teacher may assign you one). Find some information about your toxin. Your report should cover these topics.

-   According to the Environmental Protection Agency, what is the maximum contaminant level for your toxin in the drinking water supply? How is it reported?

-   If your toxin dissolves in water, how would you remove it from water?

-   If it does not dissolve in water, how would you remove it from soil?

---

# Toxins | REVIEW

## UNIT 4

Chemical changes make life possible on this planet. Through chemical reactions, our bodies process the food we eat, make new cells and tissue, and eliminate waste products. Chemical equations allow us to track these changes on an atomic level.

However, some substances cause chemical changes that affect your body in negative ways, causing damage or harm. Acids and bases can react with your skin, causing burns and destroying tissue. Metal salts can interfere with ionic substances in the bloodstream, upsetting the natural balance. Precipitates can form in your body and block passages—examples are cholesterol in the bloodstream and kidney stones in the urinary tract.

But chemical reactions can be used to address toxic reactions. Acids and bases can be dealt with through neutralization reactions or by dilution with water. Toxic substances can often be safely removed from your body through precipitation reactions. Other times, the best approach is to dissolve solid toxic substances and then remove them from the body in liquid form. These same methods also have other applications, such as purifying the drinking water that comes to your home.

Every substance on the planet has the potential to be either toxic or therapeutic. The outcome depends on the dosage or amount and the way it is administered. Health officials test substances and set limits on exposure to all the different compounds that we might encounter in our tap water, medications, atmosphere, food, or the household products we buy.

## REVIEW EXERCISES

### General Review

*Write a brief and clear answer to each question. Be sure to show your work.*

1. Balance the following chemical equations:
   a. $Zn(s) + HCl(aq) \longrightarrow ZnCl_2(aq) + H_2(g)$
   b. $KClO_3(s) \longrightarrow KCl(s) + O_2(g)$
   c. $S_8(s) + F_2(g) \longrightarrow SF_6(g)$
   d. $Fe(s) + O_2(g) \longrightarrow Fe_2O_3(s)$

2. For each of the four equations in Exercise 1, name the type of reaction it represents.

3. Using the same equations you balanced in Exercise 1, suppose that you start with 1 g of each reactant. Identify the limiting reactant and determine the mass of each product that can be made.

4. Write a balanced chemical equation for these reactions:
   a. Calcium metal reacts with water to form calcium hydroxide and hydrogen gas.
   b. A solution of hydrochloric acid reacts with solid calcium bicarbonate to produce water, carbon dioxide, and calcium chloride.

5. What is a neutralization reaction?

6. Write out the complete and net ionic equations for reactions between these compounds in solution:
   a. magnesium nitrate and calcium chloride
   b. silver nitrate and potassium iodide
   c. sodium sulfate and barium chloride

7. Does a precipitate form in any of the reactions from Exercise 6? Circle them.

8. Describe the steps you would follow to make a 1.0 M solution of silver nitrate.

9. How many grams of solute are dissolved in a 500.0 mL sample of 0.50 M sodium sulfate?

10. What is the pH of a solution with $[H^+] = 2.0 \times 10^{-8}$? What is the $[OH^-]$ of the solution?

11. This reaction is completed in a factory to create potassium phosphate:

$$\underline{\hspace{1cm}} H_3PO_4(aq) + \underline{\hspace{1cm}} KOH(aq) \longrightarrow$$
$$\underline{\hspace{1cm}} K_3PO_4(aq) + \underline{\hspace{1cm}} H_2O(l)$$

   a. What type of reaction is represented by the equation?
   b. Copy the equation and balance it.
   c. If you want to make 5.0 mol of potassium phosphate, how many moles of phosphoric acid, $H_3PO_4$, do you need?
   d. If 628.0 g of phosphoric acid are reacted with 1122.0 g of potassium hydroxide, what is the maximum amount of potassium phosphate that can be produced?
   e. Which is the limiting reactant in part d? How much (in moles and in grams) of the excess reactant is left over?

12. The reaction shown here is completed by a chemical company to create phosphoric acid, $H_3PO_4$.

$$\underline{\hspace{1cm}} K_3PO_4(aq) + \underline{\hspace{1cm}} H_2SO_4(aq) \longrightarrow$$
$$\underline{\hspace{1cm}} H_3PO_4(aq) + \underline{\hspace{1cm}} K_2SO_3(aq)$$

   a. What type of reaction is represented by the equation?
   b. Copy the equation and balance it.
   c. How many moles of potassium phosphate, $K_3PO_4$, are required to produce 6 mol of phosphoric acid?
   d. How many moles of sulfuric acid, $H_2SO_4$, are required to produce 6 mol of phosphoric acid?
   e. If 750.0 g of potassium phosphate are reacted with 0.75 L of 18.4 M sulfuric acid, what is the maximum number of moles of phosphoric acid that can be produced?
   f. Which is the limiting reactant in part e? How much (in moles and in grams) of the excess reactant is left over?

## STANDARDIZED TEST PREPARATION

### Multiple Choice

*Choose the best answer.*

1. What is a correct interpretation of the following chemical equation?

$$K_2CO_3(aq) + Mg(NO_3)_2(aq) \longrightarrow$$
$$2KNO_3(aq) + MgCO_3(s)$$

   (A) A solution of potassium nitrate is added to solid magnesium carbonate to produce a solution of potassium carbonate and magnesium nitrate.
   (B) Solutions of potassium carbonate, magnesium nitrate, and potassium nitrate are mixed to produce solid magnesium carbonate.
   (C) A solution of potassium carbonate is added to a solution of magnesium nitrate to produce a solution of potassium nitrate and solid magnesium carbonate.
   (D) Solid magnesium carbonate dissolves in potassium nitrate to form a solution of potassium carbonate and magnesium nitrate.

2. What information can be obtained from a written chemical equation?

   (A) the color change of the reactant
   (B) the speed of the overall reaction
   (C) the states of the products
   (D) the probability of an explosion

3. For the following equation, what would you expect to observe when this reaction is performed in the laboratory?

$$Na_2CO_3(aq) + 2HCl(aq) \longrightarrow$$
$$2NaCl(aq) + CO_2(g) + H_2O(l)$$

   (A) The resulting solution contains bubbles.
   (B) The resulting solution freezes.
   (C) The resulting solution changes color.
   (D) The resulting solution contains white crystals.

4. Which of the following balanced chemical equations represents a physical change?

   (A) $2H_2(g) + O_2(g) \longrightarrow 2H_2O(g)$
   (B) $N_2(l) \longrightarrow N_2(g)$
   (C) $2KClO_3(s) \longrightarrow 2KCl(s) + 3O_2(g)$
   (D) $Zn(s) + 2HCl(aq) \longrightarrow ZnCl_2(aq) + H_2(g)$

5. Which of the following correctly matches a balanced chemical reaction to its type of reaction?

   (A) single exchange:
   $$NaCl(aq) + AgNO_3(aq) \longrightarrow$$
   $$NaNO_3(aq) + AgCl(aq)$$

   (B) double exchange:
   $$2AlBr_3(aq) + 3Cl_2(g) \longrightarrow$$
   $$2AlCl_3(aq) + 3Br_2(g)$$

   (C) decomposition:
   $$Na_2CO_3(s) \longrightarrow Na_2O(s) + CO_2(g)$$

   (D) combination:
   $$2PbSe(s) + O_2(g) \longrightarrow 2PbO(s) + 2Se(s)$$

6. The $LD_{50}$ for lead toxicity is 450 mg/kg. How many milligrams would be a lethal dose for a 132 lb adult?

   (A) 0.13 mg
   (B) 50 mg
   (C) 27 g
   (D) 59 g

7. Which of the following correctly describes the comparison of one mole of copper, Cu, magnesium, Mg, and neon, Ne?

   (A) Each mole sample has the same mass.
   (B) Each mole sample has the same number of atoms.
   (C) Each mole sample has the same molar mass.
   (D) Each mole sample has a different number of atoms based on the atomic mass.

8. Which correctly describes a 500 mL bottle filled completely with water?

   (A) There are 9000 mol of water in this bottle.
   (B) There are $1.67 \times 10^{25}$ molecules of water in this bottle.
   (C) The mass of the water in the bottle is 500 mL.
   (D) The molarity of the water in this bottle is 1.00 M.

9. How many moles of $Ca(NO_3)_2$ are there in a sample with a mass of 50.0 g?

   (A) 0.305 mol
   (B) 0.333 mol
   (C) 0.490 mol
   (D) 8200 mol

10. What is the molarity of a sugar solution that has 45.0 g of sugar, $C_{12}H_{22}O_{11}$, dissolved in 250 mL of water?

    (A) 0.53 M
    (B) 0.016 M
    (C) 0.18 M
    (D) 0.0010 M

11. You want to make 100.0 mL of a 0.75 M solution of potassium sulfate. What mass of solid potassium sulfate should you use?

    (A) 430 mg
    (B) 1.3 g
    (C) 13 g
    (D) 13 kg

12. What is the pH of a $1.0 \times 10^{-5}$ M HCl solution?

    (A) −5
    (B) 5
    (C) 9
    (D) −9

13. What is the pH of a 0.001 M NaOH solution?

    (A) −3
    (B) 2
    (C) 3
    (D) 11

14. Which of the following is a balanced neutralization reaction?

    (A) $2HNO_3(aq) + CaOH_2 (aq) \longrightarrow$
    $$2H_2O(l) + Ca(NO_3)_2(aq)$$

    (B) $H_2SO_4(aq) + KOH(aq) \longrightarrow$
    $$H_2O(l) + K_2SO_4(aq)$$

    (C) $HC_2H_3O_2(aq) + NH_4OH(aq) \longrightarrow$
    $$2H_2O(l) + NH_2C_2H_3O(aq)$$

    (D) $2H_3PO_4(aq) + 3Mg(OH)_2(aq) \longrightarrow$
    $$6H_2O(l) + Mg_3(PO_4)_2(aq)$$

15. If 15.0 mL of 0.50 M HCl solution is required to neutralize 34.1 mL of NaOH, what is the molarity of the base?

    (A) 0.22 M NaOH
    (B) 0.26 M NaOH
    (C) 1.14 M NaOH
    (D) 4.6 M NaOH

16. Which solid would be best to add to a solution of $Cu(NO_3)_2$ in order to remove the $Cu^{2+}$ from the solution?

(A) $NaCl(s)$

(B) $KOH(s)$

(C) $Mg(NO_3)_2(s)$

(D) $PbCO_3(s)$

17. The reaction to form ammonia gas, $NH_3$, is shown in the chemical equation below. Which statement correctly describes this chemical reaction?

$$N_2(g) + 3H_2(g) \longrightarrow 2NH_3(g)$$

(A) The mole ratio cannot be determined because the chemical equation is not balanced.

(B) More molecules of $N_2$ are needed than $H_2$ molecules to maximize the amount of $NH_3$ produced.

(C) To produce 2 mol of $NH_3$, three times as many moles of $H_2$ are needed as moles of $N_2$.

(D) If 6 mol of $H_2$ and 2 mol of $N_2$ are used, then 2 mol of $NH_3$ will be produced.

18. Consider the reaction below. What mass of sodium chloride will be produced if 5.0 g of calcium oxalate is made?

$$CaCl_2(aq) + Na_2C_2O_4(aq) \longrightarrow$$
$$2NaCl(aq) + CaC_2O_4(s)$$

(A) 2.3 g NaCl

(B) 4.6 g NaCl

(C) 11 g NaCl

(D) 22 g NaCl

19. Consider the reaction below. How many grams of $Zn_3(PO_4)_2$ can be made with 26.5 g of $Zn(NO_3)_2$ and 19.7 g of $Na_3PO_4$?

$$3Zn(NO_3)_2(aq) + 2Na_3PO_4(aq) \longrightarrow$$
$$Zn_3(PO_4)_2(s) + 6NaNO_3(aq)$$

(A) 12.0 g $Zn_3(PO_4)_2$ because $Zn(NO_3)_2$ is the limiting reagent

(B) 18.0 g $Zn_3(PO_4)_2$ because $Zn(NO_3)_2$ is the limiting reagent

(C) 23.2 g $Zn_3(PO_4)_2$ because $Na_3PO_4$ is the limiting reagent

(D) 34.8 g $Zn_3(PO_4)_2$ because $Na_3PO_4$ is the limiting reagent

20. Consider the balanced chemical equation below. Which substance is the limiting reactant, and how many moles of $Fe_3O_4$ can be obtained by reacting 16.8 g Fe with 10.0 g $H_2O$?

$$3Fe(s) + 4H_2O(g) \longrightarrow Fe_3O_4(s) + 4H_2(g)$$

(A) Iron is the limiting reactant and 0.100 mol $Fe_3O_4$ can be formed.

(B) Water is the limiting reactant and 0.100 mol $Fe_3O_4$ can be formed.

(C) Iron is the limiting reactant and 0.902 mol $Fe_3O_4$ can be formed.

(D) Water is the limiting reactant and 0.902 mol $Fe_3O_4$ can be formed.

# Cleaning Toxins from Water

**Work Groups of 4**

### Purpose

To use the process of engineering design to plan and design a method to clean toxins from a water supply.

### Problem

Your engineering design team is charged with creating a protocol for removing toxins from the water supply in your community. This protocol involves removing an individual toxin to make the water cleaner. You will not complete the actual clean up. Instead, you will devise a method for cleaning the water.

### Review

Review the handout, *Engineering Design* (found after the Unit 1 Engineering Project), and reflect on the material that has been presented thus far in your chemistry class. For this project, you will focus on brainstorming solutions, developing a plan, and refining that plan based on feedback.

In Unit 4: Toxins, you learned about toxins, how they can be measured, how they can form solutions, influence acidity, and how they can be cleaned up. Using the principles of solutions and precipitation reactions, you can determine how to remove certain toxins from water. Make sure to consider all the topics from Unit 4 when devising your protocol. The following table shows some of the many topics and skills that were covered in Unit 4: Toxins. You may find these helpful as your work to engineer a solution to the problem:

| Chemical reactions | Making observations of change | Types of reactions |
|---|---|---|
| Conservation of Mass | Balancing equations | Molar mass |
| Toxicity | Counting by weighing | Solution concentration |
| Moles | Comparison of amounts | Acids and bases |
| Preparing solutions | Identifying solutions | Neutralization |
| Dilution | pH | |
| Precipitation reactions | Physical vs chemical changes | |

**Part 1:** Identify the Problem and Brainstorm Solutions

A sample of water has the following dissolved toxins:

lead nitrate, $PbNO_3$              methanol, $CH_3OH$             oxalic acid, $C_2H_2O_4$

1. What are some possible chemical reactions that might be used to remove these toxins from the water?

2. In your Group, brainstorm some ideas to solve this problem. Here are some questions to consider:
   • What toxin are you removing from the water?
   • What is the solubility of the toxin?
   • How does the amount of the toxin affect its toxicity?

# Engineering Project

- Can a precipitation reaction help remedy the problem?
- What considerations need to be made when using a precipitation reaction?
- What is the concentration of toxin?
- What chemical changes occur in the water due to the toxin?
- Can a chemical change be used to remediate the water?
- Can heating selectively evaporate the toxin?

**Research:** In the library and online, research possible ways to remove this toxin from the water using chemical reactions. Create a list of all of your sources to turn in with your project.

## Part 2: Create a Plan or Prototype

1. Based on your research and brainstorming, create a procedure for removing the toxin from the water.
2. Write up a description of how your procedure will work. Include appropriate chemical equations.
3. Write a list of materials and possible safety considerations needed for your procedure.
4. Submit your procedure to your teacher for review and feedback.

## Part 3: Refine and Retest

1. Ask your classmates to help you think of ways to improve your plan. What changes might you make based on their feedback?
2. Make changes to your plan based on feedback from your teacher and other students.
3. Once your experiment is refined, your teacher may ask you to present your procedure.

## Possible Presentation Options

Choose one of these options to present your procedure, or get permission from your teacher to present in a different way:

- Oral presentation describing your procedure
- Poster describing your procedure

# Unit 5 | Fire

Vadym Volodin/iStockphoto

Fire is a fascinating phenomenon and a vital source of energy on our planet.

Phil McDonald/Shutterstock

## In this unit, you will learn

- the nature of heat, energy, and fire

- how to keep track of and measure changes in energy

- about energy exchanges associated with chemical changes

- how chemical energy is transformed into work

- about energy exchanges during reactions with metals and ionic compounds

- how light energy interacts with matter

## Why Fire?

Every change that happens to matter is accompanied by a change in energy. Fire is visible evidence of the energy associated with one particular type of chemical change. When a compound burns, it is broken down into smaller, less complex substances, and heat and light are released. When fire is uncontrolled, it can be destructive. However, this same chemical reaction can also provide heat, light, and mechanical or electrical energy. This unit explores how the energy from chemical and physical change can be observed, measured, understood, and controlled.

# CHAPTER 18

# Observing Energy

## In this chapter, you will study

- energy exchange in chemical reactions

- heat, temperature, and thermal energy

- the effect of heat on different substances

- the energy associated with phase changes

*The snow on the branch is changing from the solid to the liquid phase. Energy is involved in phase changes.*

All around you is evidence of energy exchanges. Any time matter moves or changes, energy is involved. Some expressions of energy are familiar, like the heat from a campfire or the radiant energy from the Sun. But energy can be a tricky concept to define. Chapter 18 of Unit 5: Fire introduces you to energy through observations of its effects on matter.

# Energy Changes

## THINK ABOUT IT

Fire is central to our lives. The most familiar fires are wood, oil, and gas fires. When these substances are ignited, brilliant flames leap into the air and there is intense heat. The energy from fire is used to warm homes, to cook food, and to move cars. For millennia, fire has been a vital energy source for humankind.

### What reactions are sources of heat?

*To answer this question, you will explore*

⮕ Energy

⮕ Exothermic Processes

## EXPLORING THE TOPIC

### ⮕ Energy

The universe consists of matter and energy. As discussed in Unit 1: Alchemy, matter is made up of atoms of different elements. So, what is energy?

Because energy is not matter, it is not a substance. Energy does not have mass and it does not take up space. Energy is difficult to describe and define. Take a moment to examine how the word *energy* is used in the following sentences. What do these situations have in common?

- A plant needs energy from the Sun to grow.
- An athlete eats a snack bar for energy to continue running.
- A campfire provides energy to heat water.
- Energy from flowing water makes the waterwheel rotate.
- Electrical energy causes the filament in the light bulb to glow.
- Pressure from steam provides the energy to move a locomotive.

You may notice that all of the statements are about energy causing something to happen to matter. Matter moves, falls, glows, melts, breaks apart, or burns. In each case, matter changes in some way and energy causes the change to happen. So, one definition of energy is that it is about change. **Energy** is a measure of the ability to cause change to occur. Even though you cannot see energy or hold it in your hand, you can measure amounts of energy transferred to a substance or from a substance.

**ENVIRONMENTAL CONNECTION**

Lightning strikes sometimes cause wildfires. It's probable that humans first discovered fire through lightning strikes.

**LITERATURE CONNECTION**

Many myths and legends have attempted to explain the origins of fire. According to Greek mythology, Prometheus stole fire from the gods and brought it to humankind. He was punished for it by Zeus.

**HISTORY**
**CONNECTION**

The Great Fire of London occurred in 1666. The fire lasted for four days, destroying 13,200 homes. Historians credit this fire with eradicating the rat population and helping to end the Great Bubonic Plague that had killed almost 20 percent of the city's populace.

# ⬡ Exothermic Processes

The chemical reactions listed here all release energy. This is evident in the flames, the sparks, the light you see, the sounds you hear, and the heat you feel when you observe these reactions. A reaction that creates products that are hotter than the reactants is called **exothermic.** Take a moment to examine these exothermic reactions.

## HEAT, LIGHT, FLAMES

The first two equations involve molecular covalent compounds. They react with oxygen, $O_2$. The products include carbon dioxide and water. Heat, light, and flames accompany these reactions.

$$\text{Methane: } CH_4(g) + 2O_2(g) \longrightarrow CO_2(g) + 2H_2O(g)$$

$$\text{Propanol: } 2C_3H_8O(l) + 9O_2(g) \longrightarrow 6CO_2(g) + 8H_2O(g)$$

© kokopopsdave/iStockphoto

## HEAT, LIGHT, SPARKS

The second set of equations involves elemental substances. Elemental metals and many nonmetals also react with oxygen to produce heat and light. There are no flames associated with these reactions, but sometimes sparks are emitted.

$$\text{Phosphorus: } P_4(s) + 5O_2(g) \longrightarrow 2P_2O_5(s)$$

$$\text{Iron: } 4Fe(s) + 3O_2(g) \longrightarrow 2Fe_2O_3(s)$$

© Hocus Focus Studio/iStockphoto

In all of these exothermic reactions, the reactants change into products that are very hot. These hot products cool down over time as they transfer heat to the surroundings. Exothermic changes are associated with big changes to matter.

© Reuters/CORBIS

## FIRE

When you observe fire, matter is changing drastically. As a result of a fire, an entire forest can be reduced to smoke and ashes. During this change, the trees and brush are converted to carbon dioxide and water, which spread out in the atmosphere in gaseous form. Some ash and charred fuel remain after a fire. These are the result of reactants that did not react completely.

Flames are generally what distinguish fires from other exothermic reactions. Flames consist mainly of hot gases. The glowing yellow color of a wood fire is caused by small particles of carbon carried into the air by the gaseous products of this exothermic reaction.

> **Big Idea** Changes in matter are accompanied by changes in energy.

## LESSON SUMMARY

### What reactions are sources of heat?

**KEY TERMS**

energy

exothermic

Energy is not a substance, so it does not have mass and it does not take up space. However, any change in matter is accompanied by a change in energy. Many chemical reactions are exothermic, which means they result in products that are hotter than the reactants. Fire is an important example of an exothermic reaction that transfers energy in the form of light and heat. In a fire, reactants combine with oxygen to form carbon dioxide and water. The energy associated with fires has many uses such as heating your home, cooking your food, and moving cars.

## Exercises

### Reading Questions

1. In your own words, define the word *energy*.

2. Use your own words to write a short paragraph describing fire to someone who has never seen it before.

### Reason and Apply

3. Write five sentences that use the word *energy*.

4. Describe three ways in which fire is central to life on our planet.

5. Overnight, the ashes of a fire cool down. What has happened to the products and the heat?

6. **PROJECT**

   Research uses of fire on the Web. Describe how fire is used in gas heaters to heat homes.

7. **PROJECT**

   Research uses of fire on the Web. Describe how fire is used to generate electricity in a coal-burning power plant.

*Not So Hot*

# Exothermic and Endothermic

## THINK ABOUT IT

© Lon C Diehl/PhotoEdit

You just received a bad bruise on your leg playing sports. The bruise is swelling and is quite painful. Your coach pulls out a disposable cold pack from the first-aid kit. A quick twist of the package activates it, and the temperature of the cold pack suddenly decreases. It feels nice and cold on your injury. What is the source of this cold sensation?

**In what direction is heat transferred during a chemical process?**

*To answer this question, you will explore*

➲ Heat

➲ Exothermic and Endothermic Processes

➲ Kinetic Energy

## EXPLORING THE TOPIC

### ➲ Heat

Temperature drops

*When you dissolve ammonium nitrate in water, the temperature decreases. The beaker feels cold.*

Therapeutic hot and cold packs are fairly easy to make. In one type of hot pack, solid calcium chloride, $CaCl_2$, and water are separated into two different pouches. When the pack is twisted, the pouches break open and the solid $CaCl_2$ dissolves in the water, releasing heat. The temperature of the solution increases and the bag feels hot.

In contrast, in one type of cold pack, solid ammonium nitrate, $NH_4NO_3$, and water are separated into two different pouches. When you twist the pack and the $NH_4NO_3$ dissolves in the water, the temperature of the solution decreases and the bag feels cold. If you perform these two reactions in beakers, you can measure the temperature changes with a thermometer.

What the thermometer records and your hand experiences is energy transferred into or out of the products of the chemical change. This process of energy transfer is called heat. **Heat** is a transfer of energy between two objects due to temperature differences. Heat always transfers from a higher temperature to a lower temperature. So, for instance, when the beaker feels cold to your hand, it is because heat transfers from your hand to the beaker.

### SYSTEM AND SURROUNDINGS

To communicate clearly about heat transfer, it is necessary to specify where the heat is transferring to or from. The matter that you are focusing on is referred to as the **system.** Once you have defined the system, everything else is referred to as the **surroundings,** or environment. If the solution of ammonium nitrate and water is the system, then the beaker, the thermometer, the air around the beaker, your hand, and everything else in the universe constitute the surroundings. Heat transfers between a system and its surroundings.

Surroundings = thermometer, beaker, and the rest of the universe.

System

**Big Idea** Heat is a transfer of energy due to temperature differences.

CONSUMER
CONNECTION

Most household refrigerators contain a liquid called a *refrigerant* that circulates in a series of coiled tubes. As it evaporates, the refrigerant absorbs heat from inside of the refrigerator. The refrigerant is then condensed back into a liquid in tubes located in back of the refrigerator, releasing heat, and the process starts all over again.

© sciencephotos/Alamy

## ➲ Exothermic and Endothermic Processes

Chemists categorize chemical changes according to the direction of heat transfer. As you learned in the previous lesson, when heat is transferred out of the system to the surroundings, the process is exothermic. When heat is transferred from the surroundings to the system, the process is **endothermic.** (In Greek, *exo* means outside and *endo* means inside.) Exothermic processes are experienced as warm by an observer. Endothermic processes are experienced as cold by an observer.

Some examples of exothermic and endothermic changes are listed below. The word *heat* is included in the chemical equation to highlight the direction of heat transfer.

### EXOTHERMIC PROCESSES
Burning methane gas

$$CH_4(g) + 2O_2(g) \longrightarrow CO_2(g) + 2H_2O(g) + heat$$

Dissolving calcium chloride in water

$$CaCl_2(s) \longrightarrow Ca^{2+}(aq) + 2Cl^-(aq) + heat$$

Neutralization of sodium hydroxide with hydrochloric acid

$$NaOH(aq) + HCl(aq) \longrightarrow H_2O(l) + NaCl(aq) + heat$$

## ENDOTHERMIC PROCESSES

Dissolving ammonium nitrate in water

$$NH_4NO_3(s) + \text{heat} \longrightarrow NH_4^+(aq) + NO_3^-(aq)$$

Decomposition of mercury oxide

$$2HgO(l) + \text{heat} \longrightarrow 2Hg(l) + O_2(g)$$

Notice that heat transfer is not limited to chemical reactions. The process of dissolving is also exothermic or endothermic depending on the substance. This is because energy is involved every time atoms are rearranged.

## ⮕ Kinetic Energy

The shuttle launches due to the reaction of $H_2$ and $O_2$. The water vapor produced expands rapidly to thrust the shuttle upward.

Changes in temperature due to a chemical reaction are associated with changes in motion. **Kinetic energy** is the energy of motion, and temperature is a reflection of the average kinetic energy of a sample. If the products of a reaction are hotter than the reactants, they must be moving faster.

The reaction between $H_2$ and $O_2$ to produce $H_2O$ is an extremely exothermic reaction. It is the reaction used to launch the space shuttle. The water molecules produced in the reaction are so hot and they expand so rapidly that the water vapor thrusts the space shuttle up into the air. The photo here provides evidence that the product molecules are moving with explosive speed.

A great deal of the heat energy of this reaction is converted to kinetic energy in the form of rapidly moving water molecules. While the water molecules push the space shuttle up, they transfer some of their kinetic energy to the space shuttle. They also transfer some of their energy to the surroundings as heat. With less kinetic energy, the water molecules gradually become cooler and move more slowly.

## LESSON SUMMARY

### In what direction is heat transferred during a chemical process?

Heat is a transfer of energy between two objects due to temperature differences. Heat transfer accompanies all chemical changes. The heat is transferred either into the system or out of the system. Exothermic reactions are chemical processes that result in the transfer of heat *from* the products of the reaction (the system) *to* the surroundings. These reactions feel hot to the observer. Endothermic reactions are chemical processes that result in the transfer of heat *from* the surroundings *to* the products of the reaction. These reactions feel cold to the observer. Because temperature is directly related to the motion of molecules, hotter products mean faster moving molecules. Colder products mean slower moving molecules.

**KEY TERMS**

heat

system

surroundings

endothermic

kinetic energy

**Reading Questions**

1. In your own words, define exothermic and endothermic chemical changes.

2. Why does an endothermic reaction that takes place in a beaker cause the beaker to feel cold?

**Reason and Apply**

3. Methane, $CH_4(g)$, reacts with oxygen, $O_2(g)$, to produce carbon dioxide, $CO_2(g)$, and water, $H_2O(g)$. You observe flames.
   a. How does the average kinetic energy of the reactants differ from the average kinetic energy of the products?
   b. Describe what you would expect to see if you had a molecular view of the products and reactants.

4. You mix solid copper (II) sulfate, $CuSO_4(s)$, with a small amount of water, $H_2O(l)$ in a beaker. Hydrated copper (II) sulfate, $CuSO_4 \cdot 5H_2O(s)$ is produced. A thermometer inside the beaker indicates that the temperature has increased from 19 °C to 48 °C.

$$CuSO_4(s) + 5H_2O(l) \longrightarrow CuSO_4 \cdot 5H_2O(s)$$

   a. List all the substances that are part of the system.
   b. List at least four things that are part of the surroundings.
   c. Which is at a lower temperature: $CuSO_4(s)$ or $CuSO_4 \cdot 5H_2O(s)$? Explain your thinking.
   d. What will you feel if you touch the beaker?
   e. Is the process endothermic or exothermic? Explain your thinking.
   f. What evidence do you have that heat is transferred *to* the surroundings *from* the products of the reaction?

5. You mix solid hydrated barium hydroxide, $Ba(OH)_2 \cdot 8H_2O(s)$, and solid ammonium nitrate, $2NH_4NO_3(s)$, in a beaker. The reaction is shown here. A small pool of water in contact with the outside of the beaker freezes.

$$Ba(OH)_2 \cdot 8H_2O(s) + 2NH_4NO_3(s) \longrightarrow 2NH_3(g) + 10H_2O(l) + Ba(NO_3)_2(aq)$$

   a. List all the substances that are part of the system.
   b. List at least four objects that are part of the surroundings.
   c. Which is at a lower temperature: $NH_4NO_3(s)$ or $Ba(NO_3)_2(aq)$? Explain your thinking.
   d. What will you feel if you touch the beaker?
   e. Is the reaction endothermic or exothermic? Explain your thinking.
   f. What evidence do you have that heat is transferred *from* the surroundings *to* the products of the reaction?

*Point of View*

# First and Second Laws

## THINK ABOUT IT

Suppose that you have just come inside on a very cold day. There is a fire in the fireplace. If you go near the fire, you will feel the warmth. Heat is transferred to your body by the warm gases of the flames and the energy radiating from the burning wood.

**What do temperature differences indicate about heat transfer?**

*To answer this question, you will explore*

- Heat Transfer
- The First and Second Laws of Thermodynamics

## EXPLORING THE TOPIC

### ⟳ Heat Transfer

There are three main ways heat is transferred: conduction, convection, and radiation.

- Conduction takes place when a substance transfers heat to another substance or object that it is in contact with. For example, the handle on a frying pan gets hot because the frying pan is hot.

- Convection takes place when a warm substance changes location, such as when warm air rises.

- Radiation takes place when electromagnetic waves carry energy from an energy source, such as the Sun.

**CONSUMER CONNECTION**

Materials that do not conduct heat well are called *thermal insulators.* We use them in the walls and ceilings of our homes to slow down the transfer of heat from our homes in the wintertime and into our homes in the summertime. A thermos bottle is composed of several layers of insulating material, often with a vacuum in the middle, to limit heat transfer and to keep cold liquids cold and hot liquids hot.

© Ole Graf/Corbis

Conduction: molecular motion

Convection: movement of matter

Radiation: electromagnetic rays

## SYSTEM AND SURROUNDINGS

Consider each of the three illustrations. Assume water is the system. Try to identify the direction of heat transfer, the type of heat transfer, and the surroundings.

②

*Water is placed in the freezer to make ice cubes.*

①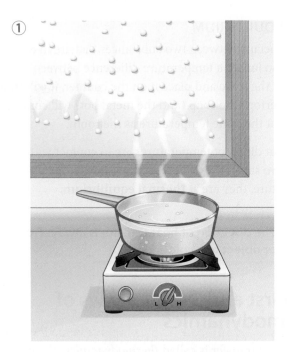

*Water boils on the stove, producing water vapor. When the water vapor comes in contact with the cold pane of glass it condenses and forms drops.*

③

*An ice cube is on a table and melts.*

① In the first illustration, water molecules are in the form of both water vapor and liquid water. For the water vapor, the surroundings include the stove, the pot, the window, and the air in the room. Heat transfers from the stove to the pot and then to the water by conduction, then to the air in the room by convection. Heat also transfers from the water vapor to the window by conduction.

② In the second illustration, water takes the form of both liquid and solid. For the liquid water in the ice cube tray, the surroundings include the ice cube tray, the freezer, and the air in the freezer. Heat transfers from the water to the surroundings by conduction. When enough heat energy transfers from the liquid water to the interior of the freezer, the water freezes and becomes solid.

③ For the ice cube on the table, the surroundings include the table and the air in the room. Heat transfers to the ice cube by conduction. When enough energy is transferred into the ice cube on the table, it melts.

## HOT AND COLD

The sensation of hot or cold depends on the direction of heat transfer. For example, if you hold an ice cube, your hand will feel cold as heat transfers from your hand to the ice cube. If you hold a cup of hot tea, your hand will feel warm as heat transfers from the hot tea to the cup and then to your hand. However, "hot" and "cold" are relative.

Just because something is at a low temperature does not mean it will feel cold. For example, if your hands are cold because you have been out in the snow all day, even holding them under cold water can feel painfully hot because the water is at a higher temperature than your hands. The experience of hot or cold depends on whether energy is being transferred to or from the observer.

## THERMAL EQUILIBRIUM

Heat transfer occurs between two substances that are in contact with each other until there is no longer a temperature difference between them. If you take a pot of hot soup off the stove and place it on the counter, it will transfer heat to the air and the countertop. The soup (and the metal pot) will continue to transfer heat until they are at the same temperature as the room.

If two objects at different temperatures are in contact with one another long enough, the two substances will reach the same temperature. When they reach the same temperature, they are at **thermal equilibrium.**

> **Big Idea** The direction of heat transfer is always from a hotter substance or object to a colder one.

## ⟳ The First and Second Laws of Thermodynamics

The study of heat transfer is called *thermodynamics.* The principles behind thermodynamics have been summarized into scientific laws that are consistent with observations of a large number of systems.

The **first law of thermodynamics** states that energy is always conserved. The heat transferred *from* a hotter object is equal to the heat transferred *to* the colder objects surrounding it. There is no loss of energy. Likewise, you cannot create energy. It can only be transferred from one place to another place, and it can change form, such as when solar energy is converted to electrical energy.

> **Big Idea** Energy is conserved. It cannot be created or destroyed.

The **second law of thermodynamics** states that energy tends to disperse, or spread out. Energy transfers from hotter objects to colder objects until the temperature evens out. It does not transfer from a cold object to a hot object because this would concentrate the energy in one place rather than spread it out. This concept helps to explain why you can sit far away from a fire and become warmed by it. The energy from the fire disperses, or spreads out, from its source.

Another way to explain this concept is to say that the **entropy** of a system tends to increase over time. This means that energy and matter have a natural tendency to become more dispersed and disordered rather than more collected and ordered. For example, when you open a bottle of perfume, the molecules disperse throughout the room. The molecules will not naturally collect back into the bottle.

> **Important to Know** There is no such thing as "cold transfer." Cold is the experience of heat transfer away from our bodies.

## RECREATION
## CONNECTION

Hypothermia occurs when the body loses heat faster than it can generate heat. A body temperature below 95 °F (35 °C) is considered dangerous. People who are outdoors in winter without enough protective clothing, or who accidentally fall into very cold water, may develop hypothermia. Hot liquids and external heat sources can help a person recover from hypothermia.

Melissa McManus/Getty Images

········································································

**Big Idea** Energy tends to disperse.

········································································

## LESSON SUMMARY

### What do temperature differences indicate about heat transfer?

**KEY TERMS**

thermal equilibrium

first law of thermodynamics

second law of
thermodynamics

entropy

The sensory experience of hot or cold depends on the direction of heat transfer. If your hand is at a higher temperature than what you are touching, the object will feel cold. If your hand is at a lower temperature, the object will feel warm. Heat is always transferred from a hotter object or to a colder one. Heat transfer is associated with phase changes. An ice cube melts because heat is transferred into it from the surroundings. If objects remain in contact with each other, heat is transferred until objects are at the same temperature. This balanced state is referred to as thermal equilibrium. Two laws of thermodynamics are related to these observations: The first law states that energy is conserved, and the second law states that energy tends to disperse, or spread out.

**Exercises**

### Reading Questions

1. Why does an ice cube feel cold to your hand?
2. What is thermal equilibrium?

### Reason and Apply

3. Is energy transferred to a liquid or from a liquid when it evaporates? When it solidifies?
4. Is energy transferred to a gas or from a gas when it condenses?
5. What is evaporation and why is it a cooling process? What cools during evaporation?
6. What determines the direction of heat transfer?
7. Do humans ever reach thermal equilibrium with the surrounding air? Explain.
8. Provide evidence to support the first law of thermodynamics.
9. Provide evidence to support the second law of thermodynamics.
10. Explain why a campfire burns out. Explain why the ashes eventually cool to the temperature of the surrounding air.
11. Could there ever be a situation in which an ice cube actually heats up something else? Explain.

# Heat Transfer

Ken Karp Photography

## Purpose

To explore the factors that affect heat transfer between two samples of water.

## Materials

- large foam cup
- 250 mL beakers (2)
- 100 mL graduated cylinder
- thermometer
- hot plate
- beaker tongs
- ice in a large beaker or plastic tub

## Procedure

1. Measure and record a volume of water. Pour it into the foam cup and record the temperature.

2. Measure and record a second volume of water. Pour it into a beaker and either heat or cool the water using a hot plate or ice. Record the temperature.

3. Add the second sample to the first. Stir and watch the thermometer closely. Record the temperature after the temperature has stabilized.

4. Repeat the experiment using different volumes and temperatures. Create a data table with results from at least three trials.

5. Convert your volume measurements to mass. The density of water is 1 g/mL.

## Analysis

1. Look for relationships in your data table. What factors affect the temperature change?

2. Chemists measure the amount of heat transfer from one sample to another in units of calories. A *calorie* is the amount of energy needed to raise the temperature of 1 g of water by 1 °C. How does this relate to your data?

3. Write an equation relating the masses of the two water samples, $m_1$ and $m_2$, and the temperature changes, $\Delta T_1$ and $\Delta T_2$, of these two masses of water.

4. **Making Sense** What happens to the motions of the water molecules if you mix hot and cold water? Use changes in molecular motion to explain how energy from the hot water transfers to the cold water.

5. **If You Finish Early** Predict the final temperature if you mix 15 g of water at 20 °C with 45 g of water at 80 °C. Show your work.

# Heat Transfer

## THINK ABOUT IT

Imagine that you want two samples of really hot water. You want a large sample to make soup and a smaller sample for a cup of hot cocoa. Will it take the same amount of heat to get both samples to the same temperature?

**What is the difference between temperature and heat?**

*To answer this question, you will explore*

➲ Particle View of Energy

➲ Measuring Heat Transfer

## EXPLORING THE TOPIC

### ➲ **Particle View of Energy**

Consider a sample of water molecules that is heated. What happens to the molecules as more energy is transferred to the sample?

*Same volume, different amounts of thermal energy*

As shown in the illustration (left), atoms and molecules move faster as they are heated. As the movement of the molecules increases, the temperature of the sample goes up.

Now consider two samples of water that contain different numbers of molecules. While the temperatures might be identical, the amount of energy represented by the two samples is not the same. In one sample, you have many quickly moving particles. In the other sample, you have fewer particles, even though they are moving just as quickly. There will be more total energy in the larger sample.

If the water samples in the large pot and the small cup shown in the illustration (bottom left) are both at 85 °C, you could say that they both have the same "degree of hotness," or temperature. However, the water in the large pot has a greater total "quantity of hotness" because it is a larger sample. This second concept is referred to as **thermal energy.** Thermal energy describes the total amount of energy in the particles of a sample. Because the temperature is the same, the gas molecules in both samples are moving with the same average kinetic energy. Because their volumes are different, the two samples have different amounts of thermal energy. So thermal energy is dependent on both the temperature of the sample *and* on the number of particles in the sample.

85 °C

85 °C

*Same temperature, different amounts of thermal energy*

## ⮂ Measuring Heat Transfer

Unfortunately, you cannot measure the thermal energy of a sample of matter directly. A thermometer gives you only an indication of the average kinetic energy. However, changes in thermal energy can be determined using heat transfer.

Imagine that you have three samples of water at different temperatures: a pitcher of water at 80 °C, a beaker of water at 80 °C, and a pitcher of water at 50 °C. Your common sense tells you that the largest 80 °C sample is the one with the greatest thermal energy. But can that energy be measured or quantified?

You can calculate the amount of thermal energy in each sample by determining how much energy can be transferred from each one. One way to do this is to use the energy in the three samples to warm identical samples of cold water.

Warming 100 mL samples of water

Notice that the large sample of water at 50 °C warms the cold water sample nearly as much as the small sample of water at 80 °C.

In each case, the mixture reaches thermal equilibrium. That is, the faster moving molecules in the hot sample slow down and the slower moving molecules in the cold sample speed up until they are all moving with the same average kinetic energy. The transfer of energy happens through collisions of the molecules as the samples mix. The temperature of the final mixture depends on the temperatures and the masses of the samples that were mixed.

## A UNIT OF ENERGY: THE CALORIE

Scientists define a unit called a **calorie** to measure heat transfer. A calorie is the amount of energy needed to raise the temperature of 1 gram of water by 1 Celsius degree. So, this unit of energy is based on heat transfer to or from water. To determine the number of calories needed, you multiply the mass of the water by the temperature increase in Celsius degrees. This can be summarized by the mathematical equation for the heat transfer, $q$, in terms of the mass of water, $m$, and the temperature change, $\Delta T$. The uppercase Greek letter delta, $\Delta$, is used to represent a change in quantity.

$$q = m(1 \text{ cal/g} \cdot {}^\circ\text{C})\Delta T$$

(Remember, 1 g of water is equivalent to 1 mL of water at room temperature.)

So, for example, to raise the temperature of 500 grams of water by 10 °C, from a temperature of 20 °C to a temperature of 30 °C, it would take 5000 calories of energy.

[For a review of this math topic, see **MATH Spotlight**: Averages on page A-7.]

## Example 1

### Calories

Suppose that you need to heat some water.

  **a.** How many calories of energy does it take to heat a 75 mL sample of water for cocoa from 25 °C to 38 °C?

  **b.** How many calories of energy does it take to heat a 2 L sample of water for a footbath from 25 °C to 38 °C?

  **c.** Which sample of water contains more thermal energy: the water for cocoa or the water for the footbath?

### Solution

Convert volume to mass in grams. Determine the change in temperature $\Delta T$ for both situations. Then use the equation for heat transfer to solve for $q$.

  **a.** 75 mL water = 75 g water

   $\Delta T = 38 - 25 = 13 \, ^\circ\text{C}$

   $q = m(1 \text{ cal/g} \cdot {}^\circ\text{C})\Delta T$

   $\quad = 75 \text{ g}(1 \text{ cal/g} \cdot {}^\circ\text{C})13 \, ^\circ\text{C}$

   $\quad = 975 \text{ calories}$

  **b.** 2 L water = 2000 g water

   $\Delta T = 38 - 25 = 13 \, ^\circ\text{C}$

   $q = m(1 \text{ cal/g} \cdot {}^\circ\text{C})\Delta T$

   $\quad = 2000 \text{ g}(1 \text{ cal/g} \cdot {}^\circ\text{C})13 \, ^\circ\text{C}$

   $\quad = 26{,}000 \text{ calories}$

  **c.** The water in the footbath represents a greater amount of thermal energy.

**Example 2**

**Conservation of Energy**

You mix 300 mL of water at 60 °C with 100 mL of water at 20 °C.

   **a.** Find the final temperature of the mixture.

   **b.** Show that the heat transferred *from* the hot water is equal to the heat transferred *to* the cold water.

*Solution*

   **a.** The final temperature will be a weighted average of the two initial temperatures. (Each temperature is "weighted" by multiplying it by the mass of each sample that is mixed.) The masses of the samples are 300 g and 100 g.

$$\frac{300 \text{ g } (60 \text{ °C}) + 100 \text{ g}(20 \text{ °C})}{400 \text{ g}} = 50 \text{ °C}$$

   The final temperature is 50 °C. It makes sense that the final temperature is closer to 60 °C, since there was more of the 60 °C water.

   **b.** The calories of energy transferred is given by this equation:

$$q = m(1 \text{ cal/g} \cdot \text{°C})\Delta T$$

Remember, 300 mL of water = 300 g of water.

   Hot water: $q = (300 \text{ g})(1 \text{ cal/g} \cdot \text{°C})(50 \text{ °C} - 60 \text{ °C}) = -3000$ calories

   Cold water: $q = (100 \text{ g})(1 \text{ cal/g} \cdot \text{°C})(50 \text{ °C} - 20 \text{ °C}) = 3000$ calories

The negative sign means that energy is transferred *from* the hot water. The positive sign for the cold water indicates that energy is transferred *to* the cold water. Based on the calculated answers, energy is conserved.

## LESSON SUMMARY

### What is the difference between temperature and heat?

Temperature and heat are different, although they are both related to the motion of atoms and molecules. The temperature of a substance is a measure of the average kinetic energy of its particles. Heat is a process of energy transfer to or from a sample of matter. Thermal energy is a term often used to describe the "energy content" of a sample. The thermal energy of a sample depends on the number of particles in a sample, whereas temperature is the same regardless of the size of the sample. A calorie is a unit of energy. It is defined as the energy required to raise the temperature of 1 g of water by 1 °C. The energy transferred is referred to as heat.

**KEY TERMS**

thermal energy

calorie

**Reading Questions**

1. If you heat up a large and a small sample of cold tap water to the same temperature, explain why it takes more energy to heat up the larger sample.

2. What is the difference between thermal energy and temperature?

*(continued)*

**Reason and Apply**

3. How many calories of energy do you need to transfer for each of the following changes?
   a. Raise the temperature of 1 g of water by 5 °C.
   b. Raise the temperature of 2 g of water by 5 °C.
   c. Raise the temperature of 9 g of water by 35 °C.

4. Which will warm a child's inflatable pool more: adding 500 g of water at 50 °C or 100 g at 95 °C? The temperature of the water in the pool is 20 °C. Explain your reasoning.

5. Suppose that two water samples are mixed. Copy the data table and fill in the amounts of heat transferred.
   a. How do the values in the last two columns compare?
   b. How does the amount of heat transferred depend on the change in temperature?
   c. How does the amount of heat transferred depend on the mass of each sample?

|  | Sample 1 mass (g) | Sample 1 temperature (°C) | Sample 2 mass (g) | Sample 2 temperature (°C) | Final temperature (°C) | Heat transfer to sample 1 (cal) | Heat transfer from sample 2 (cal) |
|---|---|---|---|---|---|---|---|
| Trial 1 | 100 g | 20 °C | 100 g | 80 °C | 50 °C | | |
| Trial 2 | 100 g | 20 °C | 100 g | 70 °C | 45 °C | | |
| Trial 3 | 100 g | 20 °C | 300 g | 80 °C | 65 °C | | |
| Trial 4 | 200 g | 20 °C | 100 g | 80 °C | 40 °C | | |

6. Suppose that you mix two water samples: 300 g of water at 20 °C and 200 g of water at 50 °C. What do you expect the final temperature of the water to be?

7. From a molecular viewpoint, explain how thermal equilibrium is reached when hot and cold water mix.

*The Heat Is On*

# Specific Heat Capacity

## THINK ABOUT IT

Imagine that you put a metal pan and a pizza in a hot oven. Both the pizza and the pan start out at room temperature. You leave them in the oven for the same amount of time. Will they heat up to the same temperature? And once they are taken out of the oven, will they cool off differently?

**How do different substances respond to heat?**

*To answer this question, you will explore*

➡ Specific Heat Capacity

➡ Bonding, Numbers, and Heat

## EXPLORING THE TOPIC

## ➡ Specific Heat Capacity

Every substance has a unique response to heat. Suppose that you place an aluminum pot of water on the stove to heat. After a minute, the metal pot will be too hot to touch, but the water will still be cool. This is because different amounts of energy are needed to raise the temperature of each type of substance.

To compare heat transfer to different substances, you can measure the energy required to raise the temperature of the same mass of each substance by the same number of degrees. The amounts of energy needed to raise the temperature of 1 g of several sample substances by 1 °C are given in the table.

| Sample | Mass (g) | $\Delta T$ (°C) | Energy required (cal) |
|---|---|---|---|
| water, $H_2O(l)$ | 1.0 g | 1 °C | 1.0 cal |
| methanol, $CH_3OH(l)$ | 1.0 g | 1 °C | 0.58 cal |
| aluminum, Al(s) | 1.0 g | 1 °C | 0.21 cal |
| copper, Cu(s) | 1.0 g | 1 °C | 0.09 cal |

Notice that fewer calories of energy are required to raise the temperature of methanol, aluminum, and copper from 20 °C to 21 °C, compared with water. This is consistent with the observation that the aluminum pot heats up and cools off faster than the water in the pot. Less energy is involved in changing the temperature of the aluminum.

The heat required to raise the temperature of 1 g of a substance by 1 °C is called the **specific heat capacity.** Every substance has a specific heat capacity. Some values are shown in the table.

| Substance | Specific heat capacity (cal/g · °C) |
|---|---|
| aluminum, Al($s$) | 0.21 |
| water, H$_2$O($l$) | 1.00 |
| copper, Cu($s$) | 0.09 |
| iron, Fe($s$) | 0.11 |
| wood (cellulose), (C$_6$H$_{10}$O$_5$)$_n$($s$) | 0.41 |
| glass, SiO$_2$($s$) | 0.16 |
| nitrogen, N$_2$($g$) | 0.24 |
| ethanol, CH$_3$CH$_2$OH($l$) | 0.57 |
| hydrogen, H$_2$($g$) | 3.34 |

The greater the specific heat capacity of a substance, the less its temperature will rise when it absorbs a given amount of energy. If a substance has a low specific heat capacity, it will change temperature easily, with a small transfer of energy. This applies whether the energy is being transferred into or out of the substance. This is why the aluminum pot heats up faster, and also cools down faster, than water. Water does not change temperature as easily as most substances.

**Big Idea** Substances with low specific heat capacities can heat up and cool down easily with only a small transfer of energy.

The specific heat capacity of a substance can be used to determine the amount of energy needed to raise the temperature of any mass by any number of degrees. The formula for heat transfer to any substance is a product of the mass, $m$, its specific heat capacity, $C_p$, and the temperature change, $\Delta T$.

$$q = mC_p \Delta T$$

## Example 1

### Cooling

Suppose that you have 15 g of methanol, CH$_3$OH, and 15 g of water, H$_2$O, both at 75 °C. You want to cool both samples to 20 °C. How much energy (in calories) do you need to remove from each sample?

### Solution

The energy transferred is a product of the mass, the specific heat capacity, and the change in temperature.

$$q = mC_p \Delta T$$

Example 1

*(continued)*

The mass of each sample is 15 g, and the change in temperature is 20 °C minus 75 °C, or −55 °C. However, the specific heat capacities of water and methanol are different. The specific heat capacity of water is 1 cal/g · °C. The specific heat capacity of methanol is 0.58 cal/g · °C.

Water:

$q = (15 \text{ g})(1 \text{ cal/g} \cdot °C)(-55 °C)$
$= -830 \text{ cal}$

Methanol:

$q = (15 \text{ g})(0.58 \text{ cal/g} \cdot °C)(-55 °C)$
$= -480 \text{ cal}$

It takes a greater transfer of energy to cool water than to cool methanol. Because of this, water is a very good insulator and retains its temperature. Note that the negative signs indicate that energy is transferred from the system to the surroundings.

Example 2

## Final Temperature

Imagine that you have 3.5 g of copper and 3.5 g of water. Both are at 25 °C. Which sample will be at a higher temperature if you transfer 150 cal to each? Show your work.

### Solution

The energy transferred is a product of the mass, the specific heat capacity, and the change in temperature.

$$q = mC_p \Delta T$$

You know the value of $q$ and the masses of the water and copper samples. You want to find the temperature change. Solve for $\Delta T$. You can rearrange the equation:

$$\Delta T = \frac{q}{mC_p}$$

You can look up the specific heat capacities of water and copper in a table like the one on page 501. The specific heat capacity of water is 1 cal/g · °C. The specific heat capacity of copper is 0.09 cal/g · °C.

Water:

$\Delta T = \dfrac{q}{mC_p} = \dfrac{150 \text{ cal}}{(3.5 \text{ g})(1 \text{ cal/g} \cdot °C)}$
$= 43 °C$

Copper:

$\Delta T = \dfrac{q}{mC_p} = \dfrac{150 \text{ cal}}{(3.5 \text{ g})(0.09 \text{ cal/g} \cdot °C)}$
$= 480 °C$

The temperature of copper increases about ten times as much as the temperature of the same quantity of water when 150 calories of energy are transferred.

To find the final temperature, add the temperature change to the initial temperature.

Water: Final temperature = 25 °C + 43 °C = 68 °C

Copper: Final temperature = 25 °C + 480 °C = 505 °C

For the same amount of heat transfer, the copper is much hotter.

© Becky Luigart-Stayner/CORBIS

## ◔ Bonding, Numbers, and Heat

It may seem strange that different amounts of energy are needed to raise the temperature of different substances. Why would you need more energy for water, compared with the same mass of methanol, aluminum, or copper to heat them all to the same temperature?

The answer is partially related to molar mass. Look back at the table of specific heat capacities on page 501. While the molar mass of aluminum is 27.0 g/mol, the molar mass of copper is 63.6 g/mol. So, there are more atoms in 1 g of aluminum than in 1 g of copper. It makes sense that more energy is required to cause a larger number of atoms to move faster.

Bonding also has an effect on the specific heat capacity of a substance. Substances that are polar tend to have high specific heat capacities. This means that molecules will have higher specific heat capacities than metals.

Finally, complicated molecules with large numbers of atoms have a variety of internal vibrational motions. The atoms within the molecule can vibrate back and forth like balls on a spring. And the entire molecule can rotate. All of these motions need to be increased to raise the temperature. So, substantially more heat is required to increase the average kinetic energy of molecules compared with individual atoms.

### HIGH SPECIFIC HEAT CAPACITY OF WATER

Water has a particularly high specific heat capacity for several reasons. First, the molar mass of water, 18.0 g/mol, is quite small. So there are a lot of water molecules to get moving in a 1 g sample. Second, because water is molecular, it has complex internal movements. Third, water consists of $H_2O$ molecules with strong intermolecular attractions called **hydrogen bonds.** A hydrogen bond is the attraction between the positive and negative dipoles of different water molecules. The hydrogen bonding restricts the motions of the molecules. Extra heat needs to be added to overcome these attractions and raise the kinetic energy. This is why the water in the metal pot heats up so much more slowly than the metal pot itself.

Hydrogen bond

## LESSON SUMMARY

### How do different substances respond to heat?

To increase the temperature of a sample of matter, you must supply energy to it. But the addition of a given amount of energy does not always result in the same rise in temperature. Substances differ in their responses to heat. Some substances, like metals, change temperature more than others in response to the transfer of the same amount of energy. This is because they have a lower specific heat capacity. Specific heat capacity is the amount of energy needed to raise the temperature of 1 g of a substance by 1 °C. The specific heat capacity of a substance depends on a number of factors, including the number of atoms or molecules in a sample and the type of bonding.

**KEY TERMS**

specific heat capacity
hydrogen bond

### Reading Questions

1. What is specific heat capacity? Give the specific heat capacity of a substance, and explain what that means in terms of that substance.

2. Why is more energy required to raise the temperature of 1.0 g of water compared with 1.0 g of aluminum?

### Reason and Apply

3. Which substance in each pair has the higher specific heat capacity? Justify your choice with an atomic view.
   **a.** Al or Pb     **b.** $H_2$ or Ar     **c.** $F_2$ or $Cl_2$

4. Imagine that you transfer the same amount of energy to all of the substances in the table of specific heat capacities on page 501. Which one would be hottest to the touch? Coolest? Explain your reasoning.

5. Use specific heat capacity to explain why metal keys left in the sun are hotter than a plastic pair of sunglasses left in the sun.

6. How much energy is required to raise the temperature of 50 g of methanol from 20 °C to 70 °C?

7. If you place 20 g of aluminum and 20 g of copper, both at 20 °C, into ice water, which will cool faster? Explain.

8. Imagine that you have 50 g of water and 50 g of methanol, each in a 100 mL beaker. You place both on a hot plate on low heat.
   **a.** Which sample will be at a higher temperature after 5 min? Explain your thinking.
   **b.** If the initial temperature of each liquid is 23 °C, what is the temperature of each after 25 cal of energy are transferred from the hot plate to each sample?

9. Would water make a good thermal insulator for a home? Explain your reasoning.

10. Why do farmers spray water on oranges to protect them from frost? Give two reasons why the water on the orange might help.

Wayne Eastep/Getty Images

# Heat and Phase Changes

## THINK ABOUT IT

Water boils at 100 °C, at 1 atmosphere pressure. Once it reaches that temperature, liquid water does not get any hotter, no matter how big the flames are under its container or how much you heat it. If the temperature of the boiling water is not changing, where is all that heat going?

**What happens to the heat during a phase change?**

*To answer this question, you will explore*

- The Heating Curve of Water
- Heat and Phase Changes

## EXPLORING THE TOPIC

### ➲ The Heating Curve of Water

Imagine that you have an ice cube that you have just removed from the freezer. The temperature of the freezer is below zero. The graph shows the changes in temperature as you heat the ice cube from below 0 °C to above 100 °C. The *x*-axis represents time. Because the water is being heated steadily, as time goes by, more and more heat is transferred to the water. This graph is the heating curve of water.

**Heating Curve of Water**

As you can see, the heating curve of water is a series of steep inclines with flat horizontal parts in between. Along the horizontal parts, energy continues to be transferred into the water, but the temperature does not change. The flat lines occur when melting or boiling is taking place, at 0 °C or 100 °C.

Note that whenever the temperature is not changing, two phases are present. As soon as only one phase—solid, liquid, or gas—is present, the temperature is able to change again. Something is happening to the energy going into the water during phase change.

## ⮕ Heat and Phase Changes

During melting and boiling, there is no doubt that energy is going into the substance. However, this energy is doing something besides making the molecules move faster. Remember, temperature is a measure of the average kinetic energy of the molecules. If the temperature is not going up, then the molecules are not moving any faster on average. Where is the energy going?

The answer lies in the intermolecular forces. These attractive forces hold particles in a solid or a liquid close to each other. When heat is transferred into a substance, like ice, the molecules in the ice move faster and the temperature rises. Once the temperature rises to 0 °C, the ice begins melting. When heat is transferred into melting ice, the average kinetic energy does not change, and the temperature does not rise. Instead, the energy transferred weakens the attractive forces between the molecules so that they can wiggle and rotate. Once the ice has melted completely, the energy transferred causes the molecules in the liquid to move faster, and the temperature rises once again.

During the boiling process, all of the energy transferred is used to completely break the intermolecular attractions so that molecules can move freely as a gas. As a result, no temperature change is observed during boiling.

> **Big Idea**   Heat transfer does not always result in a temperature change.

When a substance changes phase from solid to liquid or from liquid to gas, a certain amount of energy must be supplied to overcome the molecular or ionic attractions between the particles. This energy is changing the solid into the liquid or the liquid into a gas, without changing the temperature. For example, the heat required to change 1 gram of a substance from a liquid into a gas is called the **heat of vaporization.** The heat required to change 1 gram of a substance from a solid to a liquid is called the **heat of fusion.**

Solid                              Liquid                              Gas

When a substance changes phase from a solid to a liquid, or from a liquid to a gas, energy is transferred *to* the substance from the surroundings. When a substance changes phase from a gas to a liquid, or from a liquid to a solid, energy is transferred *from* the substance to the surroundings.

The table below shows some phase-change data for five common substances. Notice that it takes very little energy per gram to melt mercury (2.7 calories per gram), whereas it takes a great deal of energy to melt aluminum (94.5 calories per gram). Substances also require significantly more energy to change into gases than to melt.

### Phase-Change Data

| | Water | Ethanol | Mercury | Aluminum | Gold |
|---|---|---|---|---|---|
| **Heat of fusion** | 79.9 cal/g | 24.9 cal/g | 2.7 cal/g | 94.5 cal/g | 15.4 cal/g |
| **Melting point** | 0 °C | −117 °C | −39 °C | 659 °C | 1063 °C |
| **Heat of vaporization** | 539 cal/g | 205 cal/g | 70 cal/g | 2500 cal/g | 377 cal/g |
| **Boiling point** | 100 °C | 78 °C | 357 °C | 2467 °C | 2660 °C |

## Example

### Heat of Fusion

You place 11 g of ice at 0 °C into 100 g of liquid water at 20 °C in an insulated water bottle. After a few minutes, the ice melts completely. Calculate the final temperature of the water in the bottle.

© Skjold Photographs/PhotoEdit

#### Solution

The heat of fusion of water given in the table is 79.9 cal/g. This means that, to melt 11 g of ice, 79.9 cal/g · 11 g, or 879 calories, are transferred. This energy is transferred from the 100 g of liquid water in the bottle. Use the equation for heat transfer to determine how much this amount of energy changes the temperature.

$$q = mC_p \Delta T$$
$$-879 \text{ cal} = (100 \text{ g})(1 \text{ cal/g} \cdot {}^\circ\text{C})\Delta T$$
$$\Delta T = -8.8 \text{ }^\circ\text{C}$$

The final temperature of the 100 g of water is 20 °C − 8.8 °C, or 11.2 °C. But there are also 11 g melted ice at 0 °C, so you need to find the weighted average of all the water in the bottle.

$$\frac{100 \text{ g } (11.2 \text{ }^\circ\text{C}) + 11 \text{ g } (0 \text{ }^\circ\text{C})}{111 \text{ g}} = 10 \text{ }^\circ\text{C}$$

So, the final temperature of the water in your insulated water bottle will be around 10 °C.

## LESSON SUMMARY

### What happens to the heat during a phase change?

**KEY TERMS**

heat of vaporization

heat of fusion

Heat transfer does not always result in a temperature increase. During a phase change, the transfer of energy into a substance does not cause any rise in temperature. Instead, this energy is used to overcome the attractions between atoms and molecules.

The heat of fusion is the energy required to change a substance from solid to liquid. The heat of vaporization is the energy required to change a substance from liquid to gas.

### Reading Questions

1. Why doesn't the temperature of water change during boiling?

2. What is heat of vaporization? Use a substance from the phase-change data table on page 507 as an example.

### Reason and Apply

3. Use the table of phase-change data on page 507 to sketch a heating curve for mercury.

4. A jeweler pours liquid gold into a mold to make a 6 g wedding ring. How much energy is transferred from the ring when it solidifies? What temperature is the gold at the instant all of its atoms become solid?

5. Which of the following processes are exothermic? Explain your thinking.
    **(A)** boiling ethanol
    **(B)** freezing liquid mercury
    **(C)** subliming carbon dioxide

6. How much energy is required to melt these quantities of ice?
    **a.** 56 g
    **b.** 56 mol

7. If you transfer 5000 cal, how many grams of ethanol can you vaporize?

8. **a.** How much energy do you need to transfer to raise the temperature of 150 g of aluminum from 20 °C to its melting point?
    **b.** How much energy do you need to transfer to melt 150 g of aluminum?

9. You place 25.0 g of ice at 0 °C into 100 g of liquid water at 45 °C. The final temperature is 20 °C. Show that energy is conserved.

Elena Elisseeva/Shutterstock

# CHAPTER **18**

## Observing Energy

# SUMMARY

### KEY TERMS

**energy**

**exothermic**

**heat**

**system**

**surroundings**

**endothermic**

**kinetic energy**

**thermal equilibrium**

**first law of thermodynamics**

**second law of thermodynamics**

**entropy**

**thermal energy**

**calorie**

**specific heat capacity**

**hydrogen bond**

**heat of vaporization**

**heat of fusion**

# Fire Update

Energy and heat can be difficult to define. However, one thing is certain: Whenever matter changes in any way, energy is involved. Here are some generalizations about heat:

- Fire is an exothermic chemical change. Fire releases energy in the form of heat and light.

- Heat is a transfer of thermal energy. Heat always transfers from a system at a high temperature to a system at a lower temperature until both systems are at thermal equilibrium.

- In an endothermic reaction, energy is transferred from the surroundings to the system. An ice pack feels cold because your body transfers heat to the pack.

- The energy transferred by heating does not always result in temperature change. Some of that energy goes into breaking intermolecular attractions during phase change.

- A metal pot will heat up and cool off more quickly than the water inside the pot. This is because different substances have different specific heat capacities.

## REVIEW EXERCISES

1. Provide evidence to support the first and second laws of thermodynamics.

2. Why is it possible to touch a piece of aluminum foil shortly after it has been removed from a hot oven, but a piece of hot cheese pizza can burn you long after it has been removed?

Galyna Andrushko/Shutterstock

3. Thermal energy is transferred to raise the temperature of several substances as specified here. List the substances in order of increasing amount of energy that needs to be transferred. Show your work by determining the number of calories of energy transferred.
   **a.** Raise the temperature of 20 g of water by 25 °C.
   **b.** Raise the temperature of 200 g of copper by 50 °C.
   **c.** Raise the temperature of 20 g of methanol by 40 °C.

4. An ice cube tray can make 16 ice cubes. Each mold is filled with 25 mL of water, and the ice cube tray is placed in the freezer.
   **a.** As the water is freezing, is energy transferred from the freezer to the water, or from the water to the freezer? Draw a diagram of this scenario with an arrow showing the direction of heat transfer.
   **b.** Would the freezing of the water in the tray be considered endothermic or exothermic for the ice cube?
   **c.** Based upon your answer to part a, calculate the amount of heat transfer needed to freeze one ice cube.
   **d.** Based upon your answer to part a, calculate the amount of heat transfer needed to freeze all the ice cubes in the tray.

5. A 5 kg iron pot is filled with 500 mL of water, placed on the stove, and the unit is turned on. After three minutes have passed, the temperature of the iron pot has increased from 25 °C to 150 °C.
   a. How much heat was transferred to the pot?
   b. After three minutes, would you expect the water in the pot to be the same temperature as the iron pot? Or hotter or colder than the iron pot? Why?

6. A 30.4 g copper coin is heated to 400 °C with a Bunsen burner and then dropped into a beaker containing 100 mL of water at 25 °C, room temperature. The water absorbs 1000 cal of heat when this happens.
   a. What is the final temperature of the water in the beaker?
   b. Based upon your answer to part a, what will be the final temperature of the coin?
   c. What is the specific heat of the copper coin?

7. If 70 g of hot water at 85 °C are poured into 50 g of cold water at 18 °C, what is the temperature of the water mixture after it has come to thermal equilibrium?

8. A 50.0 g sample of pure water is heated from 10 °C to 120 °C.
   a. Draw and label a heating curve that represents the different stages of this process.
   b. If the specific heat capacity is 0.49 cal/g · °C for ice, 1.0 cal/g · °C for liquid water, and 0.50 cal/g · °C for steam, how much energy is added to a 50.0 g sample of water to raise its temperature from 10 °C to 120 °C?
   c. Use the heating curve and your calculations to determine which part of the process requires the most heat transfer. Explain your reasoning.

## Uses of Fire

 **PROJECT**  Pick an object that you might purchase, and consider different ways that fire might have been used to make it, harvest it, transport it, and so on. Describe step by step how fire may be involved in getting that object from its natural beginnings to you.

# CHAPTER 19

# Measuring Energy

*In this chapter, you will study*

- substances that will and will not burn
- experimental methods for measuring heat
- how to compare different fuels
- quantifying energy

*When charcoal burns in a barbecue grill, energy is released, allowing us to cook a tasty meal.*

One way to make use of the energy from chemical change is to burn a fuel. In homes, methane gas is burned in water heaters and kitchen stoves. In most automobiles, gasoline or some other fuel is burned. However, not all substances burn, and not all fuels are equally productive from an energy standpoint. Chapter 19 of Unit 5: Fire introduces you to ways that energy from chemical change can be observed and measured.

*You're Fired!*

# Combustion

## THINK ABOUT IT

An automobile engine burns gasoline or ethanol. In a fireplace, you burn paper and wood. In a barbecue grill, you burn charcoal, lighter fluid, and an occasional marshmallow. While many substances burn, others—like sodium chloride, table salt, and water—do not.

### What types of substances burn?

*To answer this question, you will explore*

- Substances That Burn
- Combustion Reactions

## EXPLORING THE TOPIC

### Substances That Burn

A wide variety of substances can burn, or combust. Solids, such as wood and coal, liquids, such as methanol or kerosene, and gases, such as methane or hydrogen, all combust. Something besides phase must determine whether or not a material can burn.

Use the table to compare substances that combust with those that do not combust.

**HISTORY**
**CONNECTION**

From the 1880s until the 1930s, photographers would ignite a sample of magnesium to produce the flash needed to take a picture. For safety and simplicity, flash bulbs, which contained magnesium wire or thin sheets of magnesium foil, were then invented. In the 1980s, these bulbs were replaced with electronic flashes.

© Steve Hamblin/Alamy

**Combustion and Bond Type**

| Substance | Chemical formula | Type of bond | Does it combust? |
|---|---|---|---|
| sodium chloride | $NaCl(s)$ | ionic | no |
| iron (III) oxide | $Fe_2O_3(s)$ | ionic | no |
| nickel | $Ni(s)$ | metallic | yes |
| magnesium | $Mg(s)$ | metallic | yes |
| water | $H_2O(l)$ | molecular covalent | no |
| cellulose (paper) | $C_6H_{12}O_6(s)$ | molecular covalent | yes |
| octane | $CH_3(CH_2)_6CH_3(l)$ | molecular covalent | yes |
| carbon dioxide | $CO_2(g)$ | molecular covalent | no |
| hydrogen | $H_2(g)$ | molecular covalent | yes |
| silicon dioxide (sand) | $SiO_2(s)$ | network covalent | no |

Using the information in the table, it is possible to make some generalizations about whether or not a substance will combust.

- Most ionic compounds do not combust.
- Most molecular covalent compounds do combust, especially those that contain carbon and hydrogen.
- Most elemental metals combust.
- Water and carbon dioxide do not combust.

## ➲ Combustion Reactions

**Combustion** is a chemical reaction in which an element or a compound reacts with oxygen and releases energy in the form of heat and light. When molecular covalent substances containing mainly carbon and hydrogen burn, they produce carbon dioxide and water. When metallic substances combust, they produce a metal oxide.

These are the chemical equations for the combustion of methane and of magnesium.

$$CH_4(g) + 2O_2(g) \longrightarrow CO_2(g) + 2H_2O(g) + \text{heat} + \text{light}$$
$$2Mg(s) + O_2(g) \longrightarrow 2MgO(s) + \text{heat} + \text{light}$$

Oxygen is always a reactant in a combustion reaction. The other reactant in a combustion reaction is usually called a *fuel*. In these reactions, methane and magnesium are considered the fuels.

Notice that both reactions have heat and light on the product side of the equation. Remember, heat is not a substance, but chemists sometimes include it in the equation to keep track of energy changes.

If a compound is already combined with oxygen, it is not likely to combust and combine with *more* oxygen. This is why some molecular covalent compounds like water, $H_2O$, and carbon dioxide, $CO_2$, do not burn. Other examples include sulfur trioxide, $SO_3(g)$, and nitrogen dioxide, $NO_2(g)$.

### FIRE—A SPECIAL TYPE OF COMBUSTION

The photo shows the flame produced when methane burns. When we call something a fire, it is because a flame is present during the combustion reaction. The combustion of methane produces a fire.

Not all combustion reactions produce fire. The burning of magnesium produces a bright light but no flame. The light produced is so bright that it can damage your eyes. Most elemental metals are combustible but do not produce fire.

Ash is often left over after a fire. For molecular compounds, ash is composed of unburned or partially burned reactants. After combustion of metallic substances, the product may look like ash, but it is actually the metal oxide product of the reaction.

## What types of substances burn?

The process of burning is also referred to as combustion. Combustion is the combining of compounds or elements with oxygen through a chemical reaction that produces energy in the form of heat and light. Most ionic compounds do not combust. Most molecular covalent compounds are combustible. When molecular substances that are made mostly of hydrogen and carbon combust, they produce carbon dioxide and water. When metals combust, they form metal oxides, producing heat and light but no flame.

**KEY TERM**

combustion

## Exercises

### Reading Questions

1. What should you look for in a chemical formula to decide if a compound or element will not combust?

2. Name three things that are true of every combustion reaction.

### Reason and Apply

3. Which of the substances listed below will combust? Explain your reasoning.
   - **a.** $Na_2SO_4(s)$
   - **b.** $Cu(s)$
   - **c.** $MgO(s)$
   - **d.** $Na(s)$
   - **e.** $C_2H_6O(l)$
   - **f.** $Ar(g)$

4. Balance the following equations for combustion reactions. Circle the fuel for each reaction.
   - **a.** $C_2H_6(g) + O_2(g) \longrightarrow CO_2(g) + H_2O(l)$
   - **b.** $C_6H_{12}O_6(s) + O_2(g) \longrightarrow CO_2(g) + H_2O(l)$
   - **c.** $C_2H_5OH(l) + O_2(g) \longrightarrow CO_2(g) + H_2O(l)$
   - **d.** $C_{21}H_{24}N_2O_4(s) + O_2(g) \longrightarrow CO_2(g) + H_2O(l) + NO_2(g)$
   - **e.** $C_2H_5SH(l) + O_2(g) \longrightarrow CO_2(g) + H_2O(l) + SO_2(g)$

5. What are the products of these combustion reactions? Write balanced chemical equations for each reaction.
   - **a.** $C_7H_6O(l) + O_2(g) \longrightarrow$
   - **b.** $CH_3COCH_3(l) + O_2(g) \longrightarrow$
   - **c.** $H_2C_2O_4(s) + O_2(g) \longrightarrow$
   - **d.** $Ca(s) + O_2(g) \longrightarrow$
   - **e.** $Li(s) + O_2(g) \longrightarrow$
   - **f.** $Si(s) + O_2(g) \longrightarrow$

Ken Karp Photography

**SAFETY
Instructions**

Safety goggles must be worn at all times. Be sure to know the location of the fire blanket and fire extinguisher.

# Calorimetry

## Purpose

To determine which snack food, cheese puffs or toasted corn snacks, transfers the most heat in a combustion reaction.

## Materials

- cheese puffs, toasted corn snacks
- ring stand with ring and clamp
- Bunsen burner
- wire mesh, paper clips, safety pins, corks, modeling clay, rubber bands, tape
- matches
- crucible
- thermometer
- 150 mL beaker
- balance
- water
- empty tuna fish can, empty soft drink cans with tabs, aluminum foil

## Procedure

1. Design an experiment using any of the materials above to determine which snack food will transfer more energy when burned. You must provide measurements and calculations to support your conclusion.

2. Write a detailed materials list of what you will use and a step-by-step procedure. You must get teacher approval before beginning your test.

3. When your teacher approves your procedure, conduct your experiment and record your observations and data.

## Analysis

1. List all the types of data your team collected.

2. How did you decide what data were important for your experiment?

3. Why do you need to know the mass of the snack food burned?

4. **Making Sense** Which snack food supplies more energy to your body, cheese puffs or toasted corn snacks? What evidence do you have to support your answer?

5. **If You Finish Early** Repeat the experiment to determine whether the volume of water or the mass of fuel used makes a difference in the outcome.

*Now We're Cooking*

# Calorimetry

## THINK ABOUT IT

Foods are sources of energy, and certain foods are better sources of energy than others. What are the energy differences between a granola bar, a handful of nuts, and a few potato chips? Burning dry food items is one way to discover the possible energy potential of each type of food.

### How are food Calories measured?

*To answer this question, you will explore*

- Foods as Fuel
- Calorimetry

## EXPLORING THE TOPIC

### Foods as Fuel

The foods you eat are sources of energy. There are no fires in your stomach, but your body processes the food you eat to generate energy. Burning two foods and comparing the energy transfer that results is a way to measure the energy of each food.

#### FOOD CALORIES

Food labels contain a wide variety of nutritional data, including Calorie content, which is a measure of energy. Food Calories, or Calories with a capital C, are kilocalories of energy, abbreviated kcal or Cal. This product label tells you that 1 ounce of potato chips has 160 Cal. In comparison, 1 ounce of walnuts has 190 Cal.

### HEALTH CONNECTION

A healthy diet for the average adult consists of an intake of about 2000 food Calories per day. One fast-food double cheeseburger contains about 750 Calories, a medium soda is about 210 Calories, and a medium order of French fries is 450 Calories. If you ate this for lunch, you would have only 600 Calories left for the rest of the day.

© 68/Ocean/Corbis

**Nutrition Facts**
Serving Size 1 oz.

Amount Per Serving

**Calories** 160    Calories from fat 80

| | % Daily Value* |
|---|---|
| **Total Fat** 10g | 16% |
| Saturated Fat 1g | 5% |
| Polyunsaturated Fat 3g | |
| Monounsaturated Fat 6g | |
| Trans Fat 0g | |
| **Cholesterol** 0mg | 0% |
| **Sodium** 160mg | 7% |
| **Potassium** 340mg | 10% |
| **Total Carbohydrate** 14g | 5% |
| Dietary Fiber 1g | 4% |
| Sugars 0g | |
| **Protein** 2g | |

| | | |
|---|---|---|
| Vitamin A 0% | • | Vitamin C 10% |
| Calcium 0% | • | Iron 0% |
| Vitamin E 8% | • | Thiamin 2% |
| Niacin 4% | • | Vitamin B$_6$ 6% |
| Phosphorus 4% | | |

* Percent Daily Values are based on a 2,000 calorie diet. Your daily values may be higher or lower depending on your calorie needs:

| | | Calories | 2,000 | 2,500 |
|---|---|---|---|---|
| Total Fat | Less than | | 65g | 80g |
| Sat Fat | Less than | | 20g | 25g |
| Cholesterol | Less than | | 300mg | 300mg |
| Sodium | Less than | | 2,400mg | 2,400mg |
| Potassium | | | 3,500mg | 3,500mg |
| Total Carbohydrate | | | 300g | 375g |
| Dietary Fiber | | | 25g | 30g |

Calories per gram:
Fat g    • Carbohydrate 4    • Protein  4

---

**Example**

### Comparing Calories per Gram

Are there more Calories in 1.0 g of potato chips or 1.0 g of walnuts? (28.3 g = 1.0 oz)

*Solution*

There are 160 Cal in 1.0 oz of potato chips. There are 190 Cal in 1.0 oz of walnuts. An ounce is the same as 28.3 g.

*(continued)*

> Potato chips: 160 Cal/28.3 g = 5.6 Cal/g
>
> Walnuts: 190 Cal/28.3 g = 6.7 Cal/g
>
> So, there are more Calories available in 1.0 g of walnuts than 1.0 g of potato chips.

According to these calculations, an ounce of walnuts contains more energy than an ounce of potato chips. However, the number of Calories per gram is fairly close. In contrast, raisins contain 3.0 Cal per gram, and raw carrots contain 0.4 Cal per gram.

## ⮞ Calorimetry

To measure the Calories in food, you have to find out how much energy is released when the food reacts with oxygen. One way to determine the amount of heat transfer during a combustion reaction is to use the reaction between the fuel and oxygen to heat a measured amount of some other substance, such as water. You can then use a thermometer to measure the change in temperature of the water. The science of measuring the energy released or absorbed in a chemical reaction or physical change is called **calorimetry.** (In Latin, *calor* means heat, and *meter* comes from the Greek word for measure.) Chemists and nutritionists alike use a bomb calorimeter to figure out the calorie content of combustible substances.

The illustration shows a bomb calorimeter. The combustion reaction takes place in an inner reaction chamber called a bomb, which is completely surrounded by water. The sample is ignited by electrical energy, and the energy, due to the burning of the fuel, is transferred to the water. A stirring rod makes sure the heat is uniform in the water. If you determine the change in temperature of the water and you know the quantity of water in the calorimeter, you can determine the heat transferred during the combustion reaction.

Science & Society Picture Library/Getty Images

Electrical leads for igniting sample

+  −

Thermometer

Motorized stir rod

Insulated outer container filled with water

Water

Bomb (reaction chamber)

## How are food Calories measured?

There are a number of different ways to compare the energy transferred when food reacts with oxygen, as it does in the overall digestion process. Some foods can be burned and then their energies can be measured and compared. Calorimetry is the science of measuring the energy released or absorbed by a chemical or physical change. You can use a bomb calorimeter to measure the amount of heat transferred to the surroundings in a combustion reaction.

**KEY TERM**

calorimetry

**Exercises**

### Reading Questions

1. What do Calories have to do with combustion?
2. Why does a bomb calorimeter have an inner chamber and an outer chamber?

### Reason and Apply

3. Can you measure the heat energy of a combustion reaction by placing a thermometer directly into the flames? Explain.
4. What is calorimetry?
5. Most school laboratories do not have bomb calorimeters.
   a. Design an experiment that would allow you to compare whether a cheese puff snack or a roasted corn snack transfers more energy when burned.
   b. In your experiment, does it matter if you burn the same mass of cheese puff and roasted corn snack? Why or why not?
   c. Why is it important to measure the mass of each snack food at the end of your experiment?
   d. What are some possible sources of error that could occur in your experiment?

6. **Lab Report** Write a lab report for the Lab: Calorimetry. In the procedure section, describe your procedure in detail and include a sketch of your setup. In your conclusion, include a discussion of how well your setup worked and whether or not you would change anything.

(Title)

Purpose: (Explain what you were trying to find out.)

Procedure: (List the steps you followed.)

Observations: (Describe your observations.)

Analysis: (Explain what you observed during the experiment.)

Conclusion: (What can you conclude about what you were trying to find out? Provide evidence for your conclusions.)

# Calorimetry Calculations

## THINK ABOUT IT

Heating a beaker of water with a burning potato chip seems like an unusual way to determine the number of Calories in this snack food. However, this is very close to the actual procedure used by nutritionists.

### How does a calorimetry experiment translate into Calories?

*To answer this question, you will explore*

⟳ Calorimetry in the Lab

⟳ Calorimetry Calculations

## EXPLORING THE TOPIC

### ⟳ Calorimetry in the Lab

Suppose that you wanted to measure the amount of energy transfer from combustion of a cashew nut. Because the digestive processes of the human body cannot be duplicated in your classroom, the next best approach is to determine the energy that can be transferred when food reacts directly with oxygen. To do this, you burn the food item and transfer the energy released due to combustion to another substance. Many substances can be used, but the most common one to use is water.

*Building a shield of some sort helps to direct the heat from burning.*

One feature of this type of experiment that is difficult to control in the lab is heat transfer to the surrounding air and equipment. Not all of the energy from burning is transferred directly to the water. Some thermal energy is transferred to the container, to the air, and even to you. One way to control this loss of energy is to burn the food sample very close to the bottom of the water container. Another way is to create some sort of shield to keep the energy transfer focused on the water. All of these details must be taken into account when designing a procedure.

### ⟳ Calorimetry Calculations

There are three measurements and one property that help you to determine the energy transferred during a calorimetry procedure.

Data needed:

1. Mass of water heated in grams; 1 mL = 1 g

2. Temperature change of the water in degrees Celsius, $\Delta T$ (delta $T$)

3. Mass of fuel burned in grams

4. Specific heat capacity of water = 1.00 cal/g °C at 25 °C and 1 atm pressure

With the first two pieces of data, you can calculate how many calories of thermal energy are transferred to the water. Recall that the equation for heat transfer is

$$q = mC_p \Delta T$$

where $C_p$ is the specific heat capacity of the substance being heated.

The number of calories of energy transferred is dependent on the mass of the water and the temperature change.

The illustration shows that if you multiply the change in temperature by the mass of the water, you can calculate the number of calories of heat transferred to the water. This is because the specific heat capacity of water is 1.00 cal/g · °C.

## Example 1

### Calories from a Cashew

These data were collected from the burning of one cashew. How much thermal energy was transferred?

$$\text{Initial } T \text{ of water} = 19.0\,°C$$
$$\text{Final } T \text{ of water} = 34.5\,°C$$
$$\text{Volume of water} = 30.0\text{ mL}$$

*Solution*

First, figure out the temperature change.

Subtract the initial temperature from the final temperature.

$$\Delta T = 34.5\,°C - 19.0\,°C = 15.5\,°C$$

Since 1 mL of water is equal to 1.0 g of water, the mass of the water heated is 30.0 g.

Use the equation for heat transfer.

Substitute values and solve.

$$q = mC_p \Delta T = 30.0\text{ g}(1.00\text{ cal/g} \cdot °C)15.5\,°C = 465\text{ cal}$$

One cashew transferred 435 cal of energy to the water. To compare different foods, it is necessary to figure out the calories *per gram* of food burned.

**Example 2**

### Calories per Gram of Cashew

How much energy per gram was released by the combustion of the cashew?

$$\text{Initial mass of cashew} = 0.66 \text{ g}$$

$$\text{Final mass of cashew} = 0.06 \text{ g}$$

*Solution*

Figure out the mass of the cashew burned.

Subtract final mass from initial mass.

$$\text{Mass of cashew burned} = 0.66 \text{ g} - 0.06 \text{ g} = 0.60 \text{ g}$$

Determine the number of calories per gram.

$$465 \text{ cal}/0.60 \text{ g} = 775 \text{ cal/g}$$

The number of calories per gram is also called the food energy density. Cashews have an energy density of 775 cal/g of cashew burned. Remember, these are chemist's calories. To convert them to food Calories, you must divide by 1000 to get 0.775 Cal/g.

The actual amount of energy a human obtains after the digestive processes are completed is about 85% of the Calorie content listed on nutrition labels. Fats have high energy densities, around 9 Cal/g. Sugars and proteins are around 4 Cal/g.

### COMPARING FOODS

In the real world, nutritionists use calorimeters to figure out the Calorie content of foods. The word *content* is misleading. The food does not actually contain the Calories. The Calorie content represents the energy transferred specifically by *combustion* of these compounds.

Bomb calorimeters minimize the heat transfer to the surroundings, and the results are much more precise than our lab experiments. The method of calculation is very similar to the one used here. If a bomb calorimeter is used, a cashew would be found to have closer to 5500 chemists' calories per gram, or 5.5 Cal/g. Our experimental results came up with 775 cal/g. There is a lot of error in the classroom procedure.

### LESSON SUMMARY

#### How does a calorimetry experiment translate into Calories?

Different fuels transfer different amounts of energy during a combustion reaction. Data about the temperature change, fuel mass, and water volume from a calorimetry procedure can be converted into Calories. Using the specific heat capacity of the substance that was heated by the reaction allows you to calculate the exact number of calories transferred by the combustion of the fuel. This provides you with a measure of the amount of thermal energy "contained" in a fuel. The number of calories transferred from a substance that burns depends on the identity of the substance and its mass.

## Exercises

### Reading Questions

1. Name three possible sources of error in a calorimetry experiment.
2. Why is it important to know the specific heat of the substance being heated by the combustion of a fuel?

### Reason and Apply

3. Could you use ethanol instead of water in a calorimetry experiment? Explain.
4. The experimental value for the Calorie content of a cashew is 0.775 Cal/g. The actual value provided on the package label is 5.5 Cal/g. Determine the percent error.
5. A cereal flake is burned under a beaker containing 25 mL of water. If the water temperature goes up 6 °C, how many calories of energy were transferred to the water?
6. Fuel pellets are used in modern energy-saving wood stoves. If the pellets used for these stoves release 742 cal/g, how many calories of energy will be released by combustion of an entire 40 lb sack of pellets?
7. Examine the illustration on page 520. Use a kinetic molecular view to explain why it takes more calories of heat to raise the water temperature in the third beaker by 5 °C than it does to raise the temperature of the water in the second beaker by 7 °C.
8. The calorie content of a peanut is measured by burning it beneath a can of water and measuring the temperature change of the water. Which of these is a possible source of error?
   (A) The initial mass of the peanut is measured incorrectly.
   (B) Some of the heat of combustion is transferred to the air.
   (C) Some unburned remnants of the peanut are lost before finding the final mass.
   (D) All of the above.
   (E) None of the above.

# Heat of Combustion

## THINK ABOUT IT

Combustion reactions are used to power vehicles, heat water and homes, and create electricity. But not all combustion reactions provide the same amount of energy. Some fuels heat water or power furnaces better than others.

### How do different fuels compare?

*To answer this question, you will explore*

➥ Burning Fuels

➥ Heat of Reaction

➥ Comparing Fuels

## EXPLORING THE TOPIC

### ↻ Burning Fuels

There are a number of different choices of fuels to power race cars. Ideally, you want to choose a fuel that transfers the most energy per gram. That way, the fuel does not unnecessarily add to the weight of the car. You can compare the energy from combustion of a variety of fuels using calorimetry to see which one would be best for a race car.

#### ETHANOL VERSUS METHANOL

Both ethanol, $C_2H_6O$, and methanol, $CH_4O$, have been used to power race cars. Suppose that you put the two fuels into alcohol burners and use them to heat two identical 50 mL samples of water by 30 °C.

**INDUSTRY
CONNECTION**

When pure alcohols, such as ethanol and methanol, combust, the flame is almost colorless and therefore dangerous because it is invisible. In 2005, the fuel for the cars racing in the Indianapolis 500 was changed from methanol to ethanol to promote safety. The ethanol now used contains 2% gasoline, which creates a visible yellow-orange flame when it is burning.

21 °C · 50 mL water · Ethanol
51 °C · 50 mL water · Ethanol
21 °C · 50 mL water · Methanol
51 °C · 50 mL water · Methanol

| Before | After | Before | After |
|---|---|---|---|
| *Burner plus ethanol = 128.0 g* | *Burner plus ethanol = 126.8 g* | *Burner plus methanol = 128.0 g* | *Burner plus methanol = 119.4 g* |

The chemical equations for the combustion of each type of fuel are shown here.

Combustion of ethanol: $C_2H_6O(l) + 3O_2(g) \longrightarrow 2CO_2(g) + 3H_2O(l)$

Combustion of methanol: $2CH_4O(l) + 3O_2(g) \longrightarrow 2CO_2(g) + 4H_2O(l)$

|  | Mass of fuel burned (g) | Mass of water heated (g) | $\Delta T$ for the water (°C) | Energy transferred to water (cal) | Energy per gram of fuel (cal/g) |
|---|---|---|---|---|---|
| **Ethanol** | 1.2 g | 50 g | 30 °C | 1500 cal | 1250 cal/g |
| **Methanol** | 1.6 g | 50 g | 30 °C | 1500 cal | 938 cal/g |

The experimental results show that the combustion of ethanol delivers more energy per gram of fuel consumed, compared with the combustion of methanol. So you would want to use ethanol, not methanol, in a race car if weight were a concern.

## Heat of Reaction

The amount of energy transferred during a chemical reaction is called the **heat of reaction.** It is often designated with the symbol $\Delta H$. The heat of reaction for a combustion reaction is usually referred to as the **heat of combustion.** Heat of reaction, $\Delta H$, is expressed as a negative number when the reaction is exothermic. This is because energy is transferred from the system. When a reaction is endothermic, the system gains energy. Heat of reaction, $\Delta H$, is expressed as a positive number when the reaction is endothermic.

Scientists have done carefully controlled experiments to measure the heat of combustion for many different fuels. Accurate measurements for the combustion of methanol result in a heat of reaction that is much larger in magnitude than the experimental value reported earlier.

Combustion of methanol: $2CH_4O + 3O_2 \longrightarrow 2CO_2 + 4H_2O$
$\Delta H = -5400$ cal/g

Combustion of ethanol: $C_2H_6O(l) + 3O_2(g) \longrightarrow 2CO_2(g) + 3H_2O(l)$
$\Delta H = -7100$ cal/g

### UNITS

The metric unit that scientists use to express energy is the joule, J. A **joule** is a unit of energy that is 4.184 times larger than a calorie. The heat of reaction can be expressed in kilocalories per gram, kilocalories per mole, kilojoules per gram, or kilojoules per mole.

1 joule = 0.239 calorie
1 calorie = 4.184 joules

### ENVIRONMENTAL CONNECTION

A biofuel is any plant substance that can be used as a fuel. For example, some people have customized their cars to burn vegetable oils. These people can collect the oil that would have been discarded after deep frying by restaurants for use in their fuel tanks.

© Gary Houlder/CORBIS

**Important to Know** The heat of combustion is reported with a negative sign to indicate that energy is released by the reaction.

## ⟳ Comparing Fuels

One way to compare fuels is to calculate the number of kilojoules of energy per mole (kJ/mol) of fuel. Simply multiply the kilojoules per gram (kJ/g) by the molar mass (g/mol) of each fuel.

This table provides the heats of combustion for several fuels in both kilojoules per gram and kilojoules per mole.

### Heats of Combustion

|          | Fuel          | Chemical formula | $\Delta H$ (kJ/g) | $\Delta H$ (kJ/mol) |
|----------|---------------|------------------|-------------------|---------------------|
| elements | hydrogen      | $H_2(g)$         | $-121.3$          | $-242$              |
|          | carbon (coal) | $C(s)$           | $-32.6$           | $-393$              |
| alkanes  | methane       | $CH_4(g)$        | $-55.6$           | $-891$              |
|          | ethane        | $C_2H_6(g)$      | $-51.9$           | $-1560$             |
|          | butane        | $C_4H_{10}(l)$   | $-49.6$           | $-2882$             |
|          | hexane        | $C_6H_{14}(l)$   | $-48.1$           | $-4163$             |
|          | octane        | $C_8H_{18}(l)$   | $-47.7$           | $-5508$             |
| alcohols | methanol      | $CH_4O(l)$       | $-22.6$           | $-724$              |
|          | ethanol       | $C_2H_6O(l)$     | $-29.7$           | $-1368$             |
|          | butanol       | $C_4H_{10}O(l)$  | $-36.0$           | $-2669$             |
| sugar    | glucose       | $C_6H_{12}O_6(s)$ | $-15.9$          | $-2828$             |

Notice that when the number of carbon atoms increases, the energy per mole increases. Also, the energy transferred per mole is smaller for the alcohols compared with the alkanes. This makes sense because combustion is a reaction in which a substance combines with oxygen, and the alcohols already contain an oxygen atom. Also notice that elemental hydrogen releases the highest energy per gram. This is why hydrogen is used as the fuel to launch a space shuttle. You get a lot of energy without adding much weight.

## LESSON SUMMARY

### How do different fuels compare?

One way to compare different fuels is to measure the amount of energy they transfer as a result of combustion. Calorimetry is the most common method of measuring this energy. The energy transferred can be expressed in kcal/mol, kcal/g, kJ/mol, or kJ/g of substance. The amount of energy transferred per gram or per mole during a chemical change is commonly called the heat of reaction, $\Delta H$. For combustion reactions, the heat of reaction is negative and is also referred to as the heat of combustion.

**KEY TERMS**

heat of reaction

heat of combustion

joule

**Reading Questions**

1. How can you use water to compare two fuels?

2. What is heat of combustion?

3. Explain why heat of combustion is expressed as a negative number.

**Reason and Apply**

4. The combustion of 4.0 g of propanol transfers energy to 150 mL of water. The temperature of the water rises from 20 °C to 45 °C.
   a. How many calories of energy are transferred to the water?
   b. How many kilocalories of energy are transferred to the water?
   c. How many kilojoules of energy are transferred to the water?
   d. What temperature increase do you expect for 300 mL of water?
   e. What temperature increase do you expect for 2.0 g of propanol and 150 mL of water?

5. Write balanced chemical equations for the combustion of the fuels in the heats of combustion table on page 525. What generalization can you make about the amount of oxygen required for combustion of hydrocarbons?

6. Explain why the combustion of glucose provides a large energy per mole but a rather small energy per gram.

7. PROJECT    In the heats of combustion table, hydrogen has the highest energy per gram of fuel that is combusted. Research to find three reasons why it is challenging to use hydrogen as a fuel for your automobile.

# CHAPTER 19

## Measuring Energy

## SUMMARY

### KEY TERMS

combustion

calorimetry

heat of reaction

heat of combustion

joule

# Fire Update

Combustion can be defined as a reaction in which a substance combines with oxygen, releasing energy. The process of combustion is commonly referred to as burning, and the substance that burns is called a fuel. Some fuels are better sources of energy than others. To measure the energy from a fire, or any chemical reaction, you must measure it indirectly by measuring its effect on matter. Energy is usually measured in units of calories or joules.

Calorimetry is the procedure used to measure the transfer of energy. The amount of energy transferred during a chemical reaction is called the heat of reaction, $\Delta H$. The energy released by the combustion of a fuel is called the heat of combustion.

## REVIEW EXERCISES

1. Describe an experimental setup that would help you compare two fuels. What would you measure?

2. A chemist wants to find out how many Calories are in a bag of peanuts. He sets up a calorimeter and burns a peanut underneath it. This is his data table.

| Initial mass of peanut (g) | Mass of water (g) | $\Delta T$ of water (°C) | Final mass of peanut (g) |
|---|---|---|---|
| 3.75 g | 100 g | 10 °C | 1.20 g |

   a. Calculate the amount of energy, in calories, that can be transferred from the combustion of one peanut.
   b. Calculate the amount of calories per gram of peanut.
   c. If a bag of peanuts is 500 g, how many food Calories does it contain?

3. If iron nails are left outside, they will rust according to the reaction: $4Fe(s) + 3O_2(g) \longrightarrow 2Fe_2O_3(s)$. Why do you think this reaction does not appear to release light and heat?

4. Propane, $C_3H_8(g)$, is used as fuel in small gas stoves and barbecue grills.
   a. Write the balanced equation for the combustion of propane.
   b. The heat of combustion of propane is $-2220$ kJ/mol. How many kJ/g would combusting propane release?
   c. According to the table on page 525, is the heat of combustion per gram of propane greater than the heat of combustion per gram of methane? Per mole? Can you explain why?

*Alan R Harris/Getty Images*

# CHAPTER **20**

# Understanding Energy

*Early trains were powered by steam engines. A steam engine burns coal or wood to boil water, then uses the pressure from the water vapor to drive the engine. Modern trains run on diesel or electricity instead.*

### *In this chapter, you will study*

- bond energies and heat of reaction

- reversing chemical reactions

- rates of chemical reactions

- how energy is converted to work

When chemical change occurs, new substances with new properties are produced. This means some chemical bonds must break and new chemical bonds must form. The energy transferred during chemical change is related to bond making and bond breaking. People have figured out how to make productive use of the energy that accompanies chemical changes. Chapter 20 of Unit 5: Fire focuses on the source of energy in chemical change. It also explores how easily or how quickly a reaction might happen.

Ken Karp Photography

# Make It or Break It

## Purpose

To explore bond energies and calculate the net energy exchange for several reactions.

## Materials

- ball-and-stick model kit

## Questions

1. List at least two patterns you see in the table of average bond energies.

### Average Bond Energies (per mole of bonds)

| Bond | H—H | C—H | C—C | C=C | O—H | C—O | C=O | O=O | O—O |
|---|---|---|---|---|---|---|---|---|---|
| Bond energy (kJ/mol) | 432 | 413 | 347 | 614 | 467 | 358 | 799 | 495 | 146 |

2. Use ball-and-stick models to create the reactants for this reaction.

   Burning methane: $CH_4(l) + 2O_2(g) \longrightarrow CO_2(g) + 2H_2O(g)$

3. Rearrange the reactant models into carbon dioxide and water. Count how many of each type of bond you must break when this reaction takes place. Record the information in a table. Use the average bond energies to determine the amount of energy transferred to break all the bonds.

4. Count how many of each type of bond you must make to form the products. Record this information in your table. Use the average bond energies to determine the amount of energy transferred out when the new bonds are formed.

5. What is the net energy you expect to be transferred to the surroundings by this reaction?

6. The energy of reaction for exothermic reactions is normally expressed as a negative number. Why do you think it is negative?

7. **Making Sense** When a substance combusts, energy is transferred to the surroundings as heat and light. Where does that energy come from?

8. **If You Finish Early** What is the energy of reaction for 1 g of methane burned, in kilocalories per gram?

*Make It or Break It*

# Bond Energy

## THINK ABOUT IT

In a chemical reaction, bonds in the reactants are broken and the atoms rearrange to form new bonds in the products. Flames associated with a combustion reaction indicate that energy is released in the form of heat and light. All combustion reactions are exothermic.

**Where does the energy from an exothermic reaction come from?**

*To answer this question, you will explore*
- Bond Breaking and Bond Making
- Energy of Change
- Energy Exchange Diagrams

## EXPLORING THE TOPIC

### Bond Breaking and Bond Making

In chemical compounds, atoms are bonded together. When the compound methane burns, it combines with oxygen to make new products. To make carbon dioxide and water from molecules of methane and oxygen, the atoms must rearrange as shown. This means that covalent bonds in the methane and oxygen molecules must be broken and new covalent bonds must be formed.

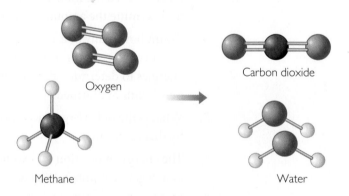

Oxygen

Methane

Carbon dioxide

Water

Energy is required to break bonds. Atoms that are bonded together are attracted to each other, so it takes effort to get these atoms apart. So bond breaking is an endothermic process.

In contrast, when new bonds are formed, energy is released. In terms of energy, bond making is the opposite of bond breaking. Because energy is conserved, the amount of energy released when a bond is made is assumed to be equal to the amount of energy required to break that same bond.

During a combustion reaction, energy is required to break the bonds in the reactant molecules and energy is released when the bonds form in the product molecules. The net energy released is observed as heat and light.

Energy must be absorbed to break bonds.

Energy is released when bonds form.

**Big Idea** Bond breaking requires energy. Bond making releases energy.

### BOND ENERGY

The amount of energy it takes to break a specific bond, such as a carbon-hydrogen bond, is called the **bond energy**. Bond energy is considered a measure of the strength of that bond. However, not all C—H bonds have the same bond energy. For example, it takes 435 kJ/mol to break the first C—H bond in a methane molecule, as compared to 339 kJ/mol to break the last C—H bond.

For this reason, bond energies are reported as averages. The bond energy of an average carbon-hydrogen bond is approximately 413 kJ/mol (or 99 kcal/mol). This means that it takes 413 kilojoules of energy to break one mole of carbon-hydrogen bonds.

This table shows the bond energies of some common bonds.

### Average Bond Energies (per mole of bonds)

| Bond | H—H | C—H | C—C | C=C | C≡C | O—H | C—O | C=O | O=O | O—O |
|------|-----|-----|-----|-----|-----|-----|-----|-----|-----|-----|
| Bond energy (kJ/mol) | 432 | 413 | 347 | 614 | 811 | 467 | 358 | 799* | 495 | 146 |

*C=O in $CO_2$: 799, C=O in organic molecules: 745

**Example**

### Methane Molecule

How much energy would it take to break all the bonds in 1 mol of methane molecules?

*Solution*

Each molecule of methane has four C—H bonds. Therefore, each mole of methane molecules has 4 mol of carbon-hydrogen bonds.

$$413 \text{ kJ/mol} \cdot 4 \text{ mol C—H bonds} = 1652 \text{ kJ}$$

In reality, all of the bonds of the reactants do not necessarily break to form the products in a chemical reaction. However, this is a useful model to explain energy exchanges related to a chemical reaction.

# ➲ Energy of Change

To estimate the energy of an entire chemical reaction, you can consider the reaction as if it takes place in two parts—*energy in* for bond breaking and *energy out* for bond making.

## COMBUSTION OF METHANE

Consider the reaction for the combustion of 1 mol of methane molecules with 2 mol of oxygen molecules. What is the net energy exchange?

$$CH_4 + 2O_2 \longrightarrow CO_2 + 2H_2O$$

First, figure out how much energy is required to break all the bonds in the reactants.

**Bond Breaking—Energy In**

413 kJ  413 kJ  495 kJ  495 kJ

413 kJ  413 kJ

1 mol of $CH_4$   2 mol of $O_2$

Total energy input = 4(413 kJ) + 2(495 kJ) = 2642 kJ

Notice that the energy required to break the bonds of the reactants is positive because energy is added to the system from the surroundings. Now figure out how much energy is released by the formation of new bonds.

**Bond Making—Energy Out**

−799 kJ  −799 kJ  −467 kJ  −467 kJ  −467 kJ  −467 kJ

1 mol of $CO_2$   2 mol of $H_2O$

Total energy output = 2(−799 kJ) + 4(−467 kJ) = −3466 kJ

Notice that the energy required to make the bonds of the reactants is negative because energy leaves the system and goes to the surroundings. The net energy exchange is equal to the sum of the energy input and the energy output.

$$\text{Net energy exchange} = (2642 \text{ kJ}) + (-3466 \text{ kJ})$$
$$= -824 \text{ kJ}$$

The magnitude of energy out is greater than the magnitude of energy in, so this reaction is exothermic. Net energy is expressed as a negative number, because the system loses energy to the surroundings. This value is very close to the actual value for the heat of combustion, $\Delta H$, for this reaction.

# Energy Exchange Diagrams

An energy exchange diagram is a way to keep track of the energy changes during a chemical reaction. The arrows pointing up represent the energy going into the system to break the bonds. The arrows pointing down represent the energy released when new bonds are made. The net energy change determines if a reaction is exothermic or endothermic.

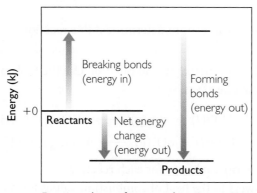

*Energy exchange for an exothermic reaction*

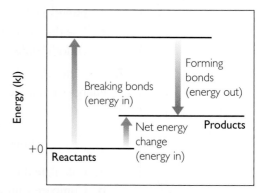

*Energy exchange for an endothermic reaction*

## LESSON SUMMARY

### Where does the energy from an exothermic reaction come from?

**KEY TERM**

bond energy

When reactions take place, bonds between atoms are broken and new bonds are formed. Energy must be transferred *to* a reactant to break a bond. The average energy required to break a certain type of bond is called its bond energy. Conversely, when new bonds are made, energy is released. The heat transferred during a chemical reaction is equal to the difference in energy between the bond-breaking process and the bond-making process. An energy diagram can be used to keep track of energy exchanges in a chemical reaction.

## Exercises

### Reading Questions

1. How can you determine the net energy exchange of a chemical reaction using average bond energies?

2. Use the energies involved in bond breaking and bond making to explain why combustion reactions are exothermic.

### Reason and Apply

3. Place these bonds in order of increasing energy required to break the bond: H—H, C—C, C=C, O—O, O=O.

4. Place these bonds in order of increasing energy released when the bond is formed: H—H, C—H, O—H, C=O.

5. What is the total energy required to break all the bonds in 1 mol of ethanol, $C_2H_6O$?

*(continued)*

6. Consider the combustion of hydrogen. Use the table of average bond energies to answer the questions.

$$2H_2(g) + O_2(g) \longrightarrow 2H_2O(l)$$

a. What is the total energy required to break the bonds in the reactant molecules?
b. What is the total energy released by the formation of the bonds in the product molecules?
c. Use the values from (a) and (b) to determine if this reaction is exothermic or endothermic.

7. Consider the combustion of methanol and butanol. Use the table of average bond energies to answer these questions.

$$\text{Methanol: } 2CH_4O(l) + 3O_2(g) \longrightarrow 2CO_2(g) + 4H_2O(l)$$
$$\text{Butanol: } C_4H_{10}O(l) + 6O_2(g) \longrightarrow 4CO_2(g) + 5H_2O(l)$$

a. Calculate the net energy exchange for each reaction.
b. Compare your answers to the heat of combustion values for methanol and butanol in Lesson 103: Fuelish Choices. What did you discover?

8. Calculate the net energy change for this reaction:

$$C_2H_6(g) \longrightarrow C_2H_4(g) + H_2(g)$$

9. In exothermic reactions,
(A) It takes more energy to make all the new bonds in the products than to break all the bonds in the reactants.
(B) There are more bonds made than broken.
(C) There are more bonds broken than made.
(D) It takes more energy to break all the bonds in the reactants than to make all the new bonds in the products.

# Reversing Reactions

## THINK ABOUT IT

So far, you have been shown chemical reactions that proceed in one direction. This means that reactants are converted to products until the reaction is complete. Some reactions can be reversed. If a reaction is run in the reverse direction, the products of the reaction are converted back into reactants.

### What is the energy associated with reversing reactions?

*To answer this question, you will explore*

➥ Reversing Reactions

➥ Kinetic and Potential Energy

➥ Activation Energy

## EXPLORING THE TOPIC

### ➥ Reversing Reactions

*Hydrogen combustion*

If you place a lit candle under a balloon filled with hydrogen, you can observe the reaction between hydrogen and oxygen. When hydrogen combines with oxygen to form water there is a large "bang."

$$2H_2(g) + O_2(g) \longrightarrow 2H_2O(g) \qquad \Delta H = -482 \text{ kJ}$$

The energy released by this reaction is the heat of reaction, $-482$ kJ. This energy corresponds to 2 mol of $H_2(g)$ reacting with 1 mol of $O_2(g)$ to form 2 mol of $H_2O(g)$. This converts to $-241$ kJ per mole of $H_2(g)$, the value given in the heat of combustion table in Lesson 103: Fuelish Choices.

Notice that the decomposition of water into molecules of hydrogen and oxygen is the reverse of the combustion of hydrogen to form water. What is the heat of reaction for this change?

$$2H_2O(g) \longrightarrow 2H_2(g) + O_2(g) \qquad \Delta H = +482 \text{ kJ}$$

The first law of thermodynamics states that energy is conserved. So, the energy exchange of the reaction in the forward direction is equal and opposite to the energy exchange required for the reverse of the reaction. Notice that when the forward reaction is exothermic, the reverse reaction must be endothermic.

*Decomposition as the reverse of hydrogen combustion*

Photo on left:

*Formation of water in a balloon filled with hydrogen*

| Important to Know | Heat of reaction values are determined experimentally by calorimetry. You can use average bond energies to estimate the heat of reaction. |

Diagrams showing the energy exchange for the formation of water and the decomposition of water are provided here. These energy diagrams are similar to the ones in Lesson 104: Make It or Break It.

*Combustion of hydrogen is exothermic.*      *Decomposition of water is endothermic.*

## ⇌ Kinetic and Potential Energy

The energy in a system is a combination of kinetic energy and potential energy. Kinetic energy is the energy of motion. **Potential energy** is energy that is stored within a physical system and can be converted into another type of energy, such as kinetic energy.

Potential energy can also be defined as the energy associated with the composition of a substance or the position or location of an object in space. For example, a ball at the top of a hill has high potential energy. This potential energy is converted into kinetic energy if the ball rolls down the hill. Likewise, some molecules have a high potential energy. For an exothermic reaction, this potential energy is converted into kinetic energy by converting reactants into products.

### CONSERVING ENERGY

These diagrams show the changes in potential energy for an exothermic reaction and an endothermic reaction.

*Decrease in potential energy = increase in kinetic energy*      *Increase in potential energy = decrease in kinetic energy*

For an exothermic reaction, the potential energy of the system decreases in converting from reactants to products. Because energy is conserved, the sum of the kinetic energy and the potential energy must remain the same, so the kinetic energy increases. The products are hotter than the reactants, because they have a higher average kinetic energy. To reach thermal equilibrium with the surroundings, the hot products transfer energy to the surroundings.

For an endothermic reaction, the potential energy of the system increases. This means that kinetic energy must be converted into potential energy to conserve the total energy. The products are colder than the reactants because they have a lower kinetic energy. To reach thermal equilibrium with the surroundings, energy is transferred from the surroundings to the cold products.

Publiphoto/Science Source

# ➲ Activation Energy

A fire can be started with a match. The combustion reaction in an automobile engine is started with a spark plug. All chemical reactions, not just combustion reactions, require some energy input to get started. This is called the **activation energy.**

Examine the energy diagram for an exothermic reaction. The potential energy of the products is lower than that of the reactants. However, notice that the potential energy first increases on the pathway from reactants to products before it is converted to kinetic energy.

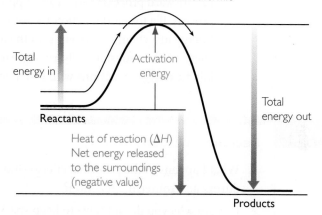

The potential energy between the reactants and products for this reaction is called the activation energy. Energy must be supplied to the reactants to get over this energy barrier. This is why reactions often require a spark to get started. Once an exothermic reaction is started, the reaction itself can provide the necessary energy for more molecules to react. This is why a single spark can cause an entire forest to burn down.

**Big Idea** Some reactions are energetically favored over other reactions.

In contrast, energy must be supplied continuously to cause the reverse of an exothermic reaction. You can see from the diagram that the reverse of the exothermic reaction is an endothermic reaction. Notice also that the activation energy, or energy barrier, is much greater in the reverse direction.

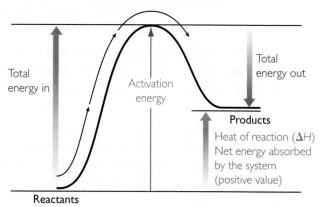

The net energy change from reactants to products of this reaction does not provide the necessary energy to activate more molecules to react. This is why you need to continuously supply energy to decompose water to hydrogen and oxygen. Endothermic reactions *generally* are less likely to occur than exothermic reactions.

## LESSON SUMMARY

### What is the energy associated with reversing reactions?

**KEY TERMS**

potential energy

activation energy

Energy is conserved in chemical processes. This means that the net energy exchange in a forward process is equal and opposite to the net energy exchange in the reverse process. If a forward reaction is exothermic, the reverse reaction is endothermic. Energy in a chemical system is in the form of either kinetic energy or potential energy. When substances burn, a great deal of potential energy is converted into kinetic energy. Reactions with lower activation energies are easier to get started.

## Exercises

### Reading Questions

1. What happens to the potential energy and the kinetic energy of a chemical system during an exothermic reaction?

2. Explain why you do not have to keep sparking a fire once it is started.

### Reason and Apply

3. Consider the combustion of ethane, $C_2H_6(g)$.
   a. Write a balanced equation for the reaction.
   b. Draw an energy diagram for the forward reaction. Label the reactants and the products. Use the table of heats of combustion from Lesson 103: Fuelish Choices to label the heat of reaction.
   c. What is the net energy exchange of the reverse reaction? Explain.

4. Consider the energy diagrams below for the combustion of one mole of methane and the reverse reaction.

   a. Which diagram represents an endothermic reaction?
   b. Which substances have the lowest potential energy?
   c. When methane reacts with oxygen to form carbon dioxide and water, the potential energy decreases. What happens to the kinetic energy?
   d. Explain why the reverse reaction requires a constant input of energy.
   e. Use the table of average bond energies from Lesson 104: Make It or Break It to estimate the heat of reaction. How does your value compare to the heat of reaction given in the diagram?

*(continued)*

**5.** Copy the three energy diagrams shown here.

    **a.** Label the heat of reaction on each.
    **b.** Label the activation energy on each.
    **c.** Which diagram represents the most exothermic reaction?
    **d.** Which reactions require energy to get started?

*Speed Things Up*

# Rate of Reaction

## THINK ABOUT IT

© Roger Ressmeyer/CORBIS

The Golden Gate Bridge in San Francisco is a national landmark. However, it is slowly eroding because of a chemical reaction in which iron in the steel beams combines with oxygen in the air to create iron oxide. This iron oxide falls away from the metal underneath. Is there a way to slow down the rate of this reaction?

**How can you control the speed of a reaction?**

*To answer this question, you will explore*

⮕ Reaction Rates

⮕ Catalysts

*Looking north from high up on the Golden Gate Bridge*

## EXPLORING THE TOPIC

## ⮕ **Reaction Rates**

Some reactions proceed quickly and others proceed slowly. When a balloon full of hydrogen is ignited, it reacts rapidly, exploding into flame. The formation of iron oxide on steel bridges is a slower process called rusting.

Combustion of hydrogen occurs rapidly.  $2H_2(g) + O_2(g) \longrightarrow 2H_2O(g)$

Rusting of iron occurs slowly.  $4Fe(s) + 3O_2(g) \longrightarrow 2Fe_2O_3(s)$

Both of these reactions are exothermic, yet the rates of each reaction are very different.

You might notice that the reactants are both gases for the quicker reaction. Gases mix on a molecular scale, so there is a lot of contact between the reactants. In contrast, the slow reaction involves a solid. The reactants are in contact with one another only at the surface of the solid. There are many more iron atoms on the interior of the solid that are not exposed to oxygen and therefore do not react. When a layer of the iron oxide, rust, flakes off, the underlying iron is then exposed to oxygen and will react.

### COLLISION THEORY

For a reaction to occur, the reactant particles must collide with enough energy to cause reactant bonds to break and product bonds to form. This energy is the activation energy of the reaction. Also, the reactant particles must collide in the correct orientation with one another. When this happens, the reactant particles are able to form a temporary arrangement called an *activated complex,* which facilitates bond breaking and bond making.

| Enough energy but incorrect orientation: | Correct orientation but not enough energy: | Enough energy and correct orientation: |
|---|---|---|
| | | |
| No products form. | No products form. | Products form. |

So, for a reaction to occur, the reactant particles must collide in the correct orientation and with enough energy. If they do not, they will bounce apart and products will not form. Various factors can increase the frequency of collisions between particles and the energy of the collisions. These factors will speed up a reaction.

### BIOLOGY CONNECTION

Enzymes could be considered nature's catalysts. They are biomolecules that speed up the rates of chemical reactions in living things. Enzymes are usually protein molecules. For example, the enzyme amylase in your saliva helps to break down sugar molecules. The enzyme lactase in the small intestine breaks down lactose. People with insufficient lactase have trouble digesting dairy products, which contain lactose.

- **Surface area:** The exposed surface area of the reactants can affect the rate of a reaction. For example, in combustion reactions, fuels with larger exposed surface area, like sawdust or twigs, burn more rapidly than fuels with smaller exposed surface area, like an entire tree or a wooden table.

- **Mixing:** When reactants are mixed well, the reaction rate increases. This is related to the increased probability that the appropriate reactants will collide.

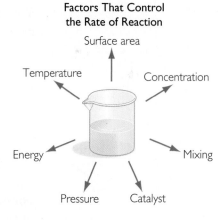

Factors That Control the Rate of Reaction

- **Temperature:** When the temperature is raised, the average kinetic energy of the reactant particles is also raised. The particles move more rapidly and more collisions between particles are possible.

- **Concentration or pressure:** Reactions usually proceed faster as the concentration of the reactants in solution increases or as the pressure of gases increases. Again, more reactants in a small volume result in more collisions.

- **Energy:** Adding energy, such as microwave radiation or light, can increase the rate of a reaction.

## Catalysts

Most of the methods of increasing the rate of reaction involve an increase in the number of collisions of the molecules or an increase in the speeds of the

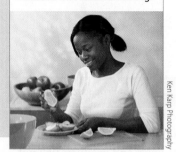
collisions. Another way to increase the rate of reaction is to add a catalyst to the reactants.

A catalyst is a substance that assists a reaction but is neither consumed nor permanently changed by the reaction. A catalyst lowers the activation energy for the reaction. In Unit 2: Smells, sulfuric acid was used to catalyze the reaction that produced sweet-smelling ester molecules.

Chemical reactions may go through a series of intermediate steps to get from reactants to products. Some catalysts work by changing the sequence of steps to shorten the reaction path. Catalysts may undergo chemical transformations during a reaction, but they always return to their original form at the end of the reaction.

Another way a catalyst works is to provide a surface on which the reaction takes place. A catalytic converter in a car provides a platinum surface where pollutants can react to form harmless gases.

The energy diagram shows the same reaction with and without a catalyst. The catalyst lowers the energy barrier for the reaction.

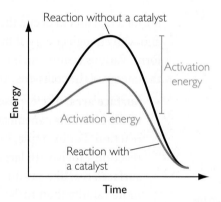

## LESSON SUMMARY

### How can you control the speed of a reaction?

For molecules to react with each other, they must collide with a sufficient amount of energy for bond breaking or bond making. There are a number of ways to increase the rate of a reaction, such as increasing the temperature of the system or changing the concentration of the reactants. These methods increase collisions between reactants. One way catalysts increase the rate of a reaction is by lowering the activation energy to get the reaction started. Catalysts are substances that speed up a reaction but are not permanently changed or consumed by the reaction.

## Exercises

### Reading Questions

1. Name three things you could do to increase the rate of a reaction between two liquids.
2. Name three things you could do to increase the rate of a reaction between two solids.
3. Name two ways you could decrease the rate of a reaction.

*(continued)*

**Reason and Apply**

4. What could you do to speed up the reactions listed here?
   a. Lighting charcoal for a barbecue
   b. Baking cupcakes
   c. Dissolving sugar in water
   d. Removing a stain from clothing

5. Which will burn faster, wood chips or a tree? Explain the difference in the rates of reaction.

6. Which will dissolve faster, a lollipop or powdered sugar? Explain the difference in the rates of reaction.

7. Which will react faster, a steel beam or powdered iron? Explain the difference in the rates of reaction.

8. Why do you think you can extinguish a flame with a carbon dioxide fire extinguisher?

9. What makes an explosion happen so quickly?

10. **PROJECT**

    Research a particular enzyme in your body. Explain how the enzyme catalyzes a reaction in your body.

11. The rate of a reaction can be affected by
    **(A)** The phase of the reactants.
    **(B)** The temperature of the reactants.
    **(C)** The presence of a catalyst.
    **(D)** All of the above.
    **(E)** None of the above.

*Make It Work*

# Work

## THINK ABOUT IT

All around us are machines doing work for us. They transport us from place to place, they help us to manufacture products, and they allow us to listen to music, wash our clothes, and watch television. The source of energy for most of these machines is chemical change.

### How can a chemical reaction be used to do work?

*To answer this question, you will explore*

⟳ Work

⟳ Converting Chemical Energy into Work

## EXPLORING THE TOPIC

### ⟳ Work

To move anything, whether it is a truck or an atom, from one place to another, requires work. **Work,** like heat, is a transfer of energy. Specifically, work is the result of a force acting through a distance. A force is a push or a pull upon an object as a result of an interaction with another object. Force is measured in newtons, N. Scientists define mechanical work as the product of force and distance.

$$\text{Work} = \text{force} \cdot \text{distance}$$
$$W = Fd$$

**LIFTING AN OBJECT**

Imagine that you pick up your baby sister, who weighs 22 pounds, or 10 kilograms, and lift her 1.2 meters. To lift her, you exert 100 newtons of force. This force is equal to the force of gravity acting on your sister, or her weight if it were expressed in newtons.

The work done picking up your 22-pound sister is equal to the force of her weight in newtons multiplied by the distance she was lifted.

$$\text{Work} \doteq 100\text{ N} \cdot 1.2\text{ m} = 120\text{ N} \cdot \text{m}$$

Work is expressed in units of newton-meters, N · m. Work and energy are related. One newton-meter is equal to one joule.

Jo unruh/Getty Images

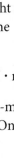
© Guy Edwardes/naturepl.com

### ENVIRONMENTAL CONNECTION

Hydroelectric power plants, windmills, and solar panels do not use chemical reactions to produce electricity. These power sources depend on physical change alone. They are good alternatives to using fossil fuels to produce electricity.

Machines (or tools) help us to do work. Levers, ramps, pulleys, wheels, and wedges are all examples of simple machines. They give you a mechanical advantage to perform work. Consider how easy it is to lift a great deal of weight using a seesaw. A claw hammer acts as both a wedge and a lever, allowing you to pull a nail out of a board, something that would be nearly impossible with your bare hands.

*A hammer acts as a lever when you use it to pull out a nail.*

### ENERGY FROM CHEMICAL REACTIONS

Chemical reactions are sources of energy. Matter has the potential to transfer energy when it is transformed into new compounds. You have seen that burning different fuels can transfer large quantities of energy in the form of heat and light. But heat is not the only useful source of energy from chemical reactions. Consider the products of a combustion reaction. Both the gases formed and the heat produced are potential sources of work.

$$CH_4(g) + 2O_2(g) \longrightarrow CO_2(g) + 2H_2O(g) + heat$$

*Two potential sources of work from a chemical reaction*

Recall from Unit 3: Weather that the pressure from expanding gases can be a powerful force. In the case of a combustion reaction, the products are in the form of heated gases that can expand and push on objects. The work done by expanding gases is often referred to as *PV* work, where *P* and *V* stand for pressure and volume. *PV* work is equal to pressure times the change in volume.

$$W = P\Delta V$$

If pressure is expressed in atmospheres and the volume is expressed in liters, *PV* work is expressed in liter-atmospheres. Liter-atmospheres can also be converted to units of energy in joules.

$$1 \text{ liter-atmosphere} = 101 \text{ joules}$$

**Important to Know** A system cannot contain or store either heat or work. So, heat and work both represent energy in transition.

## ⊃ Converting Chemical Energy into Work

Chemical systems are often paired up with machines to transform chemical change into useful work. The heat transferred, *q*, from a reaction or process can be utilized in some way, or the chemical products themselves can be used.

### STEAM ENGINE

A steam engine uses a combustion reaction to heat water in a cylinder with a piston. When the heat is transferred to the water, the liquid changes phase to a gas that expands and moves the piston.

### A Steam Engine

Work
Piston
Valve
Steam
Pump
Heat in
Cooling water
Heat out
Condenser

## HISTORY
### CONNECTION

The earliest recorded steam engine was built by an ancient Greek mathematician named Hero of Alexandria. The steam came out through two pipes, making a metal arm spin around. There is little evidence that this engine was used for anything more than amusement.

Science and Society/SuperStock

Notice that this is a two-step process as the heat from a chemical change is used to cause a physical change. The expanding gas (water vapor) does the work in this type of engine. Early steam engines were used to pump water and to power steamboats, trains, farm machinery, factories, and the first automobiles.

## INTERNAL COMBUSTION ENGINE

Internal combustion engines like those in cars also make use of expanding gases. However, this type of engine uses the gases produced in the combustion reaction, carbon dioxide and water vapor, to do the work.

The illustration shows an internal combustion engine going through a four-step process.

### Four-Step Process of an Internal Combustion Engine

Spark plug
Air/fuel mixture is sent in.
Piston
Crankshaft
Spark plug fires.
Exhaust gases are sent out.

*Fuel is added.*  *Reactants are compressed.*  *Gases force the piston down.*  *The cycle repeats.*

This four-step process is repeated at an extremely rapid rate to run the engine. The expanding carbon dioxide and water vapor from the combustion reaction move the pistons up and down, turning a large crank that ultimately makes tires, wheels, propellers, or turbines go around.

## ENTROPY

Internal combustion and steam engines convert thermal energy into work. As required by the first law of thermodynamics, the total energy is conserved in the process. However, it is impossible to completely convert all that energy into work. Some of the energy is lost to the surroundings as heat.

This loss of heat to the surroundings is related to entropy and the second law of thermodynamics. As you recall, the second law states that energy and matter tend to disperse or become more disordered. Entropy is the measure of the unavailability of a system's energy to do work due to this dispersal. So, when designing an efficient machine, scientists and engineers must account for entropy.

**Big Idea** Energy disperses; it does not collect.

## LESSON SUMMARY

### How can a chemical reaction be used to do work?

Work is force acting upon an object causing it to move in the direction of the force. So work is energy in motion. The heat transferred, $q$, from chemical reactions can be used to do work. The work done by expanding gases is called *PV* work. The combustion of fuels is one of the most common ways we transform chemical energy into work. However, some energy is always lost, due to entropy.

**KEY TERM**

work

## Exercises

### Reading Questions

1. What is the scientific meaning of the word *work*?

2. Give an example of how a chemical reaction can be used to do work.

### Reason and Apply

3. The expansion of gases pushes a piston. The volume of the gases increases from 0.5 L to 12.5 L. If the piston is pushing against a pressure of 1.2 atmospheres, how much work is done by the gases?

4. If it takes 45 N of force to lift a box 3.8 m, how much work is done to the box?

5. **PROJECT**  Research and describe three ways that a rocket engine is different from an internal combustion engine. What extra item must a rocket engine carry with it into space?

# CHAPTER 20

# Understanding Energy

# SUMMARY

## KEY TERMS

**bond energy**

**potential energy**

**activation energy**

**work**

# Fire Update

All chemical reactions involve the breaking of bonds and the making of bonds. The net energy difference between these two processes determines if a reaction is exothermic or endothermic. This is because bond breaking requires energy and bond making releases energy. By using average bond energies, you can closely approximate the heat of reaction, $\Delta H$.

Energy is required to break bonds.

Energy is released when new bonds form.

Each chemical system expresses energy in the motions of its molecules or particles. This is known as kinetic energy. In addition, each chemical system has potential, or the potential for further energy in rearranging its chemical bonds. When an exothermic reaction occurs, the system's potential energy decreases and its kinetic energy increases. The reverse is true in an endothermic reaction. An increase in kinetic energy is accompanied by an increase in temperature.

It generally takes an input of energy, called the activation energy, to get a reaction started. This is why a fire requires a match or a spark. The overall conditions and the form that the reactants are in can affect the rate of a reaction. For example, raising the temperature, reducing the surfaces of reactants, or adding a catalyst can increase the rate of a reaction.

Chemical reactions can be used to do work. For example, the internal combustion engine uses the gaseous products formed in a combustion reaction to do work on pistons.

## REVIEW EXERCISES

1. List seven ways you could speed up a chemical reaction. Explain how each method can increase the rate of reaction.

2. You have a cup of iced coffee that you are going to sweeten with a sugar cube. Name three ways that you could speed up the rate of dissolving sugar in the coffee. Explain why each of your methods works according to collision theory.

3. You are walking by a classroom, and you overhear two students having a discussion about boiling water. Whom do you agree with and why?

   *Trey:* When you're boiling water, you're breaking the bonds between the hydrogen and oxygen atoms in water. The steam you see above the boiling water is hydrogen and oxygen gas. Heating the water provides the energy to break these bonds.

   *Paige:* Boiling water is not providing nearly enough energy to decompose water. The steam you see above the boiling water is still water. You have not broken any bonds to make it. Heating the water provides the energy to vaporize the water.

4. The chemical formula for heptane is $C_7H_{16}$.
   a. Draw the molecular structure for heptane.
   b. Write the balanced equation for the combustion of heptane.
   c. Using the table of average bond energies (located on page 531) and your answers to parts a and b, calculate the heat of reaction for heptane.

5. Methane, $CH_4$, is a combustible gas used in high school and college labs to burn in Bunsen burners. Carbon dioxide, $CO_2$, is an incombustible gas used in fire extinguishers.
   a. Draw the molecular structures for methane and carbon dioxide.
   b. Using the table of average bond energies on page 531, calculate the amount of energy that would be needed to break the bonds in a mole of methane and a mole of carbon dioxide.
   c. Why do you think that methane combusts, but carbon dioxide doesn't?

6. A balloon is placed over a flask with water in it. As the flask is heated on a hot plate, the balloon expands from a volume of 0.1 L to 1.6 L at a constant 1.1 atm.
   a. Calculate the amount of work that the water molecules did on the balloon.
   b. Is there anything else that needs to be taken into account when calculating the amount of work that the water vapor did?

## Uses of Fire

PROJECT

Write a report describing where coal comes from and how the energy from coal is converted into electricity in a power plant. Include a diagram of the process in your report.

# CHAPTER 21

# Controlling Energy

## In this chapter, you will study

- exothermic reactions of metals and metal compounds

- the energy it takes to extract pure metals from compounds

- how chemical change is converted into electrical energy

- combinations of reactants that make useful batteries

*The battery in your car converts chemical reactions into electricity.*

Fires are not the only source of energy resulting from chemical change. Energy from all types of reactions can be made useful. Reactions involving metals and metal compounds are particularly valuable sources of energy. Using certain techniques, these reactions can be made portable, controllable, and safe. This chapter of Unit 5: Fire focuses on ways in which the energy from chemical change is contained and made easily available.

# Oxidation

## THINK ABOUT IT

If you examine a handful of pennies, chances are some of them will be a bright coppery color and others will be discolored and dull. These tarnished pennies have reacted with the oxygen in the air, just as a fuel does when it burns. But is this a form of combustion?

### What happens when metals react with oxygen?

*To answer this question, you will explore*

- Oxidation
- Electron Transfer

## EXPLORING THE TOPIC

### ⮕ Oxidation

No penny stays bright and shiny forever. Eventually, they all become dark and dull when the copper metal on the surface forms copper (II) oxide and other copper compounds. This reaction happens very slowly at room temperature. The reaction is also exothermic, but it happens so slowly the heat is difficult to detect.

The reaction of copper with oxygen is described by this equation.

$$2Cu(s) + O_2(g) \longrightarrow 2CuO(s)$$

On the other hand, if magnesium is held to a flame, it reacts rapidly and violently with oxygen, producing an almost blinding light. This metal is also combining with oxygen, but it is a much more exothermic reaction and happens quite rapidly once it gets started.

$$2Mg(s) + O_2(g) \longrightarrow 2MgO(s)$$

Copper and magnesium are just two examples of metals reacting with oxygen. Under the right conditions, *all* metals will react with oxygen. This process is called oxidation. Both combustion of molecules and tarnishing of metals are forms of oxidation. Compare the oxidation of two metals with a common combustion reaction:

Oxidation of copper: $2Cu(s) + O_2(g) \longrightarrow 2CuO(s)$

Oxidation of magnesium: $2Mg(s) + O_2(g) \longrightarrow 2MgO(s)$

Combustion of methane: $CH_4(g) + 2O_2(g) \longrightarrow CO_2(g) + 2H_2O(g)$

The reaction of oxygen with a metal produces a solid, whereas the reaction of oxygen with a molecular substance produces two gases. But the reactions have

**HEALTH
CONNECTION**

Antioxidants are molecules that can slow or prevent the oxidation of other molecules. Antioxidants found in fruits and vegetables can cancel out the cell-damaging effects of free radicals. This is one of the reasons why eating fruits and vegetables can lower a person's risk for heart disease, neurological disease, and some forms of cancer. Berries are high in antioxidants.

MarcQuebec/Getty Images

several characteristics in common. This table shows a comparison of the two types of reactions:

| | **Combustion of molecules** | **Oxidation of metals** |
|---|---|---|
| combines with | oxygen | oxygen |
| releases heat? | yes | yes |
| releases light? | yes | sometimes |
| produces flames? | yes | no |
| product | gaseous molecules | solid metal oxides |

Oxidation of metals transfers energy to the surroundings, just like combustion of molecular covalent compounds. This energy is in the form of heat and sometimes a bright light. However, you can see that the products of oxidation of metals and combustion of molecular compounds are different.

### FLAME VERSUS GLOW

Flames are composed largely of superheated gases. When metals oxidize, they might glow with a very bright light, but there is no flame. The bright light is due to the excitation of electrons. Recall that the colored flame tests in Unit 1: Alchemy Lesson 17 were also the result of excited electrons changing energy levels. The bright light from the oxidation of metals may resemble a flame, but it is not a flame.

## ⮕ Electron Transfer

All of the oxidation reactions discussed so far involve oxygen. However, **oxidation** is defined more broadly as any reaction in which an atom transfers electrons to oxygen or another atom. When magnesium oxide is formed, the magnesium atoms transfer electrons to the oxygen atom. When this happens, we say that the magnesium has been oxidized.

**Big Idea**  When an atom or ion is oxidized, it loses electrons.

Magnesium atom          Oxygen atom

Transfers two electrons

Magnesium ion, $Mg^{2+}$          Oxygen ion, $O^{2-}$

Ionic bond

Forms magnesium oxide

Oxidation of magnesium

The shell models show a magnesium atom being oxidized. When magnesium is oxidized, it forms $Mg^{2+}$ ions.

Recall that metal elements generally have one, two, or three electrons in their valence electron shells. Having such small number of electrons in the valence shell makes it more likely that metal atoms will lose electrons, forming cations that have a charge of $+1$, $+2$, or $+3$. Through electron transfer, metal atoms form ionic bonds with nonmetal atoms.

Molecular covalent compounds also transfer electrons to oxygen when they react. Therefore, combustion reactions are considered oxidation reactions.

## LESSON SUMMARY

### What happens when metals react with oxygen?

Metal atoms react with oxygen to form metal oxides. When metal oxides form, heat is released, and sometimes, an intense bright light is emitted. This process is called oxidation. Molecular substances can also be oxidized, as is the case in combustion. However, no flames are present when metals oxidize. This is partly what differentiates the oxidation of metals from the oxidation of molecular substances. Oxidation occurs whenever an atom, ion, or molecule transfers electrons to another atom, ion, or molecule.

**KEY TERM**

oxidation

## Exercises

### Reading Questions

1. How does oxidation of a metal compare to combustion of a covalent compound?
2. Explain what happens when magnesium metal is oxidized.

### Reason and Apply

3. **a.** Write balanced equations for the reactions of calcium, chromium, and lithium with oxygen. Identify the charge on the cation and the charge on the anion for each product.

$$Ca(s) + O_2(g) \longrightarrow CaO(s)$$
$$Cr(s) + O_2(g) \longrightarrow Cr_2O_3(s)$$
$$Li(s) + O_2(g) \longrightarrow Li_2O(s)$$

   **b.** Circle the elements that are oxidized in the chemical equations in part a.

4. Draw shell models for lithium and oxygen atoms to show the transfer of electrons during oxidation.

5. Draw shell models for silicon and oxygen atoms to show the transfer of electrons during oxidation.

*Pumping Iron*

# Heat of Formation

## THINK ABOUT IT

With all the heat and light it gives off, fire is an obvious source of energy. Less obvious is the possible use of other exothermic chemical reactions as energy sources. Because the oxidation of metals is exothermic, it may be possible to use oxidation reactions to generate power.

### How much energy is transferred during oxidation of metals?

*To answer this question, you will explore*

- Heat of Formation
- Reversing Oxidation

## EXPLORING THE TOPIC

### Heat of Formation

The oxidation of most metals is exothermic, just like the combustion of fuels. So, oxidation reactions are a potential source of energy for our daily lives. To explore the usefulness of metal oxidation as an energy source, we must examine and compare the amount of thermal energy transferred during these reactions.

### COPPER VERSUS ZINC

Copper metal and zinc metal react with oxygen to form copper (II) oxide, CuO, and zinc oxide, ZnO.

$$2Cu(s) + O_2(g) \longrightarrow 2CuO(s) \qquad \Delta H = -314 \text{ kJ}$$
$$2Zn(s) + O_2(g) \longrightarrow 2ZnO(s) \qquad \Delta H = -696 \text{ kJ}$$

Notice that $\Delta H$ is negative for both reactions. As you recall, this indicates that the reactions are exothermic and heat is transferred to the surroundings. Also notice that in both cases the heat of reaction is associated with *two* moles of metal oxide being produced. So, the formation of *one* mole of each oxide results in the release of half as much energy.

The energy released when one mole of a compound is formed from elements is called the **heat of formation,** or $\Delta H_f$. Compare the heats of formation of copper (II) oxide, CuO, and zinc oxide, ZnO:

Heat of formation of copper (II) oxide: $\Delta H_f(\text{CuO}) = -157 \text{ kJ/mol}$

Heat of formation of zinc oxide: $\Delta H_f(\text{ZnO}) = -348 \text{ kJ/mol}$

These values indicate how much heat transfer to expect when one mole of each compound is formed from its elements. The formation of 1 mol of zinc oxide

HISTORY

**CONNECTION**

Malachite, $Cu_2CO_3(OH)_2$, a green stone, was probably the first source of copper metal that was extracted through heating. Malachite may have been used to color a piece of pottery that was then fired in a very hot oven, leaving behind globs of copper.

releases over twice as much energy as the formation of 1 mol of copper (II) oxide. You can find heats of formation values of some metal oxides in the table below. These values are determined under standard conditions, at 1 atm and 25 °C. Note that standard conditions are different from STP, standard temperature and pressure, which is 1 atm and 0 °C.

### Heats of Formation

| Compound | $\Delta H_f$(kJ/mol) |
|----------|----------------------|
| $Fe_2O_3(s)$ | $-826$ |
| $Al_2O_3(s)$ | $-1676$ |
| $PbO(s)$ | $-218$ |
| $MgO(s)$ | $-602$ |
| $CuO(s)$ | $-157$ |
| $Au_2O_3(s)$ | $+81$ |

**Example 1**

### Oxidation of Mg and Fe

Which reaction will release more energy?

$$2Mg(s) + O_2(g) \longrightarrow 2MgO(s)$$
$$4Fe(s) + 3O_2(g) \longrightarrow 2Fe_2O_3(s)$$

*Solution*

The heat of formation per mole of product can be found in the table above. Multiply $\Delta H_f$ by the number of moles of each product to get the heat of the reaction, $\Delta H$:

$$(-602 \text{ kJ/mol})(2 \text{ mol MgO}) = -1204 \text{ kJ}$$
$$(-826 \text{ kJ/mol})(2 \text{ mol Fe}_2O_3) = -1652 \text{ kJ}$$

The oxidation of iron, or the formation of iron (III) oxide, $Fe_2O_3$, produces more energy than the oxidation of magnesium. Heats of formation are not limited to metal oxides. For example, the heat of formation of magnesium chloride, $MgCl_2$, is $-2686$ kJ/mol.

### ⮌ Reversing Oxidation

When an oxidation reaction for a metal is reversed, the resulting products are a pure metal and oxygen gas. This reaction is a decomposition reaction. The metal oxide decomposes and one of the products is a metal element. The decomposition of a metal oxide is one way to extract pure metals from metal compounds. But how difficult is it to accomplish energetically? Consider the decomposition of aluminum oxide, $Al_2O_3$:

$$2Al_2O_3(s) \longrightarrow 4Al(s) + 3O_2(g)$$

Energy diagrams for the oxidation of aluminum and for the decomposition of aluminum oxide are shown here. Take a moment to examine the energy changes.

$4Al(s) + 3O_2(g)$

$2Al_2O_3(s)$

Oxidation of Al
$\Delta H = -3352$ kJ
(formation of $Al_2O_3$)

$2Al_2O_3(s)$

$4Al(s) + 3O_2(g)$

Decomposition of $Al_2O_3$
$\Delta H = +3352$ kJ
(formation of Al)

The oxidation of aluminum to form aluminum oxide is very exothermic. The energy for the reverse reaction is equal and opposite. So the decomposition of aluminum oxide is highly endothermic. This means that a large amount of energy is required to extract pure aluminum from aluminum oxide.

Indeed, aluminum was not discovered and brought into use until about 200 years ago, partly because so much energy is required to extract the pure aluminum metal from aluminum ores. The simple heating of aluminum ores does not provide enough energy to decompose the aluminum compounds and extract aluminum metal.

**Example 2**

### Extracting Al

How much energy is required to extract a mole of aluminum metal from aluminum oxide, $Al_2O_3$?

*Solution*

The equation for the decomposition of aluminum oxide is

$$2Al_2O_3 \longrightarrow 4Al(s) + 3O_2(g)$$

This process is the reverse of the formation of $Al_2O_3(s)$, for which $\Delta H_f = -1676$ kJ/mol. (See the table on page 555.)

Because 2 mol of $Al_2O_3(s)$ are decomposed in the equation, the heat of reaction, $\Delta H$, is 2(1676 kJ/mol), or 3352 kJ/mol. Note that the sign is positive because the reaction is reversed.

Because 4 mol of aluminum are produced, the energy required to extract 1 mol of aluminum metal is one-fourth of the heat of reaction.

$$3352 \text{ kJ/mol}/4 = 838 \text{ kJ/mol Al}$$

### THE HISTORY OF METAL USE

It is apparent from archeological sites that gold was in common use thousands of years before iron. This makes sense from the energy data for these metals. The metals that have been in use the longest are the ones that are easiest to extract from their compounds.

*A Roman gilded-silver cup showing a scene from the Iliad.*

## INDUSTRY CONNECTION

The process of removing metals from their ores is called *smelting*. It is usually necessary to add carbon to assist in the process of removing metals from ores. Smelting is often performed at high temperatures so that the metal obtained is already molten and ready to be cast.

This table shows the energy data for the decomposition of metal oxides to metals. To extract the metals, people had to construct furnaces that transferred a sufficient amount of heat.

### Extraction of Metals from Metal Oxides

| Metal | $\Delta H$ kJ/mol | Date |
|---|---|---|
| gold, Au | −41 kJ/mol | 6000 B.C.E. |
| silver, Ag | +15 kJ/mol | 4000 B.C.E. |
| copper, Cu | +156 kJ/mol | 4000 B.C.E. |
| lead, Pb | +218 kJ/mol | 3500 B.C.E. |
| iron, Fe | +413 kJ/mol | 2500 B.C.E. |
| tin, Sn | +581 kJ/mol | 1800 B.C.E. |
| magnesium, Mg | +602 kJ/mol | 1755 C.E. |
| aluminum, Al | +838 kJ/mol | 1825 C.E. |

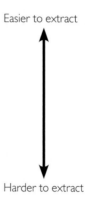

Easier to extract

Harder to extract

The difficulty of extraction of various metals from metal compounds is reflected in the chronological "discovery" and use of these metals through history. The more energy it takes to extract a metal, the later the metal was put into practical use.

Metals that are easily oxidized are hard to extract. In other words, the oxidation reaction is harder to reverse. These are also the metals that tarnish and rust easily. Gold is often found in its pure form in nature. This is because gold is not easily oxidized and is fairly unreactive.

## LESSON SUMMARY

### How much energy is transferred during oxidation of metals?

The formation of metal oxides is usually an exothermic reaction. This means oxidation reactions may be valuable sources of energy. The heat of formation of a compound is the thermal energy change associated with the formation of *one mole* of that compound from its constituent elements. The decomposition of metal oxides into pure metals and oxygen is the reverse of an oxidation reaction. These reactions are generally endothermic. This means it requires an input of energy to extract a metal from a metal compound. Metals that are easily oxidized (with high heats of formation) are more difficult to extract from their metal compounds than metals that are less easily oxidized (with low heats of formation).

**KEY TERM**

heat of formation

## Exercises

### Reading Questions

1. Of all the possible reactions in this lesson, which ones have the potential of being the best sources of energy? Explain your answer.

2. What does the heat of formation indicate about a metal in terms of its practical use as a metal?

*(continued)*

**Reason and Apply**

3. How much energy is released when 10 mol of lead (II) oxide, PbO, are formed?

4. Explain why metals that are easily oxidized are not found in nature in their pure forms. Name two of these metals.

5. Draw the energy diagrams for the oxidation of copper and its reverse reaction. Include the energy data.

6. How many kilojoules of energy does it take to extract 6 mol of tin from tin (II) oxide, SnO?

7. How many kilojoules of energy does it take to extract 10 g of copper from copper (II) oxide, CuO?

8. Which would be a source of more energy, the formation of 6 mol of magnesium oxide, MgO, or the formation of 2 mol of aluminum oxide, $Al_2O_3$? Explain.

9. **PROJECT** Research how one of the metals of antiquity was extracted from metal compounds dug out of the Earth. Describe the process and draw a sketch.

    **www**

# Oxidation-Reduction

## THINK ABOUT IT

A silvery or reddish mineral called hematite, $Fe_2O_3$, is the main source of iron metal on Earth. When this iron (III) oxide is decomposed to iron and oxygen, the oxygen atoms lose electrons. This means that the iron atoms gain electrons.

### What happens to electrons during oxidation?

*To answer this question, you will explore*

➲ Oxidation-Reduction

➲ Redox Reactions

## EXPLORING THE TOPIC

## ➲ Oxidation-Reduction

When iron oxides are decomposed to iron metal, the iron cations gain electrons and the oxygen anions lose electrons.

$$2Fe_2O_3(s) \longrightarrow 4Fe(s) + 3O_2(g)$$

The oxygen anions in $Fe_2O_3(s)$ have a charge of $-2$. In the decomposition of $Fe_2O_3(s)$, each $O^{2-}$ anion loses two electrons to form neutral oxygen atoms. The oxygen atoms bond in pairs to form elemental oxygen, $O_2$ (to satisfy the octet rule). In this process, the oxygen anions, $O^{2-}$, are oxidized to elemental oxygen, $O_2$.

*Sample of hematite, $Fe_2O_3$*

The iron cations in $Fe_2O_3(s)$ have a charge of $+3$. The iron atoms in $Fe(s)$ have a charge of 0. In the decomposition of $Fe_2O_3(s)$, each $Fe^{3+}$ cation gains three electrons to form neutral iron atoms in the elemental metal. Just as there is a term for losing electrons (oxidation), there is a term for gaining electrons: **reduction.** In this process, the iron atoms, $Fe^{3+}$, are reduced to elemental iron, Fe.

Reduction—gains electrons

$$2\overset{2+}{Fe_2}\overset{2-}{O_3}(s) \longrightarrow 4\overset{0}{Fe}(s) + 3\overset{0}{O_2}$$

Oxidation—loses electrons

Whenever metals are extracted from their compounds, they are reduced. While the term *reduction* refers to the process of gaining electrons, its origins have to do with the concept of reducing metal ores to their simplest elements. Whenever you see a metal element as the product of a reaction, you can be fairly certain that it has been reduced.

Every time a substance loses electrons, another substance gains electrons and vice versa. These two processes always take place together: oxidation and reduction. **Oxidation-reduction reactions** are often referred to as **redox** reactions.

## REACTION OF ZINC AND COPPER SULFATE

A strip of zinc metal is placed in a light blue solution of copper (II) sulfate, $CuSO_4$, as shown in the illustration. Soon, a reddish coating begins to form on the zinc strip. The blue color of the solution begins to fade. What is going on?

Start with the balanced chemical equation for this reaction:

$$Zn(s) + CuSO_4(aq) \longrightarrow ZnSO_4(aq) + Cu(s)$$

You know from Unit 1: Alchemy that when salts dissolve in water, the compound dissociates into ions. Often, chemical equations are written in ionic form to specify the charges of the ions in solution as shown here.

Notice that the sulfate ion is a spectator ion. The net ionic equation is therefore

$$Zn(s) + Cu^{2+} \longrightarrow Zn^{2+} + Cu(s)$$

The next illustration shows this reaction from a particle view. Take a moment to follow the transfer of the electrons.

*The copper ions, $Cu^{2+}$, in solution are reduced to copper metal, $Cu^0$. The zinc metal atoms on the strip, $Zn^0$, are oxidized to zinc ions, $Zn^{2+}$, and enter the solution.*

### INDUSTRY CONNECTION

Barium is a metal that is easily oxidized in air to form BaO. So, it is difficult to obtain this metal in its pure form. Because barium reacts so easily with oxygen, it is sometimes used to create vacuum tubes. It can remove the last of the oxygen.

**Important to Know** The atom that is being reduced gains electrons. The atom that is oxidized loses electrons.

## Redox Reactions

Many chemical reactions are oxidation-reduction reactions. In fact, any reaction in which electrons are transferred from one reactant to another is a redox reaction. This includes most reactions that have an element as a reactant or product, which are often single exchange reactions.

Typically, double exchange reactions are *not* redox reactions. For example, in acid-base reactions, $H^+$ trades places with another cation. In precipitation reactions, an aqueous ion from one compound combines with an aqueous ion from another. So, double exchange reactions often involve transfer of ions, while single exchange reactions involve transfer of electrons.

### SINGLE EXCHANGE: REDOX REACTION

$$Mg(s) + 2HCl(aq) \longrightarrow H_2(g) + MgCl_2(aq)$$

### DOUBLE EXCHANGE: ACID-BASE REACTION

$$NaOH(aq) + HCl(aq) \longrightarrow H_2O(l) + NaCl(aq)$$

### DOUBLE EXCHANGE: PRECIPITATION REACTION

$$AgNO_3(aq) + KCl(aq) \longrightarrow AgCl(s) + KNO_3(aq)$$

Oxidation-reduction reactions are not limited to metals and ionic compounds. Molecular covalent compounds also take part in oxidation-reduction reactions.

### COMBUSTION: REDOX REACTION

$$CH_4 + 2O_2 \longrightarrow CO_2 + 2H_2O$$

In this reaction, carbon is oxidized. Electron transfer is trickier to identify in reactions with covalent compounds. But we already know that combustion is a form of oxidation. For molecules, oxidation often involves a loss of hydrogen and a gain of oxygen, and reduction often involves a gain of hydrogen and the loss of oxygen.

**Example**

### Oxidation-Reduction

Solid magnesium, Mg, is placed in a solution of iron (III) nitrate, $Fe(NO_3)_3$. Magnesium nitrate and solid iron are formed. Show the ionic equation and determine what is oxidized and what is reduced.

*Solution*

Write and balance the chemical equation for this reaction:

$$3Mg(s) + 2Fe(NO_3)_3(aq) \longrightarrow 3Mg(NO_3)_2(aq) + 2Fe(s)$$

Translate this equation into an ionic equation. Ignore the spectator ions, in this case, the $NO_3^-$ ions.

$$3Mg(s) + 2Fe^{3+}(aq) \longrightarrow 3Mg^{2+}(aq) + 2Fe(s)$$

*(continued)*

The metal losing electrons is oxidized. The metal gaining electrons is reduced.

Each magnesium atom loses two electrons, so magnesium is oxidized. Two iron ions accept three electrons each, so iron is reduced.

> **Tracking the electrons**
>
> $3Mg^0 \longrightarrow 3Mg^{2+} + 6e^-$
>
> $2Fe^{3+} + 6e^- \longrightarrow 2Fe^0$

## LESSON SUMMARY

### What happens to electrons during oxidation?

**KEY TERMS**

reduction

oxidation-reduction
   reaction (redox)

Whenever an atom or ion loses electrons, another atom or ion gains electrons. The loss of electrons is called oxidation, and the gain of electrons is called reduction. Oxidation and reduction always occur together. This process is known as oxidation-reduction, or redox. When metals react with oxygen to form metal oxides, the metal atoms are oxidized and the oxygen is reduced. To extract elemental metals, the reverse reaction must occur and metal ores must be reduced.

## Exercises

### Reading Questions

1. Explain what is oxidized and what is reduced when copper reacts with oxygen to form copper (II) oxide, $CuO$.

2. Explain what is oxidized and what is reduced when copper (II) oxide is decomposed to copper and oxygen.

### Reason and Apply

3. Describe how you might extract a metal from a metal salt solution.

4. Determine what is oxidized and what is reduced in these reactions:
   a. $CO_2(g) + H_2(g) \longrightarrow CO(g) + H_2O(g)$
   b. $SF_4(g) + F_2(g) \longrightarrow SF_6(g)$
   c. $4Ag(s) + 2H_2S(g) + O_2(g) \longrightarrow 2Ag_2S(s) + 2H_2O(g)$
   d. $C_6H_{12}O_6(aq) + 6O_2(g) \longrightarrow 6CO_2(g) + 6H_2O(l)$

5. Write the net ionic equations for these reactions:
   a. $Zn(s) + CuCl_2(aq) \longrightarrow Cu(s) + ZnCl_2(aq)$
   b. $Ba(NO_3)_2(aq) + CuSO_4(aq) \longrightarrow BaSO_4(s) + Cu(NO_3)_2(aq)$

6. Determine which of the reactions in Exercise 5 is not a redox reaction. Explain your reasoning.

7. Which of these reactions from Unit 4: Toxins can be classified as redox reactions? For each redox reaction, show what is oxidized, what is reduced, and the total number of electrons transferred.
   a. $Tl_2O(s) + 2HCl(l) \longrightarrow 2TlCl(aq) + H_2O(l)$
   b. $HgS(s) + 2HCl(aq) \longrightarrow HgCl_2(s) + H_2S(aq)$
   c. $PbCO_3(s) + 2HCl(aq) \longrightarrow PbCl_2(aq) + H_2CO_3(aq)$
   d. $Na_2C_2O_4(aq) + CaCl_2(aq) \longrightarrow CaC_2O_4(s) + 2NaCl(aq)$
   e. $Pb(s) + 2HCl(aq) \longrightarrow PbCl_2(aq) + H_2(g)$
   f. $2As(s) + 6HCl(aq) \longrightarrow 2AsCl_3(aq) + 3H_2(g)$

# Activity of Metals

## THINK ABOUT IT

Imagine that you set up three experiments. In one, a strip of copper is placed in a solution of silver nitrate. In the second, a strip of zinc is placed in a solution of copper (II) nitrate. In the third, a strip of silver is placed in a solution of zinc nitrate. Only two of these setups will result in reactions. How can you predict which reactions will occur?

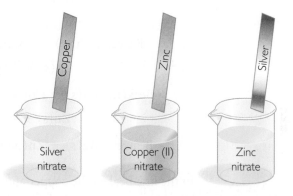

**Which metal atoms are most easily oxidized?**

*To answer this question, you will explore*

⮕ Comparing Metals

⮕ Activity Series

## EXPLORING THE TOPIC

## ⮕ Comparing Metals

When metal atoms are combined with metal ions in solution, you may or may not end up with a reaction. Some combinations of metals and metal salts result in redox reactions and others don't. It all depends on which combinations of metals are used.

Consider the three solutions shown on the next page. Each solution contains metal cations and a strip of elemental metal. If the metal atoms on the strip transfer electrons to the metal cations in solution, a reaction will occur.

### INDUSTRY
### CONNECTION

Redox reactions have been used for decades to coat metals with chromium. Chrome plating improves appearance and durability. Chromium sulfate, or chromium chloride, is often used in industrial plating solutions.

Beaker 1: Copper in silver nitrate solution

Beaker 2: Zinc in copper (II) nitrate solution

Beaker 3: Silver in zinc nitrate solution

*Silver: reduced*
*Copper: oxidized*

*Copper: reduced*
*Zinc: oxidized*

*No reaction*

$$\text{Beaker 1: } Cu(s) + 2AgNO_3(aq) \longrightarrow Cu(NO_3)_2(aq) + 2Ag(s)$$
$$Cu(s) + 2Ag^+(aq) \longrightarrow Cu^{2+}(aq) + 2Ag(s)$$
$$\text{Beaker 2: } Zn(s) + Cu(NO_3)_2(aq) \longrightarrow Zn(NO_3)_2(aq) + Cu(s)$$
$$Zn(s) + Cu^{2+}(aq) \longrightarrow Zn^{2+}(aq) + Cu(s)$$
$$\text{Beaker 3: } Ag(s) + Zn(NO_3)_2(aq) \longrightarrow \text{no reaction}$$

In the first beaker, the copper atoms transfer electrons to the silver ions. Copper atoms are oxidized and silver cations are reduced. As a result, silver metal can be seen as a coating on the copper strip. In the second beaker, zinc atoms transfer electrons to the copper ions in solution. Zinc atoms are oxidized and copper is reduced. Copper metal coats the zinc strip. However, nothing happens in the third beaker.

In the third beaker, there is no reaction. Silver does not give up its electrons in the presence of zinc ions. Chemists say that silver is not as active as zinc.

## ⮕ Activity Series

By combining different metals and metal ions, you can determine experimentally which metals are more active than other metals. The result is called an **activity series.** Working with the data we have so far, zinc, copper, and silver can be placed in order of their activity.

**Zinc**

gives up electrons to

**Copper**

which gives up electrons to

**Silver**

More active

↑

↓

Less active

Where would magnesium fit on this list? In the lab, you combined magnesium with zinc nitrate. The magnesium was oxidized and formed $Mg^{2+}$ ions.

$$Mg(s) + Zn(NO_3)_2(aq) \longrightarrow Mg(NO_3)_2(aq) + Zn(s)$$

*Because silver is easily reduced, it is relatively easy to plate metals with silver for ornamentation.*

So, magnesium belongs above zinc on the list.

**Magnesium**

gives up electrons to

**Zinc**

which gives up electrons to

**Copper**

which gives up electrons to

**Silver**

More active

Less active

You may have noticed that all the reactions in this section have been single exchange reactions in which the more easily oxidized metal displaces the other metal. In this way, the more active metal forms cations while the less active metal is reduced to a solid.

A series of experiments can reveal where other metals belong on the list. The more active atoms at the top of the list will always displace the less active ones below.

## Activity Series

Potassium — More active

Barium — Easily oxidized

Calcium

Sodium

Magnesium

Aluminum

Zinc

Chromium

Iron

Nickel

Copper

Silver

Mercury — Less active

Gold — Easily reduced

On this list, gold is at the bottom. Gold atoms are very difficult to oxidize, and gold cations are very easy to reduce. This is one reason why gold is rarely found in compounds combined with other atoms.

### A metal high in the activity series

• reacts vigorously with compounds

• readily gives up electrons in reactions to form positive ions

• is corroded easily

*A metal low in the activity series*

- does not react vigorously with chemicals
- does not readily give up electrons in reactions to form positive ions
- is not corroded easily

Example

### Activity of Lead

Lead, Pb, will give up electrons to copper cations, $Cu^{2+}$, but not to zinc cations, $Zn^{2+}$.

**a.** Write the net ionic equation for the reaction between Pb and $Cu^{2+}$.

**b.** Do you expect that Zn will transfer electrons to lead cations, $Pb^{2+}$? Explain your reasoning.

### Solution

**a.** $Pb(s) + Cu^{2+}(aq) \longrightarrow Pb^{2+}(aq) + Cu(s)$

**b.** Pb does not give up electrons to $Zn^{2+}$. This means that the reverse reaction *will* happen. Zn will give up electrons to $Pb^{2+}$. Zn is more active than Pb, so it displaces Pb in the ionic compound.

$$Zn(s) + Pb^{2+}(aq) \longrightarrow Zn^{2+}(aq) + Pb(s)$$

All of the metals on the periodic table can be placed in an activity series table. Chemists are interested in the activity of metals, because this allows them to control the outcome of different reactions. Metal compounds can be combined in different ways to form other compounds. Also, the most easily oxidized metals turn out to be a good source of electrons. You'll learn more about this in Lesson 112: Current Events.

## LESSON SUMMARY

### Which metal atoms are most easily oxidized?

Metals that are easily oxidized give up their electrons easily. They are considered more active than other metals. Comparing metals experimentally allows you to rank the metals in terms of their ease of oxidation. The activity of a metal is demonstrated when it is combined with another metal cation. The more active metal will lose electrons, displacing the other metal in the compound. The displaced metal is reduced. The ranking of metals in order of their activity is referred to as an activity series.

**KEY TERM**

activity series

Exercises

### Reading Questions

1. Explain how you would figure out where tin should be listed in the activity series.

2. Why are metals that are more active also considered less stable?

*(continued)*

**Reason and Apply**

3. If zinc nitrate, $Zn(NO_3)_2$, is paired with solid iron, what would you expect to observe? Explain your reasoning.

4. If iron (III) nitrate, $Fe(NO_3)_3$, were combined with solid zinc, what would you expect to observe? Explain your reasoning

5. Consider any reactions that occur in Exercises 3 and 4.
   a. Write chemical equations.
   b. Write the net ionic equation.
   c. Identify the more active metal.
   d. Which metal is oxidized? Which metal is reduced?

6. Platinum is more active than gold but less active than silver. Where does it belong in the activity series?

7. If cobalt ions are replaced by iron to form iron ions and solid cobalt, is cobalt above or below iron in the activity series? Explain.

8. Write net ionic equations for two reactions between magnesium and less active metals.

9. Name three combinations of ionic compounds and metals that will not react.

# Electrochemical Cell

## Purpose

To explore how to get electrical energy from a redox reaction.

## Materials

- zinc strip
- copper strip
- 250 mL beakers (2)
- 500 mL beaker
- 2 connecting wires with alligator clips
- tiny LED light bulb
- 1.0 M $CuSO_4$, 100 mL
- 1.0 M zinc sulfate solution, $ZnSO_4$, 100 mL
- saturated potassium nitrate ($KNO_3$) solution, 20 mL—for salt bridge
- filter paper approximately 1 in wide and 6 in long—for salt bridge
- gloves for handling salt bridge

## Procedure

1. Make a salt bridge. Place a piece of folded-up filter paper in the bottom of an empty 500 mL beaker. Soak it thoroughly with $KNO_3$. Set aside for later.

2. Carefully pour 100 mL of $CuSO_4$ into one beaker and 100 mL of $ZnSO_4$ into another beaker.

3. Set up the zinc strip in the $ZnSO_4$ solution and the copper strip in the $CuSO_4$ solution. Use a wire with alligator clips to connect the two metal strips. The clips should not touch the solutions.

4. Use gloves to place the salt bridge between the two beakers as shown.

## Observation and Analysis

1. What do you observe?

2. Why is it necessary to have an ionic solution in the salt bridge?

3. Which substance is being oxidized? Which substance is being reduced? How do you know?

4. Connect the tiny LED light into your circuit using both sets of alligator clips. What do you observe? What does that prove?

5. Explain how you might reverse this reaction.

6. **Making Sense** Explain where the electricity in the electrochemical cell is coming from.

Salt bridge

Copper

Zinc

Copper (II) sulfate     Zinc sulfate

# Electrochemical Cell

## THINK ABOUT IT

Batteries are used extensively in our daily lives. They are in our cell phones, automobiles, electronic toys, mobile music players, and wristwatches. A battery contains a controlled redox reaction.

**How can you use a redox reaction as an energy source?**

*To answer this question, you will explore*

➲ Electrochemical Cells

➲ Maximizing the Potential

➲ Batteries

## EXPLORING THE TOPIC

## ➲ Electrochemical Cells

The combustion of gasoline powers cars. The combustion of methane is used to cook food and heat homes. But it would be hard to imagine a cell phone or a toy running off a combustion reaction.

There is another way, besides burning, to make use of the energy of a reaction. In the case of many oxidation-reduction reactions, the transfer of electrons can be converted into electrical energy using an apparatus called an *electrochemical cell*.

### CONTROLLING REDOX REACTIONS

The first step to controlling a redox reaction is to separate the oxidation and reduction reactions. One possible way to do this is shown. Two strips of metal called *electrodes* are placed in two beakers. Both beakers are filled with aqueous salt solutions called **electrolytes.** Electrons move through a wire connecting the two metal strips, and ions move through a salt bridge connecting the two electrolyte solutions.

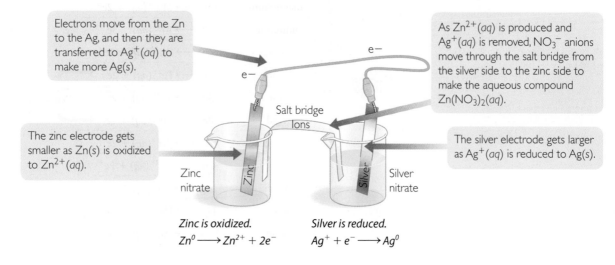

Electrons move from the Zn to the Ag, and then they are transferred to $Ag^+(aq)$ to make more Ag(s).

As $Zn^{2+}(aq)$ is produced and $Ag^+(aq)$ is removed, $NO_3^-$ anions move through the salt bridge from the silver side to the zinc side to make the aqueous compound $Zn(NO_3)_2(aq)$.

The zinc electrode gets smaller as Zn(s) is oxidized to $Zn^{2+}(aq)$.

The silver electrode gets larger as $Ag^+(aq)$ is reduced to Ag(s).

Salt bridge
Ions

Zinc nitrate

Silver nitrate

*Zinc is oxidized.*
$Zn^0 \longrightarrow Zn^{2+} + 2e^-$

*Silver is reduced.*
$Ag^+ + e^- \longrightarrow Ag^0$

© Tony Freeman/PhotoEdit

You can use the activity series table below to help you predict what will happen in the **electrochemical cell.** Because zinc is higher on the chart than silver, zinc is the more active metal and you can expect it to be oxidized. Likewise, because silver is not a very active metal, you can expect it to be reduced.

$$\mathrm{Zn}(s) + 2\mathrm{Ag}^+(aq) \longrightarrow \mathrm{Zn}^{2+}(aq) + 2\mathrm{Ag}(s)$$

If you place $\mathrm{Zn}(s)$ and $\mathrm{Ag}^+(aq)$ in the same beaker, the reaction will occur with direct transfer of electrons from the $\mathrm{Zn}(s)$ to the $\mathrm{Ag}^+(aq)$. The reaction just goes until one or both reactants are used up.

Separating the reactants allows you to control the reaction and to create an electrical current. Each half of an electrochemical cell is called a **half-cell.** The reaction in the electrochemical cell can be stopped and started when desired. It will proceed only when the circuit is closed and the two halves of the reaction are connected. The reaction that takes place in each half-cell is called a **half-reaction.**

## ⟳ Maximizing the Potential

Not every combination of metals and metal salts results in the same output of energy. Generally, the farther away two metals are from each other on the activity series list, the more energy that will be produced by a reaction between them. For example, the energy released when calcium and silver are paired is greater than the energy released when iron and lead are paired.

**Activity Series**

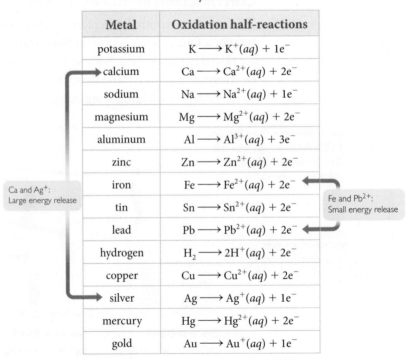

| Metal | Oxidation half-reactions |
|---|---|
| potassium | $\mathrm{K} \longrightarrow \mathrm{K}^+(aq) + 1e^-$ |
| calcium | $\mathrm{Ca} \longrightarrow \mathrm{Ca}^{2+}(aq) + 2e^-$ |
| sodium | $\mathrm{Na} \longrightarrow \mathrm{Na}^{2+}(aq) + 1e^-$ |
| magnesium | $\mathrm{Mg} \longrightarrow \mathrm{Mg}^{2+}(aq) + 2e^-$ |
| aluminum | $\mathrm{Al} \longrightarrow \mathrm{Al}^{3+}(aq) + 3e^-$ |
| zinc | $\mathrm{Zn} \longrightarrow \mathrm{Zn}^{2+}(aq) + 2e^-$ |
| iron | $\mathrm{Fe} \longrightarrow \mathrm{Fe}^{2+}(aq) + 2e^-$ |
| tin | $\mathrm{Sn} \longrightarrow \mathrm{Sn}^{2+}(aq) + 2e^-$ |
| lead | $\mathrm{Pb} \longrightarrow \mathrm{Pb}^{2+}(aq) + 2e^-$ |
| hydrogen | $\mathrm{H}_2 \longrightarrow 2\mathrm{H}^+(aq) + 2e^-$ |
| copper | $\mathrm{Cu} \longrightarrow \mathrm{Cu}^{2+}(aq) + 2e^-$ |
| silver | $\mathrm{Ag} \longrightarrow \mathrm{Ag}^+(aq) + 1e^-$ |
| mercury | $\mathrm{Hg} \longrightarrow \mathrm{Hg}^{2+}(aq) + 2e^-$ |
| gold | $\mathrm{Au} \longrightarrow \mathrm{Au}^+(aq) + 1e^-$ |

Ca and Ag⁺: Large energy release

Fe and Pb²⁺: Small energy release

Example 1

## Magnesium-Tin Cell

Imagine you want to create an electrochemical cell using magnesium, Mg, and tin, Sn.

    **a.** Which metal would be oxidized and which would be reduced?

    **b.** What ion will form and what metal will be plated out of solution as a solid?

    **c.** Write the half-reactions and the net ionic equation.

### Solution

Magnesium is higher in the activity series.

    **a.** Magnesium will be oxidized. Tin will be reduced.

    **b.** Tin will be plated out onto the tin strip. Magnesium ions, $Mg^{2+}(aq)$, will form.

    **c.** The half-reactions are $Mg^0 \longrightarrow Mg^{2+} + 2e^-$ and $Sn^{2+} + 2e^- \longrightarrow Sn^0$.
Adding these, the net reaction is $Mg(s) + Sn^{2+}(aq) \longrightarrow Mg^{2+}(aq) + Sn(s)$.

The concentrations of the reactants, as well as the identities of the metals and metal ions used, can affect the amount of energy that is produced.

### VOLTAGE

The energy that comes out of an electrochemical cell is expressed in volts. The **voltage** of a battery lets you know how much energy is "stored" in it, that is, how much potential it has to produce electricity.

Chemists can predict the voltage of an electrochemical cell by examining the two half-reactions involved. The measurement of the voltage of half-cells takes place under uniform conditions. These voltages are normally expressed as *reduction* half-reactions and are listed in a standard reduction potentials table.

**Standard Reduction Potentials at 25 °C, 1 atm, and 1 M Ion Concentration**

| Reduction Reaction | Half-cell Potential |
|---|---|
| $Li^+ + e^- \longrightarrow Li$ | $-3.05$ volts |
| $Mg^{2+} + 2e^- \longrightarrow Mg$ | $-2.37$ volts |
| $Zn^{2+} + 2e^- \longrightarrow Zn$ | $-0.76$ volt |
| $Ni^{2+} + 2e^- \longrightarrow Ni$ | $-0.26$ volt |
| $Pb^{2+} + 2e^- \longrightarrow Pb$ | $-0.13$ volt |
| $Sn^{2+} + 2e^- \longrightarrow Sn$ | $0.15$ volt |
| $Cu^{2+} + 2e^- \longrightarrow Cu$ | $0.34$ volt |
| $Ag^+ + e^- \longrightarrow Ag$ | $0.80$ volt |
| $Au^+ + e^- \longrightarrow Au$ | $1.69$ volts |

The overall voltage of an electrochemical cell can be calculated by finding the difference between the potentials of the half-reactions. This means subtracting the voltage of the oxidation process from the voltage of the reduction process.

**Example 2**

### Silver-Zinc Electrochemical Cell

Determine the voltage of a silver-zinc electrochemical cell.

*Solution*

Write the half-reactions for silver and zinc from the standard reduction potentials table on page 571. The half-reaction that is more positive will proceed as a reduction, and the half-reaction that is more negative will proceed as an oxidation (in the reverse direction).

This half-reaction proceeds as reduction.     $Ag^+ + e^- \longrightarrow Ag$     0.80 volt

This half-reaction proceeds in the opposite direction as oxidation.     $Zn^{2+} + 2e^- \longrightarrow Zn$     $-0.76$ volt

Next, subtract the reduction potential of the oxidation half-reaction from the reduction potential of the reduction half-reaction.

$$0.80 - (-0.76) = 0.80 + 0.76 = +1.56 \text{ volts}$$

This cell has a voltage of $+1.56$ volts.

## ENVIRONMENTAL CONNECTION

Batteries often contain substances that are potentially hazardous to living things. It is important to dispose of them properly so that they do not contaminate the water supply or the soil. This means they should be taken to a recycling center, not put in the trash.

© Richard T. Nowitz/Corbis

## → Batteries

Electrochemical cells have been developed into batteries, which are basically containers of chemicals just waiting to be converted to electrical energy. The chemical reaction within a battery does not take place until the two reactions are connected to one another. When you flip the switch on a flashlight, you are connecting the halves of the battery inside. If you forget and leave the flashlight on, the chemical reaction inside will continue until the battery is dead.

Some batteries are rechargeable. This means that once the battery has died, an electric current can be run through it in the opposite direction to push the reverse reaction to occur. This restores the original reactants, and the battery is good once again.

A battery that is not connected to anything still has a voltage, even though it is not yet doing any work. Most of the batteries in your household can produce voltages from about 1.25 volts to 9 volts.

## LESSON SUMMARY

### How can you use a redox reaction as an energy source?

**KEY TERMS**

electrolyte

electrochemical cell

half-cell

half-reaction

voltage

You can convert the transfer of electrons that occurs during oxidation-reduction into electrical energy. By separating the reactants into half-cells, you can control and "store" electrical energy. The activity series allows you to choose electrodes and electrolytes that will produce electricity. The voltage of an electrochemical cell can be calculated from the half-cell potentials.

### Reading Questions

1. Explain how to determine the direction of electron transfer in an electrochemical cell.

2. Explain how you might choose which reactants to put together in an electrochemical cell.

### Reason and Apply

3. Suppose that you make an electrochemical cell. One beaker has a calcium metal electrode and a solution of aqueous calcium ions, $Ca^{2+}(aq)$. The other beaker has an iron metal electrode and a solution of aqueous iron ions, $Fe^{3+}(aq)$.
   a. Write the half-reactions for each beaker.
   b. Use the activity series table on page 570 to decide which metal is oxidized and which metal cation is reduced.
   c. Make a sketch of the electrochemical cell. Show the reaction that occurs in each beaker.
   d. Write a net ionic equation for the reaction that takes place to generate electricity.

4. Repeat Exercise 3 with the pairs of metals here.
   a. copper and tin
   b. silver and aluminum
   c. copper and lead

5. Predict which combination of metals in Exercises 3 and 4 will result in the greatest voltage. Explain your prediction. Then calculate the voltages of the redox reactions. Was your prediction correct?

6. What are the pros and cons of using combustion reactions as a source of energy?

7. What are the pros and cons of using electrochemistry as a source of energy?

8. PROJECT  Research common batteries, such as AA or D cell batteries, to find out the chemistry behind them. List the metals and compounds commonly used in their construction.

9. PROJECT  Car batteries are usually 12-volt. Research car batteries and describe how such a high voltage is achieved.

10. PROJECT When car lights are left on for a long time, this can cause a "dead" battery, which will require a jump-start from another car battery. In chemical terms, what does it mean to have a "dead" car battery? What does "jump-starting" a car battery mean in terms of the overall redox reaction? Research and report on your findings.

Peter Bennett/Science Source

# CHAPTER 21

## Controlling Energy

# SUMMARY

### KEY TERMS

**oxidation**

**heat of formation**

**reduction**

**oxidation-reduction reaction (redox)**

**activity series**

**electrolyte**

**electrochemical cell**

**half-cell**

**half-reaction**

**voltage**

# Fire Update

Every chemical change is accompanied by an energy transfer. The chemical energy from reactions between metals and metal compounds can be converted to electrical energy with the appropriate apparatus.

Oxidation-reduction reactions are a particularly useful source of energy. Oxidation is the loss of electrons by an atom, ion, or molecule during a chemical change. Its counterpart, reduction, is the gaining of electrons by an atom, ion, or molecule during a chemical change. Combustion is one type of oxidation-reduction reaction in which the fuel is oxidized and the oxygen is reduced.

Metals that are easily oxidized give up their electrons easily. These metals are considered more active.

## REVIEW EXERCISES

1. Explain how an electrochemical cell works to convert energy from chemical reactions into electrical energy.

2. Why are some metal elements easier to extract from metal ores than others?

3. Based upon the value in the heats of formation table on page 555, determine the heat of reaction for the decomposition of rust into iron and oxygen gas. Is this reaction endothermic or exothermic? Explain.

$$2Fe_2O_3 \longrightarrow 4Fe(s) + 3O_2(g)$$

4. If you place a piece of aluminum foil in a beaker of 1 M $AgNO_3$, would a single exchange reaction occur? What if you place a piece of silver foil in a beaker of 1 M $Al(NO_3)_3$? How can you explain these two different results?

5. An electrochemical cell is set up with 1 M $Pb(NO_3)_2$ on one side, and 1 M $AgNO_3$ on the other.
   a. Which metal will be oxidized? Which will be reduced? Write out balanced half-reactions for these reactions.
   b. Write out the balanced overall reaction for this electrochemical cell.
   c. How many moles of electrons will flow from the anode to the cathode?
   d. Calculate the voltage for this electrochemical cell.

---

## Alternative Energy

**PROJECT** Investigate an energy source that is an alternative to fossil fuels. Create a poster promoting the benefits of the energy source and explaining how it works. Write a short paper on the pros and cons of your energy source.

---

# CHAPTER 22

# Radiating Energy

## In this chapter, you will study

- the emission, reflection, transmission, and absorption of light

- how color arises

- the path of light as a ray, as a wave, and as a particle

- how to detect electromagnetic radiation

*Bees and humans do not see the patterns on flowers in the same way. The photograph of this flower has been altered to show the way a bee sees light. The tan areas help to lead the bee to the nectar at the flower's center.*

Fire produces bright light and transfers energy. When hot gases produced in a fire come into direct contact with objects at lower temperatures, some energy is transferred. In addition, fire and extremely hot objects transfer energy as electromagnetic radiation. Fire transfers energy efficiently as infrared radiation that makes you feel warm. Chapter 22 of Unit 5: Fire focuses on models to describe light energy and its interaction with matter.

*All Aglow*

# Light Energy

## THINK ABOUT IT

Light from the Sun travels a great distance through nearly empty space to reach Earth. When the Sun's light shines on objects, those objects can reflect light into your eyes, and you are able to see the objects. Although the light from the Sun is white, the light that is reflected from objects is often colorful. What is the path of the light such that you can see white, clear, and colored objects?

---

**How can you describe light shining on matter?**

---

*To answer this question, you will explore*

- Light Emission and Light Rays
- Reflection, Transmission, and Absorption

## EXPLORING THE TOPIC

## Light Emission and Light Rays

At a very young age, children learn to draw the Sun as a circle with lines radiating outward to indicate that the Sun is glowing. These lines of light represent *light rays,* and they show that light is traveling away from the Sun. Light rays provide a useful way of indicating the path that light travels and how light interacts with matter.

### LIGHT SOURCES

Light is emitted from the Sun because the Sun is extremely hot. Nuclear reactions transfer heat to raise the temperature of the surface of the Sun to nearly 6000 K. The Sun transfers energy to Earth by the **emission** of light.

Other hot objects, such as the thin wire inside a light bulb, an electric burner on a stove, and burning wood, transfer energy as light. Creating a light source requires transfer of energy to the source. Thermal energy, electricity, exothermic chemical reactions, and striking a substance to make sparks can all generate light.

*Extremely hot objects emit light.*

As you learned in previous lessons in this unit, hot substances transfer energy to reach thermal equilibrium. So far, you have considered conduction, which relies on a hot object being in direct contact with a colder object. **Radiation** refers to energy that travels outward from a hot substance even when the substance is isolated. Visible light is one form of radiation.

**Big Idea** One way that hot substances transfer energy is by the emission of light, called radiation.

### RAY MODEL OF LIGHT

Generally, it is not possible to see the path of light from a source such as the Sun. The path of light becomes visible as a glowing sunbeam only if there are dust particles or water droplets in the air. The path of light is commonly referred to as a **light ray.** Light rays do not describe the complexities of light, but they do provide a simple model with which to track the path of light as it shines from the source onto an object and then to your eye.

*Light from this flashlight produces a bright spot on a surface. Notice that while the light is visible at the flashlight, the path of light is invisible between the flashlight and the surface. The light becomes visible when it strikes the surface.*

### WHITE LIGHT IS A MIXTURE OF COLORS

In the late 1600s, Sir Isaac Newton showed that white light produces a rainbow of colors after passing through a piece of glass in the shape of a triangular prism. At the time, other hypotheses suggested that the prism might somehow distort the light or make it less pure. However, when Newton placed a second prism in the path of the rainbow of light created by the first prism, he observed a single beam of white light. This experiment convinced Newton that the colors are already present in the light.

*Sunbeams are visible through the trees when there is a mist of water in the air. Small particles show the path of the light.*

White light — First prism — Color spectrum — Second prism — White light

*When white light moves through a prism, it spreads out into colors in a sequence: red, orange, yellow, green, blue, and purple. A second prism can recombine the colors back into white light.*

# ⟳ Reflection, Transmission, and Absorption

In a room that is dark and sealed off so that no light leaks in, you see nothing. Turn on the lights, and objects in the room become visible. The once invisible objects are now illuminated. Moreover, many of them are colorful even under light that appears white at the source, referred to as **white light.** What are the ways that white light interacts with matter to cause these perceptions of object's visibility and color?

### OPAQUE OBJECTS REFLECT LIGHT

When light shines on an object, some or all of the light is reflected. **Reflection** of light refers to light that "bounces" off a surface. When light from one source reflects from a smooth surface, all the light rays are parallel to one another. When light reflects from a rough surface, the light rays go off in many different directions. It is typical that you can see a reflection on a smooth surface such as a mirror. You can also see a faint image of a reflection in a window. Rough surfaces do not create as clear an image because the light goes off in multiple directions.

**Important to Know** Only light sources emit, or give off, their own light. Most objects only reflect light and, therefore, are not visible in a pitch-black room.

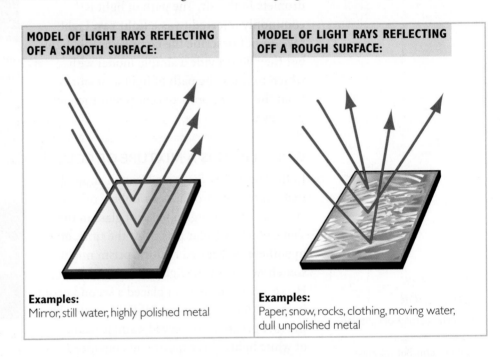

**MODEL OF LIGHT RAYS REFLECTING OFF A SMOOTH SURFACE:**

**Examples:**
Mirror, still water, highly polished metal

**MODEL OF LIGHT RAYS REFLECTING OFF A ROUGH SURFACE:**

**Examples:**
Paper, snow, rocks, clothing, moving water, dull unpolished metal

### CLEAR OBJECTS TRANSMIT LIGHT

Light passes through certain types of objects, such as glass, thin fabric, and thin sheets of plastic, by **transmission.** Objects that transmit light are **transparent.** Very little light is reflected during transmission. However, most objects do not transmit light, and are referred to as **opaque.** It is not possible to see through opaque objects.

mattjeacock/ Getty Images

*Some light moves through objects that are transparent, such as sunglasses.*

## COLORED OBJECTS PARTIALLY ABSORB LIGHT

Consider green leaves. You cannot see through leaves, so light is not transmitted. If our eyes detect a green color, then it must mean that rays of green light reflect from the leaves and enter our eyes. How is it possible that you detect green when the leaves are illuminated with white light?

For green leaves, colors other than green are taken in, or absorbed, by the leaves while green light rays are reflected. This process of taking in certain colors of light is called light **absorption.** The light that is absorbed transfers energy to the object.

*White light, which is a mixture of colors, shines on a leaf. The leaf reflects green light rays. The other colors are absorbed by the leaf.*

**Big Idea**  An object is colored because it reflects or transmits certain colors of light (these are the colors you see), while it absorbs other colors of light (these light rays transfer energy back to the object).

### Example

## Yellow Streetlamps and Car Colors

Imagine that you are walking down a street at night. The streetlamps emit yellow light.

- **a.** Explain why a blue car appears black.
- **b.** Explain why a yellow car appears yellow.

### Solution

- **a.** A blue car absorbs all colors of light except for blue light. The streetlamps emit yellow light, which is absorbed by the blue car. So, no light is reflected into your eyes, and the car appears black.

- **b.** A yellow car absorbs all colors of light except for yellow light. The streetlamps emit yellow light, which is reflected by the yellow car. So, the car appears yellow as it would during the day in sunlight.

**KEY TERMS**

emission

radiation

light ray

white light

reflection

transmission

transparent

opaque

absorption

## LESSON SUMMARY

### How can you describe light shining on matter?

Light is a form of energy emitted from sources such as the Sun and light bulbs. The light travels through space undetected until it strikes an object. The object can reflect, transmit, or absorb the light. Your eye detects reflected and transmitted light that travels toward you. A colored object reflects or transmits the color or colors you perceive, while it absorbs the other colors. Black objects absorb all colors. The energy from light that is absorbed by an object is transferred to the object.

## Exercises

### Reading Questions

1. How could you provide evidence that light is a form of energy?
2. Provide an explanation as to why we see colors.

### Reason and Apply

3. What color will you see if you shine white light through a piece of thin blue plastic?
4. What color will you see if you shine blue light through a piece of thin blue plastic?
5. What color will you see if you shine white light from a flashlight onto a green car at night?
6. What color will you see if yellow light from a streetlamp shines on a red car at night?
7. Imagine that you are wearing a pair of orange-tinted sunglasses.
   a. What color will you see when you look at an orange object? Explain your thinking.
   b. What color will you see when you look at a blue object? Explain your thinking.
8. Explain how it is possible to see a sunbeam in a room that has dust in the air. Draw a ray model to show how the light path travels from its source, to the dust, and then to your eye.

# Light Waves

## THINK ABOUT IT

Light from the Sun travels a great distance through nearly empty space to reach Earth. It is hard to imagine how the light gets from one place to another. How does it travel and with what speed?

### How is a wave model used to describe light?

*To answer this question, you will explore*

➲ The Wave Model of Light

➲ Wavelength, Frequency, and Speed

## EXPLORING THE TOPIC

### ➲ The Wave Model of Light

The ray model of light indicates the light path, but it does not reveal much about what light is and how it travels through space. The Greek philosopher Aristotle, born in 384 B.C.E., thought of light as wavelike in nature, traveling through space similar to the way ripples spread across a pond's surface. Even though this view has undergone modification and elaboration through the centuries, it remains a useful model.

### WAVES

Waves are created by a disturbance. For example, throw a pebble in a pond, and water waves spread out in circles of up-and-down motion. Or pluck a guitar string to make the entire string move back and forth.

Throw a pebble in a pond.

Pluck a guitar string.

You can also place a piece of rope on the floor and move one end back and forth while another person holds the other end fixed to create a single pulse of back-and-forth motion. This pulse travels along the rope. If you move one end back and forth in a continuous motion, the entire rope will move with a wave motion. Certain parts of the rope move left, while other parts move right.

*The pulse that you create along the rope models the movement of a wave. As seen here from above, the rope moves first to the left and then to the right.*

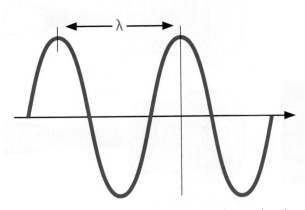

*The wave shown is traveling from left to right. The wavelength, λ, is the distance between two peaks. The frequency, f, is the number of full waves (from peak to peak) that pass by the dashed line in a given amount of time.*

Notice that when a disturbance is created on one end of the rope, a person watching from the other end can detect the disturbance once the wave moves along the rope. Even from a large distance, it is possible to detect a disturbance as a wave reaches you and transfers its energy.

**Light waves** are thought to be ripples in space created by forces that oscillate up and down at a source, such as the Sun. The oscillations move as a wave over extremely long distances to Earth.

## WAVE MEASUREMENTS

Two important measurements of waves are wavelength, $\lambda$, and frequency, $f$. The **wavelength** is the distance between two peaks or between two troughs of the wave measured in meters, m. The **frequency** is the measure of the number of waves that pass by a certain point in space per second, measured in hertz, Hz.

The wave below has different dimensions. Notice that the wavelength is shorter compared with the wave above. It also has more up-and-down motion. If you pick a place along the line and count waves that pass by, you will notice more waves per second for the wave with the shorter wavelength.

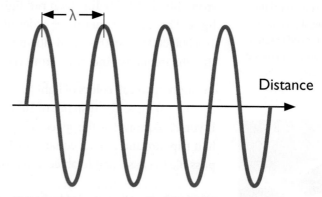

*This wave has a shorter wavelength, λ, and higher frequency, f.*

Notice that as the wavelength decreases, the frequency increases. And as the wavelength increases, the frequency decreases.

## ⤷ Wavelength, Frequency, and Speed

The bar provides the sequence of colors of the rainbow from red to violet. There are names for many of the colors, but not all of them. The six main categories are red, orange, yellow, green, blue, and violet. We tend to use labels such as red-orange and blue-green to refer to the colors in between the six main categories.

*Lowest frequency*                                                            *Highest frequency*

*Longest wavelength*                                                        *Shortest wavelength*

The wave model of light is useful in considering the differences and similarities of light of different colors. What are the dimensions of the light waves of various colors?

### WAVES OF COLORED LIGHT

The light waves of different colors have characteristic wavelengths and frequencies. The table below gives the wavelength, frequency, and **speed of light** in a vacuum for the waves of five different colors. Take a moment to notice the trends.

### Wavelengths and Frequencies of Colored Light

| Color of light | Wavelength, $\lambda$ | Frequency, $f$ | Speed of light, $c$ |
|----------------|-----------------------|----------------|---------------------|
| deep red | $7.5 \times 10^{-7}$ m | $4.0 \times 10^{14}$ Hz | $3 \times 10^8$ m/s |
| red-orange | $6.5 \times 10^{-7}$ m | $4.6 \times 10^{14}$ Hz | $3 \times 10^8$ m/s |
| yellow-green | $5.5 \times 10^{-7}$ m | $5.5 \times 10^{14}$ Hz | $3 \times 10^8$ m/s |
| blue-violet | $4.5 \times 10^{-7}$ m | $6.7 \times 10^{14}$ Hz | $3 \times 10^8$ m/s |
| deep violet | $3.9 \times 10^{-7}$ m | $7.7 \times 10^{14}$ Hz | $3 \times 10^8$ m/s |

PHYSICS
CONNECTION

Unlike light, sound does not travel through empty space—only through gases, liquids, and solids. This is because sound causes wavelike motion of the atoms and molecules in these substances. The speed of sound waves is much slower than light waves, and vehicles, like this aircraft, can reach the speed of sound.

Stocktrek Images/Getty Images

The wavelengths of visible light measured in meters, m, range from about $7.5 \times 10^{-7}$ m for deep red light to $3.9 \times 10^{-7}$ m for deep violet light. This dimension on the order of $10^{-7}$ m, or 0.0000001 m, is about the size of a large virus, one of the smallest things that can be seen using an optical microscope. Notice that the wavelength decreases from red to violet.

The frequencies, measured in inverse seconds, 1/s or $s^{-1}$ or hertz, Hz, range from $4.0 \times 10^{14}$ Hz for deep red light to $7.7 \times 10^{14}$ Hz for deep violet light. This dimension on the order of $10^{14}$ Hz is huge. It is 100 trillion or 100,000,000,000,000 up-and-down motions, or oscillations, per second. Notice that the frequency increases from red to violet.

In empty space, all colors of light travel with the same speed, called the speed of light, $c = 3 \times 10^8$ m/s. The distance of $10^8$ m is 100 million meters, or 100,000,000 meters. If you could travel at the speed of light, you would be able to make seven-and-a-half trips around Earth's equator in 1 second. This speed is extremely fast. It is not possible to cause visible objects to move at this speed!

**Big Idea** All colors of light travel with the same speed in a vacuum, called the speed of light.

A relationship exists among the wavelength, frequency, and speed of light waves:

$$\lambda \times f = c = 3 \times 10^8 \text{ m/s}$$

If you know the wavelength, you can calculate the frequency of colored light. And if you know the frequency, you can calculate the wavelength of colored light. The speed of light is the proportionality constant: frequency and wavelength are inversely proportional.

$$\text{Wavelength} = \lambda = \frac{c}{f} \qquad \text{Frequency} = f = \frac{c}{\lambda}$$

**Example**

## Wavelength of Colored Light

Imagine a light wave with the frequency of $7.5 \times 10^{14}$ Hz.

**a.** What is the wavelength of this light wave?

**b.** Use the table on page 583 to determine the approximate color of the light.

### Solution

**a.** The wavelength and frequency of colored light are related by the equation:

$$\lambda \times f = c$$

Rearrange the equation to solve for wavelength:

$$\text{Wavelength} = \lambda = \frac{c}{f}$$

Substitute values for the frequency and speed of light:

$$\text{Wavelength} = \lambda = \frac{3 \times 10^8 \text{ m}}{7.5 \times 10^{14} \text{ Hz}} = 4.0 \times 10^{-7} \text{ m}$$

**b.** The wavelength and frequency indicate that the color is at the violet end of the rainbow.

Green    Blue

| | | | |
1    2    3    4

## ANTHROPOLOGY CONNECTION

Does the language you speak influence your perception of color? Cognitive scientists have evidence that if you have a word for a specific color, you can pick that color out from among others. For example, some cultures do not distinguish between blue and green. What do you call these four colors, green or blue?

## LESSON SUMMARY

### How is a wave model used to describe light?

It is useful to characterize different colors of light with a wave model. This model treats light as a wave that travels through space. Light waves have a wavelength, $\lambda$ (pronounced *lambda*), which is the distance between the peaks of the waves. The number of waves that pass by per second is called the frequency, $f$. Wavelength is typically reported in meters, m, and frequency is reported in hertz, $\text{Hz} = 1/\text{s}$. Different colors of light have different wavelengths and frequencies. However, all colors of light travel at the same speed, called the speed of light, $c$. A relationship exists among the wavelength, frequency, and speed of light waves: $\lambda \times f = c$, where $c = 3 \times 10^8$ m/s. Assigning a wavelength to different colors of light provides a precise continuous scale for all the different colors of the rainbow.

**KEY TERMS**

light wave

wavelength

frequency

speed of light

## Exercises

### Reading Questions

1. Explain how to determine the wavelength, $\lambda$, of a wave. What would you need to know?

2. Explain how to determine the frequency, $f$, of a wave. What would you need to know?

### Reason and Apply

3. Draw sketches of two waves to show how green light and blue light differ from one another.
   a. What are the approximate wavelengths of these two waves?
   b. What are the approximate frequencies of these two waves?
   c. How fast are these waves traveling?

4. Imagine a wave of light with the frequency of $4.1 \times 10^{14}$ Hz.
   a. What is the wavelength of this light wave?
   b. Use the table on page 583 to determine the approximate color of the light.

5. Imagine a wave of light with the wavelength of $5.1 \times 10^{-7}$ m.
   a. What is the frequency of this light wave?
   b. Use the table on page 583 to determine the approximate color of the light.

6. Why might it be useful to describe colors in terms of wavelength rather than simply using words?

*Beyond What You See*

# Electromagnetic Radiation

## THINK ABOUT IT

On a dark night, it is very difficult to find your way without bumping into objects that you cannot see. Put on a pair of night vision goggles and you can detect objects fairly easily. What do the goggles detect so that you can see at night?

---

**What is the evidence for light waves that are invisible to your eyes?**

---

*To answer this question, you will explore*

⟳ Electromagnetic Radiation

⟳ Light Energy

Science Photo Library

© David J. Green—technology/Alamy

*Night vision goggles detect light that you cannot see. They convert it into light that you can see so that you can locate objects in the dark.*

### BIOLOGY
### CONNECTION

Unlike human beings, bees can see ultraviolet light. Bee vision is shifted toward the blue range, and they do not see red. Flowers have adapted to guide bees toward pollen and nectar. The center of this creeping zinnia flower is an unmistakable target for a bee, which sees two-toned petals that humans do not see. As part of its reproduction process, the zinnia advertises that it has pollen for bees to collect for the hive and deposit in other flowers.

© Dr. K. Schmitt, Weinheim Germany uvir.eu

## EXPLORING THE TOPIC

## ⟳ Electromagnetic Radiation

The human eye is able to detect a range of light waves with wavelengths from about $7.5 \times 10^{-7}$ m for red light to $3.9 \times 10^{-7}$ m for violet light. You might wonder if these are the only light waves that exist or if there are light waves with longer and shorter wavelengths outside your range of vision. The example of night vision goggles indicates that there is light even in the dark. However, human eyes cannot detect this type of light without the assistance of technology.

### A LARGE COLLECTION OF WAVES

The light that we see is a small part of a large collection of waves called **electromagnetic radiation.** As indicated in the figure, electromagnetic radiation exists with a huge range of wavelengths and frequencies. Its frequency ranges from a low frequency of less than $10^5$ Hz to a high frequency of greater than $10^{20}$ Hz. Its wavelength ranges from longer than $10^3$ m to shorter than $10^{-12}$ m. These wavelengths have dimensions that are roughly equivalent to mountains for the longest wavelengths and to smaller than atoms for the shortest wavelengths.

All of these electromagnetic waves have one important thing in common: They all travel through space with the same speed, the speed of light, $c = 3 \times 10^8$ m/s. In all cases, the frequency times the wavelength is equal to the speed of light.

$$f \times \lambda = c = 3 \times 10^8 \text{ m/s}$$

Only waves that are classified as electromagnetic radiation travel at the speed of light.

> **Big Idea** Electromagnetic waves all travel with the same speed in a vacuum, the speed of light.

The figure shows the range of wavelengths and frequencies that are observed for electromagnetic waves.

## NAMES FOR ELECTROMAGNETIC RADIATION

The various wavelength ranges are classified with names, some of which are familiar because they have entered into everyday language. Here is a description of the names for the various wavelength ranges from the longest wavelengths to the shortest wavelengths. Notice that the words *ray, wave,* and *radiation* are used in the names.

**Radio waves:** Radio waves are emitted at a radio or television station. Radio waves are all around you all the time, but your body is unable to detect them. A device such as a radio or television converts radio waves into sound and picture. Wireless routers and wireless pointers also use radio waves.

**Microwave radiation:** Microwaves are emitted by the power source in a microwave oven. Water molecules absorb microwaves. This causes the water molecules to move. The motion raises the temperature of the water contained in food, which in turn causes the temperature of the food to increase.

**Infrared radiation:** Your body senses infrared radiation as heat. For example, if you stand in direct sunlight, you will feel hotter than if you stand in the shade. In both cases, the air temperature is the same. The difference is that in direct sunlight, you are exposed to infrared radiation in addition to visible light. Infrared, or IR, radiation causes molecules to vibrate. This vibration motion also causes the temperature to increase. You may be familiar with heat lamps, such as those used to keep pet reptiles warm. These lamps use IR radiation.

**Visible light:** Your eyes detect visible light as various colors. The colors of the rainbow are in the sequence of the shortest wavelength closest to Earth (violet) to the longest wavelength at the top of the arc (red).

**Ultraviolet radiation:** If you are out in the sun too long, your skin will turn red with sunburn. This does not happen with room lights, indicating that sunlight has additional waves with wavelengths outside the visible region. The waves that cause sunburn are called ultraviolet, or UV, radiation. Sunburn is damage to your skin. The UV rays cause molecules in your skin to break apart. Sunscreens provide protection by blocking the UV rays from getting to your skin.

**X-rays:** If you have ever broken a bone, then you have probably had an x-ray photograph taken. X-rays are dangerous, so the x-ray technician keeps the exposure time short. The x-rays fly through most substances, but they are blocked by bones and teeth. By detecting the x-rays that come through your body, it is possible to capture a photograph of your bones to locate a break or a photo of your teeth to locate cavities. X-ray technicians and dentists place a lead apron on you to block x-rays from reaching parts of the body that are not under examination.

**Gamma rays:** Gamma rays have the highest frequencies and shortest wavelengths, and they are extremely dangerous. Gamma rays are emitted during a nuclear explosion, in which the nuclei of atoms are falling apart.

## ⟳ Light Energy

Consider a fire caused by a chemical reaction between wood and oxygen to produce carbon dioxide and water. The reaction is exothermic, and it transfers a lot of heat to the surroundings. How does this transfer of heat occur?

So far in this unit, you have considered energy transferred by direct contact between two objects at different temperatures. Energy is transferred by conduction until the two objects are at the same temperature.

Kevin Schafer/Getty Images

Energy is also transferred when light interacts with matter. This is referred to as transfer of energy by radiation. IR radiation is especially efficient at transferring energy to make you feel warm. Fire emits a lot of IR radiation in addition to emitting visible light. Again, while you cannot see IR radiation, your body does detect it as heat.

**Big Idea** Electromagnetic radiation is one way that energy is transferred from a source to an object.

### RELATIONSHIP BETWEEN FREQUENCY AND ENERGY

The table shows examples of values for the wavelength, frequency, and energy of various types of electromagnetic radiation. The energy is measured in joules, J. One joule is roughly equivalent to 0.24 calories. This is enough energy to raise the temperature of 1 mL of water by 0.24 degrees.

Take a moment to notice the relationships among wavelength, frequency, and energy.

**Sample Values for Types of Radiation**

| Type of EM radiation | Wavelength, λ | Frequency, $f$ | Energy, $E$ | Size of Wavelength (approx.) |
|---|---|---|---|---|
| gamma rays | $3 \times 10^{-12}$ m | $10^{20}$ Hz | $6.63 \times 10^{-14}$ J | (nucleus) |
| x-rays | $3 \times 10^{-10}$ m | $10^{18}$ Hz | $6.63 \times 10^{-16}$ J | (atom) |
| ultraviolet | $3 \times 10^{-8}$ m | $10^{16}$ Hz | $6.63 \times 10^{-18}$ J | (molecule) |
| visible light | $6 \times 10^{-7}$ m | $5 \times 10^{14}$ Hz | $3.32 \times 10^{-19}$ J | (microorganism) |
| infrared | $3 \times 10^{-5}$ m | $10^{13}$ Hz | $6.63 \times 10^{-21}$ J | (needle tip) |
| microwaves | $3 \times 10^{-2}$ m | $10^{10}$ Hz | $6.63 \times 10^{-24}$ J | (fly) |
| radio waves | $3 \times 10^{3}$ m | $10^{5}$ Hz | $6.63 \times 10^{-29}$ J | (mountain) |

The energy of light increases as the frequency increases and as the wavelength decreases. One way to remember this is by considering that to make a high-frequency wave with the rope, you made a rapid back-and-forth hand motion. It definitely required a greater input of energy to generate a higher frequency.

### PLANCK'S CONSTANT

**Planck's constant** relates the frequency of the electromagnetic wave with its energy. It is given the symbol, $h$, defined as $h = 6.63 \times 10^{-34}$ J · s. The energy of the electromagnetic wave is calculated with the formula given below. For a frequency in hertz ($s^{-1}$), the energy transferred by each wave (one up-and-down motion), is given in joules.

$$E = h \times f$$

Frequency and wavelength are related by the equation $c = f \times \lambda$. Rearranging the equation gives

$$f = c/\lambda.$$

Substituting $c/\lambda$ for the frequency in the equation $E = h \times f$ gives the relationship between energy and wavelength.

$$E = \frac{hc}{\lambda}$$

While each wave (one up-and-down motion) only transfers a tiny amount of energy, keep in mind that the waves are coming at a high frequency and with a high speed. So, the total energy transferred can be quite large.

## PHOTON MODEL OF LIGHT

So far, you have considered a ray model and a wave model to describe light. A **photon model** is a third model to describe light. It is useful to think about one up-and-down motion of the wave as a "particle," or photon of light. Instead of imagining light as a continuous light ray, the photon model suggests that a beam of light is composed of small particles.

According to the photon model, photons associated with waves of higher frequency transfer a greater amount of energy per photon. An x-ray photon transfers more energy than an infrared photon because the x-ray has a higher frequency compared with IR radiation.

Further, the photon model helps to explain the brightness of a beam of light. If the light is brighter, there should be a greater number of photons in a given area in one second. Light that is dimmer has fewer photons in the same area in one second.

Light reflected as a strong beam.

Light as a weak beam.

There are two ways of transferring more total energy in a given amount of time: (1) make the light brighter so that there are more photons, or (2) choose light of a higher frequency such that each photon transfers more energy. For example, bright red light has a greater number of photons compared with dim red light. Each photon transfers the same amount of energy, but there are more photons in the bright light. In comparison, even dim x-rays transfer a lot of energy because each photon transfers a much larger amount of energy than a photon of red light. The frequency of an x-ray is $10^{18}$ Hz, whereas the frequency of red light is only $10^{14}$ Hz. Each x-ray photon transfers about 10,000 times more energy compared with a photon of red light. This is why x-rays are dangerous.

The wave model and the photon model are both useful. Scientists refer to the *wave-particle duality* of light, but this simply means that you can choose either model. Observations of radio waves indicate that they behave most like waves. In contrast, observations of the behavior of x-rays indicate that they are more particle-like.

**Important to Know** Wave-particle duality simply indicates that both a wave model and a particle model describe electromagnetic radiation.

---

**Example**

### Energy of Colored Light

Imagine a light wave with the frequency of $10^{17}$ Hz.

  **a.** What is the energy of this electromagnetic wave?

  **b.** Use the figure on page 587 and the table on page 589 to determine the type of electromagnetic radiation.

#### Solution

  **a.** The energy and frequency of electromagnetic radiation are related by the equation:

$$E = h \times f, \text{ where } h = 6.63 \times 10^{-34} \text{ J} \cdot \text{s}$$

Substitute values for the Planck's constant, $h$, and the frequency of light:

$$\text{Energy} = E = (6.63 \times 10^{34} \text{ J} \cdot \text{s})(10^{17} \text{ Hz}) = 6.63 \times 10^{-17} \text{ J}$$

Notice that seconds cancel because Hz = 1/s, giving units of joules.

  **b.** The energy and frequency indicate that the electromagnetic radiation is x-rays.

---

### BIOLOGY
### CONNECTION

The human eye detects colored light by using three receptors that have maximum frequencies for red, green, and blue. When light strikes these receptors at the back of the eye, a signal is sent along nerves to the brain. The brain then takes the information from the three receptors to create the perception of various colors. (And unlike bees, human beings can see red.)

Human cone sensitivities

## What is the evidence for light waves that are invisible to your eyes?

**KEY TERMS**

electromagnetic radiation

radio wave

microwave radiation

infrared radiation

visible light

ultraviolet radiation

x-ray

gamma ray

Planck's constant

photon model

The light that we see is a tiny part of a large collection of waves called *electromagnetic radiation*. Electromagnetic radiation exists with a huge range of wavelengths and frequencies. The wavelength regions include radio waves, microwave radiation, infrared radiation, visible light, ultraviolet radiation, x-rays, and gamma rays. The entire collection of waves travels at the same speed, the speed of light, $c = 3 \times 10^8$ m/s. The energy transferred by electromagnetic radiation is proportional to the frequency of the wave. Electromagnetic waves of high frequency transfer more energy than waves of low frequency. High-frequency electromagnetic waves such as gamma rays and x-rays are quite dangerous because of the large amount of energy they transfer, even with a dim beam.

## Exercises

### Reading Questions

1. Summarize what you learned about electromagnetic radiation.

2. Describe the ray model, the wave model, and the photon model of light.

3. Describe the relationship between energy and frequency.

### Reason and Apply

4. What type of radiation has a wavelength of $6.12 \times 10^2$ m?

5. Identify an object that is similar in size to the wavelength of a microwave.

6. A radio station broadcasts at a frequency of 790 MHz ($7.9 \times 10^8$ s$^{-1}$). What is the wavelength of the electromagnetic radiation generated by this radio station?

7. The gamma rays emitted by $^{60}$Co are used in radiation treatment for cancer. The frequency of these gamma rays is $3 \times 10^{20}$ s$^1$. What is the energy of this radiation?

## LESSON 116

# Spectroscopy

### THINK ABOUT IT

While enjoying a day at the beach, you notice a person whose skin has turned quite red. In contrast, your skin looks the same as it did in the morning when you left your house. This is because you have applied sunscreen several times during the day. The claim on the package is that sunscreen protects against harmful UV radiation. Although you cannot see UV radiation, how is it possible to determine which substances block UV radiation effectively?

**How is it possible to measure electromagnetic radiation that you cannot see?**

*To answer this question, you will explore*

⮕ Light Detectors

⮕ Spectroscopy: Studying Matter with Light

### EXPLORING THE TOPIC

## ⮕ Light Detectors

A **light detector** is a device that can sense electromagnetic radiation. The detector provides some type of signal when it comes into contact with electromagnetic radiation. Some detectors, such as solar cells, produce an electrical signal that converts visible light into electricity. Radio antennas detect radio waves. The radio then converts the signal detected by the antenna to sound. Your skin signals exposure to UV radiation by turning red. In this lesson, you will explore what happens when paper coated with certain chemical mixtures responds to electromagnetic radiation by changing color.

Solar cells detect visible light.

The radio detects radio waves.

Your skin detects UV radiation.

## UV-SENSITIVE PAPER

For more than 150 years, people have been creating images and photos by using light-sensitive paper and film to capture light. One way to create an image this way is to place an object that blocks the incoming electromagnetic radiation onto the light-sensitive paper. The electromagnetic radiation only strikes the paper where it is not blocked by the object. This creates a silhouette of the object.

*The leaves block the UV radiation from the Sun, resulting in a white image. The background is blue due to a chemical reaction caused by the UV radiation.*

Some UV-sensitive paper is coated with a mixture of compounds. Before exposure to UV radiation, the mixture of compounds is pale green. This color indicates that one of the compounds in the mixture absorbs all colors except for green. Green light is reflected.

The green reactant compounds are soluble in water. If the paper is submerged in water, the green compounds dissolve and wash away. With the green reactants removed, the paper is no longer UV-sensitive. If you expose it to UV radiation, it will not turn blue. Instead, it will remain white.

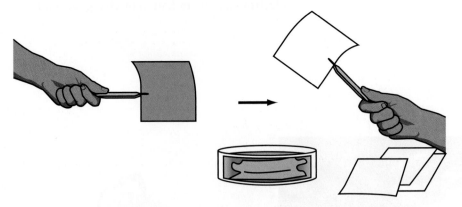

*The green coating on the UV-sensitive paper can be washed away in water to make the paper white.*

If you shine UV radiation on the green UV-sensitive paper, the energy that is transferred provides the energy needed to cause a chemical reaction to occur. The reactant that causes the green color changes into a new compound that is blue. The paper is blue because the new compound reflects blue light.

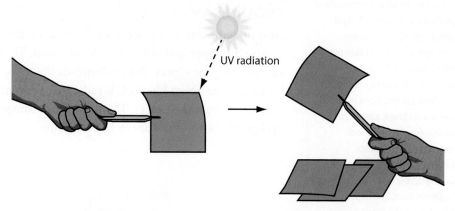

mixture of reactant compounds ⟶ mixture of product compounds

*Shining UV radiation on the UV-sensitive paper causes a chemical reaction that turns the paper from green to blue.*

The longer the paper is exposed to UV radiation, the deeper the shade of blue it turns. The reaction will continue until all the reactants are used up and the paper has the deepest blue color. The amount of the green compound is a limiting reactant. So one way to stop the reaction is to expose the paper to UV radiation until all limiting reactant is used up in the reaction.

*Longer exposure to UV radiation causes the blue color to be a deeper blue.*

Another way you could stop the reaction is related to the difference in solubility between the green reactant compound and blue product compound. The green reactant compound is soluble in water, but the blue product compound is insoluble in water. If you expose the paper to UV radiation for a very short time, only some of the reactant compound will turn blue. The paper will appear to be a pale blue color and is a mixture of green reactant compound that has not yet reacted and blue product compound. If you then submerge the partially reacted paper in water, the green reactant compound will dissolve in the water and can be washed away. The blue product compound remains attached to the paper because it does not dissolve in the water. Now, even if you were to expose the partially reacted and washed paper to UV radiation again, it would no longer change to a deeper blue. It would remain the same pale blue color because there would no longer be any green reactant compound available to change to blue product compound.

### CREATING AN IMAGE

Now imagine that you place a small object on top of the UV-sensitive paper and expose the paper and object to UV radiation. The paper turns blue in the areas surrounding the object. But underneath the object, the paper remains green. This is because the object blocks UV radiation, preventing it from shining on the paper directly beneath the object. There is a green image of the object on a blue background after exposure to UV radiation.

Shine light on paper and object.

Remove object from paper.

Image remains after washing in water.

If the paper with the green image is exposed to light, the green compounds will react and the paper will be completely blue. However, if you protect the paper with the green image from UV radiation and then wash away the green compounds, a white image of the object will remain permanently.

NASA images courtesy of Ozone Watch

*Photographers are able to make beautiful, complex images with UV-sensitive paper.*

## ➜ Spectroscopy: Studying Matter with Light

Scientists make use of light detectors to create instruments called **spectrometers** to study properties of matter. A spectrometer has four main components: a source of electromagnetic radiation, a device, such as a prism, to separate the wavelengths of light, a sample holder, and a light detector. By using electromagnetic radiation of different wavelengths, it is possible to gain a lot of information about atoms, molecules, and compounds. It is also possible to learn about the concentration of a substance and its identity depending on the choice of electromagnetic radiation. These methods for analyzing compounds are referred to as **spectroscopy.**

**ENVIRONMENTAL CONNECTION**

Ozone in the upper atmosphere protects us against UV light, which can cause serious damage to eyes and skin over time. Thinning of the ozone layer over Antarctica is caused by chemical reactions in which pollutants break apart ozone molecules. International cooperation to control pollutants has led to less stress on the ozone layer and gradual improvement. The expanse of blue here is the "ozone hole":

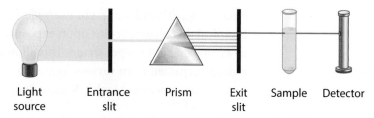

*Schematic of a basic spectrometer: light, sample, prism and detector*

### STUDYING WHAT YOU CANNOT SEE

Some substances, such as sunscreen, clothing, glass, and plastic, absorb UV radiation. Other substances, such as the atmosphere, transmit some UV radiation. You can use UV-sensitive paper as a simple UV detector to see how a substance behaves when exposed to UV light. Although it's not a spectrometer, this tool is similar to one in some ways. It allows you to study what you can't see. The light source for this tool is UV radiation, the sample is the substance you test, and the detector is the UV-sensitive paper.

First, you have to place your sample substance on a piece of UV-sensitive paper. It is best to spread the sunscreen on a thin piece of glass or plastic. Next, expose the sample on the paper to a source of UV radiation. If the substance absorbs UV radiation, the paper under the substance will remain green when exposed to the UV radiation source. If the substance transmits UV radiation, the paper under the substance will turn blue when exposed to the UV light source.

You can also use this setup to determine how different amounts of a substance block or transmit UV radiation. To do this, you need to examine samples with different concentrations of the substance being tested. For example, you might test sunscreens with the same active ingredient but different sun protection factor (SPF) ratings to see which ones block the most UV radiation. The SPF rating indicates the ability of the sunscreen to block UV radiation. Sunscreens with higher SPF ratings are more concentrated in the substance providing protection.

You could also use this tool to examine various light sources to compare the amount of UV radiation they emit. For example, the Sun emits a lot of UV radiation. In contrast, indoor lighting emits much less UV radiation. This is why you can get a sunburn when you are outside and exposed to the Sun but do not get a sunburn from indoor light sources.

> **Big Idea** Spectroscopy is a method of using electromagnetic radiation to study properties of atoms, molecules, and compounds.

## LESSON SUMMARY

### How is it possible to measure electromagnetic radiation that you cannot see?

Various sensors are used to detect electromagnetic radiation. Human eyes detect visible light as does the detector in a digital camera. It is also possible to use certain chemicals as light detectors that respond to different wavelengths of

### MEDICAL CONNECTION

Spectroscopy can be used to identify substances, such as medicines or illegal drugs that a person has ingested. Infrared spectroscopy produces a reading that shows the molecules in an unknown substance. The absorption of infrared light indicates which molecules are present and in what proportions. Shown here is the infrared spectroscopy "fingerprint" of the chemical methamphetamine.

electromagnetic radiation. For example, UV-sensitive paper is coated with a chemical that changes color upon exposure to UV light. Longer exposures result in a color that is a deeper shade. This allows you to examine the amount of UV radiation emitted from various light sources, and to determine the ability of various substances to block UV radiation.

## Exercises

### Reading Questions

1. Why does UV radiation change the color of UV-sensitive paper?
2. Make a sketch of a spectrometer and label the main parts.

### Homework

3. What happens if you leave a piece of UV-sensitive paper out in the Sun?
4. List three materials that block UV radiation.
5. What happens if you leave a piece of UV-sensitive paper out in a room with a dim red light?
6. Imagine that you want to measure two red solutions to determine if they have the same concentration of red dye or not.
   a. What type of light source do you need for the spectrometer?
   b. Sketch the placement of the light source, the detector, and the sample. Show the path of the light.
   c. What do you predict you will observe if the two solutions are different?
   d. Can you use UV-sensitive paper for the detector? Why or why not?
7. What differences do you predict if you use UV-sensitive paper to test sunglasses with UV protection compared with regular reading glass without UV protection?
8. Write a lab report for the experiment that you designed in class. In your report, include the title of the experiment, purpose, procedure, observations and conclusions.

(Title)

Purpose: (Explain what you were trying to find out.)

Procedure: (List the steps you followed.)

Observations: (Describe what you observed during the experiment.)

Conclusions: (What can you conclude about what you were trying to find out? Provide evidence for your conclusions.)

## CHAPTER 22

# Radiating Energy

# SUMMARY

### KEY TERMS

emission

radiation

light ray

white light

reflection

transmission

transparent

opaque

absorption

light wave

wavelength

frequency

speed of light

electromagnetic radiation

radio wave

microwave radiation

infrared radiation

visible light

ultraviolet radiation

x-ray

gamma ray

Planck's constant

photon model

light detector

spectrometer

spectroscopy

# Fire Update

The hot gases emitted from the reaction of wood and oxygen in a fireplace transfer thermal energy by conduction when they collide with your body. Upon collision, the faster-moving gas molecules slow down, the atoms moving more slowly in your body speed up, and you sense warmth. Conduction is one way that a fire transfers thermal energy, but this is not the only way that energy is transferred.

You feel the warmth of the Sun even though its light travels great distances through nearly empty space to reach Earth. This occurs by the emission of electromagnetic radiation. We perceive energy transfer from the Sun by radiation as visible light with our eyes, as the warmth due to infrared radiation, and, sometimes, as the sunburn caused by ultraviolet radiation. The color of a fire's flames and much of its warmth are due to energy transfer by the radiation emitted by the hot gas molecules produced by the combustion reaction.

There are a number of models that lend understanding to the nature of light. The ray model suggests that light travels in straight lines until it is reflected, transmitted, or absorbed when it strikes an object. If white light is partially absorbed, the reflected or transmitted light has a color. The wave model is used to explain that different colors are waves all traveling at the speed of light but of varying wavelength and frequency. The photon model suggests that energy is transferred in small units, called *photons*. Photon energy increases with frequency.

Electromagnetic radiation exists with a huge range of wavelengths and frequencies. The wavelength regions include radio waves, microwave radiation, infrared radiation, visible light, ultraviolet radiation, x-rays, and gamma rays. Instruments, called spectrometers, use various wavelengths of electromagnetic radiation to probe matter. In essence, spectroscopy extends our ability to see light that we would otherwise not be able to detect with only our eyes.

## REVIEW EXERCISES

1. Why are some objects colored when illuminated with white light?

2. Explain why the dentist covers you with a lead apron before taking an x-ray image of your teeth.

3. What is the wavelength of a light wave with the frequency $1.2 \times 10^6$ Hz?

4. Imagine that you want to measure two green solutions to determine if they have the same concentration of green dye or not.
   a. What type of light source do you need for the spectrometer?
   b. What do you predict you will observe if the two solutions are different?

5. What is the frequency of a light wave with energy $2.7 \times 10^{-19}$ J?

# Wavelength and Energy

- How would you answer the question: "Are black and white colors?" Debate the two questions posed below.

    1. A light bulb manufacturer might claim: "Black *is not* a color, but white *is* a color." Is this correct or not? Explain your thinking.

    2. An artist with paints might claim: "Black *is* a color, but white *is not* a color." Is this correct or not? Explain your thinking.

- Research the type of light emitted by three different light sources. Take digital images of white objects and two colored objects illuminated by these three light sources. Write a report explaining what you learned about light sources. Include a description of the experiment you carried out and an explanation of the results.

Vadym Volodin/iStockphoto

UNIT
**5**

Phil McDonald/Shutterstock

Combustion reactions are extremely useful in our daily lives. Most cars, planes, and trains move by the energy transferred from a combustion reaction. Even the digestion of our food can be considered a very controlled combustion reaction.

Calorimetry is one technique used to figure out how much energy is transferred from combustion reactions. It is possible to measure the energy transferred by measuring the temperature change of a known quantity of water. The heat required to change the temperature of 1 g of water by 1 degree, called the heat capacity of water, is needed to convert between the measured temperature change of the water and energy.

Different substances respond differently to heating. Substances that change temperature easily, with very little energy transfer, have low specific heat capacities. It takes more energy to change the temperature of substances with higher specific heat capacities.

Once the data from calorimetry experiments are collected, the data can be used to compare and classify reactions. Exothermic reactions transfer energy *to* the surroundings. Exothermic reactions are often sensed as "hot" by the observer. Endothermic reactions require energy transfer *from* the surroundings. Endothermic reactions are often sensed as "cold" by the observer.

The heat of a reaction comes from bond breaking and bond making. It takes energy to break the bonds in a compound. In contrast, bond making is a process that releases energy. An energy diagram can be used to show the energetic process of a reaction.

When the energy of making the new bonds in the products is greater than the energy that it takes to break the bonds in the reactants, energy is transferred *to* the surroundings, and the reaction is exothermic. When it takes more energy to break the bonds, energy is transferred *from* the surroundings, and the reaction is endothermic.

The net energy of a reaction is referred to as the heat of the reaction, or $\Delta H$. In the case of an exothermic reaction, $\Delta H$ is negative, because energy is transferred *from* the system. In the case of an endothermic reaction, $\Delta H$ is positive, because energy is transferred *to* the system.

The energy associated with the formation of any compound from its elements is referred to as the heat of formation, or $\Delta H_f$. Heat of formation is expressed in kilojoules per mole of product formed. There is also an energy associated with getting a reaction started, called the activation energy. For example, it takes a match or a spark to get a fire started. Once a few bonds are broken by the spark, combustion reactions proceed on their own because of the heat released by the formation of the products.

Combustion is an oxidation-reduction reaction. Oxidation is defined as the loss of electrons by an atom, an ion, or a molecule during a chemical change. Reduction is when an atom, an ion, or a molecule gains electrons in a chemical change.

Reactions are controlled efficiently in batteries. A battery keeps the reactants separated from one another, preventing the reactants from mixing together and starting a fire. When a wire connects to the battery, the reaction occurs and energy is transferred. The reaction stops when the wire is disconnected, or when the reactants are used up.

Aside from transferring energy by direct contact between a hot substance and a cold substance, called conduction, energy is transferred by radiation. Radiation refers to the generation of electromagnetic radiation. When described by the wave model, electromagnetic waves have a large range of frequency and wavelength. However, all of these waves travel at the speed of light. The energy transferred depends on the frequency of the electromagnetic waves. A fire transfers energy both by conduction and radiation.

## REVIEW EXERCISES

### General Review

*Write a brief and clear answer to each question. Be sure to show your work.*

1. What have you learned about fire in this unit?

2. Generally describe how our energy needs are satisfied by chemical changes.

3. Where does the thermal energy from an exothermic reaction come from?

4. What do the first and second laws of thermodynamics indicate about energy?

5. How many calories of energy are transferred to
   a. raise the temperature of 50 grams of water from 25 °C to boiling?
   b. lower the temperature of 100 grams of water from boiling to 25 °C?

6. The specific heat capacity of aluminum is 0.21 cal/g °C. How many calories of thermal energy must be transferred to heat up a 100 g aluminum pot from 20 °C to 80 °C?

7. For the same number of grams, does it take more or less energy to raise the temperature of aluminum metal or liquid water? Explain.

8. The heat of fusion for water is 79.9 cal/g. The heat of fusion for ethanol is 24.9 cal/g. What does this tell you about these two substances?

9. Name three ways you can increase the rate of a reaction.

10. Define work. How many joules of work are done if a 500 lb piano is lifted by a pulley to a height of 20 m?

11. Write and balance the equation for the oxidation of calcium, Ca.

12. Write and balance a chemical equation for a redox reaction.

13. Use net ionic equations to show which reactant atoms are oxidized and which atoms are reduced in the following chemical equations.

$$Zn(s) + 2MnO_2(s) \longrightarrow ZnO(s) + Mn_2O_3(s)$$

$$3CaO(s) + 2Al(s) \longrightarrow Al_2O_3(s) + 3Ca(s)$$

14. Explain how you know which atoms are oxidized and which are reduced in a redox reaction.

15. Explain the ray model, the wave model, and the photon model of light. Draw a sketch of each model showing transmission through a thin red plastic film.

16. What types of electromagnetic radiation are least harmful? Explain why in terms of wavelength, frequency, and energy.

17. Describe the main components of a spectrometer.

## STANDARDIZED TEST PREPARATION

### Multiple Choice

*Choose the best answer.*

1. Which of these is best classified as energy as opposed to matter?
   (A) safety match
   (B) potassium chlorate, $KClO_3(s)$
   (C) fireworks
   (D) light

*Use the following to answer Exercises 2 and 3.*

A solution of sodium bicarbonate is mixed with a solution of hydrochloric acid in a beaker. A solution of sodium chloride and carbonic acid then forms. A thermometer inside the beaker indicates that the temperature decreased from 24 °C to 17 °C.

$$NaHCO_3(aq) + HCl(aq) \longrightarrow$$
$$NaCl(aq) + H_2CO_3(aq)$$

2. What evidence would indicate that energy was transferred from the surroundings to the products of the reaction in the beaker?
   (A) The products are aqueous. No precipitate formed.
   (B) Both the products and reactants include acids.
   (C) The beaker would feel cold to the touch.
   (D) The beaker would feel hot to the touch.

3. Which correctly names this process?
   (A) endothermic process
   (B) exothermic process
   (C) equilibrium process
   (D) combustion process

4. Water condenses on the outside of a cold glass of lemonade on a warm day. Which of the following correctly describes this process?
   (A) When water vapor condenses, energy is transferred *from* the surroundings to the gaseous water molecules, thereby lowering the temperature.
   (B) When water vapor condenses, energy is transferred *to* the surroundings, thereby lowering the temperature of the water molecules.
   (C) Condensation is a cooling process because hydrogen and oxygen react to form water molecules.
   (D) Condensation is a cooling process because gaseous water molecules collide with one another during the phase change.

5. Choose the best description of the final temperature if 100 mL of water at 20 °C is mixed with 50 mL of water at 80 °C.
   (A) The final temperature will be 50 °C because this is the average temperature.
   (B) The final temperature will less than 50 °C because the 100 mL sample has the greater volume.
   (C) The final temperature will greater than 50 °C because the 50 mL sample has the higher temperature.
   (D) The final temperature cannot be determined with the information given.

6. Which cup of water gets the hottest, if the initial temperature is the same?

(A) 160 calories of energy are transferred to 32 mL of water.

(B) 525 calories of energy are transferred to 105 mL of water.

(C) 2375 calories of energy are transferred to 475 mL of water.

(D) Each cup of water will reach the same final temperature.

7. The specific heat capacities for several substances are listed in the following table.

| Substance | Specific heat capacity |
|---|---|
| water, $H_2O$ | 1.00 cal/g °C |
| methanol, $CH_3OH$ | 0.58 cal/g °C |
| brass, CuZn | 0.090 cal/g °C |
| aluminum, Al | 0.21 cal/g °C |

Which substance would require the most energy to be tranferred to raise the temperature from 20 °C to 30 °C?

(A) water

(B) methanol

(C) brass

(D) aluminum

8. If the specific heat capacity is 0.49 cal/g °C for ice and 1.0 cal/g °C for liquid water, how much energy must be transferred to a 32 g sample of ice to raise its temperature from −4.0 °C to 24 °C? The heat of fusion of water is 79.9 cal/g °C.

(A) 770 cal

(B) 820 cal

(C) 3300 cal

(D) 3400 cal

9. The heating curve shows what happens as heat is added to water at a constant rate, starting with a piece of ice. What is present during time D?

(A) both liquid water and ice

(B) both liquid water and water vapor

(C) only water vapor

(D) only liquid water

10. Which of the following describes a plausible combustion reaction with a balanced chemical equation?

(A) $C_6H_{14}(l) + 6\ O_2(g) \longrightarrow 6\ CO_2(g) + 7\ H_2O(l)$

(B) $C_2H_4O_2(l) + 2\ O_2(g) \longrightarrow$
$2\ CO_2(g) + 2\ H_2O(l)$

(C) $2\ Li(s) + O_2(g) \longrightarrow Li_2O(s)$

(D) $2\ MgO(s) + O_2(g) \longrightarrow 2\ MgO_2(s)$

11. A student performed a calorimetry experiment where 1.3 g of crushed potato chips were burned to warm 25.0 mL of water. If the temperature of the water increased by 6.00 °C, how much energy per gram was released by the combustion of the crushed chips?

(A) 0.416 cal/g

(B) 115 cal/g

(C) 150 cal/g

(D) 195 cal/g

12. The combustion of 2.6 g of octane transferred energy to 926 mL of water. If the temperature of the water increased from 22 °C to 54 °C, how many kilojoules of energy were transferred to the water?

(A) 0.083 kJ

(B) 0.35 kJ

(C) 29 kJ

(D) 120 kJ

Use the following table to answer Exercises 13 and 14.

### Average Bond Energies (per mole of bonds)

| Bond | H−H | C−H | C−C | C=C | C≡C | O−H | C−O | C=O | O=O | O−O |
|---|---|---|---|---|---|---|---|---|---|---|
| Bond energy (kJ/mol) | 432 | 413 | 347 | 614 | 811 | 467 | 358 | 799* | 495 | 146 |

*C=O in $CO_2$: 799, C=O in organic molecules: 745

**13.** What is the total energy required to break all the bonds in 1 mole of ethanol, $C_2H_6O$?

(A) 880 kJ/mol     (B) 1120 kJ/mol
(C) 3240 kJ/mol     (D) 4150 kJ/mol

**14.** Based on bond energies, what is the energy change for the combustion of propane?

$$C_3H_8(g) + 5\ O_2(g) \longrightarrow 3\ CO_2(g) + 4\ H_2O(g)$$

(A) −2060 kJ/mol
(B) 2060 kJ/mol
(C) −15,000 kJ/mol
(D) 15,000 kJ/mol

**15.** Which process could be represented correctly by the energy diagram?

(A) An ice cube is melting at 0 °C.
(B) $H_2O(g) \longrightarrow H_2(g) + O_2(g)$    $\Delta H = +482$ kJ
(C) $2\ C_4H_{10}(g) + 13\ O_2(g) \longrightarrow$
                 $8\ CO_2(g) + 10\ H_2O(g)$
(D) Ammonium chloride is dissolved in water. The temperature cools by 16 °C.

**16.** Which of these processes will result in an increase of the speed of the process?

(A) Use cool water to dissolve sugar crystals.
(B) Crush charcoal into smaller pieces before lighting.
(C) Decrease the concentration of detergent to remove a stain.
(D) Put a rusty nail in a small container sealed with a lid.

**17.** What would you expect to observe if you added aluminum metal to a solution of copper (II) chloride? Use the activity series diagram below to answer.

### Activity Series

Potassium        More active
Barium          Easily oxidized
Calcium
Sodium
Magnesium
Aluminum
Zinc
Chromium
Iron
Nickel
Copper
Silver
Mercury        Less active
Gold           Easily reduced

(A) Aluminum would dissolve as copper oxidized.
(B) Copper would dissolve as aluminum oxidized.
(C) Copper would precipitate as copper oxidized.
(D) Copper would precipitate as aluminum oxidized.

18. What occurs in an electrochemical cell using the two half-reactions shown below?

$$Li^+ + e^- \longrightarrow Li \qquad -3.05\ V$$
$$Zn^{2+} + 2e^- \longrightarrow Zn \qquad -0.76\ V$$

(A) Li is reduced; Zn is oxidized; cell is 2.29 V.

(B) Li is reduced; Zn is oxidized; cell is 3.81 V.

(C) Zn is reduced; Li is oxidized; cell is 2.29 V.

(D) Zn is reduced; Li is oxidized; cell is 3.81 V.

19. When visible light strikes a mirror, the light is mostly

(A) Absorbed by the mirror.

(B) Transmitted by the mirror.

(C) Absorbed and transmitted by the mirror.

(D) Reflected by the mirror.

20. Red light has a wavelength of $7.5 \times 10^{-7}$ m and a frequency of $4.0 \times 10^{14}$ Hz. Blue light has a shorter wavelength. Which statement is true about blue light?

(A) Blue light has a higher frequency compared with red light.

(B) Blue light has a greater speed compared with red light.

(C) Blue light has a lower energy compared with red light.

(D) Blue light has the same speed as red light but a lower frequency.

21. The wavelength of a radio wave is 300 m. What is the frequency of this wave?

(A) 10 Hz        (B) 1,000 Hz

(C) 1,000,000 Hz    (D) 100,000,000 Hz

22. Spectroscopy is a method that can be used to determine

(A) The speed of light.

(B) The density of a substance.

(C) The concentration of a solution.

(D) The mass of a substance.

# Calorimeter

## Work Groups of 4

### Purpose

To use the process of engineering design to plan, build, and refine a calorimeter created in your in class lab.

### Problem

In Chapter 19, Lesson 101, you designed an experiment to determine which snack food, cheese puffs or toasted corn snacks, transferred the most heat in a combustion reaction.

### Safety Notes

- Wear safety goggles at all times
- Be sure to know the location of the fire blanket and fire extinguisher

### Review

Before you begin, review the handout, *Engineering Design* (found after the Unit 1 Engineering Project), and reflect on what you have learned so far in your class from your first experiment. For this project, you will focus on, brainstorming for solutions, refining a plan and retesting.

### Part 1: Refine and Retest

1. Have your classmates and teacher provide feedback for you about your calorimeter design. Ask them to help you think of ways to improve your design. What changes might you make based on their feedback?

2. Make changes to your procedure and design based on feedback and the results of your testing. Submit your revised procedure and design to your teacher for review and approval. Include your list of resources and research material.

3. Once approved, use your procedure to conduct the experiment again.

4. Once your procedure and design are refined, your teacher may ask you to lead a small group of students in demonstrating your calorimeter.

### Possible Presentation Options

- Oral presentation describing your procedure
- Poster describing your procedure in detail with diagrams
- A protocol pamphlet describing your procedure with diagrams
- A website that outlines your project including your procedure and diagrams

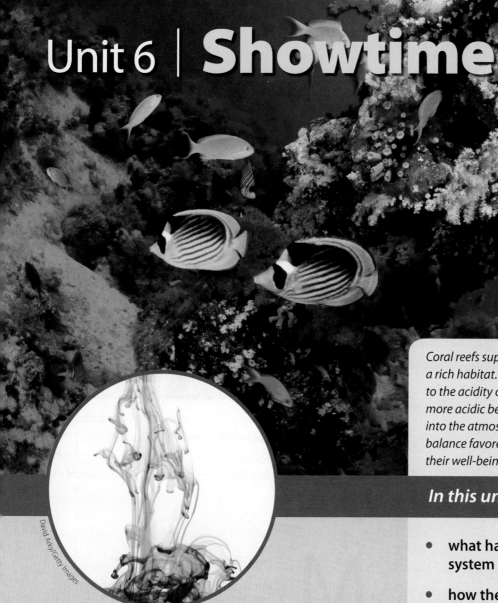

# Unit 6 | Showtime

*David Arky/Getty Images*

*Carlos Villoch/Robert Harding*

Coral reefs support a wondrous diversity of life in a rich habitat. Yet coral reefs are highly sensitive to the acidity of ocean water. As oceans become more acidic because of carbon dioxide released into the atmosphere by human activity, the balance favored by corals is thrown off, affecting their well-being.

## In this unit, you will learn

- **what happens in a chemical system at equilibrium**

- **how the balance in a reversible process is maintained**

- **the mathematical relationship between starting substances and products in a system at equilibrium**

- **about variables that affect reversible processes**

## Why Showtime?

As the *Living by Chemistry* course comes to the culminating Unit 6: Showtime, it is time for you to reflect on what you have learned and show what you know. You will explore reversible processes that have a dynamic balance between starting substances and products. This balance is known as chemical equilibrium. You will consider the behavior of molecules in the human body, the reversible processes of dissolving and precipitation, weak acids in aqueous solutions, and colorful molecules called indicators. You'll complete a research project that involves extracting colorful molecules from flowers or berries and then manipulating the solution color in a showy ending to the course.

# CHAPTER 23

# Chemical Equilibrium

## In this chapter, you will study

- reversible chemical and physical processes

- the dynamic nature of chemical equilibrium

- the mathematical relationship between products and starting substances at equilibrium

*Hot water running through the limestone (calcium carbonate) under parts of Yellowstone National Park dissolves the limestone. Hot springs leave enormous deposits above ground as the water cools and limestone precipitates as a result of this reversible process.*

**M**any chemical processes are reversible. Unlike most combustion reactions, reversible processes do not proceed all the way to the products. In reversible chemical systems, both the forward and reverse processes occur at the same time, so starting substances and products exist in the same mixture. When two opposite processes are in balance, the result is an equilibrium mixture. To the eye, the equilibrium mixture may appear unchanging, but this is not the case. Changes are continuously happening on the atomic scale in systems at equilibrium.

# Acid-Base Indicators

## THINK ABOUT IT

The same "red" cabbage plant is known by different colors in various regions, depending on the pH of the soil in which it is grown and how people describe the color in their native language. The leaves can be reddish, purple, or blue. If you cook red cabbage, it normally turns blue. But you can get the red color back by adding vinegar to the pot. What is the chemistry behind these color changes?

*The color of red cabbage depends on compounds in the soil where it is grown.*

### What do color changes indicate?

*To answer this question, you will explore*

⮕ Molecular Views of Indicator Color Changes

⮕ Chemical Symbols to Represent Indicator Color Changes

*Red cabbage turns blue when you cook it in water. It remains red if you add vinegar.*

## EXPLORING THE TOPIC

*Colors of cabbage juice solutions at different pH values*

### ⮕ Molecular Views of Indicator Color Changes

In Unit 4: Toxins, you explored cabbage juice as an **indicator** of the pH of a solution. As shown in the figure, the color of the cabbage juice indicator changes from red at pH ~3, to purple at pH ~6, and to blue at pH ~9. At higher pH values, the yellow color of a second indicator molecule is evident.

#### THE PIGMENT ANTHOCYANIN

A **pigment** is a substance that is colored because of the light it reflects. Instead of reflecting all the colors in white light from the light source, a pigment only reflects certain colors. For example, the leaves of red cabbage contain a pigment that reflects red light.

*Anthocyanins give these berries their colors.*

The pigments in red cabbage are called *anthocyanins*. Anthocyanins are found in blueberries, cranberries, raspberries, cherries, grapes, eggplant, and the flower petals of violets.

The anthocyanins that give red cabbage its color are large, complex molecules. Instead of writing out their molecular formulas, it is common to abbreviate the anthocyanin molecule as H*In*. H*In* is a weak acid, which means that it partially dissociates (breaks apart) into $H^+$ cations and $In^-$ anions when dissolved in water. The H*In* anthocyanin molecule is a red color. The $In^-$ anions that form when the anthocyanin molecules dissociate are a blue color.

Anthocyanins are called indicators because they change color when the pH of the solution they are dissolved in changes. For simplicity, all indicators in this unit will be abbreviated as H*In*. And for the purpose of this lesson, we will refer to the H*In* molecule as one form of the indicator, and the $In^-$ ion as a second form. The color of a solution that the indicator is added to depends on how much of each form of the indicator is present in that solution.

*The anthocyanin indicator is abbreviated as H*In* when in acidic solution (structure on the left) and as $In^-$ when in basic solution (structure on the right). Notice that the $In^-$ anion is missing an $H^+$.*

### CHANGES IN COLOR

When an acid-base indicator is placed in an acidic solution, mostly H*In* molecules will be present in the solution. When the indicator is placed in a basic solution, mostly $In^-$ anions will be present in the solution. The anthocyanin indicator is red in acid (H*In*) and blue in base ($In^-$). When the indicator is placed in a neutral solution, both molecules and anions, H*In* and $In^-$, are present. Because both the red and blue forms are present, the mixture appears purple. If the indicator is placed in a solution that is slightly acidic, the solution contains slightly more red H*In* molecules, and its color is red-purple. If it is placed in a solution that is slightly basic, there are more blue $In^-$ anions, and its color is blue-purple.

Molecular views of a tiny volume of solution are shown in the figure on the next page. There are 20 anthocyanin molecules in this tiny volume, either as H*In* molecules (red circles) or as $In^-$ anions (blue circles). Notice that there are no purple circles. Solutions that are varying shades of purple do not contain molecules that reflect purple light. Instead, these solutions reflect red and blue light. Our brains perceive the combination of red and blue as purple.

**ENVIRONMENTAL CONNECTION**

Besides its use as an indicator, anthocyanin is responsible for spectacular fall foliage. At the end of the growing season, when the chlorophyll in plants and trees decays, the red and purple colors of anthocyanins become visible, contributing to colorful foliage displays. Every autumn, "leaf peeping" draws many people to Vermont.

**Big Idea** The color of an indicator solution varies with pH because the solution contains varying mixtures of H*In* molecules and $In^-$ anions.

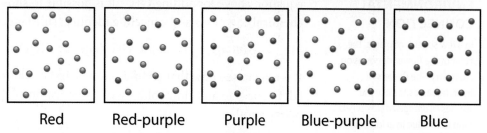

| Red | Red-purple | Purple | Blue-purple | Blue |

*Notice that the square on the left has only red circles, representing HIn. The square on the right has only blue circles, representing In⁻. The three in the middle are mixtures of HIn and In⁻. The mixtures are varying shades of purple.*

## ⟳ Chemical Symbols to Represent Indicator Color Changes

Molecular views are a good way to illustrate how the mixture of molecules in solution relates to the observation of solution color. Chemical equations are also useful for expressing the color changes that occur with indicator molecules in solution.

### ANTHOCYANIN IN A BASIC SOLUTION

When anthocyanin is added to a solution that is basic, with pH ~9, the solution turns blue because the indicator is present mostly as $In^-$ anions. If $H^+$ ions are added to the basic solution with $In^-$ ions, $H^+$ combines with $In^-$ anions to form HIn molecules. The HIn molecule is red. As blue $In^-$ anions are converted to red HIn molecules, the solution appears varying shades of purple. Once all the $In^-$ anions have combined with $H^+$, the solution turns red. The following balanced chemical equation describes the process that occurs as $H^+$ ions are added.

$$H^+(aq) + In^-(aq) \longrightarrow HIn(aq)$$
$$\phantom{H^+(aq) + {}}\text{blue}\phantom{(aq) \longrightarrow }\text{red}$$

### ANTHOCYANIN IN AN ACIDIC SOLUTION

When anthocyanin is added to a solution that is acidic, the solution turns red because the indicator is present mostly as HIn molecules. If $OH^-$ anions are then added to the acidic solution, the solution changes color from red to purple to blue. This is evidence that the number of $In^-$ anions in the solution increases as the pH increases. A reasonable conclusion to draw from these observations is that adding base to an acidic solution reverses the process described above of adding an acid to a basic solution. The following balanced chemical equation describes the process that occurs as $OH^-$ is added.

$$HIn(aq) + OH^-(aq) \longrightarrow H_2O(l) + In^-(aq)$$
$$\phantom{HIn}\text{red}\phantom{(aq) + OH^-(aq) \longrightarrow H_2O(l) + }\text{blue}$$

Although this equation does not appear to be an exact reversal of the first equation that describes adding $H^+$, it can be written as two equations: one that shows that $OH^-$ combines with $H^+$ to make $H_2O$, and another that shows the HIn molecules dissociate into $H^+$ and $In^-$ ions.

$$HIn(aq) \longrightarrow H^+(aq) + In^-(aq)$$
$$\phantom{HIn}\text{red}\phantom{(aq) \longrightarrow H^+(aq) + }\text{blue}$$

$$H^+(aq) + OH^-(aq) \longrightarrow H_2O(l)$$

These two chemical equations describe the same process as the single chemical equation above for adding OH⁻. It is possible to show this by adding the two chemical equations and canceling identical substances on both sides of the arrow.

$$HIn(aq) + \cancel{H^+}(aq) + OH^-(aq) \longrightarrow H_2O(l) + \cancel{H^+}(aq) + In^-(aq)$$

It is easy to see now that adding acid to a basic solution of anthocyanin and adding base to an acidic solution of anthocyanin are processes that are the reverse of one another.

**Anthocyanin in a Basic Solution**

$$H^+(aq) + In^-(aq) \longrightarrow HIn(aq)$$
$$\text{blue} \qquad\qquad\quad \text{red}$$

**Anthocyanin in an Acidic Solution**

$$HIn(aq) \longrightarrow H^+(aq) + In^-(aq)$$
$$\text{red} \qquad\qquad\qquad \text{blue}$$

## LESSON SUMMARY

### What do color changes indicate?

Color changes can provide information about chemical processes on a molecular level. One class of color-changing molecules is called an indicator. An indicator is a large, complex molecule that is abbreviated with the symbol H*In*. When placed in solution, indicator molecules can exist as H*In* molecules or as *In*⁻ anions. This is because the H*In* molecules can dissociate into H⁺ and *In*⁻ ions. The reverse process can also occur in which the H⁺ and *In*⁻ ions combine to form H*In* molecules. In an acidic solution, there is plenty of H⁺ such that the indicator is present as H*In* molecules. Adding base removes H⁺ to produce *In*⁻ anions. Because the H*In* molecules and *In*⁻ anions are different colors, the color of the solution indicates the acid-base properties of the solution.

**KEY TERMS**

indicator

pigment

## Exercises

### Reading Questions

1. Explain how molecules account for the purple cabbage leaves.

2. Explain how the two forms of an indicator are related to one another.

### Reason and Apply

3. Methyl red is an indicator that is red in acidic solution and yellow in basic solution. Draw three molecular views to show how the color changes from red to yellow. What color do you observe in between?

4. Write a balanced chemical equation to describe what happens as the color changes from red to yellow for methyl red indicator using the general symbols for an indicator molecule and its anion.

5. Write a balanced chemical equation to describe what happens as the color changes from yellow to red for methyl red indicator.

6. **PROJECT**  Research an acid-base indicator and explain how it works. (Do not use the universal indicators bromothymol blue, methyl red, or anthocyanin.)

# Reversible Processes

### THINK ABOUT IT

John Gollop/Getty Images

Imagine that you open a bottle of perfume. You detect a sweet smell that remains quite intense as long as the bottle is open. This indicates that molecules are attaching to receptors in your nose and causing a sensation by stimulating nerves that send signals to your brain. Once you close the bottle, the sweet smell grows fainter over time. Why don't you continue to detect the sweet smell?

**What is a reversible process?**

*To answer this question, you will explore*

⮕ Reversible Processes: The Double Arrow

⮕ Examples of Reversible Processes

⮕ Degree of Sweetness

### EXPLORING THE TOPIC

## ⮕ Reversible Processes: The Double Arrow

Based on what you have learned so far in this course, it might seem that chemical processes proceed only in the direction in which they are written as a chemical equation. For example, here is the chemical equation that describes gasoline burning.

$$2C_8H_{18}(l) + 25O_2(g) \longrightarrow 16CO_2(g) + 18H_2O(g)$$

When gasoline burns, the reaction goes in the forward direction, from reactants to products. This reaction is said to *go to completion,* which means that all reactants are completely converted to products. Although it *is* possible for gasoline to be created through chemical reactions, it takes many chemical steps and requires a great deal of energy. So this reaction, as it is written, is not reversible. Once you burn the gasoline, there is no going back.

In contrast, there are many **reversible processes** that proceed in both the forward and reverse directions. The term *process* can describe either a physical process or a chemical process (chemical reaction). In Lesson 70, you learned that chemical equations are used to represent both chemical and physical processes.

| Physical process | Chemical process |
|---|---|
| $H_2O(l) \longrightarrow H_2O(s)$ | $2Na(s) + Cl_2(g) \longrightarrow 2NaCl(s)$ |
| $Br_2(l) \longrightarrow Br_2(g)$ | $CH_4(g) + 2O_2(g) \longrightarrow CO_2(g) + 2H_2O(l)$ |
| $I_2(s) \longrightarrow I_2(g)$ | $N_2O_4(g) \longrightarrow 2NO_2(g)$ |

As written, the processes in the table appear to move in one direction only. However, some of these processes are reversible.

| Physical process | Chemical process |
|---|---|
| $H_2O(l) \rightleftharpoons H_2O(s)$ | $2Na(s) + Cl_2(g) \longrightarrow 2NaCl(s)$ |
| $Br_2(l) \rightleftharpoons Br_2(g)$ | $CH_4(g) + 2O_2(g) \longrightarrow CO_2(g) + 2H_2O(l)$ |
| $I_2(s) \rightleftharpoons I_2(g)$ | $N_2O_4(g) \rightleftharpoons 2NO_2(g)$ |

Notice that for a reversible process, a double arrow is used in the equation. The double arrow indicates that both a forward and a reverse process can occur. A reversible process can be written in either the forward or reverse direction. All phase changes are reversible, while many chemical processes are not.

**Big Idea** All physical processes and some chemical processes are reversible.

## ⮕ Examples of Reversible Processes

As is evident from the second table above, phase changes are reversible. Other common reversible processes involve processes in aqueous solution, such as dissolving and precipitation and acid-base reactions, and processes in the human body involving binding to a receptor. Reversible processes can occur in both the forward and reverse directions. Depending on the conditions, it is possible to cause these processes to proceed exclusively in one direction, or to create a mixture of starting substances and products. This is explained further with the examples below.

### PHASE CHANGES

Water going from liquid to gaseous state.

Consider water boiling in a pot on the stove. The liquid water evaporates and spreads out in the air. If the gaseous water is cooled, the reverse process occurs and water droplets form. Intermolecular forces, including hydrogen bonds and dipole-dipole attractions, account for the attraction between water molecules. Heating easily breaks these forces of attraction, and they re-form by cooling. One way to represent this reversible process of evaporating and condensing is with a double arrow as shown in the equation below.

$$H_2O(l) \rightleftharpoons H_2O(g)$$

### DISSOLVING AND PRECIPITATION

Suppose you dissolve a small amount of salt in water to create an aqueous salt solution. If you evaporate the water, the reverse process occurs and solid salt precipitates. The dominant force that accounts for the formation of the solid is cation-anion attractions. Upon dissolving, the ions are attracted to the water molecules by ion-dipole attractions. The difference in the strength of these forces is small enough such that the reaction is easy to reverse. One way to represent this

In a flask of salt solution, some of the salt remains on the bottom of the flask.

reversible process of dissolving and precipitating is with a double arrow as shown in the chemical equation below.

$$NaCl(s) \rightleftharpoons Na^+(aq) + Cl^-(aq)$$

## ACID-BASE REACTIONS

When an acid is dissolved in water, the acid molecules dissociate into $H^+$ and an anion. This process can also go in the reverse direction, in which the ions are attracted to one another to form the acid molecule, depending on the pH of the solution.

In the previous lesson, you examined a solution of an acid-base indicator, H*In*. Adding base to an aqueous solution causes the H*In* molecules to dissociate into $H^+$ and *In*$^-$ ions. Adding acid causes the reverse process in which the ions combine to form H*In* molecules.

Bromothymol blue displays as a different color for acidic (yellow), neutral (green), and basic (blue) solutions.

A solution of the indicator bromothymol blue is yellow in acidic solutions due to the presence of mostly H*In* molecules, blue in basic solutions due to the presence of mostly *In*$^-$ anions, and green at pH = 7 because both H*In* molecules and *In*$^-$ anions are present in roughly equal amounts. A double arrow, as shown in the chemical equation below, is one way to represent the reversible process of acid molecules dissociating into ions and ions attracting one another to re-form molecules.

$$HIn(aq) \rightleftharpoons H^+(aq) + In^-(aq)$$

## BINDING TO A RECEPTOR IN THE BODY

Processes involving binding to receptor sites in the human body are reversible. Prime examples include taste and smell, in which molecules attach to and reversibly release from receptors on the tongue and in the nose. The receptors are made of large protein molecules with a specific shape and sections that are polar, charged, or nonpolar.

Small molecules are attracted selectively to the protein by intermolecular attractions and how well the molecule fits. When molecules attach to the receptor, this initiates a signal sent along the nervous system to the brain. The brain receives the signal and it is perceived as a certain taste or smell. The signal is no longer detected when the molecules leave the receptor.

The general process for reversible binding between a molecule and a receptor is shown below. The blue shape represents a small molecule that is soluble in blood, saliva, or fluids in the cell. The red shape is a piece of a large protein molecule that forms the receptor. In the forward process, the molecule attaches to the receptor. In the reverse process, the molecule separates from the receptor.

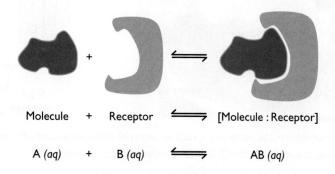

Molecule + Receptor $\rightleftharpoons$ [Molecule : Receptor]

A (aq) + B (aq) $\rightleftharpoons$ AB (aq)

There are many reversible receptor processes in the body. The smell of menthol is detected in the nose when menthol molecules are attracted to smell receptor sites. A protein in the blood called *hemoglobin* reversibly binds oxygen, $O_2$. The pain reliever ibuprofen binds reversibly to an enzyme to block it from producing molecules that cause the pain signal. The pain signal returns as the ibuprofen molecules leave the receptor. In the chemical equations below, the symbol ":" represents an attraction between the molecule and receptor.

$$menthol + receptor \rightleftharpoons [menthol : receptor]$$
$$O_2 + hemoglobin \rightleftharpoons [O_2 : hemoglobin]$$
$$ibuprofen + receptor \rightleftharpoons [ibuprofen : receptor]$$

Some of the reversible processes described above can be represented as two components, A and B, forming one component, AB, and then breaking apart as A and B. Here is a general form of a chemical equation describing this type of process:

$$A + B \rightleftharpoons AB$$

Because this describes a reversible process, the equation can also be written in the other direction:

$$AB \rightleftharpoons A + B$$

This model can be used to describe salt dissolving and precipitating, acid-base reactions, and molecules binding to and releasing from receptors. For example, for salt dissolving, AB represents the salt, and A and B are the ions that it separates into when dissolved. For sugar molecules binding to a taste receptor, A represents a sugar molecule and B represents the receptor.

**Important to Know** If a process is reversible, the equation describing it can be written in either the forward or reverse direction.

## ➲ Degree of Sweetness

According to the receptor site theory, the attraction of sugar molecules to sweet receptors in the taste buds on the tongue involves the formation of two hydrogen bonds on two adjacent carbon atoms in the sugar molecule. Recall from Unit 2: Smells that a hydrogen bond forms when an $O-H$ or $N-H$ on one molecule is attracted to an O or N atom on another molecule.

Protein chain

The receptor site is a large protein molecule consisting of a chain of C and N atoms.

*Hydrogen bonds* are between two molecules. An –OH or –NH on one molecule is attracted to an N or O atom on another molecule.

Sugar molecule

Molecules that attach to the sweet receptor have an –OH on adjacent C atoms.

Joy Brown/Shutterstock

*The molecules in substances that taste sweet, such as cake with strawberries and whipped cream, are attracted to the sweet receptor on the tip of the tongue by two hydrogen bonds formed on adjacent carbon atoms.*

The table below shows the relative sweetness for four substances. The molecules in these substances all attach to the same type of sweet receptors. Why do they vary in relative sweetness?

| Sweetener | Source | Relative sweetness |
|-----------|--------|-------------------|
| lactose | milk and other dairy products | 2 |
| maltose | beer, cereal, pasta, potatoes | 3 |
| sucrose | table sugar | 10 |
| aspartame | artificial sweetener | 1800 |

One explanation for the data is that some molecules stay in the sweet receptor longer, thereby creating a stronger sensation. This can be explained by differences in the strength of the intermolecular attractions between the molecule and the receptor, as well as a better fit. Sucrose (table sugar) is attracted more strongly to a sweet receptor than maltose (the sugar in malt) and more strongly still than lactose (the sugar in milk). The result is that, for the same number of molecules and the same number of sweet receptors, more sucrose molecules are detected because more are attached to the sweet receptors. For lactose, more molecules remain in the saliva.

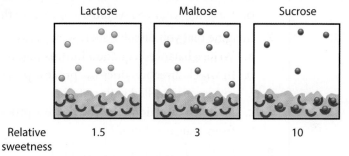

| Relative sweetness | 1.5 | 3 | 10 |

*The squares show that as more molecules bind to the sweet receptors, the relative sweetness of the substance increases. The small circles represent the sweet molecules, and the dark area at the bottom represents the area in the tongue with the sweet receptors. The white space represents saliva.*

Artificial sweeteners are about 180 times sweeter than table sugar (sucrose) per mole. This is why packets of artificial sweeteners are much smaller than packets of table sugar.

**Big Idea** Some reversible processes favor the forward direction, while others favor the reverse direction.

## LESSON SUMMARY

### What is a reversible process?

A reversible process is one that can proceed in both the forward and the reverse directions. Many physical and chemical processes are reversible given the appropriate conditions. In a chemical equation, a double arrow is used to symbolize reversible processes. Examples of reversible processes include phase

## NATURE CONNECTION

Pointed stalactites and stalagmites form when limestone alternately dissolves in and precipitates out of the dripping water in a cave. These calcite structures build slowly, at the rate of about one quarter inch to one inch per century.

Francesco Tomasinelli/Science Source

changes, dissolving and precipitation, acid-base reactions, and binding to protein receptors. In each case, it is possible to find conditions to cause these processes to proceed almost exclusively in one direction or the reverse, or to create a mixture of starting substances and products.

## Exercises

### Reading Questions

1. What evidence is there that some processes are reversible?

2. What does a double arrow in a chemical equation tell you?

### Reason and Apply

3. Provide examples to support the claim that phase changes are reversible.

4. Provide examples to support the claim that processes involving the breaking and forming of intermolecular attractions are reversible.

5. Provide examples to support the claim that processes involving breaking and making of several covalent bonds are sometimes not reversible.

6. Imagine that you take an ice cube out of the freezer and place it on a counter at room temperature.
   a. Write a balanced equation for the process that occurs.
   b. Describe what you would do to reverse the process.

7. Imagine that you dissolve sucrose, $C_{12}H_{22}O_{11}(s)$, in water.
   a. Write a balanced equation for this reversible process.
   b. In which direction does the process proceed if you place a tiny amount of sugar in a large pot filled with water?
   c. In which direction does the process proceed if you allow the water to evaporate from a sugar solution?

8. Methane, $CH_4(g)$, burns in oxygen, $O_2(g)$, to produce carbon dioxide, $CO_2(g)$, and water, $H_2O(l)$.
   a. Write a balanced equation for this process.
   b. This process is described as irreversible. What does this mean?

9. Imagine that you breathe in oxygen and it binds to hemoglobin in your blood to form $[O_2 : \text{hemoglobin}]$.
   a. Write a balanced equation for the process.
   b. The $[O_2 : \text{hemoglobin}]$ delivers oxygen to your cells. Describe this process.
   c. Provide evidence that this process is reversible.

## *How Dynamic*
# Dynamic Equilibrium

THINK ABOUT IT

A tightrope walker moves carefully and purposefully, keeping his balance as he goes. He tips back and forth while he walks, sometimes using an outstretched arm or a leg to counterbalance any movement one way or the other. He is maintaining his balance but not keeping still. Chemical systems that are at equilibrium maintain a similar type of dynamic balance.

Bettina Ritter/iStockphoto

**What is chemical equilibrium?**

*To answer this question, you will explore*
- Models of Equilibrium
- Chemical Equilibrium

EXPLORING THE TOPIC

## ⊃ Models of Equilibrium

If you put a small amount of sodium chloride, NaCl(*s*), in water, the process goes toward forming Na$^+$(*aq*) and Cl$^-$(*aq*) ion products. If you keep adding more and more NaCl(*s*), the solution will become saturated. Once this happens, no more NaCl(*s*) will dissolve. The excess NaCl(*s*) remains on the bottom of the container.

### THE VIEW AT AN ATOMIC SCALE

To gain a better understanding of why there is a limit to the amount of NaCl(*s*) that dissolves in a given quantity of water, it is useful to consider what is happening at the atomic scale. The figure on the next page shows a model of how water molecules bump into the surface of the solid and carry the ions into the solution. This is due to ion-dipole intermolecular attractions between the Na$^+$(*aq*) and Cl$^-$(*aq*) ions and the water dipoles.

It is also possible to carry out the reverse process by evaporating the water. As the water evaporates, there is less space for the Na$^+$(*aq*) and Cl$^-$(*aq*) ions to spread out, and they bump into each other more often. The forces of attraction between ions cause the salt to re-form.

H₂O
Na⁺
Cl⁻

Cl⁻ (aq)    Na⁺ (aq)

*The water molecules are attracted to Na⁺ (purple spheres) and Cl⁻ (green spheres) ions in the solid NaCl. Once surrounded by water, the ions move away from the solid and into the solution.*

## A DYNAMIC BALANCE

In a dilute aqueous solution of NaCl, the process of dissolving has gone entirely to the $Na^+(aq)$ and $Cl^-(aq)$ ion products. The ions do bump into one another, but enough of them do not cluster together to form a solid. For dilute solutions, you can represent the process of dissolving NaCl($s$) by using a chemical equation with a single arrow in the direction of forming $Na^+(aq)$ and $Cl^-(aq)$ ions.

When the concentration of $Na^+(aq)$ and $Cl^-(aq)$ ions is large enough, solid salt is visible at the bottom of the container. The solution is referred to as saturated. For an observer, it appears that nothing is happening. However, on an atomic scale, the dissolving and precipitation processes are still happening. When some $Na^+(aq)$ and $Cl^-(aq)$ ions precipitate, there is enough room for more NaCl($s$) to dissolve, and so on, back and forth. This explains why all the salt does not dissolve: some keeps precipitating. The saturated solution with excess solid NaCl is represented by a chemical equation with a double arrow to show that both the forward and reverse processes are occurring.

The saturated NaCl solution with excess solid on the bottom is at **chemical equilibrium.** Equilibrium is a dynamic state in which opposing processes occur at the same time. At equilibrium, the rate of the forward process is equal to the rate of the reverse process. Because these rates are equal, no change is visible.

In contrast, in the case of a dilute solution, the rate of the reverse process (precipitation) is never equal to the rate of the forward process (dissolution). In this case, all of the solid dissolves.

*Ken Karp Photography*

*The flask on the left shows a solution of NaCl. Because the entire solid has dissolved, the process has gone in the forward direction. The flask on the right shows a saturated NaCl solution with solid NaCl at the bottom. The solid has not dissolved completely because the process has reached equilibrium.*

**Big Idea** At equilibrium, the rate of the forward process is equal to the rate of the reverse process.

**Example**

### Evaporation of Water

Bonita Hein/iStockphoto

Consider the evaporation of water at 25° C.

**a.** Right after a rainfall, there are water puddles on the sidewalk. Once the rain stops and the sun is shining, the water puddles evaporate completely. Explain why. Provide a chemical equation to describe the process.

**b.** The water in the bottle shown in the picture does not evaporate completely. Explain why. Provide a chemical equation to describe the process.

**c.** How does chemical equilibrium explain the observation of droplets of water on the sides of the bottle above the water level?

*Solution*

**a.** The puddles evaporate and the gaseous water molecules spread out in the atmosphere. The reverse process (condensation) does not happen because the rate at which molecules attract one another is too slow to form water droplets. A chemical equation with a single arrow describes the process.

$$H_2O(l) \longrightarrow H_2O(g)$$

**b.** The water in the bottle does not evaporate completely because the closed container limits the space in which the water molecules can travel. When there are enough gaseous water molecules in the small space above the liquid water, the molecules bump into each other frequently enough to condense. This means that while water continues to evaporate, the gaseous water molecules are also condensing. So, the forward and reverse processes of evaporation and condensation are occurring simultaneously. A chemical equation with a double arrow describes the process.

$$H_2O(l) \Longleftrightarrow H_2O(g)$$

**c.** The water in the bottle is at equilibrium. This means that the rate of evaporation is equal to the rate of condensation. When the gaseous water condenses, it can form droplets on the side of the bottle above the water level, or simply fall onto the surface of the liquid water.

## Chemical Equilibrium

You can use a paper-clip model to describe an irreversible process and a reversible process. The paper-clip model is a general example of any process involving an AB pair that separates into A and B.

AB $\quad\rightleftharpoons\quad$ A + B

The paper-clip pair, AB, might represent:

- a salt such as NaCl, which dissociates to $Na^+(aq) + Cl^-(aq)$
- a strong acid such as HCl, which dissociates to $H^+(aq) + Cl^-(aq)$
- an acid-base indicator, H*In*, which dissociates to $H^+(aq) + In^-(aq)$
- a molecule bound to a receptor, which separates into a molecule and the empty receptor
- a molecule such as $N_2O_4$, which breaks apart into two molecules of $NO_2$

It is useful to examine the features of the model for both irreversible and reversible processes.

### IRREVERSIBLE VERSUS REVERSIBLE PROCESSES

The two graphs below show what happens for two different starting materials, one labeled XY and the second labeled AB. Imagine that the first graph represents a collection of chocolate pieces, XY, with each piece broken into two smaller pieces, X and Y. The second graph might represent a collection of paper-clip pairs, AB, of two different colors that are pulled apart into single paper clips, A and B, and put back together again.

Take a moment to examine the graphs and describe the similarities and differences.

Here are some similarities that you might notice:

- In both cases, the starting substance is breaking apart into two pieces.
- At the start of the process, there are 50 XY pairs and 50 AB pairs.
- The number of XY pairs decreases as the number of X or Y pieces increases. The number of AB pairs decreases as the number of A or B pieces increases.

Here are some differences that you might notice:

- The number of XY pairs goes to 0 over time. Once you break the chocolate, you cannot put it back together again. The process is irreversible.
- The number of AB pairs does not go to 0. There are 15 AB pairs remaining, even after a long time. This is because the reverse reaction makes new AB

pairs. At equilibrium, there is a mixture of 15 AB pairs, 35 individual A pieces, and 35 individual B pieces.

If you examine the graph on the right, you might notice that the amounts of AB, A, and B in the mixture no longer change after about 80 tries. This indicates that equilibrium is reached: the amounts remain fixed.

**Example**

---

### Strong versus Weak Acid

Consider the observations reported here for two aqueous solutions: one with hydrochloric acid, HCl, and a second with formic acid, HCOOH.

**a.** The process of dissolving hydrochloric acid, HCl, in water results in a solution of $H^+(aq)$ and $Cl^-(aq)$. If there are any HCl molecules remaining, the numbers are so tiny that they are not detected. Explain why. Base your explanation on the rates of the forward and reverse processes.

**b.** The process of dissolving formic acid, HCOOH, in water results in a solution of HCOOH, $H^+(aq)$, and $HCOO^-(aq)$. There are many HCOOH molecules in the solution compared with ion products. Explain why. Base your explanation on the rates of the forward and reverse processes.

*Solution*

**a.** Only the forward process occurs at a high rate. Once $H^+(aq)$ and $Cl^-(aq)$ form, they do not stay together long enough to produce a measurable amount of HCl($aq$). (See the graph for XY.) So nearly all of the HCl is converted to products. This process is often represented with a single arrow. Because very nearly all the HCl molecules dissociate, HCl is called a strong acid.

$$HCl(aq) \longrightarrow H^+(aq) + Cl^-(aq)$$

**b.** The process does not go to completion to become products because the reverse process also occurs at a high rate. Once $H^+(aq)$ and $HCOO^-(aq)$ form, they do go back together again to make HCOOH. (See the graph for AB.) At first, the rate of the forward reaction is fast and the rate of the reverse reaction is slow. The rate of the forward reaction slows down as more HCOOH molecules dissociate. The rate of the reverse reactions speeds up as the number of $H^+$ and $HCOO^-$ ions increases. Over time, the rates become equal and the process reaches equilibrium. At equilibrium, the amounts of HCOOH($aq$), $H^+(aq)$, and $HCOO^-(aq)$ no longer change. This reversible process is represented with a double arrow. Because very few of the HCOOH molecules dissociate, HCOOH is called a weak acid.

$$HCOOH(aq) \rightleftharpoons H^+(aq) + HCOO^-(aq)$$

---

### STARTING WITH ALL PRODUCTS

The two graphs below compare the same reversible processes but for different starting amounts. In the graph on the left, there are 50 AB pairs at the start but no A or B pieces. In the graph on the right, there are 50 A and 50 B pieces at the start but with no AB pairs. Take a moment to examine the graphs. What do you notice about the amounts at equilibrium?

 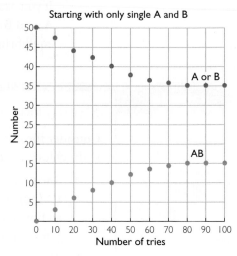

After 100 tries, the amounts of AB, A, and B are the same in both cases. This indicates that equilibrium is a property of the mixture. It does not matter if you start with the substances on the left or right of the double arrow; the resulting mixture is the same.

## LESSON SUMMARY

### What is chemical equilibrium?

Equilibrium is a dynamic process. The word *dynamic* means that things are constantly changing. Equilibrium is a balance that is reached between the forward and reverse processes in a reversible change. To an observer, it seems that things are no longer changing. In reality, however, changes are continuously occurring at an atomic scale, but the changes balance each other out. When a system is at chemical equilibrium, the rate of the forward reaction is equal to the rate of the reverse reaction. Equilibrium is a property of the mixture, meaning that it does not matter how you start the process.

**KEY TERM**

chemical equilibrium

## Exercises

### Reading Questions

1. What does it mean when a system is in a state of dynamic equilibrium?
2. Why is there a limit as to how much salt you can dissolve?

### Reason and Apply

3. Explain these observations in terms of the amounts of starting substances (AB) and products (A + B) in an equilibrium mixture.
   a. Artificial sweetener is sweeter than the same amount of sucrose.
   b. Much more sodium chloride, NaCl, dissolves in 1 L of water compared with silver chloride, AgCl.

*(continued)*

4. When phosphorus pentachloride, $PCl_5(g)$, is placed in a sealed container, it breaks apart reversibly. The concentrations of the starting substance and products are monitored. Use the data in the table to create a graph showing the concentrations of the starting substance and products over time.

$$PCl_5(g) \rightleftharpoons PCl_3(g) + Cl_2(g)$$

| Time (s) | $[PCl_5]$ (mol/L) | $[PCl_3]$ (mol/L) | $[Cl_2]$ (mol/L) |
|----------|-------------------|-------------------|------------------|
| 0        | 0.00              | 1.00              | 1.00             |
| 20       | 0.10              | 0.90              | 0.90             |
| 40       | 0.20              | 0.80              | 0.80             |
| 60       | 0.25              | 0.75              | 0.75             |
| 80       | 0.29              | 0.71              | 0.71             |
| 100      | 0.29              | 0.71              | 0.71             |
| 120      | 0.29              | 0.71              | 0.71             |

a. What are the equilibrium concentrations of the starting substance and the products?
b. At what time did the mixture reach equilibrium? How do you know?
c. At 20 seconds, which process is faster, the forward or the reverse process? At 100 seconds?
d. Why is the amount of starting substance, $PCl_5$, not equal to the amounts of each of the products, $PCl_3$ and $Cl_2$, at equilibrium? Support your answer with evidence.
e. Why are the amounts of $PCl_3$ and $Cl_2$ always equal to one another?

5. The graph below shows the reversible separation of a molecule bound to a receptor.

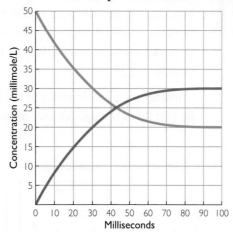

a. The process begins with all the molecules attached to receptors. Write a chemical equation to represent the process.
b. Label the two curves as either starting substance (with the molecule attached to the receptor) or products (molecule or empty receptor).
c. What are the equilibrium concentrations of the starting substance and the two products?
d. How long did it take for this process to reach equilibrium?
e. What is "equal" at equilibrium?

*How Favorable*

# Equilibrium Constant *K*

## THINK ABOUT IT

In a reversible chemical process at equilibrium, the starting substances and the products are both present as a mixture. In some processes at equilibrium, there are more products than starting substances present in the mixture. In other processes, there are more starting substances than products in the equilibrium mixture. Is there some way to predict whether more starting substances or products will be present when a chemical process reaches equilibrium?

**How can you predict if products are favored in a reversible reaction?**

*To answer this question, you will explore*

➔ Products versus Starting Substances

➔ Equilibrium Constant *K*

➔ Dissociation of Strong and Weak Acids into Ions

## EXPLORING THE TOPIC

➔ ## Products versus Starting Substances

Certain pain relievers act by attaching reversibly to receptors to block molecules that cause the pain signals from being sent to the brain. The strength of the binding between the pain reliever and its receptor site influences how effective the pain reliever will be. The stronger the binding, the longer the pain reliever stays in the receptor site, thereby blocking the molecules that cause the pain signal.

Consider the particle views of the binding of three different pain relievers. What do you think the *equilibrium constant K* indicates about the pain reliever?

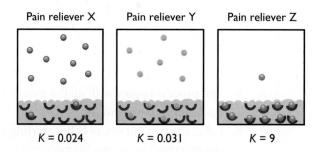

Pain reliever X     Pain reliever Y     Pain reliever Z

*K* = 0.024     *K* = 0.031     *K* = 9

pain reliever + receptor ⇌ [pain reliever : receptor]

A + B ⇌ AB

*The figures show particle views of the binding for three different pain relievers. The pain reliever molecules are represented as spheres, and the receptors are shown as half circles. The square represents a very tiny volume of solution.*

---

**HEALTH**
**CONNECTION**

Sneezing? When the body releases histamine into the nasal passages, the well-known "runny" nose occurs. Histamine is important as the immune system fights allergens. So histamine isn't just a nuisance. It is a neurotransmitter involved in regulating many bodily responses. We can take antihistamine medicine to block the binding of histamines to shut down the "runny" nose response to the flood of histamines released.

Iancu Cristian/Shutterstock

The figure shows that the equilibrium constant $K$ is smallest for pain reliever X and greatest for pain reliever Z. For pain reliever X, 8 out of 10 pain reliever molecules are unbound. For pain reliever Z, only 1 molecule out of 10 pain reliever molecules is unbound. Although other factors might be important, it is likely that pain reliever Z is more effective at relieving pain per mole because more molecules are bound to receptors. It is reasonable to conclude that a small value of $K$ suggests that fewer pain reliever molecules are bound, and therefore, the pain reliever is less effective.

## Equilibrium Constant $K$

As you have learned, many processes in the human body involve the attachment of molecules to receptors. For example, molecules bind reversibly to receptors to regulate sleep, memory, mood, muscle contraction, and pain. In studying these processes in the human body, the concepts of equilibrium and reversible processes are relevant.

The **equilibrium constant $K$** is a measure of whether starting substances or products are favored in an equilibrium mixture. A large value of $K$ indicates a large amount of product in the equilibrium mixture. A small value of $K$ indicates very little product in the equilibrium mixture. A value of $K$ can be determined experimentally for each reversible process. Because $K$ describes the concentrations of starting substances and products in the equilibrium mixture, it is a valuable concept in making predictions about equilibrium mixtures.

**Big Idea** The equilibrium constant $K$ is a measure of whether starting substances or products are favored in an equilibrium mixture.

**Example 1**

### Pain Relievers and $K$

Explain how the value of $K$ relates to the particle views for pain relievers X, Y, and Z shown on page 626.

*Solution*

The smaller the value of $K$ is, the more the starting substance is favored in the equilibrium mixture. In this case, the starting substances are individual molecules and empty receptor sites. For the larger value of $K$, the products are favored in the equilibrium mixture. In this case, the product is the pain reliever attached to the receptor.

## Dissociation of Strong and Weak Acids into Ions

The molecules in substances called acids both *dissolve* in water and *dissociate* into $H^+$ and an anion. The word *dissolve* indicates that water molecules surround the acid molecules, and the acid molecules spread out in the solution. While salts dissolve and dissociate into ions, most molecular substances *dissolve*, without dissociating, like sugar. Nothing further happens.

Acids are special because even though they are molecular substances, they *dissociate* into cations and anions when they dissolve. This is shown in the particle views for tiny volumes of hydrochloric acid, HCl, solution. The solution is mostly $H^+$ and $Cl^-$ ions, with no measurable number of HCl molecules.

Because acid molecules *dissolve* in an aqueous solution, and they also *dissociate*, you might wonder what solution molarity refers to in these cases. For example, the particle view of 0.010 M HCl does not have any molecules of HCl. Instead, the solution consists of 0.010 M $H^+$ and 0.010 M $Cl^-$ because all of the molecules are *dissociated* into ions.

Formic acid, HCOOH, also *dissolves* and *dissociates*. However, as shown in the particle view, the solution is a mixture of HCOOH molecules, $H^+$ ions, and $HCOO^-$ ions. This is because the reaction is reversible. For this solution, the concentration of acid molecules, HCOOH, remains close to the solution molarity of 0.010 M because so few molecules *dissociate*. The $H^+$ concentration is only 0.0013 M. Therefore, for the same solution molarity, the pH of the HCl solution is lower than the pH of the HCOOH solution.

**Important to Know** Hydrochloric acid is referred to as a strong acid, which means a strongly dissociated acid. Formic acid is called a weak acid, which means a weakly dissociated acid.

HCl (aq) ⟶ $H^+$ (aq) + $Cl^-$ (aq)
K = 1,300,000
pH = 2.0

HCOOH (aq) ⇌ $H^+$ (aq) + $HCOO^-$ (aq)
K = 0.00018
pH = 2.9

## Example 2

### The Value of K

A student prepares two aqueous solutions with a total acid concentration of 0.10 M. One solution is prepared with chlorous acid, HOClO, and the other with hypochlorous acid, HOCl. The equilibrium constant for HOClO is $K = 0.011$. The equilibrium constant for HOCl is $K = 0.000000029$. Which solution will have a lower pH value? Explain your thinking.

### Solution

A larger value of $K$ indicates more products and a greater degree of dissociation. This means that there will be more $H^+$ in the HOClO solution, and the pH will be lower.

**KEY TERM**

equilibrium constant $K$

## LESSON SUMMARY

### How can you predict if products are favored in a reversible process?

Chemical processes are characterized by an equilibrium constant $K$. The value of $K$ is a measure of the degree to which the starting substances are converted to products. A large value of $K$ indicates a large amount of product in the equilibrium mixture. A small value of $K$ indicates very little product in the equilibrium mixture. Strong acids have a value of $K$ that is very large, so products are favored. This means that very nearly all of the molecules are dissociated into ions. Weak acids have a value of $K$ that is very small, which means that reactants are favored. This means that at any given time, only a small number of molecules are dissociated. The equilibrium constant $K$ provides information about what is present in the solution and whether starting substances or products are favored in reversible processes.

## Exercises

### Reading Questions

1. Why is it useful to know the value of the equilibrium constant $K$ for a reversible process?
2. How does the equilibrium constant $K$ distinguish between strong and weak acids?

### Reason and Apply

3. Hemoglobin is a large molecule in red blood cells that transports $O_2$ from the lungs to cells in the human body. Consider the two reversible processes shown below involving oxygen, $O_2$, and carbon monoxide, CO, attaching to hemoglobin.

$$\text{hemoglobin} + O_2 \rightleftharpoons [\text{hemoglobin} : O_2]$$
$$\text{hemoglobin} + CO \rightleftharpoons [\text{hemoglobin} : CO]$$

The binding of CO to hemoglobin is more than 200 times greater than the binding of $O_2$ to hemoglobin.

a. Which reversible process has the larger equilibrium constant? Explain your thinking.
b. Sketch a particle view for both processes.
c. Explain why CO is extremely toxic to humans.

*(continued)*

4. Use the table to answer the questions below. The brackets, [ ], indicate concentrations in the equilibrium mixture.

| # | Solution | Formula | Solution molarity | [HA] | [H$^+$] | [A$^-$] | K |
|---|----------|---------|------------------|------|--------|--------|---|
| 1 | nitric acid | HNO$_3$ | 1.0 M | ~0 M | 1.0 M | 1.0 M | 40 |
| 2 | nitric acid | HNO$_3$ | 0.50 M | ~0 M | 0.50 M | 0.50 M | 40 |
| 3 | nitric acid | HNO$_3$ | 0.10 M | ~0 M | 0.10 M | 0.10 M | 40 |
| 4 | nitrous acid | HNO$_2$ | 1.0 M | 0.98 M | 0.022 M | 0.022 M | 0.00056 |
| 5 | benzoic acid | C$_6$H$_5$COOH | 1.0 M | 0.99 M | 0.0079 M | 0.0079 M | 0.000063 |

a. Describe how the H$^+$ concentration for Solutions 1, 2, and 3 relates to the solution molarity.

b. Explain how the value of *K* depends on the specific acid and the solution molarity.

c. Explain why [H$^+$] = [A$^-$].

d. For a weak acid, describe a particle view of the equilibrium mixture.

e. Determine the pH for all five solutions.

f. If spilled, Solution 3 will cause more harm than Solution 5. Use this information to explain why pH, as opposed to molarity, is important in determining how dangerous an acidic solution is.

# LESSON 121
# Equilibrium Calculations

## THINK ABOUT IT

One reason that aqueous acidic solutions can be dangerous is because of the sensitivity of the human body to high concentrations of $H^+$, or low pH. Because weak acids only dissociate to a small degree, the molarity of a weak acid solution alone does not provide information about the concentration of $H^+$. So far, you have considered that the equilibrium constant $K$ is a measure of the degree of dissociation. But is there a more specific relationship between $K$ and the solution molarity?

**How does $K$ describe the concentrations in an equilibrium mixture?**

*To answer this question, you will explore*

- The Equilibrium-Constant Equation for a Weak Acid
- The General Form of the Equilibrium-Constant Equation

## EXPLORING THE TOPIC

### The Equilibrium-Constant Equation for a Weak Acid

Many parts of the human body can only tolerate a specific range of pH values. For example, the normal pH range of your skin is 4 to 6, due to the presence of lactic acid from perspiration and of the acids that make up the outer skin layer. When you wash your hands with soaps with pH values that are much higher than that of your skin, this can cause skin irritation and harm beneficial microorganisms that reside in the skin. For this reason, small amounts of weak acids such as citric acid are added to hand soaps to adjust their pH.

Because weak acids do not dissociate completely, you might wonder how to determine the pH of a solution for a certain concentration of a weak acid. One way is to prepare solutions of various molarities and measure the pH. Another way is to calculate the pH by using the equilibrium constant $K$ for a specific acid and the **equilibrium-constant equation.**

**RELATIONSHIP BETWEEN $K$ AND EQUILIBRIUM CONCENTRATIONS**

In the previous lesson, $K$ was introduced as a number that describes what is present in the equilibrium mixture. The equation below defines $K$ more specifically for an acidic solution. The equation relates the equilibrium constant $K$, the concentrations of an acid, HA, and the ion products, $H^+$ and $A^-$, in an

Alexander Raths/Shutterstock

equilibrium mixture of an acid dissolved in water. The brackets, [ ], indicate the concentrations at equilibrium in moles per liter.

$$HA(aq) \Longleftrightarrow H^+(aq) + A^-(aq)$$

$$K = \frac{[H^+][A^-]}{[HA]}$$

For a weak acid solution, $K$ is a very small number. There are very few ions in the solution because very few acid molecules dissociate. It is possible to determine the value of $K$ for a specific weak acid by making a solution of known molarity and measuring the solution pH.

- $[H^+]$ at equilibrium is equal to $10^{-pH}$. This relationship is determined by rearranging $pH = -\log[H^+]$. For the dissociation of a weak acid, $[H^+]$ is a tiny number.
- $[A^-] = [H^+]$ for a weak acid dissolved in water. For every $H^+$ ion, there is one $A^-$ ion.
- $[HA] \approx$ solution molarity. The equilibrium concentration of HA is approximately equal ("$\approx$") to the solution molarity because such a small amount dissociates.

**Example 1**

### Determination of the Value of $K$

A student prepares a 0.10 M solution of acetic acid. The pH of the solution is 2.9. Determine the equilibrium constant $K$ for lactic acid.

*Solution*

$$[H^+] = 10^{-pH} = 10^{-2.9} = 1.3 \times 10^{-3} = 0.0013 \text{ M}$$

$$[A^-] = [H^+] = 0.0013 \text{ M}$$

$$[HA] \approx [HA]_{total} \approx 0.10 \text{ M}$$

Substituting the values in the equilibrium-constant equation gives:

$$K = \frac{[H^+][A^-]}{[HA]} \approx \frac{(0.0013)(0.0013)}{0.10} \approx 0.000017 \approx 1.7 \times 10^{-5}$$

Note that all these calculations are approximate because they use the solution molarity to approximate the concentration of HA at equilibrium. Since very little HA dissociates, the equilibrium concentration of HA does not differ significantly from the solution molarity. In this case, the concentration of HA is ~75 times the concentration of $H^+$.

### USING $K$ TO CALCULATE THE pH OF A WEAK ACID SOLUTION

Values of $K$ for various weak acids are listed in the table. If you look up the value of $K$ for a specific weak acid, you can use it to determine the solution pH. (The temperature is assumed to be 25° C.)

| Acid | Formula | $K$ |
|---|---|---|
| hydrochloric acid | HCl | 1,300,000 |
| nitric acid | $HNO_3$ | 40 |
| nitrous acid | $HNO_2$ | 0.00056 |
| formic acid | HCOOH | 0.00018 |
| benzoic acid | $C_6H_5COOH$ | 0.000063 |
| acetic acid | $CH_3COOH$ | 0.000018 |
| hydrogen sulfide | $H_2S$ | 0.00000010 |
| hypochlorous acid | HOCl | 0.000000030 |

First, it is necessary to rearrange the equilibrium-constant equation to solve for [H$^+$].

$$K = \frac{[H^+][A^-]}{[HA]}$$

For an acid dissolved in water: $[H^+] = [A^-]$

Substituting for [A$^-$] gives

$$K = \frac{[H^+]^2}{[HA]}$$

Solving for [H$^+$] gives

$$[H^+]^2 = K[HA]$$

$$[H^+] = \sqrt{K[HA]}$$

$$pH = -\log[H^+]$$

The next step is to find the value of $K$ in a reference table and decide on a solution molarity.

**Big Idea** The equilibrium constant $K$ is useful in predicting the pH of a weak acidic solution.

## Example 2

### Determination of the pH of Artificial Lemon Juice

Citric acid is often added to juice to give it a sour taste. However, if the pH is below 3, many people will find that the juice is too sour to drink. Given that $K = 7.2 \times 10^{-4} = 0.00072$ for the dissociation of citric acid, determine if the solutions listed below will have a pH that is below 3.

   **a.** 1.0 M

   **b.** 0.50 M

   **c.** 0.10 M

*(continued)*

*Solution*

**a.** $[H^+] = \sqrt{K[HA]} \approx \sqrt{(0.00072)(1.0)} \approx 0.027$

$pH = -\log[H^+] \approx -\log(0.027) \approx 1.6$

**b.** $[H^+] = \sqrt{K[HA]} \approx \sqrt{(0.00072)(0.50)} \approx 0.019$

$pH = -\log[H^+] \approx -\log(0.019) \approx 1.7$

**c.** $[H^+] = \sqrt{K[HA]} \approx \sqrt{(0.00072)(0.10)} \approx 0.0084$

$pH = -\log[H^+] \approx -\log(0.0084) \approx 2.0$

All three solutions have a pH less than 3, so they may all be too sour to drink.

## ⟳ The General Form of the Equilibrium-Constant Equation

So far, the discussion has centered on reversible processes involving dissociation of a substance into two pieces. However, the form of the equilibrium-constant equation changes depending on the balanced chemical equation for the reversible process. The table shows how the equilibrium-constant equation changes with different types of chemical processes. Examine the table to look for patterns.

| Reaction | Equilibrium-constant equation |
|---|---|
| $A \rightleftharpoons B$ | $K = \dfrac{[B]}{[A]}$ |
| $2A \rightleftharpoons B$ | $K = \dfrac{[B]}{[A]^2}$ |
| $A + B \rightleftharpoons C$ | $K = \dfrac{[C]}{[A][B]}$ |
| $C \rightleftharpoons A + B$ | $K = \dfrac{[A][B]}{[C]}$ |
| $A_2 + B_2 \rightleftharpoons 2AB$ | $K = \dfrac{[AB]^2}{[A_2][B_2]}$ |

You might notice two important patterns.

- The products are always in the numerator, and the starting substances are always in the denominator.

- The coefficients in the chemical equation show up in the equilibrium-constant equation as exponents.

These patterns can be generalized as shown below.

$$aA + bB \rightleftharpoons cC + dD$$

$$K = \frac{[C]^c[D]^d}{[A]^a[B]^b}$$

## PURE SOLIDS AND LIQUIDS

The equilibrium-constant equation keeps track of concentrations that vary. This includes concentrations of substances in solution and concentrations of gases, which vary as gases expand and contract. However, pure solids and pure liquids do not appear in the equation. They are assigned a value of 1 because they don't have a "concentration" that changes.

Here are two examples.

$$MgCO_3(s) \Longleftrightarrow MgO(s) + CO_2(g) \qquad K = [CO_2]$$

$$PbCl_2(s) \Longleftrightarrow Pb^{2+}(aq) + 2Cl^-(aq) \qquad K = [Pb^{2+}][Cl^-]^2$$

## LESSON SUMMARY

### How does *K* describe the concentrations in an equilibrium mixture?

A system at equilibrium reaches a balance between starting substances and products characterized by the equilibrium constant *K*. The concentrations adjust until the rates of the forward and reverse processes are equal. At equilibrium, there is a specific relationship between *K* and the concentrations of starting substances and products in an equilibrium mixture, called the *equilibrium-constant equation*. For a dissociation reaction, the relationship is:

$$HA(aq) \Longleftrightarrow H^+(aq) + A^-(aq) \qquad K = \frac{[A][B]}{[AB]}$$

**KEY TERM**

equilibrium-constant equation

If *K* is known for a reversible process at a given temperature, it can be used in calculations for various equilibrium concentrations for this same system.

## Exercises

### Reading Questions

1. How can you use the equilibrium constant *K* to determine the pH of a weak acid solution?

2. Explain how to write the equilibrium-constant equation for a reversible process.

### Reason and Apply

3. Use the information provided in the table to determine the equilibrium constant *K* for these weak acid solutions.

| Weak acid solution | [H$^+$] | *K* |
|---|---|---|
| 0.10 M chloroacetic acid | 0.0118 M | |
| 0.10 M formic acid | 0.00424 M | |
| 0.10 M acetic acid | 0.00134 M | |
| 0.050 M acetic acid | 0.000949 M | |

*(continued)*

4. Use the information provided in the table to determine the $H^+$ concentration and the pH for these weak acid solutions.

| Weak acid solution | $K$ | $[H^+]$ | pH |
|---|---|---|---|
| 0.10 M benzoic acid | $6.3 \times 10^{-5}$ | | |
| 0.10 M *p*-chlorobenzoic acid | $1.0 \times 10^{-4}$ | | |
| 0.10 M *p*-nitrobenzoic acid | $3.7 \times 10^{-4}$ | | |
| 0.050 M *p*-nitrobenzoic acid | $3.7 \times 10^{-4}$ | | |

5. Write equilibrium-constant equations for each reversible process.
   a. $NH_3(aq) + H_2O(l) \rightleftharpoons NH_4^+(aq) + OH^{2-}(aq)$
   b. $Na_2SO_4(s) \rightleftharpoons 2Na^+(aq) + SO_4^{2-}(aq)$
   c. $C_2H_4(g) + H_2(g) \rightleftharpoons C_2H_6(g)$

# CHAPTER 23

## Chemical Equilibrium

## SUMMARY

### KEY TERMS

indicator

pigment

reversible process

chemical equilibrium

equilibrium constant $K$

equilibrium-constant equation

# Showtime Update

Many chemical processes proceed in the forward and reverse directions. Colorful indicator molecules dissolved in solutions of varying acidity can provide evidence of what is happening on the atomic scale. The color of the solution shows conditions that favor the forward process and other conditions that favor the reverse process.

Unlike combustion reactions, which go to completion to become products, in reversible processes both the forward and reverse processes occur simultaneously. Reversible processes are written with a double arrow to show that both the forward and reverse processes occur at the same time.

When both the forward and reverse processes occur at equal rates, the mixture is at chemical equilibrium. Examples of equilibrium mixtures include the back-and-forth evaporation of water in a closed container or the dissociation of an acid such as hydrofluoric acid, HF into ions in solution:

$$HF(aq) \rightleftharpoons H^+(aq) + F^-(aq)$$

At equilibrium, the rate of the forward process is equal to the rate of the reverse process, but this does not mean that the amounts of starting substances and products are necessarily equal. While there is change on an atomic scale, the amounts of starting substances and products do not change once equilibrium is reached.

There is a mathematical relationship between the amounts of starting substances and products in an equilibrium mixture. This relationship can be described by the equilibrium-constant equation. The equilibrium constant $K$ is the same for a specific reversible process at a specified temperature.

## REVIEW EXERCISES

1. Suppose that you have a 0.10 M HCl solution and a 0.10 M HF solution. The HCl solution has an $H^+$ concentration that is 0.10 M. The HF solution has an $H^+$ concentration that is 0.010 M.
   a. Hydrochloric acid is a strong acid. Write a chemical equation to describe what happens in solution.
   b. Draw a particle view showing what happens to 10 HCl molecules when they are dissolved in a tiny volume of water.
   c. For the same concentration of acid, the $H^+$ concentration is smaller. Write a chemical equation to describe what happens in solution.
   d. Draw a particle view showing what happens to 10 HF molecules when they are dissolved in a tiny volume of water.

2. Suppose that you are studying an aqueous solution with the ammonium ion, $NH_4^+$. The chemical equation describing the equilibrium mixture is

$$NH_4^+(aq) \rightleftharpoons H^+(aq) + NH_3(aq)$$

   a. A 0.10 M solution has a pH of 2.9. Is the ammonium ion, $NH_4^+$, a strong acid, a strong base, a weak acid, or a weak base? Explain how you know.
   b. List all the molecules and ions that would be present in a 0.1 M solution of the ammonium ion.
   c. Write the equilibrium-constant equation for the ammonium ion ($K = 1.8 \times 10^{-5}$).
   d. If $[NH_4^+] = 0.50$ M at equilibrium, what is the pH of the solution?
   e. If you add ammonia, $NH_3$, to an acidic solution, the pH increases. Explain why.

3. Write the equilibrium-constant equations for these reversible processes.
   a. $A(aq) + B(aq) \rightleftharpoons C(aq) + D(aq)$
   b. $COCl_2(g) \rightleftharpoons CO(g) + Cl_2(g)$
   c. $PbCl_2(s) \rightleftharpoons Pb^{2+}(aq) + 2Cl^-(aq)$

4. For the reaction $N_2(g) + 3H_2(g) \rightleftharpoons 2NH_3(g)$, complete these exercises:
   a. Write the equilibrium-constant equation.
   b. Solve for $K$ if the equilibrium concentrations are $[N_2] = 0.1$ M, $[H_2] = 0.2$ M, $[NH_3] = 0.1$ M.
   c. Based on your answer to part b, do you think that this reaction produces an equilibrium mixture that favors starting substances, products, or both about equally?

## Neurotransmitters

**PROJECT** Do Web research on one neurotransmitter. What is the function of the neurotransmitter in the body? What type of receptors are involved and what happens when the neurotransmitter binds to those receptors? Write a report.

Your report should include

- a description of the structure of the neurotransmitter molecule.
- a description of the receptors for the neurotransmitter and where they are found in the body.
- an explanation of what happens when neurotransmitter molecules bind to the receptor.
- a description of how the binding process is reversed.
- a discussion of what happens if the body does not produce enough of the neurotransmitter.

# CHAPTER 24

# Changing Conditions

### In this chapter, you will study

- changing conditions for an equilibrium mixture

- how to use Le Châtelier's principle to explain how processes respond to stresses

- the chemical equilibrium of acid-base indicators

- equilibrium constants for different types of reactions

*If starting substances or products are added to an equilibrium mixture, the composition adjusts to maintain equilibrium. Sometimes, this change is visible as a color change.*

When a system is at equilibrium, there is a dynamic balance between the forward process and the reverse process. At equilibrium, the rate of the forward process is equal to the rate of the reverse process. However, conditions do not always stay the same for systems at equilibrium. When conditions change, an equilibrium mixture will adjust in a predictable way to those changes so as to reduce the effect of the change and maintain equal rates.

*How Pushy*

# Le Châtelier's Principle

## THINK ABOUT IT

Human blood has a certain pH. When people exercise, they produce a lot of carbon dioxide, which in turn builds up carbonic acid in their bodies. A build-up of carbonic acid can alter the blood's pH, which can put stress on the body. How does the body maintain a healthy equilibrium?

**What happens to a system at equilibrium when conditions change?**

*To answer this question, you will explore*
- Changing Conditions
- Le Châtelier's Principle

## EXPLORING THE TOPIC

## Changing Conditions

When a mixture is at equilibrium, the forward and reverse processes are dynamically balanced, going backward and forward at equal rates. For example, if an indicator is dissolved in water, the indicator, H*In*, dissociates in the forward process and the ions, $H^+$ and *In*$^-$, attract one another in the reverse process. Although equilibrium is dynamic, once equilibrium is established, the concentrations of H*In*, $H^+$, and *In*$^-$ in the aqueous solution do not change over time, as long as the conditions stay the same.

However, conditions do not always stay the same for an equilibrium mixture. Something new might be added to the mixture or the mixture might be heated or cooled. For example, HCl might be added to the indicator solution described above. Such changes disturb the equilibrium, which changes in a predictable way by adjusting to new concentrations to counteract the disturbance.

Here are three types of changes that can be made to an equilibrium mixture:
- A change in the concentrations—this applies when more of the starting substances or products are added to an equilibrium mixture.
- A change in external pressure—this applies to processes involving gases when the container volume is increased or decreased.
- A change in temperature—heating or cooling an equilibrium mixture changes the value of *K*.

Changing the conditions for an equilibrium mixture will disturb the system. The examples here demonstrate how each of the three types of changes (or disturbances) affects a reversible process at equilibrium.

## COBALT CHLORIDE EQUILIBRIA

The dissociation of tetrachlorocobalt (II) ions, $CoCl_4^{2-}(aq)$, to aqueous cobalt (II) ions, $Co^{2+}(aq)$ and chloride ions, $Cl^-(aq)$, is especially useful in demonstrating changes in equilibrium because it involves a striking color change.

$$CoCl_4^{2-}(aq) \rightleftharpoons Co^{2+}(aq) + 4Cl^-(aq) \qquad K = 4000$$
blue        pink

Cobalt chloride solutions can be pink, blue, or various shades of purple. This figure shows molecular views of tiny volumes of a pink and blue cobalt solution. Notice that there are more $Co^{2+}$ ions (pink circles) than $CoCl_4^{2-}$ ions (blue circles) in the pink solution, but more $CoCl_4^{2-}$ ions in the blue solution. Notice also that the blue solution has a larger number of $Cl^-$ ions (gray circles).

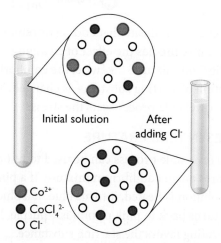

● $Co^{2+}$
● $CoCl_4^{2-}$
○ $Cl^-$

*There are more $Co^{2+}$ ions in the pink solution, but more $CoCl_4^{2-}$ ions in the blue solution.*

> **Important to Know** Equilibrium mixtures can have various concentrations of starting substance and products. For example, equilibrium mixtures of cobalt chloride can be pink, blue, or various shades of purple.

It is possible to predict how a disturbance, such as the addition of $Cl^-$ ions to the pink cobalt solution, will change the equilibrium mixture.

## CHANGES IN CONCENTRATION

Imagine that you have a pink sample of a cobalt chloride solution. The pink color indicates that the sample has more $Co^{2+}$ ions (one of the products) than $CoCl_4^{2-}$ ions (starting substance). If you change the conditions by adding more $Cl^-$ ions to the mixture, the sample turns blue. So adding more products results in the formation of more starting substance.

Add $Cl^-$ ion

Before

After

*When one of the product ions, $Cl^-$, is added, more starting substance, $CoCl_4^{2-}$, forms and the color shifts to blue.*

If you take this blue sample and add silver ions, $AgNO_3(aq)$, to remove chloride ions, $Cl^-$, by precipitating $AgCl(s)$, the mixture changes back to pink. So removing

Cl⁻ ions (one of the products) results in the formation of more products ($Co^{2+}$ and $Cl^-$ ions). The illustration shows the results of these two changes.

*When one of the product ions, $Cl^-$, is removed, more of both products, $Co^{2+}$ and $Cl^-$ ions, form and the color shifts to pink.*

The color changes indicate that adding or removing one of the products in an equilibrium mixture has the effect of changing the final concentrations of the new equilibrium mixture. Likewise, the equilibrium mixture also changes if the concentration of starting substance is changed. The change depends on which side of the reversible process gets the stress.

## CHANGES IN TEMPERATURE

This same reversible process can be used to test the effects of changing temperatures on an equilibrium mixture. If a pink starting sample of cobalt chloride solution is heated, it turns blue. If a blue sample of this mixture is cooled, it turns back to pink or pinkish-purple. In this particular reversible process, heating favors the starting substance.

For this process, heating produces more starting substance, $CoCl_4^{2-}$, and the color shifts to blue.

For this process, cooling produces more product ions, $Co^{2+}$ and $Cl^-$, and the color shifts to pink.

Heating does not always cause more starting substance to form. The direction of the change depends on whether the process is exothermic or endothermic. For an endothermic process, heating produces more products. For an exothermic process, cooling produces more products. The reversible cobalt process is exothermic in the directions the chemical equation is written, so heating causes the formation of more starting substance.

## CHANGES IN PRESSURE

Pressure changes are particularly important when considering equilibrium mixtures that are in the gas phase. Recall from Unit 3: Weather that it is common to use a cylinder and piston as a flexible container for a gas. Pushing down on the

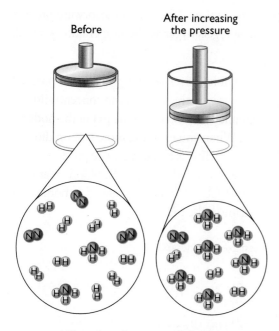

Before | After increasing the pressure

When the volume is made smaller, more ammonia forms, decreasing the number of moles of gas.

piston to decrease the gas volume increases the external pressure on a gaseous equilibrium mixture. A change will occur in the direction that decreases the number of moles of gas and relieves some of the pressure. Recall that the pressure decreases as the number of moles decreases for the same temperature and volume.

For example, consider the process for the formation of ammonia gas from nitrogen and hydrogen gases.

$$N_2(g) + 3H_2(g) \rightleftharpoons 2NH_3(g)$$

Decreasing the volume of the container increases the pressure on the gas mixture. The system will change so that more ammonia forms. This is because there are four moles of gas on the starting substance side of the equation and only two moles of gas on the product side.

The creation of more ammonia relieves the stress of the added pressure by reducing the total number of moles of gas in the mixture.

## ⟳ Le Châtelier's Principle

These very predictable changes to equilibrium mixtures were first noted around 1885 by a French chemist named Henry Le Châtelier. He observed that when the conditions of an equilibrium mixture are modified, the system responds in such a way as to reduce the effect of the change. This idea is known as **Le Châtelier's principle.**

> Le Châtelier's Principle
> When conditions change for a system at equilibrium, the system will respond by reducing the effect of the change.

Le Châtelier's principle allows you to predict qualitatively how the system responds to changes in concentration, pressure, or temperature.

### AN EXPLANATION OF LE CHÂTELIER'S PRINCIPLE

If more of one of the products is added to the equilibrium mixture of starting substances and products, this momentarily increases the rate of the reverse process. The concentrations in the mixture adjust until the rates of the forward and reverse processes are equal. This adjustment results in a new equilibrium mixture that has more starting substance compared with that of the initial solution. The value of the equilibrium constant $K$ is the same for both the adjusted mixture and the initial solution.

A second change is to raise or lower the solution's temperature, which also changes the composition of the equilibrium mixture of starting substances and products. In this case, the changes in composition result in a new value of $K$ depending on if the forward process is exothermic or endothermic. Because there is a new value of $K$, the concentrations of starting substances and products change to satisfy the new equilibrium constant equation. If the forward process is exothermic (heat is released),

**WEATHER CONNECTION**

On a humid day, there is a lot of water vapor in the air. When you exercise in humid weather, your sweat does not evaporate because the air is already nearly saturated with water vapor. Beads of sweat accumulate on your body. In contrast, when you exercise in a dry climate, you lose a lot of water because your sweat evaporates. This is why it is especially important to drink a lot of water in dry climates.

zorani/Getty Images

GIPhotoStock/Science Source

Robert Llewellyn/Getty Images

the value of $K$ gets smaller as the temperature is raised. Likewise, if the forward process is endothermic (heat is required), the value of $K$ gets larger as the temperature is raised.

## BLOOD pH AND EXERCISE

The human body is a large collection of reversible processes. For processes to proceed at the appropriate rates, and for substances to stay in the appropriate concentrations within our bodies, a balance must be maintained. For example, the pH of the body's blood is normally around 7.4, slightly basic. When you exercise, more carbon dioxide is produced, and it dissolves in the bloodstream. This is a natural waste product that you normally exhale from the lungs. However, carbon dioxide in the bloodstream also leads to the formation of carbonic acid, $H_2CO_3$. A large build up of carbonic acid would increase the acidity of the blood and lower the pH.

**Stress:** Exercise adds $CO_2$.

$$H_2CO_3(aq) \rightleftharpoons CO_2(aq) + H_2O(l)$$

**Shift:** The process shifts toward the starting substance.

One of the ways your body deals with a change in the blood's pH is to exhale more rapidly, removing $CO_2$ from the bloodstream and shifting the process back toward the right. This is one reason why we breathe more heavily when exercising—to rid the body of excess carbon dioxide and increase the blood's pH.

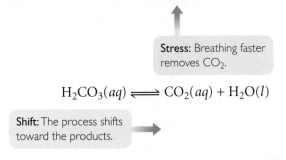

**Stress:** Breathing faster removes $CO_2$.

$$H_2CO_3(aq) \rightleftharpoons CO_2(aq) + H_2O(l)$$

**Shift:** The process shifts toward the products.

---

**Example**

## Decomposition of $SO_3$

Gaseous sulfur trioxide, $SO_3$, decomposes to form sulfur dioxide, $SO_2$, and oxygen, $O_2$. The process is endothermic. Predict how the changes listed here will affect the process.

$$2SO_3(g) \rightleftharpoons 2SO_2(g) + O_2(g)$$

**a.** Increasing the pressure on the system     **b.** Adding more $O_2$

**c.** Removing $O_2$ from the system     **d.** Adding heat

*Solution*

**a.** In the forward process, two moles of gas decompose to form three moles of gas. The system can minimize the effect of a pressure increase by converting some of the products back to $SO_3$ to decrease the number of moles of gas molecules present. This will push the process more toward the starting substances.

*(continued)*

**b.** Adding more $O_2$ to the system will cause more starting substance to form. By making more $SO_3$, some of the excess $O_2$ is removed, relieving the stress on the system.

**c.** Removing $O_2$ from the system has the opposite effect; it causes the formation of more products. This change has the effect of restoring some of the $O_2$ that was removed.

**d.** Because energy is absorbed in the forward process, heating will favor products.

## LESSON SUMMARY

### What happens to a system at equilibrium when conditions change?

When conditions are changed for a system at equilibrium, the system will respond by reducing the effect of the change. This is known as Le Châtelier's principle. Changing conditions (or stresses) include changes to temperature, concentration, or pressure. A system regains equilibrium by rebalancing the concentrations of starting substances and products. The exact nature of the effects on a system are predictable because they depend on the type of change in condition that the system encounters. A temperature change affects the value of $K$. A pressure change or a change in concentration stresses the system, but does not change the value of $K$.

**KEY TERM**

Le Châtelier's principle

### Reading Questions

1. Name two things that you can do to form more starting substances in an equilibrium mixture.

2. Based on the evidence provided in this lesson, explain how you can tell if a process is exothermic as written.

### Reason and Apply

3. If you add more ammonia, $NH_3(aq)$, to the equilibrium mixture described by the chemical equation given below, what will happen? Explain your thinking.

$$Cu(NH_3)_6{}^{2+}(aq) \rightleftharpoons Cu^{2+}(aq) + 6NH_3(aq)$$

4. If you remove ammonia, $NH_3(g)$, from the equilibrium mixture described by the equation here, what will happen? Explain your thinking.

$$N_2(g) + 3H_2(g) \rightleftharpoons 2NH_3(g)$$

5. If you heat the equilibrium mixture for this endothermic process, what will happen? Explain your thinking.

$$2NiO(s) \rightleftharpoons 2Ni(s) + O_2(g)$$

6. Suppose that you want to dissolve more copper (II) hydroxide, $Cu(OH)_2(s)$, in an aqueous solution. Would you add acid or base? Explain your thinking.

$$Cu(OH)_2(s) \rightleftharpoons Cu^{2+}(aq) + 2OH^-(aq)$$

7. For each reversible process listed, describe *two* things you can do to change the concentration of one of the products.
   **a.** $2NO_2(g) \rightleftharpoons N_2O_4(g)$
   **b.** $HF(aq) \rightleftharpoons H^+(aq) + F(aq)$
   **c.** $PbCl_2(s) \rightleftharpoons Pb^{2+}(aq) + 2Cl^-(aq)$
   **d.** $CaCO_3(s) \rightleftharpoons CaO(s) + CO_2(g)$

*How Colorful*

# Applying Le Châtelier's Principle

## THINK ABOUT IT

Acid-base indicators, such as red cabbage juice, change color as the $H^+$ concentration in a solution is changed. The pH at which the color changes is different for different indictors. For example, cabbage juice changes color from red to blue at pH ~7, whereas phenolphthalein changes color from colorless to pink at pH ~9. How do equilibrium considerations predict the pH for the color change?

### How do acid-base indicators work?

*To answer this question, you will explore*

⮕ Applying Le Châtelier's Principle to Acid-Base Indicator Solutions

⮕ Relating the Value of $K$ and the pH to the Indicator Color

⮕ Types of Equilibrium Constants

## EXPLORING THE TOPIC

### ⮕ Applying Le Châtelier's Principle to Acid-Base Indicator Solutions

Acid-base indicators are weak acids, chosen for the bright colors of the H*In* molecule and *In*⁻ ion. Like other weak acids, the dissociation process of the indicator molecules to $H^+$ and an anion in aqueous solutions is reversible.

$$\text{H}In(aq) \rightleftharpoons \text{H}^+(aq) + In^-(aq)$$

**COLORS AT DIFFERENT pH VALUES**

Each indicator has a specific value for the equilibrium constant $K$ and a certain color that depends on the pH of the solution. Examine the information in the table to look for a relationship between $K$, pH, and color.

| Indicator | H*In* color | *In*⁻ color | $K$ | Color at different pH values | | |
|---|---|---|---|---|---|---|
| | | | | pH = 4 | pH = 7 | pH = 10 |
| methyl orange | red | yellow | $3.4 \times 10^{-4}$ | orange | yellow | yellow |
| litmus | red | blue | $3.2 \times 10^{-7}$ | red | purple | blue |
| phenolphthalein | colorless | pink | $3.2 \times 10^{-10}$ | colorless | colorless | pink |

Here are some things you might notice:

- **The value of $K$:** The equilibrium constants are all much less than 1, indicating that the concentrations of the products are quite small. This is because these

## HISTORY
## CONNECTION

Litmus is a mixture of water-soluble organic compounds that are extracted from certain lichens. Often this mixture is infused into filter paper to make litmus paper. Lichens themselves are species of fungus that exist in a close symbiotic relationship with species of photosynthetic algae.

indicators are very weak acids that do not dissociate very much. The value of $K$ for methyl orange is the largest, so, of these three indicators, it dissociates the most. Phenolphthalein, with the smallest $K$ value, dissociates the least. The value of $K$ for litmus is in between the values for methyl orange and phenolphthalein.

$$K = \frac{[H^+][In^-]}{[HIn]}$$

A large denominator indicates less dissociation. $K$ is smaller.

A large numerator indicates more dissociation. $K$ is larger.

- **The color at pH = 7**: A methyl orange solution is yellow at pH = 7, showing that the yellow $In^-$ ion is present in relatively large concentration. A phenolphthalein solution is colorless, showing that the colorless $HIn$ molecules is present in relatively large concentration. A litmus solution, being purple, is a mixture of red $HIn$ molecules and blue $In^-$ ions in roughly equal concentrations. Notice that the methyl orange solution, with the largest value of $K$, has the most product ions. The phenolphthalein solution has the smallest value of $K$ and the least number of product ions.

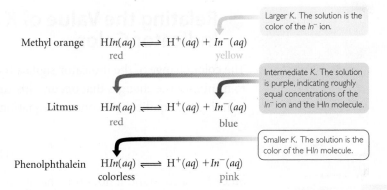

Methyl orange    $HIn(aq) \rightleftharpoons H^+(aq) + In^-(aq)$
red              yellow

Larger $K$. The solution is the color of the $In^-$ ion.

Litmus    $HIn(aq) \rightleftharpoons H^+(aq) + In^-(aq)$
red           blue

Intermediate $K$. The solution is purple, indicating roughly equal concentrations of the $In^-$ ion and the $HIn$ molecule.

Phenolphthalein    $HIn(aq) \rightleftharpoons H^+(aq) + In^-(aq)$
colorless           pink

Smaller $K$. The solution is the color of the $HIn$ molecule.

- **The color at different pH values**: As the pH value increases, the solution color changes from the color of $HIn$ in acidic solutions to the color of $In^-$ in basic solutions. The illustration below shows the color changes for the three indicators in the table. Notice that the color change occurs at higher pH values as the equilibrium constant $K$ gets smaller.

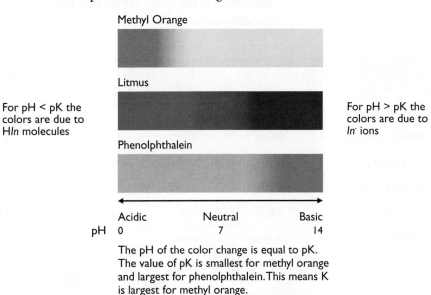

Methyl Orange

Litmus

For pH < pK the colors are due to $HIn$ molecules

For pH > pK the colors are due to $In^-$ ions

Phenolphthalein

Acidic     Neutral     Basic
pH   0        7        14

The pH of the color change is equal to pK. The value of pK is smallest for methyl orange and largest for phenolphthalein. This means K is largest for methyl orange.

**Important to Know** For these acid-base indicator solutions, an acid or base is added to change the value of the pH. For this reason, the concentration of $H^+$ is not equal to the concentration of $In^-$.

## MINIMIZING CHANGES AT EQUILIBRIUM

One way to understand the color differences for indicators in acidic and basic solutions is to apply Le Châtelier's principle. When an acid is added to a solution with an indicator, $[H^+]$ increases. Because $H^+$ is a product of the dissociation of the indicator, the system responds to minimize the change by forming $HIn$ molecules. Alternatively, if a base is added to a solution with an indicator, $[H^+]$ decreases. The system responds by making more product ions, $H^+$ and $In^-$.

**Stress:** Add $H^+$ ions.

$HIn \rightleftharpoons H^+(aq) + In^-(aq)$

**Shift:** The process shifts toward starting substances.

**Stress:** $H^+$ ions are removed.

$HIn \rightleftharpoons H^+(aq) + In^-(aq)$

**Shift:** The process shifts toward the products.

## ⟳ Relating the Value of $K$ and the pH to the Indicator Color

The color change of the indicator signals the shift that has occurred in the mixture to minimize the changes that occur. This can also be seen mathematically if you examine the equilibrium-constant equation.

$$K = \frac{[H^+][In]}{[HIn]}$$

**Important to Know** In pure water, $[H^+] = [In^-]$. However, this is no longer true if a different acid or base is added to change $[H^+]$.

Since $K$ is a constant, it does not change. If any variable in the equation changes, the other variables must adjust. For example, if $[H^+]$ increases, $[In^-]$ must decrease and $[HIn]$ must increase to keep $K$ the same.

### pH RANGE FOR COLOR CHANGE

Take a moment to examine the information for the color change for several indicators in the table. Notice that the color change occurs at different pH values for different indicators.

| Indicator | $HIn$ color | $In^-$ color | $K$ | $[H^+]$ when $[HIn] = [In^-]$ | pH range for color change |
|---|---|---|---|---|---|
| methyl orange | red | yellow | $3.4 \times 10^{-4}$ | $3.4 \times 10^{-4}$ | 3.2–4.4 |
| bromocresol green | yellow | blue | $1.3 \times 10^{-5}$ | $1.3 \times 10^{-5}$ | 3.8–5.4 |
| methyl red | red | yellow | $1.0 \times 10^{-5}$ | $1.0 \times 10^{-5}$ | 4.8–6.0 |
| bromothymol blue | yellow | blue | $5.0 \times 10^{-8}$ | $5.0 \times 10^{-8}$ | 6.0–7.6 |
| phenol red | yellow | red | $1.0 \times 10^{-8}$ | $1.0 \times 10^{-8}$ | 6.6–8.0 |
| phenolphthalein | colorless | pink | $3.2 \times 10^{-10}$ | $3.2 \times 10^{-10}$ | 8.2–10.0 |
| alizarin yellow | yellow | red | $1.0 \times 10^{-11}$ | $1.0 \times 10^{-11}$ | 10.0–12.0 |

The color change for an indicator occurs when the $H^+$ ion concentration is equal to $K$. This happens when the concentration of H$In$ molecules is equal to the concentration of the $In^-$ ions. Rearranging the equilibrium-constant equation proves this to be true.

$$K = \frac{[H^+][In^-]}{[HIn]}$$

$$= [H^+]\frac{[In^-]}{[HIn]}$$

When $[In^-] = [HIn]$, then $K = [H^+]$

When $[H^+] > K$, then $[HIn] > [In^-]$. The solution has the color of H$In$ molecules. When $[H^+] < K$, then $[HIn] < [In^-]$. The solution has the color of $In^-$ ions. When $[H^+] = K$, then $[HIn] = [In^-]$. The solution color is a mixture of the color of H$In$ molecules and $In^-$ ions.

## Example 1

### The Color of Congo Red

Determine the pH value at which the indicator Congo red changes color. The equilibrium constant for the dissociation of the indicator is $K = 10^{-4}$.

### Solution

The color change happens when the concentration of H$In$ equals the concentration of $In^-$. If $[HIn] = [In^-]$, then $K = [H^+] = 10^{-4}$.

$$pH = -\log[H^+]$$
$$= -\log[10^{-4}]$$
$$= 4$$

So the indicator changes color at pH = 4.

## ⮕ Types of Equilibrium Constants

The equilibrium constant $K$ can be used to solve a variety of equilibrium problems. In every case, the equilibrium-constant equation can be used to figure out the concentrations of one or more of the compounds in the equilibrium mixture. The table shows some specific types of chemical equilibrium constants. Each of these constants is designated by a unique subscript.

| Equilibrium constant | Chemical equation | Equilibrium-constant equation |
|---|---|---|
| Solubility product constant $K_{sp}$ | $MX(s) \rightleftharpoons M^+(aq) + X^-(aq)$ | $K_{sp} = [M^+][X^-]$ |
| Acid dissociation constant $K_a$ | $HA(aq) \rightleftharpoons H^+(aq) + A^-(aq)$ | $K_a = \dfrac{[H^+][A^-]}{[HA]}$ |
| Base dissociation constant $K_b$ | $B(aq) + H_2O(l) \rightleftharpoons BH^+(aq) + OH^-(aq)$ | $K_b = \dfrac{[BH^+][OH^-]}{[B]}$ |
| Ionization of water constant $K_w$ | $H_2O(l) \rightleftharpoons H^+(aq) + OH^-(aq)$ | $K_w = [H^+][OH^-]$ |

Notice that the equilibrium-constant equations are the same as those that you might have written, except that they have a subscript attached to the symbol $K$. As discussed previously, liquids and solids are not included in the equilibrium-constant equation because their concentrations do not change. So, equilibrium-constant equations can be used to determine the concentrations of molecules and ions in solution and the concentrations of gaseous molecules in systems at equilibrium.

The equilibrium constant $K_{sp}$ can be used to determine the solubility of an ionic compound.

## Comparing Solubility

The solubility product constants $K_{sp}$ for several alkaline earth metal sulfates at 25 °C are given in the table.

| Metal sulfate | Solubility product constant $K_{sp}$ |
|---|---|
| magnesium sulfate, $MgSO_4(s)$ | 4.7 |
| calcium sulfate, $CaSO_4(s)$ | $4.9 \times 10^{-5}$ |
| barium sulfate, $BaSO_4(s)$ | $1.1 \times 10^{-10}$ |

a. Put the three salts in order of increasing solubility from least soluble to most soluble. Explain your thinking.

b. Use the periodic table to predict whether strontium sulfate, $SrSO_4(s)$, is more soluble than calcium sulfate, $CaSO_4(s)$. Explain your thinking.

c. Suppose you add sodium sulfate, $Na_2SO_4(s)$, to a saturated solution of calcium sulfate, $CaSO_4(aq)$. Use Le Châtelier's principle to explain why a small amount of $CaSO_4(s)$ precipitates from the solution.

### Solution

a. The chemical equation for the solubility of an alkaline earth metal sulfate, $MSO_4(s)$, is

$$MSO_4(s) \rightleftharpoons M^{2+}(aq) + SO_4^{2-}(aq) \qquad K_{sp} = [M^{2+}][SO_4^{2-}]$$

If the solubility product constant $K_{sp}$ is large, products are favored (the salt is more soluble). Therefore, the solubility from least soluble to most soluble is $BaSO_4 < CaSO_4 < MgSO_4$.

b. All of the metals are in Group 2A, the alkaline earth metals. The solubility decreases as you go down the group. So, you expect strontium sulfate, $SrSO_4(s)$, to be less soluble than calcium sulfate, $CaSO_4(s)$, but more soluble than barium sulfate, $BaSO_4(s)$.

c. Adding sodium sulfate, $Na_2SO_4(s)$, adds more sulfate ions. Because $SO_4^{2-}(aq)$ is a product of dissolving calcium sulfate, $CaSO_4(aq)$, the system will reduce the stress by forming more reactants. This is why some $CaSO_4(s)$ precipitates.

The equilibrium constant $K_a$ can be used to determine the strength of an acid or a base.

## Strength of an Acid

Imagine that you prepare a 0.10 M solution of HCl and a 0.10 M solution of HOCl. The concentrations of the ions in equilibrium are given in this table.

The general chemical equation for acid dissociation is:

$$HA(aq) \rightleftharpoons H^+(aq) + A^-(aq)$$

| Acid | [HA] | [H⁺] | [A⁻] |
|---|---|---|---|
| hydrochloric acid, HCl($aq$) | ~0 | 0.10 M | 0.10 M |
| hypochlorous acid, HOCl($aq$) | 0.099994 M | $6.0 \times 10^{-5}$ M | $6.0 \times 10^{-5}$ M |

a. Determine $K_a$ for hypochlorous acid, HOCl($aq$).

b. Explain why $K_a$ for HCl($aq$) is a very large number.

c. Which acid is stronger? Explain your thinking.

*Solution*

a. $HOCl(aq) \rightleftharpoons H^+(aq) + OCl^-(aq)$

$$K_a = \frac{[H^+][OCl^-]}{[HOCl]}$$

$$= \frac{(6.0 \times 10^{-5})(6.0 \times 10^{-5})}{0.099994}$$

$$= 3.6 \times 10^{-8}$$

b. The value [HCl] is extremely small since it is approximately zero. Dividing by such a small number gives a very large number, so $K$ for HCl must be extremely large.

So $K_a$ for this acid is very large.

c. The very large $K_a$ for HCl indicates that HCl fully dissociates in water. The tiny $K_a$ for HOCl indicates that HOCl does not dissociate as much as HCl. So HCl is a stronger acid than HOCl.

## How do acid-base indicators work?

Acid-base indicators are weak acids or bases that change color depending on pH. The concentrations of an indicator molecule and its anion depend on the concentration of $H^+$ ions in solution. Therefore, the color of an indicator is a measure of the acidity of the solution. The color of an indicator changes when the concentrations of the H*In* molecules and *In*$^-$ anions are identical. This occurs when $K = [H^+]$. The lessons learned from applying equilibrium ideas to indicators are general, and they can be applied to other types of equilibria.

## Exercises

### Reading Questions

1. Explain how an acid-base indicator works to show the pH of a solution.

2. Does the value of $K$ for an indicator change when the indicator is added to an acid or a base solution? Explain.

### Reason and Apply

3. The solubility product constants $K_{ap}$ at 25 °C for several chlorides are provided in the table.

| Metal chloride | Solubility product constant $K_{sp}$ |
|---|---|
| thallium (I) chloride, TlCl($s$) | $1.86 \times 10^{-4}$ |
| copper (I) chloride, CuCl($s$) | $1.72 \times 10^{-7}$ |
| silver (I) chloride, AgCl($s$) | $1.77 \times 10^{-10}$ |

   a. Put the three salts in order of increasing solubility—from least soluble to most soluble. Explain your thinking.
   b. Suppose that you add sodium chloride, NaCl($s$), to a saturated solution of copper chloride, CuCl($aq$). Use Le Châtelier's principle to explain what happens. Assume that NaCl is very soluble.

4. Imagine you prepare a 0.10 M solution of nitric acid, HNO$_3$, and a 0.10 M solution of hydrofluoric acid, HF. The concentrations of the ions in equilibrium are given in the table.

   General equation for acid dissociation: HA($aq$) $\rightleftharpoons$ H$^+$($aq$) + A$^-$($aq$)

| Acid | [HA] | [H$^+$] | [A$^-$] |
|---|---|---|---|
| nitric acid, HNO$_3$($aq$) | ~0 | 0.10 M | 0.10 M |
| hydrofluoric acid, HF($aq$) | 0.0974 M | $2.6 \times 10^{-3}$ M | $2.6 \times 10^{-3}$ M |

   a. Determine $K$ for hydrofluoric acid, HF.
   b. Explain why $K$ for nitric acid, HNO$_3$, is a very large number.
   c. Which acid is stronger? Explain your thinking.
   d. Which solution has the higher concentration of H$^+$($aq$), 0.10 M HNO$_3$ or 0.10 M HF? Explain your thinking.

*(continued)*

5. Use the table on page 648 to predict the color of these indicators if you dissolve them in water. Explain your thinking.
   a. alizarin yellow
   b. bromocresol green
   c. methyl red
   d. phenol red

6. Use Le Châtelier's principle to explain why bromocresol green is blue if you dissolve it in a base solution, such as 0.10 M sodium hydroxide, NaOH.

7. Determine the pH value at which thymolphthalein changes color. The equilibrium constant for the dissociation of the indicator is $K = 10^{-10}$.

8. Methyl violet changes color from the yellow H*In* molecule to the blue-violet *In*⁻ ion at a pH of 1. What is the equilibrium constant for the dissociation of methyl violet?

9. **PROJECT**

   Look up the source, structure, and properties of a particular indicator. Write a short description of the molecule.

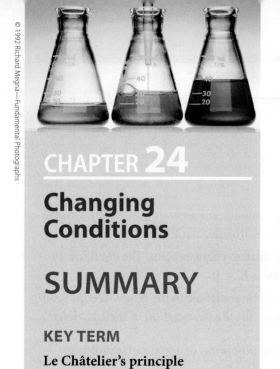

## CHAPTER 24

# Changing Conditions

# SUMMARY

**KEY TERM**

Le Châtelier's principle

# Showtime Update

Le Châtelier's principle states that when conditions change for an equilibrium mixture, the amounts of starting substances and products will adjust to reduce the effect of the change.

If you add more of a starting substance to an equilibrium mixture, the amounts will shift in the direction of forming products. If you remove a starting substance from an equilibrium mixture, the amounts will shift in the direction of forming more starting substances. The same is true for adding or removing products.

Changing temperature is another way to place a stress on an equilibrium mixture. If the process is exothermic, heating will shift the amounts in the direction of forming starting substances. For an endothermic process, heating will shift the amounts in the direction of forming products.

For equilibrium mixtures with starting substances and products that are gases, changes in volume that cause changes in pressure place a stress on the system at equilibrium. If the pressure increases, a change will occur in the direction that decreases the total number of moles of gas to reduce the pressure.

## REVIEW EXERCISES

1. The indicator called litmus, H$In$, is red in acidic solution and blue in basic solution.
   a. Write a chemical equation to describe the equilibrium mixture.
   b. What color is $In^-$? Explain how you know.
   c. What happens if you add acid to a red solution of litmus?
   c. What happens if you add base to a red solution of litmus?
   d. Draw a particle view of a purple litmus solution.

2. Hydrofluoric acid, HF, is a weak acid with $K = 6.3 \times 10^{-4}$. The chemical equation below describes the dissociation of HF in water:

$$HF(aq) \rightleftharpoons H^+(aq) + F^-(aq)$$

   a. Write the equilibrium-constant equation for the dissociation of HF.
   b. At equilibrium, is $[H^+]$ greater than, less than, or equal to $[F^-]$? Explain your answer.
   c. At equilibrium, is $[H^+]$ greater than, less than, or equal to $[HF]$? Explain your answer.
   d. How does $[HF]$ change if $H^+$ is added to the equilibrium mixture?

3. The dissolution of calcium chloride, $CaCl_2$, in water is used to create hot packs for foot and hand warmers. The chemical equation below describes a saturated solution.

$$CaCl_2(s) \rightleftharpoons Ca^{2+}(aq) + 2Cl^-(aq)$$

   a. Describe what is meant by a saturated solution. How do you know the solution is saturated?

**b.** Describe what happens if you add more $Ca^{2+}$ to a saturated solution.
   **c.** Describe what happens if you add more $Cl^-$ to a saturated solution.
   **d.** Describe what happens if you raise the temperature.
   **e.** Describe what happens if you lower the temperature.

**4.** Nitrogen, $N_2$, and oxygen, $O_2$, react in a car engine to produce nitric oxide, NO, as described by the chemical equation given below:

$$N_2(g) + O_2(g) \rightleftharpoons 2NO(g)$$

   **a.** In an engine at 1800 °C, $K = 1.0 \times 10^{-4}$. Write the equilibrium-constant equation for this reaction at 1800 °C.
   **b.** Are the starting substances or products favored in an equilibrium mixture at 1800 °C? Explain your thinking.
   **c.** If you drive the car to a mountaintop where the concentration of oxygen is much lower, will there be more or less NO in the equilibrium mixture? Explain your thinking.
   **d.** At 25 °C, $K = 1.0 \times 10^{-31}$. Write the equilibrium constant equation for this reaction at 25 °C.
   **e.** Do you expect to find much NO in the atmosphere due to the reaction of $N_2$ and $O_2$ at 25 °C? Why or why not?
   **f.** Given that the value of $K$ is much smaller at lower temperatures, is the reaction exothermic or endothermic?
   **g.** If the volume of the container with the equilibrium mixture is expanded, do you expect a change in the equilibrium mixture? Why or why not?

**5.** Suppose this chemical reaction is in a state of dynamic equilibrium in a sealed container:

$$COCl_2(g) \rightleftharpoons CO(g) + Cl_2(g) \qquad K = 4.63 \times 10^{-3}$$

   **a.** Write the equilibrium-constant equation for this reaction.
   **b.** At equilibrium, what would $[COCl_2]$ be if $[Cl_2] = [CO] = 2.0 \times 10^{-2}$ M?
   **c.** In which direction would the equilibrium shift if some additional CO were added to the system? Explain why.
   **d.** Suppose that the CO added in part c brought $[COCl_2]$ to 4.0 M at equilibrium. What would $[Cl_2]$ and $[CO]$ be?

## It's Showtime!

**PROJECT** ✔

It is time for you to show what you have learned both in Unit 6: Showtime and in the entire *Living by Chemistry* course. For this project, you will do research and design your own lab. The goal of this project is to examine the acid-base indicator properties of a naturally occurring pigment.

### Research

- Begin by doing literature research to identify a pigment molecule that gives color to a common substance and behaves as an indicator in solution. Choose a substance that is safe to handle and that you can obtain easily, such as flowers, leaves, fruits, berries, vegetables, spices, bark, or roots.

- Find out how it might be possible to safely extract this indicator molecule, so you can prepare a solution. For example, you might crush rose petals in rubbing alcohol or boil beets in water.

### Design a Lab Experiment

- Based on your research, write a detailed lab procedure that explains how to prepare a solution of the pigment.

- Next, write a lab procedure for how to test the acid-base properties of the pigment with common substances. You might want to refer to Unit 4: Toxins, Lesson 84, when preparing this procedure.

- Finally, write a lab procedure for testing a small piece of fabric, such as wool, cotton, or nylon, to determine if the pigment will stain the fabric and if the stain is permanent.

- Be sure to include a materials list and any safety considerations.

### Teacher and Peer Review

- Bring your proposed lab experiment to class and have your classmates review and critique your project.

- Submit your research and lab experiment to your teacher for review and approval.

- With teacher supervision, conduct your experiment in your school lab.

- Make changes based on feedback and the results of your testing. Once your experiment is a success, your teacher may ask you to lead a small group of students in carrying out the lab, or ask you to do a presentation to show what you learned about pigments, extraction, acid-base properties, and staining fabric.

# Showtime | REVIEW

UNIT
**6**

Carlos Villoch/Robert Harding

Many chemical processes are reversible. These processes do not convert all of the starting substances to products. Instead, the result is an equilibrium mixture of starting substances and products. In reversible chemical processes, the forward and reverse processes continue. When the forward and reverse processes occur at equal rates, these systems are said to be at equilibrium. Reversible processes are written with a double arrow to show that the forward and reverse processes occur at the same time.

Common equilibrium mixtures involve dissociation processes in which a starting substance breaks apart into two or more pieces. The back-and-forth evaporation and condensation of water on the planet is a form of phase equilibrium. Salts dissolve in water and dissociate into cations and anions, and the cations and anions combine to re-form the salt. Weak acids in water dissociate into $H^+$ and an anion, and they recombine to form the acid molecule. Molecules in the body attach and detach from receptor sites. In all of these cases, there is a dynamic balance between the starting substances and products.

There is a mathematical relationship between the products and starting substances in an equilibrium mixture. The equilibrium constant $K$ and the equilibrium-constant equation can be used to solve problems involving systems at equilibrium.

$$a\text{A} + b\text{B} \rightleftharpoons c\text{C} + d\text{D}$$

$$K = \frac{[\text{C}]^c[\text{D}]^d}{[\text{A}]^a[\text{B}]^b}$$

In general, a large value of $K$ indicates that products are favored in the equilibrium mixture, and a small value of $K$ indicates that starting substances are favored.

Le Châtelier's principle states that when conditions are changed for an equilibrium mixture, the system will respond by reducing the effect of the change. When conditions such as temperature, pressure, and concentration are changed in a system, the system regains equilibrium by rebalancing the concentrations of starting substances and products.

## REVIEW EXERCISES

### General Review

*Write a brief and clear answer to each question. Be sure to show your work.*

1. Write the reversible reactions for these processes.

   a. Chlorine gas becomes liquid chlorine.
   b. Nitrogen dioxide becomes dinitrogen tetroxide.

   c. Silver chloride precipitates from aqueous solution. Water evaporates.

2. Write the equilibrium-constant equations for the processes in Exercise 1.

3. Write two statements that are always true of a system at equilibrium.

4. When a dissolving and precipitation process is at equilibrium, the solution is saturated. Explain why this is true.

5. What is the concentration of copper ion, $Cu^{2+}$, in an aqueous solution of copper (II) carbonate, $CuCO_3$, if $K_{sp} = 1.4 \times 10^{-10}$ at 25 °C?

6. What does an acid-base indicator have to do with equilibrium?

7. The chemical equation for the dissolution of lead chloride, $PbCl_2(s)$, in water is given here.

$$PbCl_2(s) \longleftrightarrow Pb^{2+}(aq) + 2Cl^-(aq)$$

Determine the concentration of $Pb^{2+}$ ions in a saturated $PbCl_2$ solution if $K_{sp} = 1.7 \times 10^{-5}$.

8. Solid lead (II) sulfate, $PbSO_4$, is dissolved in water to form a solution with excess undissolved solute at the bottom of the container. The value of $K_{sp}$ for lead sulfate is $2.5 \times 10^{-8}$.

   a. How can you tell that this solution is saturated?

   b. Is the solution at equilibrium? How do you know?

   c. Write the chemical equation for the reversible change that is taking place.

   d. What is the equilibrium-constant equation for this equilibrium?

   e. What is the equilibrium concentration of each ion?

   f. Dissolving lead (II) sulfate is an endothermic change. Predict what will happen to the value of $[Pb^{2+}]$ and $[SO_4^{2-}]$ if the solution is heated.

   g. What would happen if you added more sulfate ions to the solution?

9. The value of $K_{sp}$ for cadmium sulfide, CdS, is $8.0 \times 10^{-27}$. Compare this value with the $K_{sp}$ for lead (II) sulfate given in Exercise 8.

   a. Which compound will have a higher concentration of ions in solution at equilibrium? Explain your reasoning.

   b. Calculate the concentration of cadmium ions in solution at equilibrium.

10. For the acid-base indicator litmus, the molecule, $HIn(aq)$, is red and the anion, $In^-(aq)$, is blue.

    a. Write a reaction for the dissociation of the $HIn$ molecule in solution.

    b. Why is litmus purple when you dissolve it in water? Explain your reasoning.

    c. What color do you think litmus is in a basic solution? Use Le Châtelier's principle to explain your thinking.

## STANDARDIZED TEST PREPARATION

### Multiple Choice

*Choose the best answer.*
Use the following to answer Exercises 1–3.

$$H^+(aq) + In^-(aq) \rightleftharpoons HIn(aq)$$
$$\text{yellow} \qquad \text{red}$$

1. Which correctly identifies the acid and base forms of the indicator?

   (A) $H^+$ is the acid form of the indicator; $HIn^-$ is the base form of the indicator.

   (B) $In^-$ is the acid form of the indicator; $HIn$ is the base form of the indicator.

   (C) $HIn$ is the acid form of the indicator; $In^-$ is the base form of the indicator.

   (D) $HIn$ is the acid form of the indicator; $H^+$ is the base form of the indicator.

2. Which statement correctly describes the composition of the solution containing the indicator when it is orange?

   (A) Only the $H^+$ ions are present in the solution.

   (B) Only $In^-$ anions are present in the solution.

   (C) Only the $HIn$ molecules are present in the solution.

   (D) There is a mixture of both the $In^-$ anions and $HIn$ molecules in solution.

3. What will happen to the reaction as $H^+$ ions are added to the solution?

   (A) The color of the solution will change from red to yellow.

   (B) The color of the solution will change from yellow to red.

   (C) The color of the solution will remain yellow.

   (D) The color of the solution will remain red.

**4.** For the cabbage juice indicator, what might be the pH of the solution when the color is red?

**(A)** 2

**(B)** 7

**(C)** 9

**(D)** 12

**5.** Which of these is a reversible process?

**(A)** lighting a safety match

**(B)** burning a log on a fire

**(C)** water freezing to become ice

**(D)** baking a cake

**6.** Carbon dioxide binds to hemoglobin, Hb, in your blood to form [$CO_2$ : hemoglobin] in a reversible process. Which of the following correctly represents a balanced chemical equation for this process?

I. [$CO_2$ : hemoglobin] $\rightleftharpoons$ $CO_2$ + hemoglobin

II. $CO_2$ + hemoglobin $\rightleftharpoons$ [$CO_2$ : hemoglobin]

III. $CO_2$ + hemoglobin $\longrightarrow$ [$CO_2$ : hemoglobin]

**(A)** I only

**(B)** II only

**(C)** III only

**(D)** Both I and II

**7.** The particle views below show 10 molecules of each of three sweeteners with their relative sweetness. Which statement correctly describes sucrose?

| Lactose | Maltose | Sucrose |
| --- | --- | --- |

Relative sweetness    1.5      3      10

**(A)** Sucrose molecules are detected as sweeter because more are attached to the receptors.

**(B)** Sucrose molecules have weaker intermolecular attractions with receptors than maltose molecules.

**(C)** More sucrose molecules remain in the saliva than either maltose or lactose molecules.

**(D)** Sucrose molecules are the sweetest because they bind to all available receptors.

**8.** An octane molecule is bound reversibly and weakly to a smell receptor in the nose. What might be responsible for reversibly binding octane to smell receptors?

**(A)** hydrogen bonds

**(B)** covalent bonds

**(C)** dipole-dipole interactions

**(D)** intermolecular forces

**9.** Which correctly describes a reversible process?

**(A)** The forward and reverse reactions can occur at the same time.

**(B)** A reversible process proceeds in both the forward and reverse directions.

**(C)** In a reversible chemical reaction system, the products and the starting substances are both present.

**(D)** All of the above are true of a reversible process.

**10.** Which chemical process might this graph represent?

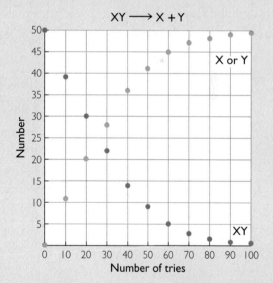

**(A)** a salt such as NaCl, which dissolves partially to $Na^+(aq)$ + $Cl^-(aq)$ to form a saturated solution

**(B)** a strong acid such as HBr, which separates to $H^+(aq)$ + $Br^-(aq)$

**(C)** an acid-base indicator, H$In$, which separates to $H^+(aq)$ + $In^-(aq)$

**(D)** a molecule bound to a receptor, which separates to a molecule and the empty receptor

**11.** Consider this chemical reaction: $AB \Longleftrightarrow A + B$. In a rigid container, 25 mol of AB decompose to reach equilibrium. The amounts of A and B are measured over time, and the values are recorded in the table below. At 240 s, which statement is correct?

| Time (s) | 30 | 60 | 90 | 120 | 150 | 180 | 210 | 240 | 270 |
|---|---|---|---|---|---|---|---|---|---|
| Amount of A (mol) | 4 | 6 | 8 | 12 | 14 | 15 | 15 | 15 | 15 |
| Amount of B (mol) | 4 | 6 | 8 | 12 | 14 | 15 | 15 | 15 | 15 |

(A) The concentrations of AB, A, and B are equal.

(B) The concentration of A is greater than the concentration of AB.

(C) The concentration of AB is greater than the concentration of B.

(D) The system is not at equilibrium.

Use this graph to answer Exercises 12–14. The graph represents the chemical reaction:

$$A_2(g) + 3B_2(g) \Longleftrightarrow 2AB_3(g)$$

**12.** Given that the initial concentration of $A_2$ is 1.0 M and the initial concentration of $B_2$ is 3.0 M, what are the equilibrium concentrations of the starting substances and the products?

(A) $[A_2] = 0.37$ M  $[B_2] = 1.1$ M  $[AB_3] = 1.2$ M

(B) $[A_2] = 0.40$ M  $[B_2] = 1.5$ M  $[AB_3] = 0.50$ M

(C) $[A_2] = 1.0$ M  $[B_2] = 1.5$ M  $[AB_3] = 0.50$ M

(D) $[A_2] = 1.2$ M  $[B_2] = 1.1$ M  $[AB_3] = 0.37$ M

**13.** Which correctly describes the reaction process at equilibrium?

(A) The rate of the forward reaction is equal to the rate of the reverse reaction.

(B) The concentration of the starting substances equals the concentration of the products.

(C) The rate of the forward reaction is faster than the rate of the reverse reaction.

(D) The concentrations of the starting substances are greater than the concentration of the products.

**14.** How long did it take for this reaction to reach equilibrium?

(A) 6 s

(B) 12 s

(C) 25 s

(D) 50 s

Use this table to answer Exercises 15–18.

| Solution | Formula | Solution molarity | [HA] | [H$^+$] | [A$^-$] | K |
|---|---|---|---|---|---|---|
| nitric acid | $HNO_3$ | 1.0 M | ~0 M | 1.0 M | 1.0 M | 40 |
| nitrous acid | $HNO_2$ | 1.0 M | 0.97 M | 0.023 M | 0.023 M | 0.00056 |
| benzoic acid | $C_6H_5COOH$ | 1.0 M | 0.99 M | 0.0079 M | 0.0079 M | 0.000063 |

**15.** Which is the correct chemical equation for the dissociation of nitrous acid, $HNO_2$?

(A) $HNO_2(aq) \Longleftrightarrow H^+(aq) + NO_2^-(aq)$

(B) $H^+(aq) + NO_2^-(aq) \Longleftrightarrow HNO_2(aq)$

(C) $HNO_2(aq) \longrightarrow H^+(aq) + NO_2^-(aq)$

(D) $H^+(aq) + NO_2^-(aq) \longrightarrow HNO_2(aq)$

**16.** Which correctly describes the dissociation of benzoic acid at equilibrium?

(A) There are equal concentrations of $C_6H_5COO^-$ and $H^+$ ions.

(B) There are more $C_6H_5COOH$ molecules than $C_6H_5COO^-$ and $H^+$ ions.

(C) The pH of the solution will be between 2 and 3.

(D) All of the above are true statements.

17. Which correctly describes the three reactions based on the value of their equilibrium constants?

    (A) Both nitrous acid and benzoic acid dissolve and partially dissociate in water.

    (B) Both nitrous acid and benzoic acid dissolve and completely dissociate in water.

    (C) Both nitrous acid and nitric acid dissolve and partially dissociate in water

    (D) Both nitrous acid and nitric acid dissolve and completely dissociate in water.

18. Which of the following is the correct equilibrium-constant expression for the dissociation of benzoic acid?

    (A) $K = \dfrac{[C_6H_5COO^-]}{[C_6H_5COOH]}$

    (B) $K = \dfrac{[C_6H_5COOH][H^+]}{[C_6H_5COO^-]}$

    (C) $K = \dfrac{[C_6H_5COO^-][H^+]}{[C_6H_5COOH]}$

    (D) $K = \dfrac{[C_6H_5COO^-][H^+]^2}{[C_6H_5COOH]}$

19. The dissociation of alizarin yellow indicator, a weak acid, is given below.

    $$HIn(aq) \rightleftharpoons H^+(aq) + In^-(aq) \quad K = 3.4 \times 10^{-11}$$

    What is the pH value of a 1.0 M solution of alizarin yellow, H$In$?

    (A) pH = 3.3

    (B) pH = 5.2

    (C) pH = 10.5

    (D) pH = 10.7

20. Which is the correct equilibrium-constant expression for the following reaction?

    $$(NH_4)_3PO_4(s) \rightleftharpoons 3NH_4^+(aq) + PO_4^{3-}(aq)$$

    (A) $K = [(NH_4)_3PO_4]$

    (B) $K = [NH_4^+]^3 [PO_4^{3-}]$

    (C) $K = \dfrac{[NH_4^+]^3[PO_4^{3-}]}{[(NH_4)_3PO_4]}$

    (D) $K = \dfrac{[(NH_4)_3PO_4]}{[NH_4^+]^3[PO_4^{3-}]}$

# It's Showtime

## Work Groups of 4

### Purpose
To use the process of engineering design to refine a procedure.

### Problem
At the end of Unit 6: Showtime, you designed a lab experiment to examine the acid-base indicator properties of a naturally occurring pigment. The goal of this project is to use engineering design to refine and improve your procedure.

### Safety Notes
- Wear safety goggles at all times
- Have baking soda and citric acid on hand to neutralize acid and base spills

### Review
Before you begin, review the handout, *Engineering Design* (found after the Unit 1 Engineering Project), and reflect on what you have learned so far in your class from your first experiment. For this project, you will focus on, brainstorming for solutions, refining a plan and retesting.

### Part 1: Refine and Retest

1. Have your classmates and teacher provide feedback for you about your procedure. Ask them to help you think of ways to improve your design. What changes might you make based on their feedback?

2. Make changes to your procedure and design based on feedback and the results of your testing. Submit your revised procedure and design to your teacher for review and approval. Include your list of resources and research material.

3. Once approved, use your procedure to conduct the experiment again.

4. Once your procedure and design are refined, your teacher may ask you to lead a small group of students in demonstrating your procedure.

### Possible Presentation Options:
- Oral presentation describing your procedure
- Poster describing your procedure in detail with diagrams
- A protocol pamphlet describing your procedure with diagrams
- A website that outlines your project including your procedure and diagrams

# Appendix A: Math Spotlights

| Quantity | Unit (abbreviation) |
|---|---|
| length | meter (m) |
| mass | kilogram (kg) |
| time | second (s) |
| temperature | Kelvin (K) |
| amount of substance | mole (mol) |

## SI Units of Measure

Scientists rely on repeatable measurements as they study the physical world. It is important that they use consistent units of measure worldwide. In 1960 an international council standardized the metric system, creating the *Système International d'Unités* (International System of Units), abbreviated as SI.

In the table at left, there are the basic SI units used in chemistry. Other units such as density and volume are combinations of these.

SI units are based on powers of 10. Larger and smaller units get their names by combining standard prefixes with these basic units. For example, the word *centimeter* is a combination of *centi-* and *meter*. A centimeter is one one-hundredth of a meter. These are the prefixes used in the SI, along with their abbreviations and their meanings. (A kilogram is the only basic SI unit that has a prefix as part of its name.)

| Prefix | Multiple | Scientific Notation | Prefix | Multiple | Scientific Notation |
|---|---|---|---|---|---|
| tera- (T-) | 1,000,000,000,000 | $10^{12}$ | pico- (p-) | 0.000 000 000 001 | $10^{-12}$ |
| giga- (G-) | 1,000,000,000 | $10^{9}$ | nano- (n-) | 0.000 000 001 | $10^{-9}$ |
| mega- (M-) | 1,000,000 | $10^{6}$ | micro- (μ-) | 0.000 001 | $10^{-6}$ |
| kilo- (k-) | 1,000 | $10^{3}$ | milli- (m-) | 0.001 | $10^{-3}$ |
| hecto- (h-) | 100 | $10^{2}$ | centi- (c-) | 0.01 | $10^{-2}$ |
| deka- (da-) | 10 | $10^{1}$ | deci- (d-) | 0.1 | $10^{-1}$ |

### Example 1

**Length Conversions**

How many meters does each of these lengths represent?

    **a.** 562 centimeters    **b.** 2.5 kilometers

*Solution*

    **a.** $562 \text{ cm} \cdot \dfrac{1 \text{ m}}{100 \text{ cm}} = 5.62 \text{ m}$    **b.** $2.5 \text{ km} \cdot \dfrac{1000 \text{ m}}{1 \text{ km}} = 2500 \text{ m}$

### Example 2

**The Mass of One Liter**

One milliliter, or cubic centimeter, of water at 4 °C weighs 1 g. How much does 1 L of water at 4 °C weigh?

*(continued)*

*Solution*

One milliliter is one one-thousandth of a liter; multiply the mass of 1 mL by 1000 to get the mass of 1 L.

$$1000 \cdot 1 \text{ g} = 1000 \text{ g} = 1 \text{ kg}$$

**Practice Exercises**

Convert these measurements to the indicated units.

1. 7 m = _____ cm
2. 3200 mL = _____ L
3. 20,012 cm = _____ km
4. 0.003 kg = _____ g
5. 16 m² = _____ cm²
6. 2 m³ = _____ dm³

*Answers*

1. 700 cm   2. 3.2 L   3. 0.20012 km   4. 3 g   5. 160,000 cm²   6. 8000 dm³

## Accuracy, Precision, and Significant Digits

There are two kinds of numbers in the world—exact and inexact. For instance, counting is exact because you can safely say that there are exactly 12 eggs in a dozen. However, no measurement with a ruler, a balance, or a graduated cylinder is ever exact. So, when a measurement or the average of several measurements comes out extremely close to the actual true value, we say that the measurement or average is *accurate*.

If you measure something several times and get very similar answers each time, your measurements are *precise*.

The ability to make precise measurements depends partly on the equipment used. For example, a graduated cylinder is more precise than a beaker. Precision also depends on how carefully a measurement was made. For example, a measurement of 23.76 mL is more precise than a measurement of 24 mL.

To understand the difference between precision and accuracy, imagine a lab experiment to measure the boiling point of water at sea level. Several readings are taken: 97.2 °C, 97.0 °C, and 97.1 °C. The measurements are close to each other; repeating the experiment would likely give similar results, so they are precise. However, they are not accurate; the boiling point of water at sea level is known to be 100 °C. Perhaps the thermometer was faulty, or the person taking the measurements consistently read the thermometer incorrectly.

Using significant digits in a measurement allows you to indicate the degree of certainty in the measurements. In general, the last digit of any measurement is uncertain. For example, suppose you use a meterstick, marked in millimeters, to measure the length of an object. If you record a measurement of 24.33 cm, the last digit is an estimate based on the closest millimeter markings and is not certain. Another person might measure the length as 24.34 cm or 24.32 cm. However, you will both agree on the 24.3 because you can read the meterstick accurately to the millimeter, or 0.1 cm. The measurement 24.33 has four significant digits.

The rules for determining the number of significant digits are complicated. Nonzero digits always count. Zeros sometimes count depending on where they are. Zeros might be leading, trapped, or trailing.

- *Leading zeros,* such as those in 0.004728, never count as significant digits.

- *Trapped zeros,* as in 1.08, always count as significant digits.

- *Trailing zeros,* or those at the end of a number, count only when there is a decimal point. In the numbers 20.0, 300.00, and even 50., the zeros count as significant digits. If there is no decimal point, as in 500, then it isn't possible to tell whether the zeros are significant. Sometimes you can deduce that zeros are significant based on the instrumentation. For example, a thermometer generally measures to the nearest degree. You can use scientific notation to avoid ambiguity. The measurement $5.00 \times 10^2$ has three significant digits, whereas the measurement $5 \times 10^2$ has only one significant digit.

As you do calculations involving numbers with different numbers of significant digits, follow these two rules.

- Adding or subtracting: Your final answer will have only as many *decimal places* as the measurement with the fewest decimal places.

- Multiplying or dividing: The result can have only as many significant digits as the number with the fewest significant digits.

As you add, multiply, or combine measurements in other ways, you will often have to round the result of your calculation to obtain the correct number of significant digits.

| Example 1 | **Rounding the Sum or Difference** |

Calculate each sum or difference and round the answer to the appropriate number of significant digits.

  **a.** $2.24 + 3.4 + 5.231$    **b.** $10.5 \text{ cm} - 3.36 \text{ cm}$

*Solution*

  **a.** First add the numbers to get the sum 10.871. Then consider the decimal places to arrive at the final answer.

| 2 decimal places | 2.24 |
| 1 decimal place | 3.4 |
| 3 decimal places | + 5.231 |
| | 10.871 |

Because 3.4 has only one decimal place, the sum should be rounded to 10.9. Always complete the calculation before rounding.

  **b.** First subtract the numbers.

| 1 decimal place | 10.5 cm |
| 2 decimal places | − 3.36 cm |
| | 7.14 cm |

Because 10.5 has only one decimal place, round the final answer to one decimal place: 7.1 cm.

**Example 2**

## Rounding a Product

Calculate each product and round to the appropriate number of significant digits.

**a.** 12.34 · 1.6   **b.** 4.71 m · 5.28 m

*Solution*

**a.** First multiply the two decimals.

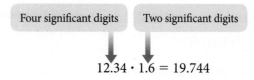

$$12.34 \cdot 1.6 = 19.744$$

Because there are only two significant digits in 1.6, you can have only two significant digits in your answer. You must round the product to 20. You might write this answer as 2.0 · 10 to make it clear that both digits are significant.

**b.** First multiply: 4.71 m · 5.28 m = 24.8688 m².

Both factors have three significant digits, so the answer must be rounded to three significant digits: 24.9 m².

### Practice Exercises

1. How many significant digits are in each of these numbers?
   **a.** 20.1   **b.** 300.0   **c.** 0.0031   **d.** 0.03010

2. Complete these calculations.
   **a.** 25.14 + 3.4 + 15.031   **b.** 100.04 cm − 7.362 cm   **c.** 3005 · 45.20

3. What is the volume of a box 14.5 cm by 15.9 cm by 21.1 cm?

4. The density of copper is 8.92 g/cm³. What is the mass of 24 cm³ of copper?

> *Answers*
>
> **1a.** 3   **1b.** 4   **1c.** 2   **1d.** 4   **2a.** 43.5   **2b.** 92.68   **2c.** 135,800
> **3.** 4860 cm³   **4.** 210 g

## Solving Equations

Solving chemistry problems sometimes involves solving a math equation. When the quantity you are looking for is isolated, or alone on one side of the equation, all you need to do is complete the calculations. A simple example is finding the Fahrenheit equivalent of 28 °C using the equation

$$F = \frac{9}{5}C + 32°$$

Sometimes the quantity that answers your question is not alone on one side of the equation. To solve these problems, isolate the variable you are looking for on one side of the equation, so that calculations that will lead to your answer are on the other side of the equation. For example, to find a temperature in degrees Celsius, rearrange the equation in terms of *C*.

Start with the known relationship

$$F = \frac{9}{5}C + 32°$$

Subtract 32° from both sides

$$F - 32° = \frac{9}{5}C$$

Multiply both sides by $\frac{5}{9}$.

$$\frac{5}{9}(F - 32°) = C$$

Now that $C$ is isolated, you can substitute any Fahrenheit temperature and carry out the calculations to find the corresponding Celsius temperature.

**Example 1**

## Solving for $x$

Solve these equations for $x$.

**a.** $0.1x + 12 = 2.2$    **b.** $\dfrac{12 + 3.12x}{3} = -100$

*Solution*

**a.** Original equation. $\qquad\qquad\qquad\qquad 0.1x + 12 = 2.2$

Subtract 12 from both sides. $\qquad 0.1x + 12 - 12 = 2.2 - 12$

$$0.1x = -9.8$$

Divide both sides by 0.1. $\qquad\qquad\qquad\qquad x = -98$

**b.** Original equation. $\qquad\qquad\qquad \dfrac{12 + 3.12x}{3} = -100$

Multiply both sides by 3. $\qquad\qquad 12 + 3.12x = -300$

Subtract 12 from both sides. $\quad -12 + 12 + 3.12x = -300 - 12$

$$3.12x = -312$$

Divide both sides by 3.12. $\qquad\qquad\qquad\qquad x = -100$

**Example 2**

## Solving for $V_2$

Solve this equation for $V_2$.

$$\frac{V_1 P_1}{T_1} = \frac{V_2 P_2}{T_2}, \text{ for } V_2$$

*Solution*

Original equation.

$$\frac{V_1 P_1}{T_1} = \frac{V_2 P_2}{T_2}$$

Multiply both sides by $T_2$; remove the factor of 1.

$$\frac{T_2 V_1 P_1}{T_1} = \frac{\cancel{T_2} V_2 P_2}{\cancel{T_2}}$$

Divide both sides by $P_2$; remove the factor of 1.

$$\frac{T_2 V_1 P_1}{T_1 P_2} = \frac{V_2 \cancel{P_2}}{\cancel{P_2}}$$

$$V_2 = \frac{T_2 V_1 P_1}{T_1 P_2}$$

The equation is now expressed in terms of $V_2$. Substituting in values for the other five variables will give a value for $V_2$.

**Practice Exercises**

1. Solve these equations. Indicate the action you take at each stage.

   **a.** $144x + 33 = 45$     **b.** $\frac{1}{6}x + 2 = 8$     **c.** $5(x - 7) = 15 + 5^2$

2. Solve these equations for the variable indicated.

   **a.** $d = rt$, for $t$     **b.** $P = 2(l + w)$, for $w$     **c.** $A = \frac{1}{2}h(a + b)$, for $h$

---

*Answers*

**1a.** $x = \frac{1}{12}$ (On both sides, subtract 33 and divide by 144.)

**1b.** $x = 36$ (On both sides, subtract 2 then multiply by 6.)

**1c.** $x = 15$ (Combine 15 and 25 on the right side, divide both sides by 5, and add 7 to both.)

**2a.** $t = \dfrac{d}{r}$     **2b.** $l = \dfrac{P}{2} - w$     **2c.** $h = \dfrac{2A}{a} + b$

---

## Order of Operations

Math expressions and equations often involve several operations. For example, to convert from Celsius to Fahrenheit, you first multiply the number of Celsius degrees by the fraction $\frac{9}{5}$ and then add 32 degrees. To convert from Fahrenheit to Celsius, you subtract 32 degrees from the number of Fahrenheit degrees and then multiply the result by $\frac{5}{9}$. A rule called the *order of operations* is used to write math expressions clearly so that anyone seeing the formula or equation would know whether multiplication was the first step or the second step.

---

*Order of Operations*

1. Evaluate all expressions within parentheses.

2. Evaluate all terms with exponents.

3. Multiply and divide from left to right.

4. Add and subtract from left to right.

---

**Example 1**

### Temperature Conversions

Convert these temperatures.

   **a.** 37 °C to degrees Fahrenheit     **b.** 48 °F to degrees Celsius

*Solution*

Substitute the known value into each equation and then solve using the order of operations.

   **a.** Substitute 37° into the equation.     $F = \dfrac{9}{5} \cdot 37° + 32°$

   Multiply.     $F = 66° + 32°$

   Add.     $F = 98\ °F$

*(continued)*

**b.** Substitute 48° into the equation.    $C = \dfrac{5}{9}(48° - 32°)$

Subtract.    $C = \dfrac{5}{9}(16°)$

Multiply.    $C = 9\,°C$

---

**Example 2**

## Parentheses, Exponents, and Fractions

Evaluate these expressions.

**a.** $\dfrac{5}{9}(96 - 15)$    **b.** $\dfrac{(70 - 64)^2}{2 \cdot 5} + 12$

*Solution*

Evaluate the expression in the parentheses first.

**a.** Original expression    $\dfrac{5}{9}(96 - 15)$

Subtract the numbers within the parentheses.    $= \dfrac{5}{9}(81)$

Multiply $\frac{5}{9}$ by 81.    $= 45$

**b.** The fraction line acts like parentheses. In fact, when the expression is entered into a calculator, parentheses are required around the $2 \cdot 5$.

Original expression.    $\dfrac{(70 - 64)^2}{2 \cdot 5} + 12$

Evaluate the expressions above and below the fraction line.    $= \dfrac{6^2}{10} + 12$

Divide 36 by 10 and then add 12.    $= 15.6$

### Practice Exercises

1. Evaluate these expressions.
   **a.** $3 \cdot 24 \div 8$    **b.** $3 + 24 \cdot 8$    **c.** $3 - 24 + 8$
   **d.** $(3 + 24) \cdot 8$    **e.** $(3 + 21) \div 8$    **f.** $3 - (24 + 8)$

2. Calculate the value of each expression.
   **a.** $-2 + 5 - (-8)$    **b.** $(-5^2) - (-3)^2$    **c.** $-0.3 \cdot 20 + 15$

3. Insert parentheses as needed to make each equation true.
   **a.** $15 \div 3 + 7 - 4 = -48$    **b.** $15 \div 3 + 7 - 4 = 8$    **c.** $-4^2 + -3^2 = -7$

---

*Answers*

**1a.** 9    **1b.** 195    **1c.** $-13$    **1d.** 216    **1e.** 3    **1f.** $-29$

**2a.** 11    **2b.** $-34$    **2c.** 9

**3a.** $(15 \div 3 + 7)(-4) = -48$    **3b.** $15 \div 3 + 7 - 4 = 8$

**3c.** $-4^2 + (-3)^2 = -7$

# Averages

Scientists are often interested in the typical result of a repeated experiment. The average, also called the *mean,* is one way to determine a typical value. The average is calculated by totaling the data values and dividing by the number of values.

**Example 1**

### The Mean

Find the average of these numbers: 14, 23, 10, 21, 7, 80, 32, 30, 92, 14, 26, 21, 38, 20, 35, 21.

*Solution*

$$\frac{14 + 23 + 10 + 21 + 7 + 80 + 32 + 30 + 92 + 14 + 26 + 21 + 38 + 20 + 35 + 21}{16} = 30.25$$

The average is 30.25.

Sometimes the data values are not of equal importance in contributing to the average. You need to weight the values differently.

**Example 2**

### The Weighted Average

On a chemistry quiz, 1 student got 100%, 7 students got 95%, 12 students got 90%, 1 student got 85%, 5 students got 80%, 3 students got 75%, and 1 student got 70%. What was their average (mean) score?

*Solution*

One solution would be to change the list so that 95% appeared 7 times, 90% appeared 12 times, and so on. However, it is more efficient to use a weighted average as shown here.

$$(1)(100) + (7)(95) + (12)(90) + (1)(85) + (5)(80) + (3)(75) + (1)(70) = 2625$$

Thirty students took the quiz, so divide 2625 by 30.

The average score was 87.5%.

**Example 3**

### Average Atomic Mass

The element silver, Ag, has two naturally occurring isotopes. Approximately 52% of all silver consists of atoms with 60 neutrons, and 48% consists of atoms with 62 neutrons. Calculate the average atomic mass of silver atoms.

*(continued)*

## Solution

Silver atoms have 47 protons. The atoms with 60 neutrons have masses of $47 + 60 = 107$ amu. The atoms with 62 neutrons have masses of $47 + 62 = 109$ amu. If you have a sample of 100 silver atoms, 52 atoms will have a mass of 109 amu and 48 atoms will have a mass of 109 amu. Use a weighted average. The total mass is $52(107 \text{ amu}) + 48(109 \text{ amu}) = 10{,}796$ amu. The average mass is 10,796 amu/100 = 107.96. This is close to the atomic mass of 107.9 amu shown on the periodic table.

### Practice Exercises

1. Find the average of these numbers: 52.3, 18.91, 35.66, 4.35.

2. A student had these scores on chapter tests: 87%, 90%, 95%, 92%. He got 92% on one unit test and 86% on the other. His final exam score was 91%. Unit tests count twice as much as chapter tests, and the final exam counts four times as much as a chapter test. What is his average score?

3. About 76% of all chlorine atoms have an atomic mass of 35.00 amu; about 24% have an atomic mass of 37.00 amu. Use this information to calculate the average atomic mass of 100 chlorine atoms.

### Answers

**1.** 27.8     **2.** 90%     **3.** 35.48 amu

# Graphing

Many chemical experiments involve changing one variable and seeing how another variable changes as a result. The results of experimental procedures can be listed in a table, but it is often helpful to look on a graph for any trends in the data.

Data consisting of two variables, such as temperature and volume measurements, can be graphed using a coordinate plane. Each point on the coordinate plane can be identified by a pair of numbers $(x, y)$ called *coordinates*. The first number, the $x$-coordinate, describes how far the point is to the right or left of the origin; the second number, the $y$-coordinate, describes how far up or down the point is.

In the example below, a coordinate pair represents one (temperature, volume) data point. The quantity that the experimenter is changing, in this case temperature, is called the *independent variable*. It is graphed on the horizontal axis, the $x$-axis. The resulting measurement, or the *dependent variable*, is graphed on the vertical axis, or $y$-axis. The dependent variable in this example is the volume. If the experiment had been set up to measure changes in temperature as a result of changes in volume, the independent variable would have been volume, and temperature would have been the dependent variable.

Normally, you graph the dependent variable versus the independent variable. (*Versus* means "compared with" and is abbreviated "vs.")

## Example 1

### Graphing Coordinate Pairs

Graph these results for the heating of a gas:

| Temperature (°C) | Volume (mL) | Temperature (°C) | Volume (mL) |
|---|---|---|---|
| 0 | 465 | 15 | 491 |
| 5 | 475 | 20 | 498 |
| 10 | 481 | 25 | 509 |

#### Solution

The temperature is being changed, so temperature is the independent variable and should be graphed on the *x*-axis. The volume changes as a result, so it is the dependent variable and should be graphed on the *y*-axis. To graph the data, you need to decide on the scale for each axis. The *x*-axis can start at 0 °C and go up to 30 °C. To save space, you don't need to show the *y*-axis starting at 0; it can go from 450 to 520. Label the axes "Temperature (°C)" and "Volume (mL)." Graph the coordinate pairs. Title the graph. In the title, the dependent variable is usually named first.

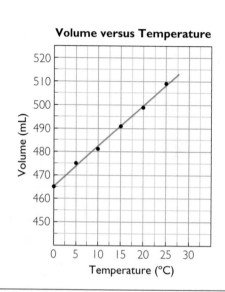

Notice that the points in the graph of volume versus temperature lie nearly in a straight line. Small errors in measurement account for the variation. The points are close enough to a line to indicate that the relationship between temperature and volume is linear. As one quantity increases, the other increases at a constant rate. So the quantities vary directly, or are *directly proportional*.

## Example 2

### Volume versus Pressure

Graph the results of an experiment in which a gas's volume is measured as pressure is applied. The temperature of the gas is kept from changing. This table shows the data collected.

| Pressure (atm) | Volume (L) | Pressure (atm) | Volume (L) |
|---|---|---|---|
| 1.0 atm | 10.0 L | 3.0 atm | 3.3 L |
| 1.5 atm | 6.7 L | 3.5 atm | 2.9 L |
| 2.0 atm | 5.0 L | 4.0 atm | 2.5 L |
| 2.5 atm | 4.0 L | | |

*(continued)*

### Solution

An appropriate scale for the independent variable on the *x*-axis is 0 atm to 4.5 atm. The scale for the dependent variable on the *y*-axis can go from 0 L to 12 L. Label the axes and indicate the units: "Pressure (atm)" and "Volume (L)." The coordinate pairs to graph are (1.0, 10), (1.5, 6.7), (2.0, 5.0), and so on. Title the graph *Volume versus Pressure*.

The points in the graph of volume versus pressure do not lie in a line, but you can draw a smooth curve through the points to show the trend. The shape of the graph is typical of relationships that are inversely proportional.

### Practice Exercises

1. The examples all used coordinate pairs with positive values. But points can also have negative coordinates. The coordinates of point *A* are $(-7, -4)$. Name the $(x, y)$ coordinates of each point pictured.

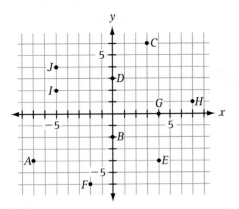

| Temperature (°C) | Pressure (atm) |
|---|---|
| 22.0 °C | 3.20 atm |
| 35.0 °C | 3.34 atm |
| 45.0 °C | 3.45 atm |
| 60.0 °C | 3.61 atm |

2. A sealed rigid container is heated and the pressure in the container is measured. Describe the graph of the data (your scales, axes titles, graph title, plotted points, and shape of the curve through the points).

3. These coordinate pairs represent measurements of volume and mass of copper: (1.1, 9.8), (2.3, 20.5), (2.7, 24.1), (4.5, 40.1). If these are graphed, what will be the shape of the graph? Are mass and volume directly proportional or inversely proportional?

# Ratios and Proportions

As chemists study matter, they are often working with ratios. The four statements here all express some sort of ratio.

Density is the ratio of mass to volume: $D = m/V$.

The units of molar mass are g/mol.

The mole ratio of $H_2$ to $O_2$ is 2:1.

The reaction gave a 78% yield.

Ratios can be written as fractions. For instance, 78% is $\frac{78}{100}$. The ratio 2:1 can be written $\frac{2}{1}$. Ratios can be reduced like fractions, or they can be added once they have a common denominator. Unit fractions, which equal 1 because the numerator is equivalent to the denominator, can be used to convert between units or to create equivalent fractions.

Whenever you are working with ratios, use the rules for working with fractions.

- *To create equivalent fractions,* multiply a fraction by a unit fraction. In a unit fraction, the numerator is equivalent to the denominator. A unit fraction is equal to 1 and can be used to convert between units or create equivalent fractions.

Unit fraction

$$\frac{1}{2} = \frac{1}{2} \cdot \frac{5}{5} = \frac{5}{10}$$

Unit fraction

$$300 \text{ m} = 300 \text{ m} \cdot \frac{1 \text{ km}}{1000 \text{ m}} = 0.3 \text{ km}$$

- *To add or subtract fractions,* find equivalent fractions with a common denominator, then add or subtract the numerators and put that number over the common denominator.

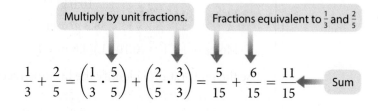

Multiply by unit fractions. Fractions equivalent to $\frac{1}{3}$ and $\frac{2}{5}$

$$\frac{1}{3} + \frac{2}{5} = \left(\frac{1}{3} \cdot \frac{5}{5}\right) + \left(\frac{2}{5} \cdot \frac{3}{3}\right) = \frac{5}{15} + \frac{6}{15} = \frac{11}{15} \leftarrow \text{Sum}$$

- To *multiply fractions,* multiply the two numerators, and multiply the two denominators. You may want to reduce the fraction so that there are no common factors between the numerator and the denominator.

$$\frac{1}{3} \cdot \frac{3}{5} = \frac{1 \cdot 3}{3 \cdot 5} = \frac{3}{15} = \frac{1}{5}$$

- To *divide fractions,* multiply by the reciprocal of the divisor.

Divisor      Reciprocal of divisor

$$\frac{1}{6} \div \frac{1}{2} = \frac{1}{6} \cdot \frac{2}{1} = \frac{2}{6} = \frac{1}{3}$$

- To *reduce a fraction,* divide the numerator and denominator by any factor they have in common; this is often referred to as *canceling.*

$$\frac{6}{15} = \frac{2 \cdot 3}{3 \cdot 5} = \frac{2}{5} \qquad \frac{300 \,\cancel{m} \cdot 1 \,km}{1000 \,\cancel{m}} = 0.3 \,km$$

Proportions are equations that state two ratios are equal. Each of the equations shown here is a proportion.

$$\frac{P_1 V_1}{T_1} = \frac{P_2 V_2}{T_2} \qquad \frac{3.20 \,atm}{22 \,°C} = \frac{P_2}{50 \,°C}$$

$$\frac{1 \,mol \,H_2O}{18.0 \,g \,H_2O} = \frac{x \,mol \,H_2O}{51.2 \,g \,H_2O}$$

Proportions can be solved by multiplying (or dividing) both sides of the equation by the same value. As you solve chemistry problems, you will often use ratios and proportions.

## Example 1

### Limiting Reactant

Imagine you start with 10.0 g of $Mg(OH)_2$ and 100.0 mL of 4.0 M HCl. How much $MgCl_2$ will be produced?

### Solution

Write the balanced chemical equation.

$$Mg(OH)_2(s) + 2HCl(aq) \longrightarrow MgCl_2(aq) + 2H_2O(l)$$

Determine the molar masses of each compound.

Reactants: $Mg(OH)_2$ = 58.3 g/mol      HCl = 36.5 g/mol
Products: $MgCl_2$ = 95.2 g/mol      $H_2O$ = 18.0 g/mol

Determine the number of moles of each reactant that you have.

$$Moles \,of \,Mg(OH)_2 = \frac{10.0 \,g}{58.3 \,g/mol} = 0.17 \,mol$$

Moles of HCl = (0.100 L)(4.0 mol/L) = 0.40 mol

Use the mole ratio to identify the limiting reactant.

**Example**

(continued)

The reactants combine in a 1:2 ratio. So you need 0.17 mol · 2 = 0.34 mol of hydrochloric acid, HCl, to react with the 0.17 mol of magnesium hydroxide, $Mg(OH)_2$.

You have 0.40 mol of HCl, which is plenty, so the limiting reactant is $Mg(OH)_2$. When the reaction is complete, there will be 0.06 mol of HCl left over.

Use the limiting reactant to determine the maximum amount of product.

For every 1 mol of $Mg(OH)_2$, 1 mol of $MgCl_2$ is produced.

So 0.17 mol of $MgCl_2$ is produced. Use the molar mass of $MgCl_2$ to determine the mass of $MgCl_2$.

$$\frac{95.2 \text{ g MgCl}_2}{1 \text{ mol}} = \frac{x \text{ g MgCl}_2}{0.17 \text{ mol}}$$

$$0.17 \text{ mol} \cdot \frac{95.2 \text{ g}}{1 \text{ mol}} = x$$

$$x = 16 \text{ g MgCl}_2$$

**Example 2**

## Percent Yield

Suppose you ran the reaction from Example 1 again, this time starting with 35.0 g $Mg(OH)_2$ and the same amount of HCl as before. In the laboratory, 15.4 g of $MgCl_2$ are produced. What is the percent yield of your reaction?

### Solution

First, determine the number of moles of each reactant you have. This will make it possible to identify the limiting reactant.

$$\frac{58.3 \text{ g}}{1 \text{ mol Mg(OH)}_2} = \frac{35.0 \text{ g}}{x \text{ mol Mg(OH)}_2}$$

$$x = 0.60 \text{ mol Mg(OH)}_2$$

$$\frac{y \text{ mol HCl}}{0.100 \text{ L}} = \frac{4.0 \text{ mol}}{1 \text{ L}}$$

$$y = 0.40 \text{ mol HCl}$$

The mole ratio of reactants is 1:2, so you need 1.2 mol of HCl to react with 0.60 mol of $Mg(OH)_2$. You have only 0.40 mol of HCl available, so this time HCl is the limiting reagent.

It takes 2 mol of HCl to produce each mole of $MgCl_2$, so you can expect to produce 0.20 mol of $MgCl_2$: $\left(\frac{2}{1} = \frac{0.40}{0.20}\right)$.

Next, use the molar mass of $MgCl_2$ to calculate the theoretical yield.

$$\frac{x \text{ g MgCl}_2}{0.20 \text{ mol}} = \frac{95.2 \text{ g}}{1 \text{ mol}}$$

$$x = \frac{95.2 \text{ g}}{1 \text{ mol}} \cdot 0.20 \text{ mol}$$

$$x = 19 \text{ g}$$

*(continued)*

Now you can calculate the percent yield for the reaction you ran. Percent yield is yield per 100, or yield/100.

$$\frac{\text{yield}}{100} = \frac{\text{actual yield}}{\text{theoretical yield}}$$

$$\text{yield} = \frac{15.4\text{ g}}{19\text{ g}} \cdot 100$$

Percent yield = 81%

Your percent yield was significantly lower than 100%.

### Practice Exercises

1. What ratios and proportions appear in Examples 1 and 2?

2. Write each expression as a fraction.
   a. 68%
   b. 3:4
   c. the ratio of volume to temperature
   d. grams per cubic centimeter

3. Complete these calculations.

   a. $\dfrac{2}{3} + \dfrac{1}{5}$   b. $\dfrac{4}{5} - \dfrac{1}{6}$   c. $\dfrac{1}{3} \cdot \dfrac{9}{11}$   d. $\dfrac{3}{8} \div \dfrac{1}{4}$

---

*Answers*

1. $\dfrac{10\text{ g}}{58.3\text{ g/mol}}$, 1:2, $\dfrac{95.3\text{ g MgCl}_2}{1\text{ mol}} = \dfrac{x\text{ g MgCl}_2}{0.17\text{ mol}}$,

   $\dfrac{58.3\text{ g}}{1\text{ mol Mg(OH)}_2} = \dfrac{35.0\text{ g}}{x\text{ mol Mg(OH)}_2}$   $x = 0.60\text{ mol Mg(OH)}_2$, $\dfrac{15.4\text{ g}}{19.4\text{ g}}$

2a. $\dfrac{68}{100}$   2b. $\dfrac{3}{4}$   2c. volume/temperature, or $\dfrac{V}{T}$   2d. $\dfrac{\text{g}}{\text{cm}^3}$

3a. $\dfrac{13}{15}$   3b. $\dfrac{19}{30}$   3c. $\dfrac{3}{11}$   3d. $\dfrac{3}{2}$, or $1\dfrac{1}{2}$

---

## Scientific Notation

The mass of a hydrogen atom is 0.00000000000000000000000014 g. The number of atoms in 1 mol of carbon is 602,000,000,000,000,000,000,000. To make these numbers easier to read, compare, and use in calculations, scientists use scientific notation. The long numbers above can be written as $1.4 \times 10^{-24}$ and $6.02 \times 10^{23}$.

Each number is written as a decimal with one digit to the left of the decimal point times a power of 10. A number written in scientific notation has the form $a \times 10^n$, where $1 \le a < 10$ or $-10 < a \le -1$, and $n$ is an integer. In other words, if $a$ is a positive number, it is greater than or equal to 1 and less than 10. If $a$ is a negative number, it is less than or equal to $-1$ and greater than $-10$.

Use the properties of exponents as you combine numbers written in scientific notation.

- For addition and subtraction, convert all the numbers so that the powers of 10 you are combining are the same. (The converted numbers might not be in scientific notation.)

- For multiplication, add the exponents. For any values of $b$, $m$, and $n$, $b^m \cdot b^n = b^{m+n}$.

- For division, subtract the exponents. For any nonzero value of $b$, and any values of $m$ and $n$, $\frac{b^m}{b^n} = b^{m-n}$.

## Example 1

### Billions

Write these numbers in scientific notation.

    **a.** 2,110,000,000    **b.** 0.0074

*Solution*

    **a.** Move the decimal point so that there is only one digit to the left of it, in this case 2. The decimal point moved nine places. The number in scientific notation is $2.11 \times 10^9$. This means that to get the number in standard notation you start with 2.11 and move the decimal nine places to the right.

    **b.** Move the decimal after the 7 to get 7.4. From here, the decimal will need to move four places to the left, so the number in scientific notation is $7.4 \times 10^{-4}$.

## Example 2

### Operations with Scientific Notation

Calculate these values.

    **a.** $(3.2 \times 10^5)(4.0 \times 10^{-8})$    **b.** $\dfrac{1.4 \times 10^3}{2.8 \times 10^5}$    **c.** $(3.2 \times 10^5) + (4.0 \times 10^4)$

*Solutions*

    **a.** $1.28 \times 10^{-2}$. Multiplying the decimals gives 12.8. The product of the powers of 10 is $10^{-3}$. To put this into scientific notation, divide 12.8 by 10; to keep things balanced, increase the power of 10 by a factor of 10 to $-2$.

    **b.** $5.0 \times 10^7$. The division of the numbers gives 0.50. To divide the powers of 10, subtract the exponents: $3 - (-5)$, or 8. To rewrite in scientific notation, multiply 0.50 by 10, so subtract 1 from the power of 10.

    **c.** $3.6 \times 10^5$. Change the second number to $0.4 \times 10^5$ before you add.

### Practice Exercises

    **1.** Use scientific notation to write the speed of light, 299,792,458 meters per second, accurate to five digits.

    **2.** How many zeros would follow the final 2 if $6.022 \times 10^{23}$ were written without scientific notation?

    **3.** Calculate these values.

        **a.** $(3.0 \times 10^{14})(4.0 \times 10^{-4})$    **b.** $\dfrac{2.8 \times 10^{-31}}{7 \times 10^{-28}}$

        **c.** $(1.21 \times 10^{-4})(4.18 \times 10^4)$    **d.** $(3.61 \times 10^7) - (2.5 \times 10^6)$

# Dimensional Analysis

As you work on problems that involve numbers with units of measurement, it is convenient to consider the units (or dimensions) as factors. For example, you might want to change 0.002 kilometer to millimeters. You can always multiply by 1 without changing a value, so to convert between units you can multiply by a unit fraction. A unit fraction's value is 1 because its numerator and denominator are equivalent. Some unit fractions are shown here.

$$\frac{1000 \text{ m}}{1 \text{ km}} = 1 \qquad \frac{1000 \text{ mm}}{1 \text{ m}} = 1 \qquad \frac{1 \text{ m}}{1000 \text{ mm}} = 1$$

To convert from kilometers to millimeters, use the unit fraction with meters in the numerator and the equivalent kilometers in the denominator and a unit fraction with millimeters in the numerator and equivalent meters in the denominator.

$$0.002 \text{ km} \cdot \frac{1000 \text{ m}}{1 \text{ km}} \cdot \frac{1000 \text{ mm}}{1 \text{ m}} = 2000 \text{ mm}$$

You can see that the unit fractions have been chosen so that most of the units cancel out. The answer is then in millimeters as they are the only remaining units.

Dimensional analysis is also helpful when you need to convert between two different systems of measurement. For example, 1 inch is equivalent to 2.54 centimeters. To convert from centimeters to inches, use the unit fraction $\frac{1 \text{ in}}{2.54 \text{ cm}}$. To convert from inches to centimeters, use the unit fraction $\frac{2.54 \text{ cm}}{1 \text{ in}}$.

## Example 1

### Calculating Volume

What is the volume of a box 15 cm by 10 cm by 5.12 in.?

*Solution*

The volume is the product of the three dimensions, but you need to include a factor to convert the inches to centimeters.

$$15 \text{ cm} \cdot 10 \text{ cm} \cdot 5.12 \text{ in.} \cdot \frac{2.54 \text{ cm}}{1 \text{ in.}} \approx 1950 \text{ cm}^3$$

If the problem specified the answer in cubic inches, you would have used this product.

$$15 \text{ cm} \frac{1 \text{ in.}}{2.54 \text{ cm}} \cdot 10 \text{ cm} \cdot \frac{1 \text{ in.}}{2.54 \text{ cm}} \cdot 5.12 \text{ in} \approx 119 \text{ in}^3$$

**Example 2**

### Speed

A radio-controlled car travels 30 feet across the room in 1.6 seconds. How fast is it traveling in miles per hour?

*Solution*

From the information, the rate of the car is $\frac{30 \text{ ft}}{1.6 \text{ s}}$. Multiply by unit fractions such as $\frac{60 \text{ s}}{1 \text{ min}}$ to go from seconds to minutes to hours and from feet to miles. These fractions are chosen so that the final units are miles per hour.

$$\frac{30 \text{ ft}}{1.6 \text{ s}} \cdot \frac{60 \text{ s}}{1 \text{ min}} \cdot \frac{60 \text{ min}}{1 \text{ h}} \cdot \frac{1 \text{ mi}}{5{,}280 \text{ ft}} = \frac{108{,}000 \text{ mi}}{8{,}448 \text{ mi}} \approx \frac{12.8 \text{ mi}}{1 \text{ h}}$$

or 12.8 miles per hour

### Practice Exercises

1. Show how you would use dimensional analysis to convert 6 cm/s to km/h.

2. Use dimensional analysis to change
   a. 50 meters per second to kilometers per hour
   b. 0.025 day to seconds
   c. the speed 60 miles per hour to kilometers per hour. (1609 meters = 1 mile)

3. The equation for the universal gas law is $PV = nRT$. Suppose the pressure $P$ is in pascals (Pa), the volume $V$ is in cubic meters ($m^3$), the amount of substance $n$ is in moles (mol), and the temperature $T$ is in kelvins (K). What must be the units of the universal gas law constant $R$?

*Answers*

1. $\frac{6 \text{ cm}}{s} \cdot \frac{1 \text{ m}}{100 \text{ cm}} \cdot \frac{1 \text{ km}}{1000 \text{ m}} \cdot \frac{60 \text{ s}}{1 \text{ min}} \cdot \frac{60 \text{ min}}{1 \text{ hr}} = 0.216 \text{ km/hr}$

**2a.** 180 km/h  **2b.** 2160 s  **2c.** 97 km/h  **3.** $\dfrac{m^3 \cdot Pa}{mol \cdot K}$

## Logarithms

Logarithms can be used to solve problems involving exponential functions, such as the half-life of a radioactive substance or the pH of a solution.

A *logarithm* is an exponent and is abbreviated *log*. For example, $\log_{10} x$ is the exponent you put on 10 to get $x$. So, if $x = 100$, then $\log_{10} x = 2$ because $10^2 = 100$. In the same way, $\log_{10} 1000 = 3$, $\log_{10} 10 = 1$, $\log_{10} 1 = 0$, $\log_{10} \frac{1}{10} = -1$. Ten is a common base for logarithms, so log $x$ is called a *common logarithm* and is shorthand for $\log_{10} x$. Check out these log equivalents on your calculator, and use your calculator to determine the logs of numbers that are not multiples of 10.

Because logarithms are exponents, they follow the properties of exponents.

| Product property | $a^m \cdot a^n = a^{m+n}$ | $\log xy = \log x + \log y$ |
| Quotient property | $a^m/a^n = a^{m-n}$ | $\log x/y = \log x - \log y$ |
| Power property | | $\log x^n = n \log x$ |

In some problems you know what the logarithm is, but you want to find the number that has that logarithm. What you are looking for is called the *antilogarithm*, abbreviated antilog. Use the 10x key on your calculator to find an antilog.

## Example 1

### Comparing Logarithms

Find the values *a–f.* Which values are equal?

$a = \log 18 \quad b = \log 71 \quad c = a + b \quad d =$ antilog $c \quad e = 18 \cdot 71 \quad f = \log(8 \cdot 71)$

*Solution*

$a \approx 1.255; \quad b \approx 1.851; \quad c \approx 3.106; \quad d = 1278; \quad e = 1278; \quad f \approx 3.106$

$d = e$, or antilog $(\log 18 + \log 71) = 18 \cdot 71$

$c = f$, or $\log 18 + \log 71 = \log(8 \cdot 71)$

## Example 2

### pH

What is the pH of a solution in which $[H^+]$ is $5.3 \times 10^{-4}$?

*Solution*

| Start with the definition of pH. | $pH = -\log[H^+]$ |
| Substitute the value of $[H^+]$. | $= -\log[5.3 \times 10^{-4}]$ |
| Find the log, then change the sign. | $= 3.35$ |

### Practice Exercises

1. Find the log.

   **a.** 456    **b.** $\dfrac{x}{15}$    **c.** $3.4 \times 10^{-6}$

2. Find the antilog

   **a.** 2    **b.** $-3$    **c.** 3.4

3. The pH of a solution is 8.25. What is the concentration of hydrogen ions in the solution?

---

*Answers*

**1a.** 2.659    **1b.** $\log x - 1.176$    **1c.** 5.469

**2a.** 100    **2b.** $\dfrac{1}{1000}$    **2c.** 2512    **3.** $5.62 \times 10^{-9}$

# Appendix B: Connecting Chemical Concepts

## Chemical Names and Formulas

Recall from Unit 1: Alchemy that all matter is made up of elements that can combine with one another to form compounds. Chemists have created an organized system of chemical names and formulas that specify what elements are present in a given compound. In Unit 2: Smells, chemical names and formulas were used to predict the smell of different substances. So knowing the chemical name and formula of a compound can often be useful in predicting properties of substances. Understanding some basic patterns in naming and writing formulas is also useful because it allows chemists from different parts of the world to communicate chemical information easily. Examine the table that contains binary ionic compounds and binary covalent compounds.

**Binary Compounds**

| Chemical name | Chemical formula | Type of compound |
|---|---|---|
| carbon dioxide | $CO_2$ | molecular covalent |
| calcium chloride | $CaCl_2$ | ionic |
| carbon monoxide | $CO$ | molecular covalent |
| dinitrogen tetroxide | $N_2O_4$ | molecular covalent |
| sodium iodide | $NaI$ | ionic |

Here are some patterns you might notice.

- Each compound is made up of two different elements.
- The name of each compound ends in the suffix *-ide*.
- The names of the molecular covalent compounds include prefixes: *di-*, *mono-*, *tri-*, and *tetr-* correspond to the number of atoms of each element in the formula.
- The names of the ionic compounds do not include prefixes.

The compounds in the table are called *binary compounds* because they are made up of only two different elements. Binary ionic compounds are made up of metal and nonmetal atoms and binary covalent compounds are made up of nonmetals only. Notice that with binary ionic compounds, the name alone does not tell you how many atoms of each element are in the compound. However, with binary covalent compounds, the prefixes in the name indicate how many of each atom are present in the formula.

Be careful not to confuse binary compounds with elements that exist as diatomic molecules such as hydrogen, $H_2$, nitrogen, $N_2$, and oxygen, $O_2$. These diatomic molecules are made up of only one type of element. The elements that exist in nature as diatomic molecules include hydrogen, nitrogen, oxygen, and the halogens.

## NAMES OF BINARY COVALENT COMPOUNDS

If you know the formula for a binary covalent compound, some simple rules can help you to name it correctly.

**1.** Write the name of the first element in the formula. Use a prefix in front of the element name to indicate if there are two or more atoms of that element in the formula.

**2.** Write the name of the second nonmetal in the formula and change the ending of the name to the suffix *-ide*. Use a prefix to indicate how many atoms of that element are in the formula, even if there is only one.

The table lists common prefixes and the number of atoms the prefix represents in the formula for a binary covalent compound.

### Prefixes for Binary Covalent Compounds

| Prefix | Number |
|--------|--------|
| mono- | one |
| di- | two |
| tri- | three |
| tetra- | four |
| penta- | five |
| hexa- | six |
| hepta- | seven |
| octa- | eight |
| nona- | nine |
| deca- | ten |

The first molecular covalent compound in the binary compounds table on page B-1 is $CO_2$. The compound is named *carbon dioxide* following the rules for naming binary covalent compounds. Notice that this name helps to distinguish $CO_2$ from carbon monoxide, CO, which has one oxygen atom in the formula. Notice also that the prefix *mono-* is used in carbon monoxide to indicate that there is only one oxygen atom in that compound but that the prefix is not used in front of carbon in the name of either compound.

## FORMULAS OF BINARY COVALENT COMPOUNDS

If you know the name of a binary molecular compound, you can determine the correct chemical formula.

**1.** Write the symbol for the first element in the name. Use a subscript to indicate the number of atoms of that element in the formula, based on the prefix in the name.

**2.** Write the symbol for the second element. Use a subscript to indicate the number of atoms of the second element, based on the prefix in the name.

**3.** Remember that, in formulas, the numeral 1 is not written as a subscript.

For example, the compound dinitrogen tetroxide contains nitrogen, N, and oxygen, O. The *di-* prefix indicates that there are two nitrogen atoms in the formula and the *tetra-* prefix indicates that there are four oxygen atoms in the formula. (Note that in this case, the "a" is dropped.) So the correct formula is $N_2O_4$.

### Practice Exercises

1. Write the name for these binary covalent compounds.
   **a.** $CO_2$   **b.** $S_2O$   **c.** $PCl_3$   **d.** $N_2O$   **e.** $SO_2$   **f.** $NF_3$

2. Write the formula for these binary covalent compounds.
   **a.** dinitrogen tetrafluoride     **b.** dinitrogen pentoxide
   **c.** carbon monoxide             **d.** diphosphorus trisulfide
   **e.** carbon tetrabromide          **f.** sulfur trioxide

---

*Answers*

**1a.** carbon dioxide      **1b.** disulfur monoxide  **1c.** phosphorus trichloride
**1d.** dinitrogen monoxide  **1e.** sulfur dioxide      **1f.** nitrogen trifluoride
**2a.** $N_2F_4$  **2b.** $N_2O_5$  **2c.** CO  **2d.** $P_2S_3$  **2e.** $CBr_4$  **2f.** $SO_3$

---

## NAMES OF BINARY IONIC COMPOUNDS

If you know the formula for a binary ionic compound, some simple rules can help you to name it correctly. Recall that a binary ionic compound is a combination of a positively charged metal cation and a negatively charged nonmetal anion.

1. Write the element name of the cation followed by the element name of the anion.

2. Change the ending of the element name of the anion to the suffix *-ide*.

For example, NaCl is the formula for sodium chloride, and $CaCl_2$ is the formula for calcium chloride. Notice that no prefixes are used in naming binary ionic compounds.

## FORMULAS OF BINARY IONIC COMPOUNDS

If you know the name of a binary ionic compound, you can use it to determine the correct chemical formula.

1. Write the symbol for the cation, including the charge.

2. Write the symbol for the anion, including the charge.

3. Use subscripts in the formula so the net charge of the compound is zero.

4. Remember not to use a subscript to indicate the numeral 1, and do not include charges in the formula.

For example, the compound strontium chloride is made up of a combination of the cation strontium, $Sr^{2+}$, and the anion chloride, $Cl^-$. For the net charge to be zero, the formula for strontium chloride requires one strontium cation and two chloride anions:

$$1(+2) + 2(-1) = (+2) + (-2) = 0$$

So the formula for strontium chloride is $SrCl_2$.

**Practice Exercises**

1. Write the name for the following binary ionic compounds.
   **a.** KBr     **b.** MgO     **c.** $Na_3P$     **d.** NaCl     **e.** $Ca_3N_2$     **f.** $LiS_2$

2. Write the formula for the following binary ionic compounds.
   **a.** calcium sulfide     **b.** beryllium nitride     **c.** aluminum oxide
   **d.** potassium oxide     **e.** strontium sulfide     **f.** barium bromide

---

*Answers*

**1a.** potassium bromide     **1b.** magnesium oxide     **1c.** sodium phosphide
**1d.** sodium chloride     **1e.** calcium nitride     **1f.** lithium sulfide
**2a.** CaS     **2b.** $Be_3N_2$     **2c.** $Al_2O_3$     **2d.** $K_2O$     **2e.** SrS     **2f.** $BaBr_2$

---

## NAMES OF IONIC COMPOUNDS WITH TRANSITION METALS

Unlike main group atoms, most transition metals can have more than one ion charge. When naming ionic compounds that contain transition metals, Roman numerals are used to indicate the charge on the cation. If you know the formula for an ionic compound that contains a transition metal, you can determine its name.

1. Follow the same basic rules for naming binary ionic compounds. Name the cation first, followed by the anion, changing the ending of the element name of the cation to the suffix -*ide*.

2. To indicate the charge on the cation, use a Roman numeral in parentheses. Use the rule of zero charge and the charge on the anion to determine the charge on the cation.

3. Note that if the compound contains silver, Ag, or zinc, Zn, there is no need to use a Roman numeral in the name. Silver typically has a charge of +1 and zinc has only one cation with a charge of +2.

For example, the formula for $Fe_2O_3$ contains three oxide anions, each with a charge of −2. The formula shows that there are two Fe ions. For the net charge on the compound to be zero, each iron cation must have a charge of +3:

$$2(+3) + 3(-2) = (+6) + (-6) = 0$$

So, the name of this compound is *iron (III) oxide*.

## FORMULAS OF IONIC COMPOUNDS WITH TRANSITION METALS

If you know the name of an ionic compound that contains a transition metal, you can use the name to determine the correct formula.

1. Write the symbol for the metal cation. The Roman numeral in parentheses indicates the charge on the cation. Remember that silver has a charge of +1 and zinc has a charge of +2.

2. Write the symbol for the anion, including the charge.

3. Use subscripts in the formula to indicate that the net charge of the compound is zero.

**4.** Remember not to use a subscript to indicate the numeral 1, and do not include charges in the formula.

For example, the compound chromium (III) oxide contains a chromium cation, $Cr^{3+}$ as indicated by the Roman numeral III in parentheses. The oxide anion has a charge of $(-2)$. Using the rule of zero charge, you can determine how many of each ion is in one formula unit of chromium oxide.

$$2(+3) + 3(-2) = (+6) + (-6) = 0$$

So the formula contains two chromium cations and three oxygen anions: $Cr_2O_3$.

### Practice Exercises

**1.** Write the name for the following binary ionic compounds.
   **a.** $CuF_2$   **b.** $AgCl$   **c.** $FeCl_3$   **d.** $PbS$   **e.** $SnO_2$   **f.** $Mn_3P_2$

**2.** Write the formula for the following binary ionic compounds.
   **a.** cobalt (III) chloride   **b.** chromium (IV) sulfide   **c.** copper (I) nitride
   **d.** zinc oxide   **e.** cobalt (II) iodide   **f.** lead (IV) sulfide

---

*Answers*

**1a.** copper (II) fluoride   **1b.** silver chloride   **1c.** iron (III) chloride
**1d.** lead (II) sulfide   **1e.** tin (IV) oxide   **1f.** manganese (II) phosphide
**2a.** $CoCl_3$   **2b.** $CrS_2$   **2c.** $Cu_3N$   **2d.** $ZnO$   **2e.** $CoI_2$   **2f.** $PbS_2$

---

### NAMES OF IONIC COMPOUNDS WITH POLYATOMIC IONS

Some ionic compounds contain ions made up of two or more elements called polyatomic ions.

**1.** Follow the basic rules for naming binary ionic compounds.

**2.** Insert the name of any polyatomic ions in the formula. There is a list of common polyatomic ions in the reference tables at the end of your textbook.

For example, $Ca(NO_3)_2$ contains more than two elements, so it's not a binary ionic compound. It's a compound that contains the polyatomic ion nitrate, $(NO_3)^-$. The metal cation is calcium, so the name of this compound is *calcium nitrate*.

### FORMULAS OF IONIC COMPOUNDS WITH POLYATOMIC IONS

If you know the name of an ionic compound containing a polyatomic ion, you can use it to determine the correct chemical formula. Remember that, in the formula, the charge on the polyatomic ion is for the entire group of atoms as a unit. When there is more than one of the same polyatomic ion in a formula, the ion is enclosed in parentheses and a subscript number indicates how many ions are in the compound. Follow the same rules for writing formulas for other types of compounds, but make sure polyatomic ions are treated as a unit.

**1.** Write the symbol for the cation, including the charge.

**2.** Write the symbol for the anion, including the charge.

**3.** Use subscripts in the formula so the net charge of the compound is zero.

For example, in the compound magnesium hydroxide, the cation is magnesium, $Mg^{2+}$, and the anion is hydroxide, $(OH)^-$. For the charge on the compound to be zero, the formula must contain one magnesium ion and two hydroxide polyatomic ions:

$$1(+2) + 2(-1) = (+2) + (-2) = 0$$

So the formula for magnesium hydroxide is $Mg(OH)_2$. Notice that parentheses and a subscript are needed to show that there are two hydroxide units in the formula. An ionic compound could contain both a transition metal ion and a polyatomic ion. If that's the case, follow the rules for both transition metal ions and polyatomic ions.

### Practice Exercises

1. Write the name for the following ionic compounds with polyatomic ions.
   **a.** $Ca(NO_3)_2$ **b.** $Ag_2SO_4$ **c.** $Ba(OH)_2$
   **d.** $CuCO_3$ **e.** $Ca_3(PO_4)_2$ **f.** $Pb(HCO_3)_4$

2. Write the formula for the following ionic compounds with polyatomic ions.
   **a.** calcium bromate **b.** ammonium hydroxide **c.** potassium nitrate
   **d.** nickel (II) cyanide **e.** iron (III) sulfate **f.** copper (II) phosphate

---

*Answers*

**1a.** calcium nitrate **1b.** silver sulfate **1c.** barium hydroxide
**1d.** copper (II) carbonate **1e.** calcium phosphate
**1f.** lead (IV) hydrogen carbonate

**2a.** $Ca(BrO_3)_2$ **2b.** $NH_4OH$ **2c.** $KNO_3$ **2d.** $Ni(CN)_2$
**2e.** $Fe_2(SO_4)_3$ **2f.** $Cu_3(PO_4)_2$

---

### NAMES OF ACIDS

In Unit 4: Toxins, you learned that acids could be defined as molecules that break apart, or dissociate, in solution to form at least one hydrogen cation, $H^+$, and an anion. Examine the table of acids. Try to find patterns in the names, formulas, and ions in solution.

#### Acids

| Acid name | Chemical formula | Cations in solution | Anions in solution |
|---|---|---|---|
| sulfuric acid | $H_2SO_4$ | 2 hydrogen ions, $H^+$ | 1 sulfate ion, $SO_4^{2-}$ |
| sulfurous acid | $H_2SO_3$ | 2 hydrogen ions, $H^+$ | 1 sulfite ion $SO_3^{2-}$ |
| hydroiodic acid | $HI$ | 1 hydrogen ion, $H^+$ | 1 iodide ion, $I^-$ |
| hydrochloric acid | $HCl$ | 1 hydrogen ion, $H^+$ | 1 chloride ion, $Cl^-$ |
| chloric acid | $HClO_3$ | 1 hydrogen ion, $H^+$ | 1 chlorate ion, $ClO_3^-$ |
| perchloric acid | $HClO_4$ | 1 hydrogen ion, $H^+$ | 1 perchlorate ion, $ClO_4^-$ |
| hypochlorous acid | $HClO$ | 1 hydrogen ion, $H^+$ | 1 hypochlorite ion, $ClO^-$ |
| chlorous acid | $HClO_2$ | 1 hydrogen ion, $H^+$ | 1 chlorite ion, $ClO_2^-$ |

Here are some patterns you might notice.

- The formula for each acid contains the element hydrogen, and each acid dissociates to form $H^+$ ions in solution.
- When dissociated in solution, some of the anions are polyatomic ions, while others are monatomic ions.
- The names of acids containing only two elements begin with "hydro."
- The names of the acids with polyatomic ions start with the root of the name of the polyatomic ion.
- Some acids have a name that ends *-ic acid*; others have names that end in *-ous acid*.
- There is no net charge on the compound before it dissociates, and the net charge of the ions in solution is zero.

### NAMES OF BINARY ACIDS

In the table, some of the acids are made up of covalent molecules that contain only two elements. These compounds dissociate into hydrogen ions and nonmetal anions that are not polyatomic ions. To name these binary acids, you need to know the root of the name of the anion. For example, the root of the chloride ion is *chlor-*, and the root of the iodide ion is *iod-*.

1. Add the prefix *hydro-* to the root name of the anion.
2. Add the suffix *-ic* to the root of the anion name.
3. Add the word *acid* to the name.

For example, the molecule HBr is an acid that dissociates into $H^+$ and $Br^-$ ions in solution. The root of the anion name is *brom-*, so the name of this acid is *hydrobromic acid*. Likewise, HF is hydrofluoric acid.

### FORMULAS OF BINARY ACIDS

Acids have formulas that begin with hydrogen, H. Binary acids are named using the root of the anion that forms in solution, so you can work backward from the root to determine the name.

1. Identify the root of the anion name by removing the prefix *hydro-* and the suffix *-ic* from the root of the anion name. Rename the anion using the suffix *-ide*.
2. Write the symbol for the anion, including the charge.
3. Write the symbol for the hydrogen ion, including the charge.
4. Use subscripts to write a chemical formula for a molecule that has no charge and when dissociated in solution has no net charge.

For example, if you remove the prefix *hydro-* and the suffix *-ic* from hydrobromic acid, you are left with the root, *brom-*. Adding back the *-ide* ending indicates that the anion is bromide, $Br^-$. This acid is a molecule with no charge that dissociates into hydrogen ions, $H^+$, and $Br^-$ anions so the formula can be determined using the charges of the ions in solution, such that they add to zero charge.

$$1(+1) + 1(-1) = 0$$

This means that when one molecule of this acid dissociates, one hydrogen ion and one bromide ion form in solution, so the compound is the molecule, HBr. Consider hydroselenic acid. The root of the anion name is *selen-*, which corresponds to the selenide ion, $Se^{2-}$. For the solution to have no net charge, the molecule must dissociate into 2 hydrogen ions, $H^+$ and 1 selenide ion, $Se^{2-}$.

$$2(+1) + 1(-2) = (+2) + (-2) = 0$$

So the formula for this acid is $H_2Se$.

## NAMES OF ACIDS THAT CONTAIN POLYATOMIC ANIONS WITH OXYGEN

In the table, some of the acids are made up of covalent molecules that dissociate into hydrogen ions and polyatomic anions that contain oxygen. To name these acids, you use the root of the name of the polyatomic anion. For example, the root of the sulfate ion is *sulf-* and the root of the chlorite ion is *chlor-*.

1. Identify the root of the name of the polyatomic anion.
2. If the polyatomic anion ends in *-ate*, add the suffix *-ic* to the root of the polyatomic anion name.
3. If the polyatomic anion ends in *-ite*, add the suffix *-ous* to the root of the polyatomic anion name.
4. Add the word *acid* to the name.

For example, HClO dissociates into one hydrogen ion, $H^+$, and the polyatomic ion hypochlorite, $ClO^-$, which has the root, *hypochlor-*. Since the polyatomic anion ends in *-ite*, the suffix *-ous* is added to the root, and the word *acid* is added to form hypochlorous acid. As another example, $HClO_2$ dissociates into one hydrogen $H^+$ ion and the polyatomic ion chlorite, with the root, *chlor-*. Because the anion name ends in *-ite*, this acid is called *chlorous acid*. Likewise, $HClO_4$ with the perchlorate ion is perchloric acid, and $HClO_3$ with the chlorate ion is chloric acid.

## FORMULAS OF ACIDS THAT CONTAIN POLYATOMIC ANIONS WITH OXYGEN

Remember that these acids are molecules with no charge that dissociate into hydrogen ions and polyatomic anions that contain oxygen in solution. They are named using the root of the polyatomic anion that forms in solution. So work backward from the root of the anion to determine the name.

1. Identify the name and charge of the polyatomic ion from the root of the anion name.
2. Write the symbol for the polyatomic anion, including the charge.
3. Write the symbol for the hydrogen ion, including the charge.
4. Use subscripts to write a chemical formula for a molecule that has no charge and when dissociated in solution has no net charge.

For example, sulfuric acid does not have the prefix *hydro-*, so it must contain a polyatomic ion. Both sulfite, $(SO_3)^{2-}$ and sulfate $(SO_4)^{2-}$ have the root *sulf-*, but the name of this acid ends in *-ic*. The correct anion is sulfate, because its name ends in *-ate*. The sulfate ion has a charge of $+2$, so when it dissociates, the original molecule must break apart into two hydrogen ions, $H^+$, and one sulfate ion, $(SO_4)^{2-}$.

$$2(+1) + 1(-2) = (+2) + (-2) = 0$$

So the formula for this acid contains two hydrogen atoms and one sulfate ion, $H_2SO_4$.

### Practice Exercises

1. Write the name for these acids.
   **a.** HBr    **b.** $H_2SO_4$    **c.** $HNO_3$    **d.** HCl    **e.** HBrO    **f.** $H_3PO_3$
2. Write the formula for these acids.
   **a.** carbonic acid    **b.** hydrofluoric acid    **c.** phosphoric acid
   **d.** sulfurous acid    **e.** perbromic acid    **f.** nitrous acid

---

*Answers*

**1a.** hydrobromic acid    **1b.** sulfuric acid    **1c.** nitric acid
**1d.** hydrochloric acid    **1e.** hypobromous acid    **1f.** phosphorous acid
**2a.** $H_2CO_3$   **2b.** HF   **2c.** $H_3PO_4$   **2d.** $H_2SO_3$   **2e.** $HBrO_4$   **2f.** $HNO_2$

---

## NAMES AND FORMULAS FOR ARRHENIUS BASES

According to the Arrhenius theory of acids and bases, a base is defined as a substance that adds $OH^-$ to an aqueous solution. Examine the table of Arrhenius bases.

### Arrhenius Bases

| Acid name | Chemical formula | Cations in solution | Anions in solution |
|---|---|---|---|
| potassium hydroxide | KOH | 1 potassium ion, $K^+$ | 1 hydroxide ion, $OH^-$ |
| calcium hydroxide | $Ca(OH)_2$ | 1 calcium ion, $Ca^{2+}$ | 2 hydroxide ions, $OH^-$ |
| aluminum hydroxide | $Al(OH)_3$ | 1 aluminum ion, $Al^{3+}$ | 3 hydroxide ions, $OH^-$ |

Here are some patterns you might notice.

- They each contain the hydroxide polyatomic ion.
- They each dissociate into metal cations and hydroxide polyatomic anions.
- There are as many hydroxide ions as the charge on the metal.
- The names all contain the word *hydroxide*.

When naming and writing formulas for Arrhenius bases, you simply follow the rules for naming ionic compounds that contain a polyatomic ion. For these bases, the polyatomic ion is the hydroxide ion, $OH^-$.

### Practice Exercises

1. Write the name for these bases.
   **a.** $Mg(OH)_2$      **b.** NaOH      **c.** $Sr(OH)_2$
2. Write the formula for these bases.
   **a.** potassium hydroxide
   **b.** barium hydroxide
   **c.** lithium hydroxide

---

*Answers*

**1a.** magnesium hydroxide    **1b.** sodium hydroxide    **1c.** strontium hydroxide
**2a.** KOH   **2b.** $Ba(OH)_2$   **2c.** LiOH

---

# Percent Composition

Elements combine in specific ratios to form compounds, so the ratio of one element to another in a compound is always the same. For example, water, $H_2O$, is made up of the elements hydrogen and oxygen in a 2:1 ratio. If the ratio of hydrogen to oxygen were different, the compound would not be water. For example, hydrogen peroxide, $H_2O_2$, is also made up of hydrogen and oxygen atoms, but the ratio of hydrogen to oxygen in this compound is 1:1. In one mole of water, there are always two moles of hydrogen atoms and one mole of oxygen atoms, as indicated by the ratios of elements in the chemical formula.

Examine the table that shows the composition of water in terms of moles, mass, and percent mass of each element in the compound.

**Composition of Different Samples of Water, $H_2O$**

| Sample size | Moles of $H_2O$ | Moles of hydrogen atoms | Moles of oxygen atoms | Mass of sample | Mass of hydrogen in sample | Mass of oxygen in sample | Percent by mass of hydrogen | Percent by mass of oxygen |
|---|---|---|---|---|---|---|---|---|
| 1 mol $H_2O$ | 1 mol | 2 mol | 1 mol | 18.02 g | 2.02 g | 16.00 g | 11.2% | 88.8% |
| 5 mol $H_2O$ | 5 mol | 10 mol | 5 mol | 90.10 g | 10.1 g | 80.00 g | 11.2% | 88.8% |
| 18.02 g $H_2O$ | 1 mol | 2 mol | 1 mol | 18.02 g | 2.02 g | 16.00 g | 11.2% | 88.8% |
| 200 g $H_2O$ | 11.1 mol | 22.2 mol | 11.1 mol | 200 g | 22.4 g | 177.6 g | 11.2% | 88.8% |

Notice that for each sample size of water, the percent of hydrogen in the sample and the percent of oxygen in the sample are always the same. This is because these elements always combine in the same molar ratio to form water. The percent composition of elements in a compound is the percent by mass of each element in the compound. A percent is an amount per 100. You can calculate a percent by taking part of a whole amount, dividing it by the whole amount, and multiplying by 100%.

$$\text{percent} = \left(\frac{\text{part}}{\text{whole}}\right) \cdot 100\%$$

In terms of percent composition of elements in a compound, the "part" is the mass of one of the elements in a sample of a compound and the "whole" is the mass of the entire sample of the compound. If you add the masses of each element present in the sample of the compound, the result is the mass of the compound. If you add the percent by mass of each element in a compound the total should be 100%.

## DETERMINING PERCENT COMPOSITION USING THE CHEMICAL FORMULA

If you know the chemical formula for a compound, you can determine the percent composition of each element in it by using the mass of one mole of the compound and the mass of each element that would be found in one mole of the compound.

Example 1

## Percent Composition of CO

What is the percent composition of carbon monoxide, CO?

*Solution*

Assume that you have one mole of CO.

**Step 1:** Use the periodic table to calculate the molar mass of CO.

$$\text{molar mass of CO} = 12.01 \text{ g/mol} + 16.00 \text{ g/mol} = 28.01 \text{ g/mol}$$

One mole of CO is 28.01 g.

**Step 2:** Determine the percent mass of each element in the compound.

Divide the mass of each element in one mole of the compound by the mass of one mole of the compound and multiply by 100%. The molar ratio of C:O in the compound is 1:1, so there is one mole of each element in one mole of CO.

The mass of one mole of carbon, C, is 12.01 g. The mass of one mole of oxygen, O, is 16.00 g. Use these values to determine the percent mass of each compound in one mole of the compound.

$$\text{percent by mass of C} = \left(\frac{12.01 \text{ g}}{28.01 \text{ g}}\right) \cdot 100\% = 42.88\%$$

$$\text{percent by mass of O} = \left(\frac{16.00 \text{ g}}{28.01 \text{ g}}\right) \cdot 100\% = 57.12\%$$

**Step 3:** Check your answer by adding the percent by mass of each element. The total should be 100%.

42.88% + 57.12% = 100.00% (or 100% when rounded to the nearest whole number)

Example 2

## Percent Mass of Oxygen in KClO₃

What is the mass of oxygen that can be produced from the decomposition of 150.0 g of potassium chlorate, $KClO_3$? (*Hint:* Find the percent composition of oxygen in the compound to figure out the answer.)

*Solution*

When potassium chlorate, $KClO_3$, decomposes, the oxygen produced will be the amount present in the 150.0 g sample of the compound.

**Step 1:** Calculate the mass of one mole of potassium chlorate, $KClO_3$.

$$\begin{aligned}\text{molar mass of } KClO_3 &= 1(39.10 \text{ g/mol}) + 1(35.45 \text{ g/mol}) + 3(16.00 \text{ g/mol}) \\ &= 39.10 \text{ g/mol} + 35.45 \text{ g/mol} + 48.00 \text{ g/mol} \\ &= 122.55 \text{ g/mol}\end{aligned}$$

One mole of potassium chlorate, $KClO_3$, is 122.55 g.

*(continued)*

**Step 2:** Determine the percent mass of oxygen, O, in one mole of the compound.

The mass of one mole of oxygen is 16.00 g. In one mole of potassium chlorate, $KClO_3$, there are three moles of oxygen, O.

$$\text{mass of O in one mole of compound} = 3(16.00 \text{ g})$$
$$= 48.00 \text{ g}$$

Divide the mass of oxygen, O, in one mole of potassium chlorate, $KClO_3$, by the molar mass of the potassium chlorate, $KClO_3$, and multiply by 100%.

$$\text{percent by mass O} = \left(\frac{48.00 \text{ g}}{122.55 \text{ g}}\right) \cdot 100\% = 39.17\%$$

**Step 3:** Use the percent composition of oxygen, O, to determine the amount of oxygen produced in the decomposition reaction.

The mass of oxygen produced when 150.0 g of $KClO_3$ decomposes is 39.17% of 150.0 g. Convert the percentage to a decimal and multiply by 150.0 g.

$$0.3917 \times 150.0 \text{ g} = 58.75 \text{ g}$$

The decomposition of 150.0 g of potassium chlorate, $KClO_3$, produces 58.75 g of oxygen, O.

**Practice Exercises**

1. What is the percent composition of HCl?
2. What is the percent composition of ammonium phosphate $(NH_4)_3PO_4$?
3. What is the mass of aluminum needed to produce 298 g of $Al_2(SO_4)_3$?

*Answers*

1. H = 2.7%, Cl = 97.3%
2. N = 28.19%, H = 8.113%, P = 20.77%, O = 42.93%. The percent by mass values add to 100.10% or 100% if rounded to the nearest whole number.
3. 47.0 g Al

## Molecular and Empirical Formulas

In Unit 2: Smells, you learned that the chemical formula for a molecular covalent compound is also called the *molecular formula*. The molecular formula indicates the number of atoms of each element present in one molecule of a compound. Examine the table of formulas. Notice that for each compound, there is a molecular formula and what is called the *empirical formula*.

## Molecular and Empirical Formulas

| Name | Molecular formula | Empirical formula |
|---|---|---|
| ethyne | $C_2H_2$ | CH |
| benzene | $C_6H_6$ | CH |
| carbon monoxide | CO | CO |
| carbon dioxide | $CO_2$ | $CO_2$ |
| tetraphosphorus decaoxide | $P_4O_{10}$ | $P_2O_5$ |

Here are some patterns you might notice.

- For each compound, the molecular formula and the empirical formula both have the same type of atoms.

- Some compounds have a molecular formula that is different from the empirical formula.

- Some compounds have a molecular formula that is the same as the empirical formula.

- The molecular formula and the empirical formula are related mathematically.

- If the empirical formula is different from the molecular formula, you can multiply the subscripts in the empirical formula by a single number to get the values for subscripts in the molecular formula. For example, if you multiply the subscripts for CH by 6, you get the molecular formula for benzene, $C_6H_6$. You can multiply the subscripts in $P_2O_5$ by 2 to get the molecular formula $P_4O_{10}$.

The empirical formula is the simplest formula for a compound and shows the smallest whole-number ratio of the atoms present in that compound. There is a relationship between molecular formula and empirical formula that can be seen in the patterns of the subscripts. If you multiply the subscripts in the empirical formula by the number of empirical units, you get the molecular formula. Remember if there is no subscript, the value is 1.

$$(\text{empirical formula})_{(\text{number of empirical units})} = \text{molecular formula}$$

ethyne: $(CH)_2 = C_2H_2$

benzene: $(CH)_6 = C_6H_6$

carbon monoxide: $(CO)_1 = CO$

carbon dioxide: $(CO_2)_1 = CO_2$

If the molecular formula is different from the empirical formula, you can determine the empirical formula by dividing each subscript by the largest whole-number factor they have in common. For example, you can divide subscripts 4 and 10 in $P_4O_{10}$ by the number 2 to get the empirical formula $P_2O_5$.

**Example 1**

## Determining Empirical Formula from Molecular Formula

Write the empirical formula for these compounds.

   a. $C_6H_{12}O_2$
   b. $C_4H_8O_2$
   c. $H_2SO_4$

*Solution*

   a. The largest whole-number factor the subscripts have in common is 2. You can divide each subscript by 2 to determine that the empirical formula is $C_3H_6O$.

   b. The largest whole-number factor the subscripts have in common is 2. You can divide each subscript by 2 to determine that the empirical formula is $C_2H_4O$.

   c. The largest whole-number factor the subscripts have in common is 1. You can divide each subscript by 1 to determine that the empirical formula is $H_2SO_4$, which is the same as the molecular formula.

If you know the amount in grams of each element present in a sample of a compound, or you know the percent composition for a compound, you can determine its empirical formula. If you also know the molar mass of the compound, you can use it along with the empirical formula to determine the molecular formula of the compound.

**Example 2**

## Determining the Molecular Formula Using Mass Data

A 500.0 g sample of hexanoic acid contains 310.2 g of carbon, 52.1 g of hydrogen, and 137.8 g of oxygen. The molar mass of this compound is 116.16 g/mol. Determine the empirical and molecular formulas for this compound.

*Solution*

**Step 1:** The empirical and molecular formulas show the relative numbers of atoms in an element, not the relative amount of mass in each element. So the first step is to convert the amount of each element in grams to the amount in moles. You can do this using the molar mass of the element from the periodic table.

$$\text{moles of C} = \left( \frac{310.2 \text{ g}}{12.01 \text{ g/mol}} \right) = 25.83 \text{ mol}$$

$$\text{moles of H} = \left( \frac{52.1 \text{ g}}{1.008 \text{ g/mol}} \right) = 51.69 \text{ mol}$$

$$\text{moles of O} = \left( \frac{137.8 \text{ g}}{16.00 \text{ g/mol}} \right) = 8.62 \text{ mol}$$

*(continued)*

**Step 2:** The molecular formulas show the relative amounts of moles in whole-number ratios. To convert the amounts of moles to whole numbers, divide each value by the lowest number of moles.

$$\text{moles of C} = \frac{25.83 \text{ mol}}{8.62} = 2.997 \text{ mol, which rounds to 3 mol}$$

$$\text{moles of H} = \frac{51.69 \text{ mol}}{8.62} = 5.997 \text{ mol, which rounds to 6 mol}$$

$$\text{moles of O} = \frac{8.62 \text{ mol}}{8.62} = 1 \text{ mol}$$

Use the ratio of moles, 3:6:1, to write the empirical formula, $C_3H_6O$.

**Step 3:** You can calculate the number of empirical units by dividing the molar mass of the compound by the empirical molar mass. The empirical molar mass is the sum of the molar masses of the elements in the empirical formula.

$$\text{empirical molar mass} = 3(12.01 \text{ g/mol}) + 6(1.008 \text{ g/mol}) + (16.00 \text{ g/mol})$$
$$= 58.08 \text{ g/mol}$$

$$\text{number of empirical units} = \frac{\text{molar mass}}{\text{empirical molar mass}}$$
$$= \frac{116.16 \text{ g/mol}}{58.08 \text{ g/mol}}$$
$$= 2 \text{ units}$$

$$(\text{empirical formula})_{(\text{number of empirical units})} = \text{molecular formula}$$
$$(C_3H_6O)_2 = C_6H_{12}O_2$$

The molecular formula for the compound is $C_6H_{12}O_2$.

**Example 3**

### Determining the Molecular Formula Using Percent Composition Data

A compound contains 71.65% Cl, 24.27% C, and 4.07% H. The molar mass of the compound is 98.96 g/mol. Determine the empirical formula and molecular formula for this compound.

*Solution*

**Step 1:** When amounts in mass are not provided, assume that you have 100 g of the compound. Find the mass of each element in grams by converting the percent values to decimals and multiplying by the total mass.

$$\text{mass of Cl} = 0.7165 \cdot (100 \text{ g}) = 71.65 \text{ g of Cl}$$
$$\text{mass of C} = 0.2427 \cdot (100 \text{ g}) = 24.27 \text{ g of C}$$
$$\text{mass of H} = 0.0407 \cdot (100 \text{ g}) = 4.07 \text{ g of H}$$

*(continued)*

**Step 2:** Convert the amounts in grams to moles.

$$\text{moles of Cl} = \left(\frac{71.65 \text{ g}}{35.45 \text{ g/mol}}\right) = 2.02 \text{ mol}$$

$$\text{moles of C} = \left(\frac{24.27 \text{ g}}{12.01 \text{ g/mol}}\right) = 2.02 \text{ mol}$$

$$\text{moles of H} = \left(\frac{4.07 \text{ g}}{1.008 \text{ g/mol}}\right) = 4.04 \text{ mol}$$

**Step 3:** Convert the amounts in moles to whole numbers by dividing by the lowest value.

$$\text{moles of Cl} = \left(\frac{2.02 \text{ mol}}{2.02}\right) = 1 \text{ mol}$$

$$\text{moles of C} = \left(\frac{2.02 \text{ mol}}{2.02}\right) = 1 \text{ mol}$$

$$\text{moles of H} = \left(\frac{4.04 \text{ mol}}{2.02}\right) = 2 \text{ mol}$$

The empirical formula is $CH_2Cl$.

**Step 4:** Determine the empirical molar mass.

$$\text{empirical molar mass} = (12.01 \text{ g/mol}) + 2(1.008 \text{ g/mol}) + (35.45 \text{ g/mol})$$
$$= 49.48 \text{ g/mol}$$

**Step 5:** Calculate the number of empirical units.

$$\text{number of empirical units} = \frac{\text{molar mass}}{\text{empirical molar mass}}$$
$$= \frac{98.96 \text{ g/mol}}{49.48 \text{ g/mol}}$$
$$= 2 \text{ units}$$

**Step 6:** Multiply the subscripts in the empirical formula by the number of empirical units to determine the molecular formula.

$$(\text{empirical formula})_{(\text{number of empirical units})} = \text{molecular formula}$$
$$(CH_2Cl)_2 = C_2H_4Cl_2$$

The molecular formula for the compound is $C_2H_4Cl_2$.

**Practice Exercises**

1. What is the molecular formula for a compound with an empirical formula of $C_4H_9$ and a molar mass of 114.22 g/mol?

2. A compound has a mass percent composition of 5.927% hydrogen and 94.073% oxygen. The molecular mass of this compound is 34.016 g/mol. Find both the empirical and molecular formulas.

3. A compound has a mass percent composition of 40.00% carbon, 6.72% hydrogen, and 53.28% oxygen. The molecular mass of this compound is 150.13 g/mol. Determine both the empirical and molecular formulas.

4. A compound contains 0.384 g carbon, 0.048 g hydrogen, and 0.5674 g chlorine. The molecular mass of this compound is 125 g/mol. Calculate the empirical and molecular formulas.

---

**Answers**

1. $C_8H_{18}$

2. empirical formula: HO, molecular formula: $H_2O_2$

3. empirical formula: $CH_2O$, molecular formula: $C_5H_{10}O_5$

4. empirical formula: $C_2H_3Cl$, molecular formula: $C_4H_6Cl_2$

---

# More Stoichiometry Practice

Recall from Unit 4: Toxins that stoichiometry involves calculations based on the quantitative relationship between amounts of reactants and products in a chemical reaction. Using mole ratios from a balanced chemical equation, it's possible to determine amounts of reactants and products that will be used or formed during a reaction.

When the calculations involve amounts in mass, you need to convert given amounts in mass to moles. To do this, you use the molar mass for each substance, because the molar mass is the amount in grams of exactly one mole of that substance.

$$\text{number of moles} = \frac{\text{mass in g}}{\text{molar mass in g/mol}}$$

$$\text{mass in g} = (\text{molar mass in g/mol}) \cdot (\text{number of moles})$$

Once you convert amounts to moles, you can use mole ratios from the balanced chemical equation to relate reactants to products. The "mole tunnel" is a way to help you remember these steps.

Reactant A + Reactant B ⟶ Product C + Product D

**Step 1** Convert grams of reactants to moles of reactants.

$$\text{moles} = \frac{\text{grams}}{\text{molar mass}}$$

Grams

Moles

Mole ratio

Moles

Grams

**Step 3** Convert moles of products to grams of products.

$$\text{grams} = (\text{molar mass})(\text{moles})$$

Mole tunnel

**Step 2** Relate moles of reactants to moles of products using mole ratios.

If you are solving a problem that involves a particular volume of gas, you also need to convert to moles. Recall from Unit 3: Weather that at standard temperature and pressure, STP, one mole of gas occupies a volume of 22.4 liters.

$$\text{number of moles} = \frac{\text{volume at STP in L}}{22.4 \text{ L/mol}}$$

$$\text{volume at STP in L} = \text{number of moles} \cdot (22.4 \text{ L/mol})$$

If the reaction is not carried out under STP conditions, you can use the ideal gas law, $PV = nRT$, to solve for the number of moles, $n$, as long as you are given the temperature, volume, and pressure.

$$n = \frac{PV}{RT}$$

$$V = \frac{nRT}{P}$$

In the above equations, the number of moles is represented by $n$, pressure by $P$, volume by $V$, temperature by $T$, and $R$ is the ideal gas constant.

It is also possible to solve stoichiometry problems in terms of the numbers of particles of each substance because one mole of a substance = $6.02 \times 10^{23}$ particles (such as atoms, molecules, ions, or formula units).

$$\text{number of particles} = 6.02 \times 10^{23} \text{ particles/mol} \cdot (\text{number of moles})$$

$$\text{number of moles} = \frac{\text{number of particles}}{6.02 \times 10^{23} \text{ particles/mol}}$$

**Example 1**

## Volume of Gas Produced in a Reaction at STP

What volume of $H_2O$ vapor is produced when 0.35 mol of $O_2$ gas reacts with an excess of $H_2$ gas at STP? How many molecules of $H_2O$ is this?

*Solution*

**Step 1:** Write the balanced equation for this reaction.

$$2H_2(g) + O_2(g) \longrightarrow 2H_2O(g)$$

**Step 2:** The oxygen gas, $O_2$, is already reported in moles. There is an excess of hydrogen, $H_2$, so oxygen, $O_2$, is the limiting reactant. Use mole ratios to determine the amount of product in moles.

The mole ratio of $O_2$ to $H_2O$ is 1:2.

$$0.35 \text{ mol } O_2 \cdot \left( \frac{2 \text{ mol } H_2O}{1 \text{ mol } O_2} \right) = 0.70 \text{ mol } H_2O$$

So 0.70 mole of $H_2O(g)$ is produced in this reaction.

**(continued)**

**Step 3:** Convert moles of $H_2O(g)$ to volume of $H_2O(g)$. This reaction takes place at standard temperature and pressure, STP.

> volume of $H_2O$ at STP $=$ number of moles $\cdot$ (22.4 L/mol)
> volume of $H_2O$ at STP $=$ 0.70 mol $H_2O \cdot$ (22.4 L/mol)
> volume of $H_2O$ at STP $=$ 15.68 L, rounded to two significant figures, 16 L

**Step 4:** Convert moles of $H_2O$ (g) to number of molecules.

> number of molecules $H_2O = 6.02 \times 10^{23}$ particles/mol $\cdot$ (number of moles)
> number of molecules $H_2O = 6.02 \times 10^{23}$ particles/mol $\cdot$ (0.70 mol $H_2O$)
> $= 4.21 \times 10^{23}$ molecules $H_2O$, rounded to two
> significant figures, $4.2 \times 10^{23}$ molecules $H_2O$

Recall from Unit 3: Weather that if there is a mixture of nonreacting gases, the pressure exerted by one gas in the mixture is called the *partial pressure*. Dalton's law of partial pressures states that the partial pressures of all the gases add up to the total pressure exerted by that mixture of gases. This information can also be helpful in solving stoichiometry problems involving gases.

Dalton's Law of Partial Pressures

$$P_{total} = P_1 + P_2 + P_3 \dots$$

In the above equation, the numbered subscripts refer to each individual gas in the mixture.

**Example 2**

## Stoichiometry with Partial Pressures and the Ideal Gas Law

A student carried out the experiment from the illustration with an excess of hydrochloric acid. The student added a step to bubble the hydrogen gas formed through a closed container of water to collect it in the space above the water. When finished, a 75 mL mixture of water vapor and hydrogen gas was above the water. If the temperature in the container was 30 °C, the total pressure of the gas mixture was 1.015 atm, and the pressure of water vapor was 0.042 atm, what mass of chromium must have reacted?

Solid chromium disappears. Hydrogen gas bubbles form and escape. A blue solution of chromium (II) chloride forms.

Hydrochloric acid
HCl(*aq*)

Hydrogen
$H_2$(*g*)

Chromium (II) chloride
$CrCl_2$(*aq*)

Chromium
Cr(*s*)

Hydrochloric acid is added to solid chromium.

All the hydrogen has escaped. The beaker now contains only aqueous chromium (II) chloride.

*(continued)*

*Solution*

**Step 1:** Start by writing the balanced chemical equation for the reaction.

$$Cr(s) + 2HCl(aq) \longrightarrow H_2(g) + CrCl_2(aq)$$

**Step 2:** To determine the mass of chromium reacted, you need to determine how much hydrogen gas was formed. The problem states the total pressure of the final mixture of gas, the total volume of the mixture of gas, and the partial pressure of water vapor under these conditions. You can use Dalton's law of partial pressures to determine the pressure of the hydrogen gas in the mixture and then the ideal gas law to determine the amount of hydrogen in moles.

$$P_{total} = P_{H_2O} + P_{H_2}$$
$$P_{H_2} = P_{total} - P_{H_2O}$$
$$P_{H_2} = 1.015 \text{ atm} - 0.042 \text{ atm}$$
$$P_{H_2} = 0.973 \text{ atm}$$

**Step 3:** Use the ideal gas law to calculate the number of moles of hydrogen gas produced.

$$T = 30\,°C + 273 = 303 \text{ K}$$
$$75 \text{ mL} = 0.075 \text{ L}$$
$$n = \frac{PV}{RT}$$
$$n = \frac{0.973 \text{ atm} \cdot 0.075 \text{ L}}{\left(0.082\dfrac{\text{L} \cdot \text{atm}}{\text{mol} \cdot \text{K}}\right) \cdot 303\text{K}}$$
$$= 2.9 \times 10^{-3} \text{ mol H}_2$$

**Step 4:** Use mole ratios to work backward to determine the number of moles of chromium that reacted.

The mole ratio of Cr to $H_2$ is 1:1

$$2.9 \times 10^{-3} \text{ mol H}_2 \cdot \left(\frac{1 \text{ mol Cr}}{1 \text{ mol H}_2}\right) = 2.9 \times 10^{-3} \text{ mol Cr}$$

**Step 5:** Convert moles of Cr to grams.

$$\text{mass of Cr} = (52.00 \text{ g/mol}) \cdot (2.9 \times 10^{-3} \text{ mol})$$
$$= 0.15 \text{ g Cr}$$

### BASIC STEPS FOR SOLVING STOICHIOMETRY PROBLEMS

In general, you can follow some basic steps to solve stoichiometry problems.

1. If not given, write a balanced chemical equation.
2. Convert amounts given in mass, volume, or number of particles to moles.
3. If needed, determine the limiting reactant using mole ratios.
4. Use mole ratios to relate reactants to products.
5. Convert amounts in moles back to mass, volume, or number of particles.

Sometimes, you have to incorporate other chemistry concepts, as in Example 2, before you can solve the problem. So before you begin each problem, determine:

- What information is given in the problem?
- What exactly are you being asked to calculate or determine?
- Does the problem contain all the necessary information or should you think about other chemistry concepts that are related to the question?
- What unit conversions will you have to perform?
- What mathematical relationships or formulas are relevant?

The problems that follow give you an opportunity to practice these problem-solving skills.

## Practice Exercises

1. Using the balanced equation from Example 1, determine the volume of $O_2$ gas required to produce 5.50 moles of $H_2O$ vapor at STP.

2. Consider this balanced chemical equation.

$$2C_2H_2(g) + 5O_2(g) \longrightarrow 4CO_2(g) + 2H_2O(g)$$

   a. Find the mole ratio of $C_2H_2$ to $O_2$.
   b. Find the mole ratio of $C_2H_2$ to $CO_2$.
   c. Suppose you have 2.0 L of ethyne, $C_2H_2$. What volume of oxygen gas would be needed to react completely with 2.0 L of ethyne at STP?
   d. What volume of carbon dioxide would be produced from 2.0 L of ethyne at STP?

3. Consider this balanced chemical equation.

$$Mg(s) + 2HCl(aq) \longrightarrow MgCl_2(aq) + H_2(g)$$

   A 120 mL sample of hydrogen gas was collected over water at 24 °C. The vapor pressure of water at 24 °C is 0.030 atm. The total pressure of gas collected was 1.010 atm. What mass of magnesium must have reacted to produce this amount of hydrogen gas with excess hydrochloric acid?

4. Consider this balanced chemical equation.

$$2KClO_3(s) \longrightarrow 2KCl(s) + 3O_2(g)$$

   a. How many moles of $KClO_3$ are needed to produce 0.60 mol of $O_2$?
   b. What mass of $KClO_3$ is needed to produce 0.20 mol of KCl?
   c. How many molecules of $O_2$ are produced from 0.50 mol $KClO_3$?

5. Consider this balanced chemical equation.

$$2Mg(s) + O_2(g) \longrightarrow 2MgO(s)$$

   a. What volume of oxygen gas is needed to burn 0.50 g of magnesium ribbon at STP?
   b. What mass of MgO will be produced from the 0.50 g magnesium ribbon?

6. Consider this balanced chemical equation.

$$4Al(s) + 3O_2(g) \longrightarrow 2Al_2O_3(s)$$

   What mass of aluminum would be used in the production of 10.2 g of aluminum oxide if aluminum reacted directly with unlimited oxygen?

7. Consider this balanced chemical equation.

$$C(s) + O_2(g) \longrightarrow CO_2(g)$$

   a. If 6.00 g of carbon were burned to form $CO_2$, what mass of $CO_2$ would be expected?
   b. If the actual yield of $CO_2$ was measured to be 19.0 g, what is the percent yield?

8. Consider this balanced chemical equation.

$$CH_4(g) + 2O_2(g) \longrightarrow CO_2(g) + 2H_2O(g)$$

   a. If 50.0 g of methane, $CH_4$, were burned, what mass of $CO_2$ would be expected?
   b. If the actual yield of $CO_2$ was measured to be 98.0 g, what is the percent yield?

9. Consider this balanced chemical equation.

$$3Fe(s) + 4H_2O(g) \longrightarrow Fe_3O_4(s) + 4H_2(g)$$

   a. How many moles of $Fe_3O_4$ can be obtained by reacting 16.8 g Fe with 10.0 g $H_2O$?
   b. Which substance is the limiting reactant?
   c. Which substance is in excess?

10. The following reaction was performed with 36.0 g Mg and 0.25 L of a 2.0 M HCl.

$$Mg(s) + 2HCl(aq) \longrightarrow MgCl_2(aq) + H_2(g)$$

   a. Which compound is the limiting reactant?
   b. What mass of hydrogen gas was produced?
   c. What mass of magnesium chloride was produced?

---

*Answers*

1. 61.6 L $O_2$

2a. $\dfrac{2 \text{ mol } C_2H_2}{5 \text{ mol } O_2}$     2b. $\dfrac{1 \text{ mol } C_2H_2}{2 \text{ mol } CO_2}$     2c. 5.0 L $O_2$     2d. 4.0 L $CO_2$

3. 0.12 g Mg

4a. 0.40 mol $KClO_3$     4b. 24.5 g $KClO_3$     4c. 4.52 × $10^{23}$ molecules $O_2$

5a. 0.23 L $O_2$     5b. 0.83 g MgO

6. 5.40 g Al

7a. 22.0 g $CO_2$     7b. 86.4%

8a. 137 g $CO_2$     8b. 71.5%

9a. 0.100 mol $Fe_3O_4$     9b. iron, Fe     9c. water, $H_2O$, is in excess

10a. HCl     10b. 0.50 g $H_2$

10c. 23.8 g $MgCl_2$, rounded to two significant figures, 24 g $MgCl_2$

# Appendix C: Answers to Selected Exercises

## Unit 1: Alchemy

### Lesson 1

1. Possible answer: We use glass because substances in a glass container are visible and glass containers are relatively easy to clean and reuse. Tempered glass containers can be heated over flames without shattering.

3. Possible answers: Know the location of safety equipment. Read lab instructions carefully. Check to be sure that you are using the right chemicals and equipment.

5. Possible answers: Put all equipment in its proper place. Clean your work area. Make sure all bottles and containers holding chemicals are closed and stored properly.

### Lesson 2

1. Alchemists developed some of the first laboratory tools and chemistry techniques. They classified substances into categories and experimented with mixing and heating different substances to create something new.

5. Observable changes that involve chemistry often involve an alteration in the appearance of matter. Examples include metal rusting, cookies baking, and ice cubes forming. Changes that do not involve chemistry only involve matter moving to a different location. Examples include the Sun going down, objects falling, and hands moving on a clock.

### Lesson 3

1. Mass is the amount of material in an object. Volume is the amount of space the object takes up.

5. Possible answer: Examples of things that are matter: a car, a tree, a person, water. Each of these objects has mass and volume. Examples of things that are not matter: sound, movement, feelings, energy. These objects do not have mass or volume.

### Lesson 4

1. Measure the dimensions of the object and calculate its volume using a geometric formula, or if it does not float or dissolve, measure the amount of liquid that it displaces when it is submerged.

3. Yes, the volume of an object is the amount of space it fills. You can usually see how much space an object fills and estimate its volume based on its dimensions. However, for objects that have an irregular shape, are very thin, or have a surface with lots of holes or pits, determining the volume of the object by sight may be difficult.

5. Yes, the mass of the rubber band is the same because only the shape of the rubber band changes, not the amount of matter in it.

11. Possible answer: The balance or scale that was used to measure the two different cubes may not be able to measure accurately to a hundredth of a gram. The two cubes could have exactly the same mass because the measurement error is greater than 0.03 g.

### Lesson 5

1. Possible answer: Density is the mass of an object divided by its volume.

3. The density of aluminum is less than the density of gold. More matter is present in a given volume of gold than in the same volume of aluminum.

5. Possible answer: The object that has a density of 2.7 $g/cm^3$ has the larger volume. The two objects have the same mass, but the mass is packed into a smaller space in the denser object.

7. a. 2.87 $g/cm^3$    b. 111 g, 38.7 $cm^3$, 2.87 $g/cm^3$
   c. The density is the same.

### Chapter 1 Review Exercises

1. Possible answer: Determine the volume of a powdered solid or of a liquid by pouring the substance into a graduated cylinder or beaker and reading the markings on the side. Determine the volume of a rock by submerging the rock in a graduated cylinder partially filled with water and then reading how much the water level changes.

3. Density is no help in determining which object will displace more water. A large object will displace more water than a small object no matter how dense the two objects are.

## Lesson 6

1. Elements are the building materials of all matter. A compound is matter that is made up of two or more elements combined in a specific ratio. An element cannot be broken down into simpler substances by chemical means, but a compound can be broken down into elements.

3. Sodium nitrate contains three elements: sodium, nitrogen, and oxygen.

5. The chemical formula for cubic zirconia is $ZiO_2$. The chemical formula for a diamond is C. The stone cannot be a diamond because it has a different chemical formula than a diamond.

## Lesson 7

1. Possible answer: A chemical reaction is a change leading to the final substance or substances being different from the original substance or substances. Some of the signs that a new substance has formed include color changes, formation of a new solid or gas, and the release of energy as heat or light.

3. Possible answer: A chemical reaction combined the zinc with part of the dissolved copper compound. The resulting zinc compound was also dissolved in water. The zinc was still present, but not as a pure element.

5. a. The baking soda is a solid, the vinegar is a liquid, the clear colorless liquid is a liquid, and the $CO_2$ is a gas.
   b. Yes, the production of carbon dioxide gas is evidence that a chemical change has occurred after the original liquid and solid substances were mixed.
   c. Before the change, the sodium is in the solid baking soda. After the change, the sodium must be in the clear colorless liquid.

## Lesson 8

1. Possible answer: The chemical names and symbols indicate what compounds were combined in each step. Because matter is conserved, the products must contain the same elements as the original compounds, which enables you to make a reasonable guess about the products. For example, when sulfuric acid combines with copper oxide, it is likely that copper sulfate and water are the products.

5. The solution would be yellow because the combination of nickel, Ni, and hydrochloric acid, HCl, can only produce a solution containing compounds with nickel, hydrogen, and chlorine. Nickel chloride, $NiCl_2$, is a possible product. Nickel sulfate, $NiSO_4$ is not.

## Lesson 9

1. Three useful properties for sorting elements are reactivity, formulas of their compounds, and atomic mass.

5. a. CaS
   b. The compound with sulfur will have more mass for a given amount of calcium. The two compounds have the same number of atoms, but the atomic mass of sulfur is 32, and the atomic mass of oxygen is only 16.

## Lesson 10

1. Within Group 1A, the elements tend to get more reactive as you move from the top of the column to the bottom.

5. **(B)** titanium   **(C)** lead   **(E)** potassium   **(F)** silicon

7. Elements copper and mercury are the least reactive. On the periodic table the least reactive elements (aside from the noble gases) are the transition metals that are located in the center of the table. The other elements listed are from more reactive groups near the edge of the table: alkali metals (potassium and rubidium), alkaline earth metals (barium), and halogens (chlorine).

## Chapter 2 Review Exercises

1. Gold, represented by the symbol Au, is a transition metal that is a solid at room temperature. It has an atomic number equal to 79 and an average atomic mass of 196.97. Gold is nonreactive, a good conductor of heat, and a good conductor of electricity. It has properties similar to those of copper and silver.

3. Possible answer: A chemical formula is a symbol that represents a compound. The chemical formula shows what elements are in the compound and the ratio in which the elements combined. It can also show what physical form the compound is in.

## Lesson 11

1. When Thomson zapped atoms with electricity, he found that a negatively charged particle was removed. Because the solid sphere model does not allow for particles splitting off atoms, he created the plum pudding model.

3. Bohr revised the nuclear model of the atom when he noticed different atoms giving off different colors of light when exposed to flame or electric fields. Because the nuclear model fails to account for this process, he created the solar system model.

5. Possible answer: The two types of atoms could have different amounts of positive fluid and different numbers of electrons. Each atom would still have a net charge of zero.

11. No, the Greeks were not correct. It is now known that atoms are made up of smaller particles such as electrons, protons, and neutrons. Each of these particles has a mass and a volume, so they are matter.

## Lesson 12

1. The atomic number indicates the number of protons in the nucleus of an atom.

3. magnesium

5. The atomic mass is the sum of the number of protons and the number of neutrons in the nucleus of an atom. Although boron and carbon each have six neutrons, carbon has six protons while boron has only five protons.

## Lesson 13

1. Atomic number refers to the number of protons in the nucleus of an atom. Atomic mass, when expressed in amu, is the sum of the number of protons and the number of neutrons.

3. Possible answer: The isotopes differ from one another in the number of neutrons in their nuclei. The isotopes are $^{39}_{19}K$, $^{40}_{19}K$, $^{41}_{19}K$.

5. **a.** 58 amu **b.** $^{58}_{26}Fe$

7. 35.48 amu

9. **(B)** Because nitrogen has an atomic number of 7 on the periodic table, any possible isotopes must have an atomic number of 7.

## Lesson 14

1. Possible answer: An element is a fundamental building block of matter. An atom is the smallest possible unit of an element. An atom is the smallest unit of an element that still has the same characteristics as the element.

3. Oxygen has three stable isotopes. Neodymium has five stable isotopes. Copper has two stable isotopes. Tin has ten stable isotopes.

5. The diagonal line on the graph represents isotopes that have equal numbers of protons and neutrons, because the line passes through points that have the same x- and y-coordinates.

7. No, because no isotope is indicated on the graph of isotopes of the first 95 elements that has 31 protons and 31 neutrons.

11. no

## Lesson 15

1. A nuclear reaction is a change in the nucleus of an atom.

3. Possible answer: Gamma radiation is the most harmful because it has the most power to penetrate living tissues and cause damage.

5. The mass number of the atom does not change during beta decay because the nucleus loses one electron, which has a mass that is only a tiny fraction of the total mass of the nucleus.

7. **a.** calcium-42, $^{42}_{20}Ca$✓ **b.** xenon-131, $^{131}_{54}Xe$✓
   **c.** cobalt-52, $^{52}_{27}Co$ **d.** magnesium-24, $^{24}_{12}Mg$✓

## Lesson 16

1. Possible answers: In alpha decay, a nucleus emits a particle consisting of two protons and two neutrons. The atomic number decreases by 2 and the atomic mass decreases by 4. In beta decay, a nucleus emits an electron. The atomic number increases by 1 as one of the neutrons becomes a proton. The atomic mass does not change. In nuclear fission, a nucleus splits apart to form the nuclei of two or more lighter elements. In nuclear fusion, two nuclei combine to form the nucleus of a heavier element.

3. $^{141}_{58}Ce \longrightarrow {}^{0}_{-1}e + {}^{141}_{59}Pr$

5. Possible answer: $^{50}_{24}Cr + {}^{4}_{2}He \longrightarrow {}^{54}_{26}Fe$

## Chapter 3 Review Exercises

1. Possible answer: An atom changes identity when the number of protons in its nucleus changes. Processes in which this can occur are radioactive decay, fission, and fusion. In radioactive decay, emission of an alpha particle decreases the atomic number by 2, while emission of a beta particle increases the atomic number by 1. Nuclear fission is the process in which a nucleus breaks apart, forming two or more smaller nuclei. In nuclear fusion, two nuclei join to form one larger nucleus.

## Lesson 17

1. The color of the flame produced during a flame test is a characteristic of particular metallic elements. When a compound containing one of the metallic elements is heated, its atoms emit light of a specific color.

5. Possible answer: Red fireworks could contain lithium chloride, and purple fireworks could contain a mixture of lithium chloride and copper chloride. Red fireworks could contain lithium sulfate, and purple fireworks could contain a mixture of lithium sulfate and copper sulfate.

7. No, the flame color of each nitrate compound is different and matches the flame color of the metal in the compound. This indicates that the nitrate is not responsible for the color of the flame.

9. **a.** yellow-orange   **b.** green   **c.** pink-lilac
   **d.** pink-lilac   **e.** green

## Lesson 18

1. For main group elements, the number of shells containing electrons is equal to the period number. All of the shells, except the highest, are completely filled. The group number indicates the number of electrons in the outermost shell.

3. Possible answer: Beryllium, magnesium, and calcium are all alkaline earth metals. They are located in Group 2A, so their atoms all have two valence electrons.

5. The number of core electrons does not change across a period.

7. **a.** Element number 17 is chlorine. It has the chemical symbol Cl and is located in Group 7A. This information comes directly from square 17 on the periodic table.
   **b.** The nucleus contains 17 protons. The number of protons is equal to the atomic number.
   **c.** Possible answer: The nucleus can contain either 18 or 20 neutrons. This information is given in the graph of isotopes of the first 95 elements.
   **d.** The number of electrons in a neutral atom of chlorine is 17. In a neutral atom, the number of electrons equals the number of protons.
   **e.** Chlorine has 7 valence electrons. The number of valence electrons is equal to the group number.
   **f.** Chlorine has 10 core electrons. The number of core electrons equals the difference between the total number of electrons and the number of valence electrons, or $17 - 7$.
   **g.** Possible answer (any three elements): Fluorine, bromine, iodine, and astatine all have the same number of valence electrons. All of these elements are in the same Group, 7A, as chlorine.

## Lesson 19

1. A cation is an ion that has a positive charge. An anion is an ion that has a negative charge.

3. 2 electrons, 3 protons, and either 3 or 4 neutrons

7. The noble gas closest to sulfur is argon. A sulfur atom gains two electrons to have an electron arrangement similar to that of argon.

9. Possible answers: $S^{2-}$, $Cl^-$, $K^+$, $Ca^{2+}$

11. $Ti^{4+}$

13. Elements on the right side of the table gain electrons to have a noble gas arrangement. They do not tend to lose electrons because the charge would be too large.

## Lesson 20

1. The number of valence electrons can predict whether an atom will form a cation or an anion, as well as the size of the charge on the ion. Ionic compounds form between cations and anions in a ratio so that the charges are balanced.

3. **a.** $+1$   **b.** $-3$
   **c.** Lithium nitride has three lithium ions with charge $+1$ and one nitride ion with charge $-3$.
   $3(+1) + (-3) = 3 + (-3) = 0$
   **d.** 8 valence electrons

5. **a.** KBr has one potassium ion with charge $+1$ and one bromide ion with charge $-1$.
   $+1 + (-1) = 0$
   **b.** CaO has one calcium ion with charge $+2$ and one oxide ion with charge $-2$.
   $+2 + (-2) = 0$
   **c.** $Li_2O$ has two lithium ions with charge $+1$ and one oxide ion with charge $-2$.
   $2(+1) + (-2) = 2 + (-2) = 0$
   **d.** $CaCl_2$ has one calcium ion with charge $+2$ and two chloride ions with charge $-1$.
   $+2 + 2(-1) = 2 + (-2) = 0$
   **e.** $AlCl_3$ has one aluminum ion with charge $+3$ and three chloride ions with charge $-1$.
   $+3 + 3(-1) = 3 + (-3) = 0$

7. **a.** $NaCl_2$ does not form because it has a net charge of $-1$. The sodium ion has a charge of $+1$ and each chloride ion has a charge of $-1$.
   **b.** CaCl does not form because it has a net charge of $+1$. The calcium ion has a charge of $+2$ and the chloride ion has a charge of $-1$.
   **c.** AlO does not form because it has a net charge of $+1$. The aluminum ion has a charge of $+3$ and the oxide ion has a charge of $-2$.

## Lesson 21

1. For main group elements, the group number shows the number of valence electrons. Metal atoms lose all of their valence electrons when they form an ion, adding a positive charge for each electron lost. Nonmetal atoms gain enough electrons to have eight valence electrons, adding one negative charge for each electron gained.

3. **a.** LiCl is possible because the total of the charges on the ions equals zero. The lithium ion has a charge of $+1$ and the chloride ion has a charge of $-1$.

**b.** $LiCl_2$ is not possible because the total of the charges on the ions equals $-1$. **c.** MgCl is not possible because the total of the charges on the ions equals $+1$. Magnesium ions have a charge of $+2$. **d.** $MgCl_2$ is possible because the total of the charges on the ions equals zero. **e.** $AlCl_3$ is possible because the total of the charges on the ions equals zero. Aluminum ions have a charge of $+3$.

7. **a.** $AlBr_3$, aluminum bromide
   **b.** $Al_2S_3$, aluminum sulfide
   **c.** AlAs, aluminum arsenide
   **d.** $Na_2S$, sodium sulfide
   **e.** CaS, calcium sulfide
   **f.** $Ga_2S_3$, gallium sulfide

## Lesson 22

1. A polyatomic ion is an ion that consists of two or more elements.

3. **a.** ammonium chloride   **b.** potassium sulfate
   **c.** aluminum hydroxide   **d.** magnesium carbonate

5. $-1$

## Lesson 23

1. The Roman numeral indicates the charge on the transition metal cation in the compound.

3. **a.** $+2$, mercury (II) sulfide **b.** $+2$, copper (II) carbonate
   **c.** $+2$, nickel (II) chloride **d.** $+3$, cobalt (III) nitrate
   **e.** $+2$, copper (II) hydroxide **f.** $+2$ iron (II) sulfate

5. $Co_3(PO_4)_2$

## Lesson 24

1. Electron subshells are divisions within a specific electron shell of an atom.

3. As the number of electrons in an element increases, they are added in a specific sequence that is illustrated by the position of the element on the periodic table. Each section of the table corresponds to a particular subshell of electrons.

5. 15 subshells

7. 4p

9. **a.** $1s^2 2s^2 2p^6 3s^2 3p^1$ **b.** Element 13 has 3 valence electrons because it has three electrons in the outer shell, $n = 3$. **c.** Element 13 has 10 core electrons because it has two shells filled completely: $n = 1$ with 2 electrons and $n = 2$ with 8 electrons.

11. **a.** $1s^2 2s^2 2p^4$, [He] $2s^2 2p^4$
    **b.** $1s^2 2s^2 2p^6 3s^2 3p^5$, [Ne] $3s^2 3p^5$
    **c.** $1s^2 2s^2 2p^6 3s^2 3p^6 4s^2 3d^6$, [Ar] $4s^2 3d^6$
    **d.** $1s^2 2s^2 2p^6 3s^2 3p^6 4s^2$, [Ar] $4s^2$

**e.** $1s^2 2s^2 2p^6 3s^2$, [Ne] $3s^2$
**f.** $1s^2 2s^2 2p^6 3s^2 3p^6 4s^2 3d^{10} 4p^6 5s^2 4d^9$, [Kr] $5s^2 4d^9$
**g.** $1s^2 2s^2 2p^6 3s^2 3p^2$, [Ne] $3s^2 3p^2$
**h.** $1s^2 2s^2 2p^6 3s^2 3p^6 4s^2 3d^{10} 4p^6 5s^2 4d^{10} 5p^6 6s^2 4f^{14} 5d^{10}$, [Xe] $6s^2 4f^{14} 5d^{10}$

13. **a.** chromium   **b.** silicon   **c.** nitrogen   **d.** cesium
    **e.** lead   **f.** silver

## Chapter 4 Review Exercises

1. Valence electrons are important because they are in the outermost electron shell of an atom and will interact with other atoms. This interaction is what determines the properties of an element.

3. As the number of electrons in an element increases, they are added in a specific sequence that is illustrated by the position of the element on the periodic table. Each section of the table corresponds to a particular subshell of electrons.

5. **a.** The anion is $Mg^{2+}$ and has a charge of $+2$. The cation is $Cl^-$ and has a charge of $-1$.
   **b.** The anion is $Ca^{2+}$ and has a charge of $+2$. The cation is $NO_2^-$ and has a charge of $-1$.

## Lesson 25

1. A substance is insoluble if it fails to dissolve in a particular solvent.

3. Possible answer: The substance is most likely an ionic compound. Many compounds dissolve in water, but electrical conductivity is a characteristic of compounds that separate into ions, such as ionic compounds.

5. No, though ionic compounds generally do not conduct electricity as solids, they do conduct electricity as aqueous solutions.

## Lesson 26

1. The atoms that make up substances are held together by chemical bonds. The bond is an attraction between the positively charged nuclei of atoms and the valence electrons of other atoms.

3. **a.** metallic   **b.** molecular covalent   **c.** ionic

5. $NO_2$ is a gas that is made up of nonmetal atoms, so it has molecular covalent bonds.

7. Possible answer: Drop the mixture into a beaker of water. The sodium chloride will dissolve in the water, but the carbon will not.

9. Carbon is a solid because many carbon atoms are held together in a large array by network covalent bonds.

## Lesson 27

1. Possible answers: finding pure metals in nature, heating ionic compounds to separate the metal, extracting metals with electricity.

3. Attach the coated object to the positive terminal of a battery in an electroplating circuit.

5. The copper sulfate solution is composed of cations ($Cu^{2+}$) and anions ($SO_4^-$) dissolved in water. The only thing that is added to the solution during the experiment is a stream of electrons. When the electrons are added to the copper ions, copper atoms (Cu) are formed on the metal strip. This indicates that the $Cu^{2+}$ ions are simply copper ions that are missing electrons.

7. Possible answer: Nickel cannot change into copper unless the number of protons in the nucleus changes. The plating apparatus only adds electrons to the nickel strip, causing the plating to occur. Adding electrons does not change the nucleus, so it cannot change the identity of the atoms.

## Chapter 5 Review Exercises

1. Possible answer: While it is not practical to try to make gold, many substances that are quite valuable can be made through chemistry.

## Unit 1 Review Exercises

## General Review

1. Possible answer: An element is the basic building block of compounds. An element has only one type of atom, while a compound has at least two types of atoms held together by chemical bonds.

3. **a.** Lithium, Li, has atomic number 2 and is in Group 2A. It has 2 protons and 2 electrons. **b.** Bromine, Br, has atomic number 35 and is in Group 7A. It has 35 protons and 35 electrons. **c.** Zinc, Zn, has atomic number 30 and is in Group 2B. It has 30 protons and 30 electrons. **d.** Sulfur, S, has atomic number 16 and is in Group 6A. It has 16 protons and 16 electrons. **e.** Barium, Ba, has atomic number 56 and is in Group 2A. It has 56 protons and 56 electrons. **f.** Carbon, C, has atomic number 6 and is in Group 4A. It has 6 protons and 6 electrons.

5. An isotope is an atom of an element with a specific number of neutrons in its nucleus. Predict the most common isotope by rounding the average atomic mass of the element to the nearest whole number.

7. Cations are ions that have lost electrons, causing them to have a positive charge. Anions are ions that have gained electrons, causing them to have a negative charge.

9. **a.** Magnesium chloride has ionic bonding and conducts electricity in solution only. **b.** Oxygen has molecular covalent bonding and does not conduct electricity. **c.** Silver (I) hydroxide has ionic bonding and conducts electricity in solution only. **d.** Platinum has metallic bonding and conducts electricity.

11. A material that does not dissolve in water and does not conduct electricity is held together by network covalent bonds or molecular covalent bonds.

## Standardized Test Preparation

| | | |
|---|---|---|
| 1. C | 3. C | 5. B |
| 7. A | 9. D | 11. C |
| 13. D | 15. A | 17. B |
| 19. B, C | 21. B | |

# Unit 2: Smells

## Lesson 28

1. Possible answer: Scientists classify smells by placing similar types of smells in a category. This allows scientists to talk about smells in a consistent way.

7. **a.** Methyl octenoate probably smells sweet because it has two oxygen atoms and twice as many hydrogen atoms as carbon atoms in each molecule. The chemical name also ends in "-ate." **b.** Monoethylamine probably smells fishy because it has one nitrogen atom and no oxygen atoms in each molecule. **c.** Ethyl acetate probably smells sweet because it has two oxygen atoms and twice as many hydrogen atoms as carbon atoms in each molecule. The chemical name also ends in "-ate."

## Lesson 29

1. Structural formulas show what atoms are present in a molecule and how the atoms are bonded to one another.

3. Yes, the structural formula shows all of the atoms in a molecule, enabling use of the structural formula to determine the molecular formula.

5. **a.** $C_3H_6O_2$   **b.** $C_5H_{10}O$   **c.** $C_4H_{10}O$
**d.** $C_4H_8O_2$   **e.** $C_4H_{11}N$   **f.** $C_2H_4O_2$

## Lesson 30

1. The HONC 1234 rule describes the bonding patterns in molecules. Hydrogen forms one bond with other atoms. Oxygen forms two bonds with other atoms.

Nitrogen forms three bonds with other atoms. Carbon forms four bonds with other atoms.

3. Possible answers:

a.
$$H-O-\overset{\overset{\displaystyle H}{|}}{C}-\overset{\overset{\displaystyle H}{|}}{\underset{\underset{\displaystyle H}{|}}{C}}-\overset{\overset{\displaystyle H}{|}}{\underset{\underset{\displaystyle H}{|}}{C}}-O-H$$

b. (structure with N branch)

c. (four-carbon chain structure)

d. (five-carbon chain with -O-H)

5. Possible answer: Atoms on the molecules might react with smell receptors inside the nose. Different receptors may detect different types of molecular structures, leading to different sensations of smell.

## Lesson 31

1. Possible answer: In an ionic bond, a valence electron is transferred from one metal atom to a nonmetal atom. In a covalent bond, two nonmetal atoms share a pair of electrons. Ionic and covalent bonds are similar because both involve the valence electrons of the atoms that are bonded together. The result of the bond is that the two atoms have an outer shell that is filled. In both types of bonds, electrical forces hold the atoms together.

3. a.   K·   In·   ·Pb·   ·Bi·   ·Te:   :I:

   Group   1   3   4   5   6   7

   b. two covalent bonds: Te, one covalent bond: I, does not form covalent bonds: K, In, Pb, Bi

5. a. :Cl:  b. :I:H  c. :Br:
      :Te:Cl:           :As:Br:
         ··               :Br:

   d. :F:Si:F:  e. :F:F:
         :F:

## Lesson 32 Exercises

1. Possible answer: Nitrogen has three unpaired electrons, as shown in the Lewis dot structure. Hydrogen atoms have one unpaired electron, so three hydrogen atoms can form bonds with one nitrogen, giving the nitrogen an octet of electrons and each hydrogen two electrons. In $NH_2$, nitrogen would have only seven electrons, and in $NH_4$, an extra hydrogen atom is left over after all the unpaired electrons in the nitrogen atom have formed bonds.

   ·N·   ·N:H   H:N:H   H:N:H·
                 H       H

3. Possible answer: hydrogen, chlorine, or another fluorine

5. a. :F:        b. H        c. H
      :F:C:F:       H:C:Cl:      :Cl:Si:Cl:
      :F:           H            H

   d. H          e. H:O:Cl:   f. H H
      H:C:O:H                    H:C:N:
      H                          H H

9. $CH_4$ would form a stable compound because all of the atoms in the molecule are surrounded by the most stable number of valence electrons—eight for carbon, two for hydrogen. In $CH_3$, the carbon atom would have only seven electrons in its outer shell.

## Lesson 33

1. A functional group is a portion of a molecular structure that is the same in all molecules of a certain type.

3. $C_2H_4$ has fewer hydrogen atoms than $C_2H_6$ because the two carbon atoms are held together by a double bond. That means there are only two unpaired electrons in each carbon atom available to form bonds with hydrogen.

5. a. $C_6H_{14}O$

   b. $H-\overset{\overset{\displaystyle H}{|}}{\underset{\underset{\displaystyle H}{|}}{C}}-\overset{\overset{\displaystyle H}{|}}{\underset{\underset{\displaystyle H}{|}}{C}}-\overset{\overset{\displaystyle H}{|}}{\underset{\underset{\displaystyle H}{|}}{C}}-\overset{\overset{\displaystyle H}{|}}{\underset{\underset{\displaystyle H}{|}}{C}}-\overset{\overset{\displaystyle H}{|}}{\underset{\underset{\displaystyle H}{|}}{C}}-\overset{\overset{\displaystyle H}{|}}{\underset{\underset{\displaystyle H}{|}}{C}}-O-H$

   c. The structural formula is more useful because it shows the −OH functional group that is characteristic of alcohols, and alcohols have a characteristic smell.

11. a. $C_3H_6O_2$   b. $C_5H_{10}O$   c. $C_5H_{10}O$
    d. $C_4H_8O_2$   e. $C_4H_{11}N$   f. $C_2H_4O_2$

Appendix C: Answers to Selected Exercises

## Lesson 34

1. Combine the alcohol and acid with a strong acid and heat the mixture.

3. Possible answer: When butyric acid is heated with methanol, the two compounds react to form a new compound, which is a sweet-smelling ester. The butyric acid, which has a foul smell, is no longer present.

5. Possible answer: Combine the acetic acid with an alcohol and sulfuric acid and then heat the mixture. The hydrogen atom of the acid is replaced by the part of the alcohol molecule that is attached to the $-OH$ functional group.

## Lesson 35

1. Possible answer: Converting a carboxylic acid into an ester is a way to make a new compound with more desirable properties. For example, eliminating a putrid odor is one possible goal of converting a carboxylic acid into an ester.

3. A catalyst is a chemical that is added to a reaction mixture to help get the reaction started, but is not consumed by the reaction.

5. a. $C_8H_{16}O_2$   b. $C_7H_{14}O_2$   c. $C_9H_{18}O_2$   d. $C_6H_{12}O_2$

## Chapter 6 Review Exercises

1. a. ester   b. ketone   c. carboxyl   d. amine

3. a. 4 lone pairs   b. 12 lone pairs   c. no lone pairs

5. Possible answer: The smell of a compound is strongly related to the functional group of the molecule.

## Lesson 36

1. Possible answer: A structural formula shows the arrangement of atoms and the types of bonds within a molecule in a two-dimensional form. A ball-and-stick model adds information about the arrangement of the atoms in three dimensions.

3. 10 black balls (carbon), 18 white balls (hydrogen), 1 red ball (oxygen), 30 connectors (chemical bonds)

7. Possible answer: The three molecules geraniol, menthol, and fenchol each have ester functional groups in their structural formulas. However, each of the three molecules has a distinctive smell.

## Lesson 37

1. A tetrahedral shape has four single bonds spaced equally around one central atom.

5. Possible answer: $CH_4$, $SiCl_4$, $CF_4$

7. $AsH_3$ will have a pyramidal shape because the three bonded pairs and the one lone pair on the arsenic atom

form a tetrahedron. Therefore, the three single bonds form a pyramid shape with the arsenic atom at its top.

9.

Methane         Ammonia         Water

## Lesson 38

1. If a molecule has three electron domains, its shape will be trigonal planar.

3. $Cl_2$ and $CO_2$ are linear and $H_2O$ is bent.

5. A molecule with two atoms will be linear. A molecule with three atoms can be either linear or bent. A molecule with four atoms can be linear, bent, trigonal planar, or pyramid-shaped. A molecule with five atoms can be linear, bent, tetrahedral, or trigonal planar or pyramidal with the fifth molecule attached to one of the triangle's or pyramid's vertices.

7. a.
$$\text{H:}\overset{\text{H}}{\underset{\text{H}}{\text{C}}}\text{:}\overset{\text{H}}{\underset{\text{H}}{\text{C}}}\text{:}\overset{\text{H}}{\underset{\text{H}}{\text{C}}}\text{:}\overset{\text{H}}{\underset{\text{H}}{\text{C}}}\text{:H}$$
   b. 13

   c. Possible answer: a crooked, zigzag shape

   d. Each carbon atom is surrounded by four electron domains. Therefore, the shape of the molecule is a series of linked tetrahedra.

## Lesson 39

1. A space-filling model is a more accurate model of how the atoms of the molecule are arranged in space. Space-filling models give a better picture of the shape of the molecule than ball-and-stick models.

3. Possible answer:

$$\text{H}-\overset{\overset{\text{H}}{|}}{\underset{\underset{\text{H}}{|}}{\text{C}}}-\overset{\overset{\text{H}}{|}}{\underset{\underset{\text{H}}{|}}{\text{C}}}-\overset{\overset{\text{H}}{|}}{\underset{\underset{\text{H}}{|}}{\text{C}}}-\overset{\overset{\text{H}}{|}}{\underset{\underset{\text{H}}{|}}{\text{C}}}-\overset{\overset{\text{H}}{|}}{\underset{\underset{\text{H}}{|}}{\text{C}}}-\overset{\overset{\text{H}}{|}}{\underset{\underset{\text{H}}{|}}{\text{C}}}-\overset{\overset{\text{H}}{|}}{\underset{\underset{\text{H}}{|}}{\text{C}}}-\overset{\overset{\text{H}}{|}}{\underset{\underset{\text{H}}{|}}{\text{C}}}-\overset{\overset{\text{H}}{|}}{\underset{\underset{\text{H}}{|}}{\text{C}}}-\text{O}-\text{H}$$

$C_9H_{20}O$ is long and stringy, with a zigzag shape.

5. Possible answer:

$C_{10}H_{16}O$ has a ring-shaped structure of carbon atoms with a "handle." The overall shape of the molecule resembles a frying pan.

## Lesson 40

1. The molecular formula of a compound can provide some information about its smell but is not useful in most cases because it fails to give information about functional groups. For alkanes and amines, the chemical formula is a good indicator of smell.

3. Possible answer: The minimum information required to determine that a molecule smells sweet is the shape and the functional group or, if it is an ester, just the functional group.

## Lesson 41

1. According to the receptor site theory, the nose is lined with sites that match the shapes of molecules that have smells. When a molecule fits into this site, it stimulates nerves to send a message to the brain. The brain interprets the message as a smell.

3. Possible answer: The compound would most likely have a minty smell. The nose detects the minty smell when the molecule shown fits into a receptor site in the nose that detects minty smells. In order for this to happen, some of the rub must change phase and become a gas.

## Chapter 7 Review Exercises

1. **a.** $C_5H_{10}O_2$

   **b.**

   H H H    :O  H
   H:C:C:C:O:C:C:H
   H H H        H

   **c.** ester
   **d.** The molecule probably has a sweet smell.

3. The shape of a molecule is determined by the number of electron domains around its atoms. The electron domains are all negatively charged, so they are positioned as far apart as possible.

5. Possible answer: Although the functional group appears to be a main factor that determines the smell of a molecule, its shape also affects the smell because it helps determine which receptors will be stimulated to send a signal to the brain.

## Lesson 42

1. Possible answer: A polar molecule is a molecule in which the charge is not evenly distributed around the molecule. This means that different portions of the molecule will have partial electric charges.

5. Hexane would not be expected to dissolve in water because the information given indicates that it is a nonpolar compound. Compounds with nonpolar molecules do not tend to dissolve in water.

## Lesson 43

1. Possible answer: In a polar covalent bond, the electron is attracted more by one atom than by the other, so one of the atoms has a partial negative charge and the other has a partial positive charge. In a nonpolar covalent bond, in which the two atoms share the electrons equally, there are no partial charges.

3. Although the two carbon-oxygen bonds of carbon dioxide are polar, the molecule itself has a linear shape, so the partial negative charges are on opposite sides of the molecule. These charges balance one another, so there is no dipole.

## Lesson 44

1. Electronegativity values help determine the polarity of the bond between two atoms because they can be used to determine the tendency of an electron to be attracted to one atom rather than to another atom. The greater the difference in the electronegativities of the two atoms, the more polar the bond that forms between the atoms.

3. **a.** Li—F, Na—F, K—F, and Rb—F (same polarity), Cs—F; For an alkali metal bonding with fluorine, the polarity increases from the top of the group to the bottom.

   **b.** P—S, N—F, Al—N, Mg—O, K—Cl; The polarity of the molecules increases as the distance between the atoms on the periodic table increases.

5. No, hydrogen will have a partial positive charge only when it is bonded to an atom that has a greater electronegativity. If the hydrogen atom is bonded to an atom with a smaller electronegativity, such as boron, then the hydrogen atom will have a partial negative charge. If a hydrogen atom is bonded with another hydrogen atom, the bond will be nonpolar.

7. Possible answer: O—F, 0.54; C—H, 0.45; S—F, 1.40

9. To say that bonding is on a continuum means that the type of bonding changes gradually as the difference in electronegativity between atoms increases. There is no sharp distinction between polar covalent and ionic bonds.

## Lesson 45

1. To determine whether a molecule is polar, use a Lewis dot structure or structural diagram to figure out the shape of the molecule. Then look at whether the

individual bonds are polar. If the molecule is asymmetrical and has polar bonds, then it will be a polar molecule.

3. **a.**

H
$\cdot\cdot$
: Se : H
$\cdot\cdot$

$H_2Se$ is a bent molecule. Because its bonds are polar and it is asymmetrical, it is likely to have a smell.

**b.** H : H

$H_2$ is a linear molecule. Its bond is nonpolar. Because it is symmetrical and has a nonpolar bond, it is not likely to have a smell.

**c.** : Ar :
$\cdot\cdot$

Because Ar consists of a single atom, it is not polar and is not likely to have a smell.

**d.**

: F :
$\cdot\cdot$
: O : H
$\cdot\cdot$

HOF is a bent molecule. Its bonds are polar and it is asymmetrical, so it is likely to have a smell.

**e.**

H
$\cdot\cdot$ $\cdot\cdot$
: F : C : Cl :
$\cdot\cdot$ $\cdot\cdot$
: F :
$\cdot\cdot$

$CHClF_2$ is a tetrahedral molecule and it has polar bonds. Its shape is symmetrical, but because the atoms at each point in the tetrahedron are different, the bonds are not symmetrical. It is likely to have a smell.

**f.**

H
$\cdot\cdot$
: C : : O :
$\cdot\cdot$
H

$CH_2O$ is a trigonal planar molecule. Its bonds are polar and it is asymmetrical, so it is likely to have a smell.

5. Possible answer: Water does not have a smell because there are no receptors in the nose that are sensitive to water molecules. If the nose had receptor sites that were sensitive to water, they would always be filled by the water in the mucous membrane.

## Lesson 46

3. Decanol has a smell because it is a medium-sized molecular compound and because it is polar. Lead does not have a smell because it is a metal. Iron oxide does not have a smell because it is an ionic compound. Potassium chloride does not have a smell because it is an ionic compound.

5. Possible answer: If a substance can become a gas under normal conditions, you should be able to smell it as long as receptor sites in the nose can detect it. In general, small nonpolar molecules are odorless.

7. Possible answer: When the T-shirt comes out of the clothes dryer, it is so warm that molecules from the detergent and fabric softener are likely to have changed phase and become gases. When they reach your nose, you are able to smell them.

## Chapter 8 Review Exercises

1. **a.** Br   **b.** Li   **c.** Au

3. **a.** $^{\delta+}$H — Cl $^{\delta-}$

**b.**

$\delta+$
H
|
$\delta+$H — $^{\delta-}$C — H$^{\delta+}$
|
H
$\delta+$

**c.**

H
$\delta-$O   $^{\delta+}$
$\setminus$
H $^{\delta+}$

HCl and $H_2O$ are polar molecules because they are asymmetrical and have polar bonds. $CH_4$ is not a polar molecule because its polar bonds are symmetrically arranged around the carbon atom.

## Lesson 47

1. The mirror image of the letter "D" can be superimposed on the original as long as the mirror is placed horizontally with respect to the letter. Then the mirror image and the original image are identical.

3. Examples of objects that look different in a mirror include a glove, a written sentence, a pair of scissors. Examples of objects that look the same in a mirror include a spoon, an empty glass, and a pencil with no writing on it.

9. **a.** citronellol: $C_{10}H_{20}O$; geraniol: $C_{10}H_{18}O$

**b.** Citronellol has two distinct smells because the structural formulas have mirror-image isomers that interact with different receptors in the nose. Citronellol has mirror-image isomers because the third carbon atom from the right in the structural formula is bonded to four different groups.

**c.** The geraniol molecule has only one smell because the molecule and its mirror image are identical. None of the carbon atoms in the structure are bonded to four different groups.

## Lesson 48

1. Amino acids are the building materials for many of the structures in the body, especially those related to the functioning of cells. An amino acid has a carbon backbone and two functional groups, an amine group and a carboxylic acid group.

## Chapter 9 Review Exercises

1. Possible answer: Two molecular models represent mirror-image isomers if they are mirror images of each other and one cannot be rotated in space to be superimposed on top of the other. When the models have a carbon atom that is attached to four different groups, then they will be mirror-image isomers.

3. Possible answer: A molecule fits into its receptor site only when the receptor has the same "handedness" as the molecule. This is similar to the way that each of your feet will fit into only one shoe of a pair of shoes, even though the shoes are mirror images of one another.

## Unit 2 Review Exercises

### General Review

1. Isomers are molecules that have the same molecular formula but different structural formulas.

3. A molecular formula provides the elements and number of atoms per element in a compound. A structural formula shows the way in which these atoms are bound to each other. A ball-and-stick model provides a three-dimensional image of the bonds. A space-filling model shows the actual amount of space each atom takes up, with atoms that are sharing electrons overlapped in space.

5. **a.** yes  **b.** $C_8H_8O_3$
   **c.** The molecule has a flat section with the ester functional group attached to its side, similar to the shape of a frying pan.
   **d.** The bond is bent because there are four electron domains. The oxygen atom has two bonded pairs and two lone pairs. The shape that separates the four domains as much as possible is tetrahedral, which causes the two bonds to be bent.

### Standardized Test Preparation

| | | |
|---|---|---|
| 1. D | 3. B | 5. D |
| 7. C | 9. C | 11. C |
| 13. C | 15. D | 17. B |
| 19. B | | |

# Unit 3: Weather

## Lesson 49

1. Possible answer: To have weather, a planet must have a layer of gases surrounding its surface.

3. A physical change is a change in the form of a substance that does not change the identity of the substance.

## Lesson 50

1. Possible answer: One way to determine the volume of 3 cm of water in the rain gauge is to use the graph. Find 3 cm on the *x*-axis, move up to the line on the graph, and then read the corresponding *y*-value. A second way is to use a proportionality constant. Examining the data in the table shows that the value of the volume is always twice the value of the height.

3. Possible answer: Both the volume and height of water in the rain gauge will keep track of the increase because they both increase by the same proportion as additional rain falls into the gauge.

5. Possible answer: The three containers would have different volumes of water in them after the storm because their bases have different areas. The greatest volume of water will be in the washtub and the least will be in the graduated cylinder. However, the height of water in each container will be the same.

## Lesson 51

1. You can convert the volume of snow in a snowpack to a volume of water by using the mathematical equation for density, $D = \frac{m}{v}$. If you know the volume and density of snow, you can determine the mass of the snow in the snowpack. If the snow melts, the mass of the water will be the same as the mass of the snow. Then you can calculate the volume of water by dividing the mass by the density of water, which is equal to 1 g/mL.

7. **a.** 16.1 g  **b.** 17.5 g

9. **a.** 4.3 g of iron occupies a greater volume because volume increases as density decreases, and iron is less dense than lead.
   **b.** 2.6 mL of lead has a larger mass because mass increases as density increases, and lead is more dense than iron.

## Lesson 52

1. Possible answer: Liquids usually expand when they are heated and contract when they are cooled. At any given temperature the volume of a fixed amount of liquid will always be the same. If the liquid is in a tube, the height of the column of liquid will change in a predictable way with changes in temperature.

5. $-40\,°F$

7. Normal body temperature is about 98.6 °F. A body temperature of 40 °C is the same as 104 °F, which indicates fever and probable illness.

**9. a.**

**b.** 50 °F   **c.** 13 °C

## Lesson 53

1. Possible answer: Absolute zero is the temperature at which the volume of a gas would be equal to zero. It is considered a theoretical temperature because a real gas would condense into a liquid and then a solid before reaching absolute zero, making a volume of zero impossible.

3. Possible answer: The kinetic theory of gases defines the temperature of a gas as the average kinetic energy of the particles of the gas.

5. The smallest unit is 1 °F because there are 180 Fahrenheit degrees between the freezing temperature of water and the boiling temperature of water, while there are only 100 Celsius degrees or kelvins between the freezing temperature of water and the boiling temperature of water.

7. **a.** −173 °C   **b.** 333 K   **c.** −23 °C   **d.** 298 K
   **e.** 27 °C   **f.** 173 K   **g.** 127 °C

9. **a.** 308 K   **b.** 450 K   **c.** 258 K

11. B

## Lesson 54

1. Determine the proportionality constant, *k*, for a gas by dividing the volume of the gas by its temperature measured on the Kelvin temperature scale.

3. 689 mL

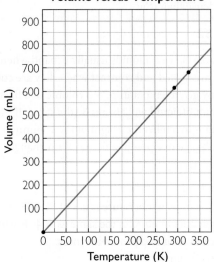

The answer is reasonable because 325 K corresponds to about 700 mL.

5. 186 K

## Lesson 55

1. When air gets warmer, its volume increases and its density decreases. Air that is less dense than the surrounding air will rise.

3. A

## Chapter 10 Review Exercises

1. 283 K

3. 760 mL

5. **a.** 3.3 mL **b.** 2700 mL. The volume of gas can be calculated using the density equation. The mass of the carbon dioxide does not change when it changes phase and becomes a gas.

## Lesson 56

1. The density of a gas is much lower than the density of a solid.

3. Possible answer: Sublimation occurs when a substance changes from the solid phase directly to the gas phase without becoming a liquid. Evaporation occurs when a substance changes from the liquid phase to the gas phase.

5. Possible answer: Ice floats in water so it must have a slightly lower density than water.

7. 4.1 mL

9. 12 g

## Lesson 57

3. Because the air inside the balloon exerts pressure on the inside wall but there is no pressure balancing it on the outside wall, the balloon will expand and most likely pop.

## Lesson 58

1. As you push down on the plunger of the syringe, the gas pressure inside the syringe increases and the pressure on the scale increases. The pressure inside the syringe is equal to the pressure you are exerting on the plunger, which can be read from the scale. To calculate the pressure of the gas, divide the weight on the scale by the surface area of the plunger. Then add atmospheric pressure.

5. Possible answer: The relationship between gas pressure and gas volume is different from the relationship

between gas volume and gas temperature because an increase in pressure causes the volume to decrease, while an increase in temperature causes the volume to increase.

## Lesson 59

1. Possible answer: You can change the pressure, the volume, or the temperature of a gas sample.

3. 720 L

5. a. The volume of air inside the bottle stays the same because the bottle is a rigid container.
   b. Yes, the air inside the bottle will lose energy until it has the same temperature as its surroundings because glass does not insulate against temperature changes.
   c. Pressure and temperature are directly proportional. The pressure of the air inside the bottle will decrease as its temperature decreases.
   d. 0.93 atm

## Lesson 60

1. According to the kinetic theory of gases, increasing the gas volume decreases the gas pressure because the molecules of gas have more room in which to travel and therefore they strike the walls of the container less often.

3. Possible answer: An increase in temperature probably caused the change in pressure because pressure and temperature are proportional and the volume of a gas cylinder doesn't change.

5. The volume of a gas cannot decrease to zero because the particles of the gas themselves occupy space.

7. a. The volume of your lungs at 3.5 atm is much lower than the volume at sea level.
   b. Holding your breath when you ascend quickly is dangerous because the pressure decreases rapidly as you rise. If you hold your breath, your lungs will act as sealed containers, making the pressure inside the lungs much greater than the pressure outside the lungs, possibly causing your lungs to rupture.

## Lesson 61

1. The combined gas law is a mathematical equation that relates gas pressure, temperature, and volume. This law applies when all three variables can change at the same time.

3. a. The volume of the gas will decrease because pressure and volume are inversely proportional when the temperature and the amount of gas remain unchanged.
   b. $PV = k$    c. 0.49 L

5. a. The volume of the balloon will increase because, based on the values given, the change in pressure has a greater effect on the final volume than the change in temperature.
   b. $\dfrac{PV}{T} = k$    c. 1.6 L

## Lesson 62

1. Possible answer: High-pressure areas are generally associated with clear skies. Low-pressure areas are generally associated with cloudy weather and precipitation.

## Chapter 11 Review Exercises

1. The pressure of a gas is proportional to the temperature of the gas. As a gas becomes warmer, its molecules move faster and collide with each other and other objects more frequently.

3. 0.80 atm

5. 2.0 atm

## Lesson 63

1. Possible answer: Air pressure decreases with increasing altitude because the number of molecules of air in a given volume decreases. This means that there are fewer collisions of air molecules with one another and with other objects, and therefore the air exerts less pressure.

5. Possible answers: Increase the amount of air in the tire. Increase the temperature of the air in the tire. Press on the outside of the tire to decrease its volume.

7. When a high-pressure system moves into a region, the air pressure increases. When the air pressure on the open end of the barometer increases, it pushes on the mercury harder than it did before. Mercury is pushed higher into the closed end of the barometer until the downward pressure caused by the weight of the column of mercury is equal to the atmospheric pressure.

## Lesson 64

1. Possible answer: Chemists invented the mole so as to have a sufficiently large unit with which to count the enormous number of particles in a sample of gas.

3. a. Each gas sample contains 1 mol of atoms.
   b. 0.0446 mol/L c. The xenon sample has the largest mass because the mass of each xenon atom is greater than the mass of each neon atom or argon atom.
   d. the xenon sample

5. 8.0 g of helium

7. The two balloons hold the same volume. The volumes of the balloons are the same because the number of moles of He is equal to the number of moles of $O_2$.

## Lesson 65

1. The ideal gas law is an equation that relates the variables pressure, volume, number of moles, and temperature for any sample of any gas. The equation is $PV = nRT$.

3. 0.13 mol

5. 3.90 atm

7. Possible answers: Reduce the number of particles to 0.5 mol. Reduce the temperature to 137 K. Increase the volume to 44.8 L.

## Lesson 66

1. Humidity is a measure of the number density of molecules of water vapor in the air.

3. about 1.7 mol per 1000 L

5. 2.2 mol per 1000 L

7. 0.43 mol per 1000 L

11. **a.** No, if the amount of water vapor in the air remains constant, it will still be less than the maximum possible vapor.
    **b.** Yes, the amount of water vapor in the air cannot remain constant, because the maximum possible vapor at 20 °C is less than the original amount of water vapor. Therefore, some water vapor will condense and form fog.

## Lesson 67

1. Possible answer: For a hurricane to form, the temperature at the ocean's surface must be at least 80 °F and the air must contain a lot of moisture.

## Chapter 12 Review Exercises

3. D

5. 68%

## Unit 3 Review Exercises
## General Review

1. The ice has a lower mass because it has a lower density than liquid water. The density of water is 1.0 g/mL. Mass can be calculated by multiplying the density by the volume, and because the samples have the same volume, the ice has less mass.

3. According to the kinetic theory of gases, gas particles move faster when they have more energy. When a gas is heated, its particles gain energy, move faster, and

scatter farther apart, causing the gas to expand. When the gas is cooled, its particles lose energy, causing them to move more slowly and remain closer together, so the gas contracts.

7. 2.05 atm

9. **a.** 273 K, 1.0 atm    **b.** 1 mol
   **c.** 2 mol               **d.** 0.0446 mol/L

11. 12 mol

## Standardized Test Preparation

| | | |
|---|---|---|
| 1. A | 3. D | 5. B |
| 7. D | 9. C | 11. A |
| 13. D | 15. A | 17. C |
| 19. B | | |

# Unit 4: Toxins

## Lesson 68

1. Reactants are the substances that are present before a chemical reaction, and products are the substances that are formed during the reaction.

3. Possible answer: Toxic substances cause harm to living organisms.

5. **a.** Solid mercury chloride is added to a solution of EDTA to produce a solution containing a compound of mercury with EDTA and hydrochloric acid.
   **b.** If the two reactants are mixed, the solid will disappear and a mixture of two liquids remains.

## Lesson 69

1. Possible answer: When sugar melts, it changes from the solid phase to the liquid phase. When sugar dissolves in water, the sugar molecules spread throughout the water.

3. **a.** The solid magnesium will react and disappear into an aqueous compound, and gas bubbles will appear.
   **b.** Gas bubbles will appear in the solution.
   **c.** A solid will form when the two solutions are mixed.

5. In both cases, the mixture of two clear solutions produces gas bubbling out of the resulting solution.

## Lesson 70

1. During a physical change, substances change form but new chemical substances are not made. During a chemical change, chemical substances are changed into other chemical substances.

3. In each case, the material changes its appearance or breaks apart, but it does not change identity. Examples include melting ice, breaking glass, grinding pepper, dissolving sugar, and bending metal.

5. 1: chemical change; 2: physical change; 3: chemical change; 4: chemical change; 5: physical change; 6: chemical change; 7: chemical change

## Lesson 71

1. Possible answer: The law of conservation of mass states that matter is not created or destroyed during a chemical reaction. The atoms involved in the reaction are rearranged but are not created or destroyed.

## Lesson 72

1. Possible answer: You need to balance chemical equations to show how many molecules, atoms, or formula units of each substance take part in the reaction or are produced by the reaction.

## Lesson 73

1. Combination reactions and decomposition reactions are opposites because combination reactions put two or more atoms or compounds together to make a new compound, while decomposition reactions break a compound into two or more atoms or compounds.

3. **a.** double exchange reaction
   $NaOH(aq) + HNO_3(aq) \longrightarrow$
   $\qquad\qquad NaNO_3(aq) + H_2O(l)$
   **b.** combination reaction
   $C_2H_4(g) + Cl_2(g) \longrightarrow C_2H_4Cl_2(g)$
   **c.** single exchange reaction
   $Cl_2(g) + MgBr_2(s) \longrightarrow Br_2(s) + MgCl_2(s)$

5. $SO_3(g) + H_2O(l) \longrightarrow H_2SO_4(aq)$
   Because sulfur trioxide is a gas, it could enter the body through the nose and mouth and travel to the lungs. Sulfuric acid is an aqueous liquid that could enter the body by ingestion or through the skin.

## Chapter 13 Review Exercises

1. Possible answer: Identifying physical or chemical changes through observations alone is difficult because often no change occurs that can be detected by sight or smell.

3. **a.** Each reactant has ionic bonds because they both consist of metal atoms bonded to nonmetal atoms or polyatomic ions. **b.** double exchange reaction **c.** chemical change **d.** A balanced reaction shows that matter is conserved because all of the atoms in the reactants are accounted for in the products. **e.** calcium phosphate

## Lesson 74

1. Possible answer: Toxicities of most substances are measured by exposing laboratory animals to the substance in different dosages.

5. **a.** 90,600 mg   **b.** 227,000 tablets

## Lesson 75

1. Possible answer: If you know the mass of an individual object in a large group of objects with the same mass, you can use it and the total mass to calculate the total number of objects.

5. 740 marbles will have a greater mass than 740 tiny plastic beads because the mass of a single marble is greater than the mass of a single bead.

7. 333 g of red jelly beans

9. 0.000000000000000000000327 g

## Lesson 76

1. The average atomic mass of atoms of an element is given on the periodic table in units of amu.

5. **a.** 1 mol carbon   **b.** 1 mol iron
   **c.** 1 mol gold   **d.** 1 mol gold

## Lesson 77

1. To determine the molar mass of sodium chloride, add the atomic mass of sodium and the atomic mass of chlorine to determine the mass of one unit of sodium chloride. The atomic mass in amu is the same as the molar mass in moles.

3. Ne(g): 20.2 g/mol, Ca(s): 40.1 g/mol, $CO_2(g)$: 44.0 g/mol, $CaCO_3(s)$: 100.1 g/mol, $CH_4O(l)$: 32.0 g/mol, $C_2H_6O(l)$: 46.1 g/mol, $Fe_2O_3(s)$: 159.7 g/mol

5. **a.** 10.0 g calcium, Ca
   **b.** 5.0 g sodium fluoride, NaF
   **c.** 2.0 g iron oxide, FeO

## Lesson 78

1. To convert between moles of a substance and grams of the substance, use the formula: mass (g) = molar mass · moles.

3. 12.0 mol

5. 169 g $BaO_2$

7. **a.** 1.0 mol $Cu_2O(s)$
   **b.** 1.0 mol $Cu_2O(s)$
   **c.** $Cu_2O(s)$ represents the best deal for the company because you get twice as much copper for the same price.

## Lesson 79

1. Possible answer: A comparison of the amount of each compound needed to make a soft drink taste sweet provides evidence that aspartame is sweeter than fructose. It takes many more molecules of fructose than aspartame to sweeten the drink.

3. **a.** about 360 cans    **b.** 0.00072 mol/kg

## Chapter 14 Review Exercises

3. Possible answer: Scientific notation is a useful tool for chemists because it is often necessary to use very large or very small numbers in calculations. Scientific notation simplifies these calculations and also makes comparison of the values easier.

5. AgCl, NaCl, LiCl

## Lesson 80

1. Possible answer: "Uniform throughout" means that the components of the solution are so well mixed that two samples taken from anywhere in the solution will be identical.

5. In order of increasing molarity, the solutions are c, a, b.

7. In order of increasing molarity, the solutions are a, 0.56 M, c, 0.58 M, and b, 1.1 M.

## Lesson 81

1. Concentration refers to the amount of a substance divided by the volume in which the substance is dissolved. Two solutions of different volumes can have the same concentration if the proportion of the substance to the volume is the same.

5. **a.** The first two solutions have the same number of molecules. Each of these solutions has more molecules than the third solution. The number of molecules in each sample is the product of the molarity and the volume. The first two solutions have 1 mol of molecules, and the third solution has 0.5 mol of molecules. **b.** All three solutions have the same concentration, 1.0 M. Molarity is a measure of concentration. **c.** The 1.0 L sucrose solution has the greatest mass because each sucrose molecule has more mass than each glucose molecule. Therefore, one mole of sucrose has more mass than one mole of glucose. Also, one mole of sucrose has more mass than half a mole of sucrose.

9. **a.** 0.25 L    **b.** 1.5 L    **c.** 6.0 L    **d.** 300 L

## Lesson 82

1. Possible answer: To prepare a 0.25 molar solution of sugar, you would calculate the mass of 0.25 mol of sugar, then measure the amount of sugar and dissolve it in enough water to make 1 L of solution.

3. **a.** 29 g    **b.** 220 g    **c.** 60 g
5. **a.** 10 L    **b.** 0.33 L    **c.** 0.18 L
7. 1.0-liter solution with 20 g of glucose

## Lesson 83

1. A substance dissolved in the water adds mass to the solution. A sample of contaminated water should have a greater density than a sample of pure water. If this difference can be measured, the greater density indicates that the water is not pure.

3. The 0.10 M copper chloride ($CuCl_2$) solution has a greater density than the 0.10 M potassium chloride (KCl) solution. Copper chloride has a molar mass of 134.5 g and potassium chloride has a molar mass of 75.6 g. So if you have equal volumes of each solution there would be more mass in the copper chloride solution. And if the mass is greater, the density is greater.

7. The sodium bromide (NaBr) solution would weigh the most because it has more solute added than the sodium hydroxide (NaOH) solution. The KCl weighs about half as much as the other two samples because a 500 mL aqueous solution will always weigh less than a 1 L aqueous solution.

## Chapter 15 Review Exercises

1. Possible answer: Knowing exactly how many moles or grams of contaminant are in a water sample determines whether the sample is toxic because toxicity is based on dosage.

3. **a.** 0.00002 mol $Pb^{2+}$    **b.** 0.0066 g Pb $(NO_3)_2$
   **c.** 0.0041 g $Pb^{2+}$

## Lesson 84

1. Possible answer: Acids have a sour taste and they can burn the skin in concentrated form. Bases have a bitter taste and a slippery feel. Bases can also cause skin burns in concentrated form. Most acidic and basic solutions are colorless and odorless.

## Lesson 85

1. According to the Arrhenius theory, an acid is a molecule that dissociates to form hydrogen ions in solution and a base is a substance that dissociates to form hydroxide ions in solution.

7. Washing soda, $Na_2CO_3$, forms a basic solution because it accepts a proton from water, thereby forming a hydroxide ion in solution.

## Lesson 86

1. The pH of a solution is equal to the negative logarithm of the $H^+$ concentration. Therefore, the pH increases by 1 as the hydrogen ion concentration decreases by a factor of 10.

7. pH $= -0.40$

## Lesson 87

1. When you add water to an acidic solution, the pH of the solution increases to a maximum of 7.

5. Because the concentration of hydrogen ions contributed by the acid is less than the concentration of hydrogen ions in pure water, the addition of $1 \times 10^{-9}$ moles of HCl does not affect the pH of pure water.

7. **a.** 0.075 mol    **b.** 0.068 M    **c.** pH $= 1.2$

## Lesson 88

1. Two ways to make a strong acid solution safer are dilution with water and neutralization with a base.

5. Each mole of sulfuric acid forms two moles of hydrogen ions in solution, while each mole of sodium hydroxide forms one mole of hydroxide ions. Therefore, the final solution has more hydrogen ions than hydroxide ions, so it is acidic and the pH is less than 7.

## Lesson 89

3. 1000 mL

5. **a.** The final solution is acidic because the solution has more $H^+$ ions than $OH^-$ ions. **b.** The final solution is neutral because the solution has equal numbers of $H^+$ ions and $OH^-$ ions. **c.** The final solution is basic because the solution has fewer $H^+$ ions than $OH^-$ ions.

## Chapter 16 Review Exercises

1. According to the Arrhenius theory, an acid is a substance that dissociates to form hydrogen ions in solution and a base is a substance that dissociates to form hydroxide ions in solution. According to the Brønsted-Lowry theory, an acid is a substance that donates protons in solution while a base is a substance that accepts protons. This is similar to the Arrhenius theory, but it accounts for bases that do not dissociate to form hydroxide ions.

5. **a.** Sodium bicarbonate can be classified as a base because it accepts a proton from water, forming hydroxide ions. **b.** Sodium bicarbonate was available in case sulfuric acid spilled. The sodium bicarbonate could have been used to neutralize any acid that was spilled.

## Lesson 90

1. A precipitate is an insoluble compound that forms as a product of a chemical reaction in solution.

3. Possible answer: A spectator ion is an ion that is present in both the reactants and the products of a chemical reaction. Spectator ions do not participate in the reaction.

## Lesson 91

1. Possible answer: Coefficients are important in chemical equations because they represent the proportion in which reactants combine to form products.

5. **a.** 0.10 mol    **b.** 5.8 g

## Lesson 92

1. Possible answer: You need to convert between grams and moles to use mole ratios to find the right mass proportions of reactants.

3. **a.** 12.6 g    **b.** 19.8 g NaCl, 47.0 g $PbCl_2$

## Lesson 93

5. **a.** $AgNO_3(aq) + NaCl(aq) \longrightarrow$ $AgCl(s) + NaNO_3(aq)$
   **b.** 0.037 mol $AgNO_3$; 0.077 mol NaCl
   **c.** silver nitrate, $AgNO_3$
   **d.** 5.3 g AgCl; 3.1 g $NaNO_3$
   **e.** 2.3 g NaCl

## Chapter 17 Review Exercises

1. Possible answer:

$$2NaOH(aq) + NiCl_2(aq) \longrightarrow$$
$$2NaCl(aq) + Ni(OH)_2(s)$$
$$2Na^+(aq) + 2OH^-(aq) + Ni^{2+}(aq) + 2Cl^-(aq) \longrightarrow$$
$$2Na^+(aq) + 2Cl^-(aq) + Ni(OH)_2(s)$$
$$2OH^-(aq) + Ni^{2+}(aq) \longrightarrow Ni(OH)_2(s)$$
$$K_2CO_3(aq) + NiSO_4(aq) \longrightarrow$$
$$K_2SO_4(aq) + NiCO_2(s)$$
$$2K^+(aq) + CO_3^{2}(aq) + Ni^{2+}(aq) + SO_4^{2+}(aq) \longrightarrow$$
$$2K^+(aq) + SO_4^{2+}(aq) + NiCO_3(s)$$
$$CO_3^{2+}(aq) + Ni^{2+}(aq) \longrightarrow NiCO_3(s)$$
$$2Li_3PO_4(aq) + 3NiCl_2(aq) \longrightarrow$$
$$6LiCl(aq) + Ni_3(PO_4)_2(s)$$
$$6Li^+(aq) + PO_4^{3+}(aq) + 3Ni^{2+}(aq) + 6Cl^-(aq) \longrightarrow$$
$$6Li^+(aq) + 6Cl^-(aq) + Ni_3(PO_4)_2(s)$$
$$2PO_4^{3+}(aq) + 3Ni^{2+}(aq) \longrightarrow Ni_3(PO_4)_2(s)$$

## Unit 4 Review Exercises

### General Review

1. **a.** $Zn(s) + 2HCl(aq) \longrightarrow ZnCl_2(aq) + H_2(g)$
   **b.** $2KClO_3(s) \longrightarrow 2KCl(s) + 3O_2(g)$
   **c.** $S_8(s) + 24F_2(g) \longrightarrow 8SF_6(g)$
   **d.** $4Fe(s) + 3O_2(g) \longrightarrow 2Fe_2O_3(s)$

3. **a.** limiting reactant: HCl; 1.87 g $ZnCl_2$ made
   **b.** limiting reactant: $KClO_3$; 0.608 g KCl made
   **c.** limiting reactant: $F_2$; 1.28 g $SF_6$
   **d.** limiting reactant: Fe; 1.43 g $Fe_2O_3$

5. A neutralization reaction is a chemical reaction in which an acid and a base combine to form a salt and water.

9. 36 g

### Standardized Test Preparation

| | | |
|---|---|---|
| **1.** C | **3.** A | **5.** C |
| **7.** B | **9.** A | **11.** C |
| **13.** D | **15.** A | **17.** C |
| **19.** B | | |

# Unit 5: Fire

## Lesson 94

1. Possible answer: Energy is a measure of the ability to cause a change in matter.

5. Possible answer: Most of a fire is hot gases that become part of the atmosphere. Ashes are leftover material that was not converted to light and heat by a fire. The heat from the ashes transfers into the air and other materials around the site of the fire as it cools.

## Lesson 95

1. Possible answer: Exothermic chemical changes are changes that transfer heat into the environment, and endothermic chemical changes are changes that absorb heat from the environment.

3. **a.** The average kinetic energy of the products is greater than the average kinetic energy of the reactants because the reaction is exothermic.
   **b.** In a molecular view, the gas particles in the reactant would be moving more slowly than the particles in the products.

5. **a.** barium hydroxide, $Ba(OH)_2 \cdot 8H_2O$; ammonium nitrate, $NH_4NO_3$; ammonia, $NH_3$; water, $H_2O$; barium nitrate, $Ba(NO_3)_2$
   **b.** Possible answer: the beaker, the pool of water outside the beaker, air, my hand

**c.** The $Ba(NO_3)_2$ is at a lower temperature than the $NH_4NO_3$ because the water freezing outside the beaker shows the temperature decreased after the reactants were mixed.
**d.** The beaker will feel cold.
**e.** The process is endothermic because the temperature decreases during the reaction. Energy was transferred from the surrounding environment into the reaction.
**f.** A chemical reaction must have occurred for a change in energy to occur. The reaction must have been the cause of the water outside the beaker turning into ice.

## Lesson 96

1. The ice cube feels cold to your hand because heat is being transferred from your hand to the ice cube, which is at a lower temperature.

3. When a liquid evaporates energy is transferred to it, and when it solidifies it transfers energy to the surroundings.

5. Evaporation is the process of a liquid changing into a gas. It is a cooling process because the substance absorbs energy during the phase change, so that energy must be transferred from the surrounding environment. Objects and substances in the surroundings of the evaporating liquid are cooled.

9. Possible answer: When an object is warmer than its environment, the heat spreads into the environment. For example, a fire will eventually cause air in all parts of a room to feel warmer.

11. Yes, if the ice cube is warmer than the air around it or an object that is touching it, heat will transfer from the ice cube to its surroundings or into the other object.

## Lesson 97

1. Because the larger sample has more particles, it takes more energy to increase the average kinetic energy of that sample by the same amount as the smaller sample.

3. **a.** 5 cal   **b.** 10 cal   **c.** 315 cal

## Lesson 98

1. Possible answer: The specific heat capacity of a substance is the amount of thermal energy needed to raise the temperature of 1 g of the substance by 1 °C. For example, the specific heat capacity of aluminum is 0.21 cal/g °C, which means that 0.21 cal of thermal energy is needed to increase the temperature of 1 g of aluminum by 1 °C.

5. Possible answer: Because metals have a lower specific heat capacity than plastic, they experience a greater

change in temperature when they absorb a specific amount of energy from sunlight. Plastics are composed of large molecules that require more energy to cause a rise in temperature than a metal does.

7. The copper will cool faster because it has a lower specific heat capacity. Copper does not need to transfer as much thermal energy as aluminum does to cool to the temperature of the water.

## Lesson 99

1. Water temperature doesn't change during boiling because the energy added to the water doesn't increase the kinetic energy of molecules.

3.

5. B.

7. 24.4 g

## Chapter 18 Review Exercises

3. methanol (460 cal), water (500 cal), copper (900 cal)

7. 57 °C

## Lesson 100

1. If a compound is ionic, or a substance contains water or carbon dioxide, it is unlikely to combust.

3. a. $Na_2SO_4$ will not combust because it is an ionic compound.
   b. Cu will combust because it is a metallic element.
   c. MgO will not combust because it is an ionic compound.
   d. Na will combust because it is a metallic element.
   e. $C_2H_6O$ will combust because it is a molecular substance containing carbon and hydrogen.
   f. Ar will not combust because it is a noble gas.

## Lesson 101

1. Possible answer: Calories are a measure of heat. The heat that is produced by the reaction of the substance with oxygen during combustion can be measured in calories.

3. No, placing a thermometer in the flames measures the temperature of the combustion reaction, but it

cannot measure the total heat energy given off by the reaction.

## Lesson 102

5. 150 cal

7. Temperature is a measure of the average kinetic energy of the particles of water in the beakers. It takes more energy to increase the temperature of the third beaker because there are many more particles of water, making the amount of thermal energy needed to increase the temperature greater even though there is less temperature change.

## Lesson 103

3. Heat of combustion is expressed as a negative number because energy transfers out of the system and into the environment.

## Chapter 19 Review Exercises

3. Possible answer: The reaction does not appear to release light and heat because it occurs slowly, so the energy is lost to the environment without being noticed.

## Lesson 104

3. O—O, C—C, H—H, O=O, C=C

5. 3237 kJ

7. a. methanol: −660 kJ/mol; butanol: −2509 kJ/mol
   b. The calculated values are slightly higher (slightly less negative) than the values given in the table.

## Lesson 105

1. During an exothermic reaction, the potential energy of the system decreases and the kinetic energy of the system increases. Some of the potential energy in the system is converted into kinetic energy.

3. a. $2C_2H_6(g) + 7O_2(g) \longrightarrow 4CO_2(g) + 6H_2O(g)$
   b.

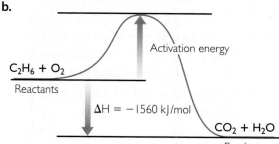

   c. The net energy exchange of the reverse reaction is 1560 kJ/mol. Conservation of energy requires that the amount of energy used to make ethane and oxygen from carbon dioxide and water is equal to the amount of energy released when ethane and oxygen are converted into carbon dioxide and water.

5. **a.** and **b.**

**c.** the second diagram

**d.** All of the reactions require energy to get started.

## Lesson 106

5. Wood chips will burn faster because they have a much greater surface area than a tree, allowing the molecules in the wood chips more contact with the oxygen in the air.

7. The powdered iron oxidizes faster than the steel beam because the small particles have much more total surface area than the beam. Because a greater percentage of the iron is exposed to the oxygen in the air, the reaction is faster.

9. During an explosion, a large amount of energy is released very quickly from chemical bonds in molecules. This energy acts as activation energy to other molecules and causes the explosion to occur rapidly.

## Lesson 107

1. Work is the result of a force acting through a distance. It is a transfer of energy, like heat.

3. $14 \text{ atm} \cdot \text{L}$

## Chapter 20 Review Exercises

3. Paige's answer is correct. The water will evaporate before any hydrogen-oxygen bonds are broken.

## Lesson 108

3. **a.** $2Ca(s) + O_2(g) \longrightarrow 2CaO(s); Ca^{2+}, O^{2-}$
$4Cr(s) + 3O_2(g) \longrightarrow 2Cr_2O_3(s); Cr^{3+}, O^{2-}$
$4Li(s) + O_2(g) \longrightarrow 2Li_2O(s); Li^+, O^{2-}$

**b.** Calcium, Ca, is oxidized. Chromium, Cr, is oxidized. Lithium, Li, is oxidized.

## Lesson 109

1. Possible answer: The oxidation reactions are better sources of energy than the reverse reactions because they

are almost always exothermic. The metals that are the hardest to extract have the lowest (most negative) heats of formation, which means that they release the most energy into the environment. The oxidation of aluminum is the best source of energy listed in the lesson.

3. 2180 kJ

5.

7. 24.5 kJ

## Lesson 110

1. In this reaction, copper is oxidized because the neutral copper atom loses the electrons and becomes a positively charged copper ion. The oxygen atom is reduced because it gains electrons and becomes a negatively charged oxygen ion.

5. **a.** $Zn(s) + Cu^{2+}(aq) \longrightarrow Cu(s) + Zn^{2+}(aq)$
**b.** $Ba^{2+}(aq) + SO_4{}^{2-}(aq) \longrightarrow BaSO_4(s)$

7. **a.** not a redox reaction **b.** not a redox reaction **c.** not a redox reaction **d.** not a redox reaction **e.** redox reaction, lead is oxidized and hydrogen is reduced, 2 electrons are transferred **f.** redox reaction, arsenic is oxidized and hydrogen is reduced, 6 electrons are transferred

## Lesson 111

3. No change would be observed because iron is less active than zinc. Therefore, iron cannot be oxidized by the zinc nitrate solution and the iron strip would not be affected.

5. **a.** No reaction occurs in Exercise 3. The chemical equation for the reaction in Exercise 4 is
$2Fe(NO_3)_2(aq) + 3Zn(s) \longrightarrow$
$$2Fe(s) + 3Zn(NO_3)_2(aq)$$
**b.** $2Fe^{3+}(aq) + 3Zn(s) \longrightarrow 2Fe(s) + 3Zn^{2+}(aq)$
**c.** zinc
**d.** Zinc is oxidized, and iron is reduced.

7. Cobalt is below iron in the activity series because the iron is oxidized by the cobalt ions, making iron the more active metal.

## Lesson 112

1. In an electrochemical cell, electrons are transferred from the more active metal to the less active metal.

3. **a.** $Ca \longrightarrow Ca^{2+} + 2e^-$
$Fe \longrightarrow Fe^{3+} + 3e^-$
**b.** Calcium is oxidized and iron is reduced.

**c.**

$$Ca \longrightarrow Ca^{2+} \qquad Fe^{3+} \longrightarrow Fe$$

**d.** $Ca(s) + Fe^{3+}(aq) \longrightarrow Ca^{2+}(aq) + Fe(s)$

## Chapter 21 Review Exercises

1. Possible answer: An electrochemical cell works by separating the two parts of a redox reaction. As electrons are transferred from one metal to the other, they travel through a wire connecting the two half-reactions. The electrical energy is the stream of moving electrons. The chemical energy is the voltage potential of the redox reaction.

3. In the extraction of metals from metal oxides table, the heat of reaction of iron (III) oxide decomposing into pure iron and oxygen gas is listed as $-413$ kJ/mol. Because this value is positive, the reaction is endothermic. Energy must be continuously added into the system for the reaction to occur.

## Lesson 113

1. Light shining on matter raises its temperature.

3. Blue

5. Green

7. **a.** Orange, glasses transmit orange light, so an orange object would look orange.
   **b.** Black, because blue light would not be transmitted.

## Lesson 114

1. $\lambda = c/f$; you need to know the frequency of the light, $f$, and the speed of light, $c$

5. **a.** $5.9 \times 10^{14}$ Hz   **b.** Green

## Lesson 115

3. $E = hf$; energy and frequency are proportional to one another

5. Possible answer: an insect

7. $2 \times 10^{-13}$ J

## Lesson 116

1. A chemical reaction happens.

5. It remains green. No reaction occurs.

7. The sunglasses with UV protection will block UV light such that the UV-sensitive paper remains green. The regular reading glasses will transmit more UV light, thereby exposing the UV-sensitive paper and turning it blue.

## Chapter 22 Review Exercises

1. Colored objects reflect the color we see and absorb other colors.

3. $2.5 \times 10^2$ m

5. $4.1 \times 10^{14}$ Hz

## Unit 5 Review Exercises

### General Review

3. The thermal energy of an exothermic reaction is the result of bond making releasing more energy than is required for bond breaking for the reaction.

5. **a.** 3800 cal
   **b.** $-7500$ cal

7. More energy is required to raise the temperature of water than to raise the temperature of aluminum by the same amount because water has a higher specific heat capacity than aluminum.

13. $Zn(s) + 2Mn^{4+}(aq) \longrightarrow Zn^{2+}(aq) + 2Mn^{3+}(aq)$
    Zinc loses electrons, so it is oxidized. Magnesium gains electrons, so magnesium ions are reduced.
    $2Al(s) + 3Ca^{2+}(aq) \longrightarrow 2Al^{3+}(aq) + 3Ca(s)$
    Aluminum loses electrons, so it is oxidized. Calcium gains electrons, so calcium ions are reduced.

### Standardized Test Preparation

| | | | | | |
|---|---|---|---|---|---|
| 1. | D | 3. | A | 5. | B |
| 7. | A | 9. | B | 11. | B |
| 13. | C | 15. | C | 17. | D |
| 19. | D | 21. | C | | |

# Unit 6: Showtime

## Lesson 117

1. Both H*In* (red) and *In*$^-$ (blue) are present.

5. $H^+ + In^- \longrightarrow HIn$

## Lesson 118

1. Possible answer: evaporation and condensation of water

3. freezing and melting, evaporation and condensing

7. **a.** $C_{12}H_{22}O_{11}(s) \rightleftharpoons C_{12}H_{22}O_{11}(aq)$
   **b.** forward   **c.** reverse

## Lesson 119

1. Dynamic equilibrium means that the rate of a forward process and the rate of its reverse process are equal.

3. **a.** Artificial sweetener is sweeter than the same amount of sucrose. Answer: Many more artificial sweetener molecules bind to the sweet receptors than do the same number of sucrose molecules. So AB is more favored for the artificial sweetener compared with sucrose. **b.** Much more sodium chloride, NaCl, dissolves in 1 L of water compared with silver chloride, AgCl. Answer: More products form for NaCl, whereas AgCl remains mostly as the starting substance. So AB is more favored for AgCl compared with NaCl.

## Lesson 120

1. A large value of $K$ indicates that there are more products in the equilibrium mixture, while a small value of $K$ indicates that more starting substance is present.

3. **a.** The binding of CO to hemoglobin has the larger equilibrium constant because [hemoglobin:CO] is greater than [hemoglobin:$O_2$]. **b.** Each particle view should show a hemoglobin binding to a particle. The view with oxygen should have more free oxygen and hemoglobin particles than the view for carbon monoxide. **c.** CO is toxic because it takes up binding sites in hemoglobin so that they are not available to oxygen. Because the [hemoglobin:CO] product is favored, fewer free hemoglobin molecules are available to bind oxygen.

## Lesson 121

1. You can rearrange the equilibrium constant equation to solve for hydrogen ion concentration. Once you have the hydrogen ion concentration, you can solve for pH. $[H^+] = \sqrt{K[HA]}$, pH $= -\log[H^+]$

3. Solution 1: $1.4 \times 10^{-3}$, Solution 2: $1.8 \times 10^{-4}$, Solution 3: $1.8 \times 10^{-5}$, Solution 4: $1.8 \times 10^{-5}$

## Chapter 23 Review Exercises

1. **a.** $HCl(aq) \longrightarrow H^+(aq) + Cl^-(aq)$ **b.** The particle view should show 10 $H^+$ and 10 $Cl^-$ ions and no HCl molecules. **c.** $HF(aq) \rightleftharpoons H^+(aq) + F^-(aq)$ **d.** The particle view should show mostly HF molecules as well as a few $H^+$ and $F^-$ ions.

3. **a.** $K = [C][D]/[A][B]$ **b.** $K = [CO][Cl_2]/[COCl_2]$ **c.** $K = [Pb^{2+}][Cl^-]^2$

## Lesson 122

1. Add more of one of the products. Remove some of the starting substance.

3. More $Cu(NH_3)_6{}^{2+}(aq)$ forms to reduce the concentration of $NH_3$.

5. Since this is an endothermic process, heating favors products. Therefore more products will form to relieve the stress.

## Lesson 123

1. Indicator molecules have a bright color that changes to a different color when the H*In* molecules dissociate into $H^+$ and *In*$^-$. When the solution is more acidic, the color of H*In* molecules is observed. In basic solution, the solution displays the color of *In*$^-$.

3. **a.** In order of increasing solubility, AgCl, CuCl, TlCl, the most soluble is TlCl. A small $K_{sp}$ indicates mostly undissolved solid. **b.** CuCl($s$) will precipitate.

5. **a.** yellow, due to the large amount of H*In* **b.** blue, due to the large amount of *In*$^-$ **c.** yellow, due to large amount of *In*$^-$ **d.** orange, due to presence of both H*In* and *In*$^-$

7. pH $\sim 10$

## Chapter 24 Review Exercises

3. **a.** No more solid will dissolve. Excess solid remains. In a saturated solution, no more solid can be dissolved. Excess solid will remain at the bottom of the container. **b.** More $CaCl_2$ precipitates. **c.** More $CaCl_2$ precipitates. **d.** More $CaCl_2$ precipitates. **e.** More $CaCl_2$ dissolves.

5. **a.** $K = \dfrac{[CO][Cl_2]}{[COCl_2]}$ **b.** $8.6 \times 10^{-2}\,M$ **c.** More $COCl_2$ forms to minimize the effect of adding one of the products. **d.** 0.14 M

## Unit 6 Review Exercises

### General Review

1. **a.** $Cl_2(g) \rightleftharpoons Cl_2(l)$
   **b.** $2NO_2(g) \rightleftharpoons N_2O_4(g)$
   **c.** $Ag^+(aq) + Cl^-(aq) \rightleftharpoons AgCl(s)$
   $H_2O(l) \rightleftharpoons H_2O(g)$

3. Possible answers: The reaction is reversible. The equilibrium constant does not change as the concentrations of products and reactants change.

5. $1.2 \times 10^{-5}\,M$

### Standardized Test Preparation

| | | |
|---|---|---|
| 1. C | 3. B | 5. C |
| 7. A | 9. D | 11. B |
| 13. A | 15. A | 17. A |
| 19. B | | |

# Appendix D: Reference Tables

## Table of Average Atomic Masses

| Element and symbol | Atomic number | Average atomic mass (amu) | Element and symbol | Atomic number | Average atomic mass (amu) |
|---|---|---|---|---|---|
| Actinium, Ac | 89 | (227) | Dysprosium, Dy | 66 | 162.5 |
| Aluminum, Al | 13 | 26.98 | Einsteinium, Es | 99 | (252) |
| Americium, Am | 95 | (243) | Erbium, Er | 68 | 167.3 |
| Antimony, Sb | 51 | 121.8 | Europium, Eu | 63 | 152.0 |
| Argon, Ar | 18 | 39.95 | Fermium, Fm | 100 | (257) |
| Arsenic, As | 33 | 74.92 | Flerovium, Fl | 114 | (289) |
| Astatine, At | 85 | (210) | Fluorine, F | 9 | 19.00 |
| Barium, Ba | 56 | 137.3 | Francium, Fr | 87 | (223) |
| Berkelium, Bk | 97 | (247) | Gadolinium, Gd | 64 | 157.3 |
| Beryllium, Be | 4 | 9.012 | Gallium, Ga | 31 | 69.72 |
| Bismuth, Bi | 83 | 209.0 | Germanium, Ge | 32 | 72.64 |
| Bohrium, Bh | 107 | (272) | Gold, Au | 79 | 197.0 |
| Boron, B | 5 | 10.81 | Hafnium, Hf | 72 | 178.5 |
| Bromine, Br | 35 | 79.90 | Hassium, Hs | 108 | (277) |
| Cadmium, Cd | 48 | 112.4 | Helium, He | 2 | 4.003 |
| Calcium, Ca | 20 | 40.08 | Holmium, Ho | 67 | 164.9 |
| Californium, Cf | 98 | (251) | Hydrogen, H | 1 | 1.008 |
| Carbon, C | 6 | 12.01 | Indium, In | 49 | 114.8 |
| Cerium, Ce | 58 | 140.1 | Iodine, I | 53 | 126.9 |
| Cesium, Cs | 55 | 132.9 | Iridium, Ir | 77 | 192.2 |
| Chlorine, Cl | 17 | 35.45 | Iron, Fe | 26 | 55.85 |
| Chromium, Cr | 24 | 52.00 | Krypton, Kr | 36 | 83.80 |
| Cobalt, Co | 27 | 58.93 | Lanthanum, La | 57 | 138.9 |
| Copernicium, Cn | 112 | (285) | Lawrencium, Lr | 103 | (262) |
| Copper, Cu | 29 | 63.55 | Lead, Pb | 82 | 207.2 |
| Curium, Cm | 96 | (247) | Lithium, Li | 3 | 6.941 |
| Darmstadtium, Ds | 110 | (281) | Livermorium, Lv | 116 | (292) |
| Dubnium, Db | 105 | (268) | Lutetium, Lu | 71 | 175.0 |

The values in parentheses are the average atomic mass of the longest-lasting isotope of the element at the time of writing.

| Element and symbol | Atomic number | Average atomic mass (amu) | Element and symbol | Atomic number | Average atomic mass (amu) |
|---|---|---|---|---|---|
| Magnesium, Mg | 12 | 24.31 | Roentgenium, Rg | 111 | (280) |
| Manganese, Mn | 25 | 54.94 | Rubidium, Rb | 37 | 85.47 |
| Meitnerium, Mt | 109 | (276) | Ruthenium, Ru | 44 | 101.1 |
| Mendelevium, Md | 101 | (258) | Rutherfordium, Rf | 104 | (267) |
| Mercury, Hg | 80 | 200.6 | Samarium, Sm | 62 | 150.4 |
| Molybdenum, Mo | 42 | 95.94 | Scandium, Sc | 21 | 44.96 |
| Moscovium, Mc | 115 | (289) | Seaborgium, Sg | 106 | (271) |
| Neodymium, Nd | 60 | 144.2 | Selenium, Se | 34 | 78.96 |
| Neon, Ne | 10 | 20.18 | Silicon, Si | 14 | 28.09 |
| Neptunium, Np | 93 | (237) | Silver, Ag | 47 | 107.9 |
| Nickel, Ni | 28 | 58.69 | Sodium, Na | 11 | 22.99 |
| Nihonium, Nh | 113 | (286) | Strontium, Sr | 38 | 87.62 |
| Niobium, Nb | 41 | 92.91 | Sulfur, S | 16 | 32.07 |
| Nitrogen, N | 7 | 14.01 | Tantalum, Ta | 73 | 180.9 |
| Nobelium, No | 102 | (259) | Technetium, Tc | 43 | (98) |
| Oganesson, Og | 118 | (294) | Tellurium, Te | 52 | 127.6 |
| Osmium, Os | 76 | 190.2 | Tennessine , Ts | 117 | (293) |
| Oxygen, O | 8 | 16.00 | Terbium, Tb | 65 | 158.9 |
| Palladium, Pd | 46 | 106.4 | Thallium, Tl | 81 | 204.4 |
| Phosphorus, P | 15 | 30.97 | Thorium, Th | 90 | 232.0 |
| Platinum, Pt | 78 | 195.1 | Thulium, Tm | 69 | 168.9 |
| Plutonium, Pu | 94 | (244) | Tin, Sn | 50 | 118.7 |
| Polonium, Po | 84 | (209) | Titanium, Ti | 22 | 47.87 |
| Potassium, K | 19 | 39.10 | Tungsten, W | 74 | 183.8 |
| Praseodymium, Pr | 59 | 140.9 | Uranium, U | 92 | 238.0 |
| Promethium, Pm | 61 | (145) | Vanadium, V | 23 | 50.94 |
| Protactinium, Pa | 91 | 231.0 | Xenon, Xe | 54 | 131.3 |
| Radium, Ra | 88 | (226) | Ytterbium, Yb | 70 | 173.0 |
| Radon, Rn | 86 | (222) | Yttrium, Y | 39 | 88.91 |
| Rhodium, Rh | 45 | 102.9 | Zinc, Zn | 30 | 65.39 |
| Rhodium, Rh | 45 | 102.9 | Zirconium, Zr | 40 | 91.22 |

# Table of Electronegativities

* Electronegativity values for the lanthanides and actinides range from about 1.10 to 1.50.

# Bonding Continuum

Electronegativity difference

| Nonpolar covalent | Polar covalent | Ionic |
|---|---|---|

| 0 | 0.5 | 0.6 | 1.9 | 2.1 | 3.1 | 3.3 |
| $H_2$ | | HI | HF | | NaF | |

## Average Bond Energies (per mole of bonds)

| Bond | H—H | C—H | C—C | C=C | C≡C | O—H | C—O | C=O | O=O | O—O |
|---|---|---|---|---|---|---|---|---|---|---|
| Bond energy (kJ/mol) | 432 | 413 | 347 | 614 | 811 | 467 | 358 | 799* | 495 | 146 |

*C=O in $CO_2$: 799, C=O in organic molecules: 745

## Heats of Combustion

| | Fuel | Chemical formula | $\Delta H$ (kJ/mol) | $\Delta H$ (kJ/g) |
|---|---|---|---|---|
| Elements | hydrogen | $H_2(g)$ | −243 | −121.3 |
| | carbon (coal) | $C(s)$ | −393 | −32.6 |
| Alkanes | methane | $CH_4(g)$ | −891 | −55.6 |
| | ethane | $C_2H_6(g)$ | −1560 | −51.9 |
| | butane | $C_4H_{10}(l)$ | −2882 | −49.6 |
| | hexane | $C_6H_{14}(l)$ | −4163 | −48.1 |
| | octane | $C_8H_{18}(l)$ | −5508 | −47.7 |
| Alcohols | methanol | $CH_4O(l)$ | −724 | −22.6 |
| | ethanol | $C_2H_6O(l)$ | −1368 | −29.7 |
| | butanol | $C_4H_{10}O(l)$ | −2669 | −36.0 |
| Sugar | glucose | $C_6H_{12}O_6(s)$ | −2828 | −15.9 |

## Solubility Trends

| Cations | | | Anions | | | | | | |
|---|---|---|---|---|---|---|---|---|---|
| | | | $NO_3^-$ | $Cl^-$ | $OH^-$ | $SO_4^{2-}$ | $CO_3^{2-}$ | $C_2O_4^{2-}$ | $PO_4^{3-}$ |
| Cations | Most alkali metals, such as $Li^+$, $Na^+$, $K^+$, $NH_4^+$ | | S | S | S | S | S | S | S |
| | Most alkaline earth metals, such as $Mg^{2+}$, $Ca^{2+}$, $Sr^{2+}$ | | S | S | N | S | N | N | N |
| | Some Period 4 transition metals, such as $Fe^{3+}$, $Co^{3+}$, $Ni^{2+}$, $Cu^{2+}$, $Zn^{2+}$ | | S | S | N | S | N | N | N |
| | Other transition metals, such as $Ag^+$, $Pb^{2+}$, $Hg^{2+}$ | | S | N | N | N | N | N | N |

S = very soluble, N = not very soluble or insoluble

# Polyatomic Ions

## Anions

### Charge: $1^-$

| Name | Formula |
|---|---|
| acetate | $CH_3CO_2^-$ |
| chlorate | $ClO_3^-$ |
| chlorite | $ClO_2^-$ |
| hypochlorite | $ClO^-$ |
| perchlorate | $ClO_4^-$ |
| cyanide | $CN^-$ |
| formate | $HCO_2^-$ |
| hydroxide | $OH^-$ |
| nitrate | $NO_3^-$ |
| nitrite | $NO_2^-$ |
| permanganate | $MnO_4^-$ |
| thiocyanate | $SCN^-$ |

### Charge: $2^-$

| Name | Formula |
|---|---|
| oxalate | $C_2O_4^{2-}$ |
| carbonate | $CO_3^{2-}$ |
| chromate | $CrO_4^{2-}$ |
| sulfate | $SO_4^{2-}$ |
| sulfite | $SO_3^{2-}$ |

### Charge: $3^-$

| Name | Formula |
|---|---|
| phosphite | $PO_3^{3-}$ |
| phosphate | $PO_4^{3-}$ |

## Cations

### Charge: $1^+$

| Name | Formula |
|---|---|
| ammonium | $NH_4^+$ |

### Charge: $2^+$

| Name | Formula |
|---|---|
| hexaaquocobalt | $Co(H_2O)_6^{2+}$ |

## Specific Heat Capacities at 20 °C

| Substance | Specific heat capacity (cal/g °C) | Specific heat capacity (J/g °C) |
|---|---|---|
| aluminum | 0.21 | 0.88 |
| water | 1.00 | 4.18 |
| brass | 0.09 | 0.38 |
| iron | 0.11 | 0.46 |
| rubber | 0.41 | 1.72 |
| glass | 0.16 | 0.67 |
| sand | 0.19 | 0.95 |
| mercury | 0.03 | 0.13 |
| hydrogen | 3.34 | 13.97 |
| copper | 0.09 | 0.38 |
| wood | 0.41 | 1.72 |
| silver | 0.05 | 0.21 |
| concrete | 0.20 | 0.84 |
| air (dry, sea level) | 0.24 | 1.00 |
| zinc | 0.09 | 0.38 |
| ethanol | 0.57 | 2.38 |
| lead | 0.03 | 0.13 |
| helium | 1.24 | 5.19 |
| silver | 0.05 | 0.23 |
| tin | 0.05 | 0.21 |
| gold | 0.03 | 0.13 |

# SI Units and Conversion Factors

### Length

| SI unit: meter (m) |
|---|
| 1 meter = 100 cm = 1.0936 yards |
| 1 inch = 2.54 centimeters (exactly) |
| 1 kilometer = 1000 m = 0.62137 mile |

### Mass

| SI unit: kilogram (kg) |
|---|
| 1 kilogram = 1000 grams = 2.2046 pounds |
| 1 pound = 453.59 grams = 0.45359 kilogram = 16 ounces |
| 1 atomic mass unit = $1.66054 \times 10^{-27}$ kilogram |

### Volume

| SI unit: cubic meter (m³) |
|---|
| 1 liter = 1000 milliliters = 0.001 cubic meter |
| 1 cubic centimeter = 1 milliliter |

### Pressure

| SI unit: pascal (Pa) |
|---|
| 1 Pa = 1 newton/m² = 0.01 millibar |
| 1 atmosphere = 101.325 kilopascals |
| = 760 torr (mm Hg) |
| = 14.70 pounds per square inch |
| = 1.01 bar |

### Energy

| SI unit: joule (J) |
|---|
| 1 joule = 0.23901 calorie |
| 1 calorie = 4.184 joules |

## Physical Constants

| Constant | Symbol | Value |
|---|---|---|
| atomic mass unit | amu | $1.66054 \times 10^{-24}$ g |
| Avogadro's number | $N_A$ | $6.02214 \times 10^{23}$ particles/mole |
| charge of an electron | $e$ | $-1.60218 \times 10^{-19}$ coulomb |
| universal gas constant | $R$ | 8.31451 J/K · mol <br> $0.08206 \dfrac{\text{L} \cdot \text{atm}}{\text{mol} \cdot \text{K}}$ |
| mass of an electron | $m_e$ | $9.10938 \times 10^{-28}$ g <br> $5.48580 \times 10^{-4}$ amu |
| mass of a neutron | $m_n$ | $1.67493 \times 10^{-24}$ g <br> 1.00866 amu |
| mass of a proton | $m_p$ | $1.67262 \times 10^{-24}$ g <br> 1.00728 amu |
| density of water (at 4 °C) | — | $1.0000$ g/cm$^3$ |
| mass of carbon-12 atom | — | 12 amu (exactly) |
| ideal gas molar volume at STP | $V_m$ | $22.4 \dfrac{\text{L}}{\text{mol}}$ |

## Symbols and Abbreviations

| | | | | |
|---|---|---|---|---|
| $m$ | mass | | $F$ | force |
| $V$ | volume | | $A$ | area |
| $D$ | density | | $h$ | height |
| $\alpha$ | alpha particle | | $LD_{50}$ | lethal dose 50% |
| $\beta$ | beta particle | | [ ] | concentration in moles per liter |
| $\gamma$ | gamma ray | | $q$ | heat |
| $\delta^+ \, \delta^-$ | partial ionic charge | | $E$ | energy |
| $n$ | number of moles | | $C_p$ | specific heat capacity |
| $T$ | temperature | | $\Delta H$ | heat of reaction (change in enthalpy) |
| $\Delta T$ | change in temperature | | $w$ | work |
| $P$ | pressure | | $K_{eq}$ | equilibrium constant |
| STP | standard temperature and pressure | | $K_{sp}$ | solubility product constant |
| $R$ | universal gas constant | | $K_a, K_b$ | acid/base dissociation constant |

## Proportional Relationships

| Mathematical relationship | Variables | Proportionality constant |
|---|---|---|
| $m = D \cdot V$ <br> $D = \dfrac{m}{V}$ | mass, $m$ <br> volume, $V$ | density, $D$ |
| $n = M \cdot V$ <br> $M = \dfrac{n}{V}$ | moles, $n$ <br> volume, $V$ | molarity, $M$ |
| $m = \text{molar mass} \cdot n$ <br> $\text{molar mass} = \dfrac{m}{n}$ | mass, $m$ <br> moles, $n$ | molar mass (in g/mol) |
| $F = P \cdot A$ <br> $P = \dfrac{F}{A}$ | force, $F$ <br> area, $A$ | pressure, $P$ |
| $V = A \cdot h$ <br> $A = \dfrac{V}{h}$ | volume, $V$ <br> height, $h$ | area, $A$ |

## Proportional Relationships in the Gas Laws

| Law | Variables | Mathematical relationships | Variables held constant |
|---|---|---|---|
| Charles's law | volume, $V$ <br> temperature, $T$ | $V = k \cdot T$ <br> $k = \dfrac{V}{T}$ | pressure, $P$ <br> number of moles, $n$ |
| Gay-Lussac's law | pressure, $P$ <br> temperature, $T$ | $P = k \cdot T$ <br> $k = \dfrac{P}{T}$ | volume, $V$ <br> number of moles, $n$ |
| Boyle's law | pressure, $P$ <br> volume, $V$ | $k = P \cdot V$ <br> $p = \dfrac{k}{V}$ | temperature, $T$ <br> number of moles, $n$ |
| combined gas law | pressure, $P$ <br> volume, $V$ <br> temperature, $T$ | $PV = kT$ <br> $k = \dfrac{PV}{T}$ | number of moles, $n$ |
| ideal gas law | pressure, $P$ <br> volume, $V$ <br> temperature, $T$ <br> moles, $n$ | $PV = nRT$ <br> $R = \dfrac{PV}{nT}$ | none |

## Other Important References

| Table | Location |
|---|---|
| Isotopes of the Elements | Unit 1: Alchemy, Chapter 3, Lesson 14 |
| Four Models of Bonding | Unit 1: Alchemy, Chapter 5, Lesson 26 |
| Water Vapor Density Versus Temperature | Unit 3: Weather, Chapter 12, Lesson 66 |
| Lethal Doses | Unit 4: Toxins, Chapter 14, Lesson 74 |
| Activity Series | Unit 5: Fire, Chapter 21, Lesson 112 |
| Standard Reduction Potentials | Unit 5: Fire, Chapter 21, Lesson 112 |

## Some Key Formulas

| | |
|---|---|
| **Density** | $D = \dfrac{m}{V}$ |
| **Ideal gas law** | $PV = nRT$ |
| **Heat** | $q = C_p m \Delta T$ |
| **Acidity** | $\text{pH} = -\log [\text{H}^+]$ |
| **Basicity** | $\text{pOH} = -\log [\text{OH}^-]$ |

# Glossary

## English/Inglés

### A

**absolute zero** The temperature defined as 0 K on the Kelvin scale and $-273.15\ °C$ on the Celsius scale. Considered to be the lowest possible temperature that matter can reach. (p. 279)

**absorption** Of light, not being transmitted because certain colors are removed by an object. Absorption transfers energy back to the object. (p. 579)

**acid** A substance that adds hydrogen ions, $H^+$, to an aqueous solution; a substance that donates a proton to another substance in solution. (p. 430)

**actinides** A series of elements that follow actinium in Period 7 of the periodic table and that are typically placed separately at the bottom of the periodic table. (p. 44)

**activation energy** The minimum amount of energy required to initiate a chemical process or reaction. (p. 537)

**activity series** A table showing elements in order of their chemical activity, with the most easily oxidized at the top of the list. (p. 564)

**actual yield** The amount of a product obtained when a reaction is run (as opposed to the theoretical yield). (p. 473)

**air mass** A large volume of air that has consistent temperature and water content. (p. 288)

**alkali metals** The elements in Group 1A on the periodic table, except for hydrogen. (p. 44)

**alkaline earth metals** The elements in Group 2A on the periodic table. (p. 44)

## Spanish/Español

**cero absoluto** La temperatura definida como 0 K en la escala Kelvin y $-273.15\ °C$ en la escala Celsius. Se considera como la temperatura más baja que puede alcanzar la materia. (p. 279)

**absorción** De la luz: cuando esta no se transmite porque ciertos colores han sido eliminados (por un objeto). La absorción transfiere energía al objeto. (p. 579)

**ácido** Sustancia que cede iones de hidrógeno, $H^+$, a una solución acuosa; una sustancia que dona un protón a otra sustancia en la solución. (p. 430)

**actínidos** Serie de elementos que están después del actinio en el séptimo período de la tabla periódica y que usualmente, aparecen en la parte de abajo de la tabla. (p. 44)

**energía de activación** La mínima cantidad de energía que se necesita para iniciar una reacción o un proceso químico. (p. 537)

**serie de actividad** Una tabla que muestra elementos ordenados de acuerdo con la actividad química de cada uno de dichos elementos, empezando por aquellos que se oxidan con mayor facilidad. (p. 564)

**rendimiento real** La cantidad de un producto que se obtiene cuando se ejecuta una reacción (al contrario del rendimiento teórico). (p. 473)

**masa de aire** Gran volumen de aire cuya temperatura y contenido de agua son constantes. (p. 288)

**metales alcalinos** Los elementos del grupo 1A de la tabla periódica, con excepción del hidrógeno. (p. 44)

**metales alcalinotérreos** Los elementos del grupo 2A de la tabla periódica. (p. 44)

**alpha decay**  A nuclear reaction in which an atom emits an alpha particle consisting of two protons and two neutrons. Alpha decay decreases the atomic number of an atom by 2 and the mass number by 4. (p. 72)

**alpha particle**  A particle made of two protons and two neutrons, equivalent to the nucleus of a helium atom. (p. 72)

**amino acid**  A molecule that contains both an amine ($-NH_2$) and a carboxylic acid ($-COOH$) functional group. (p. 247)

**anion**  An ion that has a negative charge. (p. 96)

**aqueous solution**  A solution in which water is the dissolving medium or solvent. (p. 25)

**Arrhenius theory of acids and bases**  An acid-base theory that defines an acid as a substance that adds hydrogen ions, $H^+$, to an aqueous solution, and a base as a substance that adds hydroxide ions, $OH^-$, to an aqueous solution. (p. 434)

**atmosphere (atm)**  A unit of measurement for gas pressure. One atmosphere is equivalent to 14.7 pounds of pressure per square inch or a barometric reading of 760 millimeters of mercury. (p. 301)

**atmospheric pressure**  Pressure exerted by the weight, or force, of the air pressing down on a surface. Air pressure is a result of air molecules colliding with the surfaces of objects on Earth. At sea level, atmospheric pressure equals approximately 1 atm or 14.7 lb/in². (p. 301)

**atom**  The smallest unit of an element that retains the chemical properties of that element and can exist as a separate particle. (p. 51)

**atomic mass**  The mass of a single atom (or isotope) of an element. (p. 38)

**atomic mass unit (amu)**  The unit used for expressing atomic mass. 1 amu $= 1.66 \times 10^{-24}$ g, the mass of one hydrogen atom. This is 1/12 the mass of a carbon-12 atom. (p. 38)

**desintegración alfa**  Reacción nuclear en la que un átomo emite partículas alfa que contienen dos protones y dos neutrones. La desintegración alfa disminuye el número atómico de un átomo en 2 y el número de masa en 4. (p. 72)

**partícula alfa**  Partícula que se compone de dos protones y dos neutrones, lo que equivale al núcleo de un átomo de helio. (p. 72)

**aminoácido**  Molécula que contiene un grupo funcional amino ($-NH_2$) y un ácido carboxílico ($-COOH$). (p. 247)

**anión**  Ion que tiene una carga negativa. (p. 96)

**solución acuosa**  Solución en la que el agua es el medio de disolución o el disolvente. (p. 25)

**teoría de ácidos y bases de Arrhenius**  Teoría ácido-base que define un ácido como una sustancia que cede iones de hidrógeno, $H^+$, a una solución acuosa; y una base como una sustancia que cede iones de hidróxido, $OH^-$, a una solución acuosa. (p. 434)

**atmósfera estándar (atm)**  Unidad de medida de la presión de los gases. Una atmósfera estándar equivale a 14.7 libras de presión por pulgada cuadrada o a la medida barométrica de 760 milimetros de mercurio. (p. 301)

**presión atmosférica**  La presión ejercida por el peso o fuerza del aire sobre una superficie. La presión del aire es el resultado del choque de las moléculas de aire con la superficie de los objetos que están sobre la Tierra. Al nivel del mar, la presión atmosférica es igual a 1 atm o 14.7 lb/pulg², aproximadamente. (p. 301)

**átomo**  La unidad más pequeña de un elemento que mantiene sus propiedades químicas y que puede existir como una partícula independiente. (p. 51)

**masa atómica**  La masa de un solo átomo (o isótopo) de un elemento. (p. 38)

**unidad de masa atómica (uma)**  La unidad que se usa para expresar la masa atómica. 1 uma $= 1.66 \times 10^{-24}$ g, la masa de un átomo de hidrógeno. Esto equivale a 1/12 de la masa de un átomo de carbono-12. (p. 38)

**atomic number** The consecutive whole numbers associated with the elements on the periodic table. The atomic number is equal to the number of protons in the atomic nucleus of an element. (p. 41)

**atomic theory** A theory that states that all matter is made up of individual particles called atoms. (p. 52)

**average atomic mass** The weighted average of the mass of the isotopes of an element. (p. 62)

**Avogadro's law** A scientific law stating that equal volumes of gases at the same temperature and pressure contain equal numbers of particles. (p. 333)

**Avogadro's number** A number equal to $6.02 \times 10^{23}$. It is the number of particles (atoms, ions, molecules, or formula units) present in one mole of a substance. (p. 332)

**número atómico** Los números enteros consecutivos asociados con los elementos de la tabla periódica. El número atómico es igual a la cantidad de protones que hay en el núcleo de un elemento. (p. 41)

**teoría atómica** Teoría que afirma que toda la materia se compone de partículas individuales llamadas átomos. (p. 52)

**masa atómica promedio** El promedio ponderado de la masa de los isótopos de un elemento. (p. 62)

**ley de Avogadro** Ley científica que establece que volúmenes iguales de gases que tienen la misma temperatura y la misma presión contienen la misma cantidad de partículas. (p. 333)

**número de Avogadro** Número igual a $6.02 \times 10^{23}$. Es el número de partículas (átomos, iones, moléculas o unidades de fórmula) que existen en un mol de una sustancia. (p. 332)

# B

**ball-and-stick model** A three-dimensional representation of a molecule that uses color-coded balls to represent atoms and sticks to represent bonds. (p. 186)

**base** A substance that adds hydroxide ions $OH^-$, to an aqueous solution, or a substance that accepts protons, $H^+$ ions, from another substance in solution. (p. 430)

**bent shape** The nonlinear shape around a bonded atom with two lone pairs of electrons. (p. 194)

**beta decay** A nuclear reaction in which a neutron changes into a proton and an electron, and the atom emits a beta particle, which is the electron. Beta decay increases the atomic number of the atom without changing the mass. (p. 72)

**beta particle** An electron emitted from the nucleus of an atom during beta decay. (p. 72)

**boiling point (boiling temperature)** The temperature at which both liquid and gas phases of a single substance are present and in equilibrium; the temperature at which equilibrium is established between a liquid and its vapor at a pressure of 1 atm. (p. 274)

**modelo de esferas y varillas** Representación tridimensional de una molécula que usa esferas de colores específicos para representar átomos y varillas para representar enlaces. (p. 186)

**base** Sustancia que cede iones de hidróxido, $OH^-$, a una solución acuosa o una sustancia que recibe protones, iones $H^+$, de otra sustancia en una solución. (p. 430)

**forma curva** La forma curvilínea alrededor de un átomo enlazado con dos pares aislados de electrones. (p. 194)

**desintegración beta** Reacción nuclear en la cual un neutrón pasa a ser un protón y un electrón, y el átomo emite una partícula beta, que es el electrón. La desintegración beta aumenta el número atómico de un átomo sin alterar su masa. (p. 72)

**partícula beta** Electrón emitido por el núcleo de un átomo durante una desintegración beta. (p. 72)

**punto de ebullición (temperatura de ebullición)** La temperatura a la cual los estados líquido y gaseoso de una misma sustancia están presentes y en equilibrio; la temperatura a la cual se establece el equilibrio entre un líquido y su vapor a una presión de 1 atm. (p. 274)

**bond energy** The amount of energy that is required to break a specific chemical bond. (p. 531)

**bonded pair** A pair of electrons that are shared in a covalent bond between two atoms. (p. 163)

**Boyle's law** The scientific law that states that the volume of a given sample of gas at a given temperature is inversely proportional to its pressure. (p. 306)

**Brønsted-Lowry theory of acids and bases** An acid-base theory that defines an acid as a substance that donates a proton to another substance, and a base as a substance that accepts a proton from another substance. (p. 434)

# C

**calorie (cal)** A unit of measurement for thermal energy. The amount of heat required to raise the temperature of 1 gram of water by 1 Celsius degree. A kilocalorie, 1000 calories, is equal to one food Calorie. (p. 497)

**calorimetry** A procedure used to measure the heat transfer that occurs as a result of chemical reactions or physical changes. (p. 517)

**catalyst** A substance that accelerates a chemical reaction but is itself not permanently consumed or altered by the reaction. (p. 181)

**cation** An ion that has a positive charge. (p. 96)

**Charles's law** The law that states that the volume of a given sample of gas is proportional to its Kelvin temperature if the pressure is unchanged. (p. 284)

**chemical bond** An attraction between atoms that holds them together. (p. 129)

**chemical change** See **chemical reaction.**

**chemical equation** A representation of a chemical reaction written with chemical symbols and formulas. (p. 179)

**chemical equilibrium** A dynamic state in which opposing processes occur at the same time and the rate of the forward process is equal to the rate of the reverse process. (p. 620)

**energía de enlace** La cantidad de energía necesaria para romper un enlace químico. (p. 531)

**par enlazado** Par de electrones que están siendo compartidos en un enlace covalente entre dos átomos. (p. 163)

**ley de Boyle** La ley científica que establece que el volumen de una cantidad dada de un gas a una temperatura dada es inversamente proporcional a la presión. (p. 306)

**teoría de ácidos y bases de Brønsted-Lowry** Teoría ácido-base que define un ácido como una sustancia que dona protones a otra, y una base como una sustancia que recibe un protón de otra. (p. 434)

**caloría (cal)** Unidad de medida de la energía térmica. La cantidad de calor que se necesita para elevar la temperatura de 1 gramo de agua en 1 grado centígrado. Una kilocaloría, 1000 calorías, es igual a una caloría alimenticia. (p. 497)

**calorimetría** Procedimiento usado para medir la transferencia de calor que se produce como resultado de las reacciones químicas o de los cambios físicos. (p. 517)

**catalizador** Sustancia que acelera una reacción química pero que no se consume o se transforma de manera permanente por dicha reacción. (p. 181)

**catión** Ion que tiene una carga positiva. (p. 96)

**ley de Charles** La ley que establece que el volumen de una cantidad dada de un gas es proporcional a su temperatura en grados Kelvin si la presión no cambia. (p. 284)

**enlace químico** Fuerza de atracción entre átomos que los mantiene unidos. (p. 129)

**cambio químico** Ver **reacción química.**

**ecuación química** Representación de una reacción química escrita con símbolos químicos y fórmulas. (p. 179)

**equilibrio químico** Estado dinámico en el que se producen simultáneamente reacciones opuestas y la velocidad de la reacción hacia delante es igual a la velocidad de la reacción inversa. (p. 620)

**chemical formula** A combination of element symbols and numbers representing the composition of a chemical compound. (p. 24)

**chemical reaction (chemical change)** A transformation that alters the composition of one or more substances such that one or more new substances with new properties are produced. (p. 28)

**chemical symbol** A one- or two-letter representation of an element. The first letter is always uppercase. If there is a second letter, it is lowercase. (p. 24)

**chemistry** The study of substances, their properties, and how they can be transformed; the study of matter and how it can be changed. (p. 8)

**coefficients** The numbers in front of the chemical formulas of the reactants and products in a balanced chemical equation. They indicate the correct ratio in which the reactants combine to form the products. (p. 372)

**combination reaction** A reaction in which two or more reactants combine to form a single product. (p. 375)

**combined gas law** The law that describes the proportional relationship among the pressure, temperature, and volume of a gas. It states that the value of $PV/T$ will be constant for a given sample of any gas. (p. 318)

**combustion** An exothermic chemical reaction between a fuel and oxygen, often producing flames; burning. (p. 513)

**complete ionic equation** A chemical equation that shows all of the soluble ionic compounds as independent ions. (p. 459)

**compound** A pure substance that is a chemical combination of two or more elements in a specific ratio. (p. 24)

**concentration** A measure of the amount of solute dissolved per unit of volume of solution, often expressed as moles of solute per liter of solution, mol/L. (p. 410)

**fórmula química** Combinación de símbolos de elementos y números que representan la composición de un compuesto químico. (p. 24)

**reacción química (cambio químico)** Transformación que altera la composición de una o más sustancias de tal manera que se producen una o más sustancias nuevas con propiedades nuevas. (p. 28)

**símbolo químico** Representación de un elemento por medio de una o dos letras. La primera letra es siempre mayúscula y la segunda, si la hay, es minúscula. (p. 24)

**química** El estudio de las sustancias, sus propiedades y cómo se pueden transformar; el estudio de la materia y cómo se puede transformar. (p. 8)

**coeficientes** Los números delante de las fórmulas químicas de los reactantes y los productos en una ecuación química balanceada. Indican la velocidad de reacción exacta a la que los reactantes se combinan para formar productos. (p. 372)

**reacción de combinación** Reacción en la que dos o más reactantes se combinan para formar un solo producto. (p. 375)

**ley combinada de los gases** La ley que describe la relación proporcional entre la presión, la temperatura y el volumen de un gas. Establece que el valor de $PV/T$ será constante para una cantidad dada de un gas. (p. 318)

**combustión** Reacción química exotérmica entre un combustible y oxígeno que suele producir llamas; fuego. (p. 513)

**ecuación iónica completa** Ecuación química que muestra todos los compuestos iónicos solubles como iones independientes. (p. 459)

**compuesto** Sustancia pura que es una combinación química de dos o más elementos en una proporción determinada. (p. 24)

**concentración** Medida de la cantidad de soluto diluido por unidad de volumen de solución, usualmente expresada en moles de soluto por litro de solución, mol/L. (p. 410)

**conductivity**  A property that describes how well a substance transmits electricity, heat, or sound. (p. 125)

**core electrons**  All electrons in an atom that are not valence electrons. (p. 93)

**covalent bonding**  A type of chemical bonding in which one or more pairs of valence electrons are shared between the atoms. (p. 130)

**conductividad**  Propiedad que describe la capacidad de una sustancia de transmitir electricidad, calor o sonido. (p. 125)

**electrones internos**  Todos los electrones de un átomo que no son electrones de valencia. (p. 93)

**enlace covalente**  Enlace químico entre átomos en el que uno o más pares de electrones de valencia son compartidos por dichos átomos. (p. 130)

# D

**daughter isotope**  An isotope that is formed as a result of a nuclear reaction. (p. 78)

**decomposition reaction**  A chemical change in which a single substance is broken down into two or more simpler substances. (p. 376)

**density**  The measure of the mass of a substance per unit of volume, often expressed as grams per milliliter, g/mL, or grams per cubic centimeter, $g/cm^3$. (p. 17)

**diatomic molecule**  A molecule consisting of two atoms. (p. 228)

**dilution**  The process of adding solvent to a solution to lower the concentration of solute. (p. 442)

**dipole**  A molecule or covalent bond with a nonsymmetrical distribution of electrical charge that makes the molecule or bond polar. (p. 224)

**dissociate**  To break apart to form ions in solution. (p. 433)

**dissolve**  To disperse a substance homogeneously into another substance at the molecular, ionic, or atomic level. (p. 125)

**double bond**  A covalent bond where four electrons are shared between two atoms. (p. 166)

**double exchange reaction (double displacement reaction)**  A chemical change in which both reactants break apart and then recombine to form two new products; chemical change where there is an exchange of ions between reactants to form new products. (p. 377)

**isótopo descendiente**  Isótopo que se forma como resultado de una reacción nuclear. (p. 78)

**reacción de descomposición**  Cambio químico en el que una sola sustancia se descompone en dos o más sustancias más simples. (p. 376)

**densidad**  La medida de la masa de una sustancia por unidad de volumen, la cual se expresa usualmente en gramos por mililitro, g/mL, o gramos por centímetro cúbico, $g/cm^3$. (p. 17)

**molécula diatómica**  Molécula compuesta por dos átomos. (p. 228)

**dilución**  El proceso de añadir solvente a una solución para bajar la concentración del soluto. (p. 442)

**dipolo**  Molécula o enlace covalente con una distribución asimétrica de la carga eléctrica que hace que dicha molécula o enlace sea polar. (p. 224)

**disociar**  Se romper para formar iones en solución. (p. 433)

**disolver**  Dispersar un sustancia de manera homogénea en otra sustancia a nivel molecular, iónico o atómico. (p. 125)

**doble enlace**  Enlace covalente en el que dos átomos comparten cuatro electrones. (p. 166)

**reacción de doble intercambio (reacción de doble desplazamiento)**  Cambio químico en el que los dos reactantes se rompen y se vuelven a combinar para formar dos nuevos productos; cambio químico en el que hay un intercambio de iones entre los reactantes para formar productos nuevos. (p. 377)

## E

**electrochemical cell** A device used for generating electrical energy from chemical reactions. The electrical current is caused by substances releasing and accepting electrons in oxidation-reduction reactions. (p. 570)

**electrolyte solution** A solution that contains ions and can conduct electricity. (p. 569)

**electromagnetic radiation** A large range of waves, some visible, some invisible, all traveling at the speed of light, in a vacuum, and at widely varying wavelengths and frequencies. (p. 586)

**electron** An elementary particle with a negative charge that is located outside of the nucleus of an atom. It has a mass of about 1/1838 amu. (p. 54)

**electron configuration** A notation for keeping track of where the electrons in an atom are distributed among the shells and subshells in an atom. (p. 117)

**electron domain** The space occupied by bonded pairs or lone pairs of valence electrons in a molecule. Electron domains affect the overall shape of a molecule. (p. 192)

**electron domain theory** A scientific theory that states that every electron domain is located as far away as possible from every other electron domain in a molecule. (p. 193)

**electronegativity** A measure of the ability of an atom in a molecular substance to attract electrons to itself. (p. 224)

**electroplating** The process by which a material is coated with a thin layer of metal using electricity passed through a suitable ionic solution. (p. 136)

**element** A unique substance that cannot be broken down into simpler substances through physical or chemical processes. Elements serve as the building materials of all matter. (p. 24)

**emission** In general, transmission of energy from a source. Of light, a transmission of energy from the Sun or other bright body. (p. 576)

**celda electroquímica** Dispositivo usado para generar energía con reacciones químicas. La corriente eléctrica se genera mediante sustancias que ceden y reciben electrones en reacciones de oxidación-reducción. (p. 570)

**solución de electrolitos** Solución que contiene iones y que es conductora de la electricidad. (p. 569)

**radiación electromagnética** Una amplia gama de ondas, algunas visibles, otras invisibles, todas viajando a la velocidad de la luz (en el vacío) y a muy diversas longitudes de onda y frecuencias. (p. 586)

**electrón** Partícula elemental con carga negativa que está fuera del núcleo de un átomo. Tiene una masa de 1/1838 uma, aproximadamente. (p. 54)

**configuración electrónica** Notación que sirve para llevar la cuenta de cómo están distribuidos los electrones de un átomo en las capas y las subcapas electrónicas. (p. 117)

**dominio del electrón** Espacio que ocupan los pares de electrones de valencia enlazados o aislados en una molécula. El dominio de los electrones afecta la forma de una molécula. (p. 192)

**teoría del dominio del electrón** Teoría científica que establece que el dominio de cada electrón se encuentra lo más lejos posible del dominio de otro electrón en una molécula. (p. 193)

**electronegatividad** Medida de la capacidad de atraer electrones que tiene un átomo en una sustancia molecular. (p. 224)

**galvanostegia** Proceso mediante el cual se cubre un material con una capa delgada de metal haciendo pasar electricidad a través de una solución iónica especial. (p. 136)

**elemento** Sustancia única que no se puede descomponer en sustancias más simples a través de procesos físicos o químicos. Los elementos son los componentes que constituyen toda la materia. (p. 24)

**emisión** Se refiere en términos generales a la transmisión de energía a partir de una fuente. De luz: transmisión de energía del sol o de otro cuerpo incandescente. (p. 576)

**endothermic** Describes a process in which heat transfers from the surroundings to the system; heat-absorbing. (p. 487)

**energy** A measure of the capacity of an object or system to do work or produce heat. (p. 483)

**entropy** The tendency of energy or matter in a system to disperse; the energy of a system that cannot be used for external work. (p. 492)

**equilibrium constant** $K$ A measure of whether starting substances or products are favored in an equilibrium mixture. A large value for $K$ indicates a large amount of product. A small value for $K$ indicates very little product. (p. 627)

**equilibrium-constant equation** A specific mathematical relationship between $K$ and the concentrations of starting substances and products in an equilibrium mixture. (p. 631)

**equivalence point** The point in an acid-base titration when the acid and base have completely neutralized each other. (p. 451)

**evaporation** The phase change from a liquid to a gas. (p. 297)

**exothermic** Describes a process in which heat is transferred from a system to the surroundings; heat-releasing. (p. 484)

**extensive property** A characteristic, such as volume or mass, that is specific to the amount of matter and therefore changes if the quantity of the substance changes. (p. 19)

## F

**first law of thermodynamics** The scientific law that states that energy is conserved; therefore it cannot be created or destroyed. (p. 492)

**fission (nuclear)** The splitting apart of an atomic nucleus into two smaller nuclei, accompanied by a release of energy. (p. 74)

**endotérmico** Describe un proceso en el que se transfiere calor del entorno al interior del sistema; absorción de calor. (p. 487)

**energía** Medida de la capacidad que tiene un objeto o un sistema de hacer trabajo o producir energía. (p. 483)

**entropía** La tendencia de la energía o la materia de un sistema a dispersarse; la energía de un sistema que no puede usarse para trabajo externo. (p. 492)

**constante de equilibrio** $K$ Medida que indica si las sustancias iniciales o los productos finales son favorecidos en una mezcla en equilibrio. Un valor grande de $K$ indica una gran cantidad de producto final. Un valor pequeño de $K$ indica muy poco producto final. (p. 627)

**ecuación de la constante de equilibrio** Relación matemática específica entre $K$ y las concentraciones de las sustancias iniciales y los productos finales en una mezcla en equilibrio. (p. 631)

**punto de equivalencia** Momento, en una valoración química ácido-base, en que el ácido y la base se han neutralizado totalmente entre sí. (p. 451)

**evaporación** El cambio de estado de líquido a gas. (p. 297)

**exotérmico** Describe un proceso en el que el calor se transfiere del interior del sistema al entorno; que emite calor. (p. 484)

**propiedad extensiva** Característica, como el volumen o la masa, que es específica a una cantidad de materia y por tanto, cambia si la cantidad de la sustancia cambia. (p. 19)

**primera ley de la termodinámica** La ley científica que establece que la energía se conserva y por tanto no se puede crear ni destruir. (p. 492)

**fisión (nuclear)** La ruptura del núcleo de un átomo en dos núcleos más pequeños, que viene acompañada de liberación de energía. (p. 74)

**flame test** A laboratory procedure used to determine the presence of certain metal atoms in a chemical sample by heating the sample in a flame and observing the resulting flame color. (p. 85)

**formula unit** The simplest chemical formula that can be used to represent network covalent or ionic compounds that shows the elements present in the smallest whole number ratio. (p. 372)

**frequency** The number of waves that pass by a certain point in space per second, abbreviated as $f$. The frequency of light is measured in hertz, Hz. (p. 582)

**functional group** A structural feature of a molecule; consists of a specific arrangement of atoms, responsible for certain properties of the compound. (p. 170)

**fusion (nuclear)** The joining of two atomic nuclei to form a larger nucleus, accompanied by a release of energy. (p. 75)

# G

**gamma ray** Electromagnetic radiation with the shortest wavelengths, less than $10^{-11}$ m. Emitted when a nucleus decays or during a nuclear explosion. Used in irradiation of food and in some cancer treatments. (p. 74)

**Gay-Lussac's law** The scientific law that states that the pressure of a given amount of gas is directly proportional to temperature, if the gas volume does not change. (p. 310)

**group** A vertical column on the periodic table, also called a family. Elements in a group have similar properties. (p. 44)

# H

**half-cell** One-half of an electrochemical cell, the site of either oxidation or reduction. (p. 570)

**half-life** The amount of time required for one-half of the radioactive atoms in a sample to decay. (p. 73)

**ensayo a la llama** Procedimiento de laboratorio que se usa para determinar la presencia de átomos de un determinado metal en una muestra química, introduciendo una muestra en una llama y observando la coloración que adquiere la llama. (p. 85)

**unidad de fórmula** La fórmula química más simple que se puede usar para representar una red de compuestos iónicos o covalentes, y que muestra la proporción simplificada de los elementos presentes con números enteros. (p. 372)

**frecuencia** Número de ondas que pasan por un punto determinado en el espacio en un segundo, su abreviatura es $f$. La frecuencia de la luz se mide en hercios, Hz. (p. 582)

**grupo funcional** Característica estructural de una molécula que consiste en una distribución específica de los átomos, y a la que se deben ciertas propiedades de un compuesto. (p. 170)

**fusión (nuclear)** La unión de dos núcleos atómicos para formar un núcleo más grande, que viene acompañada de liberación de energía. (p. 75)

**rayo gamma** Radiación electromagnética con las longitudes de onda más cortas de menos de $10^{-11}$ m. Se emiten cuando un núcleo decae o durante una explosión nuclear. Se utiliza en la irradiación de alimentos y en algunos tratamientos contra el cáncer. (p. 74)

**ley de Gay-Lussac** La ley científica que establece que la presión de una cantidad dada de un gas es directamente proporcional a la temperatura, si su volumen no cambia. (p. 310)

**grupo** Columna vertical en la tabla periódica, también denominada familia. Los elementos de un grupo tienen propiedades similares. (p. 44)

**semicelda** La mitad de una celda electroquímica, en la que puede producirse oxidación o reducción. (p. 570)

**vida media** El tiempo que se necesita para que la mitad de los átomos radiactivos de una muestra se desintegren. (p. 73)

**half-reaction** A chemical equation that represents either the oxidation or the reduction part of an oxidation-reduction reaction. (p. 570)

**halogens** The elements in Group 7A on the periodic table. (p. 44)

**heat** A transfer of energy between two substances, from a hotter body to a colder body. (p. 486)

**heat of combustion** The energy released as heat when a compound undergoes complete combustion with oxygen, usually expressed per mole of the fuel. (p. 524)

**heat of formation** The energy transferred as heat during a chemical reaction when a compound is formed from its constituent elements, expressed per mole of product. (p. 554)

**heat of fusion** The amount of heat transfer required to change a substance from a solid to a liquid, usually expressed per mole or gram. (p. 506)

**heat of reaction** The amount of energy transferred as heat during a chemical reaction. (p. 524)

**heat of vaporization** The amount of heat transfer required to change a substance from a liquid to a gas, usually expressed per mole or gram. (p. 506)

**heterogeneous mixture** A mixture whose composition is not uniform throughout. (p. 410)

**homogeneous mixture** A mixture whose composition is uniform throughout. (p. 410)

**HONC 1234 rule** A rule that states that in most molecules, hydrogen makes 1 bond, oxygen makes 2 bonds, nitrogen makes 3 bonds, and carbon makes 4 bonds. (p. 156)

**humidity** The concentration of the water vapor in the air at any given time. (p. 340)

**semirreacción** Ecuación química que puede representar la oxidación o la reducción de una reacción de oxidación-reducción. (p. 570)

**halógenos** Los elementos del grupo 7A de la tabla periódica. (p. 44)

**calor** Una transferencia de energía entre dos sustancias, de un cuerpo más caliente a uno más frío. (p. 486)

**calor de combustión** La energía liberada en forma de calor cuando un compuesto pasa por una combustión completa con oxígeno. Usualmente se expresa por mol de combustible. (p. 524)

**calor de formación** La energía que se transfiere en forma de calor durante una reacción química cuando se produce un compuesto a partir de sus elementos constituyentes. Se expresa por mol de producto. (p. 554)

**calor de fusión** La cantidad de calor que debe ser transferida para que una sustancia pase de ser un sólido a ser un líquido. Usualmente se expresa por mol o gramo. (p. 506)

**calor de reacción** La cantidad de energía que se transfiere en forma de calor durante una reacción química. (p. 524)

**calor de vaporización** La cantidad de calor que debe ser transferida para que una sustancia pase de ser un líquido a ser un gas. Usualmente se expresa por mol o gramo. (p. 506)

**mezcla heterogénea** Mezcla cuya composición no es uniforme. (p. 410)

**mezcla homogénea** Mezcla cuya composición es uniforme. (p. 410)

**regla de HONC 1234** Regla que establece que en la mayoría de moléculas, el hidrógeno hace 1 enlace, el oxígeno hace 2 enlaces, el nitrógeno hace 3 enlaces y el carbono hace 4 enlaces. (p. 156)

**humedad** La concentración de vapor de agua en el aire en un momento determinado. (p. 340)

**hydrogen bond**  An intermolecular attraction between a hydrogen atom in a molecule and an electronegative atom in another molecule (especially nitrogen, oxygen, or fluorine). The hydrogen atoms in water molecules form hydrogen bonds with the oxygen atoms in other water molecules. (p. 503)

**hypothesis**  A proposed explanation for an observation or scientific problem, which can be tested by further investigation. (p. 7)

# I

**ideal gas law**  The scientific law that relates volume, pressure, temperature, and the number of moles of a gas sample: $PV = nRT$, where $R$ is the universal gas constant. (p. 337)

**indicator (acid-base)**  A chemical compound that indicates the relative acidity or basicity of a solution through its characteristic color changes. (p. 429)

**infrared radiation**  Electromagnetic radiation with longer wavelengths than those of visible light, from $10^{-3}$ m to $10^{-6}$ m. Causes molecules to vibrate, which human beings sense as heat. Abbreviated as IR radiation. (p. 587)

**insoluble**  Unable to be dissolved in another substance. (p. 125)

**intensive property**  A characteristic, such as boiling point or density, that does not depend on the size or amount of matter and can be used to identify matter. (p. 19)

**intermolecular force**  A force of attraction that occurs between molecules. (p. 216)

**inversely proportional**  Related in such a way that when one quantity increases, the other decreases in a mathematically predictable way. The variables $x$ and $y$ are inversely proportional to each other if $y = k/x$, where $k$ is the proportionality constant. (p. 306)

**ion**  An atom or group of bonded atoms that has a positive or negative charge. (p. 95)

**enlace de hidrógeno**  Atracción intermolecular entre un átomo de hidrógeno en una molécula y un átomo electronegativo en otra (especialmente de nitrógeno, oxígeno o flúor). Los átomos de hidrógeno en las moléculas de agua forma enlaces con átomos de oxígeno en otras moléculas de agua. (p. 503)

**hipótesis**  Una explicación dada a una observación o a un problema científico, la cual se puede probar con más investigación. (p. 7)

**ley de los gases ideales**  La ley científica que relaciona el volumen, la presión, la temperatura y el número de moles de una muestra de gas: $PV = nRT$, cuando $R$ es la constante universal de los gases ideales. (p. 337)

**indicador (ácido-base)**  Compuesto químico que sirve para mostrar la acidez o la basicidad relativa de una solución a través de cambios de color característicos. (p. 429)

**radiación infrarroja**  Radiación electromagnética con longitudes de onda más largas que las de la luz visible, entre $10^{-3}$ m y $10^{-6}$ m. Causa la vibración de las moléculas, lo que los seres humanos perciben como calor. Su abreviatura es IR radiación. (p. 587)

**insoluble**  Que no se puede disolver en otra sustancia. (p. 125)

**propiedad intensiva**  Característica como el punto de ebullición o la densidad que no depende del tamaño o la cantidad de materia y que puede ser usada para identificar dicha materia. (p. 19)

**fuerza intermolecular**  Fuerza de atracción que se produce entre moléculas. (p. 216)

**inversamente proporcional**  Relacionadas de tal manera que cuando una variable aumenta, la otra disminuye de forma que se puede predecir matemáticamente. Las variables $x$ y $y$ son inversamente proporcionales si $y = k/x$, donde $k$ es la constante de proporcionalidad. (p. 306)

**ion**  Átomo, o grupo de átomos enlazados, que tiene carga positiva o negativa. (p. 95)

**ionic bonding** A type of chemical bonding that is the result of the transfer of electrons from one atom to another, typically between metal and nonmetal atoms. (p. 130)

**ionic compound** A compound that consists of positively charged metal cations and negatively charged nonmetal anions formed when valence electrons are transferred. (p. 98)

**isomers** Compounds with the same molecular formula but different structural formulas. Isomers differ in molecular structure and in chemical and physical properties. (p. 152)

**isotopes** Atoms of the same element that have different numbers of neutrons. These atoms have the same atomic number but different mass numbers. (p. 62)

**enlace iónico** Atracción entre átomos que es el resultado de la transferencia de electrones de un átomo a otro, y que es común entre átomos de metales y de no metales. (p. 130)

**compuesto iónico** Compuesto que contiene cationes de metales con carga positiva y aniones de no metales con carga negativa que se forman al transferirse los electrones de valencia. (p. 98)

**isómeros** Compuestos que tienen una misma fórmula molecular pero fórmulas estructurales diferentes. Los isómeros son distintos en su estructura molecúlar y en sus propiedades químicas y físicas. (p. 152)

**isótopos** Átomos de un mismo elemento que tienen distinta cantidad de neutrones. Estos átomos tienen un mismo número atómico pero diferente número de masa. (p. 62)

# J

**joule (J)** A unit of measurement of energy. One calorie is equal to 4.184 joules. (p. 524)

**julio (J)** Unidad de medida de la energía. Una caloría es igual a 4.184 julios. (p. 524)

# K

**Kelvin scale** A temperature scale with units in kelvins, K, that sets the zero point at $-273.15\ ^\circ$C, which is also known as absolute zero. Kelvin units are equivalent in scale to Celsius units. (p. 279)

**kinetic energy** The energy of motion. (p. 488)

**kinetic theory of gases** The scientific theory that states that gases are composed of tiny particles in continuous, random, straight-line motion and collide with each other and the walls of the container. (p. 280)

**escala Kelvin** Escala de temperatura cuyas unidades son los kelvins, K, con el punto cero en $-273.15\ ^\circ$C, temperatura también conocida como cero absoluto. Las unidades Kelvin son equivalentes en escala a las unidades Celsius. (p. 279)

**energía cinética** La energía del movimiento. (p. 488)

**teoría cinética de los gases** La teoría científica que establece que los gases están compuestos por pequeñas partículas en movimiento continuo, aleatorio y en línea recta, y que se chocan entre sí y con las paredes que las contienen. (p. 280)

# L

**lanthanides** A series of elements that follow lanthanum in Period 6 of the periodic table; they are typically placed separately at the bottom of the periodic table. (p. 44)

**law of conservation of mass** The scientific law that states that mass cannot be gained or lost in a chemical reaction and that matter cannot be created or destroyed. (p. 34)

**lantánidos** Conjunto de elementos que están después del lantano en el sexto período de la tabla periódica y que usualmente aparecen en la parte de abajo de la tabla. (p. 44)

**ley de conservación de la masa** La ley científica que establece que en una reacción química no se gana ni se pierde masa y que la materia no se puede crear ni destruir. (p. 34)

**Le Châtelier's principle** An experimental observation that, when a stress is put on a system at equilibrium, the system responds by reducing the effect of the change. (p. 643)

**Lewis dot structure** A diagram that shows a molecule's structure by using dots to represent the valence electrons. (p. 162)

**Lewis dot symbol** A diagram that uses dots to represent the valence electrons of a single atom. (p. 162)

**light detector** A device that senses electromagnetic radiation, such as a radio or a photovoltaic cell. (p. 593)

**light ray** A model of light that shows the straight-line path that light travels and how a beam of light interacts with matter. (p. 577)

**light wave** Ripples of light energy created by forces that oscillate up and down, radiating in space from a source, such as the Sun. (p. 582)

**limiting reactant** The reactant that runs out first in a chemical reaction. It is the reactant that limits the amount of product that can be produced in the reaction. (p. 464)

**linear shape** A straight-line shape found in small molecules. (p. 197)

**lone pair** A pair of unshared valence electrons that are not involved in bonding in a molecule. (p. 163)

## M

**main group elements** The elements in Groups 1A to 7A on the periodic table. (p. 44)

**mass** A measure of the quantity of matter in an object. (p. 9)

**mass number** The sum of the number of protons and neutrons in the nucleus of an atom. (p. 62)

**matter** Anything that has substance and takes up space; anything that has mass and volume. (p. 9)

**principio de Le Châtelier** Observación experimental que, cuando se somete un sistema en equilibrio a cierto nivel de perturbación externa, el sistema responde a fin de reducir el efecto de la perturbación. (p. 643)

**estructura de puntos de Lewis** Diagrama de la estructura de una molécula en el que se usan puntos para representar los electrones de valencia. (p. 162)

**símbolo de puntos de Lewis** Diagrama en el que se usan puntos para representar los electrones de valencia de un solo átomo. (p. 162)

**detector de luz** Dispositivo que detecta la radiación electromagnética, p. ej. una radio o una célula fotovoltaica. (p. 593)

**rayo de luz** Modelo de la luz que muestra la ruta en línea recta trazada por la luz y cómo un rayo de luz interactúa con la materia. (p. 577)

**onda de luz** Ondas de energía lumínica creadas por fuerzas que oscilan hacia arriba y hacia abajo; estas ondas se irradian en el espacio a partir de una fuente tal como el Sol. (p. 582)

**reactante limitador** El reactante que se acaba primero en una reacción química. Es el que limita la cantidad de producto que se puede producir en una reacción. (p. 464)

**forma recta** Forma de línea recta que aparece en moléculas pequeñas. (p. 197)

**par aislado** Par de electrones de valencia que no están siendo compartidos y que no están involucrados en el enlace de una molécula. (p. 163)

**elementos del grupo principal** Los elementos de los grupos 1A a 7A de la tabla periódica. (p. 44)

**masa** Medida de la cantidad de materia en un objeto. (p. 9)

**número de masa** La suma del número de protones y neutrones en el núcleo de un átomo. (p. 62)

**materia** Todo lo que tiene sustancia y ocupa un lugar en el espacio; todo lo que tiene masa y volumen. (p. 9)

**melting point (melting temperature)** The temperature at which both solid and liquid phases of a single substance can be present and in equilibrium. (p. 274)

**meniscus** The curvature of the top of a liquid in a container, which is the result of intermolecular attractions between the liquid and the container. (p. 12)

**metal** An element that is generally shiny and malleable and an excellent conductor of heat and electricity. Metals are located to the left of the stairstep line on the periodic table. (p. 45)

**metallic bonding** A type of bonding between metal atoms in which the valence electrons are free to move throughout the substance. (p. 130)

**metalloid** An element that has properties of both metals and nonmetals. Metalloids are located along the stair-step line of the periodic table. (p. 46)

**microwave radiation** Electromagnetic radiation with wavelengths longer than visible light and infrared radiation from 1 m to $10^{-3}$ m. Absorption of these waves cooks food. (p. 587)

**mirror-image isomer** Molecules whose structures are mirror images of each other and cannot be superimposed on one another. (p. 243)

**mixture** A blend of two or more substances that are not chemically combined. (p. 24)

**model** A simplified representation of a real object or process that facilitates understanding or explanation of that object or process. (p. 52)

**molarity** The concentration of dissolved substances in a solution, expressed in moles of solute per liter of solution. (p. 410)

**molar mass** The mass in grams of one mole of a substance. (p. 391)

**mole** A counting unit used to keep track of large numbers of particles. One mole represents $6.02 \times 10^{23}$ items. (p. 332)

**mole ratio** The ratio of the moles of one reactant or product to the moles of another reactant or product in a balanced chemical equation. (p. 463)

**punto de fusión (temperatura de fusión)** La temperatura a la cual los estados sólido y líquido de un misma sustancia están presentes y en equilibrio. (p. 274)

**menisco** La curvatura que se forma en la parte superior de un líquido que está dentro de un envase y que es el resultado de las atracciones intermoleculares entre el líquido y el envase. (p. 12)

**metal** Elemento que generalmente es brillante y maleable y un excelente conductor del calor y de la electricidad. Los metales están a la izquierda de la línea escalonada en la tabla periódica. (p. 45)

**enlace metálico** Enlace entre átomos de un metal en el que los electrones de valencia se mueven libremente por la sustancia. (p. 130)

**metaloide** Elemento que tiene propiedades tanto de los metales como de los no metales. Los metaloides están sobre la línea escalonada de la tabla periódica. (p. 46)

**radiación de microondas** Radiación electromagnética con longitudes de onda más largas que la luz visible y la radiación infrarroja, de 1 m a $10^{-3}$ m. La absorción de estas ondas cocina los alimentos. (p. 587)

**isómero especular** Moléculas cuyas estructuras se reflejan entre sí y no se pueden superponer. (p. 243)

**mezcla** Combinación de dos o más sustancias que no están combinadas químicamente. (p. 24)

**modelo** Representación simplificada de un objeto o proceso real que facilita la comprensión o explicación de dicho objeto o proceso. (p. 52)

**molaridad** La concentración de sustancias disueltas en una solución. Se expresa en moles de soluto por litro de solución. (p. 410)

**masa molar** La masa en gramos de un mol de una sustancia. (p. 391)

**mol** Unidad de conteo usada para llevar la cuenta de grandes cantidades de partículas. Un mol representa $6.02 \times 10^{23}$ partículas. (p. 332)

**razón molar** La proporción entre los moles de un reactante o producto y los moles de otro reactante o producto en una ecuación química balanceada. (p. 463)

**molecular covalent bonding** A type of chemical bonding characterized by the sharing of valence electrons between atoms, resulting in individual units called molecules. (p. 130)

**molecular formula** The chemical formula of a molecular substance, showing the identity of the atoms in each molecule and the ratios of those atoms to one another. (p. 148)

**molecule** A group of atoms that are covalently bonded together. (p. 131)

**monatomic ion** An ion that consists of only one atom. (p. 109)

## N

**net ionic equation** A chemical equation that is written without including spectator ions. (p. 460)

**network covalent bonding** A type of chemical bonding characterized by the sharing of valence electrons throughout the entire solid sample. (p. 130)

**neutralization reaction** A chemical reaction in which an acid and base react to form a salt and water. (p. 448)

**neutron** A particle that is located in the nucleus of an atom and does not have an electric charge. The mass of a neutron is almost exactly equal to that of a proton, about 1 amu. (p. 54)

**noble gases** The elements in Group 8A on the periodic table. Noble gases are known for not being reactive. (p. 44)

**nonmetal** An element that does not exhibit metallic properties. Nonmetals are often gases or brittle solids at room temperature. Nonmetals are poor conductors of heat and electricity and are located to the right of the stair-step line on the periodic table. (p. 45)

**nonpolar molecule** A molecule that is not attracted to an electrical charge. A molecule is nonpolar if each atom shares electrons equally or there is no net dipole in the molecule. (p. 215)

**enlace covalente molecular** Tipo de enlace químico caracterizado por átomos que comparten electrones de valencia y forman unidades individuales llamadas moléculas. (p. 130)

**fórmula molecular** La fórmula química de una sustancia molecular que muestra la identidad de los átomos de cada molécula y la proporción de dichos átomos con respecto a los demás. (p. 148)

**molécula** Grupo de átomos que están unidos por enlaces covalentes. (p. 131)

**ion monoatómico** Ion que contiene un solo átomo. (p. 109)

**ecuación iónica total** Ecuación química en la que no se incluyen los iones espectadores. (p. 460)

**enlace covalente encadenado** Tipo de enlace químico caracterizado porque se comparten los electrones de valencia en toda la muestra sólida. (p. 130)

**reacción de neutralización** Reacción química en la que un ácido y una base reaccionan para formar sal y agua. (p. 448)

**neutrón** Partícula que está en el núcleo de un átomo y que no tiene carga eléctrica. La masa de un neutrón es casi igual a la de un protón, 1 uma, aproximadamente. (p. 54)

**gases nobles** Los elementos del grupo 8A de la tabla periódica. Los gases nobles se conocen por no ser reactivos. (p. 44)

**no metal** Elemento que no tiene propiedades metálicas. Los no metales suelen ser gases o sólidos quebradizos cuando están a temperatura ambiente. Los no metales no son buenos conductores del calor ni de la electricidad, y están a la derecha de la línea escalonada en la tabla periódica. (p. 45)

**molécula apolar** Molécula que no es atraída por una carga eléctrica. Una molécula es apolar si cada átomo comparte electrones equitativamente o si no hay bipolaridad neta en la molécula. (p. 215)

**nuclear chain reaction**  A nuclear reaction in which neutrons emitted from the nucleus during fission strike surrounding nuclei, causing them to split apart as well. (p. 79)

**nuclear equation**  A representation of a nuclear reaction written with isotope symbols. (p. 78)

**nuclear reaction**  A process that changes the energy, composition, or structure of an atom's nucleus. (p. 71)

**nucleus**  The dense, positively charged structure composed of protons and neutrons that is found in the center of an atom. (p. 54)

**number density**  The number of gas particles per unit volume usually expressed in moles per liter or moles per cubic centimeter. (p. 328)

## O

**octet rule**  Nonmetal atoms combine by sharing electrons so that each atom has a total of eight valence electrons. After bonding, each atom resembles a noble gas in its electron arrangements. (p. 165)

**opaque**  Not capable of transmitting light and, therefore, an object or surface not possible to see through. (p. 578)

**oxidation**  A chemical reaction in which an atom or ion loses electrons. Oxidation is always accompanied by reduction. (p. 552)

**oxidation-reduction reaction**  A chemical reaction that involves electron transfer. (p. 560)

## P

**parent isotope**  A radioactive isotope that undergoes decay. (p. 78)

**partial charge**  A less than full charge on part of a molecule, created by the unequal sharing of electrons. Partial charges are represented with the symbol delta ($\delta+$ for partial positive charge and $\delta-$ for partial negative charge). (p. 215)

**partial pressure**  The pressure exerted by one gas in a mixture of nonreacting gases. The partial pressures of all the gases add up to the total pressure exerted by that mixture of gases. (p. 341)

**reacción nuclear en cadena**  Reacción nuclear en la que los neutrones emitidos desde el núcleo durante la fisión chocan con los núcleos alrededor haciendo que estos también se dividan. (p. 79)

**ecuación nuclear**  Representación de una reacción nuclear que se expresa con símbolos de isótopos. (p. 78)

**reacción nuclear**  Proceso que transforma la energía, la composición o la estructura del núcleo de un átomo. (p. 71)

**núcleo**  Estructura densa de carga positiva que está compuesta de protones y neutrones, y que se encuentra en el centro de un átomo. (p. 54)

**densidad numérica**  El número de partículas de gas por unidad de volumen que usualmente se expresa en moles por litro o moles por centímetro cúbico. (p. 328)

**regla del octeto**  Los átomos de no metales se combinan con otros átomos hasta que cada uno queda rodeado por ocho electrones de valencia. Después de enlazarse, cada átomo se asemeja a un gas noble en la distribución de sus electrones. (p. 165)

**opaco**  Que no es capaz de transmitir la luz. En el caso de superficies u objetos opacos, no es posible ver a través de ellos. (p. 578)

**oxidación**  Reacción química en la que un átomo o un ion pierde electrones. La oxidación siempre viene acompañada de reducción. (p. 552)

**reacción de oxidación-reducción**  Reacción química que involucra transferencia de electrones. (p. 560)

**isótopo precursor**  Isótopo radiactivo que sufre desintegración. (p. 78)

**carga parcial**  Una carga incompleta en un sector de una molécula que se crea porque se comparten electrones de manera desigual. Las cargas parciales se representan con el símbolo delta ($\delta+$ para cargas positivas y $\delta-$ para cargas negativas). (p. 215)

**presión parcial**  La presión ejercida por un gas en una mezcla de gases no reactivos. La suma de la presión parcial de todos los gases es igual a la presión total ejercida por esa mezcla de gases. (p. 341)

**peptide bond** The bond between two amino acids; also called an amide bond. (p. 249)

**percent error** A calculation used to find the accuracy of a measurement. The lower the percent error, the more accurate the measurement. (p. 388)

**percent yield** A calculation that expresses the success of a chemical process in terms of product yield; the ratio of the actual yield to the theoretical yield expressed as a percentage. (p. 473)

**period** The elements in a horizontal row on the periodic table. (p. 44)

**periodic table of the elements** A table with elements organized in order of increasing atomic number and grouped such that elements with similar properties are in vertical columns. (p. 39)

**pH scale** A logarithmic scale describing the concentration of hydrogen ions, $H^+$, in solution. $pH = -\log [H^+]$. (p. 430)

**phase** The physical form of matter such as the solid, liquid, or gaseous state. (p. 25)

**phase change** A transition between solid, liquid, or gaseous states of matter. (p. 260)

**photon** A particle of light; one of many separate particles in a beam of light. (p. 590)

**physical change** A change that alters the form of a substance but does not change the chemical identity of a substance. (p. 259)

**pigment** A substance that is colored because of the light it reflects. Pigments reflect only a certain color instead of all the colors in white light from the source. (p. 609)

**Planck's constant** A mathematical relationship of the frequency of the electromagnetic wave to its energy. Given the symbol, $h$, and defined as $h = 6.63 \times 10^{-34}$ J · s. (p. 589)

**polar molecule** A molecule that has a negatively charged end and a positively charged end due to electronegativity differences between the atoms and/or the asymmetry of its structure. (p. 215)

**enlace péptido** El enlace entre dos aminoácidos; también se llama enlace amino. (p. 249)

**error porcentual** Cálculo que se usa para hallar la precisión de una medición. Entre más bajo es el error porcentual, más precisa es la medición. (p. 388)

**rendimiento porcentual** Cálculo que expresa el éxito de un proceso químico en términos del rendimiento del producto; la proporción entre el rendimiento real y el rendimiento teórico, expresada como un porcentaje. (p. 473)

**período** Los elementos en una línea horizontal de la tabla periódica. (p. 44)

**tabla periódica de los elementos** Tabla con todos los elementos organizados en orden ascendente de acuerdo a su número atómico y agrupados de tal manera que los elementos con propiedades similares están en columnas verticales. (p. 39)

**escala de pH** Escala logarítmica que describe la concentración de iones de hidrógeno, $H^+$, en una solución. $pH = -\log [H^+]$. (p. 430)

**fase** La forma física de la materia como el estado sólido, líquido o gaseoso. (p. 25)

**cambio de fase** Transición entre los estados sólido, líquido y gaseoso de la materia. (p. 260)

**fotón** Partícula de luz, una de las muchas partículas individuales que viajan en un rayo de luz. (p. 590)

**cambio físico** Cambio que altera la forma de una sustancia pero que no transforma la identidad química de la sustancia. (p. 259)

**pigmento** Sustancia que cambia el color de la luz reflejada. Los pigmentos reflejan solamente un determinado color en lugar de la luz blanca emitida por la fuente. (p. 609)

**constante de Planck** Relación matemática entre la frecuencia de la onda electromagnética y su energía. Su símbolo es $h$ y se define como $h = 6.63 \times 10^{-34}$ J · s. (p. 589)

**molécula polar** Molécula que tiene un lado con carga negativa y otro con carga positiva debido a las diferencias de electronegatividad entre los átomos y/o a que su estructura es asimétrica. (p. 215)

**polyatomic ion** An ion that consists of two or more atoms covalently bonded. (p. 109)

**potential energy** Energy that is stored within a system and can be converted into other types of energy. This energy may be associated with the composition of a substance or its location. (p. 536)

**precipitate** A solid produced in a chemical reaction between two solutions. (p. 457)

**precipitation reaction** A chemical reaction that results in the formation of a solid substance (a precipitate) that separates out of a solution because it is not very soluble. (p. 457)

**pressure** Force applied over a specific area. Force per unit area. (p. 299)

**product** A substance produced as the result of a chemical reaction. (p. 180)

**property** A characteristic or quality of a substance. (p. 7)

**proportional** Related such that when one quantity increases, the other also increases. Two variables are proportional when you can multiply one variable by a constant to obtain the other. The variable $y$ is proportional to the variable $x$ if $y = kx$ where $k$ is the proportionality constant. (p. 265)

**proportionality constant** The number that relates two variables that are proportional to one another. It is often represented by $k$. (p. 265)

**protein** A large molecule made up of chains of amino acids bonded together. Typical protein molecules consist of more than 100 amino acids. (p. 249)

**proton** A positively charged particle located in the nucleus of an atom. The mass of a proton is almost exactly equal to that of a neutron, about 1 amu. (p. 54)

**pyramidal shape** The shape assumed by other bonded atoms around an atom with one lone pair of electrons. (p. 193)

# R

**radiation** Emission of energy as light either in the visible or invisible part of the spectrum. (p. 577)

**ion poliatómico** Ion que contiene dos o más átomos con enlaces covalentes. (p. 109)

**energía potencial** La energía que está almacenada en un sistema y que puede convertirse en otros tipos de energía. Esta energía se puede asociar con la composición de la sustancia o con su ubicación. (p. 536)

**precipitado** Sólido que se produce como resultado de una reacción química entre dos soluciones. (p. 457)

**reacción de precipitación** Reacción química que resulta en la formación de una sustancia sólida (un precipitado) que se separa de la solución porque no es muy soluble. (p. 457)

**presión** Fuerza aplicada sobre un área específica. Fuerza por unidad de área. (p. 299)

**producto** Sustancia que se produce como resultado de una reacción química. (p. 180)

**propiedad** Característica o cualidad de una sustancia. (p. 7)

**proporcional** Relacionadas de tal manera que cuando una cantidad aumenta, la otra también lo hace. Dos variables son proporcionales cuando una se puede multiplicar por una constante para obtener la otra. La variable $y$ es proporcional a la variable $x$ si $y = kx$ donde $k$ es la constante de proporcionalidad. (p. 265)

**constante de proporcionalidad** El número que relaciona dos variables que son proporcionales entre sí. Generalmente se representa con la letra $k$. (p. 265)

**proteína** Molécula grande hecha de cadenas de aminoácidos enlazados. Las moléculas de proteínas suelen tener más de 100 aminoácidos. (p. 249)

**protón** Partícula de carga positiva localizada en el núcleo de un átomo. La masa de un protón es casi igual a la de un neutrón, 1 uma, aproximadamente. (p. 54)

**forma piramidal** Forma que adquieren otros átomos enlazados alrededor de un átomo con un par aislado de electrones. (p. 193)

**radiación** Emisión de energía en forma de luz, ya sea en la parte visible o invisible del espectro. (p. 577)

**radio wave** Electromagnetic radiation with very long wavelengths from $10^5$ m to 1 m. Emitted at a television or radio station and used in communications satellites. (p. 587)

**radioactive decay** Spontaneous disintegration of an atomic nucleus accompanied by the emission of particles and radiation. A radioactive substance will decay with a specific half-life. (p. 71)

**radioactive isotope** Any isotope that has an unstable nucleus and decays over time. (p. 67)

**reactants** The starting materials in a chemical reaction that are transformed into products during the reaction. (p. 180)

**reactivity** The tendency of an element or compound to combine chemically with other substances, as well as the ease or speed of the reaction. (p. 37)

**receptor site theory** The currently accepted model explaining how specific molecules are detected by the nose. Molecules fit into receptor sites that correspond to the overall shape of the molecule. This stimulates a response in the body. (p. 208)

**redox reaction** Another name for an *oxidation-reduction* reaction. (p. 560)

**reduction** A chemical reaction in which an atom or ion gains electrons. Reduction is always accompanied by oxidation. (p. 559)

**reflection** The bouncing of light off a surface, allowing it to be seen. (p. 578)

**relative humidity** The amount of water vapor in the air compared to the maximum amount of water vapor possible for a specific temperature, expressed as a percentage. (p. 341)

**reversible process** A process that can proceed in both the forward direction to produce products and the opposite direction (reverse direction) to produce starting substances. (p. 613)

**ondas de radio** Radiación electromagnética con longitudes de onda muy largas, entre $10^5$ m y 1 m. Emitidas por los canales de televisión o las estaciones de radio y se utilizan en los satélites de comunicaciones. (p. 587)

**desintegración radiactiva** Desintegración espontánea del núcleo de un átomo que viene acompañada por la emisión de partículas y radiación. Una sustancia radiactiva se desintegrará en una vida media específica. (p. 71)

**isótopo radiactivo** Cualquier isótopo que tiene un núcleo inestable y que se desintegra con el tiempo. (p. 67)

**reactantes** Los materiales iniciales de una reacción química que se transforman en productos durante dicha reacción. (p. 180)

**reactividad** La tendencia de un elemento o compuesto a combinarse químicamente con otras sustancias, así como la facilidad o la velocidad de la reacción. (p. 37)

**teoría de los sitios receptores** El modelo aceptado en la actualidad que explica cómo detectar moléculas específicas mediante la nariz. Las moléculas caben dentro de sitios receptores que corresponden a la forma general de cada molécula. Esto estimula una respuesta en el cuerpo. (p. 208)

**reacción redox** Otro nombre para la reacción de *oxidación-reducción* (p. 560)

**reducción** Reacción química en la que un átomo o un ion gana electrones. La reducción siempre viene acompañada de oxidación. (p. 559)

**reflexión** Rebote de la luz contra una superficie, lo que permite que dicha superficie pueda ser vista. (p. 578)

**humedad relativa** La cantidad de vapor de agua en el aire comparada con la máxima cantidad de vapor de agua posible para una temperatura específica. Se expresa como un porcentaje. (p. 341)

**reacción reversible** Reacción que puede proceder tanto en la dirección hacia delante para producir productos finales y en las direcciones opuestas (dirección inversa) para producir sustancias iniciales. (p. 613)

**rule of zero charge** The rule that states that in an ionic compound, the positive charges on the metal cations and the negative charges on the nonmetal anions add up to zero. (p. 99)

**regla de la carga cero** La regla que establece que en un compuesto iónico, la suma total de las cargas positivas de los cationes de metales y de las cargas negativas de los aniones de no metales, es cero. (p. 99)

# S

**saturated solution** A solution that contains the maximum amount of solute that can be dissolved in a given amount of solvent at a particular temperature. (p. 411)

**solución saturada** Solución que contiene la cantidad máxima de soluto que se puede disolver en una cantidad dada de solvente a un temperatura determinada. (p. 411)

**scientific notation** A shorthand notation used for writing numbers that are very large or very small. In this notation, the number is expressed as a decimal number with one digit to the left of the decimal point, multiplied by an integer power of 10. For example, the number 890,000 is written as $8.9 \times 10^5$ in scientific notation. (p. 391)

**notación científica** Notación abreviada que se usa para escribir números muy grandes o muy pequeños. En esta notación, el número se expresa como un número decimal con un dígito a la izquierda del punto decimal, multiplicado por una potencia entera de 10. Por ejemplo, el número 890,000 se escribe $8.9 \times 10^5$ en notación científica. (p. 391)

**second law of thermodynamics** The scientific law that states that energy tends to disperse or spread out. Thermal energy is always spontaneously transferred from a hotter object to a cooler object. (p. 492)

**segunda ley de la termodinámica** La ley científica que establece que la energía tiende a dispersarse o diseminarse. La energía térmica siempre se transfiere espontáneamente de un objeto más caliente a uno más frío. (p. 492)

**single bond** A covalent bond where two electrons are shared between two atoms. (p. 156)

**enlace sencillo** Enlace covalente en el que dos átomos comparten dos electrones. (p. 156)

**single exchange reaction (single displacement reaction)** A chemical change in which an element is displaced from a compound by a more reactive element. (p. 377)

**reacción de intercambio simple (reacción de desplazamiento simple)** Cambio químico en el que un elemento es desplazado de un compuesto por un elemento más reactivo. (p. 377)

**soluble** Capable of being dissolved into another substance. (p. 125)

**soluble** Que tiene la capacidad de disolverse en otra sustancia. (p. 125)

**solute** A substance dissolved in a solvent to form a solution. (p. 409)

**soluto** Sustancia disuelta en un solvente para formar una solución. (p. 409)

**solution** A homogeneous mixture of two or more substances. (p. 409)

**solución** Mezcla homogénea de dos o más sustancias. (p. 409)

**solvent** A substance in which another substance is dissolved, forming a solution. (p. 409)

**disolvente** Sustancia en la que otra sustancia se disuelve, para formar una solución. (p. 409)

**space-filling model** A three-dimensional representation of a molecule with no space between bonded atoms, as distinct from a ball-and-stick model. (p. 200)

**modelo de espacio relleno** Representación tridimensional de una molécula en la que no hay espacio entre los átomos enlazados, a diferencia del modelo de esferas y varillas. (p. 200)

**specific heat capacity** The amount of heat required to raise the temperature of 1 gram of a substance by 1 °C. (p. 501)

**spectator ion** An ion that does not directly participate in a chemical reaction. Spectator ions appear on both sides of a complete ionic equation. (p. 459)

**spectrometer** A device for capturing electromagnetic radiation to study properties of atoms, molecules, and compounds. (p. 596)

**spectroscopy** A method of using electromagnetic radiation to study properties of atoms, molecules, and compounds as they interact and as they emit light. (p. 597)

**speed of light** The speed that light travels in a vacuum. Symbolized as $c$, the speed of light is $3 \times 10^8$ m/s. (p. 583)

**standard temperature and pressure (STP)** A standard set of conditions at which gases can be measured and compared. Standard pressure is 1 atm and standard temperature is 273 K (0 °C). (p. 332)

**stoichiometry** The quantitative relationship between amounts (usually moles) of reactants and products in a chemical reaction. (p. 468)

**strong acid (or base)** An acid (or a base) that dissociates completely in solution. (p. 435)

**structural formula** A two-dimensional drawing or diagram that shows how the atoms in a molecule are connected. Each line represents a covalent bond. (p. 152)

**sublimation** The process of changing phase from a solid to a gas without passing through the liquid phase. (p. 295)

**surroundings** Everything in the universe outside the system being investigated. (p. 487)

**synthesis** The creation of specific compounds by chemists, through controlled chemical reactions. (p. 176)

**system** The part of the universe being investigated. (p. 486)

**capacidad calorífica específica** La cantidad de calor que se necesita para aumentar la temperatura de 1 gramo de una sustancia en 1 °C. (p. 501)

**ion espectador** Ion que no participa directamente en una reacción química. Los iones espectadores aparecen a ambos lados de una ecuación iónica completa. (p. 459)

**espectrómetro** Dispositivo que se utiliza para captar la radiación electromagnética con el propósito de estudiar las propiedades de átomos, moléculas y compuestos. (p. 596)

**espectroscopia** Método que utiliza la radiación electromagnética para estudiar las propiedades de átomos, moléculas y compuestos a medida que interactúan y emiten luz. (p. 597)

**velocidad de la luz** Velocidad a la que viaja la luz en el vacío. Se simboliza como $c$, la velocidad de la luz es $3 \times 10^8$ m/s. (p. 583)

**temperatura y presión estándar (TPE)** Conjunto estándar de condiciones en las que los gases pueden ser medidos y comparados. La presión estándar es 1 atm y la temperatura estándar es 273 K (0 °C). (p. 332)

**estequiometría** La relación cuantitativa entre la cantidad (usualmente en moles) de reactantes y productos en una reacción química. (p. 468)

**ácido (o base) fuerte** Ácido (o base) que se disocia completamente en una solución. (p. 435)

**fórmula estructural** Dibujo o diagrama de dos dimensiones que muestra cómo están conectados los átomos en una molécula. Cada línea representa un enlace covalente. (p. 152)

**sublimación** El proceso de pasar de una fase sólida a una fase gaseosa sin pasar por la fase líquida. (p. 295)

**entorno** Todo el universo que está por fuera del sistema que se está investigando. (p. 487)

**síntesis** La creación de compuestos específicos por parte de científicos, a través de reacciones químicas controladas. (p. 176)

**sistema** La parte del universo que se está investigando. (p. 486)

# T

**temperature** A measure of the average kinetic energy of the atoms and molecules in a sample of matter. (p. 280)

**temperatura** Medida del promedio de energía cinética de los átomos y las moléculas en una muestra de materia. (p. 280)

**tetrahedral shape** The shape defined by the symmetrical distribution of four bonded pairs of electrons around a central atom. (p. 192)

**forma tetraédrica** La forma definida por la distribución simétrica de cuatro pares de electrones enlazados alrededor de un átomo central. (p. 192)

**theoretical yield** The maximum amount of product that could be produced in a chemical reaction when a limiting reactant is entirely consumed. The value is calculated based on a balanced chemical equation. (p. 473)

**rendimiento teórico** La máxima cantidad de producto que puede producirse en una reacción química cuando el reactante limitador se ha consumido completamente. El valor se calcula basándose en una ecuación química balanceada. (p. 473)

**thermal energy** The total kinetic energy associated with the mass and motions of the particles in a sample of matter measured as heat energy. (p. 495)

**energía térmica** La energía cinética total asociada con la masa y el movimiento de partículas en una muestra de materia medida como energía calorífica. (p. 495)

**thermal equilibrium** When two objects in contact with one another reach the same temperature, they are in thermal equilibrium. (p. 492)

**equilibrio térmico** Cuando dos objetos que están en contacto alcanzan la misma temperatura, están en equilibrio térmico. (p. 492)

**titration** An analytical procedure used to determine the concentration of an acid or a base. A measured volume of an acid or a base of known concentration is reacted with a sample in the presence of an indicator. (p. 451)

**valoración química** Procedimiento analítico usado para determinar la concentración de un ácido o de una base. En presencia de un indicador, se hace reaccionar un ácido o una base de volumen y concentración conocidos, con una muestra de la sustancia que se quiere analizar. (p. 451)

**toxicity** The degree to which a substance can harm an organism. Toxicity depends on the toxin and the dose in which it is received. (p. 381)

**toxicidad** El grado de daño que una sustancia puede hacerle a un organismo. La toxicidad depende de la toxina y de la dosis suministrada. (p. 381)

**transition elements** The elements in Groups 1B to 8B on the periodic table. (p. 44)

**elementos de transición** Los elementos de los grupos 1B a 8B de la tabla periódica. (p. 44)

**transmission** Of light, passing through certain kinds of objects. (p. 578)

**transmisión** De la luz: cuando esta pasa a través de ciertos objetos. (p. 578)

**transparent** Of an object, capable of transmitting light, meaning that it is possible to see through the object. (p. 578)

**transparente** En relación a un objeto: cuando este es capaz de transmitir la luz, lo que significa que es posible ver a través del objeto. (p. 578)

**trigonal planar shape** A flat triangular shape that is found in small molecules with three electron domains surrounding a central atom. (p. 196)

**forma plana triangular** Forma plana triangular que aparece en las moléculas pequeñas con dominio de tres electrones alrededor del átomo central. (p. 196)

**triple bond** A covalent bond in which three electron pairs are shared between two atoms. (p. 167)

**enlace triple** Enlace covalente en el que dos átomos comparten tres pares de electrones. (p. 167)

# U

**ultraviolet radiation** Electromagnetic radiation with shorter wavelengths than those of visible light, from $4 \times 10^{-7}$ m to $10^{-8}$ m. Causes sunburn and breaking of chemical bonds. Abbreviated as UV radiation. (p. 588)

**universal gas constant, *R*** A number that relates the volume, temperature, pressure and number of moles of gas in the ideal gas law. The value of *R* is dependent on the units used. One value of *R* is 0.08206 L · atm/mol · K. (p. 337)

# V

**valence electrons** The electrons located in the outermost electron shell of an atom, which participate in chemical bonding. (p. 90)

**valence shell** The outermost electron shell in an atom. (p. 90)

**visible light** Electromagnetic radiation with wavelengths that human beings can see from about $10^{-6}$ m to $4 \times 10^{-7}$ m. The wavelength increases from violet to blue to green to yellow to orange to red. (p. 588)

**voltage** The electrical potential of an electrochemical cell expressed in volts. (p. 571)

**volume** The amount of space a sample of matter occupies. (p. 9)

# W

**water displacement (method)** A method for measuring the volume of a solid object by immersing it in water. The volume of the object is equal to the amount of water displaced by the object when fully submerged. (p. 14)

**wavelength** The distance between two peaks or two troughs of a wave, usually measured in meters. The symbol for wavelength is λ (lambda). (p. 582)

**weak acid (or base)** An acid (or a base) that does not dissociate completely in solution. (p. 435)

**weather** The day-to-day atmospheric conditions such as temperature, cloudiness, and rainfall, affecting a specific place. (p. 259)

**radiación ultravioleta** Radiación electromagnética con longitudes de onda más cortas que las de la luz visible, entre $4 \times 10^{-7}$ m y $10^{-8}$ m. Produce la ruptura de las uniones químicas y quemaduras solares. Su abreviatura es UV radiación. (p. 588)

**constante universal de los gases ideales, *R*** Número que relaciona el volumen, la temperatura, la presión y el número de moles de un gas en la ley de los gases ideales. El valor de *R* depende de las unidades de medida. Un valor de *R* es 0.08206 L · atm/mol · K. (p. 337)

**electrón de valencia** Los electrones ubicados en la capa electrónica exterior de un átomo, que participan de un enlace químico. (p. 90)

**capa de valencia** La capa electrónica exterior de un átomo. (p. 90)

**luz visible** Radiación electromagnética con longitudes de onda que los seres humanos pueden ver, entre $10^{-6}$ m y $4 \times 10^{-7}$ m. La longitud de onda aumenta del violeta al azul al verde al amarillo al naranja al rojo. (p. 588)

**voltaje** El potencial eléctrico de una celda electroquímica expresado en voltios. (p. 571)

**volumen** La cantidad de espacio que ocupa una muestra de materia. (p. 9)

**desplazamiento de agua (método)** Método para medir el volumen de un objeto sólido sumergiéndolo en agua. El volumen del objeto es igual a la cantidad de agua desplazada cuando el objeto se sumerge completamente. (p. 14)

**longitud de onda** Distancia entre dos picos o dos valles de una onda, por lo general se mide en metros. El símbolo de longitud de onda es λ (lambda). (p. 582)

**ácido (o base) débil** Ácido (o base) que no se disocia completamente en una solución. (p. 435)

**tiempo atmosférico** Las condiciones atmosféricas diarias como la temperatura, la nubosidad y la precipitación, que afectan un lugar específico. (p. 259)

**white light** The mixture of all wavelengths of visible radiation (colors of the rainbow) normally seen as white rays. (p. 578)

**work** Describes what happens when a force causes an object to move in the direction of the force applied. Work can be calculated as force · distance (p. 544)

# X

**x-ray** Electromagnetic radiation with wavelengths shorter than those of visible light and ultraviolet radiation, from $10^{-8}$ m to $10^{-11}$ m. Causes electrons to be ejected from atoms and molecules. Used mainly in medical care to create images of bones and teeth. (p. 588)

**luz blanca** Mezcla de todas las longitudes de onda del espectro visible (los colores del arco iris). Comúnmente se manifiesta como rayos blancos. (p. 578)

**trabajo** Describe qué pasa cuando una fuerza hace que un objeto se mueva en la dirección en la que se aplica dicha fuerza. El trabajo se puede calcular como fuerza · distancia (p. 544)

**rayo X** Radiación electromagnética con longitudes de onda más cortas que las de la luz visible y la radiación ultravioleta, entre $10^{-8}$ m y $10^{-11}$ m. Produce la expulsión de los electrones de los átomos y las moléculas. Se utiliza principalmente en medicina para crear imágenes de los huesos y los dientes. (p. 588)

# Index

ammonium nitrate
    endothermic chemical reaction with, 488
    water and, 486
Amontons, Guillaume, 278
amu. *See* atomic mass units
aniline, 435
animals, sense of smell of, 148
anions, 96
    in acids, 433
    chlorine, 438
    of nonmetals, 99, 104
ant stings, formic acid in, 435
antacids, 430
anthocyanins
    as acids, 611–612
    as bases, 611
    as indicator, 610
    pigment, 609–610
    in solution, 611–612
anthropology connection, on language and
        color, 585
antidote, for toxins, 356
antioxidants, 551
antiperspirants, 467
aqueous solutions
    chemical symbol for, 25
    liquids as, 25
    precipitation from, 458
    reversible processes with, 614
    sugar as, 25, 359
argon
    radioactive isotopes of, 67
    valence electrons of, 94
Aristotle, 581
Arrhenius, Svante, 434
arsenic
    amu of, 390
    chelation therapy for, 460
    LD$_{50}$ of, 382
    mass of, 391
    moles of, 390
    as toxin, 420
    in water, 420
art connection
    on acid rain, 364
    on stained-glass windows, 579
artificial sweeteners
    LD$_{50}$ of, 403–404
    sweet taste of, 617
    toxicity of, 402–404
ascorbic acid, 542
aspartame
    amino acids and, 403
    molar mass of, 402–403
    toxicity of, 403
aspartic acid, 403
aspirin
    dose for, 398–399
    LD$_{50}$ of, 382
astronomy connection
    on Boomerang Nebula, 280
    on dry ice, 297
    on ethane, 157
    on light signature of elements, 86
    on nuclear fusion, 78
    on scientific notation, 392
    on weight and mass, 13
asymmetry, of polar molecules, 230, 231
atmosphere (atm)
    air pressure and, 299, 301–302
    weather and, 258, 259, 261
atmospheric pressure, 301–302
    partial pressure and, 341
atomic mass, 38, 56–57. *See also* average atomic
        mass

atomic mass units (amu), 38
    of arsenic, 390
    of carbon, 61, 390
    of hydrogen, 388, 390
    of iron, 390
    of mercury, 390
atomic model, 53
    periodic table of elements and, 58–59
atomic number
    of copper, 56
    of gold, 56
    of isotopes, 67
    in periodic table of elements, 41, 43, 56
    protons and, 56
atomic theory, 52
atoms
    atomic mass of, 56–57
    bonds of, 123
    compounds and, 51, 94
    in copper, 59
    electron cloud model for, 53
    electrons in, 52, 54, 82
    elements and, 51
    equilibrium and, 619
    in gold, 59
    isotopes and, 61–64
    matter from, 50–81
    models of, 51–54
    nuclear model for, 52
    nucleus in, 52, 53, 54
    oxidation of, 552–553
    in pennies, 57
    plum pudding model for, 52
    proton model for, 53
    reactivity of, 93
    redox and, 560
    smell and, 146
    solar system model for, 53
    solid sphere model for, 52, 53
    structural patterns of, 89–90
    subshells in, 116–118, 119
    weight of, 388
atropine, 382
average atomic mass, 38, 57
    of isotopes, 62–63
    molar mass and, 391
    of neon, 62–63
    in periodic table of elements, 43, 62
    on periodic table of elements, 391
average kinetic energy, 496
Avicenna, 187
Avogadro, Amedeo, 333
Avogadro's law, 332–333
Avogadro's number, 332
    toxins and, 390–392
Axel, Richard, 256

## B

bakeware, tempered glass in, 5
baking soda. *See* sodium bicarbonate
balance
    for density, 18
    electronic, 11, 386–387
    for mass, 3, 10, 11
    for solids, 16
balanced chemical equations, 371–374, 467
ball shape, of camphor, 205
ball-and-stick model
    for citral, 186–187
    citral for, 186–187
    for ethyl acetate, 186
    lone pairs and, 194
    for methane, 191
    for molecules, 185–188, 190
    structural formulas and, 187

balloons
    carbon dioxide in, 331, 333–334
    helium, 331, 333–334
    hot air, 283, 294, 331
    weather, 317–318, 319
barium
    activity series of, 565
    maximum contaminant level in drinking
        water, 424
    oxidation of, 560
barium hydroxide, 435
barium-140, 78
barometer, 329–3320
Bartlett, Neil, 93
base dissociation constant, 649
bases
    anthocyanin as, 611
    Arrhenius theory of, 434
    Brønsted-Lowry theory of, 434–435
    chemical burns from, 429
    chemical formulas for, 434
    dilution of, 443
    in household products, 429
    indicators for, 429–430
    neutralization reactions with, 445
    OH$^-$ in, 447
    pH of, 430–431, 439–440
    properties of, 429
    protons and, 434
    in solution, 434, 611
    theories on, 433–436
    as toxins, 428–455
    weak and strong, 435
batteries
    disposal of, 572
    as electrochemical cells, 568–572
    voltage of, 572
beakers
    for heating, 4
    measurements in, 11
    for mixing, 3
    for volume, 11
bees
    stings by, formic acid in, 435
    UV and, 586
bent shape, water with, 194
benzoic acid, 633
Berthelot, Marcellin, 517
beryllium
    electron configuration of, 117
    naturally occurring isotopes of, 65
beta decay, 72–73
beta particles, 72–73, 78
biofuels, 524
biology connection
    on amino acids, 247
    on bee vision, 586
    on color perception of humans, 591
    on diaphragm, 315
    on dissolved oxygen, 125
    on enzymes as catalysts, 541
    on equilibrium, 622
    on ethylthiol, 152
    on formic acid, 425
    on heavy metals, 464
    on hydrochloric acid, 433
    on isotopes, 63
    on octopus blood, 28
    on photosynthesis, 547
    on sense of smell of animals, 148
    on sense of smell of fish, 237
    on sense of smell of humans,
        147, 249
    on skeleton, 466
    on sleep, 627

isotopes
  atomic number of, 67
  atoms and, 61–64
  average atomic mass of, 62–63
  of carbon, 61–62, 72
  of copper, 63–64
  daughter, 78–79
  of elements, 65–67
  human-made, 66
  mass of, 62, 63
  natural abundance of, 63
  naturally occurring, 65–67
  neutrons in, 66
  nucleus and, 61–62
  parent, 78–79
  radioactive, 65–68, 72
  stable, 65–68, 72, 77
  of uranium, 79

J

jet stream
  air pressure and, 324
  on air pressure map, 323
  in weather maps, 260, 261
joules, 524
  in electromagnetic radiation, 589

K

Kelvin temperature scale, 278–280
  Charles's law and, 284
  Gay-Lussac's law and, 310
ketones, 172
  chemical names and, 205
Kevlar, 129
kidney stones, 457
kilograms, 11
kilojoules, 524
  in mole, 525
kinetic energy
  average, 496
  chemical reactions and, 488
  defined, 488, 536
kinetic theory of gases, 280
  air pressure and, 314–315, 328
krypton, 94
Kwolek, Stephanie, 129

L

lactic acid
  in circulatory shock, 644
  equilibrium constants for, 632
language
  of chemistry, 23
  color and, 585
  oil and water and, 216
lanthanides
  electronegativity value of, 227
  in periodic table of elements, 42–43, 44, 45
Laplace, Pierre, 517
Lavoisier, Antoine, 517
law of conservation of mass. See conservation of mass
LD50. See lethal dose
Le Châtelier, Henry, 643
Le Châtelier's principle
  application of, 646–652
  equilibrium mixtures and, 640–645
  indicators and, 646–648
lead
  activity series of, 566, 570
  chelation therapy for, 460
  copper and, 566
  density of, 17
  half-reaction of, 571
  heat of formation for, 557

  as metal of antiquity, 38
  molar mass of, 394–395
  paint, 395
  toxicity of, 396
  as toxin, 394–396
lead iodide, 457–458
lemon juice, 430–431
lethal dose (LD50), 381–382
  of acetylsalicylic acid, 382
  of arsenic, 382
  of artificial sweeteners, 403–404
  of aspirin, 382
  body weight and, 383, 405
  of caffeine, 382, 383
  of coffee, 382
  of fructose, 403–404
  of glucose, 382
  of salt, 382
  of snake venom, 382
  of sodium chloride, 382
  of sweeteners, 403–404
Lewis, Gilbert Newton, 162
Lewis dot structure, 162
  of carbon, 165
  of carbon dioxide, 167
  double bonds and, 166
  of fluorine, 165
  of nitrogen, 165
  octet rule and, 165
  of oxygen, 165
  of phosphine, 195
  of phosphorus trichloride, 163
Lewis dot symbols
  bonded pairs and, 162
  HONC 1234 rule and, 162
  lone pairs and, 162
  smell and, 161–164
lichens, litmus from, 647
light. See also ultraviolet radiation/light; visible
    light
  absorption of, 579
  from combustion, 552
  electricity from, 593
  emission of, 576–577
  from exothermic chemical reactions, 484, 576
  from nuclear reactions, 576
  from oxidation, 552
  photon model of, 590–591
  rays, 577
  reflection of, 578
  sources of, 576–577
  speed of, 583, 584, 587
  transmission of, 578
  wave model of, 581–583
  waves, 581–585
  white, 577, 579
light detectors
  for electromagnetic radiation, 593
  spectroscopy and, 593–596
light energy, 576–579
  as electromagnetic radiation, 588
light signatures
  of elements, 86
  of xenon, 93
lightning
  fire from, 483
  from volcanoes, 577
limestone
  calcium carbonate as, 608
  stalagmites and stalactites from, 617
  water through, 608, 617
limiting reactant (limiting reagent), toxins and, 464, 471–473
limonene, 242, 245
linear shape, double bonds and, 197

liquid thermometers, 272
liquids
  as aqueous solutions, 25
  density of, 267–270
  equilibrium-constant equation for, 635
  evaporation of, 298
  mass of, 9
  measurement of, 263–266
  molecules in, 279–280
  in periodic table of elements, 42, 45
  as solutes, 409
  as solvent, 409
  sugar as, 359
  volume of, 9, 11
  water as, 25
liter-atmospheres, 545
literature connection, on myths about fire, 483
liters, 11
  molarity and, 410
lithium
  batteries, 88, 564
  chlorine with, 88
  electron configuration of, 117
  electron shells of, 88–89
  flame color of, 85
  half-reaction of, 571
litmus
  paper, 430
  from photosynthetic algae, 647
L-limonene, 245
lock-and-key model. See receptor sites
logarithms, for $H^+$, 438
lone pairs
  ball-and-stick model and, 194
  Lewis dot symbols and, 162
  of nitrogen, 167
low pressure, 291, 321–324
  clouds and, 323
  precipitation and, 323
  in tropical depressions, 344
Lowry, Thomas, 434–435
Lundström, Johan, 513
lye. See sodium hydroxide

M

M. See molar
0.15 M solution, of salt, 418
macromolecules, 249
magnesium
  acid rain and, 452
  activity series of, 564–565, 570
  combustion of, 512, 513
  in electrochemical cells, 571
  half-reaction of, 571
  heat of formation for, 555, 557
  iron nitrate and, 561–562
  oxidation of, 551–553, 555
  reactivity of, 38
  redox of, 561–562
  tin and, 571
magnesium chloride, 100
  percent yield of, 474
  for tofu, 472
magnesium hydroxide, 110
  chemical formula for, 434
  hydrogen chloride and, 449
  in milk of magnesia, 449
magnesium nitrate, 561–562
main group elements
  in acids, 433
  ionic compounds with, 104
  on periodic table of elements, 42, 44
malachite, 554
maltose, 617
manganese dioxide, 113

manometer, 328–329

Mars, 262

mass. *See also* atomic mass units; average atomic mass; conservation of mass; molar mass
- air, 288–289, 291
- of arsenic, 391
- atomic, 38, 56–57
- average kinetic energy and, 496
- for calorimetry, 519
- comparing with volume, 14–15
- of copper, 391
- counting with, 386–388
- of cubes, 15
- cubic centimeters and, 18
- density and, 20
- of gases, 368
- of gold, 13, 14–15
- in grams, 398
- of hydrogen, 332
- of isotopes, 62, 63
- of liquids, 9
- of matter, 13–16
- measurement of, 3, 11
- moles and, 392, 395, 399–400, 419
- of neon, 332
- of pennies, 21
- of plastic, 14–15
- of reactants, 468
- of solids, 9, 16
- solutes and, 423
- of solutions, 422–424
- volume and, 14–15, 268
- of water, 9–10
- weight and, 13
- of wood, 14–15

mass number, 62

mass spectrometer, 62

mass-mole conversions, 398–400, 419
- for medicines, 398–399

matches (for fire), 513

mathematical formula, for density, 18

matter, 2–20
- in alchemy, 234
- from atoms, 50–81
- average kinetic energy of, 496
- in chemical equations, 180
- in chemistry, 8
- chemistry lab and, 3–6
- clouds as, 10
- compounds in, 24
- conservation of, 31–34
- conservation of mass and, 34, 367
- defined, 9–10
- density and, 19, 268
- elements in, 22, 24
- feelings and, 9
- gases as, 10
- heating and, 9
- lab equipment and, 3–6
- mass of, 13–16
- measurement of, 11–12
- music and, 10
- phase of, 25, 210, 268
- pressure and, 294–302
- smell and, 210
- sound and, 9
- substance in, 9, 22
- temperature and, 273
- thoughts and, 9
- viruses as, 9
- volume of, 9, 13–16
- weather and, 258–293, 326–347
- wind and, 10

maximum contaminant levels, in drinking water, 424

medical connection
- on atropine, 382
- on circulatory shock, 644
- on spectroscopy, 597

medicinal plants, for cancer, 177

medicines
- from chemical reactions, 467
- dose of, 398–399
- mass-mole conversion for, 398–399
- in milligrams, 398
- percent yield for, 475
- receptor sites and, 208
- spectroscopy for, 597

melatonin, 627

melting point (melting temperature), 274
- for aluminum, 507
- energy and, 506
- for ethanol, 507
- for gold, 507
- for mercury, 507
- of polar molecules, 222
- of sugar, 359
- temperature scale and, 275

memory, smell and, 204

Mendeleyev, Dmitri, 37, 38, 39–40, 56

meniscus, in graduated cylinders, 11, 12

menthol, 185, 188

menthone, 200–201

mercury
- activity series of, 565, 570
- amu of, 390
- boiling point for, 507
- chelation therapy for, 460
- heat of fusion for, 507
- heat of vaporization for, 507
- melting point for, 507
- as metal of antiquity, 38
- moles of, 390
- phase change of, 507
- in thermometers, 274
- as toxin, 390, 391

mercury oxide, 488

metallic bonds, 129–130, 234
- properties of, 131–132
- smell and, 235

metalloids, 42, 45–46

metals. *See also* alkali metals; alkaline earth metals; transition metal compounds; *specific metals*
- alloys of, 132
- cations of, 99, 104
- coinage, 41, 47
- combustion of, 513
- conductivity of, 127
- dissolving of, 127
- in electrochemical cells, 569–572
- electroplating of, 134–138
- extraction of, 556–557
- flame tests for, 83–87
- glow of, 552
- heat of formation and, 554–557
- heavy, 460, 464
- historic use of, 135, 556
- ionic bonds iand, 161
- ions of, 563–566
- with nonmetals, 98
- oxidation of, 551–553
- in periodic table of elements, 42, 45–46
- redox of, 559–566
- smelting of, 557
- toxins as, 357
- valence electrons of, 105, 131, 553
- in water, 462–463

metals of antiquity, 135
- copper as, 38
- gold as, 38
- iron as, 38
- lead as, 38
- mercury as, 38
- silver as, 38

meteorology connection
- on relative humidity, 340
- on weather balloons, 318

methane
- ammonia from, 371
- ball-and-stick model for, 191
- carbon in, 162–163
- combustion of, 513, 532, 551, 569
- covalent bonds in, 530
- electron domains of, 192–193
- exothermic chemical reaction with, 487
- as gas, 217
- heat of combustion for, 525
- oxygen and, 364
- smell of, 230
- tetrahedral shape of, 192

methanol
- aspartame and, 403
- calories for, 500
- combustion of, 521–522
- heat of combustion for, 524, 525
- water and, 217

methyl methanoate, 166

methyl octenoate, 200–201

methyl orange, 647
- pH of, 648

methyl red, 648

microns, 9

microwave radiation, 587
- wavelength and frequency of, 589

milk of magnesia, 449

milligrams
- $LD_{50}$ and, 405
- medicines in, 398

milliliters, 12
- cubic centimeters and, 14

minerals, in water, 418

minty smell, 148–149
- functional groups for, 170
- receptor sites and, 208
- space-filling model for, 200–201, 202

mirror-image isomers
- of amino acids, 248
- properties of, 244–245
- smell and, 242–245

mixing
- beakers for, 3
- rate of reaction and, 541
- tools for, 3

molar (M), 410

molar mass
- of aluminum, 395
- of aspartame, 402–403
- of chlorine, 394–395
- of compounds, 396
- of fructose, 402–403
- of helium, 395
- of lead, 394–395
- moles and, 419
- of oxygen, 395
- on periodic table of elements, 391, 400, 420, 468
- of salt, 395
- specific heat capacity and, 503
- stoichiometry and, 468
- of sucrose, 400
- of toxins, 394–396

molarity
- hydrogen ions and, 437
- of solutions, 410
- volume and, 419

troposphere, 261, 327
turpentine, 409
two-pan balance, 10

## U

ultraviolet radiation/light (UV), 588
  bees and, 586
  light detectors for, 593
  ozone layer and, 596
  sensitive papers for, 594–596
  spectroscopy and, 593
  wavelength and frequency of, 589
universal gas constant, 337
unpaired electrons, 162
uranium
  alpha decay of, 72
  fission of, 79
  isotopes of, 79
uranium-234, 79
uranium-235, 79
  for nuclear power, 80
uranium-238, 79
UV. *See* ultraviolet radiation/light

## V

valence electrons, 90–91
  of argon, 94
  chemical formulas and, 102
  in covalent bonds, 130
  covalent bonds and, 161
  in electron shells, 116, 122
  of fluorine, 94
  of helium, 94
  ionic compounds and, 99
  of krypton, 94
  of metals, 105, 131, 553
  of neon, 94
  of noble gases, 94
  of nonmetals, 105
  octet rule and, 165–168
  oxidation and, 553
  periodic table of elements and, 162
  of sodium, 94
valence shell, 90
valine, 247
vaporization. *See* heat of vaporization
vegetable oils, in soap, 631
vinegar, 431
viruses, 9
visible light, 577, 588
  frequency of, 589
  solar cells and, 593
  wavelength of, 584, 589
volcanoes, lightning from, 577
voltage
  of batteries, 572
  in electrochemical cells, 571, 572
volume
  altitude and, 317–318
  beakers for, 11
  combined gas law and, 318
  concentration and, 413–416
  of cubes, 15
  density and, 18, 20
  gas thermometers and, 274
  of gases, 9, 11, 282–283, 306, 336
  geometric equations for, 13
  of gold, 14
  graduated cylinders for, 3, 11
  of liquids, 9, 11
  mass and, 14–15, 268
  of matter, 9, 13–16
  measurement of, 3, 9, 11–14
  molarity and, 419
  of mole of oxygen, 395

of nucleus, 54
pressure and, 306
in rain gauges, 264–265
of solids, 9, 13–14
of solutions, 415, 420
temperature and, 18, 283
thermometers and, 273–274
of water, 12
water displacement for, 14–15
work and, 545
von Guericke, Otto, 299

## W

warm fronts, 289–291
waste containers, in chemistry lab, 5
water
  acids in, 440
  amino acids and, 248
  ammonia and, 232, 434–435
  ammonium nitrate and, 486
  arsenic in, 420
  with bent shape, 194
  boiling point for, 507
  bottled, 422
  from butyric acid and ethanol, 179
  calories for, 500
  in clouds, 297
  combustion of, 512
  conductivity in, 127
  copper chloride in, 364
  decomposition of, 535
  density of, 270, 297–298
  dilution with, 442–444
  dipoles of, 222
  drinking, 420, 424
  electron domains of, 194
  ethanol in, 364
  evaporation of, 297, 339, 621
  from fire, 484
  as gases, 25
  in graduated cylinders, 12
  hardness of, 418
  heat of fusion for, 507
  heat of vaporization for, 507
  heating curve of, 505
  hydrochloric acid in, 364
  insoluble in, 127
  intermolecular force of, 216–217
  ionization of water constant, 649
  through limestone, 608, 617
  as liquid, 25
  mass of, 9–10
  melting point for, 507
  metals in, 462–463
  from methane and oxygen, 364
  minerals in, 418
  molar mass of, 395
  molecular shape of, 197, 230
  from neutralization reactions, 448, 449
  oil and, 216–217
  oxygen from, 376
  pH of, 439–440
  phase change of, 260, 297, 507
  phase of, 257, 258, 297–298
  physical change of, 363
  as polar molecule, 215, 232
  pollution of, 424
  smell of, 233
  in snow, 269
  sodium carbonate and, 367
  soft, 418
  as solids, 25
  solubility in, 125, 127
  as solvent, 409
  specific heat capacity for, 501, 503

sugar in, 123, 358–359
toxins in, 356, 426
volume of, 12
weather and, 257, 258, 260, 261
water displacement
  for measuring volume, 14–15
  for solids, 14
water vapor, 297
  in air, 340
  in clouds, 324
  condensation and, 339, 491
  humidity and, 340
  light through, 577
  from sugar chemical reaction, 360
  sweat and, 643
  in tropical storms, 344
water-soluble dyes, 410
wave model of light, 581–583
wavelength, 582–583
  of colors, 583–584
  of electromagnetic radiation, 586
  of gamma rays, 589
  of infrared radiation, 589
  of microwave radiation, 589
  Planck's constant and, 589
  of radio waves, 589
  in visible light, 584
wave-particle duality, 591
waves
  in electromagnetic radiation, 586–587
  frequency of, 582–583
  light, 581–585
  radio, 587, 589
  of sound, 584
weak acids
  acetic acid as, 435
  anthocyanin as, 610
  dissociation of, 627–628, 646
  dissolving of, 627–628
  equilibrium constants for, 647
  equilibrium-constant equation for,
    631–634
  formic acid as, 628
  indicators as, 646
  ions from, 627–628
  pH of, 632–633
  in soap, 631–634
  strong acids versus, 623
weak bases, 435
weather, 257–347. *See also* air pressure
  atmosphere and, 258, 259, 261
  Boyle's law and, 303–307
  Charles's law and, 282–285
  clouds and, 259
  combined gas law and, 317–320
  defined, 259
  density and, 288–291
  density of liquids and solids, 267–270
  extreme physical change and, 343–346
  fronts and, 288–291
  gases and, 258
  Gay-Lussac's law and, 309–313
  humidity and, 339–342
  maps, 260–261
  matter and, 258–293, 326–347
  measuring liquids, 263–266
  physical change and, 259, 343–346
  science of, 259–261
  temperature and, 288–291
  thermometer and, 272–276
  water and, 257, 258, 260, 261
weather balloons, 317–318, 319
weather connection
  on evaporation of sweat, 643
  on precipitation, 458

# Periodic Table of the Elements

Main Group Elements

Alkali Metals*

1A

Alkaline Earth Metals
2
2A

Metals | Solids
Metalloids | Liquids
Nonmetals | Gases

Transition Elements

| | | 3 3B | 4 4B | 5 5B | 6 6B | 7 7B | 8 8B | 9 8B |

**Row 1**
1 H Hydrogen 1.008

**Row 2**
3 Li Lithium 6.941 | 4 Be Beryllium 9.012

**Row 3**
11 Na Sodium 22.99 | 12 Mg Magnesium 24.31

**Row 4**
19 K Potassium 39.10 | 20 Ca Calcium 40.08 | 21 Sc Scandium 44.96 | 22 Ti Titanium 47.87 | 23 V Vanadium 50.94 | 24 Cr Chromium 52.00 | 25 Mn Manganese 54.94 | 26 Fe Iron 55.85 | 27 Co Cobalt 58.93

**Row 5**
37 Rb Rubidium 85.47 | 38 Sr Strontium 87.62 | 39 Y Yttrium 88.91 | 40 Zr Zirconium 91.22 | 41 Nb Niobium 92.91 | 42 Mo Molybdenum 95.94 | 43 Tc Technetium (98) | 44 Ru Ruthenium 101.1 | 45 Rh Rhodium 102.9

**Row 6**
55 Cs Cesium 132.9 | 56 Ba Barium 137.3 | 71 Lu Lutetium 175.0 | 72 Hf Hafnium 178.5 | 73 Ta Tantalum 180.9 | 74 W Tungsten 183.8 | 75 Re Rhenium 186.2 | 76 Os Osmium 190.2 | 77 Ir Iridium 192.2

**Row 7**
87 Fr Francium (223) | 88 Ra Radium (226) | 103 Lr Lawrencium (262) | 104 Rf Rutherfordium (267) | 105 Db Dubnium (268) | 106 Sg Seaborgium (271) | 107 Bh Bohrium (272) | 108 Hs Hassium (277) | 109 Mt Meitnerium (276)

**Lanthanides**
57 La Lanthanum 138.9 | 58 Ce Cerium 140.1 | 59 Pr Praseodymium 140.9 | 60 Nd Neodymium 144.2 | 61 Pm Promethium (145) | 62 Sm Samarium 150.4

**Actinides**
89 Ac Actinium (227) | 90 Th Thorium 232.0 | 91 Pa Protactinium 231.0 | 92 U Uranium 238.0 | 93 Np Neptunium (237) | 94 Pu Plutonium (244)

* Group IA, except for hydrogen, H

The value in parentheses is the average atomic mass of the longest lasting isotope of the element at the time of writing.